Probability at Saint-Flour

Editorial Committee: Jean Bertoin, Erwin Bolthausen, K. David Elworthy

For further volumes:
http://www.springer.com/series/10212

Saint-Flour Probability Summer School

Founded in 1971, the Saint-Flour Probability Summer School is organised every year by the mathematics department of the Université Blaise Pascal at Clermont-Ferrand, France, and held in the pleasant surroundings of an 18th century seminary building in the city of Saint-Flour, located in the French Massif Central, at an altitude of 900 m.

It attracts a mixed audience of up to 70 PhD students, instructors and researchers interested in probability theory, statistics, and their applications, and lasts 2 weeks. Each summer it provides, in three high-level courses presented by international specialists, a comprehensive study of some subfields in probability theory or statistics. The participants thus have the opportunity to interact with these specialists and also to present their own research work in short lectures.

The lecture courses are written up by their authors for publication in the LNM series.

The Saint-Flour Probability Summer School is supported by:

– Université Blaise Pascal
– Centre National de la Recherche Scientifique (C.N.R.S.)
– Ministère délégué à l'Enseignement supérieur et à la Recherche

For more information, see back pages of the book and
http://math.univ-bpclermont.fr/stflour/

Jean Picard
Summer School Chairman
Laboratoire de Mathématiques
Université Blaise Pascal
63177 Aubière Cedex
France

Frank den Hollander • Stanislav A. Molchanov
Ofer Zeitouni

Random Media at Saint-Flour

 Springer

Frank den Hollander
Mathematisch Instituut
Universiteit Leiden
Leiden, Netherlands

Stanislav A. Molchanov
Department of Mathematics
The University of North Carolina
Charlotte, North Carolina, USA

Ofer Zeitouni
Department of Mathematics
Weizmann Institute of Science
Rehovot, Israel

Reprint of lectures originally published in the Lecture Notes in Mathematics volumes 1581 (1994), 1837 (2004) and 1974 (2009)

ISBN 978-3-642-32948-7
Springer Heidelberg New York Dordrecht London

Library of Congress Control Number: 2012949817

Mathematics Subject Classification (2010): 60K35; 60F10; 60K37; 60G50; 60-02; 82-02

Printed on acid-free paper

Springer is part of Springer Science+Business Media (www.springer.com)

Preface for Saint Flour collections

The *École d'Été de Saint-Flour*, founded in 1971, is organised every year by the *Laboratoire de Mathématiques* of the *Université Blaise Pascal* (Clermont-Ferrand II) and the *CNRS*. It is intended for PhD students, teachers and researchers who are interested in probability theory, statistics, and in applications of stochastic techniques. The summer school has been so successful in its 40 years of existence that it has long since become one of the institutions of probability as a field of scholarship.

The school has always had three main simultaneous goals:

1. to provide, in three high-level courses, a comprehensive study of 3 fields of probability theory or statistics;
2. to facilitate exchange and interaction between junior and senior participants;
3. to enable the participants to explain their own work in lectures.

The lecturers and topics of each year are chosen by the Scientific Board of the school. Further information may be found at http://math.univ-bpclermont.fr/stflour/

The published courses of Saint-Flour have, since the school's beginnings, been published in the *Lecture Notes in Mathematics* series, originally and for many years in a single annual volume, collecting 3 courses. More recently, as lecturers chose to write up their courses at greater length, they were published as individual, single-author volumes. See www.springer.com/series/7098. These books have become standard references in many subjects and are cited frequently in the literature.

As probability and statistics evolve over time, and as generations of mathematicians succeed each other, some important subtopics have been revisited more than once at Saint-Flour, at intervals of 10 years or so.

On the occasion of the 40th anniversary of the *École d'Été de Saint-Flour,* a small ad hoc committee was formed to create selections of some courses on related topics from different decades of the school's existence that would seem interesting viewed and read together. As a result Springer is releasing a number of such theme volumes under the collective name "Probability at Saint-Flour".

Jean Bertoin, Erwin Bolthausen and K. David Elworthy

Jean Picard, Pierre Bernard, Paul-Louis Hennequin
  (current and past Directors of the *École d'Été de Saint-Flour*)

September 2012

Table of Contents

Table of Contents

Lectures on Random Media

S. Molchanov
Department of Mathematics
University of Southern California
Los Angeles, CA 90089-1113

This text is based on a short course of lectures which I gave at the Summer School in Probability Theory (Saint - Flour, July, 1992). In the process of preparation of these notes for publishing, some new results, bibliography, physical discussions etc were added in hopes of making the lectures more self-contained. The publication contains a general introduction lecture 1; with a description of the basic models and three sections corresponding to the three central ideas in random media (RM) theory: homogenization, localization and intermittency. Every subsequent section contains 3 lectures. I did not try as much to describe the contours of the generalmathematical theory (such a theory does not exist) but tried to give a mathematical analysis of a few specific models,related with modern physical applications. More details on the applications in geophysics, astrophysics, oceanography, disordered solid state physics etc, can be found in my recent review [55], which can be considered as a physical introduction to the subject for mathematicians. I want to recommend this review as an independent but essential part of the text, especially for those people working in areas intermediant between mathematics and nature sciences.

At the end of every section (i.e. after lectures 4, 7 and 10) I give (without proofs) some additional information related to recent progress in the area, possible applications, and open problems.

The author thanks Professor P.L.Hennequin and P. Bernard for the invitation to the Saint-Flour summer school, students of this school for the very interesting discussions and A. Reznikova, Prof. K. Alexander (USC) and Prof. R. Carmona (UCI) for help in preparing the lectures for publication. I am especially indepted to Prof B. Simon (Cal Tech), who suggested important improvements and Prof R. Anderson (UNCC) for his great help in the final preparation of the manuscript.

Research partially supported by the NSF under Grant DMS - 9310710 and ONR under Grant N 00014-91-J-1526.

Originally published in: *École d'Été de Probabilités de Saint-Flour XXII – 1992*,
Lecture Notes in Mathematics, Vol. **1581**, 242–411, DOI: 10.1007/BFb0073874,
© Springer-Verlag Berlin Heidelberg 1994, Reprint by Springer-Verlag Berlin Heidelberg 2012

1

Lecture 1. Introduction

The purpose of my lectures is to make a pure mathematical contribution to the theory of random media (RM), one of the most popular branches of modern mathematical and theoretical physics.

The central ideas of RM theory (homogenization, localization and intermittency) are fundamental not only for pure physics, but for many natural sciences, such as oceanography, astrophysics, seismology and so on. Mathematical analysis of these ideas is a significant problem, especially for probability theory and mathematical physics.

The homogenization approach in RM theory is based on the notion that for some large scales the random fluctuations are not essential and we can "substitute" for RM a homogeneous medium with the corresponding "effective" parameters.

Localization and intermittency describe strong fluctuations of the physical fields in RM and they are expressed in a way opposite to homogenization: the random fluctuations play a key role and determine the physical properties of RM.

The theory of homogenization was conceived more than one hundred years ago (J. Maxwell, D. Rayleigh, R. Taylor). The ideas of localization and intermittency are more recent, having been discovered in the 1960s (N. Mott, P. Anderson, I. Lifshitz, Ja. Zeldovic and others). The physics literature on this subject is boundless, but many (if not the majority) of the formulas which one can find in physical monographs and journals are not correct mathematically. What are the corresponding precise conditions on RM (for example, ergodicity)? What are the boundaries of applicability of these formulas? These and related questions are very interesting and important.

All problems in RM theory are well posed mathematically. Usually (see below) they have the form of equations of classical mathematical physics (hyperbolic, parabolic, Schrödinger, etc.) with random coefficients. The qualitative or asymptotical analysis of such equations is a real challenge for mathematicians.

A few years ago, I published a review [55] based on lectures which I gave for a large group of Russian scientists (mathematicians, physicists, geophysicists, ect.). In these lectures I made an attempt to describe the connections between the basic ideas of RM theory, formulate some strict mathematical results, and describe a new mathematical technique.

The lectures [55] and another review [75] (where it is possible to find additional references) were a result of the joint activity of a group of mathematicians composed of myself and my students in collaboration with Ya. B. Zeldovich and his group

(D. Sokolov, A. Ruzmaikin). Both publications [55], [75] were near the rigorous mathematical level, but they were not mathematical papers in the strict sense.

Now there are many new mathematical results in RM theory and there is an opportunity to realize a pure mathematical program involving the transformation of RM theory into an independent branch of probability, intermediate between probability theory and mathematical physics. This area can be a very rich source of new analytical problems, limiting theorems, etc. It is not necessary to mention the application of this mathematical theory to the natural sciences. Oceanography, astrophysics and seismology are waiting for new methods (including statistical approaches) to understand real experimental data.

I'll now describe a few basic models in RM theory and formulate mathematical problems. The solution of some of those problems will be the content of further lectures. I'll discuss only the problems where we have already real results and where we can hope to have progress in the nearest future.

Basic Models

1. Heat propagation. Many physical processes related with transport in a passive scalar field (such as temperature field, salinity of water in the ocean, diffusion and so on) can be described in an isotropic homogeneous medium (such as metal) by the classical heat (parabolic) equation with diffusion coefficient

$$
\begin{cases}
\dfrac{\partial T}{\partial t} = D\Delta T = \dfrac{\sigma^2}{2}\Delta T, \\[2mm]
T = T(t,x), \quad t \geq 0, \quad x \in R^d \\[2mm]
T(0,x) = T_0(x)
\end{cases}
\tag{0.1}
$$

The solution of problem (0.1) is unique (in the class of, for example, functions bounded from below) and has a gaussian representation

$$
T(t,x) = \int_{R^d} g(t, x - y) T_0(y)\, dy
$$

$$
g(t, x - y) = \frac{\exp\left\{-\dfrac{|x-y|^2}{2\sigma^2 t}\right\}}{(2\pi\sigma^2 t)^{d/2}}
\tag{0.2}
$$

This formula has a very clear probabilistic interpretation: the operator $\dfrac{\sigma^2}{2}\Delta$ is the generator of the Markov diffusion process $x_t = x + \sigma W_t$, where W_t is a standard

Wiener process (Brownian notion) in R^d. The solution (0.2) can be written now as an integral over the trajectories:

$$T(t,x) = E \quad T_0(x + \sigma W_t) \qquad (0.3)$$

and E means integration with respect to Wiener measure on $C(R^d)$.

Real media usually are non-homogeneous and in the case of liquids or gases the diffusion mechanism is not the leading term for transport in a scalar fields: motion of the medium (for example, convection) is more essential. This gives two different models of heat propagation which are popular in the physics literature:

a) *Disordered Solid State Model*

$$\frac{\partial T}{\partial t} = \frac{\partial}{\partial x_i} \quad a_{ij}(x,\omega) \frac{\partial T}{\partial x_j} \qquad (t,x) \in (0,\infty) \times R^d T(0,x) = T_0(x) \qquad (0.4)$$

Here the diffusivity tensor (matrix of heat conductivity) is a homogeneous ergodic matrix field on R^d with a condition of uniform ellipticity (with respect to x and ω):

$$\varepsilon|\lambda|^2 \quad \leq a_{ij}(x,\omega)\lambda_i\lambda_j \quad \leq \frac{1}{\varepsilon}|\lambda|^2 \qquad (0.4')$$

for some fixed $0 < \varepsilon < 1$. The index ω is a point of RM probabilistic space (Ω_m, F_m, μ), which usually we can understand as an ensemble of all the possible realizations of the field $a_{ij}(\cdot,\omega)$. The measure μ on this ensemble is invariant with respect to all translation of the phase space R^d : $x \to x + h$; $x, \quad h \in R^d$. For the expectation $\int_{\Omega_m} X(\omega)\mu(d\omega)$ we will use the notation $\langle \quad \rangle$:

$$\langle X \rangle = \int_{\Omega_m} X(\omega)\mu(d\omega). \qquad (0.5)$$

The solution $T(t,\kappa,\omega)$ of problem (0.4), which exists even if the coefficients $a_{ij}(x,\omega)$ are only measurable is, of course, a random function on $(0,\infty) \times \Omega_m$ and the standard approach to the analysis of (0.4) is related with the calculation (or estimation) of the statistical moments (correlation functions)

$$\begin{aligned} m_1(t,x) &= \langle T(t,x,\cdot) \rangle \\ m_2(t,x_1,x_2) &= \langle T(t,x_1,\omega)T(t,x_2,\omega) \rangle \end{aligned}$$

etc.

We will not discuss this model here, but the operator $h = \frac{\partial}{\partial x_i} \quad a_{ij}(x,\omega)\frac{\partial}{\partial x_j}$ will appear in model II (wave propagation) in another context.

b) *Turbulent transport*

$$\frac{\partial T}{\partial t} = D\Delta T + (\vec{a}, \nabla)T = \frac{\sigma^2}{2}\Delta T + a_i(t, x, \omega)\frac{\partial T}{\partial x_i} \tag{0.6}$$

Here $D = \dfrac{\sigma^2}{2}$ is the coefficient of molecular diffusivity (usually it is a small parameter) and $\vec{a}(t, x, \omega)$ is a random (turbulent) drift. It is a vector field, ergodic and homogeneous in time and space. If the drift $a_i(x, \omega)$ does not depend on time t, we call this *RM stationary* (such a situation is typical for the magneto hydrodynamics and plasma physics, where the drift is generated by magnetic fields, constant or slowly varying in time). If the field $a_i(t, x, \omega)$ is incompressible, div $\vec{a} = 0$, and the time and space correlations are decreasing fast enough, we call this *RM nonstationary* (or turbulent).

Model (0.6) plays a fundamental role in modern astrophysics, oceanography, statistical hydrodynamics and so on.

Let's describe the probabilistic approach to problem (0.6) and the procedure of homogenization.

For fixed $\omega \in \Omega_m$ (and smooth enough coefficients $a_i(t, \omega)$) we can construct a Markov process x_t (the so-called Lagrangian trajectory), related with the right hnad side of (0.6) as the solution of a (Ito's) stochastic differential equation:

$$dx_t = \sigma d\omega_t + \vec{a}(t, x_t)dt, \quad x_t = x_0 + \sigma\omega_t + \int_0^t \vec{a}(s, x_s)ds \tag{0.7}$$

The solution x_t depends on two random parameters: $\omega_m \in \Omega_m$, which is the index of a given realization of our RM and $\omega \in \Omega_\omega$, which is the trajectory of a standard Wiener process W_t. $t \geq 0$. In different terms, $x_t = x_t(\omega_m, \omega)$ is a random process on the direct product $\Omega_m \times \Omega_\omega$ with probability measure $\mu \times D^\omega$ (D_ω is a standard Wiener measure) and corresponding σ-algebra. For the expectation with respect to D^ω we'll use standard notation E^ω or E_x^ω if the Wiener process is started from $x \in R^d$.

We consider next the first two statistical moments of $x_t(\omega_m, \omega)$ for fixed $\omega = \omega_m$:

$$m_t = E(x_t - x_0), B_t = E(x_t - x_0 - m_t) \otimes (x_t - x_0 - m_t), \tag{0.8}$$

i.e. the mean drift and the covariance matrix for a fixed RM. The functions m_t, B_t are random ones (on Ω_m). We can expect (and it is true in a very general situation, see below), that the following limits exist:

$$a_0 = \lim_{t \to \infty} \frac{m_t}{t}(\mu - a.s) \tag{0.9'}$$

(which is the so-called Stokes drift)

$$B_0 = \lim_{t \to \infty} \frac{B_t}{t}(\mu - a.s) \tag{0.9''}$$

(which is the so-called turbulent diffusivity, or Taylor's diffusivity). Moreover, we can apply in this case the central limit theorem ($\mu - a.s$): for a fixed domain $\mathcal{D} \subset R^d$

$$P^\omega \left\{ \frac{x_t - x_0 - a_0 t}{\sqrt{t}} \in D \right\} \underset{t \to \infty}{\longrightarrow} \int_D G_{B_0} \ (dy), \qquad (0.10)$$

where $G_{B_0}(dy)$ is a gaussian measure with zero mean and covariance B_0. Very often we have not just CLT, but a functional CLT: for $a.e$ $\omega_m \in \Omega_m$ the process

$$\frac{x_{ts} - x_0 - a_0 ts}{\sqrt{t}} = x_s^{\sim}(\omega, \omega_m), \quad s \in [0,1] \qquad (0.11)$$

converges in distribution (in $C(R^d)$) to the Wiener process ξ_s, $s \in [0,1]$ with covariance B_0:

$$E\xi_s = 0, \quad E\xi_{s_1} \otimes \xi_{s_2} = (s_1 \wedge s_2) B_0 \qquad (0.11')$$

These results ((0.9) - (0.11)) are the probabilistic version of homogenization theorems. Let's give an analytical version of the same result. Suppose that the drift $\vec{a}(t, x, \omega)$ has the property of reflection symmetry, i.e $\vec{a}$ and $-\vec{a}$ are identically distributed and the same property holds for $\vec{a}(t, x)$ and $\vec{a}(t, -x)$ (for example, the field $\vec{a}(t, x)$ is isotropic). Then, as is easy to see, the Stokes drift $\vec{a_0} \equiv 0$ (if it exists!) Normalization in CLT shows that it's natural to make the scaling $x \to \dfrac{x}{\varepsilon}$, $t \to \dfrac{t}{\varepsilon^2}$ in equation (0.6). Then this equation will have the form

$$\frac{\partial T^\varepsilon}{\partial t}(t, x) = D \Delta T^\varepsilon + \frac{1}{\varepsilon} a_i (\frac{t}{\varepsilon^2}, \frac{x}{\varepsilon}) \frac{\partial T^\varepsilon(t, x)}{\partial x_i} \qquad (0.12)$$

Suppose that the initial data in (0.12) does not depend on ε! (In the language of the initial equation (0.6) it means, that the scale of the initial data increases if $\varepsilon \to 0$). We have now the following problem

$$\frac{\partial T^\varepsilon}{\partial t}(t, x) = D \Delta T^\varepsilon + \frac{1}{\varepsilon} a_i \left(\frac{t}{\varepsilon^2}, \frac{x}{\varepsilon} \right) \frac{\partial T^\varepsilon}{\partial x_i}$$

$$T^\varepsilon(0, x) = T_0(x) \in C(R^d) \qquad (0.12')$$

The homogenization result means, that for $\varepsilon \to 0$ $\mu a.s.$, $t \in [0, t_0]$, $x \in R^d$)

$$T^\varepsilon(t, x) \to \bar{T}(t, x)$$

and

$$\frac{\partial \bar{T}}{\partial t} = \frac{1}{2}(B_0 \nabla, \nabla) \bar{T}$$

$$\bar{T}(0, x) = T_0(x) \qquad (0.13)$$

The goal of the mathematical theory is

a) to give conditions (sufficient or necessary and sufficient) for the applicability of homogenization). Very wide sufficient conditions are known, nevertheless, the situation is not clear if $d \geq 2$ (for the model of turbulent transport),

b) to give an algorithm for the calculation of Stokes drift and Taylor diffusivity

c) to give asymptotically formulas for "extremal" physical conditions (for example, if $\sigma = \sqrt{2D} \to 0$). In this direction (at least for some special cases) there is recent progress (A. Majda and M. Avellaneda, see below).

Instead of continuous models (disordered solid state and turbulent transport) introduced above, we'll discuss lattice models: random walks in random environment. It is simpler analytically and the corresponding physical information is similar.

c) Random walk in RM. Let $\mathbf{Z}^d$ be a d-dimensional lattice with basis $\{e_i, i = 1, 2, \cdots, d\}$ and $\mathcal{P}(t, x, \omega)$ a field of transition probabilities from the point x to the (for simplicity) neighboring points at moment $t \in \mathbf{Z}_+^1 = \{0, 1, \cdots\}$

$$\mathcal{P}(t, x) = \{P(t, x \pm e_i), \quad i = 1, 2 \cdots d; \sum_i P(t, x \pm e_i) = 1\}$$

The field $\mathcal{P}(t, x)$ is a homogeneous and ergodic one in space and time. If $\mathcal{P}(t, x) = \mathcal{P}(x, \omega)$, we have a stationary RM; if there is a nontrivial dependence on time, we have a nonstationary RM, i.e the model of turbulent transport. For fixed $\omega = \omega_m$, we can construct a Markov chain $x_t(\omega, \omega_m)$ on the probability space of all paths on $\mathbf{Z}^d$ with the transition probabilities

$$P^\omega\{x_{t+1} = x \pm e_i \mid x_t = x\} = p(t, x \pm e_i) \in P(t, x) \qquad (0.14)$$

As earlier, the homogenization of this model means the calculation (in asymptotics $t \to \infty$) of the first two moments of the random vector x_t and the demonstration of the CLT (maybe in functional form).

In chapter 1 (homogenization) we will describe a few specific models of this type.

II Localization of electrons and waves.

The motion of a quantum particle (say an election) is described by the Schrödinger equation (for a corresponding system of units):

$$i\frac{\partial \psi}{\partial t} = H\psi = -\Delta\psi(t, x) + V(x)\psi(t, x), \quad (t, x) \in (0, \infty) \times R^d$$
$$\psi(0, x) = \psi_0(x), \quad \psi_0(x) \in L_2(R^d), \quad \|\psi_0\| = 1 \qquad (0.15)$$

The operator H (Hamiltonian of the particle) has two parts, a kinematics one $(-\Delta)$ and a potential $V(x)$, which describes the interaction of the particle with exterior fields. The solution of (0.15), in the case when the Hamiltonian is essentially self-ajoint (say $V(x)$ bounded from below) is given by the unitary group:

$$\psi(t, x) = (\exp(itH)\psi_0)(x). \tag{0.15'}$$

Of course, $\|\psi(t, \cdot)\| = 1$ and according to the statistical interpretation of the quantum mechanics function,

$$p_t(x) = |\psi(t, x)|^2 = \psi(t, x)\bar{\psi}(t, x)$$

gives the distribution of probability for the location of the particle at the moment $t \geq 0$.

If $V(t) \equiv 0$, (free electron) the equation (0.15) has a solution of "planar waves" form (the so-called de Broile waves):

$$\psi(t, x) = \exp[i(kx - k^2 t)], \quad k \in R^d. \tag{0.16}$$

Suppose that

$$\psi_0(x) = (\pi\sigma^2)^{-d/2} \exp\{-\frac{x^2}{4\sigma^2}\}. \tag{0.17}$$

Of course (Fourier transform)

$$\psi_0(x) = \int_{R^d} 4^{d/2} \exp\{-\sigma^2 k^2\} e^{ikx} dk \cdot \frac{1}{(2\pi)^d}. \tag{0.18}$$

Using the linearity of (0.15) and (0.16), from (0.18) we have

$$\psi(t, x) = \int \frac{4^{d/2}}{(2\pi)^d} \exp\{-\sigma^2 k^2\} e^{-ikx - ik^2 t} dk$$

and after a trivial calculations we'll get

$$P_t(x) = |\psi(t, x)|^2 = \exp\left(-\frac{x^2\sigma^2}{2(\sigma^4 + t^2)}\right) (2\pi)^{-d/2}(\sigma^4 + t^2)^{-d/4} \tag{0.19}$$

From (0.19) it follows that

$$\int_{R^d} |x| P_t(x) dx \xrightarrow{t \to \infty} \text{const} \cdot t$$

i.e. we have some sort of "quantum-mechanical" diffusion of the particles (although we can't talk about a "trajectory of an electron" or a random process related with an electron and so on). The spectrum of energy for a free particle (i.e. spectrum of $-\Delta$ in $L_2(R^d)$) is, of course, $[0, \infty]$.

If the potential $V(x)$ is a periodic one, i.e. $V(x + T) \equiv V(x)$, $T \in \Gamma = $ lattice of periods), then the situation is technically more complicated but qualitative structure of solutions is the same. Hamiltonians with the periodic potential $V(x)$ appear in the classical theory of ideal (crystal) solid state in the framework of one-body approximation.

The quasiperiodic Bloch solutions of the Schrödinger equation with periodic potential have the form

$$\psi(t, x) = \exp i(kx - Et)\psi_0(x) \tag{0.20}$$

where $\psi_0(x) = \psi_0(x + T)$, $T \in \Gamma$ and the energy E is a many-valued analytical function of the "quasimomentum" $k \in R^d$. These solutions "play the role" of the de Broile waves. For additional information see an arbitrary book on solid state physics (for example, the fundamental monograph [10]) or my review [55], where it's possible to find information on homogenization for periodic Schrödinger equations).

In any case, the electron in the periodic potential is "almost free" and this fact lies at the base of the classical theory of electrical or heat conductivity for the crystal state.

If the potential $V(x)$ has the form of a potential well i.e. $V(x) \to +\infty$, $x \to \infty$, the situation is quite different. In this case, the operator $H = -\triangle + V(x)$ has a discrete spectrum $\{E_i, \quad i = 1, 2, \cdots, \quad E_i \to +\infty\}$ and the corresponding eigenfunctions $\psi_i(x) : \quad H\psi_i = E_i\psi_i$ form an (orthnormal) basis in $L_2(R^d)$. If (for the sake of simplicity) the spectrum $\{E_i\}$ has multiplicity 1, the density of the distribution for the position of the electron at moment t, i.e. $P_t(x) = |\psi(t, x)|^2$ has a weak limit

$$P_t(x) \xrightarrow{t \to \infty} P_\infty(x) = \sum_{i=1}^{\infty} \alpha_i |\psi_i(x)|^2$$

where $\alpha_i = |(\psi_0 \cdot \psi_i)|^2$ are the squares of the Fourier coefficients ot the initial data $\psi_0(x)$. In different terms, we have a statistical stabilization of the quantum particle for $t \to \infty$, i.e., a localization. Of course this fact is obvious physically, since "the height of the potential barriers is equal to infinity".

If we consider a disordered solid state (crystal with impurities, polycrystal, alloy and so on), the corresponding models must be described in the language of RM theory. The potential $V(x, \omega)$ will be a homogeneous ergodic random field (with some very weak conditions of boundness from below, to guarantee essential self ajointness for $H = -\triangle + V(x, \omega)$) for $\mu - a.e.$ ω. In many cases the potential will be bounded or it will be a small perturbation of a periodic potential. The solution of the Schrödinger equation

$$\frac{\partial \psi}{\partial t} = -\triangle \psi + V(x, \omega)\psi$$
$$\psi(0, x) = \psi_0(x) \in L^2, \quad ||\psi_0|| = 1 \tag{0.21}$$

has a simple probabilistic sense (mentioned above) : $P_t(x) = |\psi(t, x)|^2$ is the density of the distribution for the electron at moment $t > 0$, $P_0(x) = |\psi_0(x)|^2$ is an initial distribution.

Asymptotically the behavior of the ψ-function for $t \to \infty$ depends principally on the structure of the spectrum of $H = H(\omega)$, i.e. on the solutions of the spectral problem

$$-\Delta\psi + V(x,\omega)\psi = E\psi \tag{0.22}$$

(In the case of periodic $V(x)$ the spectrum was absolutely continuous, i.e. $SpH = Sp_{a.c.}H$, but for the potential well, $SpH = Sp_{d.}H$).

The fundamental discovery of P. Anderson and N. Mott (see the original papers [6], [60] or the recent monographs [49], [14]) was that localization is "typical' for disordered systems. In other terms, the spectrum of $H = H(\omega)$ for a random Hamiltonian with "good" mixing properties must either be pure point (p.p.) ($d = 1$) or, at least, contain some (p.p.) component. The difference between the discrete spectrum and the p. p. spectrum (or, better to say, dense p.p. spectrum) is simple: the eigenvalues of the p.p. spectrum (levels) are not separated (as in the case of a potential well) but are distributed, like the rational numbers on some interval. In this case we may speak about a density of states etc.

We will study the continuous model (0.22) (or the more general model, where $H = \frac{1}{\rho}\frac{\partial}{\partial x}(\frac{1}{a}\frac{\partial}{\partial x})$) only for $d = 1$, where the picture is now more or less clear. For $d > 2$, we will study only the lattice (the so-called Anderson's) model in $L_2(\mathbf{Z}^d)$:

$$H = \Delta + \sigma V(x,\omega), \quad x \in \mathbf{Z}^d, \quad \omega \in (\Omega_m, \mathcal{F}_m, \mu)$$

Here, $\Delta\psi(x) = \sum_{|x'-x|=1} \psi(x')$ or $\Delta\psi(x) = \sum_{|x'-x|=1} (\psi(x') - \psi(x))$ is one of the forms of the lattice Laplacian, the field $V(x,\omega)$ is "random enough" and the coupling constant σ is "large enough" (case of large disorder). In both cases I'll give short proofs of very general results which contain almost all localization theorems published earlier. This part of the lectures will be based on the recent papers [2], [53], [54].

For the wave equation in RM we can introduce similar models and find similar results. We will discuss in Chapter II the wave equation for a random string, i.e. the equation

$$\frac{\partial^2 u}{\partial t^2} = \frac{1}{\rho(x,\omega)}\frac{\partial}{\partial x}\left(\frac{1}{a(x,\omega)}\frac{\partial u}{\partial x}\right)$$
$$u(0,x) = \phi_0(x) \in C_0, \quad t \geq 0, \quad x \in R^1$$
$$\frac{\partial u}{\partial t}(0,x) = \phi_1(x) \in C_0 \tag{0.23}$$

The coefficient $\rho(x,\omega)$ has the meaning of local density of the string and $a(x,\omega)$ is an elasticity coefficient. The two-component process (ρ, a) is a homogeneous, ergodic and uniformly non-degenerate one, i.e. for all $\omega \in \Omega_m$ and some $\varepsilon > 0$)

$$\varepsilon \leq \rho \leq \frac{1}{\varepsilon}, \quad \varepsilon \leq a \leq \frac{1}{\varepsilon}.$$

Under some additional mixing conditions, $\mu - a.s$ the operator $H = -\dfrac{1}{\rho}\dfrac{\partial}{\partial x}\left(\dfrac{1}{a}\dfrac{\partial}{\partial x}\right)$, which is essentially self-ajoint in the Hilbert space $L_2(R^1,\rho)$ with the inner product $(\psi_1,\psi_2)_\rho = \int_{R^1}\psi_1(x)\psi_2(x)\rho(x,\omega)dx$ and non-negative, has a $p.p.$ spectrum $SpH = \{E_i\}$. $\{E_i\} \in (0,\infty)$, and the corresponding eigenfunctions decrease exponentially (exponential localization). As in the case of the Schrödinger operator with the $p.p.$ spectrum, it means, that the elastic waves can not propagate along the random string. For example, if $\phi_1 \equiv 0$ in (0.23) the solution $u(t,x)$ has form

$$u(t,x,\omega) = \sum_{i=1}^{\infty} \cos \sqrt{E_i}t \ \ \psi_i(x,\omega)(\phi_0,\psi_i)_\rho \tag{0.24}$$

and (as in the case of potential well, see above) one can just prove, that for every $\delta > 0$ we can find $R = R(\delta,\phi_0,\omega)$, such that for all $t \geq 0$

$$\int_{|x|>R} u^2(t,x)\rho dx \leq \delta$$

i.e. the energy of elastic waves uniformly in time concentrate "not very far" from the initial fluctuation of the string.

For the wave equation, we can apply the homogenization approach. After scaling $x \to \dfrac{x}{\varepsilon}$, $t \to \dfrac{t}{\varepsilon}$, the equation will have form

$$\frac{\partial^2 u^\varepsilon}{\partial t^2} = \frac{1}{\rho(x/\varepsilon,\omega)} \ \frac{\partial}{\partial x}\left(\frac{1}{a(x/\varepsilon,\omega)}\frac{\partial u^\varepsilon}{\partial x}\right) \tag{0.23'}$$

If the initial data do not $depend$ on ε, i.e. $u^\varepsilon(0,x) = \phi_0(x)$, $\dfrac{\partial u^\varepsilon}{\partial t}(0,x) = \phi_1(x)$, then (according to well-known and basic results, see the discussion and the references in [55]) we have asymptotically for $\varepsilon \to 0$, $t \in [0,t_0]$, $x \in R^1$,

$$u^\varepsilon(t,x) \to \bar{u}(t,x) \tag{0.25}$$

$$\frac{\partial^2 \bar{u}}{\partial t^2} = C^2\frac{\partial^2 \bar{u}}{\partial x^2}, \quad \bar{u}(0,x) = \phi_0, \quad \frac{\partial \bar{u}}{\partial t}(0,x) = \phi_1$$

where the constant C (the effective spreed of propagation for elastic waves in the string) is given by the simple formula

$$C = (\langle\rho\rangle\langle a\rangle)^{-1/2} \tag{0.25'}$$

The last result contradicts the phenomenon of localization for a random string. I have discussed this "dialectical" contradiction in the review [55], where it's possible to find other similar examples. Some information about the relation between homogenization and localization for the random string will be described in the conclusion of Ch. II.

III Reaction-diffusion equations and parabolic Anderson model

This model is close to the Schrödinger equation discussed before but has the form of a parabolic equation with random potential:

$$\frac{\partial u}{\partial t} = \kappa \triangle u + \xi_t(x)u,$$
$$u(0,x) = u_0(x), \quad t \geq 0, \quad x \in \mathbf{Z}^d \tag{0.26}$$

For $u_0(x)$ we have two different variants: $u_0(x) \equiv$ const (say, $u_0(x) = 1$) or $u_0(x) = \delta_0(x)$. Of course, $\triangle$ is the lattice Laplacian, i.e. bounded operator in $L^2(\mathbf{Z}^d)$. The potential $\xi_t(x)$ can be time independent (stationary RM) or homogeneous not only in space, but in time also. Time and space correlations will decrease fast enough. A typical example will be the case when $\xi_t(x,\omega)$ are independent for different $x \in \mathbf{Z}^d$ and are white noises in time for fixed x (i.e. $\langle \xi_t(x,\omega) \rangle = 0$, $\langle \xi_t(x,\omega)\xi_t,(x',\omega) \rangle = \delta(x-x')\delta(t-t'))$. Of course in this latter case equation (0.26) has the meaning of a stochastic partial differential equation (for details, see Ch. III). Equation (0.26) is closely related to chemical kinetics. Let's discuss the simplest form of a model for a system of non-interacting Brownian particles on the lattice $\mathbf{Z}^d$. The dynamics of the system of particles is given by two functions $\xi^+(t,x)$ and $\xi^-(t,x)$. They represent, respectively the rates at which a particle at time t at $x \in \mathbf{Z}^d$ splits into two identical particles or dies. One also needs to know the diffusion constant κ which determines the holding time at each point of the lattice; a particle at time t at $x \in \mathbf{Z}^d$ jumps during the time interval $[t, t + dt)$ to a neighboring point x' such that $|x' - x| = 1$ with probability κdt. We denote by $n(t,x)$ the number of particles at $x \in \mathbf{Z}^d$ at time t. We compute its moment generating function. For $s < t, x, y \in \mathbf{Z}^d$ and $\alpha > 0$ we set:

$$M_\alpha(s,x,t,y) = E\{e^{-\alpha n(t,y)} \mid n(s,x) = \delta_y(x)\}. \tag{0.27}$$

It is then easy to derive for (t,y) fixed the *backward Kolmogorv's equation*:

$$0 = \frac{\partial M_\alpha}{\partial s} + \kappa \triangle_x M_\alpha + \xi^+(s,x)M_\alpha^2 - (\xi^+(s,x) + \xi^-(s,x))M_\alpha + \xi^-(s,x) \tag{0.28}$$

which is satisfied for $s \leq t$ and which is to be complemented with the *initial* or *boundary* condition:

$$M_\alpha(t,x,t,y) = e^{-\alpha}\delta_y(x) + (1 - \delta_y(x)).$$

Unfortunately this equation may not have a unique solution. In any case, if we denote by $m_1(s,x,t,)$ the first moment of $n(t,x)$ under the same condition, i.e.

$$m_1(s,x,t,y) = E\{n(t,y) \mid n(s,x) = \delta_y(x)\}$$

then, using the fact that:

$$m_1(s, x, t, y) = -\frac{\partial M_\alpha}{\partial \alpha}\Big|_{\alpha=0}$$

one gets the following equation for m_1:

$$0 = \frac{\partial m_1}{\partial s} + \kappa\Delta_x m_1 = (\xi^+(s, x) - \xi^-(s, x))m_1 \tag{0.29}$$

with the initial-boundary condition:

$$m_1(t, x, t, y) = \delta_y(x).$$

One gets similar equations for the higher moments by taking more derivatives. For example the equation for the second moment:

$$m_2(s, x, t, y) = E\{n(t, y)^2 \mid n(s, x) = \delta_y(x)\}$$

is obtained by taking one extra derivative. It reads:

$$0 = \frac{\partial m_2}{\partial s} + \kappa\Delta_x m_2 + (\xi^+(s, x) - \xi^-(s, x))m_2 - \xi^+(s, x)m_1^2 \tag{0.30}$$

with the same initial-boundary condition:

$$m_2(t, x, t, y) = \delta_y(x).$$

The first part of the right hand side is the same as for m_1. The only difference is the term $\xi^+(s, x)m_1^2(s, x, t, y)$. Note that this last term can be viewed as a source term if the equation for the first moment has already been solved. As we have explained, similar equations can be derived for the higher moments:

$$m_p(s, x, t, y) = (-1)^p \frac{\partial^p M_\alpha}{\partial \alpha^p}\Big|_{\alpha=0}.$$

These equations are not very easy to deal with because of the *boundary condition* which assume knowledge of the number of particles at the terminal time t. Moreover, this condition states that there is a single particle at a specific point. In particular, this is far from the desirable condition:

$$n(s, x) \equiv 1.$$

This is the main reason for an investigation of the more complicated *forward equation* which we derive now. In order to do so we need to consider the number of particles $n(s, x)$ as a function $n(s)$ of $x \in \mathbf{Z}^d$. If $\alpha = \alpha(x)$ is a function with compact support (i.e. $\alpha(x) = 0$ except for finitely many x) then the duality:

$$< \alpha, n(s, \cdot) > = \sum_x \alpha(x)n(s, x)$$

makes sense and the moment generating function M_α can be defined as a Laplace transform on a functional space by the formula:

$$M_\alpha(s,t) = E\{e^{-<\alpha,n(t)>}|n(s)\}. \tag{0.31}$$

We use the same arguments as before to derive the Kolmogorv's equation. But this time we give details because the results seem to be less standard. We consider the three possible transitions for the number of particles in an infinitesimal time interval $[t, t + dt)$.

- $n(t,y) \to n(t,y) + 1$ with probability

$$\xi^+(t,y)n(t,y)dt + \sum_{|y'-y|=1} \kappa n(t,y')dt + O(dt^2)$$

- $n(t,y) \to n(t,y) - 1$ with probability

$$\xi^-(t,y)n(t,y)dt + 2d\kappa n(t,y)dt + O(dt^2)$$

- $n(t,y) \to n(t,y)$ with probability

$$1 - \xi^+(t,y)n(t,y)dt - \sum_{|y'-y|=1} \kappa n(t,y')dt - \xi^-(t,y)n(t,y)dt$$
$$-2d\kappa n(t,y)dt + O(dt^2)$$

Next we notice that, up to terms of order $O(dt^2)$, the $n(t+dt,y) - n(t,y)$'s many be regarded as independent for different sites y. Consequently, one easily checks that:

$$M_\alpha(s,t+dt) - M_\alpha(s,t) = -E\{e^{-<\alpha,n(t)>} \sum_y \left(e^{-\alpha(y)} - 1\right) [\xi^+(t,y)n(t,y)$$

$$+ \sum_{|y'-y|=1} \kappa n(t,y')] + \sum_y \left(e^{\alpha(y)} - 1\right) [\xi^-(t,y)n(t,y) + 2d\kappa n(t,y)]\}dt + O(dt^2).$$

If one notices that:

$$\frac{\partial M_\alpha}{\partial \alpha(y)} = -E\{e^{-<\alpha,n(t)>}n(t,y)\}$$

one immediately gets the (formal) equation for M_α :

$$\frac{\partial M_\alpha}{\partial t} = \sum_y \left(e^{-\alpha(y)} - 1\right) [\xi^+(t,y)\frac{\partial M_\alpha}{\partial \alpha(y)}$$

$$+ \sum_{|y'-y|=1} \kappa \frac{\partial M_\alpha}{\partial \alpha(y')}] + \sum_y \left(e^{\alpha(y)} - 1\right)(\xi^-(t,y) + 2d\kappa)\frac{\partial M_\alpha}{\partial \alpha(y)} \tag{0.32}$$

with the initial condition $M_\alpha(s,s) = e^{\alpha(y)}$ if one wants to consider the initial condition $n(s,y) = \delta_x(y)$, or:

$$M_\alpha(s,s) = \exp\{\sum_x \alpha(x)\}$$

if one wants to consider the initial condition $n(s,y) \equiv 1$. In any case, equation (0.32) makes it possible to derive equations for the moments and the correlation coefficients of the field $n(t, y)$. Let us consider for example the case $s = 0$ and $n(0, x) \equiv 1$. Then we have:

$$m_1(t, y) = < n(t, y) >= -\frac{\partial M_\alpha}{\partial \alpha(y)}|_{\alpha=0}$$

and:

$$m_2(t, x, y) = < n(t, x)n(t, y) >= \frac{\partial^2 M_\alpha}{\partial \alpha(x)\partial \alpha(y)}_{\alpha=0}$$

and consequently, we have for the first moment:

$$\frac{\partial m_1}{\partial t} = \kappa \triangle m_1 + [\xi^+(t, y) - \xi^-(t, y)]m_1 \tag{0.33}$$

with initial condition $m_1(0, y) \equiv 1$. Notice that equation (0.33) is exactly the fundamental equation (0.26) introduced above. For the highest moments we can find similar equations. We'll discuss this equation from the point of view of intermittency theory at the beginning of Ch. III. For a detailed analysis of the parabolic Anderson model see [15].

IV Cell-dynamo model

The last model which we are going to discuss in our lectures will be the vector version of the Anderson parabolic problem:

$$\frac{\partial \vec{H}}{\partial t} = k\triangle \vec{H} + \xi_t(x)\vec{H} \tag{0.34}$$

$$\vec{H}(0, x) = H_0(x), \quad x \in \mathbf{Z}^d$$

Here $\xi_t(x) = \{\dot{W}_{ij}(t, x)\}$ is a $N \times N$ matrix of white noise with isotropic independent components:

$$\langle \xi_t(x) \rangle = 0, \quad \langle \xi_{ij}(t, x)\xi_{i'j'}(t', x') \rangle =$$

$$= \delta(i - j')\delta(j - j')\delta(t - t')\delta(x - x').$$

Of course, it is only a simple generalization of the previous (model III) Scalar Anderson model. The real interest in the equation (0.34) and its non-linear variant

$$\frac{\partial \vec{H}}{\partial t} = \kappa \triangle \vec{H} + \xi_t(x)\vec{H} - \varepsilon \vec{H}|\vec{H}|^\beta, \quad \beta > 0 \tag{0.35}$$

is based on the connection of these equations with the problem of the generation of a magnetic field by the random (turbulent) motion of a conduction fluid (the so-called dynamo), in its original classical form. The dynamo is a magnetohydrodynamic problem. It concerns the time asymptotic behavior of the magnetic field in a flow of a conducting fluid. The evolution of the divergenceless vector magnetic field, $H_i(t, x), i, j = 1, 2, 3; \nabla_i H_i = 0$, is described by the induction equation

$$\frac{\partial H_i}{\partial t} + (v_j \nabla_j)H_i = \eta \triangle H_i + (H_j \nabla_j)v_i,$$

$$\text{div } \vec{H} = \text{ div } \vec{v} = 0 \tag{0.36}$$

$$H_i(0, x) = H_{0i}(x)$$

where η is the magnetic diffusivity (the inverse conductivity of the fluid) and $v_i(t, x)$ is the hydrodynamic velocity of the fluid, usually considered as incompressible, i.e. $\nabla_i v_i = 0$. The term $(v_j \nabla_j)H_i$ describes the advection of the field, $\eta \triangle H_i$ is diffusion term, and the last term in the right side of the equation

$$(h_j \nabla_j)v_i = H_j \frac{\partial v_i}{\partial x_j} = V_{ij} H_j$$

may be interpreted as a matrix potential. It is this term which is mainly responsible for the generation (asymptotic growth) of the magnetic field. In applications the parameter η is typically very small, and the fluid is turbulent so the velocity may be described by a random ergodic field on some probability space. If we assume that the velocity is independent of the magnetic field (the kinematic approximation) then the problem is linear in H_i and the rates of growth (the Lyapunov exponents) can be estimated.

The probabilistic form of the solution of Eq. (0.36) can be expressed in terms of Lagrangian paths and a multiplicative matrix integrals. The Lagrangian path for (0.36) is given by the equation

$$d\xi_i(s) = (2\eta)^{1/2} dw_i(s) - v_i(t - s, x_s)ds \tag{0.37}$$

where $w_i(s)$ is a standard 3-dimensional Wiener process, $0 \le s \le t$. The solution has the form of a multiplicative integral

$$H_i(t, x) = E_\xi[\prod_{s=0}^t (\delta_{ij} + V_{ij}(t - s, \xi_j(s))ds]H_{0i}(\xi(t)) \tag{0.38}$$

where E_ξ is the expectation with respect to Wiener paths starting at $x \in R^3$. The statistical moments of the magnetic field are defined as follows:

$$M_1(t, x) = \langle H_i(t, x) \rangle,$$
$$M_2(t, x_1, x_2) = \langle H_i(t, x_1)H_j(t, x_2) \rangle,$$

etc. When the velocity field is Gaussian and has a very short time correlations we may consider it as a "white noise" in time, i.e. defined by the correlation function

$$\langle v_i(t_1, x_1)v_j(t_2, x_2) \rangle = \delta(t_1 - t_2)U_{ij}(x_1 - x_2) \tag{0.39}$$

where $U_{ij}(x_1 - x_2)$ is assumed to be a smooth function with rapidly decreasing spacial correlation. For small enough η the moments of the field, M_k, grow exponentially in time. However this growth is non-uniform with respect to k: the fourth moment grows faster than the squared second moment, etc. Physically, this means that the growing magnetic field is distributed intermittently, i.e. appears in the form of strong concentrations superposed on a relatively weak background.

The growth of the field eventually breaks the kinematic approximation. In other words, the kinematic dynamo describes only the initial stage of the magnetic field evolution. The back - reaction of the field on the velocity may, in principle, be described by the Navier-Stokes equations with a quadratic magnetic force. However the non-linear character of these equations and the randomness makes the solution of the problem practically impossible. At the present level of knowledge only some models of this back action effect are considered such as to prescribe some form of the non-linear dependence of the velocity on the magnetic field. The model (0.34) (for $N = 3$) and non-linear model (0.35) contain all the essentials from the point of view of the physical features of the dynamo process: a matrix potential as a generator of an increasing magnetics field, intermittency (i.e. progressive growth of the statistical moments), a non-linear mechanism of suppressing a very large magnetic field and so forth. We'll discuss this model, which was introduced in the article [66] in Ch. III. Our consideration will be based on the recent paper [58]..

Chapter I. Homogenization

Lecture 2. General principles of the homogenization. Examples.

In this section we'll discuss the asymptotical properties of the Lagrangian trajectories for the models of diffusion in RM or random walks in random environment, i.e. Stokes drift, turbulent diffusivity etc. As it was mentioned above (lecture 1), in probabilistic terms homogenization means the proof of the CLT for Lagrangian trajectories ($\mu - a.s$). All the known results of homogenization theory for classical (second order) differential equation in RM are based on the following idea. Under some conditions (but not always!), a Lagrangian trajectory $x_t(w, \omega)$ is a process with stationary increments on the probabilistic space $\Omega_m \times \Omega_w$ (see model I in the introduction for details). The increments $x_{n+1} - x_n = \Delta_n$ form a homogeneous sequence. Now we can use the CLT for weakly-dependent random vectors, i.e. estimate the mixing coefficient for $\{\Delta_n\}$ in some sense, calculate variance of $\sum \Delta_n$ and so forth. But a more powerful method is based on the transformation of the Lagrangian trajectory into a martingale with square-integrable ergodic increments followed by a utilization of the classical Billingsley CLT for martingale-differences [11] (or some of its generalizations). This approach requires the construction of special solutions (harmonic coordinates and invariant densities) for the random elliptical operator in the right hand part of the parabolic (or wave) equation. I think that the first mathematical paper to use this idea was S. Kozlov's publication [41]. For additional details see [42], [73].

Let's consider two examples which will illustrate some features of the martingale methods in homogenization theory.

Example 1. Homogenization of heat transport in a stationary periodic medium. Let's consider the heat equation

$$\frac{\partial u}{\partial t} = D\Delta u + (\vec{a}(x), \nabla)u = Lu, \quad (t, x) \in (0, \infty) \times R^d \tag{1.1}$$

$$u(0, x) = u_0(x), \qquad\qquad D = \frac{\kappa^2}{2}$$

in the incompressible (div $\vec{a}(x) = 0$) periodic medium. This means that the drift $\vec{a}(x)$ is periodic with respect to some lattice of periods: $\vec{a}(x + T) \equiv \vec{a}(x)$, $T \in \Gamma$. For the sake of simplicity we'll discuss the simplest case $\Gamma = \mathbf{Z}^d$ (d- dimensional cubic lattice with unit scale) and in addition take $\int \vec{a}(x)dx = 0$ for the cell of periodicity. The field $\vec{a}(x)$ is not random, but we can include it into the ensemble (Ω_m, μ) according to the formula

$$\vec{a}(x, \omega) = \vec{a}(x + \omega) \tag{1.2}$$

The vector $\omega = \omega_m$ will be uniformly distributed inside one cell $S^d = R^d/\mathbf{Z}^d$. The cell S^d (our RM probabilistic space) is a torus with Lebesgue measure μ as a probabilistic measure on S^d.

The operator L in the right part of (1.1) generates the Lagrangian trajectory $x_t(w, \omega)$ as a solution of the SDE

$$dx_t = \kappa dw_t + \vec{a}(x_t + \omega)dt \qquad (1.3)$$

or

$$x_t = \omega + \kappa w_t + \int_0^t a(x_s + \omega)ds \qquad (1.3')$$

Let $x_t^* = x_t \pmod 1$ be a projection of Lagrangian trajectory on the fixed cell S^d (containing the initial point ω). Since the generator L of x_t commutes with $\mathbf{Z}^d$, the process x_t^* is a diffusion process on the compact manifold $\Omega_m = S^d$.

Because of the condition div $\vec{a} = 0$, the adjoint operator is given by the formula $L^* = D\triangle - (\vec{a}(x + \omega), \nabla)$. As a result an invariant measure π for x_t^* exists and its corresponding density $\pi(x) \equiv 1$. The process x_t^* has a strictly positive transition probability density
$\overset{*}{p}(t, x, y), x, y \in S^d$, i.e. for x_t^* we have the best possible mixing conditions (for example, Döblin's condition).

Now we can "lift" x_t^* to R^d, but it's better to use a different strategy. Let $y(x)$ be a system of almost linear harmonic functions. This means that $y = (y(x), \cdots, y_d(x))$, $Ly_k(x) = 0$ and $y_k(x) = x_k + o(|x|)$, $x \to \infty$. We will construct such functions in a very special form: $y_k(x) = x_k + h_k(x)$, where h_k is a periodic function. For h_k we have the equation

$$Lh_k = -Lx_k = -a_k(x)$$

$$D\triangle h_k + (\vec{a}, \nabla)h_k = -a_k(x) \qquad (1.4)$$

This equation is valid not only in R^d, but on S^d too. The operator L (or L^*) has a simple eigenvalue $\lambda_0 = 0$ (because of the Döblin condition. Let's note that the total spectrum of L in $L_2(S^d)$ is not real, because $L \neq L^*$!). According to Fredholm, the alternative equation (1.4) has an unique solution if

$$\int_{S^d} \pi(x)(-a_k(x))dx = -\int_{S^d} a_k(x)dx = 0$$

Due to our condition, this relation holds, i.e. we have proved the existence of harmonic coordinates. According to Ito's formula,

$$y(x_t) = y(x_0) + k \int_0^t \nabla y(x_s^*) dw_s + \int_0^t Ly(x_s) ds =$$

$$= \kappa \int_0^t \nabla y(x_s^*) dw_s + y(x_0) =$$

$$= \kappa w_t + \int_0^t \nabla h(x_s^*) dw_s + y(x_0)$$

i.e. $y(x_t) - y(x_0)$ is a martingale over process x_s^*. Now, it's easy not only to prove CLT for

$$\xi_t = \frac{y(x_t)}{\sqrt{t}}, \quad t \to \infty$$

but to find the effective tensor of diffusivity (turbulent diffusion):

$$B_0 = \lim_{t \to \infty} \frac{\operatorname{var}(y(x_t) - y(x_0)) \otimes (y(x_t) - y(x_0))}{t} = \{b_{ij}\}, \tag{1.5}$$

$$b_{ij} = D\{\delta_{ij}\} + \{\int_{S^d} (\nabla h_i(x) \cdot \nabla h_j(x)) dx\}$$

Let's observe that $B_0 > D \cdot I$ (it is a consequence of the incompressibility) and the Stokes (mean) drift is equal to zero. But $y(x_t) = x_t + 0(1)$, i.e. the same asymptotic result holds not only for $y(x_t)$, but for the initial Lagrangian trajectory x_t. We have proved the following.

Theorem 1.1. Let's consider the parabolic problem

$$\frac{\partial T^\epsilon}{\partial t} = D \triangle T^\epsilon + \frac{1}{\epsilon}(\vec{a}(\frac{x+\omega}{\epsilon}), \nabla) T^\epsilon \tag{1.6}$$

$$T^\epsilon(0, x) = T_0(x)$$

where the drift $\vec{a}(x)$ is periodic with respect to the lattice $\mathbf{Z}^d$, smooth-enough, div $\vec{a} = 0$ and $\int_{S^d} a(x) dx = 0$. Then, for $\epsilon \to 0$ and $t \in [0, t_0]$

$$T^\epsilon(t, x) \longrightarrow \bar{T}(t, x).$$

The limiting function $\bar{T}$ is the solution of the problem

$$\frac{\partial \bar{T}}{\partial t} = (B_0 \nabla, \nabla)\bar{T}, \quad \bar{T}(0, x) = T_0(x) \tag{1.6'}$$

The effective diffusivity tensor B_0 has a representation (1.5) which depends on the solutions $h_k(x)$ of the equations (1.4).

If the integral of the drift $\vec{a}(x)$ over the cell of periodicity is not equal to 0, the Lagrangian trajectory has a non-trivial Stokes drift $\vec{a}_0$. To find this drift, we can construct a martingale transformation of x_t, using a function $y(t, x)$ of two variables. The process $y(t, x_t)$ will be a martingale if

$$\frac{\partial y}{\partial t} + Ly = 0. \tag{1.7}$$

If $y = (y_1, (t, x), \cdots y_d(t, x))$ has form

$$y_k(t, x) = x_k + h_k(x) - \alpha_k t,$$

then for $h_k(x)$ the equation

$$Lh_k(x) = -(a_k(x) - \alpha_k) \tag{1.8}$$

and the condition of solvability gives

$$\alpha_k = \int_{S^d} a_k(x) dx. \tag{1.8'}$$

As is easy to see, this means that

$$\lim_{t \to \infty} \frac{x_t}{t} = \vec{a}_0 = \int_{S^d} \vec{a}(x) dx.$$

For turbulent diffusivity B_0 we have the same formula (1.5), but $h_k(x)$ is given by equations(1.8), (1.8').

I suppose that the method discussed above (at least its probabilistic interpretation) first appeared in M. Friedlin's paper [25].

This method is very general. It works for an arbitrary parabolic equation with periodic coefficients

$$\frac{\partial T^\varepsilon}{\partial t} = \frac{\partial}{\partial x_j} a_{ij}(\frac{x}{\varepsilon}) \frac{\partial}{\partial x_j} + \frac{1}{\varepsilon} b_i(\frac{x}{\varepsilon}) \frac{\partial}{\partial x_i} \tag{1.9}$$

of course, the problem of how to construct (or estimate) the invariant measure for x_t^* on R^d/Γ or harmonic almost linear coordinates $y_k(x)$ is not solved. In our special case ($a_{ij} = D\delta_{ij}$, div $\vec{a} = 0$), if $D \to 0$ it's possible to get some asymptotical results, using general ideology of the small perturbations of dynamical systems (Wentzell and Freidlin, [26]).

Example 2. CLT for an almost periodic additive functional of symmetrical random walk.

We had seen already, that the problem of homogenization can be reduced in some cases to the analysis of additive functionals of ergodic Markov chains. In Example 1, this chain(or more precisely, diffusion process) was excellent: positive transition densities, exponential decreasing of correlations in the strongest form, etc. In the general case, the Markov process related with a homogenization problems is not so good. The following general approach to the CLT for Markov chains is well-known to the specialist. We'll discuss now only the discrete time case (but general phase space).

Let $(X, \mathcal{F})$ be a measurable space and $p(x, dy)$ be a transition probability on $(X, \mathcal{F})$. It generates two stochastic operators, one acting on bounded $\mathcal{F}$-measurable functions and the other on measures μ:

$$Pf(x) = \int_X p(x, dy)f(y)$$

$$\mu P(dy) = \int_X \mu(dx)P(x, dy) \tag{1.10}$$

Suppose that there exists a finite (probabilistic) invariant measure π : $\pi P = \pi$ and this measure is ergodic in the standard sense: for every $\Gamma \in \mathcal{F}$, $\pi(\Gamma) > 0$ and π-a.e initial point x

$$\frac{1}{n} \sum_{k=1}^{n} P(k, x, \Gamma) \to \pi(\Gamma) \text{ (in measure } \pi) \tag{1.11}$$

It's possible to consider ergodicity in the L_2 sense: if $f \in L_2(X, \pi)$, then

$$l.i.m \frac{1}{n} \sum_{k=1}^{n} P^k f(x) = \int_X f(x)\pi(dx). \tag{1.11'}$$

In all interesting cases these conditions are equivalent; we'll suppose that it is true in our case.

If $\{x_k(\omega), k = 0, 1, \cdots\}$ is a homogeneous Markov chain with transition probabilities $p(x, dy)$ and initial distribution π, then this chain will be ergodic and for bounded $f(x) \in B(X)$,

$$\frac{1}{n} \sum_{k=0}^{n} f(x_k) \xrightarrow{n \to \infty} (f, \pi) = \int_X f(x)\pi(dx) \ (\pi - a.s)$$

(law of large numbers for the additive functional $S_n = \sum_{k=0}^{n} f(x_k)$). If $(f, \pi) = 0$, we can expect (under some additional conditions, which guarantee square-integrability of S_n and "good" mixing), that

$$\frac{S_n}{\sqrt{n}} \xrightarrow{\text{dist}} N(0, \sigma^2).$$

As earlier, we'll try to reduce S_n to a martingale with stationary increments (with respect to the filtration $\Phi_k = \sigma(x_0, \cdots, x_k)$). To do this, consider the "homological equation"

$$(g - Pg) = f(x), \quad (f, \pi) = 0. \tag{1.12}$$

Suppose that for a given (maybe vector) function $f \in L^2(X, \pi)$ equation (1.12) has a solution $g(x) \in L^2$. Solution will be unique under the additional condition $(g, \pi) = 0$ (why ?).

Now we can write down our functional in the form

$$S_n = f(x_0) + \cdots + f(x_n) = [g(x_0) - Pg(x_n)] +$$

$$+ (g(x_1) - Pg(x_0)) + \cdots + (g(x_n) - Pg(x_{n-1}))$$

$$= R_n + \triangle_1 + \cdots + \triangle_n = R_n + \tilde{S}_n$$

It's easy to see (because of (1.12)) that $E(\triangle_k \mid \Phi_{k-1}) = 0(a.e)$, i.e. $\tilde{S}_n$ is a martingale with respect to $\{\Phi_n\}$. Its increments $\triangle_k$ are stationary ergodic and square integrable. Now we can calculate the variance:

$$B_0 = \lim_{t \to \infty} \text{Var} \frac{S_n}{\sqrt{n}} = \lim_{t \to \infty} \text{Var} \frac{\tilde{S}_n}{\sqrt{n}} =$$

$$= \text{Var} \ (g(x_1) - Pg(x_0)) \otimes (g(x_1) - Pg(x_0)) \qquad (1.13)$$

$$= \int_x g(x) \otimes g(x) \pi(dx) - \int_x (Pg \otimes Pg)(x) \pi(dx)$$

CLT follows from the Billingsley theorem [11] mentioned above.

Now we'll return to Example 2. Let $S^1 = [0, 1)$ with group operation $(x + y)$ mod 1 be an one-dimensional torus. The operator P (i.e. transition probabilities) is given by the formula

$$Pf(x) = \frac{f(x + \alpha) + f(x - \alpha)}{2}. \qquad (1.14)$$

It is obvious that $\pi(dx) = dx$ (Lebesgue measure) is invariant, the operator P is symmetrical in $L_2(S^1, dx)$ and the basic functions of L_2, $e_k(x) = \exp\{2\pi i k x\}$, are the eigenfunctions of P:

$$Pe_k(x) = \cos 2\pi \alpha k \cdot e_k(x). \qquad (1.15)$$

As a result, as was clear a priori, $\|P\|_{L_2} \le 1$.

Suppose that α is an irrational number, i.e. $|\cos 2\pi \alpha k| < 1$ if $k \ne 0$.

Let's check the ergodicity conditions (1.11), (1.11'). First of all

$$\frac{1}{n} \sum_{k=0}^{n} P^k e_{i_0}(x) = \frac{1}{n} \sum_{k=0}^{n} \cos^k 2\pi \alpha i_0 \cdot e_{i_0}(x) = \frac{1}{n} \frac{1 - \cos^{n+1} 2\pi \alpha i_0}{1 - \cos 2\pi \alpha i_0} e_{i_0}(x) \xrightarrow[n \to \infty]{} 0$$

for every fixed $i_0 \ne 0$,

$$\frac{1}{n} \sum_{k=0}^{n} P^k e_0(x) = \frac{1}{n} \sum_{k=0}^{n} P^k 1 = 1.$$

Both these facts together with the contractivity of P give immediately (1.11'). Because $I_\Gamma(x) \in L_2$ we have

$$\pi_n(x, \Gamma) = \frac{1}{n} \sum_{k=0}^{n} P(k, x, \Gamma) \longrightarrow \pi(\Gamma) = \int_\Gamma dx$$

in the L_2 sense and, consequently, in measure. It gives us (1.11).

Let $f(x) \in L^2(S^1, dx)$ be a smooth function and $\int_0^1 f(x)dx = 0$. Suppose additionally that the irrational number α is not Liouvillian, i.e. for some $c, \delta > 0$ and all integers p and q

$$\left| \alpha - \frac{p}{q} \right| \geq \frac{c}{q^{2+\delta}}. \tag{1.16}$$

The homological equation

$$g - Pg = f$$

(under the condition $\int_0^1 g dx = 0$) can be solved in Fourier representation: if $f(x) = \sum_{n \neq 0} a_n e_n(x) = \sum a_n \exp\{2\pi i n x\}$ then

$$\begin{aligned}
g(x) &= \sum_{n \neq 0} \frac{a_n}{1 - \cos 2\pi \alpha n} \exp\{2\pi i n x\} \\
&= \sum_{n \neq 0} \frac{a_n}{2\sin^2 \pi \alpha n} \exp\{2\pi i n x\}
\end{aligned} \tag{1.17}$$

From (1.16) if follows that

$$|\sin \pi \alpha n| \geq \frac{c_1}{n^{1+\delta}}$$

If $f(x) \in C^{2+\delta_1}$ (i.e. $a_n = o(\frac{1}{n^{2+\delta_2}})$) then $g(x) \in C^{\delta_2} \subset L_2$ (for some $\delta_2 > 0$) so we have proved the following.

Theorem 1.2. If $\{x_n\}$ is a homogeneous Markov chain on $S^1[0,1]$ with transition operator $Pf(x) = \frac{f(x+\alpha) + f(x-\alpha)}{2}$ (addition mod 1!), α is an irrational number with Diophantine condition (0.16), $f(x) \in C^{2+\delta}(S^1)$, $\int_0^1 f(x)dx = 0$, then

$$\lim_{n \to \infty} \frac{\sum_{k=0}^{n} f(x_k)}{n} = 0 \quad (a.s) \tag{1.18}$$

$$\frac{\sum_{k=0}^{n} f(x_k)}{\sqrt{n}} \xrightarrow{\text{dist}} N(0, \sigma^2) \tag{1.18'}$$

and

$$\sigma^2 = \sum_{k \neq 0} \frac{a_n^2(1 - \cos^2 2\pi n\alpha)}{(1 - \cos 2\pi n\alpha)^2} = \sum_{n \neq 0} a_n^2 ctg^2 \pi n\alpha. \tag{1.18''}$$

(The initial distribution may be arbitrary!). Of course, this method works not only for S^1, but for S^d and, after some modifications, for all compact Lie groups. A very interesting and, I think, open problem is to find the limiting distributions in the case when the function $f(x)$ is not smooth. In this situation the Fourier coefficients a_n are not small enough to compensate for the denominators $(1 - \cos 2\pi n\alpha)$. Standard normalization $1/\sqrt{n}$ does not work in this case.

Especially important is to obtain a nonstandard (i.e. not infinite-divisible) limiting distributions for this model (or related ones).

If α is a Liouvillian number, I can prove for some (smooth!) $f(x)$ the existence of such a non-trivial distribution.

Now we'll return to homogenization. I'll formulate the general homogenization theorems for random walk in a random environment. These theorems give an algorithm for the calculation of the Stokes drift and turbulent diffusivity. Let $p(x, y, \omega)$ be the transition probability of the homogeneous Markov chain on the lattice $\mathbf{Z}^d$, $\quad p(x, y, \omega) = 0, \quad |x - y| \neq 1$. As earlier, the parameter $\omega = \omega_m$ (the state of the RM) is an element of $(\Omega_m, \mathcal{F}_m, \mu)$. RM is statistically homogeneous and ergodic. This means that we can introduce a dynamical system $T_x, \quad x \in \mathbf{Z}^d$ on $(\Omega_m, \mathcal{F})$ preserving measure μ and ergodic. The functions $p(x, y, \omega)$ themselves are the realization of random variables on the dynamical system. Namely, let

$$p_z(\omega) = p(0, z, \omega), \quad |z| = 1, \quad \omega \in \Omega_m. \tag{1.19}$$

Then

$$p(x, x + z, \omega) = p_z(T_x \omega) \tag{1.20}$$

The random variables $p_z(\omega)$ are the probabilities of a transition from the origin $0 \in \mathbf{Z}^d$ to one of the neighbor points $z, \quad |z| = 1$. Of course, $p_z(\omega) \geq 0, \quad \sum_{|z|=1} p_z = 1$.

Fixing a state $\omega \in \Omega$ of the RM, we can introduce the Markov chain $X_n(\omega)$ with the transition probabilities $p(x, x + z) = p_z(T_x \omega), \quad X_0 = 0$. Now we'll construct (as in Example 1) some "projection" of $X_n(\omega)$ on $(\Omega_m, \mathcal{F}, \mu)$. Instead of observing the particle from the origin $X_0 = 0$ it is convenient to follow it along the lattice $\mathbf{Z}^d$ and change the state of the environment at each step. More formally: let

$$\omega_0 = \omega, \quad \omega_n = T_{X_n(\omega)} \omega_0$$

and (more general)

$$\omega_n = T_{X_{n-k}(\omega_k)} \omega_k, \quad 0 \leq k \leq n.$$

Of course, ω_n is a Markov chain on the measurable space $(\Omega_m, \mathcal{F})$ with very singular (discrete) transition probabilities. The process ω_n describes the state of the environment from the point of view of an observer moving with a Markov particle. To reconstruct the original trajectory $X_n(\omega)$, our observer must now register the jumps of the particle. We come to a Markov chain $x_n = (z_n, \omega_n)$ on $\mathbf{Z}^d \times \Omega_m$, where $z_n(\omega_0) = (X_{n+1} - X_n)(\omega_0) = z_1(\omega_n)$.

The transition operators for the chains ω_n, $x_n = (z_n, \omega_n)$ are given by the formulas

$$(Pf)(w_k) = E(f(\omega_{k+1})|\omega_k) = \sum_{|z|=1} p_z(\omega_k) f(T_z \omega_k) \tag{1.21}$$

$$\tilde{P}g(x_k) = E(g(x_{k+1})|x_k)) = E(g(z_{k+1}, \omega_{k+1})|z_k, \omega_k)$$

$$= \sum_{|z'|=1} g(z', T_{z_k}\omega_k) p_{z'}(T_{z_k}\omega_k). \tag{1.22}$$

The probability space for x_n is a skew product of Ω_m by finite set $\{z : |z| = 1\}$. Since this space "is not very large", i.e. has finite measure and so forth, we can expect ergodicity of x_n, at least under some conditions (compare with example 2). The difficulties are connected with "singularity" of the chain ω_n, the absence of Döblin type conditions etc.

Hypotheses about transition probabilities $p(x, x + z, \omega)$:

α) *Uniform ellipticity.* For some $\varepsilon > 0$, all $\omega \in \omega_m$ and z ($|z| = 1$)

$$p_z(\omega) \geq \varepsilon, \text{ i.e. } p(x, x + z, \omega) \geq \varepsilon \tag{1.23}$$

In this case, the random walk $X_n(\omega)$ has only one closed class, uniformly (in n) connected and so on. Condition α) is a technical one.

The following condition β) is principal.

β) **Hypothesis I.** *Absolute continuity of the invariant measure.* The Markov chain ω_n has an invariant measure of the form

$$\Pi(d\omega) = \pi(\omega)\mu(d\omega).$$

Of course, $\pi(\omega) \geq 0$ and $\int_{\Omega_m} \pi(\omega)\mu(d\omega) = \langle \pi \rangle = 1$

It means that

$$P^*\pi(\omega) = \pi(\omega) \text{ or } \pi(\omega) = \sum_{|z|=1} p_{-z}(T_z\omega)\pi(T_z\omega) \tag{1.24}$$

In the language of the initial transition probability $p(x, x + z, \omega)$ it mean that the function

$$\Pi(x, \omega) = \pi(T_x\omega)$$

is a homogeneous ergodic nonnegative solution of the equation

$$\Pi(x) = \sum_{|z|=1} \Pi(x - z)p(x - z, x, \omega) \tag{1.25}$$

It is not difficult to prove that under conditions $\alpha), \beta)$ the chain ω_n will be an ergodic and stationary one (if $\pi(\omega)\mu(dw)$ is an initial distribution). The solution of (1.24) (or 1.25) is unique in $L_1(\Omega, \mu)$ and $\mu\{\omega : \pi(\omega) > 0\} = 1$; for details see [42], [43].

The same properties hold for the chain $x_n = (z_n, \omega_n)$.

Lemma 1. Under conditions $\alpha), \beta)$, the chain x_n has an invariant measure with density

$$\pi(z, \omega) = p_z(\omega)\pi(\omega)$$

on every sheet (z, Ω_m) with respect to $\mu(dw)$ on every sheet).

If $\pi(z, \omega)\mu(dw)$ is the initial distribution for x_n, than $\{x_n, n = 0, 1, \cdots\}$ is an ergodic Markov process.

Theorem 1.3 If conditions $\alpha)\beta)$ hold then $\mu - a.s$ there exists

$$\lim_{n \to \infty} \frac{x_n(\omega)}{n} = \lim_{n \to \infty} \frac{\sum\limits_{k=0}^{n} z_n(\omega)}{n} = \sum_{|z|=1} \langle p_z(\omega)\pi(\omega)\rangle z = \vec{a}_0 \qquad (1.26)$$

for $P_X(\omega)$-almost all paths $X \in \mathbf{Z}^d$.

Of course, $\vec{a}_0$ is a Stokes drift.

Training example. Suppose that transition probabilities are symmetric: $p(x, y, \omega) = p(y, x, \omega)$ and $\sum_z p(x, x+z, \omega)|z| < \infty$ uniformly in x and ω (it means that we consider a general random walk with arbitrary jumps and some moment restrictions). Then

$$\pi(\omega) \equiv 1 \qquad (1.27)$$

and

$$\vec{a}_0 = \sum_{z \in \mathbf{Z}^d} \langle p_z(\omega)z \rangle = (\sum_{z \in \mathbf{Z}^d} p(0, z, \omega)z) = 0 \qquad (1.27')$$

The same answer ($\pi \equiv 1$) is valid in more general situation of double-stochastic probabilities:

$$\sum_{y \in \mathbf{Z}^d} p(x, y, \omega) = \sum_{x \in \mathbf{Z}^d} p(x, y, \omega) = 1$$

In the continuous case, the condition (1.27") is equivalent to incompressibility of the drift $\vec{a}$ (for the operator $L = D\Delta + (\vec{a}, \nabla)$). For the CLT we have to postulate the existence of almost linear harmonic coordinates. Probably these exist some relations between hypothesis I on the invariant density $\pi(\omega)$ and hypothesis II on harmonic coordinates, but I only have some "physical" ideas about these relations.

Hypothesis II. There exists a (vector) function $\vec{X}(\omega, x)$, $\omega \in \Omega_m$, $x \in \mathbf{Z}^d$ such that

$$\alpha) \tilde{P} \vec{X} = \vec{X} \quad \text{(harmonicity)}$$

$$\beta) \vec{X}(\omega, \vec{x}) = \vec{x} + \vec{h}(\omega, x)$$

and the field $\vec{h}(\omega, x)$ has homogeneous stationary ergodic increments on $\mathbf{Z}^d$,

$$\langle \vec{h} \rangle = 0, \quad \langle |\vec{h}|^2 \rangle < \infty$$

It is an assumption of almost linearity. Using the Billingsley theorem [11] or its non-homogeneous version (Brown's theorem, see book by R. Durrett [22], which contains both these theorems and further generalizations) we obtain now the following.

Theorem 1.4. Under hypotheses I, II, the Stokes drift $a_0 = 0$ and

$$\frac{x_n}{\sqrt{n}} \xrightarrow{\text{dist}} N(0, B_0)$$

For the turbulent diffusivity B_0, we have formula

$$B_0 = \{b_{ij}\} = \frac{1}{2} \{ (\pi(\omega) \sum_{|z|=1} p_z(\omega) X_i(\omega, z) X_j(\omega, z)) \} \tag{1.28}$$

For details of the calculations see [42].

This approach works in more a general situations. Of course (as mentioned above) we can consider random walks with arbitrary jumps. More interesting is an application of this method to Markov chains on $\mathbf{Z}^d$ with continuous time. The corresponding generator (in the simplest case of transitions to the neighboring points) has the form

$$Lf = (L(\omega)f)(x) = \sum_{|z|=1} \lambda(x, x + z)(f(x + z) - f(x))$$

If $0 < \varepsilon \leq \lambda(x, x + z, \omega) \leq \frac{1}{\varepsilon} < \infty$ and the field $\Lambda(x, \omega) = \{\lambda(x, x + z, \omega), \ |z| = 1\}$ is an ergodic and homogeneous we can formulate the obvious analogues of the conditions I, II, Theorems 1.3, 1.4 etc. The existence theorems for invariant density π and harmonic coordinates $\vec{X}$ can be found in [42]. In the one-dimensional case it's possible to give a complete analysis. It will be the subject of the following lecture.

Lecture 3. Example of homogenization

The first example will be simple. I give it only for the goal of training.

Example 1. Let $r(x, \omega)$, $x \in \mathbf{Z}^d$, $\omega \in \Omega_m$ be a homogeneous ergodic field on the lattice $\mathbf{Z}^d$. Suppose that

$$\mu\{\omega : \varepsilon \leq r(x, \omega) < \frac{1}{2d}\} = 1$$

for some fixed $\varepsilon > 0$. Consider the transition probabilities $p(x, x + z, \omega)$:

$$p(x, x + z, \omega) = 0, \quad |z| > 1$$

$$p(x, x + z, \omega) = r(x, \omega), \quad |z| = 1 \tag{1.29}$$

$$p(x, x, \omega) = 1 - 2dr(x, \omega), \quad z = 0$$

The transition operator can be written down in the form

$$Pf(x) = \sum_{|z|=1} f(x + z)r(x, \omega) + (1 - 2dr(x))f(x)$$

$$= f(x) + r(x)\Delta f(x). \tag{1.29'}$$

where, of course

$$\Delta f(x) = \sum_{|x'-x|=1} (f(x') - f(x))$$

is a lattice Laplacian. The adjoint operator has form

$$P^*g(x) = g(x) + \Delta(r(x)g(x)) \tag{1.29''}$$

(because $\Delta = \Delta^*$).

The equation for the invariant density $\pi(\omega)$ or, more precisely, for the function $\Pi(x, \omega) = \pi(T_x\omega)$ has form

$$P^*\Pi = \Pi, \quad \text{i.e.} \quad \Delta(r(x)\Pi) = 0 \tag{1.30}$$

and an obvious solution is

$$\Pi(x) = \frac{c}{r(x)}. \tag{1.30'}$$

Normalization $\langle \Pi(\cdot) \rangle$ gives

$$\Pi(x) = \frac{1}{r(x)} \left\langle \frac{1}{r(x)} \right\rangle^{-1} \tag{1.30''}$$

From (1.29') it is clear that $\vec{X}(x) \equiv \vec{x}$ are harmonic coordinates. From the general formula (1.28) it follows that

$$B_0 = \frac{1}{2} \left\langle \frac{1}{r(x)} \right\rangle^{-1}$$

i.e. the limiting diffusion equation has form

$$\frac{\partial \bar{T}}{\partial t} = \frac{1}{2} \left\langle \frac{1}{r(o,\omega)} \right\rangle^{-1} \Delta \bar{T} \tag{1.31}$$

For a similar result (and its version for random walks with continuous time and generator $\lambda(x.\omega)\Delta$) see in [42].

This model has some physical interest and is related with propagation of light through a medium with homogeneous and random optical density. The following example will be central to this lecture.

Example 2. Let $p(x,\omega), x \in \mathbf{Z}^1, \omega \in \Omega_m$ be a homogeneous ergodic field on the lattice $\mathbf{Z}^1$ (*one-dimensional case*) and

$$\mu\{\omega: \quad \varepsilon \le p(x,\omega) \le 1 - \varepsilon\} = 1$$

(ellipticity condition). Consider the transition probabilities

$$p(x, x+1, \omega) = p(x, \omega)$$
$$p(x, x-1, \omega) = q(x, \omega) = 1 - p(x, \omega). \tag{1.32}$$

Of course, $\varepsilon \le q(x) \le 1 - \varepsilon$ and

$$\frac{\varepsilon}{1-\varepsilon} \le \frac{p(x)}{q(x)} \le \frac{1-\varepsilon}{\varepsilon}, \quad \left\langle \ln \frac{p(x)}{q(x)} \right\rangle < c(\varepsilon). \tag{1.33}$$

Suppose in addition that the homogeneous random process $\xi(x) = \ln\frac{p(x)}{q(x)}$ has "good" mixing properties. It's enough that

$$\alpha(R) = o(R^{-1-\delta}), \quad \delta > 0, \quad R \to \infty, \tag{1.34}$$

where

$$\begin{array}{ll} \alpha(R) = & \sup |\mu(A_1 A_2) - \mu(A_1)\mu(A_2)| \\ & A_1 \in \mathcal{F}_{\le 0} \\ & A_2 \in \mathcal{F}_{\ge R} \end{array} \tag{1.34'}$$

is the strong mixing coefficient.

Under condition (1.34), the correlation function

$$R(\tau) = Cov(\xi(0), \xi(\tau)) = \langle (\xi(0) - \alpha_0)(\xi(\tau) - \alpha_0) \rangle$$

$$\alpha_0 = \langle \xi(x) \rangle = \langle \ln p(x) - \ln q(x) \rangle; \tau, x \in \mathbf{Z}^1$$

is integrable and the corresponding spectral measure $\hat{R}(\phi), \phi \in [-\pi, \pi]$ is continuous (for details see the monographs [22] or [35]).

Ja Sinai has discovered [70] a very interesting phenomenon. In a very natural case (maybe the most natural!), when the random variables $p(x)$ (or $\xi(x) = ln\dfrac{p(x)}{q(x)}$) are independent and $\langle \xi(x) \rangle = 0$, the random walk in the random environment (1.32) has no homogenization.

He proved that, first of all, the typical deviations of the random walk $x_t(w, \omega_m)$ have order not $\sqrt{t}$, but only $ln^2 t$.

Second, after the normalization

$$x_t^* = x_t(w, \omega_m)/ln^2 t$$

the process x_t^* has the following localization property. There exists a random sequence $a_t(\omega_m) \in R^1$ such that for every $\delta > 0$ and $t \to \infty$

$$P_{\xi(\omega)}\{|x_t^*(w, \omega) - a_t(\omega)| > \delta\} \to 0$$

for $\mu - a.e$ realizations of ω. It means, of course, that the limiting distribution for x_t^* does not exist.

The process x_t^* has many other "pathological" properties (for additional information see [70]).

Ja Sinai's approach was based on the functional form of the CLT for the sums of $i.i.d.r.v.'s$ $\xi(x)$ and the properties of Wiener excursions. Some background of his approach will be clear in the future.

Sinai's result shows that, at least in the one-dimensional case, homogenization of the random walk in the RM assumes some "inner symmetry" of RM.

We will formulate now several general results, to begin with, at the "physical level". Technical details (mixing conditions, moment conditions etc.) will be formulated below.

α) Suppose that $\alpha_0 = \left\langle ln\dfrac{p_x}{q_x} \right\rangle \neq 0$. Then, in a very general situations, the random walk has non-trivial Stokes drift and CLT holds (after corresponding normalization). We will give some formulas for the mean drift $\alpha_0 \neq 0$ and turbulent diffusivity B_0

β) Suppose that $\alpha_0 = \left\langle ln\dfrac{p_x}{q_x} \right\rangle = 0$ (as in Sinai's case). Then we have one of two situations:

β_1) $\hat{R}(0) > 0$. This condition (see [22], [35]) has two equivalent formulations:

1a) $\sum\limits_{\tau \in Z^1} R(\tau) > 0$

1b) $\quad var\left(\sum\limits_{x=0}^{n}\xi(x)\right) \sim_{n\to\infty} c\cdot n, \quad c > 0$ $\hspace{2cm}$ (1.35)

In this case we have the same behavior of $x_t(w,\omega)$ as in Sinai's model (absence of homogenization and so forth).

$\beta_2)\hat{R}(0) = 0$. This condition is equivalent to the following statements:

1a. $\quad \sum\limits_{\tau\in\mathbf{Z}^1} R(\tau) = 0$

1b. $\quad var\left(\sum\limits_{x=0}^{n}\xi(x)\right) = O(1)$ $\hspace{4cm}$ (1.36)

1c. The process $\xi(x)$ can be represented in the form

$$\xi(x) = \eta(x) - \eta(x-1) \hspace{2cm} (1.37)$$

where $\eta(x)$ is another homogeneous ergodic process.

In the case of β_2 we have homogenization for typical cases (Stokes drift, of course, equal to 0!). The formula for the effective diffusivity depends on the process $\eta(\kappa)$, $\quad \kappa \in \mathbf{Z}^1$ in the representation (1.37).

Calculations. Part I. We first suppose that $\alpha_0 = \langle\frac{p_x}{q_x}\rangle \neq 0$ and, for definiteness, $\alpha_0 < 0$. Let's construct an invariant density, i.e. a homogeneous solution of the equation

$$P^*\Pi = \Pi \hspace{2cm} (1.38)$$

or

$$\Pi(x-1)p(x-1) + \Pi(x+1)q(x+1) = \Pi(x) \hspace{2cm} (1.38')$$

We will be looking for the function $\Pi(x)$ in the form

$$\Pi(x) = \alpha(x)h(x) + \alpha(x)\alpha(x-1)h(x-1) + \cdots$$
$$= \alpha(x)h(x) + \alpha(x)\Pi(x-1) \hspace{2cm} (1.39)$$

Then

$$\Pi(x+1) = \alpha(x+1)h(x+1) + \alpha(x+1)\alpha(x)h(x) + \alpha(x+1)\alpha(x)\Pi(x-1) \hspace{1cm} (1.39')$$

A substitution of (1.39), (1.39') in (1.38') and a comparison of the coefficients of $\Pi(x-1)$ and the "free terms" give us two equations

$$p(x-1) + \alpha(x)\alpha(x+1)q(x+1) = \alpha(x)$$
$$\alpha(x+1)h(x+1)q(x+1) + \alpha(x+1)\alpha(x)h(x)q(x+1) = \alpha(x)h(x) \hspace{1cm} (1.40)$$

The first one of these equations has an obvious solution

$$\alpha(x) = \frac{p(x-1)}{q(x)} \hspace{2cm} (1.41)$$

and for the second we can take

$$h(x) = \frac{1}{p(x-1)} \tag{1.41'}$$

It gives the following (formal) solution for $\Pi(x)$:

$$\Pi(x) = \frac{1}{q(x)} + \frac{p(x-1)}{q(x)q(x-1)} + \frac{p(x-1)p(x-2)}{q(x)q(x-1)q(x-2)} +$$

$$= \frac{1}{q(x)} \left\{ 1 + \sum_{n=1}^{\infty} \Pi_{k=1}^{n} \left(\frac{p(x-k)}{q(x-k)} \right) \right\} = \tag{1.42}$$

$$= \frac{1}{q(x)} \left\{ 1 + \sum_{n=1}^{\infty} \exp \left\{ \sum_{k=1}^{n} \xi(x-k) \right\} .$$

For fixed x, (say, $x = 0$) and $\mu - a.c$

$$\frac{S_n}{n} = \frac{\sum_{k=1}^{n} \xi(-k)}{n} \to \alpha_0 < 0$$

(ergodic theorem) so the last expression in (1.42) makes sense.

Nevertheless, to prove the existence of the invariant density

$$\pi(\omega) = \frac{\Pi(0,\omega)}{\langle \Pi(0,\omega) \rangle} \tag{1.43}$$

we have to check the integrability of $\Pi(0,\omega)$, i.e. to prove that

$$\langle \Pi(0,\omega) \rangle < \infty \tag{1.44}$$

The last inequality is very close to the statement

$$\langle \exp S_n \rangle = \left\langle \exp \left(\sum_{k=1}^{n} \xi(+k) \right) \right\rangle =$$

$$\tag{1.44'}$$

$$= \left\langle \prod_{k=1}^{n} \frac{p_k}{q_k} \right\rangle \leq c \exp(-\gamma n), \quad \gamma > 0$$

It is a Cramer-type condition for the sum of weakly dependent r.v. $\xi(k)$, which guarantees not only CLT but the large-deviation exponential estimation. Because r.v. $\xi(k)$ are uniformly bounded, the inequality (1.44) holds in the case when $\xi(k)$ is a Markov chain with a Döblin condition, or a component of such a Markov chain. Such estimations are typical for Gibbs fields in the high temperature region and so forth.

I don't know conditions in terms of the mixing coefficient which can guarantee (1.44)

We will suppose in the future that the inequality (1.44) holds (Hypothesis I). If $\langle \Pi(0,\omega) \rangle < \infty$, then $\pi(\omega) = \dfrac{\Pi(0,\omega)}{\langle \Pi(0,\omega) \rangle} \in L_1(\Omega_m, \mu)$ and Stokes drift $\vec{a}_0$ exists. According to Theorem 1.3 of the previous lecture,

$$a_0 = \langle (p(0,\omega) - q(0,\omega))\pi(\omega) \rangle$$

and simple calculations using the representation (1.42) give the following elegant formula

$$a_0 = -\frac{1}{\langle \Pi(0,\omega) \rangle} \tag{1.45}$$

If the invariant density exists and $\alpha_0 = \left\langle ln\dfrac{p(0,\omega)}{q(0,\omega)} \right\rangle < 0$, then $a_0 < 0$.

The following problem is open: suppose that $\alpha_0 < 0$, but $\langle \Pi(0,\omega) \rangle = \infty$. This situation is possible even in the case when $p(x)$ (or $\xi(x)$) is a Markov chain with a finite number of states. Is this true that in this case the Stokes drift $a_0 = 0$? What is the limiting distribution of the chain $x_t(w,\omega)$? What is the "perfect" normalization in this case?

Now we'll prove CLT in the first case, $\alpha_0 = \left\langle ln\dfrac{p(x)}{q(x)} \right\rangle \neq 0$. Of course, we can not use directly the general Theorem 1.4 of the previous lecture. As the Stokes drift $a_0 \neq 0$, almost linear coordinates $X(x,\omega))$ for the equation

$$PX(x) = p(x)X(x+1) + q(x)X(x-1) = X(x) \tag{1.46}$$

do not exist. A direct analysis of this equation is based on its representation in the form

$$X(x) - X(x-1) = \frac{p(x)}{q(x)}(X(x+1) - X(x)) \tag{1.46'}$$

shows, that together with solution $X_1(x) \equiv 1$ equation (1.46) has a second linear independent solution $X_2(x)$ which increases exponentially if $x \to +\infty$ and decreases exponentially if $x \to -\infty$.

In this case of not-trivial mean drift, we have to construct a time-dependent martingale $y_t = X(t, x_t)$. For the function $X(t,x)$ we take a special form

$$X(t,x) = x - kt + h(x).$$

The function $X(t,x)$ must be harmonic in space and time. More precisely, the following equation must hold:

$$PX(t+1,x) = X(t,x) \tag{1.47}$$

or

$$p(x)h(x+1) + q(x)h(x-1) + (p(x) - q(x) - k) = h(x) \tag{1.48}$$

Let's take $k = a_0$ (Stokes drift). In this case function $z(x) = p(x) - q(x) - a_0$ will be orthogonal to the invariant density $\pi(\omega)$:

$$\langle \pi(\omega) \cdot z(0, \omega) \rangle = \langle \pi(\omega)(p(0) - q(0) - a_0) \rangle = 0$$

and we can expect the existence of a solution $h(x, \omega)$ which increases not very fast as $x \to \infty$, i.e. $h(x, \omega) = o(x)$, $x \to \infty$, $\mu - a.s.$

Put $\Delta(x) = h(x + 1) - h(x)$. We can rewrite (1.48) in the form

$$\Delta(x - 1) = \frac{z(x)}{q(x)} + \frac{p(x)}{q(x)} \Delta(x)$$

The iteration together with the fact that $\sum\limits_{x=1}^{n} \left(\dfrac{p(x)}{q(x)} \right) = o(\exp(-\delta n))$ (our hypothesis I) give us immediately

$$\Delta(x - 1) = h(x) - h(x - 1) = \frac{z(x)}{q(x)} + \frac{p(x)z(x + 1)}{q(x)q(x + 1)} \cdots$$

or, after transformations,

$$h(x) - h(x-1) = -1 - a_0 \left(\frac{1}{q(x)} + \frac{p(x)}{q(x)q(x + 1)} + \frac{p(x)p(x + 1)}{q(x)q(x + 1)q(x + 2)} + \cdots \right) \quad (1.49)$$

Let's introduce the notation for the ergodic process which is a factor of a_0:

$$Y(x) = \frac{1}{q(x)} + \frac{p(x)}{q(x)q(x + 1)} + \cdots \quad (1.50)$$

If we remember formulas (1.45) and (1.42), we'll obtain

$$h(x) - h(x - 1) = \frac{Y(x, \omega) - \langle Y(x, \omega) \rangle}{\langle \Pi(\omega) \rangle} \quad (1.49')$$

It means, at last, that the increments of the process $h(x, \omega)$ are stationary ergodic with zero mean. For the process

$$\frac{x_t - a_0 t}{\sqrt{t}} \underset{\sim}{\text{dist}} \frac{X(t, x_t)}{\sqrt{t}}$$

we can use Theorem 1.4 (if the second moments exist!). For the coefficient of turbulent diffusion we have expression

$$B_0 = \frac{1}{2} \langle \pi(0) a_0^2 [p(0, \omega) Y^2(1) + q(0, \omega) Y^2(-1)] \quad (1.51)$$

where $Y(x)$ is given by (1.50); $\pi(0), a(0)$ are given by the formulas (1.42), (1.45).

To guarantee the finiteness of B_0 we have to suppose more than hypothesis I (inequality (1.44)). The following result is a central one in our analysis (part I) of the one-dimensional random walks in a random non-symmetrical environment.

Theorem 1.5. Let $p(x,\omega), q(x) = 1 - p(x,\omega)$ be transition probabilities on $\mathbf{Z}^1$, where the process $p(x,\omega)$ is homogeneous and ergodic, $\mu\{\omega : \quad \varepsilon \leq p(x,\omega) \leq 1 - \varepsilon\} = 1$ for some $\varepsilon > 0$.

Suppose, that

1) $\left\langle ln\dfrac{p(x)}{q(x)}\right\rangle = \alpha_0 \neq 0$

2) The strong mixing coefficient for $p(x,\omega)$ is decreasing fast enough (see (1.34))

3) If $\alpha_0 < 0$, then for all $0 < z < 3$, $\left\langle \left(\sum\limits_{x=1}^{n} \dfrac{p(x)}{q(x)}\right)^z \right\rangle < 1$ (Hypothesis I', which is stronger than (1.44), but has the same form).

Then after the normalization

$$\frac{x_{[nt]} - a_0 t}{\sqrt{t}} = x_i^*(w,\omega)$$

the process x_t^* converges on every interval $t \in [0, t_0]$ in distribution to the Wiener process with generator

$$\bar{L} = B_0 \frac{\partial^2}{\partial x^2}, \quad x \in R^1$$

Parameters a_0 (Stokes drift) and B_0 (turbulent diffusivity) are given by the formulas (1.42), (1.45), (1.50), (1.51).

Part II. In this part we'll consider "symmetrical" RM, i.e. the case when

$$a_0 = \left\langle ln\frac{p(x)}{q(x)}\right\rangle = 0$$

The typical example is a reflection symmetrical RM, where $p(x,\omega) \overset{dist}{=} q(x,\omega)$. In different terms,

$$p(x) = \frac{1}{2} + \phi(x,\omega)$$

$$q(x) = \frac{1}{2} - \phi(x,\omega)$$

and r.v. $\phi(x,\omega)$ are symmetrically distributed on $\left[-\frac{1}{2} + \varepsilon, \frac{1}{2} - \varepsilon\right]$ for some $\varepsilon > 0$.

As earlier, we'll suppose that the process $p(x,\omega)$ is homogeneous and has fast decreasing mixing coefficient (see (1.34), (1.34')). Under this condition, we can apply the functional CLT for weakly dependent r.v.'s $\xi(x) = ln\dfrac{p(x)}{q(x)}$, if, of course, var $S_n = \text{var} \left(\sum\limits_{x=0}^{n} \xi(x)\right) \sim \sigma^2 n, \quad n \to \infty$ and $\sigma^2 > 0$. The last statement means ([22], [35]) that $\hat{R}(0) > 0$, where $\hat{R}(\phi)$, $\phi \in [0.2\pi]$ it the spectral density of the process $\xi(x)$, $x \in \mathbf{Z}^1$. It's known that $\sigma^2 = 2\pi\hat{R}(0)$. Functional CLT means that

$$\frac{S_{[nt]}}{\sqrt{n}} \xrightarrow[n\to\infty]{dist} \sigma w_t, \quad \sigma > 0(!) \tag{1.52}$$

on every interval $[-t_0, t_0]$. The Wiener process for negative t we understand as an independent copy of the standard Wiener process, which goes in the inverse direction of time.

We will prove now that if (1.52) holds, the random walk $x_t(w, \omega)$ has exactly the same pathological properties as in Ja Sinai's paper [70]: there is no homogenization, $x_t(w, \omega) = 0(ln^2 t)$ etc. (see discussion at the beginning of the lecture).

This means that *homogenization* of $x_t(w, \omega)$ in the symmetrical case ($\alpha_0 = 0$!) *implies the equality* $\hat{R}(0) = 0$! In different terms, process $\xi(x)$ must be the first difference ("derivative") of some another homogeneous square integrable process: $\xi(x) = \eta(x + 1) - \eta(x)$.

In this part we will follow our short paper [13], where one can find additional information and technical details. Let

$$\tau_N = \{\min t : \quad |x_t(w, \omega)| = N\}$$

and

$$Q_N(\omega) = P_\omega\{x_{\tau_N}(w, \omega) = +N | x_0 = 0\}$$

i.e. $Q_N(\omega)$ is the probability of the first exit from $[-N, N]$ through right boundary. Of course, $Q_N(\omega)$ is a *r.v.*

It is not difficult to prove that homogenization of $x_t(w, \omega)$, i.e. weak convergence of the distributions $P^*(\omega)$ of $x_t^*(w, \omega) = \frac{x_{[Nt]}}{\sqrt{N}}$, $N \to \infty$ to the distribution of some non-degenerated Wiener process $\sqrt{2B_0}W_t$, implies the following fact

$$Q_N(\omega) \xrightarrow[n \to \infty]{\text{dist}} 1/2 \text{ or } Q_N(\omega) \xrightarrow{\mu} 1/2 \tag{1.53}$$

(convergence with respect to the measure μ!).

Now we'll show, that

$$Q_N(\omega) \xrightarrow[n \to \infty]{\text{dist}} \Theta(\omega), \quad \mu\{\Theta = 1\} = \mu\{\Theta = 0\} = 1/2. \tag{1.54}$$

Contradiction between (1.53), (1.54) will show the *absence of homogenization*.

For the random probability

$$Q_N(x, \omega) = P_\omega\{x_{\tau_N} = +N | x_0 = x\}, \quad |x| < N$$

we have the equation

$$Q_N(x + 1)p(x) + Q_N(x - 1)q(x) = Q_N(x)$$

$$Q_N(N) = 1, \quad Q_N(-N) = 0 \tag{1.55}$$

As earlier, this equation can be solved explicitly, using transformation

$$(Q_N(x+1) - Q_N(x))\frac{p(x)}{q(x)} = Q_N(x) - Q_N(x-1)$$

The answer is

$$
Q_N(0) = \frac{\sum\limits_{k=-N}^{-1}\left[\prod\limits_{x=k+1}^{0}\left(\frac{p(x)}{q(x)}\right)\right]}{\sum\limits_{k=-N}^{-1}\left[\prod\limits_{x=k+1}^{0}\left(\frac{p(x)}{q(x)}\right)\right] + \sum\limits_{k=0}^{N}\left[\prod\limits_{x=1}^{K}\left(\frac{q(x)}{p(x)}\right)\right]}
$$

$$
= \frac{\sum\limits_{k=-N}^{-1}\exp\left(\sum\limits_{x=k+1}^{0}\xi(x)\right)}{\sum\limits_{k=-N}^{-1}\exp\left(\sum\limits_{x=k+1}^{0}\xi(x)\right) + \sum\limits_{K=0}^{N}\exp\left(-\sum\limits_{x=1}^{K}\xi(x)\right)}
$$

(1.56)

$$
= \frac{\sum\limits_{k=-N}^{-1}\exp\left(\sqrt{N}W_N\left(\frac{k+1}{N}\right)\right)}{\sum\limits_{k=-N+1}^{-1}\exp\left(\sqrt{N}W_N\left(\frac{k+1}{N}\right)\right) + \sum\limits_{k=0}^{N}\exp\left(\sqrt{N}W_N\left(\frac{k}{N}\right)\right)}
$$

Distributions of $W_N(t)$, $t \in [-1, +1]$ come close to Wiener distribution on $C_{[-1,1]}$ if $N \to \infty$. Let's consider, in the space $C_{[-1,+1]}$ under the additional condition $f(0) = 0$ and for fixed $\delta > 0$, two sets:

$$
A_1^\delta = \{f: \quad \max_{x \in [-1,0]} f(x) \;\geq\; \max_{x \in [0,1]} f(x) + \delta\}
$$

$$
A_2^\delta = \{f: \quad \max_{x \in [-1,0]} f(x) \;\leq\; \max_{x \in [0,1]} f(x) - \delta\}
$$

For standard Wiener measure on [-1, 1] (under condition $f(0) = 0$)

$$
P^W\left(A_1^\delta\right) \xrightarrow[\delta \to 0]{} 1/2, \quad P^W\left(A_2^\delta\right) \xrightarrow[\delta \to 0]{} \frac{1}{2}
$$

But on the set A_1^δ (meaning, $W_M(\cdot) \in A_1^\delta$) expression (1.56) is near to 1 for large N. If $W_N(\cdot) \in A_2^\delta$, the probability (1.56) is close to 0 and we proved (1.54).

Analysis of Sinai's paper [70] shows that he used only the invariance principal for $\{\xi(x)\}$, but not the independence of $\{\xi(x)\}$. This means that in our case all principal statements of [21] hold.

Part III. Suppose, that $\alpha_0 = \left\langle \ln\frac{p(x)}{q(x)}\right\rangle = 0$ and process of $\xi(x) = \ln\frac{p(x)}{q(x)}$ is homological to zero:

$$\xi(x) = \eta(x+1) - \eta(x) \text{ or } \hat{R}(0) = 0.$$

(1.57)

We'll suppose in addition, that (*Hypothesis II*)

$$|\eta(x)| \leq \text{const} \quad \mu - a.s. \tag{1.57'}$$

In this case we have homogenization. Simple calculations give (in terms of (1.57))

$$p(x) = \frac{\exp(\eta(x+1))}{\exp(\eta(x+1)) + \exp\eta(x)}, \quad q(x) = \frac{\exp\eta(x)}{\exp(\eta(x+1)) + \exp(\eta(x))} \tag{1.58}$$

The equation for invariant density $\Pi(x,\omega)$, which has a form

$$\Pi(x-1)p(x-1) + \Pi(x+1)q(x+1) = \Pi(x),$$

has (in view of (1.58)) a simple solution

$$\Pi(x,\omega) = \frac{\exp(\eta(x+1)) + \exp\eta(x)}{2\langle \exp\eta(x)\rangle} \tag{1.59}$$

The transition matrix P has in this case almost linear (time independent) harmonic coordinates $X(x,\omega)$, i.e. a solution of the equation

$$p(x)X(x+1) + q(x)X(x-1) = X(x) \tag{1.60}$$

Trivial computation gives us

$$X(x) = \sum_{k=0}^{x} \frac{\exp(-\eta(k))}{\langle \exp(-\eta(k))\rangle}, \quad x \geq 0$$

$$X(x) = -\sum_{k=-x}^{0} \frac{\exp(-\eta(k))}{\langle \exp(-\eta(k))\rangle}, \quad x < 0$$

Now we can give the very simple expressions for effective diffusivity B_0. Instead of boundness of $\eta(x)$ (condition (1.57')) it's enough to suppose only (as in part 1) $\langle \exp(|\eta(x)|)\rangle < \infty$. Namely:

$$B_0 = \frac{< e^{-\eta(0)} >}{< e^{\eta(0)} >^3}.$$

Lecture 4. Random walks in non-stationary RM.

We had seen in the previous lecture, that (at least in one-dimensional case) homogenization of a random walk in stationary (time-independent) RM imposes strong restrictions on the transition probabilities. The physical explanation of this fact is very simple. Realizations of such RM contain a system of (large enough) random "traps". For the case of Z^1, every such trap is an interval $[a_n, b_n]$, where the local drift goes to the center of interval. A Markov particle will leave this interval in a time which is the exponential of the "size of the trap". We will discuss this possible mechanism of the absence of homogenization in the end of this chapter, even in the multidimensional case.

If the RM is non-stationary (turbulent), we can expect that homogenization is a "generic case". We'll discuss now the simplest non-stationary model (with discrete space and time). Nevertheless, this model contains all the qualitative features of the realistic hydrodynamical models. We will be trying to understand (in the framework of this model) a few physical effects, related with heat propagation in a turbulent flow. These effects are essential for oceanography. (See [55], [75], [56], [57]).

Consider random nearest-neighbor transition probabilities on Z^d, which have form

$$p(t, x; t+1, x+n, \omega) = \begin{cases} \varepsilon, & n \neq n_0(t, x) \\ 1 - (2d-1)\varepsilon, & n = n_0(t, x) \end{cases} \tag{1.62}$$

where field (random current) $n_0(t, x, \omega)$ has independent values for different $(t, x) \in Z^1_+ \times Z^d$ and uniformly distributed (isotropy):

$$P\{n_0 = \pm e_i, \quad i = 1, 2, \cdots d\} = \frac{1}{2d} \tag{1.63}$$

Here, of course, $\{e_i\}$ is a basis of Z^d. Small parameter ε plays the role of the molecular diffusion.

Random transition probabilities $p(t, x; t+1, x+n, \omega_m)$ generate a random walk $X_0, X_1, \cdots, X_t, \cdots$ in non-stationary environment (1.62), (1.63). Suppose that $X_0 = 0$. Let

$$p^{(t)}(0, 0, t, x) = P\{X_t = x\}.$$

Of course $p^{(t)}(\cdots)$ are random variables on (Ω, μ).

Let's introduce moments

$$m_1(t, x) = \langle p^{(t)}(0, 0, t, x) \rangle$$

$$m_2(t, x_1, x_2) = \langle p^{(t)}(0, 0, t, x_1) p^{(t)}(0, 0, t, x_2) \rangle \tag{1.64}$$

and find the equations for these moments. It's obvious that

$$p^{(t+1)}(0,0,t+1,x) = \sum_{|z-x|=1} p^{(t)}(0,0,t,z)p(t,z,t+1,x) \qquad (1.65)$$

and the first and the second factors are independent. It gives us

$$m_1(t+1,x) = \frac{1}{2d} \sum_{x':|x'-x|=1} m_1(t,x') \qquad (1.66)$$

i.e. the equation of the transition probabilities of symmetrical r.w. in $\mathbf{Z}^d$.

Central limit theorem in local form gives us immediately, for $|x| = 0(\sqrt{t})$

$$m_1(t,x) \underset{t\to\infty}{\sim} C \; \frac{\exp\left(-\dfrac{|x|^2}{2t}\right)}{t^{d/2}}, \quad |x| \equiv t \bmod 2$$

i.e. gaussian asymptotics.

Let's find an equation for the second moment. We have

$$p^{(t+1)}(0,0,t+1,x_1)p^{(t+1)}(0,0,t+1,x_2)$$

$$= \sum_{\substack{z_1:|z_1-x_1|=1 \\ z_2:|z_2-x_2|=1}} p^{(t)}(0,0,t,z_1)p^{(t)}(0,0,t,z_2)$$

$$\cdot p(t,z_1;t+1,x_1)p(t,z_2,t+1,x_2) \qquad (1.67)$$

$$= \sum_{z_1=z_2} \cdots + \sum_{z_1\neq z_2} \cdots$$

Let's take the mathematical expectation of both parts in (1.67) . If $z_1 \neq z_2$, then $p(t,z_1,t+1,x_1)$ is independent from $p(t,z_2,t+1,x_2)$ and we will obtain

$$m_2(t+1,x_1,x_2) = \sum_{\substack{|z_1-x_1|=1 \\ |z_2-x_2|=1 \\ z_1\neq z_2}} \frac{1}{(2d)^2} m_2(t,z_1,z_2)$$

$$+ \sum_{\substack{z:|z-x_1|=1 \\ |z-x_2|=1}} m_2(t,z,z) \langle p(t,z,t+1,x_1)p(t,z,t+1,x_2)\rangle$$

In the last factor we have two possibilities:

a) $x_1 = x_2 = x$, then

$$\langle p(t,z,t+1,x)\, p(t,z,t+1,x)\rangle =$$

$$= \frac{1}{2d}\left((1-(2d-1)\varepsilon)^2 + \left(\frac{2d-1}{2d}\right)\varepsilon^2\right) =$$

$$= \frac{1}{2d} - \frac{(2d-1)\varepsilon}{d} + o(\varepsilon^2).$$

b) $x_1 \neq x_2$, then

$$\langle p(t, z, t+1, x_1)\, p(t, z, t+1, x_2)\rangle =$$

$$= \frac{2}{2d}(1 - (2d-1)\varepsilon) \cdot \varepsilon + \left(\frac{2d-2}{2d}\right)\varepsilon^2 = \frac{\varepsilon}{d} + o(\varepsilon^2)$$

This means that

$$m_2(t+1, x_1, x_2) = \sum_{\substack{|z_1 - x_1| = 1 \\ |z_2 - x_2| = 1}} m_2(t, z_1, z_2)\, q(z_1, z_2, x_1, x_2) \tag{1.68}$$

and

$$q(z_1, z_2; x_1, x_2) = \begin{cases} \dfrac{1}{(2d)^2} & z_1 \neq z_2 \\[2mm] \dfrac{1 - 2(2d-1)\varepsilon}{2d} + 0(\varepsilon^2), & \begin{array}{l} z_1 = z_2 \\ x_1 = x_2 \end{array} \\[2mm] \dfrac{2\varepsilon}{2d} + 0(\varepsilon^2), & \begin{array}{l} z_1 = z_2 \\ x_1 \neq x_2 \end{array} \end{cases} \tag{1.68'}$$

Non-random transition probabilities

$$q(z_1, z_2; x_1, x_2); \quad |z_1 - x_1| = 1, \quad |z_2 - x_2| = 1$$

generate a random walk in $\mathbf{Z}^d \times \mathbf{Z}^d$. This random walk is a direct sum of two independent symmetrical random walks $X_t^{(1)}, X_t^{(2)}$ outside the "diagonal" $z_1 = z_2$ and has a new behavior (delay) at diagonal $z_1 = z_2$.

Namely, our $r.w$ for $z_1 = z_2$ will stay in the diagonal with probability

$$1 - 2(2d-1)\varepsilon + 0(\varepsilon^2)$$

and will leave diagonal with probability

$$2(2d-1)\varepsilon + 0(\varepsilon^2).$$

All the points inside diagonal are equivalent, and we have the same situation outside it.

Theorem 1.6. For fixed $\varepsilon > 0$ and $t \to \infty$, uniformly in $x_1, x_2 \in \mathbf{Z}^d$; $|x_1|, |x_2| \leq$ const $\sqrt{t}$

$$m_2(t, x_1, x_2) \sim m_1(t, x_1) m_1(t, x_2) \tag{1.69}$$

This means that for $|x| \leq$ const $\sqrt{t}$

$$\frac{p(0, 0, t, x)}{m_1(t, x)} \xrightarrow[t \to \infty]{\mu} 1, \tag{1.69'}$$

i.e. the Markov chain X_t, which was introduced in the beginning of the lecture, has homogenization (in the sence of convergence in measure μ). The result of homogenization is a symmetrical random walk on $\mathbf{Z}^d$ (equation (1.66)).

I will not present the complete proof of Theorem 1.6, which is sufficiently technical. I'll only explain the idea of the proof and the probabilistic sense of the construction.

First of all, we have to do Laplace transform of the "parabolic" equation (1.68) in time, i.e. introduce for $|\lambda| < 1$, a generating function (Green function)

$$G_\lambda((x_1, x_2)) = \sum_{n=0}^{\infty} \lambda^n m_2(n, (x_1, x_2))$$

For the function $G_\lambda(x_1, x_2)$ we have an "elliptical" equation

$$\sum_{\substack{|z_1-x_1|=1 \\ |z_2-x_2|=1}} q(x_1, x_2; z_1, z_2) G_\lambda(z_1, z_2) = \lambda G_\lambda(x_1, x_2) - \delta(x_1)\delta(x_2) \qquad (1.70)$$

It is natural to introduce the new variables

$$x_1^* = \frac{x_1 + x_2}{2}, \quad x_2^* = \frac{x_1 - x_2}{2}, \quad z_1^* = \frac{z_1 + z_2}{2}, \quad z_2^* = \frac{z_1 - z_2}{2}$$

(because the kernel $q(x_1, x_2; z_1, z_2)$ depends on the differences $x_1 - x_2, z_1 - z_2$).

Fourier transform with respect to x_1^* will reduce problem (1.70) to the following one:

$$\sum_{z_2^* \in \mathbf{Z}^d} q^*(z_2^* - x_2^*) G_\lambda^*(z_2^*, \phi_1) + \delta(x_2^*) \sum_{|z|=1} G_\lambda^*(z, \phi_1) r(\varepsilon) + H(\phi_1) G_\lambda^*(x_2^*, \phi_1) =$$

$$\qquad (1.71)$$

$$= \lambda G_\lambda^*(x_2^*, \phi_1) - \delta(x_2^*)$$

where

$$G_\lambda^*(x_2^*, \phi_1) = \sum_{x_1^* \in \mathbf{Z}^d} G_\lambda^*(x_2^*, x_1^*) e^{i(\phi_1, x_1^*)},$$

$q^*(z_2^* - x_2^*)$ are the transition probabilities of the symmetrical walk in the new coordinates, $H(\phi_1)$ is a Fourier transform of the Laplacian (with respect to x_1^*), $r(\varepsilon)$ are the transition probabilities from the point 0 described by (1.68').

The main term in (1.71) is a potential $\delta(x_2^*)$ with corresponding factor. This term describes the delay in the point $x_2^* = 0$ (i.e. on the diagonal $x_1 = x_2$!).

Equations of the form (1.71) are typical for many problems in the RM theory (see, for example [15], where one can find the full analysis of similar equations). It can be solved explicitly in terms of Fourier transform in x_2^*.

This means that we can reconstruct the Fourier transform of $G_\lambda^*(x_1^*, x_2^*)$ or $G_\lambda(x_1, x_2)$. After this, we have to use the asymptotical information about this Fourier transform

$\hat{G}_\lambda(\phi_1, \phi_2)$ for $\phi_1, \phi_2 \approx 0$ and Tauberian theorems for $\lambda \to 1$. Corresponding analysis is standard for the limit theorems in the theory of random walks (see [71]), but, of course, it is sufficiently cumbersome. Let's give different (probabilistic) interpretation of the same result, which is easy to transform into the rigorous mathematical proof, at least, for $d \geq 3$. Transition probabilities (1.68') generate on $\mathbf{Z}^d \times \mathbf{Z}^d$ the random walk $X_t^{(1)}, X_t^{(2)}$. We have to study

$$p(t, (0,0), (x_1, x_2)) = P\{X_t^{(1)} = x_1, X_t^{(2)} = x_2 | X_0^{(1)} = X_0^{(2)} = 0\} =$$

$$P\{X_t^{(1)} = 0, X_t^{(2)} = 0 | X_0^{(1)} = x_1, X_0^{(2)} = x_2\} = p(t, (x_1, x_2), (0,0))$$

Let's introduce $r.v$ $\tau = \min(t : X_t^{(1)} = X_t^{(2)})$, i.e. first entrance time on the diagonal $\{x_1 = x_2\}$. It $d \geq 3$, this $r.v.$ is equal to ∞ with positive probability, because $X_t^{(1)} - X_t^{(2)}$ is a transient Markov chain. According to the strong Markovian property,

$$p(t, (x_1, x_2), (0,0)) = \sum_{k=1}^n \sum_{z \in \mathbf{Z}^d} f(k, (x_1, x_2), (z, z)) \cdot p(n - k, (z, z), (0, 0)),$$

where

$$f(k, (x_1, x_2), (z, z)) = P\{X_\tau^{(1)} = X_\tau^{(2)} = z, \quad \tau = k | X_0^{(1)} = x_1, X_0^{(2)} = x_2\}$$

It is essential that $f(k, (x_1, x_2), (z, z))$ has the same structure for the walk $(X_t^{(1)}, X_t^{(2)}$ and for the symmetrical random walk (see (1.68'')). After moment τ, the random walk $(X^{(1)}, X^{(2)})$ can make several cycles, i.e. jump out from the diagonal $(X^{(1)} = X^{(2)})$ and to return to the same diagonal. For the walk $X_t^{(1)} - X_t^{(2)}$, the cycles are related to returns to 0. Symmetrical random walk on $\mathbf{Z}^d \times \mathbf{Z}^d$ has $a.s.$ only a finite number of such cycles on the infinite time interval. Let ν be a corresponding $r.v.$ The only difference between symmetrical random walk and our random walk is the following one: symmetrical random walk leaves the diagonal with probability $1 - \left(\dfrac{1}{2d}\right)$, but for the process $X^{(1)}, X^{(2)}$ this probability has an order ε (it is the reason of delay on the diagonal, mentioned above).

During one cycle, the random walk spends a random time on the diagonal. This time has a geometrical distribution with the mean of order const (depending on d) in the symmetrical case and order ε^{-1} for the walk $X^{(1)}, X^{(2)}$. Total time on the diagonal is a random sum (ν terms) of independent geometrical $r.v.$ This time has some exponential moment (because ν has this property). This total time T generates some random shift along the diagonal, because the process $X_t^{(1)}, X_t^{(2)}$ (on the diagonal $x_1 = x_2$) is a symmetrical random walk.

Now it is easy to obtain the following result: for $d \geq 3$, on the probabilistic space of the random walk $X^{(1)}, X^{(2)}$ we can define a symmetrical random walk $Y^{(1)}, Y^{(2)}$ in such a way, that

$$|X_t^{(1)} - Y_t^{(1)}| + |X_t^{(2)} - Y_t^{(2)}| \le Z \tag{1.72}$$

and r.v. Z has the exponential moments. Moreover, for the random variable Z we can use a very simple explicit construction in the terms of $Y^{(1)}, Y^{(2)}$.

But for the symmetrical random walk Theorem 1.6 is an obvious consequence of the local CLT. Using (1.72) we not only can prove Theorem 1.6 for the process $X^{(1)}, X^{(2)}$, but study the remainder term.

In the case of small dimensions ($d = 1, 2$ especially in the case $d = 1$) this approach works only after essential modifications.

Essential remark. Asymptotical equivalence in the Theorem 1.6 is not uniform in ε. The random variable in (1.72) has order $\varepsilon^{-1/2}$, if $\varepsilon \to 0$, and the *scale of homogenization must be essentially bigger than* ε^{-1}. This is clear from the previous considerations. Let's recall them in a slightly different form.

If $X_0^{(1)} = X_0^{(2)} = z$, i.e. $X^{(1)} - X^{(2)} = 0$, then $P\{X_t^{(1)} - X_t^{(2)} = 0, \quad t = 0, 1, \cdots k\} = (1 - 2\varepsilon)^k$, i.e. for $t = o(\varepsilon^{-1})$ there is an essential difference between the symmetrical random walk $Y_t^{(1)}, Y_t^{(2)}$ and the random walk $X_t^{(1)}, X_t^{(2)}$. (This difference is connected with the fact that $X_t^{(1)} - X_t^{(2)} = o(1)$ and $Y_t^{(1)} - Y_t^{(2)} = o(\varepsilon^{-1/2})$).

Physical discussion. We had started with the random walk X_t in the random non-stationary medium (1.62). For small ε, this walk includes a motion along integral lines of the random "velocity field" $n_0(t, x, w_m)$ (see 1.63) and a small diffusion (with "molecular diffusivity" ε) around these lines.

Integral lines are by themselves the trajectories of the symmetrical random walk (on the probabilistic space $(\Omega_m, \mathcal{F}_m, \mu)$). Now it is not surprising that turbulent diffusivity does not depend on ε. Homogenization of the RM is given by the equation (1.66) for the first moment $m_1(t, x)$.

But the molecular diffusivity ε is very important. It defines the time scale of homogenizations, which has an order ε^{-1}. If we consider for fixed ω_m (i.e. "velocity field" $n_0(t, x, \omega_m)$) two independent trajectories $X_t(w, \omega_m), \quad X_t(w', \omega_m)$ and $X_0(w, \omega_m) = X_0(w', \omega_m)$, then on a time interval of order $o(\varepsilon^{-1})$ these two trajectories will coincide.

This is the explanation of a well known physical effect (see [56], [57]): the temperature fluctuations in the ocean (so-called temperature spots) have a very long lifetime. The reason is that the heat conductivity of the water is very small. This does not contradict the large value of the turbulent diffusivity, because this diffusivity includes not only dissipation of the heat energy ("cooling" of the high temperature fluctuation), but the motion of the fluctuations along random drift field $n_0(t, x, \omega_m)$.

Of course, the fluctuation must have the same (or smaller) space scale as the field $n_0(t, x)$.

Continuous theory of this type is more complicated analytically and includes some additional physical effect. This theory has not been realized on a rigorous mathematical level, nevertheless, at the physical level, for the case of a δ-correlated drift field, we understand the situation sufficiently well ([56], [57], [64]).

To understand the difference between stationary RM and non-stationary RM, we'll now briefly discuss the model similar to the central model (1.62), (1.63) of this lecture, but with a time-independent drift.

Let $n_0(x, \omega_m)$ be a system of unit arrows. We'll suppose the existence isotropy:

$$\mu\{n_0(x, \omega) = \pm e_i\} = \frac{1}{2d} \tag{1.73}$$

where e_i, $i = 1, 2, \cdots d$ is a basis of the lattice $\mathbf{Z}^d$, and independence for different $x \in \mathbf{Z}^d$. As earlier, let

$$p(x, z, \omega) = \begin{cases} 0, & |x - z| \neq 0 \\ \varepsilon, & |x - z| = 1, \\ 1 - (2d - 1)\varepsilon, & x - z = n_0(x, \omega) \end{cases} \quad x - z \neq n_0(x) \tag{1.74}$$

be transition probabilities in the *stationary* RM with the "drift" $n_0(x, \omega)$ and small molecular diffusivity ε.

If $d = 1$, it is a Sinai-type model (see lecture 3) and the process of "trapping" is very strong. Here we have no homogenization, pathological properties of the random walk $X_n(w, \omega)$ etc.

I can't prove the homogenization theorem for this model, although homogenization, probably, has place at least for $d \geq 3$. But the turbulent diffusivity must degenerate, if $\varepsilon \to 0$. It is a consequence (at the physical level) of the following pure mathematical fact.

Theorem 1.7. Let $X^{(x)}(t, \omega)$ be a trajectory of the dynamical system generated by random field $n_0(t, x)$ and started from the initial point x, i.e.

$$X^{(x)}(0, \omega) = x$$

$$X^{(x)}(t + 1, \omega) = X^{(x)}(t, \omega) + n_0(X^{(x)}(t, \omega), \omega)$$

Then

$$\mu\{\omega : \sup_t |X^{(x)}(t, \omega) - x| < \infty\} = 1 \tag{1.75}$$

and moreover, for some constant ρ : $0 < \rho < 1$}

$$\mu\{\omega \quad \sup_t |X^{(x)}(t) - x| \geq m\} \leq \rho^m \tag{1.75'}$$

Proof is simple. Let $\eta(\omega)$ be the number of bonds of the path $X^{(x)}(t, \omega)$ before the first self-intersection. It is obvious that

$$\mu\{\eta \geq m\} \leq \left(\frac{2d-1}{2d}\right)^m.$$

If $X(\eta, \omega) = X(t_1, \omega)$, $0 \leq t_1 < \eta$, then the path $X(s, \omega)$, $t_1 \leq s \leq \eta$ is a closed loop and after moment η, trajectory X will be moving periodically along this loop.

We proved Theorem 1.7 with the estimation $\rho = \dfrac{2d-1}{2d}$, but, of course, we can obtain better estimations, using number $N_d(n)$ of the self avoiding paths on the lattice $\mathbf{z}^d$ and well known fact that

$$\lim_{n \to \infty} \sqrt[n]{N_d(n)} = c(d)$$

(the so-called coordination number of the lattice $\mathbf{Z}^d$). For the const $c(d)$ we can find in the literature upper and lower estimations with good accuracy (at least, for $d = 2$, $d = 3$).

In any case

$$\mu\{\eta \geq m\} \leq \left(\frac{c(d) + \delta}{2d}\right)^m$$

for every $\delta > 0$ and $m \geq m_0(\delta)$.

The "velocity field" $n_0(x, \omega)$ can not be a model of incompressible flow. The dynamical system $X(t, \omega)$ has stable states ("traps"). This is a closed loop such that for every point z on this loop and for all directions, except $n_0(x, \omega)$, we have entering arrows. If molecular diffusivity ϵ is very small, the random walk $X_t(w, \omega)$ generated by transition probabilities (1.73), (1.74) will stay a very long time inside these traps. Nevertheless, for $d \geq 3$ these traps can't be sufficiently large to destroy the homogenization.

Probabilistic analysis of this stationary model (as many others models of this type) is an open problem (see the review at the end of section I).

Conclusion to Ch. I "Homogenization" (review of the physical literature, remarks, open mathematical problems).

This part of the lectures corresponds to seminars which I gave in Saint-Flour. Such general discussion will accompany also Chapters II and III.

The fundamental problem in homogenization theory (which is still far enough from the complete solution) is the problem of turbulent diffusivity (which I'll explain again in continuous variant only).

Let

$$\frac{\partial T}{\partial t} = D\triangle T + (\vec{a}, \triangledown)T$$

$$T(0, x) = T_0(x), \quad (t, x) \in (0, \infty) \times R^d$$

(1.76)

be a heat equation with molecular diffusivity D and random drift $\vec{a}$, which can be time-independent (stationary RM) or time-dependent (turbulent RM).

What are the general conditions of homogenization for the equation after standard scaling $x \to \dfrac{x}{\varepsilon}$, $t \to \dfrac{t}{\varepsilon^2}$, $\varepsilon \to 0$? What are the conditions guaranteeing, at least, a non-trivial Stokes drift? How can one find or estimate the turbulent diffusivity?

Today we have a more or less full solution of this problem only for $d = 1$ and stationary RM (see lecture 3). But even in this case, there are very interesting open questions.

The continuous version of results obtained in lecture 3 has the following form (for the general Fokker-Plank equation $\dfrac{\partial T}{\partial t} = \dfrac{\partial}{\partial x}\left(\sigma(x, \omega)\dfrac{\partial T}{\partial x}\right) + a(x, \omega)\dfrac{\partial T}{\partial x})$. The answer depends, first of all, on the expectation

$$\alpha_0 = \left\langle \frac{a(x, \omega)}{\sigma(x, \omega)} \right\rangle$$

If $\alpha_0 \neq 0$, we can construct harmonic coordinates and an invariant measure, but under some restriction on the exponential moments of the following type. Let $\alpha_0 < 0$, then for some $\gamma > 0$ we must have an exponential estimation

$$\left\langle \exp \int_0^t \frac{a(s, \omega)}{\sigma(s, \omega)}ds \right\rangle \leq \text{ const } \exp(-\gamma t)$$

(1.77)

(compare hypothesis I in lecture 3). Suppose that this estimation doesn't hold. What can one say about Stokes drift? If we have (1.77), but higher exponential moments (of the order 2 or 3) do not exist, we have a mean drift, but probably no CLT, at least in the standard normalization. If $\alpha_0 = 0$ and the spectral density of the process $\dfrac{a}{\sigma}(x,\omega)$ is non-degenerate at the point $\omega = 0$ (very long waves), we have no homogenization and Lagrangian trajectories have the same behavior as in Sinai's case (see the discussion in lecture 3).

If the spectral density is degenerated at the point $\phi = 0$ (and $\alpha_0 = 0$) we have homogenization, but *again under some conditions on the exponential moments.* More precisely, in this case process $\dfrac{a}{\sigma}$ must have the form: $\dfrac{a}{\sigma}(x,\omega) = \dfrac{d\Phi(x,\omega)}{dx}$, where $\Phi(x,\omega)$ is a new homogeneous process (potential). Homogenization is related with exponential moments of the process Φ.

I am sure that in both cases ($\alpha_0 \neq 0$, $\alpha_0 = 0$) we can obtain non-trivial limit theorems, if only the conditions on the exponential moments fail. Probably, the problem of homogenization for the equation $\dfrac{\partial T}{\partial t} = D\dfrac{\partial^2 T}{\partial x^2} + \delta a(x,\omega)\dfrac{\partial T}{\partial x}$ has bifurcations with respect to parameter δ for many natural drift coefficients $a(x,\omega)$.

We'll finish our deviation from the main line by the following lattice one-dimensional example (M. Rost), which can be studied in all details. We analyzed this model with my student L. Bogachev, but the corresponding publication still isn't ready.

Let $\{x_n, \quad n = 0, \pm 1, \cdots\}$ be a homogeneous renewal process on the lattice $\mathbf{Z}^1$, i.e. differences $x_{n+1} - x_n = \Delta_n$ are *i.i.d.* positive *r.v.* Random transition probabilities (to the nearest neighbor points) are given by the formula

$$p(x, x+1, \omega) = 1, \quad x \in \{x_n\}$$
$$p(x, x-1, \omega) = 0 \tag{1.78}$$

$$p(x, x+1, \omega) = p, \quad x \notin \{x_n\}$$
$$p(x, x-1, \omega) = q \tag{1.79}$$
$$p, q > 0, \quad p + q = 1$$

The corresponding random walk X_t has a positive drift in the sense, that $\mu - a.s$

$$p^\omega\{X_t \xrightarrow[t \to \infty]{} +\infty\} = 1.$$

In physical language, this model describes so-called "gate conductivity" for one-dimensional semi-conductors.

Random walk X_t has (if $p < q$) drift in the negative direction between "gates" and, as result the transition from the point x_n to the point x_{n+1} can be extremely long. I am going to formulate only the final result.

1. If $r.v.\{\Delta_n\}$ have exponential moments of any order, i.e. characteristic function $\phi(\lambda) = \langle \exp i\lambda \Delta_n \rangle$ is entire, then the random walk X_t has $\mu - a.s$ non-trivial Stokes drift $a_0 = \lim\limits_{t \to \infty} \dfrac{X_t}{t}$ ($p^\omega - a.e$) and non-trivial gaussian limit distribution with diffusivity B_0. Both these parameters are analytical functions of p, $0 < p < 1$.

2.   If the random variables $\{\Delta_n\}$ have a distribution with exponential tails, the model has bifurcation with respect to parameter p. There exists $P_r < \dfrac{1}{2}$, such that for $p > P_r$ we have the situation of the previous item. But if $p < P_r$ (i.e. left drift between "gates" $\{x_n\}$ is strong enough), the behavior of X_t is absolutely different: the Stokes drift $a_0 \equiv 0$ and the limit distribution isn't a gaussian one.

3.   If $\{\Delta_n\}$ have no exponential moments and $p < 1/2$, the situation is similar to the second part of the previous item.

4.   Of course, if $p > \dfrac{1}{2}$ we have the CLT under very weak conditions on the $r.v.\Delta_n$.

Now we will return to the main line: the multidimensional problem of turbulent diffusivity. Physical literature in this area is boundless, see, for example, the recent review [36], which contains the full bibliography. My opinion is that many of the physical "results" about turbulent transport, based very often on the computer experiments, are not correct (not only mathematically, but physically too!).

Mathematical progress in this problem is connected with the series of papers by A. Majda and M. Avellaneda [8], [9] which had studied in details one of the models of this type: so-called random shear flow. In this case,

$$d = 2, \quad \vec{a}(t,x,\omega) = (0, \quad h(t,x_1,\omega))$$

and $h(t,x,\omega)$ is a gaussian process with zero mean and realistic time-space energy spectrum (Kolmogorv's spectrum). Of course, this flow is an incompressible one: div $\vec{a} = 0$.

In addition, Majda and Avellaneda [8] had proved the existence of homogenization for the general *incompressible* drift $\vec{a}$ under very weak moments conditions. This results contains previous results of S. Kozlov [42] and Varadhan and Papanicolaou [73].

If the condition div $\vec{a} = 0$ doesn't hold, our knowledge is very restricted. I'll try to describe the physical scenario based on the numerical experiments and fragmentary mathematical results.

In the non-stationary case $\vec{a} = \vec{a}(t,x,\omega)$ for the fast decreasing time and space mixing coefficients, homogenization must be a generic case without any additional restrictions (of the type div $\vec{a} = 0$, curl $\vec{a} = 0$ etc). I don't know general results of this type (for $d \geq 2$).

In the stationary case the worst situation, probably, is the potential flows:

$$\vec{a}(x,\omega) = \text{ grad } A(x,\omega)$$

where scalar potential A is just a process with a stationary (homogeneous) increments.

If the potential $A(x, \omega)$ is homogeneous by itself, the situation is simpler. In this case, the function $\pi(x, \omega) = \exp A(x)$ will be (after normalization) the invariant measure for our random process. Of course, this requires the integrability of $\exp A(x)$ (as in lecture 3 for $d = 1$). Existence of the invariant measure guarantees the existence of the Stokes drift. For the lattice analogy of this case, S. Kozlov proved [42] the CLT (i.e. constructed harmonic coordinates). However, I would like to emphasize that in general potential case, the process $A(x, \omega)$ need not be a homogeneous one. Local maxima (peaks) of the potential A are generating "traps" for the flow $\vec{a} = \text{grad } A$. If this "traps" are large enough, they can produce effects, similar to one-dimensional Sinai's model.

It is not difficult to construct examples of this type, even for homogeneous Poisson type potentials (shot noise)

$$A(x) = \sum \phi \left(\frac{x - x_i}{\theta_i} \right)$$

where ϕ is an elementary fast decreasing non-negative potential, $\{x_i\}$ is a point Poisson set and $\{\theta_i\}$ are the scaling factors.

I am sure that for the potential of this type we can observe even a bifurcation with respect to the parameter of disorder δ in the parabolic equation

$$\frac{\partial T}{\partial t} = D \triangle T + \delta (\text{grad } A, \nabla T).$$

I have to note that the potential flows are essential for some astrophysical applications. They describe the hydrodynamics of the so-called "self-gravitating" media and are closely related with the three dimensional Bürger equation (see the recent mathematical article [4], where it's possible to find physical references).

Thus, the situation with homogenization isn't clear for the general potential case or for the flows which are the sum of incompressible and potential components. But in any case, *if div $\vec{a} = 0$* (i.e. for real liquids) *the fact of homogenization is known* for all physically interesting currents (gaussian, shot noise type and so forth).

Suppose, that $\langle \vec{a} \rangle = 0$, then $a_0 = 0$, i.e. we have no mean drift. What can one say about the turbulent diffusivity $B_0 = B_0(D)$, at least for $D \to 0$? The last case is especially interesting in applications to oceanography.

According to the physical literature, the two cases $d = 2$ and $d \geq 3$ are essentially different. If $d = 2$, the incompressible flow has a scalar "potential", which called a stream function:

$$\vec{a}(x,\omega) = \text{curl } \psi(x,\omega) = \left(-\frac{\partial \psi}{\partial x_2}, \frac{\partial \psi}{\partial x_1} \right) \qquad (1.80)$$

For small D the integral lines of the flow $\vec{a}$ i.e. the solutions of ODE will give the main contribution to the turbulent diffusivity.

$$\dot{x}_t = \vec{a}(x_t,\omega), \quad x|_{t=0} = x_0 \qquad (1.81)$$

For smooth $\vec{a}$ with non-degenerated distribution for a.e x_0 these lines will be locally simple curves. Branching of these lines occurs only for critical points of the stream function ψ.

It obvious that these lines will be the level lines of the function ψ, i.e. within parametrization, Lagrangian path (1.81) is given by equation

$$\psi(x_t,\omega) = \psi(x_0,\omega) = h$$

There is just a countable set of levels h for which the corresponding level lines have self-intersection branching. The last statement holds, for example, for a gaussian (non degenerated) function $\psi(x,\omega)$, for shot noise stream functions etc.

Physicists expect, that in "typical situations"

1. For arbitrary $h \in R^1$ level lines $L_n = \{x : \psi(x,\omega) = h\}$ have just bounded connected components (which are simple for a.e $h \in R^1$).

2. It's possible to introduce the distribution of the "typical" component or some functionals of these components.

For example, let N_r be a number of the connected components in the circle $|x| \leq R$ and $l_1, l_2, \cdots, l_r$ are their lengths. The following limit exists

$$F_h(x) = \lim_{R \to \infty} \frac{\#\{l_i \leq x\}}{N(R)}$$

and we can call this limit the distribution function for the length of the typical connected component of the level line L_h.

The area of typical connected component, its diameter and so forth have a similar meaning.

3. There exists a critical level h_{cr}, such that for $h \neq h_{cr}$ and some $z = z(h) > 0$

$$\int_0^\infty e^{zx} dF_h(x) < \infty$$

(i.e. length of the typical connected component has an exponential moment). For $h = h_{cr}$ the tail of distribution $F_h(x)$ is decreasing slowly, i.e., for example,

$$\int_0^\infty x dF_{h_{cr}}(x) = +\infty$$

The distribution of the area or diameter of the typical component have similar properties.

4. Level lines have a "scaling behavior" if $h \to h_{cr}$. For example

$$m_1(h) = \int_0^\infty h dF_h(x) \sim \frac{c}{|h - h_{cr}|^\alpha}.$$

Of course, all these hypotheses have prototypes in classical percolation theory on the lattice $\mathbf{Z}^2$ (bond or site problems), see [38].

This complex of assumptions predicts (on physical level of considerations), that

$$B_0(D) \sim_{D \to 0} \text{const } D^\gamma, 1 > \gamma > 0$$

(see discussion in [36]). The critical exponent γ depends on the critical indices of the percolation problem for level lines and $h \to h_{cr}$.

I don't know when conditions 1-4 hold, but I know that there are many models in R^2 (or $\mathbf{Z}^2$), which are natural physically (smooth realization of the field $\psi(x, \omega)$, fast decreasing of the correlations ect.), however the conditions 1-4 are false for these models.

We (with K. Alexander) recently constructed [5] a series of examples, where, for the level lines on R^2, we can have all the spectrum of possibilities: uniformly bounded components, unbounded components for $|h| \leq h_1$, models with unbounded statistical moments for $|h| \leq h_{cr}$ (i.e. instead of one critical point h_{cr} we have in this case critical strip and so forth). I'll now explain the idea of the last construction for the simplest variant.

Let $\mathbf{Z}^2_\triangle$ be a perfect triangle lattice on the plane R^2 and let $\phi(r)$, $r \geq 0$ be a smooth decreasing function with finite support $[0, 1/2 + \varepsilon]$. Here $1/2$ is a half of the scale of our lattice and ε is sufficiently small. If $\{\varepsilon_n\}$, $n \in \mathbf{Z}^2_\triangle$ are i.i.d.r.v, $p\{\varepsilon_n = 1\} = p\{\varepsilon_n = -1\} = 1/2$, we can define a smooth random field $\psi(x, \omega)$, $x \in R^2$ by the formula

$$\psi(x, \omega) = \sum_{n \in \mathbf{Z}^2_\triangle} \varepsilon_n \phi(|x - n|)$$

For small ε the supports of the different elementary potentials $\phi(|x - n|)$ have only pair-wise intersections. If n, $n' \in \mathbf{Z}^2_\triangle$, $|n - n'| = 1$ are two neighbors and

$\varepsilon_n = \varepsilon_{n'} = 1$, the field $\psi(x,\omega)$ has a saddle point for $x = \dfrac{n+n'}{2}$ and the height of this point is $h_{cr} = 2\phi\left(\dfrac{1}{2}\right)$. If $\varepsilon_n = \varepsilon_{n'} = -1$, the field has a negative saddle-point and for $\varepsilon_n \cdot \varepsilon_{n'} = -1$ it has a zero-line, which intersects the line $[n, n']$.

As the bond percolation problem for triangle lattice has critical probability exactly equal to $1/2$ and because there are no saddle-points with height inside interval $(0, h_{cr})$ we can prove easily that for $|h| < h_{cr} = 2\phi(1/2)$, the connected components of the level line $L_n = \{x : \psi(x,\omega) = h\}$ have the same structure as the connected components of the bond problem on $\mathbf{Z}^2_\triangle$ with critical probability $p = p_{cr} = 1/2$.

I assume, that results of [5] contradict with the "universal" picture of [36]. For $d = 2$, structure of turbulent diffusivity $B_0(D)$ can be, probably, essentially different for the different "physical" models, even in the simplest case of gaussian or shot noise flows $\vec{a}(x,\omega)$.

Anyway, this subject is very interesting and can be basis of the further interesting theory.

In the multidimensional case $d \geq 3$, we have no mathematical results. The standard physical idea (R. Sagdejev's hypothesis) is the following one.

If div $\vec{a}(x,\omega) = 0$, $d \geq 3$, then Lagrangian trajectories x_t : $\dot{x}_t = \vec{a}(x_t,\omega)$ can be divided into two classes: stable (i.e. bounded), which corresponds to the "islands of stability" for the dynamical system $\dot{x} = \vec{a}(x,\omega)$ and unstable, which have diffusion behavior for $t \to \infty$. If this is true, the turbulent diffusivity $B_0(D)$ must have non-degenerate limit $B_0^+ = \lim\limits_{D\to 0} B_0(D) > 0$. Unfortunately, we don't even have examples supporting this "statement". This problem is very complicated; its lattice version is more or less equivalent to the problem of the asymptotical behavior of the self-avoiding random walks on the lattice $\mathbf{Z}^d$, $d \geq 3$.

In the conclusion, I have to note once again that the problem of the molecular diffusivity $B_0(D)$ for small D (in both cases: stationary and non-stationary RM) and gaussian or shot noise flows $\vec{a}$ with div $\vec{a} = 0$ is a real challenge for mathematicians. After the papers of Majda and Avellaneda we can expect that it is an area where mathematicians can generate real progress in classical statistical hydrodynamics.

Chapter II Localization
Lecture 5. The general introduction

In this chapter we'll mainly study the spectral theory of the lattice Schrödinger equation with a random potential. At the beginning we will introduce some notions and definitions.

Let Γ be a countable graph with a bounded number of neighbors for every point $x \in \Gamma$ (i.e. bounded index of branching). Typical examples are: lattices $\mathbf{Z}^d$ (index of branching equal to $2d$), groups with the finite number of generators, homogeneous (Bethe) tree (every point has $(d+1)$ neighbors: d in the "future" and 1 in the "past").

Let $L_2(\Gamma)$ is a space of the square-integrable functions $f(x) : \Gamma \to C$ with a standard dot product and norm

$$(f, g) = \sum_{x \in \Gamma} f(x)\bar{g}(x), \quad \|f\|^2 = \sum_{x \in \Gamma} |f(x)|^2. \tag{2.1}$$

If $d(x, y)$ is an obvious metric on Γ (the minimal number of edges in the class of paths $\gamma : x \to y$) the volume of the ball, centered point $x_0 \in \Gamma$, can not increases faster than an exponential:

$$|B_R(x_0)| = |\{x : d(x, x_0) \leq R| \leq N^{R+1},$$

where N is a bound for the index of branching.

The Laplacian in the space $L_2(\Gamma)$ is given by a standard formula

$$\triangle f(x) = \sum_{x' : d(x, x') = 1} f(x') \tag{2.2}$$

and, of course,

$$\|\triangle\| = \sup_{f : \|f\| = 1} \|\triangle f\| \leq N, \tag{2.2'}$$

i.e. $\triangle$ (lattice Laplacian!) is a bounded operator.

The Schrödinger operator (Hamiltonian) has form

$$H = \triangle + \sigma V(x) \tag{2.3}$$

where $V(x)$ (potential) is an arbitrary function (multiplication operator) and σ is a coupling parameter. In the sequel the potential V will be a random one, $V = V(x, \omega)$, $\omega \in (\Omega_m, \mathcal{F}, \mu)$.

The Hamiltonian H (for arbitrary V, which is a difference between lattice and continuous cases) is symmetrical

$$(H f_1, f_2) = (f_1, H f_2), \quad f_1, f_2 \in D(H)$$

and essentially self ajoint (see the detail in the monographs [2], [14], [49], [67]). This means that operator H has an unique spectral decomposition

$$H = \int \lambda E(d\lambda), \tag{2.4}$$

where $E(d\lambda)$ is the resolution of the identity corresponding to the Schrödinger operator H. For given $f \in L_2(\Gamma)$ we can introduce the spectral measure of the element f :

$$\nu_f(d\lambda) = (E(d\lambda)f, f).$$

If $\|f\| = 1$, the measure ν_f is probabilistic.

Element f_0 has a maximal spectral type, if for arbitrary $f \in L_2(\Gamma)$ the measure $\nu_f(d\lambda)$ is absolutely continuous with respect to $\nu_{f_0}(d\lambda)$. In many different senses the "majority" of the elements have a maximal spectral type.

The spectrum $\sum(H)$ of H (as a set) has many equivalent definitions. It is :

1. The set of energies E such, that resolvent operator $(H - E)^{-1}$ is not bounded (this set is closed, its complimentary set $R^1 \backslash \sum(H)$ has the property that the operator $(H - E)^{-1}$ is bounded and gives a one-to-one mapping $L_2(\Gamma) \to L_2(\Gamma)$.

2. $\sum(H)$ is the support of the maximal spectral type, i.e. the minimal closed set which supports a measure $\nu_{f_0}(d\lambda)$.

3. Spectrum $\sum(H)$ can be divided into two parts: the discrete spectrum $\sum_d$ i.e. the set of isolated points and the essential spectrum $\sum_e = \sum \backslash \sum_d$ (a closed set without isolated points).

Of course, for every $E \in \sum_d$ we can construct at least one eigenfunction $\psi_E(x) \in L_2(\Gamma)$: $\quad H\psi = E\psi$.

4. More detailed classification of the spectrum depends on the properties of the maximal spectral type. The spectral measure $\nu_{f_0}(d\lambda)$ can be decomposed into three parts:

$$\nu_{f_0}(d\lambda) = \nu_{ac}(d\lambda) + \nu_{sc}(d\lambda) + \nu_{pp}(d\lambda) \tag{2.5}$$

(absolutely continuous spectrum, singular continuous spectrum, pure point spectrum). By the definition, operator H has a p.p. spectrum $(\sum(H) = \sum_{pp}(H))$ if $\nu_{ac} = \nu_{sc} = 0$, has singular spectrum, if $\nu_{ac} = 0$ etc.

As is easy to see, $\sum(H) = \sum_{pp}(H)$ if there exists a complete orthonormal system of eigenfunctions $\psi_{E_i}(x)$: $\quad H\psi_{E_i} = E_i\psi_{E_i}, \quad i = 1, 2, \cdots$

Of course $\sum_d \subset \sum_{pp}$, but usually in localization theory $\sum_d = \phi$ and $\sum_e = \{\bar{E}_i\}$. This means that the eigenvalues $\{E_i\}$ are dense inside the essential spectrum $\sum_e(H)$ which is very often a closed interval $[E_{\min}, E_{\max}] = \sum_e(H) = \sum(H)$.

How can we find $\sum(H), \sum_e(H)$? How can we prove that, for example, $\sum(H) = \sum_{ac}(H)$ or $\sum(H) = \sum_{pp}(H)$? General theory gives some (very poor) information of this type, but for the random operators there exist specific methods.

Proposition 2.1. An energy $E \in \sum_e$ if we can construct a family of orthogonal "almost eignenfunctions", i.e.

$$\exists \{f_n \in L_2(\Gamma): \quad \|f_n\| = 1, \quad (f_n, f_m) = \delta_{m,n},$$

$$\|Hf_n - Ef_n\| \to 0, \quad n \to \infty\}. \tag{2.6}$$

Corollary 2.1. Let $\Gamma = \mathbf{Z}^d$, $d \geq 1$ and $v(x) = v(x, \omega)$ be $i.i.d.r.v.$ It follows from the general (and elementary) theory that $\sum H(\omega)$ does not depend on $\omega \in \Omega_m$, $\mu - a.s$ (L. Pastur, [49], see also [14]) and we can write $\sum H(\omega) = \sum(H)$ (This fact holds not only for $i.i.d.r.v$, but for arbitrary ergodic homogeneous potential $V(x, \omega)$). But in the special case of $i.i.d.r.v.$, which is known as an *Anderson model [6], [60]* this *spectrum can be calculated explicitly*:

$$\sum(H) = \text{Supp } P_v \oplus [-2d, 2d] \quad (\mu - a.s). \tag{2.7}$$

Here $P_v(d\lambda) = \mu\{V(x, \omega) \in (\lambda, \lambda + d\lambda)\}$ is the distribution of the potential at a given point $x \in \mathbf{Z}^d$, Supp P_v is the support of this distribution and $\oplus$ is a sign for the algebraic sum of two sets. Formula (2.7) is our Corollary 2.1.

The proof of (2.7) is based on Proposition (2.1) and is a simple one. First of all, for arbitrary operator of multiplication $V : f \to V(x)f(x)$ we have

$$\sum(V) = \overline{\{V(x), \quad x \in \mathbf{Z}^d\}}. \tag{2.8}$$

Using the ergodic theorem (or the law of the large numbers) we can check that in our case

$$\sum(V(\omega)) = \text{Supp } P_v \quad (\mu - a.s).$$

But inequality $\|\Delta\| \leq 2d$ and the elementary facts of the general spectral analysis give us, that

$$\sum(H(\omega)) = \sum(\Delta + V(x, \omega)) \subset \text{Supp } P_v \oplus [-2d, 2d]$$

Now, suppose that $E \in \text{Supp } P_v \oplus [-2d, 2d]$. Of course,

$$E = a + b, \quad a \in \text{Supp } P_v, \quad b \in [-2d, 2d].$$

It follows from the Borel-Cantelli lemma that $\mu - a.s$ we can find a system of non-intersected balls $B_n = \{x : |x - x_n| \leq R_n\}$ such that

1. $x_n \to \infty$, $R_n \to \infty$, $n = 1, 2, \cdots$.

2. $B_n B_m = \phi, \quad n \neq m$.

3. $\sup |V(x,\omega) - a| \xrightarrow{n \to \infty} 0, \quad x \in B_n$

Because $b \in [-2d, 2d]$, we can find parameters $\phi_1, \cdots, \phi_d \in S^1 = [-\pi, \pi]$ for which

$$b = 2\sum_{i=1}^{d} \cos \phi_i,$$

i.e. b is a generalized eigenvalue of the operator $\triangle$ and the corresponding generalized eigenfunction has form

$$\psi_b(x) = \exp(i(\vec{\phi}, \vec{x})). \tag{2.9}$$

This means that $\triangle \psi_b(x) = b\psi_b(x)$. Now we can construct the "almost eigenfunctions" of the random Hamiltonian $H(\omega) = \triangle + V(x, \omega)$ in the following form:

$$f_n(x) = \exp\{i(\phi, x)\} \cdot I_{B_n} \cdot |B_n|^{-1/2}.$$

Using (2.6), we can easily check, that

$$\|Hf_n - Ef_n\| \to 0 \, (P - a.s), \text{ i.e. } E \in \Sigma_c(H).$$

We have proved the formula (2.7).

Proposition 2.2 For an arbitrary graph Γ the Schrödinger operator $H = \triangle + v(x)$ has a discrete spectrum (i.e. $\Sigma(H) = \Sigma_d(H)$) iff

$$v(x) \to \infty, \quad x \to \infty.$$

I'll give the proof of this proposition, because it isn't simple to find it in the literature. In the continuous case the situation isn't so obvious.

At the beginning, suppose that $v(x) \to \infty, \quad x \to \infty$. Then for every given energy interval $\triangle = [\alpha, \beta]$ and its neighborhood $\tilde{\triangle} = [\alpha - 2d - 1, \beta + 2d + 1]$ there exists only a finite set Γ of points $x : v(x) \in \tilde{\triangle}$. Let $\tilde{H} = \triangle + \tilde{v}(x)$, and $\tilde{v}(x) = v(x), \quad x \notin \Gamma, \quad \tilde{v}(x) = \text{const} \notin \tilde{\triangle}$, if $x \in \Gamma$. The resolvent of the operator $\tilde{v}$ is analytical inside $\tilde{\triangle}$ and, as a result, the resolvent of H is analytical inside $\triangle$ (because $\|\triangle\| = 2d$).

This means that the operator $\tilde{H}$ has no eigenvalues (or spectrum) inside the interval $\triangle$. But the difference $\tilde{H} - H$ is an operator of finite rank, i.e. H has (in the same interval $\triangle$) not more than finite number of eigen values.. We have proved that $\Sigma = \Sigma_d$.

Inverse statement. If the potential $v(x)$ does not tend to infinity, we can find a constant $C > 0$ and a sequence of points $x_n \to \infty$ such that

$$|V(x_n)| \leq c, \quad |x_n - x_m| > 1, n \neq m$$

On the linear span $L = \{\sum_n a_n \delta_{x_n}\}$ the quadratic form $(H\Psi, H\Psi) = (H^2\Psi, \Psi)$ is bounded. Indeed, simple calculations show that if $\Psi = \sum_n a_n \delta_{x_n}$ and $\sum_n |a_n|^2 = 1$ then

$$(H\Psi, H\Psi) \le 2N + C^2.$$

Courant"s variational principle gives immediately that the interval $[0, 2N + C^2]$ contains "infinitely many eigenvalues of H^2, or more perciously, non-empty essential spectrum of H^2. But this last fact contradicts the discretness of the spectrum of H (or H^2).

Example 2.1. Consider (following [30]) the random Schrödinger operator on Z^d, $d \ge 1$ with a strongly oscillating potential

$$H(\omega) = \triangle + |x|^\alpha v(x, \omega), \quad \alpha > 0, \tag{2.10}$$

where $v(x, \omega)$ are *i.i.d.r.v.*, uniformly distributed on [0,1] (more general distributions can be considered, using the same method).

Proposition 2.3. (see [30]) 1. Operator $H(\omega)$ has $\mu - a.s$ discrete spectrum if $\alpha > d$. 2. If $\dfrac{d}{k+1} < \alpha \le \dfrac{d}{k}$, $k = 1, 2, \cdots$, then $\sum_e = [c(k), +\infty)$, $\mu\{\sum_d \ne \phi\} > 0$, but $\sum_d$ can have only one limit point $E_\ell = c(k)$. For the const $c(k), -2d < c(k)$ and we can give (see below) some combinatorial expression, $c(k) \to -2d$, $k \to \infty$.

The proof of the first part of the proposition is based on the general criterion, Proposition 2.2. Consider the system of the events $B_x^c = \{\omega : |x|^\alpha v(x, \omega) < c\}$, $x \ne 0$ where $c > 0$ is a const. Of course

$$\mu\{B_x^c\} = \frac{c}{|x|^\alpha}$$

and the Borel-Cantelli lemmas (we can use both of them, because events B_x^c are independent for different $x \in Z^d$) give us immediately that for $\alpha > d$ ($\mu - a.s.$) $v(x) \to \infty$ and for $\alpha \le d$ the set $\{v(x, \omega)\}$ has non-trivial limit points, i.e. $\sum_e \ne \phi$.

Second part. Suppose, that $\dfrac{d}{k+1} < \alpha \le \dfrac{d}{k}$, $k = 1, 2, \cdots$. Let A_k be the set of all connected sets (of k points) in Z^d, for which the minimal point (in the sense of lexicographic order) coincides with $0 \in Z^d$. This set (the set of so-called k-animals) is finite. Let's introduce the events

$$B_a^{\varepsilon,c}(x) = \{\omega : V(y)|y|^\alpha \in [c, c + \varepsilon], y \in x + a\},$$

for fixed const $c > 0$, $\varepsilon > 0$ and "animal" $a \in A_k$. It is easy to see, that

$$\mu\{B_a^{\varepsilon,c}(x)\} \sim \frac{\text{const } (\varepsilon)}{|x|^{k\alpha}}, \quad |x| \to \infty$$

and the first Borel-Cantelli lemma together with the fact that ε is arbitrarily small, shows that $\mu - a.s.$ there exists a sequence $\{x_n(\omega)\}$, $x_n \to \infty$, such that

$$v(y) - c \to 0, \quad y \in x_n + a$$

If $E_a, f_a(x)$ are the minimal eigenvalue and corresponding eigenfunction of the problem

$$\triangle f = Ef, \quad x \in a$$

$$f(x) = 0, \quad x \in a$$

and $f_a^{(n)}(x) = f_a(x - x_n)$, then (using proposition 2.1) we can prove the following fact: for arbitrary $c > 0$ and $a \in A_k$

$$c + E_a \in \sum_e(H(\omega)), \quad \mu - a.s.$$

This means that

$$\sum_e(H(\omega)) \subseteq [c(k)\infty),$$

where

$$c(k) = \min_{a \in A(k)} E_a. \tag{2.11}$$

We will not finish the proof of the second statement $(\sum_e(H(\omega)) \supseteq [c(k), \infty))$ but note only that the second Borel-Cantelli lemma shows $(\mu - a.s)$ that the "walls" around our k-animals $x_n + a$ must tend to ∞, if $x_n \to \infty$. See in [30] technical details of the proof, together with asymptotical analysis of the const $c(k)$, defined by formula (2.11). The most important result of [30] is the statement that the spectrum of the operator (2.10) is $p.p.$ (for all $\alpha > 0$). We'll obtain this result later as a consequence of the general localization theorem.

We used earlier only the so-called direct methods of spectral analysis (test functions, the variational approach and so forth). But more deep analysis is based on the study of the resolvent $R_E = (H - E)^{-1}$ in the complex domain $Im E \neq 0$.

In this case the resolvent is a bounded operator and corresponding kernel (Green function) $R_E(x, y)$, $x, y \in \Gamma$ is symmetrical : $R_E(x, y) = R_E(y, x)$. As a function of x, the Green kernel is the solution of the equation

$$\triangle R_E(x, y) + v(x)R_E(x, y) - ER_E(x, y) = \delta_y(x) \tag{2.12}$$

and (again in the case $Im E \neq 0$)

$$\sum_{x \in \Gamma} |R_E(x, y)|^2 = ||R_E(\cdot, y)||^2 \leq$$

$$||R_E - I||^2 \cdot ||\delta_y(\cdot)||^2 \leq \frac{1}{|Im E|^2}.$$

It's easy to give an "exact" formula for $R_E(x,y)$, at last for $|ImE| \gg 1$. This is the so-called cluster (or path) expansion. We can get this formula either by the iteration of the relation

$$R_E(x,y) = \frac{\sum\limits_{d(x',x)=1} R_E(x',y)}{E - v(x)} - \frac{\delta_y(x)}{E - v(x)}$$

or by direct substitution into equation (2.12) of

$$R_E(x,y) = \sum_{\gamma:x\to y} \left(\prod_{z\in\gamma} \left(\frac{1}{E - v(z)} \right) \right), \quad x \neq y$$

$$\tag{2.13}$$

$$R_E(y,y) = -\frac{1}{E - v(y)} + \sum_{\gamma:y\to y} \left(\prod_{z\in\gamma} \left(\frac{1}{E - v(z)} \right) \right).$$

Here $\gamma : x \to y$ is an arbitrary path from x to y, i.e. sequence

$$z_0 = x, \quad z_1, z_2, \cdots, \quad z_n = y$$

$$d(z_i, z_{i+1}) = 1, \quad i = 0, 1, \cdots, n - 1 \tag{2.13'}$$

$$|\gamma| = n.$$

Because $\left| \prod\limits_{z\in\gamma} \frac{1}{(E - v(z))} \right| \leq \left(\frac{1}{|ImE|} \right)^{|\gamma|+1}$ and $\#\{\gamma : |\gamma| = n\} \leq N^n$ the series (2.13) are absolutely convergent if $|ImE| > N$. For all $E : ImE > 0$, we have to use analytic continuation with respect to the complex argument E (x and y are fixed).

We will use the expansion (2.13) and some of its generalizations in lecture 7. The resolvent kernel is a very complicated functional of all values $v(x)$, $x \in \Gamma$ of our potential, nevertheless this dependence has a special form.

Proposition 2.4. (Rank 1 perturbation formula or Krein formulas).

Let $\tilde{H}_a$ be an operator

$$\tilde{H}_a = \Delta + \tilde{V}_a(x), \tag{2.14}$$

where

$$\tilde{V}_a(x) = \begin{cases} V(x), & x \neq a \\ 0, & x = a, \end{cases}$$

and $\tilde{R}_{E,a}(x,y) = \left((\tilde{H}_a - E)^{-1}\delta_y, \delta_x \right)$ is the corresponding resolvent kernel. Then

$$R_E(x,y) = \tilde{R}_{E,a}(x,y) - \frac{(V(a) - E)\tilde{R}_{E,a}(x,a)\tilde{R}_{E,a}(a,y)}{1 + (V(a) - E)\tilde{R}_{E,a}(a,a)}. \tag{2.15}$$

The proof is based on the calculations, which make sense for $ImE \neq 0$.

Equation

$$(H - E)R_E(x,y) = \delta_y(x)$$

can be represented in the form

$$(\tilde{H}_a - E)R_E(x,y) = \delta_y(x) - \delta_a(x)(V(a) - E)R_E(a,y)$$

i.e.

$$R_E(x,y) = \tilde{R}_{E,a}(x,y) - (V(a) - E)R_E(a,y)\tilde{R}_{E,a}(x,a).$$

For $x = a$ we'll obtain

$$R_E(a,y) = \frac{\tilde{R}_E(a,y)}{1 + (V(a) - E)\tilde{R}_E(a,a)}.$$

Now (returning to the previous formula)

$$R_E(x,y) = \tilde{R}_{E,a}(x,y) - \frac{(V(a) - E)\tilde{R}_{E,a}(x,a)\tilde{R}_{E,a}(a,y)}{1 + (V(a) - E)\tilde{R}_{E,a}(a,a)}.$$

In the special case, $a = x$ this formula is especially simple:

$$R_E(x,y) = \frac{\tilde{R}_{E,x}(x,y)}{1 + (V(x) - E)\tilde{R}_{E,x}(x,x)}. \tag{2.15'}$$

At last, for $y = a = x$ we have

$$R_E(x,x) = \frac{1}{V(x) - E + \tilde{R}_{E,x}^{-1}(x,x)}. \tag{2.15''}$$

Formula (2.15") is the basis of Wegners estimation for the density of states [74], which played a critical role in the original and fundamental papers on localization theory [27], [21] (for additional references see [14]).

Formula (2.15') or the similar perturbation considerations is the basis of Simon-Wolff theorem [69]. In our recent paper with M. Aizenman [2] all the rank-one (sometimes rank-two) perturbation formulas; (2.15), (2.15'), (2.15"), together with Simon-Wolff theorem [69], are again a central element of the construction.

Formula (2.15) show that

$$R_E(x,y) = \frac{\alpha V(a,\omega) + \beta}{\gamma V(a,\omega) + \delta} \tag{2.16}$$

where functions $\alpha, \beta, \gamma, \delta$ depend on $\tilde{V}_a(x)$, i.e. do not depend of $V(a)$ (for fixed a, x, y). This fractional-linear dependence of $R_E(x,y)$ on every given value $V(a)$ of the potential is very important (see following lecture 6).

The especial role of the resolvent in the spectral analysis is related to the following fact: if $E \notin \Sigma(H)$

$$R_E(x,y) = \int_{\Sigma(H)} \frac{(dE(\lambda)\delta_x, \delta_y)}{\lambda - E}, \tag{2.17}$$

in particular

$$R_E(x,x) = \int_{\Sigma(H)} \frac{\mu_{\delta_x}(d\lambda)}{\lambda - E}.$$

This means that $R_E(x,x)$ is the Hilbert transform of the spectral measure of the element $f(\cdot) = \delta_x(\cdot)$.

In the theory of random Schrödinger operators, resolvent analysis has many important applications.

One of them is the formula for the density of states. Let's consider (following [49] or [14]) the Schrödinger operator on $\mathbf{Z}^d$

$$H = \triangle + V(x,\omega)$$

with the homogeneous ergodic potential $V(x,\omega)$. Let H_n be the restriction of H to the cube $Q_n = \{x : \quad |x_1| \leq n, \cdots, |x_d| \leq n\}$ with zero boundary condition on ∂Q_n. Operator H_n has a discrete (in fact, finite) spectrum $\lambda_i^{(n)}$, $i = 1, 2, \cdots, \quad |Q_n|$. We can introduce the counting function

$$N_n(\lambda) = \#\{\lambda_i^{(n)} < \lambda\}.$$

Then (see [49], [14]) for $n \to \infty$, $\quad \mu - a.s.$

$$N_n^*(\lambda) = \frac{N_n(\lambda)}{|Q_n|} \xrightarrow[n \to \infty]{} N(\lambda). \qquad (2.18)$$

The limit is a nonrandom continuous distribution function, $N(-\infty) = 0$, $N(+\infty) = 1$, the so-called integral density of states. It is one of the central notions in the physics of disordered state.

If $\psi_i^{(n)}$ are the (orthonormal) eigenfunctions, corresponding to eigenvalues $\lambda_i^{(n)}$ of H_n, then

$$\left(R_E^{(n)} \delta_x, \delta_x\right) = \sum_{i=1}^{|Q_n|} \frac{(\delta_x, \psi_i^{(n)})^2}{\lambda_i^{(n)} - E}$$

and

$$\sum_{x \in Q_n} \left(R_E^{(n)} \delta_x, \delta_x\right) = \sum_{i=1}^{|Q_n|} \sum_{x \in Q_n} \frac{(\delta_x, \psi_i^{(n)})^2}{\lambda_i^{(n)} - E}.$$

But δ_x, $x \in Q_n$ in an orthonormal system and Parseval's identity gives

$$\frac{1}{|Q_n|} \sum_{x \in Q_n} \left(R_E^{(n)} \delta_x, \delta_x\right) = \int \frac{dN_n^*(\lambda)}{\lambda - E}.$$

The random variables $(R_E^{(n)} \delta_x, \delta_x)$ are "almost identically distributed" and in the limit $Q_n \uparrow \mathbf{Z}^d$ we obtain the following formula.

Proposition 2.5. For all E, $\quad Im E \neq 0$

$$\int \frac{dN(\lambda)}{\lambda - E} = \langle R_E(0,0) \rangle. \tag{2.19}$$

(For technical details see [49], [14]). In fact, these considerations show, that for $\mu - a.e.$ realizations ω of the random potential $V(\cdot, \omega)$ the spectrum $\sum(H(\omega))$ does not depend on ω and coincides with the set of all growth points of the function $N(\lambda)$.

Wegner [74] proved (using formulas (2.19) and (2.15")) that for $i.i.d.r.v$ $V(x)$ with common bounded distribution density

$$p = p(v) = \frac{d}{dv} \mu\{V(\cdot) < v\}, \quad p(v) \leq c,$$

there exists bounded density of states:

$$n(\lambda) = \frac{dN(\lambda)}{d\lambda} \quad \text{and} \quad n(\lambda) \leq c. \tag{2.20}$$

I mentioned already that this Wegner's theorem is the basis of all the initial articles on localization theory (Frölich, Spencer, Martinelli and others).

I will not prove the estimation (2.20) here, but our further considerations will in fact contain the generalization of the basic idea of Wegner [74].

A second application is a very useful criterion of Simon-Wolff for pure point spectrum. We'll give this criterion in a special form convenient for further applications. For the abstract form of the Simon-Wolff theorem see [69].

Proposition 2.6 Let Γ be an arbitrary graph with the bound N for the index of branching, let $H = \Delta + V(x, \omega)$ be a random hamiltonian, and suppose the random variables $V(x, \cdot)$ for different $x \in \Gamma$ are independent and have absolutely continuous (a.c) distributions.

Suppose, that for a.e. $E \in \Delta \subset R^1$ (very often interval Δ coincides with R^1) and all $x \in \Gamma$

$$\lim_{\varepsilon \to 0} \sum_{y \in \Gamma} |R_{E+i\varepsilon}(x,y)|^2 < \infty, \quad \mu - a.s. \tag{2.21}$$

Then $\mu - a.s.$ the operator H has in Δ $\quad p.p.$ spectrum.

If for some $\gamma > 0$

$$\lim_{\varepsilon \to 0} \sum_{y \in \Gamma} |R_{E+i\varepsilon}(x,y)|^2 \exp(\gamma d(x,y)) < \infty, \mu - a.s, \tag{2.21'}$$

then the eigenfunctions of the point spectrum are not only square-integrable, but exponentially decrease (more precisely, exponentially integrable with exponential weight). It means, in H we have not only localization, but exponential localization.

Remark. For fixed E

$$\sum_{y \in \Gamma} |R_{E+i\varepsilon}(x,y)|^2 = \int \frac{\mu_{\delta_x}(d\lambda)}{|\lambda - E - i\varepsilon|^2}$$

and, as easy to see, this form is monotone function of ε, i.e. the limit in (2.21) exists.

Lecture 6. Localization Theorems for large disorder. Examples

In this lecture I'll give the proofs of the localization theorems for general graphs and Hamiltonians with independent values of the potential. I will follow our paper with M. Aizenman [2], but for the sake of simplicity our restrictions on the density of the distributions will be stronger than in [2].

The following result will be a prototype of more general constructions.

Theorem 2.1. Consider the discrete Schrödinger operator $H = H(\omega)$, $\omega \in (\Omega_m, \mathcal{F}, \mu)$ on $L_2(\Gamma)$

$$H = \triangle + V_o(x) + \sigma V(x, \omega), \qquad (2.22)$$

where the potential consists of the some specified (non-random) background $V_0(x)$ and random part $\sigma V(x, \omega)$. Random variables $V(x, \omega), x \in \Gamma$, are independent and uniformly distributed on $[0, 1]$. Parameter σ (the coupling constant) is "the measure of disorder".

There exists const $\sigma_0 = \sigma_0(\Gamma)$ such that for all $|\sigma| > \sigma_0$ and $\mu - a.s$, the operator H has p.p. spectrum with exponential estimation of eigenfunctions (exponential localization).

For $\Gamma = \mathbf{Z}^d$, $d \geq 1$ and $V_0(x) \equiv 0$ model (2.22) is an initial version of Anderson's model [6].

Proof of Theorem 2.1. According to Proposition 2.6 it's enough to prove (2.21'). Instead of $\mu - a.s$ considerations (which usually are complicated) we will be working with the moments. Using trivial inequality: for all complex numbers $z_1, \cdots, z_n$ and every $0 < s < 1$

$$|z_1 + \cdots + z_n|^s \leq |z_1|^s + \cdots + |z_n|^s, \qquad (2.23)$$

we can reduce (2.21') to a different (moment) inequality which is simpler;

$$\sum_{y \in \Gamma} \left\langle |R_E(x, y)|^{2s} \right\rangle \cdot \exp(s\gamma d(x, y)) \leq C(x, s), \qquad (2.24)$$

uniformly in $Im\, E = \varepsilon \neq 0$. (Reduction means that from (2.24) follows (2.21').)

Now we have to estimate the moments of the resolvent kernel. The following simple statement is a variant (in our special case) of Wegner's basic estimation [74].

Lemma 2.1. There exists constant C_0, such that for all $x \in \Gamma$, $ImE \neq 0$ and $0 < s < 1$

$$\langle |R_E(x, x)|^s \rangle \leq \frac{C_0}{|\sigma|^s (1 - s)} \qquad (2.25)$$

The proof is an elementary one. According to (2.15)

$$R_E(x,x) = \frac{1}{\tilde{R}_{E,x}^{-1}(x,x) + \sigma V(x,\omega) + V_0(x) - E}$$

$$= \frac{1}{\sigma V(x,\omega) + \zeta(x,\omega)}$$

$$(2.15')$$

where $\zeta(x,\omega)$ is a random variable, independent of $V(x,\omega)$. To prove (2.25) it's enough to check that for arbitrary (complex!) a,

$$E\frac{1}{|\sigma V + a|^s} \leq \frac{C_0}{|\sigma|^s(1-s)}$$

$$(2.25')$$

and to use an integration with respect to distribution of ζ. But

$$E\frac{1}{|\sigma V + a|^s} = \int_0^1 \frac{dv}{|\sigma v + a|^s} =$$

$$= \frac{1}{|\sigma|^s}\int_0^1 \frac{dv}{|v + a'|^s} \leq \frac{1}{|\sigma|^s}\int_0^1 \frac{dv}{|v + Re\, a'|^s}.$$

The last integral has order $|Re\, a'|^{-s}$, if $Re\, a' \to \infty$. For $|Re\, a'| < 2$, we can use the direct integration, since the singularity is not strong. If $s \uparrow 1$, we have an asymptotic of the order $\frac{c}{1-s}$. Of course, it's possible to give expression for the constant C_0.

Let's note that for $s = 1$ and $a' \in [0,1]$

$$\int_0^1 \frac{dv}{|v - a'|} = \infty.$$

This shows that the resolvent kernel $R_E(x,x,\omega)$ for real E has, roughly speaking, Cauchy tails of distribution (i.e finite moments of order $s < 1$, but infinite expectation).

Of course, for the expectation (2.25') we can give a double-sided estimation:

$$\frac{c^-(s)}{|\sigma|^s + |a|^s} \leq E\frac{1}{|\sigma v + a|^s} \leq \frac{c^+(s)}{|\sigma|^s + |a|^s}.$$

$$(2.26)$$

This is the conception of the following lemma, which is the fundamental point of [2].

Lemma 2.2. There exists constant $C_0 = C_0(s,k)$ such that for all $0 < s < \frac{1}{k+1}$, $k = 1, 2, \cdots$ and complex numbers $a_i, b_i, c_i, d_i, i = 1, 2 \cdots, k$ and v'

$$\int_0^1 \frac{dv}{|v-v'|^s} \left|\frac{a_i v + b_i}{c_i v + d_i}\right|^s \cdots \left|\frac{a_k v + b_k}{c_k v + d_k}\right|^s dv \leq$$

$$\leq C_0(s,k) \int_0^1 \left|\frac{a_i v_i + b_i}{c_i v + d_i}\right|^s \cdots \left|\frac{a_k v + b_k}{c_k v + d_k}\right|^s dv$$

$$(2.27)$$

Proof. Because both parts in (2.27) are the homogeneous functions of a_i, c_i, we need only consider the case $a_i = c_i = 1$, $i = 1, 2, \cdots k$. To prove (2.27), it's now sufficient to verify that

$$\int_0^1 \left|\frac{v + b_1}{v + c_1}\right|^s \cdots \left|\frac{v + b_k}{v + c_k}\right|^s dv \asymp \frac{(1 + |b_1|^s) \cdots (1 + |b_k|^s)}{(1 + |c_1|^s) \cdots (1 + |c_k|^s)}$$

$$\int_0^1 \frac{1}{|v - v'|_s} \left|\frac{v + b_1}{v + c_1}\right|^s \cdots \left|\frac{v + b_k}{v + c_k}\right|^s dv \asymp \frac{1}{1 + |v'|^s} \cdot \frac{(1 + |b_1|^s) \cdots (1 + |b_k|^s)}{(1 + |c_1|^s) \cdots (1 + |c_k|^s)}$$

(2.28)

As usually, $F \asymp G$ means, that

$$c_1 \leq \frac{F}{G} \leq c_2. \tag{2.28'}$$

In our case (2.28), the constants in these "rough" equalities must depend only on s and k.

Let's check, for example, the first part of (2.28). We can divide the complex parameters $b_i, c_i, i = 1, 2, \cdots k$ into the following groups:

$$b_i(\text{ or } c_i) \in \quad G_1 \quad iff \quad |b_i| \leq 2$$

$$b_i \in \quad\quad\quad G_2 \quad iff \quad |b_i| \geq 2$$

If $b_i \in G_2$, in the integrand of (2.28) we have:

$$|v + b_i|^s \asymp |b_i|^s \asymp 1 + |b_i|^s \tag{2.29}$$

This means that it is sufficient just to verify the following inequality

$$\int_0^1 \frac{|v + b_1|^s \cdots |v + b_{i_1}|^s}{|v + c_1|^s \cdots |v + c_{i_2}|^s} dv \asymp \frac{(1 + |b_1|^s) \cdots (1 + |b_{i_1}|^s)}{(1 + |c_1|^s) \cdots (1 + |c_{i_2}|^s)}$$

under condition $0 \leq i_1 \leq k$, $0 \leq i_2 \leq k$, $|b_i| \leq 2$, $|c_i| \leq 2$, i.e. $b_i, c_i \in G_1$. But in this case the left part is a positive continuous function of all parameters, because for $s < \frac{1}{k + 1} < \frac{1}{k}$ we have the uniform integrability in the left part.

Since the parameters have values in a compact set, both parts are roughly equivalent to const, and we proved (2.28).

In the special case $k = 1$, $s < \frac{1}{2}$ we the have inequality

$$\int_0^1 \frac{dv}{|v - v'|^s} \left|\frac{av + b}{cv + d}\right|^s dv \leq c(s) \int_0^1 \left|\frac{av + b}{cv + d}\right|^s dv, \tag{2.30}$$

or

$$E \frac{1}{|V - v'|^s} \left|\frac{aV + b}{cV + d}\right| \leq c(s) E \left|\frac{aV + b}{cV + d}\right|^s, \tag{2.30'}$$

where V is a r.v. uniformly distributed on $[0, 1]$.

Now everything is ready for the estimation of the moments for the Green function. For complex E $(ImE \neq 0)$, consider the resolvent kernel $R_E(x,y)$ for the operator (2.22) as a function of x for fixed $y \in \Gamma$. If $x \neq y$, we can write down

$$\triangle R_E(x,y) + (V_0(x) + \sigma V(x,\omega) - E)R_E(x,y) = 0$$

or

$$R_E(x,y) = \frac{1}{(E - V_0(x) + \sigma V(x,\omega))} \sum_{x':d(x',x)=1} R_E(x',x).$$

Using (2.23), we have

$$\langle |R_E(x,y)|^s \rangle \leq \sum_{x':d(x',y)=1} \left\langle \frac{|R_E(x',y)|^s}{(|E - V_0(x) - \sigma V(x,\omega)|^s} \right\rangle.$$

We can apply the rank-one perturbation formula (2.16) and after trivial transformations obtain

$$\langle |R_E(x,y)|^s \rangle \leq \sum_{x':d(x,y)=1} \frac{1}{|\sigma|^s} \left\langle \frac{\left| \dfrac{\alpha(x')V(x) + \beta(x')}{\gamma(x')V(x) + \delta(x')} \right|^s}{|V(x) - v(x')|^s} \right\rangle.$$

Here $\alpha, \beta, \gamma, \delta, v$ are very complicated r.v, depending on x' and the realization of the potential V outside point x, i.e. they are **independent** from $V(x,\omega)$. This is the situation of Lemma 2.2 in the special case (2.30'). From (2.30') it follows that (for $s < \frac{1}{2}$)

$$\langle |R_E(x,y)|^s \rangle \leq \frac{C_0(s)}{|\sigma|^s} \sum_{d(x',x)=1} \langle |R_E(x',y)|^s \rangle. \tag{2.31}$$

We have, in addition,

$$\langle |R_E(y,y)|^s \rangle \leq \frac{C_1(s)}{\sigma^s} \qquad \text{(Lemma 2.1)}$$

and

$$\langle |R_E(x,y)|^s \rangle \leq \frac{1}{|ImE|^s}$$

(because $\|R_E(\cdot,\cdot)\| \leq \frac{1}{|ImE|}$).

Let's put $\langle |R_E(x,y)|^s \rangle = h(x)$. For the function $h(x)$, $y \neq x$ we have an inequality of "superharmonicity":

$$h(x) \leq \varepsilon \triangle h(x), \quad \varepsilon = \frac{c_0(s)}{|\sigma|^s} \tag{2.32}$$

and additional restrictions

$$h(y) \leq \frac{c_1(s)}{|\sigma|^s}, \quad 0 \leq h(x) \leq \frac{1}{|ImE|^s}. \tag{2.33}$$

Lemma 2.3. Every solution of the system of inequalities (2.32), (2.33) admits for $\varepsilon N < 1$ the estimation

$$\langle |R_E(x,y)|^s \rangle = h(x) \leq \frac{c_2(s,N)}{|\sigma|^s}(N\varepsilon)^{d(x,y)}. \tag{2.34}$$

Proof In the beginning consider the function $h(x)$ in the "ball" $B_R\{x : d(x,y) \leq R\}$ without its center y. Iteration of the inequality (2.32) gives us a "path expansion" (compare (2.13) or probabilistic representation for the λ-Green function of the random walk):

$$h(x) \leq h(y) \sum_{\gamma : x \to y} \varepsilon^{|\gamma|} + \sum_{x' \in \partial B_R} h(x') \sum_{\gamma : x \to x'} \varepsilon^{|\gamma|}. \tag{2.35}$$

But $\#\{\gamma : x \to y, \quad |\gamma| = n\} \leq N^n$, i.e.

$$\sum_{\gamma : x \to x_0} \varepsilon^{|\gamma|} \leq \sum_{k=d(x,x_0)} \varepsilon^k \cdot N^k \leq \frac{(\varepsilon N)^{d(x,y)}}{1 - \varepsilon N} \tag{2.36}$$

(if, of course, $\varepsilon N < 1$). For the second term in (2.35), we have

$$\sum_{x' \in \partial B_R} h(x') \sum_{\gamma : x \to x'} \varepsilon^{|\gamma|} \leq$$

$$\leq \sum_{x' \in \partial B_R} h(x') \frac{(\varepsilon N)^{d(x,x')}}{1 - \varepsilon N} \leq$$

$$\leq \sup_{x \in \partial B_R} h(x') \sum_{k=R-d(x,y)}^{\infty} \varepsilon^k \#\{\gamma : x \to \partial B_R, \quad |\gamma| = k\} \tag{2.36'}$$

$$\leq \sup_{x \in \partial B_R} h(x') \sum_{k=R-d(x,y)}^{\infty} (\varepsilon N)^k \leq C(\varepsilon N, ImE) \cdot (\varepsilon N)^R (\varepsilon N)^{-d(x,y)}.$$

Combination of (2,35), (2.36) and (2.36') gives in the limit $R \to \infty$ the statement (2.34).

Of course, if we have the graph where the volume of the "ball" B_R is only a power function of R (for example, lattice $\mathbf{Z}^d$) in Lemma 2.3 instead of condition $\varepsilon N < 1$ it's enough to assume weaker conditions.

We can finish now the proof of Theorem 2.1. Consider the inequality (2.24) and rewrite it, using the result of Lemma 2.3

$$\sum_{x \in \Gamma} \langle |R_E(x,y)|^{2s} \rangle e^{s\gamma d(x,y)} \leq$$

$$\leq c_1(s) \sum_{k=0}^{\infty} \left(\frac{c_0(s)N}{|\sigma|^s} \right)^k e^{s\gamma k}.$$

This series is finite if

$$\frac{c_0(s)N}{|\sigma|^s}e^{s\gamma} < 1, \quad 0 < s < \frac{1}{2}. \tag{2.37}$$

If, to say, $s = 1/3$, for given N we can find $\sigma_0 = \sigma_0(s, N)$ to satisfy (2.37) for $|\sigma| > \sigma_0$. Theorem 2.1 is proved.

Our goal now is to generalize Theorem 2.1 for the wide class of distributions. The key elements of the proof were the Lemmas 2.1 and 2.2. Integration over $[0, 1]$ in these Lemmas was connected with the fact that r.v. $V(x, \omega)$ has a uniform distribution on $[0, 1]$. Let's introduce a more general class of distributions $\mathcal{K}_0 = \mathcal{K}_0(A, a, H, h)$.

By definition, a density $p(v)$ is an element of this class, if

1.

$$p(v) \leq H I_{[0,A]}(v). \tag{2.38'}$$

2. There exists interval $[x_0, x_0 + a] \subset [0, A]$ of the length a and

$$p(v) \geq h I_{[x_0, x_0 + a]}(v). \tag{2.38''}$$

We can now formulate Lemma 2.2 (or its especially important consequence (2.30')) in the following form.

Lemma 2.4. If r.v $V(\cdot, \omega)$ has a density from the class $\mathcal{K}_0(A, a, H, h)$, then for $s < \frac{1}{2}$ and arbitrary complex numbers $v; \alpha, \beta, \gamma, \delta$

$$\left\langle \frac{1}{|\sigma V - v'|^s} \cdot \left|\frac{\alpha V + \beta}{\gamma V + \delta}\right|^s \right\rangle \leq c(s, A, a, \frac{H}{h})\frac{1}{|\sigma|^s} \cdot \left\langle \left|\frac{\alpha V + \beta}{\gamma V + \delta}\right|^s \right\rangle. \tag{2.38}$$

Only const is different; it is essential that this const includes not H, h but only their ratio. The proof is trivial. To obtain the upper estimation of expectation, we will replace $p(v)$ in the integral representation of $\langle \cdots \rangle$ by $H \cdot I_{[0,A]}(v)$, for the lower estimation we will use (2.38').

In fact, the same proof works in essentially more general situations.

We will call $p(v) \in \mathcal{K}_1(A, a, \rho)$, $0 < \rho < 1$, if it is possible to cover line R^1 by the intervals Δ_i, $i = 0, \pm 1, \cdots$ of the length A and to find inside every Δ_i the subinterval $\Delta_i' = [x_i', x_i' + a]$ of the length a in such a way, that

$$\sup_{v \in \Delta_i} p(v) = H_i, \quad \inf_{v \in \Delta_i'} p(v) = h_i \text{ and } \frac{h_i}{H_i} \geq \rho.$$

Class $\mathcal{K}_1$ is very wide. It includes all densities which are monotone at infinity and many others. Densities $p(v) \notin \mathcal{K}_1$ exist, but are pathological, such a density must include infinite many more and more thin "bumps", moreover the distances between these "bumps" are to became longer and longer. The proof of the statement (2.38) and of more general statements, similar to Lemmas 2.1 and 2.2, are the same as in Lemma 2.4.

As a result, we have proved the following generalization of the Theorem 2.1.

Theorem 2.1'. Suppose that the Schrödinger operator on the graph Γ with bounded index of branching (bound equal to N) has form

$$H = H(\omega) = \Delta + V_0(x) + \sigma V(x, \omega)$$

where $V_0(x)$ is an arbitrary non-random function and r.v. $V(x, \omega)$ for different $x \in \Gamma$ are independent, have a.c. distributions and corresponding densities are elements of the fixed class $\mathcal{K}_1(A, a, \rho)$ (but may be different!).

Then, we can find the constant $\sigma_0 = \sigma_0(N, A, a, \rho)$, such that for all $|\sigma| > \sigma_0$ the operator $H(\omega)$ has $\mu - a.s$ the property of the exponential localization.

Very important feature of the Theorem 2.1' (or 2.1) is the absence of the conditions on the function V_0 (background potential). It can be for example, the realization of arbitrary random field $\xi(x)$, $x \in \Gamma$, independent of V. Fubini argument shows that the result of Theorem 2.1' holds with the probability 1 (in the product space $\Omega_1 \times \Omega_m$) for the potential of the form:

$$\Delta + v_0(x) + \xi(x, \omega_1) + \sigma v(x, \omega),$$

where $v_0(x)$ is an arbitrary non-random function, $\xi(x, \omega_1)$ is an arbitrary random field on the space $(\Omega_m, \mathcal{F}_m, \mu)V$, is independent of $\xi(\cdot)$ and has the properties mentioned in Theorem 2.1' (class $\mathcal{K}_1(\cdots)$ etc.).

Example. Consider the operator

$$H = \Delta + \sigma V(x, \omega)$$

on $\mathbf{Z}^d$, $d \geq 1$ with homogeneous gaussian potential. If $\langle V \rangle = 0$, and $\Gamma(x_1 - x_2) = \langle V(x_1)V(x_2) \rangle$ is the corresponding correlator, then it can be represented in terms of spectral measure $F(d\lambda)$ on the torus $S^d = [-\pi, \pi]^d$:

$$\Gamma(x) = \frac{1}{(2\pi)^d} \int\limits_{S^d} \exp(i(x, \lambda)) F(d\lambda). \tag{2.40}$$

If $F(d\lambda)$ dominates Lebesgue measure $\sigma^2 d\lambda$ on S^d, then the field $V(x, \omega)$ can be decomposed:

$$V(x, \omega) = \sigma \eta(x, \omega) + V_1(x, \omega) \tag{2.41}$$

where $\eta(x,\omega)$ are $i.i.d.$ $N(0,1)r.v$ and $V_1(x,\omega)$ is homogeneous; as a result, we proved exponential localization for large disorder in the class of homogeneous gaussian potentials on $\mathbf{Z}^d$, for which

$$F(d\lambda) \geq \text{const } d\lambda. \tag{2.41'}$$

Because the spectral measure of the field V_1 can be "very bad", the correlator $\Gamma(x_1-x_2)$ may not tend to 0, the process $V(x,\omega)$ may be even non-ergodic etc.

Now we'll develop Theorem 2.1' in two directions. First of all, we'll replace the Laplacian $\triangle$ by a very general non-local operator. Second, we'll consider the case when, roughly speaking, the coupling constant σ can be "small" for a finite number of points $x \in \Gamma$ and "large" for all others.

Theorem 2.2. Consider the Hamiltonian of the form

$$H = T + v_0(x) + \sigma v(x,\omega) \tag{2.42}$$

where $T(x,y) = \bar{T}(y,x)$ is a selfadjoint bounded operator in $L_2(\Gamma)$. In addition, we'll suppose (without loss of generality)

1. $T(x,x) = 0, \quad x \in \Gamma. \tag{2.43}$

Diagonal terms of T can be included in background potential $V_0(x)$.

2. For some $s < 1$ and all $x,y \in \Gamma$

$$|T(x,y)|^s \leq h(d(x,y))$$
$$\text{and } \sum_{k=1}^{\infty} h(k) \cdot \#\{x : d(x,y) = k\} < \infty. \tag{2.43'}$$

If for different $x \in \Gamma$ random variables $v(x,\omega)$ are independent and have (maybe different) densities $p_x(v)$ from the fixed class $\mathcal{K}_1(A,a,\rho)$, then (as earlier) we can find $\sigma_0(s,T,A,a,\rho)$ such that $\mu - a.s$ $\Sigma(H(\omega)) = \Sigma_{pp}(H(\omega))$ for all $|\sigma| > \sigma_0$.

The only difference in comparison with the previous localization theorems is that generally speaking, we have no exponential localization, even for $\sigma \to \infty$, if the matrix elements are decreasing slowly. Typical examples of the application of the Theorem 2.2 are related with the fractional powers of Laplacian $\triangle$ on $\mathbf{Z}^d$. More precisely, we can consider

$$T_\alpha = -(2d + \triangle)^\alpha, \quad 0 < \alpha < 1, \tag{2.44}$$

Matrix elements of T_α have power asymptotics at the infinty:

$$T_\alpha(x,y) \sim \frac{c_\alpha}{|x-y|^{d+2\alpha}}. \tag{2.44'}$$

Another important applicaiton (see later) is connected with the operators in half-space with random boundary conditions.

Proof of Theorem 2.2. We will use the same idea as earlier, but because $s < 1$ ($s < 1/2$ in the proofs of the theorems 2.1, 2.1'!) calculations will be just slightly different.

The resolvent kernel $R_E(x, y)$, $x \neq y$ and $Im E \neq 0$, is a bounded solution of the equation

$$\sum_{z \neq x} T(x, z) R_E(z, y) = R_E(x, y)(E - \sigma v(x, \omega) - v_0(x))$$

and for $s < 1$ given in condition (2.43')

$$|R_E(x, y)|^s \quad |E - \sigma v(x, \omega) - v_0(x)|^s \leq$$

$$\leq \sum_{z \neq x} h(d(z, x))|R_E(z, y)|^s. \tag{2.45}$$

Using, as earlier, (2.15') we can represent the expectation of the left part as

$$\left\langle \left| \frac{\alpha}{v(x) + \beta} \right|^s \cdot |\sigma v(x) + \delta|^s \right\rangle \tag{2.46}$$

where α, β, δ are the functions, independent on $V(x, \omega)$. Using the method of Lemma 2.2 we can check that

$$\left\langle \left| \frac{\alpha}{v(x) + \beta} \right|^s \quad |\sigma v(x) + \delta|^s \right\rangle \asymp$$

$$\asymp |\sigma|^s \frac{(1 + |\alpha|^s)(1 + |\delta|^s)}{(1 + |\beta|^s)}, \quad \text{i.e.}$$

for some const c_1, depending only on the parameters of the class $\mathcal{K}_1$ and s

$$\left\langle \left| \frac{\alpha}{v(x) + \beta} \right|^s \cdot |\sigma v(x) + \delta|^s \right\rangle \geq$$

$$\geq c_1 \left\langle \left| \frac{\alpha}{v(x) + \beta} \right| \right\rangle |\sigma|^s. \tag{2.47}$$

From (2.75), (2.47) it follows that for $H(x) = \langle |R_E(x, y|^s \rangle$ and $c_2 = \dfrac{1}{c_1}$

$$\frac{c_2}{|\sigma|^s} \sum_{z \neq x} h(d(z, x)) H(z) \geq \quad H(x). \tag{2.48}$$

Because $\sum_{k=1}^{\infty} h(k) \cdot \#\{z : d(x, z) = k\} < \infty$, we can find σ_0 is such a way that

$$\frac{c_1 \sum_{k=1}^{\infty} h(k) \cdot \#\{z : d(x, z) = k\}}{|\sigma|^s} = \varepsilon < 1. \tag{2.49}$$

Let's put $p(x,y) = \dfrac{h(d(x,y))}{\sum\limits_{k} h(k) \cdot \#\{y : d(x,y) = k\}}$. Of course, $p = \{p(x,y)\}$ is a stochastic matrix on Γ and basic inequality (2.48) has a form

$$H(x) \leq \varepsilon(PH)(x), \quad x \neq y. \tag{2.48'}$$

Let's remember that $H(y) = \langle |R_E(y,y)|^s \rangle$ is uniformly bounded in E (Lemma 2.1, which, of course, holds for arbitrary operators H).

Now we have to prove the analogue of Lemma 2.3 in more general situations.

Lemma 2.3'. For fixed $y \in \Gamma$ the bounded solution $H(x)$ of the inequality (2.48') is square-integrable uniformly in E, $\quad Im\, E \neq 0$.

Of course, the proof of Theorem 2.2 is an immediate consequence of Lemma 2.3'.

Proof of Lemma 2.3'. We will be using the language of Markov chain theory. Let $x(t)$ be a Markov chain on Γ with transition matrix P and let τ_y be the first entry time to the state y (maybe $p_x\{\tau_y = \infty\} > 0$). If we consider from the beginning the inequality (2.48') in the ball B_R and use the probabilistic representation for the (bounded for $Im\, E \neq 0$) solution of the equality

$$\tilde{H}(x) = \varepsilon(PH)(x), \quad x \in B_R(y)$$

$$\tilde{H}(y) = H(y), \quad \tilde{H}(z) = H(z), \quad z \notin B_R \tag{2.48''}$$

after letting $R \to \infty$ we'll obtain as earlier the inequality (uniform in E!):

$$H(x) \leq H(y) E_x(\varepsilon^{\tau_y}) = \tilde{H}_y(x) =$$

$$= H(y) \sum_{k=1}^{\infty} \varepsilon^k f(k,x,y) \leq c_0 \sum_{k=1}^{\infty} \varepsilon^k f(k,x,y)$$

where $f(k,x,y) = P_x\{\tau_y = k\}$. But

$$f(k,x,y) \leq P(k,x,y)$$

i.e.

$$\left\| \sum_{k=1}^{\infty} \varepsilon^k f(k,x,\cdot) \right\|_{L_2} \leq$$

$$\leq \sum_{k=1}^{\infty} \varepsilon^k \| p(k,x,\cdot) \|_{L_2} \leq \sum_{k=1}^{\infty} \varepsilon^k < \infty.$$

(We have used the following obvious fact: $\sum_x p^2(k,x,y) = \sum_x p^2(k,y,x) \leq \sum_x p(k,y,x) = 1$).

Remark. If in the condition (2.43') of Theorem 2.2 we have exponential convergence

$$\sum_{k=1}^{\infty} h(k) \cdot \#\{y : d(x,y) = k\} e^{\gamma k} < \infty, \quad \gamma > 0$$

then the resolvent kernel will be $\mu - a.s$ square-integrable with exponential weight and under the other conditions of Theorem 2.2 we have not just localization, but an exponential localization.

In the case $\Gamma = \mathbf{Z}^d$, it means that for large disorder we have an **exponential localization** for the random operator (2.42) if

$$|T(x,y)| \leq \text{const } \exp(-\gamma'|x-y|), \quad \gamma' > 0.$$

Some details, additional discussions and generalizations of the result from this remark can be found in [2].

Now we'll consider one of important applications of "non-local" localization theorem 2.2 to the Schrödinger operator with the random boundary condition. For the sake of simplicity, we will be analyzing the models with homogeneous boundary conditions.. Such models appear naturally in the solid state physics, when considering disordered surface effects.

Description of the model. Consider the Laplacian Δ (without potential) in the upper half space $\mathbf{Z}_+^D = \mathbf{Z}^{d-1} \times [-1, \infty)$ with the boundary condition

$$\psi(x, -1) = \sigma v(x, \omega)\psi(x, 0), x \in \mathbf{Z}^{d-1}. \qquad (2.50)$$

Hamiltonian H can be defined alternatively by the formula

$$H\psi(x,z) = \Delta\psi(x,z), \quad z \geq 1$$

$$H\psi(x,0) = \psi(x,1) + \sum_{|x'-x|=1} \psi(x',0) + \sigma v(x,\omega)\psi(x,0). \qquad (2.51)$$

Suppose, that $v(x,\omega)$ are $i.i.d.r.v$ with common density $p(v)$ of the class $\mathcal{K}_1(\cdots)$. Our goal is to study the spectral problem

$$H\psi = E\psi \quad \text{in} \quad L_2(\mathbf{Z}_+^d).$$

First of all, we will find the spectrum $\sum(H)$, which $\mu - a.s$ does not depend of ω. Using Fourier transform in x we can change the operator H to an isometric operator $\hat{H}$ in the Hilbert space $L_2(S^{d-1}) \times L_2(\mathbf{Z}_+^1)$

$$\hat{H}\hat{\psi}(\phi,z) = \hat{\psi}(\phi,z+1) + \hat{\psi}(\phi,z-1) + \Phi(\phi)\hat{\psi}(\phi,z)$$

$$\hat{H}\hat{\psi}(\phi,0) = \hat{\psi}(\phi,1) + \Phi(\phi)\hat{\psi}(\phi,0) + \sigma(V\psi)(\phi,0). \qquad (2.52)$$

Here $\Phi(\phi) = 2\sum_{i=1}^{d-1} \cos\phi_i$, $|\Phi(\phi)| \leq 2d-2$ is a symbol of the Laplacian (with respect to x) and $\hat{}$ is a sign of the Fourier transform.

The general solution of the equation $\hat{H}\hat{\psi} = E\hat{\psi}$ (without boundary condition) has form

$$\hat{\psi}(\phi, z) = c_1(\phi)\lambda_1^z(\phi, E) + c_2(\phi)\lambda_2^z(\phi, E),$$

where λ_1, λ_2 are the roots of the characteristic equation

$$\lambda + \lambda^{-1} + \Phi(\phi) = E. \tag{2.53}$$

If $|\Phi(\phi) - E| \leq 2$, this equation has two roots of the form $\lambda_{1,2} = \exp(\pm\theta)$, $\theta \in R^1$ and corresponding bounded solutions correspond to planar waves, i.e. generalized eigenfunctions of the Laplacian. Such E, of course, are the elements of the spectrum of $H(\omega)$.

Inequality $|\Phi(\phi) - E| \leq 2$ has solutions if $E \in [-2d, 2d] = \sum(\Delta)$, i.e. $[-2d, 2d] \in \sum(H)$ Physically it is obvious, because boundary effects are not important when the "electron" is sufficiently far from the boundary. However equation (2.53) may have the real roots λ_1, λ_2, $\lambda_1 \cdot \lambda_2 = 1$ if $|\Phi(\phi) - E| > 2$. Just the root λ_1, $|\lambda_1| < 1$, has a special sense (the generalized eigenfunctions can not increase exponentially!).

It means that $E \in \sum(H)$ iff $\hat{\psi}(\phi, z) = c_1(\phi)\lambda_1^z(\phi, E)$, $c_1(\phi) \in L_2(S^{d-1})$ and λ_1 is the smaller root of the equation

$$\lambda + \frac{1}{\lambda} + (\Phi(\phi) - E) = 0, \quad |\Phi(\phi) - E| > 2.$$

We have to note that function $c_1(\phi)$ must be equal to 0 on the set $\{\phi : |\Phi(\phi) - E| \leq 2\}$.

The second part of the equation (2.52) gives additional information about $c_1(\phi)$:

$$c_1(\phi)\lambda_1(\phi, E) + (\Phi(\phi) - E)c_1(\phi) + \sigma v\hat{\psi}(\phi, 0) = 0$$
$$\tag{2.52'}$$
$$c_1(\phi) = \hat{\psi}(\phi, 0).$$

If $\sigma v(x) = a = $ const, equation (2.52') has the trivial form

$$c_1(\phi)[\lambda_1(\phi, E) + (\Phi(\phi) - E) + a] = 0.$$

If for some $\phi_0 \in S^{d-1}$

$$\lambda_1(\phi_0, E) + (\Phi(\phi_0) - E) + a = 0$$

then equation (2.52') has solution

$$c_1(\phi) = \delta_{\phi_0}$$

corresponding to the generalized eigenfunction $\psi(x, z)$ which is const in x and exponentially decreasing in z. As easy to see it means that all $E \in [-2d+1+a, 2d-1+a] \in \sum(H_a)$.

But in the realization of the random field $V(x, \omega)$, $x \in \mathbb{Z}^{d-1}$ we can find (for arbitrary $a \in \sigma$ supp $p(\cdot)$ and arbitrary $\varepsilon > 0$) a sequence of balls $B_{R_n}(x_n)$, $R_n \to \infty$, $x_n \to \infty$ such that

$$|\sigma v(x, \omega) - a| \leq \varepsilon, \quad x \in B_{R_n}(x_n).$$

As earlier, this gives (compare with Corollary 2.1 in the lecture 5) that $\mu - a.s$

$$\sum(H(\omega)) \supseteq \{[-2d + 1, 2d - 1] \oplus \sigma \text{ supp } p(\cdot)\} \cup [-2d, 2d].$$

The reader must check that for all other values of E the equation $H\psi = E\psi$ only has an exponentially increasing solution and we have proved.

Proposition 2.7. Operator $H(\omega)$ given by expression (2.51) has $\mu - a.s$ the spectrum

$$\sum(H(\omega)) = [-2d, 2d] \cup \{-2d + 1, 2d - 1] \oplus \sigma \text{ supp } p(v)\}. \tag{2.54}$$

If supp $p(v) = R^1$ (a typical example is the gaussian $N(0, 1)$ distribution), then $\sum(H(\omega)) = R^1$.

We will now prove the following theorem, which gives some information about p.p. spectrum of the operator (2.51) with random boundary conditions. Possibly it is the first physically interesting and natural random Hamiltonian where we can guarantee an existence of the continuous spectrum for small coupling const σ.

Theorem 2.3. 1.　For arbitrary $\delta > 0$, we can fine $\sigma_0 = \sigma_0(\delta, d, \mathcal{K}_1)$ such that $\mu - a.s$ for $|\sigma| > \sigma_0$
$$\sum(H(\omega)) \cap \overline{[-2d - \delta, 2d + \delta]} = \sum_{pp}(H(\omega)) \cap \overline{[-2d - \delta, 2d + \delta]}. \tag{2.55}$$
The corresponding eigenfunctions are decreasing exponentially.

2.　For fixed σ it's possible to find $E_0 = E_0(d, \sigma, \mathcal{K}_1)$ such that
$$\sum(H(\omega)) \cap \overline{[-E_0, E_0]} = \sum_{pp}(H(\omega)) \cap \overline{[-E_0, E_0]}. \tag{2.56}$$

3.　Suppose that r.v $v(x, \omega)$ are uniformly bounded, i.e. supp $p(\cdot) \in [-a, a]$. Then, for arbitrary $\delta > 0$, we can find $\sigma_0 = \sigma_0(d, a)$, such that for $|\sigma| < \sigma_0$
$$\sum_{pp}(H(\omega)) \cap [-2d + \delta, 2d - \delta] = \phi. \tag{2.57}$$

Proof. If $|E| > 2d$, the previous considerations show that the solution of the equation $\Delta \psi = E\psi$, $z \geq 1$ can be expressed in the terms of $\psi(x, 0)$. Namely,

$$\hat{\psi}(\phi, z) = c_1(\phi)\lambda_1^z(\phi, E) = \hat{\psi}(\phi, 0)\lambda_1^z(\phi, E) \tag{2.58}$$

and $\lambda_1(\phi, E)$ is an analytical function on the torus S^{d-1}. This means that

$$\psi(x, z) = \sum_{y \in \mathbb{Z}^{d-1}} D_E(x - y, z)\psi(y, 0) \tag{2.59}$$

and

$$\hat{D}_E(\phi, z) = \lambda_1^z(\phi, E).$$

The Dirichlet kernel $D_E(x - y, z)$ is decreasing exponentially (because $\lambda_1(\phi, E)$ is analytic) and corresponding exponent $\gamma(E)$ becomes larger and larger if $|E|$ increases. Equation (2.51) for eigenfunctions of H (with the condition $|E| > 2d$) has now the form

$$E\psi(x, 0) = H\psi(x, 0) = \psi(x, 1) + \triangle_{d-1}\psi(x, 0) + \sigma v(x, \omega)\psi(x, 0)$$

$$= \sum_{y \in \mathbf{Z}^{d-1}} D_E(x - y, 1)\psi(\phi, 0) + \triangle_{d-1}\psi(x, 0) + \sigma v(x, \omega)\psi(x, 0)$$

i.e.

$$E\psi = T\psi(x, 0) = D_E(\cdot, 1)\psi(x, 0) + \triangle_{d-1}\psi(x, 0) + \sigma v(x, \omega)\psi(x, 0). \tag{2.60}$$

This is a $(d - 1)$-dimensional equation for eigenfunctions with exponentially decreasing kernel $D_E(x - y, 1)$. The only difference from Theorem 2.2 is that the kernel $D_E(x - y, 1)$ depends on E. But for fixed $E : |E| > 2d$, we can prove, using the same arguments, that the resolvent $(T - E)^{-1}$, $|E| > 2d + \delta$ has, for $|\sigma| > \sigma_0(\delta)$, an exponentially decreasing symbol (on the lattice $\mathbf{Z}^{d-1}$!). Then formula (2.58) will show that the same exponential decreasing occurs for the resolvent $R_E(x - y, z)$ of the initial operator H (on $\mathbf{Z}_d^+$).

Now we have to use Simon-Wolff's theorem, however not in the simplest form mentioned in the previous lecture, but in the original form [69]. The reason is that we have now the random potential only on the boundary $\partial \mathbf{Z}^{d-1}$.

From [69] it follows that we have $\mu - a.s\ p.p$ spectrum in some energy interval $\triangle$ if, first of all, for every $E \in \triangle$ the resolvent kernel $R_E(x, y)$, $(x, y \in \triangle)$ is square-integrable for $x \in X$, $\mu - a.s$ and, second, the direct sum of the cyclic supspaces generated by H and δ_x, $x \in X$ is equal to $L^2(\Gamma)$.

In our case $X = \partial \mathbf{Z}_+^d$ and solvability of the Dirichlet problem in $L^2(\mathbf{Z}_+^d)$ of the (for $|E| > 2d$) guarantees the second condition. This means that the first statement of Theorem 2.3 is proved.

The proof of the second part of this theorem is the same. We just have to notice that for $E \to \infty$

$$\sum_{x \in \mathbf{Z}^{d-1}} |D_E(x, 1)|^s \longrightarrow 0$$

for every fixed $s < 1$. After this we can apply the arguments of the theorem for fixed σ and sufficiently large energy E.

Let's prove the last (and most interesting) statement. From (2.50) and (2.58) it follows that

$$\hat{\psi}(\phi,0)\lambda_1^{-1}(\phi,E) = \overbrace{(\sigma v(\cdot,\omega)\hat{\psi}(\cdot,0)}(\phi)$$

(if $\psi(x,z)$ is an eigenfunction with the eigenvalue $E \in [-2d, 2d]$), or

$$\hat{\psi}(\phi,0)\lambda_2(\phi,E) = \sigma\overbrace{(v\hat{\psi}})(\phi). \qquad (2.61)$$

Let's remember, that $\lambda_2(\phi,E) = 0$ if $|\Phi(\phi) - E| \leq 2$, i.e. the support of $\lambda_2(\phi,E)$ is given by the inequality $|\Phi(\phi) - E| = \left|2\sum_{i=1}^{d-1}\cos\phi_i - E\right| > 2$.

If $E \in [-2d + \delta, 2d - \delta]$ and $\psi_E(x,z) \in L^2(\mathbf{Z}^d)$ we have, under the normalization condition

$$\sum_{x\in\mathbf{Z}^{d-1}}|\psi(x,0)|^2 = \int_{S^{d-1}}|\hat{\psi}(\phi,0)|^2 dp = \int_{S^{d-1}\cap\{|\Phi(\phi)-d|>2\}}|\hat{\psi}(\phi,0)|^2,$$

the following inequalities:

$$1 = \int_{S^{d-1}\cap\{\Phi(\phi)-E|>2\}}|\hat{\psi}(\phi,0)|^2 \leq \int_{S^{d-1}\cap\{|\Phi(\phi)-E|>2\}}|\hat{\psi}(\phi,0)\lambda_2(\phi,E|^2 d\phi \leq |\sigma|\sup|\xi|.$$

If $|\sigma|$ is sufficiently small, we have a contradiction, which proves part 3 of Theorem 2.3.

In the final part of this lecture, I'll give a proof of the localization theorem for random potentials, increasing in probability for $|x| \to \infty$. This subject is closely related with the paper [30], where we analyzed the spectral properties of the operator

$$H = \triangle + \sigma(1 + |x|^\alpha)\xi(x) \qquad (2.61')$$

for *i.i.d.r.v.* $\xi(\cdot)$, $x \in \mathbf{Z}^d$. (The main goal of [30] was not the theorem on the p.p. spectrum, but phenomena of bifurcation of the essential spectrum of (2.61) with respect to the parameter α).

Theorem 2.4. Let $H = \triangle + \sigma D(x)v(x,\omega)$ $x \in \Gamma$ be a random Schrödinger operator on $L^2(\Gamma)$ with the random potential $\tilde{v}(x,\omega) = D(x)v(x,\omega)$, where $v(x,\omega)$ are *i.i.d.r.v* with common density of the class $\mathbf{K}_1$ (as in Theorems 2.1' and 2.2) and amplitude $D(x)$, $D(x) \neq 0$ is an increasing function: $\lim_{x\to\infty} D(x) = \infty$.

Then, for arbitrary σ, with μ-probability 1 the operator H has p.p. spectrum and corresponding eigen function tend to 0 superexponentially.

The proof is simple, modulo the following lemma of Wegner's type, which is a deep and nontrivial generalization of the Lemma 2.1. My proof of this lemma is slightly different from our paper with M. Aizenman [2], but the principal idea is the same.

Lemma 2.5. Let $H = \Delta + \sigma v(x, \omega) + v_0(x)$ and the r.v $v(x, \omega)$ are independent and have uniformly bounded densities of the distributions $p_x(v) : \|p_x(\cdot)\|_\infty \le c < \infty$. Then, for all x, $y \in \Gamma$, $\sigma > 0$, $E \in R^1$ and $0 < s < 1$

$$\langle |R_E(x, y)|^s \rangle \le \frac{c_0(s) \cdot c}{|\sigma|^s}. \tag{2.62}$$

To begin let's prove Theorem 2.4 using Lemma 2.5.

For arbitrary σ_1 we can find a finite volume v_{σ_1}, such that

$$|\sigma D(x)| > \sigma_1, \quad x \notin v_{\sigma_1}.$$

We can estimate now the moment of the resolvent kernel by the following expression : if $x_0 \in v_{\sigma_1}$, then

$$\langle |R_E(x_0, x)|^s \rangle \le \sup_{y \in V_{\sigma_1}} \langle |R_E(x_0, y)|^s \rangle \exp\{-\varepsilon \min_{y \in V_{\sigma_1}} d(x, y)\}, \quad \varepsilon = \varepsilon(\sigma_1). \tag{2.63}$$

Using the fact that $\varepsilon(\sigma_1) \to 0$, $\sigma_1 \to \infty$ and Lemma 2.5, we can easily obtain both statements of Theorem 2.4.

Now we have to prove Lemma 2.5. In this case it is necessary to use rank-two perturbations.

For given $v(x)$, $x \in \Gamma$ consider the following "perturbed" potential

$$\tilde{v}(z) = \begin{cases} v(z), & z \neq x, \quad y \\ 0, & z = x, \quad z = y \end{cases}$$

(of course, we consider only the case, when $x \neq y$. If $x = y$, the situation is simpler and has been discussed earlier).

If $\tilde{R}(z, y) = (\Delta + \tilde{v} - E)^{-1}$ is the perturbation of the resolvent (for some fixed E, $Im E \neq 0$) then

$$R(z, y) = \tilde{R}(z, y) - v(x)R(x, y)\tilde{R}(z, x) -$$
$$-v(y)R(y, y)\tilde{R}(z, y). \tag{2.64}$$

Let's put $z = x$, $z = y$. We'll obtain the system of two equations for $R(x, y)$ and $R(y, y)$ in terms of $\tilde{R}$:

$$R(x, y)(1 + v(x)\tilde{R}(x, x)) + R(y, y)v(y)\tilde{R}(x, y) = \tilde{R}(x, y)$$
$$R(x, y)v(x)\tilde{R}(x, y) + R(y, y)(1 + v(y)\tilde{R}(y, y)) = \tilde{R}(y, y) \tag{2.65}$$

and elementary algebra gives

$$R(x, y) = \frac{\tilde{R}(x, y)}{(1 + v(x)\tilde{R}(x, x))(1 + v(y)\tilde{R}(y, y)) - v(x)v(y)\tilde{R}^2(x, y)}$$

or

$$R(x,y) = \frac{\dfrac{\tilde{R}(x,y)}{\Delta}}{\left(v(x) + \dfrac{\tilde{R}(x,x)}{\Delta}\right)\left(v(y) + \dfrac{\tilde{R}(y,y)}{\Delta}\right) - \left(\dfrac{\tilde{R}(x,y)}{\Delta}\right)^2}, \qquad (2.66)$$

where $\Delta = \tilde{R}(x,x)\tilde{R}(y,y) - \tilde{R}^2(x,y)$. Let's introduce notations $\xi_1 = v(x) + \dfrac{\tilde{R}(x,x)}{\Delta}$, $\xi_2 = v(y) + \dfrac{\tilde{R}(y,y)}{\Delta}$, $b = \dfrac{\tilde{R}(x,y)}{\Delta}$. Random variables $v(x)$, $v(y)$ and $\tilde{R}$ are independent and (under condition $\mathcal{F}_{\neq x, \neq y}$) r.v ξ_1, ξ_2 have densities bounded by $c/|\sigma|$, where $c = \sup_x \|p_x(\cdot)\|_\infty$. To prove Lemma 2.5 it's enough to check now that uniformly in b,

$$\left\langle \left|\frac{b}{\xi_1, \xi_2 - b^2}\right|^s \right\rangle \leq \frac{c_0(s)}{|\sigma|^s}.$$

Consider the following three events:

$$A_1: \ |\xi_1| < \frac{b}{2}, \quad |\xi_2| < \frac{b}{2}$$
$$A_2: \ |\xi_1| > \frac{b}{2}$$
$$A_3: \ |\xi_2| > \frac{b}{2}$$

Of course, $A_1 \cup A_2 \cup A_3 = \Omega_m$. But

$$\left\langle \left|\frac{b}{\xi_1, \xi_2 - b^2}\right|^s I_{A_1} \right\rangle \leq \frac{2}{b^s} P\{|\xi_1| < b, \ |\xi_2| < b\}$$

$$\leq \frac{2}{|b|^s} \cdot \left(\frac{2bc}{\sigma} \wedge 1\right)^2 \leq \frac{2}{|\sigma|^s} \max_{y \in R_+^d}\left(\frac{1}{ys} \cdot \min(2cy^2, 1)\right) \leq \frac{c_1(s)}{|\sigma|^s}$$

and, for example,

$$\left\langle \left|\frac{b}{\xi_1 \xi_2 - b^2}\right| I_{A_2} \right\rangle =$$

$$= \left\langle \left\langle \left|\frac{\dfrac{b}{\xi_1}}{\xi_2 - \dfrac{b^2}{\xi_1}}\right|^s I_{A_2} \,\Big|\, \xi_1 \right\rangle \right\rangle \leq$$

$$\leq \left\langle \frac{c_0(s)\left|\dfrac{b}{\xi_1}\right|^s}{\sigma^s} I_{A_2} \right\rangle \leq \frac{c_1(s)}{\sigma^s}.$$

At the last step we have used the rank-one perturbation formula (or a Lemma 2.1 type formula).

Remark. Lemma 2.5 means that generalized eigenfunction of the random Schrödinger operator are, roughly speaking, bounded in probability.

Lecture 7. One-dimensional localization.

In the previous lecture we had proved a very general form of localization theorem (non-local Laplacian, general graph Γ and so forth) but under essential additional assumption of the **large disorder**. In the situation of the small coupling constant σ, physicists expect an existence of the spectral bifurcations (so-called "mobility edge"). We'll be discussing this problem in the conclusion of the chapter.

One-dimensional case (lattice and continuous) plays a special role in the localization theory, because it admits special direct methods based on the phase formalism. The first mathematical results in this area were related with the random one-dimensional Schrödinger (or Shturm-Liouville) operators [28], [52].

In the case of homogeneous ergodic potentials the central achievement was a series [44] - [47] of S. Kotani articles, where he had discovered a deep connection between Ljapunov exponents $\gamma(E)$, the prediction properties of the random potential $V(x,\omega)$, $x \in R^1(\mathbf{Z}^1)$, absolute-continuous spectrum of H etc.

As a result, he proved localization theorem for arbitrary coupling constant and for a wide class of the "non-deterministic potentials". Probabilistic approach of the papers [44] - [47], however didn't clear up the central physical idea of one-dimensional localization: absence of resonances between quantum particle with a given admissible energy E and some (rich enough) family of the "blocks" of the potential $V(\cdot)$.

It's natural to attempt formulating the direct geometrical conditions of localization for the individual (may be, non-random) potential. The first and a very important step in this direction was made in the paper by Simon-Spencer [68]. Among many others, they proved the following result:

Theorem. If $H = \triangle + v(x)$, $x \in \mathbf{Z}^1$ is an one-dimensional lattice Hamiltonian and

$$\lim_{x \to +\infty} \sup |v(x)| = \lim_{x \to -\infty} \sup |v(x)| = \infty,$$

then

$\sum_{ac}(H) = \phi$, i.e. $\sum(H) = \sum_{sing}(H)$.

This theorem (and its generalizations, [68]) will be an instructing example for our future activity, but we will try substituting $\sum_{pp}$ instead of $\sum_{sing}$. Some of the results of this lecture (in the weaker form) were published in [40], but I'll use here the materials of my lectures [53].

Let's start with the following result.

Theorem 2.4. Consider on $L^2(\mathbf{Z}_+^1)$, $\mathbf{Z}_+^1 = (0, 1, \cdots)$ the Schrödinger operator

$$H^\theta = \triangle + v(x), \quad x \in \mathbf{Z}_+^1 \tag{2.67}$$

with boundary condition

$$\psi(-1)\cos\theta + \psi(0)\sin\theta = 0, \tag{2.67'}$$

depending on parameter $\theta \in [0, \pi]$ ("boundary phase").

Assume, that for some energy interval $I \subseteq R^1$ and almost all (in Lebegusgue sense) $E \in I$ there exists the sequence of "non-resonant blocks" $[x_n, y_n](I)$, $n = 1, 2, \cdots$, that is the sequence of the points $0 < x_1 \leq y_1 < x_2 \leq y_2 \cdots$, such that

$$|R_E^{(n)}(x_n, y_n)| \leq \delta_n, \quad \delta_n \to 0, \quad n \to \infty. \tag{2.68}$$

Here $R_E^{(n)} = (H^{(n)} - E)^{-1}$ be a resolvent of "block" operator

$$H^{(n)}\psi(x) = \triangle\psi(x) + v(x)\psi(x), \quad x_n \leq x \leq y_n,$$

$$\psi(x_n - 1) = \psi(y_n + 1) = 0. \tag{2.69}$$

Let for some nondecreasing sequence of the constants $A_n \geq 1$, $n = 1, 2, \cdots$ and const $c \geq 1$

$$\sum_n \sqrt{A_n}\delta_n < \infty, \quad y_{n+1} \leq A_1 \cdots A_n c^n. \tag{2.68'}$$

Then for a.e. $\theta \in [0, \pi]$

$$\Sigma(H^\theta) \cap I = \Sigma_{pp}(H^\theta) \cap I.$$

Remark 1. If we have only $\delta_n \to 0$, $n \to \infty$, than $\Sigma(H^\theta) = \Sigma_{sing}(H^\theta)$ for all $\theta \in [0, \pi]$.

Remark 2. The case $A_n \equiv 1$ is especially important: conditions $\sum_n \delta_n < \infty$, $x_n \leq c^n$ (for some $c > 1$) are sufficient for the full localization in I for almost all $\theta \in [0, \pi]$.

Proof of the theorem 2.4. As earlier, the basic idea is connected with analysis of the resolvent kernel. The following lemma is, of course, the special case of the Simon-Wolff's theorem. But historically it goes back to S. Kotani's paper [45], which stimulated more general results [69].

Lemma 2.6. Let $H^0 = \triangle + V, \psi(-1) = 0$ be operator (2.67) with zero boundary condition, and denote by

$$R_{E+i\epsilon}^0(0, x) = (H^0 - \lambda - i\epsilon)^{-1}(0, x), x \in \mathbf{Z}_+^1, E \in R^1, \epsilon > 0$$

its resolvent kernel. If for almost all $E \in I$

$$\limsup_{\epsilon \to 0} \sum_0^\infty |R_{E+i\epsilon}^0(0, x)|^2 < \infty \tag{2.70}$$

then for almost all $\theta \in [0, \pi]$

$$\Sigma(H^\theta) \cap I = \Sigma_{pp}(H^\theta) \cap I.$$

Lemma 2.6 admits the estimations of the resolvent kernel for complex energies $E + i\varepsilon$. Sometimes it's more convenient to work in the real domain. The following version of the lemma 2.6 will be useful in the future.

Lemma 2.6'. Suppose, that $R^0_{n,E}(x, y)$ is a resolvent kernel of the operator $H = \Delta + V$ on $[0, L_n]$, $L_n \uparrow \infty$ with zero boundary conditions. If a.e. $E \in I$

$$\lim_{n \to \infty} \sup \sum_{x=0}^{L_n} |R^0_{n,E}(0, x)|^2 < C(E), \qquad (2.71)$$

then for almost all $\theta \in [0, \pi]$

$$\Sigma(H^{(\theta)}) \cap I = \Sigma_{pp}(H^{(\theta)}) \cap I.$$

Of course, the result of the lemma 2.6' can be found in the S. Kotani's papers [44]-[47] in some (indirect) form. I'll give the short proof of the statement.

Suppose, that $E \in I'$ $\operatorname{mes} I' = \operatorname{mes} I$ and I' does not include the poles of the resolvent $R^0_{n,\cdot}(\cdot, \cdot)$ and the point spectrum of H^0 on $\mathbf{Z}^1_+$ (both these sets at most countable). Then (in the obvious notations)

$$R^{(0)}_{n,E}(0, x) = \sum_{k=1}^{L_n} \frac{\psi_{k,n}(0)\psi_{l,n}(x)}{E - E_{n,k}} \qquad (2.72)$$

and

$$\|R^0_{n,E}\|^2_{L^2} = \sum_{x=0}^{L_n} |R^0_{n,E}(o, x)|^2 =$$

$$= \sum_{k=1}^{L_n} \frac{\psi^2_{k,n}(0)}{|E - E_{n,k}|^2} = \int \frac{\mu_n(d\lambda)}{|E - \lambda|^2}. \qquad (2.73)$$

Of course,

$$\|R^0_{n,E+i\varepsilon}(0, \cdot)\|^2_{L_2} \xrightarrow{\varepsilon \to 0} \|R^0_{n,E}\|^2_{L_2}.$$

According to (2.71), it means that for all $\varepsilon > 0$

$$\|R^0_{n,E+i\varepsilon}(0, \cdot)\|^2_{L_2} \leq C(E).$$

But

$$\int \frac{\mu_n(d\lambda)}{|E + i\varepsilon - \lambda|^2} \xrightarrow[n \to \infty]{} \int \frac{\mu(d\lambda)}{|E + i\varepsilon - \lambda|^2},$$

where $\mu_n(d\lambda)$ is a spectral measure of the operator H^0 on $[0, L_n]$ (with zero boundary conditions), and $\mu(d\lambda)$ is a spectral measure of H^0 on $\mathbf{Z}^1_+$. I.e.

$$\int \frac{\mu(d\lambda)}{|E + i\varepsilon - \lambda|^2} \leq C(E), \limsup_{\varepsilon \to 0} \|R^0_{E+i\varepsilon}(0, \cdot)\|^2_{L_2} \leq C(E)$$

and we proved (2.70).

Now we'll introduce the cluster expansion of the resolvent in the form which is convenient for us. Let's define the family of block operators, related with the points $(x_n, y_n, n = 1, 2, \cdots)$, from the main condition of the theorem 2.4.

Blocks $[x_0 = 0, y_1] = (0), \quad [x_1, y_2] = (1), \cdots [x_n, y_{n+1}] = (n) \cdots$ will be called "main" blocks, and their intersections $(0) \cap (1) = [x_1, y_1] = (0, 1), \quad (1) \cap (2) = [x_2, y_2] = (1, 2) \cdots$ will be "boundary" or nonresonant blocks.

We introduce the main block operator

$$H^{(n)}(x) = \Delta + V(x), x_n \leq x \leq y_{n+1}, \psi(x_n - 1) = \psi(y_{n+1} + 1) = 0. \qquad (2.74)$$

The boundary operators $H^{(n-1,n)}$ was introduced above.

Denote by $R_E^{(n)}(x, y)$, $x, y \in (n), R_E^{(n-1,n)}(x,y)$, $x, y \in (n-1, n)$ corresponding resolvent kernels.

Definition: Elementary path γ on the lattice Z_+^1 is a finite sequence of points

$$\gamma = (Z_0, \cdots, Z_n), |Z_k - Z_{k+1}| = 1, \quad 0 \leq k \leq n - 1).$$

By $\gamma : Z_0 \to Z_e$ we denote arbitrary elementary paths between points Z_0 and Z_e. It is a well-know fact that for initial resolvent kernel $R_\lambda^0(0, x)$ and for the resolvents of our blocks, there exists a formal path representation of the form

$$R_E^{(n)}(x, y) = \sum_{\substack{\gamma : x \to y \\ \gamma \subset (n)}} \prod_{z \in \gamma} \left(\frac{1}{E - V(z)} \right) \qquad (2.75)$$

$$R_E^0(0, x) = \sum_{\substack{\gamma : 0 \to x \\ \gamma \subset Z_+^1}} \prod_{\varepsilon \in \lambda} \left(\frac{1}{E - V(z)} \right). \qquad (2.75')$$

(see lecture 5).

Remark 3. If $x = y$, it is necessary to include the path of the length zero, which gives contribution $\dfrac{1}{\lambda - V(x)}$. The number of paths beginning at the fixed point and having the length k is less than or equal to 2^k. It follows that formal path expansion absolutely converges if $|Im(\lambda)| > 2$.

Path decomposition describes physically all the ways that a quantum particle with energy λ can go from x to y.

Let us introduce graph Γ whose sites are main blocks $(0), (1), \cdots, (n)$, and the bonds connect only the neighbor blocks: $(0) \to (1); (1) \to (0), (1) \to (2)$ etc.

Let's consider some elementary path $\gamma : 0 \to x$ (from the expansion (2.72')). This path make the sequential transitions between main blocks, starting from the block $(0) \ni 0$ and finishing in the block $(k) \ni x$.

We understand this transition in the following sense. Path γ begins from the point $0 \in (0)$, in some moment reaches the point y_1 and after (first time) jumps to the point $y_1 + 1$. This moment of the first entrance to the point $y_1 + 1$, by the definition, is the moment of the transition $(0) \to (1)$ on the graph Γ. The first exit time from the block (1) is the first moment, when path γ will reach either point $x_1 - 1$ (which corresponds to transition $(1) \to (0)$ or point $y_2 + 1$ (transition $(1) \to (2)$) and so forth.

It means, that for every $\gamma : 0 \to x$ we have defined the new path $\tilde{\gamma}$ on $\Gamma : \tilde{\gamma} : (0) \to (k)$. Of course, every $\tilde{\gamma}$ represents many possible elementary paths, corresponding set we will denote $\{\tilde{\gamma}\}$.

Let's

$$R_E^{(\tilde{\gamma})}(0, x) = \sum_{\substack{\gamma \subset \{\tilde{\gamma}\} \\ \gamma \ 0 \to x}} \prod_{z \in \gamma} \frac{1}{(E - V(z))} \tag{2.76}$$

of course,

$$R_E(0, x) = \sum_{\tilde{\gamma} : (0) \to (k) \ni x} R_E^{(\tilde{\gamma})}(0, x) \tag{2.77}$$

Using simple combinatorial calculations we can express $R_E^{(\tilde{\gamma})}(0, x)$ in terms of the resolvents of the "main" and "boundary" blocks. In order to avoid the bulky notations, I'll formulate the following lemma only for specific case. Nevertheless, the structure of the general formula is obvious.

Lemma 2.7. Let $\tilde{\gamma} = (0) \to (1) \to (2) \to (1) \to (2) \to \cdots \to (k+1) \to (k)$ is some (specific) path on Γ. Then

$$R_E^{(\tilde{\gamma})}(0, x) = R_E^{(0)}(0, x_1 - 1) R_E^{(0,1)}(x_1, y_1) R_E^{(1)}(y_1 + 1, x_2 - 1) \cdot$$

$$R_E^{(1,2)}(x_2, y_2) R_E^{(2)}(y_2 + 1, y_2 + 1) R_E^{(2,1)}(y_2, x_2) \cdot$$

$$R_E^{(1)}(x_2 - 1, x_2 - 1) R_E^{(1,2)}(x_2, y_2) R_E^{(2)}(y_2 + 1, \cdot) \cdots \tag{2.78}$$

$$\cdots R_E^{(k+1)}(\cdot, y_{k+1} + 1) R_E^{(k+1,k)}(y_{k+1}, x_{k+1}) \cdot$$

$$\cdot R_E^{(k)}(x_{k+1} - 1, x).$$

Let's remark that for the resolvent of the boundary block we have in the formula (2.75) only one variant for the arguments: $R^{(i-1,i)}(x_i, y_i) = R^{(i,i-1)}(x_i, y_i)$. For the resolvents of the main blocks we have four possible versions:

$$R^{(i)}(y_i + 1, \, x_{i+1} - 1), \quad R^{(i)}(y_i + 1, \, y_i + 1),$$

$$R^{(i)}(x_{i+1} - 1, \, x_{i+1} - 1), \quad R^{(i)}(x_{i+1} - 1, \, y_i + 1)$$

(except, of course, the last factor in (2.75), which depends on $x \in (k)$).

Formula (2.78) includes the factor of two types: boundary resolvents $R_E^{(i,i+1)}(\cdot, \cdot)$, which are small (according to conditions of the theorem) and resolvents of the main blocks $R_E^{(i)}(\cdots, \cdots)$. We will prove now that this factors are not too large. It's interesting, that we have no any information about the structure of the potential inside the "main" blocks! The following simple lemma gives the central idea of the proof (see also, [53], [40], the article [40] contains this lemma in a slightly weaker form).

Lemma 2.8. Consider the resolvent $R_E^{(n)}(a, b), a, b \in (n)$ and define

$$L_n = |y_{n+1} - x_n| + 1 = |(n)|.$$

Then for any $M > 0$ and any $(a, b) \in (n)$
a)

$$\mu\{E \in I : |R_E^{(n)}(a, b)| > M|\} \leq \frac{C_0}{M} \qquad (2.79)$$

where μ is Lebesgue measure, C_0-absolute constant (in fact $C_0 = 4$),
b)

$$\mu\{E \in I : \| R_E^{(n)}(a, \cdot)\|_{L_2}^2 = \sum_{b=x_n}^{y_{n+1}} |R_E^{(n)}(a, b)|^2 > M\} \leq 4\sqrt{\frac{L_n}{M}}. \qquad (2.80)$$

Proof: The main difficulty is that we don't know anything about block potentials of the block operators $H^{(n)}$. Nevertheless
a)

$$R_E^{(n)}(a, b) = \sum_{\kappa=1}^{L_n} \frac{\psi_{\kappa,n}(a)\psi_{\kappa,n}(b)}{E - E_{\kappa,n}},$$

where $E_{\kappa,n}$ are eigenvalues and $\psi_{\kappa,n}(x), x \in (n)$ are orthonormal eigenfunction of our operator $H^{(n)}$.

Let us remark that

$$\sum_{\kappa=1}^{L_n} |\psi_{\kappa,n}(a)|^2 = \sum_{\kappa=1}^{L_n} (|\delta_a \cdot \psi_{\kappa,n})|^2 = \|\delta_a\|^2 = 1$$

and also

$$\sum_{\kappa=1}^{L_n} |\psi_{\kappa,n}(a)| \cdot |\psi_{\kappa,n}(b)| \leq 1.$$

If $\psi_{\kappa,m}(a) \cdot \psi_{\kappa,n}(b) = \alpha_\kappa$, then we can write

$$R_E^{(n)}(a,b) = \int \frac{d\mu(\lambda)}{E - \lambda} = \sum_{\alpha_k^+ > 0} \frac{\alpha_k^+}{E - \lambda_{k,n}} - \sum_{-\alpha_k^- < 0} \frac{\alpha_k^-}{E - \lambda_{k,n}} = R_{E,+}^{(n)} + R_{E,-}^{(n)}.$$

It's obvious that

$$\mu\{E \in I : |R_E^{(n)}(a,b)| > \mu\} \leq \mu\{E \in I : |R_{E,+}^{(n)}| > \mu\}$$

$$+ \mu\{E \in I : |R_{E,-}^{(n)}(\cdot,\cdot)| > \mu\}.$$

It is interesting, that $\mu\{E \in I : |R_{E,+}^{(n)}| > \mu\}$ can be calculated explicitly. In fact, consider equations

$$\sum_{k=1}^{n'} \frac{\alpha_k^+}{E - \lambda_k} = \pm M. \tag{2.81}$$

It can be represented in the form

$$\pm M \cdot \prod_{k=1}^{n'} (E - \lambda_k) = \sum_{i=1}^{n'} \alpha_i^+ \prod_{k \neq i} (E - \lambda_k),$$

or

$$\pm M \cdot E^{n'} \mp M \left(\sum_{k=1}^{n'} \lambda_k\right) E^{n'-1} + \cdots =$$

$$= \left(\sum_{k=1}^{n'} \alpha_k^+\right) E^{n'-1} + \cdots$$

Algebraic equation (2.81) has n' roots E_k^+ located for large M near points λ_k. Elementary theorem gives

$$\sum_{k=1}^{n'} E_k^+ = \sum_{k=1}^{n'} \lambda_k + \left(\sum_{k=1}^{n'} \alpha_k^+\right) M^{-1}$$

$$\cdot \sum_{k=1}^{n'} E_k^- = \sum_{k=1}^{n'} \lambda_k - \left(\sum_{k=1}^{n'} \alpha_k^+\right) M^{-1}.$$

But in the case when $\alpha_k^+ > 0$

$$\mu\{E : \left|\sum_{k=1}^{n'} \frac{\alpha_k^+}{E - \lambda_k}\right| > M\} = \sum_{k=1}^{n'} (\lambda_k^+ - \lambda_k^-)$$

$$= \frac{2}{M} \left(\sum_{k=1}^{n'} \alpha_k^+\right).$$

(Last step was based on (2.82)). It means, at last, that

$$\mu\{E : |R_E^{(n)}(a,b)| > \mu\} \leq \frac{4}{M},$$

i.e. we proved (2.79).

This result goes back to the old Kolmogorv's paper about divergent Fourier series.
b) We begin with the formula (2.72) mentioned above:

$$h_2(E) = \|R_E^{(n)}(a, \cdot)\|_{L_2}^2 = \sum_{k=1}^{L_n} \frac{\beta_k}{(E - E_{n,k})^2}, \tag{2.83}$$

where $L_n = |y_{n+1} - x_n| + 1$, $\beta_k = |\psi_{n,k}^2(a)|$, $\sum_{k=1}^{L_n} \beta_k = 1$. Let's introduce the intervals

$$\triangle_{k,n} = \left\{ E : |E - E_{n,k}| \le \sqrt{\frac{\beta_k}{M}} \right\}.$$

Then

$$\sum_k |\triangle_{k,n}| = \frac{2}{\sqrt{M}} \sum_{k=1}^{L_n} \sqrt{\beta_k} \le \frac{2\sqrt{L_n}}{\sqrt{M}},$$

(Cauchy-Schwarz inequality).

Outside the set $\cup_k \triangle_{k,n}$, the function $h_2(E)$ has no singularities and

$$\int_{E \backslash \cup_k \triangle_{k,n}} h_2(E)dE \le \sum_{k=1}^{L_n} 2 \int_{\sqrt{\frac{\beta_k}{M}}} \frac{\beta_k dE}{E^2} = 2\sqrt{M} \sum_{k=1}^{L_n} \sqrt{\beta_k} \le$$

$$\le 2\sqrt{M} L_n.$$

Chebyshev inequality gives us

$$\mu\{E : E \notin \cup_k \triangle_{k,n}, h_2 > M\} \le \left(\int_{I \backslash \cup_k \triangle_{k,n}} h_2(E)dE \right) / M$$

$$\le 2\sqrt{\frac{L_n}{M}}.$$

This fact together with estimation of $\cup_k \triangle_{k,n}$ is equivalent to (2.80).

Corollary 2.1. Suppose that M_n and N_n are such sequences ($M_n \to \infty, N_n \to \infty$)
that

$$\sum \frac{1}{M_m} < \infty, \sum \sqrt{\frac{L_n}{N_n}} < \infty.$$

Lemma 2.8 implies that

$$\sum \mu\{E \in I : |R_E^{(n)}(:\,:)| \ge M_n\} \le \sum \frac{4C_0}{M_n} < \infty$$

where : : means the four possible variants of the arguments in Lemma 2.7. Similarily

$$\sum \mu\{E \in I : \|R_E^{(n)}(\cdot,\cdot)\|_{L^2}^2 > N_n\} \le \sum \sqrt{\frac{L_n}{N_n}} < \infty,$$

point in the argument means one of the two possible variants of argument. Borel-Cantelli Lemma shows that for almost all $E \in I$ there exists number $n_0(E)$ such that for $n \ge n_0(E)$ we will have

$$|R^{(n)}(: \ :)| \le M_n, \|R_E^{(n)}(\cdot,x)\|_{L^2}^2 \le N_n. \tag{2.85}$$

So we showed that resolvent kernels are not too "big". Now everyting is ready to finish the proof of the theorem 2.4.

Let us put $M_n = \dfrac{c_0}{\sqrt{A_n}\delta_n}$, $N_n = A_1, A_2, \cdots, A_n c_1^n$. If the constant c_1 is sufficiently large, c_0 is small enough, then (according to condition (2.68') of the theorem 2.4) the following three series are convergent:

$$\sum_n \frac{1}{M_n} < \infty, \quad \sum_n \sqrt{\frac{L_n}{N_n}} < \infty, \quad \sum_n \frac{c_0^n N_n}{A_0 \cdot A_n} < \infty \tag{2.86}$$

and $M_n \delta_n < \dfrac{1}{3}$. $n = 0, 1, \cdots$.

Suppose at the beginning that the random index $n_0(E)$ in the corollary 2.1 is equal to 0. Then we can estimate $R_E^{(\bar{\gamma})}(0,x), x \in (k)$ (see (2.78)) by the expression

$$|R_E^{(\bar{\gamma})}(0,x)| \le (M_0\delta_0)\cdots(M_k\delta_k)\left(\frac{1}{3}\right)^{|\bar{\gamma}|-k-1}|R_E^{(k)}(\cdot,x)|.$$

It follows immediately from the (2.77) that

$$|R_E(0,x)| \le \prod_{i=0}^{n}(M_i\delta_i)\left(\sum_{j=0}^{\infty}\left(\frac{2}{3}\right)^j\right)|R_E^{(k)}(\cdot,x)| \le$$

$$c|R_E^{(k)}(\cdot,x)| \cdot \prod_{i=0}^{k}(M_i\delta_i)| \le c \cdot \frac{c_0^{k+1}}{\sqrt{A_0 \cdots A_{k+1}}}|R_E^{(k)}(\cdot,x)|.$$

Then

$$\sum_{x \in (k)}|R_E^{(0)}(0,x)|^2 \le c_1^2 \frac{c_0^{2k}}{A_0 A_1 \cdots A_k}\|R_E^{(k)}(\cdots,\cdots)\|_{L^2}^2 \le$$

$$\le c_1^2 \frac{c_0^{2k}}{A_0 A_1 \cdots A_k} N_k,$$

i.e. (see (2.86))

$$\|R_E^0(0,\cdot)\|^2 \le \sum_k \sum_{x \in (k)}|R_E(0,x)|^2 \le$$

$$\le \sum_k \frac{c_0^{2k} N_k}{A_0 A_1 \cdots A_k} < \infty.$$

If $n_0(E) > 0$, we can construct (using the same estimation) the resolvent of the hamiltonian H on the half-axes $x \geq L_0^{(n)}$ with zero boundary condition at the point $L_0 - 1$. It will give us for $x > L_0$ a square-integrable solution of the equation

$$H\psi = E\psi.$$

We can extend this solution for all $x \geq 0$. If $E \notin \sum_{pp}(H^0)$ then this solution (with the suitable factor) will give us $R_E^0(0, x)$. Theorem is proved.

Remark 4. If in the conditions of the theorem 2.4 we have only $\delta_n \to 0$, then for arbitrary $\{x_n, y_n\}$, $n = 1, 2, \cdots$ and arbitrary $\theta \in [0, \pi]$

$$I \cap \left(\sum \left(H^{(\theta)} \right) \right) = \left(\sum_{sing}(H^\theta) \right) \cap I.$$

This fact can be proved using the same claster expansion, but in the neighborhood of the point $E = i$. In fact this statement is contained (non-directly) in [68] and represents, probably, the strongest known criteria of the singular spectrum. We will not discuss this subject in details (see [68]).

Let's describe now some special cases of the theorem 2.4 and then apply this theorem to the random potentials $v = v(x, \omega)$, $x \in \mathbf{Z}_+^1$.

Case 1. High unique barriers
Suppose that for some sequence of points $x_n, n \geq 1, |v(x_n)| = h_n \to \infty, x_n < C^n$, and $v(y)$ is arbitrary for $y \neq x_n, n \geq 1$. In this case any boundary block consists of exactly one point from the set $\{x_n\}$ and it follows from Theorem 2.4 that the condition

$$\sum \frac{1}{h_n} < \infty \tag{2.87}$$

is sufficient for $p.p.$ spectrum of H^θ for a.e. $\theta \in [0, \pi]$.

Case 2. Barriers of the fixed length.
Suppose that for fixed $d > 1$ there exist sequence $x_n \to \infty$ such that $x_n \leq c^n$ for some $c > 1$ and

$$|v(x_n), \quad |v(x_n + 1)|, \cdots |v(x_n + d - 1)| \geq h_n.$$

If

$$\sum_n \frac{1}{h_n^d} < \infty \tag{2.88}$$

we can again apply the theorem 2.4 and prove the localization a.e. $\theta \in [0, \pi]$.

Case 3. Very long barriers (bumps).
Suppose that for arbitrary $E \in R^1$ we can find the sequence of the blocks $[x_n, y_n]$ such that

1. $|v(x) - E| \geq c > 2, \quad x \in [x_n, y_n]$

2. $l_n = |x_n - y_n| + 1 \geq c_1 ln\, n$ (2.89)

and $c_1 > \left(ln\frac{c}{2}\right)^{-1}$, i.e. series $\sum_n \left(\frac{2}{c}\right)^{ln}$ is convergent. Then $\sum(H^\theta) = \sum_{pp}(H^\theta)$ for a.e. $\theta \in [0, \pi]$.

Case 4. Periodic support (PS) (polycrystals).

Suppose that there exists a finite number of periodic potentials $v_i(x)$, $x \in \mathbf{Z}^1$ with periods T_i, $i = 1, 2 \cdots N$ with the following properties

1. $\cup_i G_i = R^1$, where G_i is an open set of all gaps in the spectrum of the periodic Hamiltonian $H_i = \Delta + v_i$. These gaps include no more than $T_i - 1$ bounded gaps and two unbounded gaps outside $\sum_{ess}(H_i)$.

2. For every $i = 1, 2, \cdots n$ we can find the system of the blocks $[x_n^i, y_n^i]$ such that

2a. $\lim\limits_{n \to \infty} \dfrac{|y_n^i - x_n^i|}{ln\, n} = +\infty$ (2.90)

2b. $\sup\limits_{x \in [x_n^i, y_n^i]} |v(x) - v_i(x)| \xrightarrow[n \to \infty]{} 0.$ (2.90')

Then $\sum(H^\theta) = \sum_{pp}(H^\theta)$ for a.e. $\theta \in [0, \pi]$.

The proof of this result is based on the positivity of the Ljapunov exponents in the gaps of periodic Hamiltonian and is elementary.

By the definition, periodic functions $v_i(x)$ such that

$$\sup\limits_{x \in [x_n, y_n]} |v(x) - v_i(x)| \xrightarrow[n \to \infty]{} 0$$

and $|x_n - y_n| \to \infty$ for some sequence of blocks $[x_n, y_n]$ are the elements of the periodic support PS of the potential $v(x)$, $x \in \mathbf{Z}^1$.

If the periodic support is sufficiently rich (i.e. $\cup_i G_i = R^1$) we can guarantee (without condition (2.90)!) that

$$\sum(H^\theta) = \sum_{sing}(H^\theta)$$

for all $\theta \in [0, \pi]$. This statement is an equivalent (or slightly stronger) of the main result of the S. Kotani's paper [46].

Theorem 2.4 is very general and includes the majority of the known results about localization on the half-axis. Let's discuss situation of the full axis $\mathbf{Z}^1$.

Consider in $L^2(\mathbf{Z}^1)$ Schrödinger operator

$$H = \Delta + v(x).$$

Assume that for given energy interval I (I may be all line R^1), conditions of the Theorem 2.4 are true separately for $x > 0$ and $x < 0$, that is, there exists a system of nonresonant blocks $[X_n, Y_n], n \in \mathbf{Z}^1$ for every $E \in I$ with corresponding constants δ, ℓ_n, L_n. The same consideration shows that a.e. in $E \in I$

$$R_E(0, x) = (H - E)^{-1}(0, x) \in L^2(\mathbf{Z}^1). \tag{2.91}$$

In order to glue together solutions $\psi_E^{\pm}(x)$ which decay when $x \to \pm\infty$ it is necessary to introduce some additional random parameter. In the case of $\mathbf{Z}_+^1$ it was a boundary phase θ uniformly distributed in $[0, \pi)$. For the case of $\mathbf{Z}^1$ this parameter must be included in potential v. The following Lemma is contained in paper of Souillard at all [19] and (in different situation) in [69] (Simon-Wolff).

Lemma 2.9. Let $V(x) = V_0(x) + V_1(x, \omega), \omega \in (\Omega, \mathcal{F}, \mu)$ and random part of the potential has finite support: $V_1(x, \omega) \equiv 0, x \notin [a, b]$. Consider for given $E \in I$ evolution of the phase of some solution $\psi_\lambda(x)$ of the equation $(H - E)\psi = 0$ in the $[a, b]$ given by lattice Riccaty equation in $[0, \pi]$;

$$\theta_E(x + 1) = \text{ arc ctg } (V(x) - E + \text{ ctg } \theta_\lambda(x)). \tag{2.92}$$

Assume that for fixed $\theta_E(a) = \phi$ the random value $\theta_E(b, \omega) = \theta_E(b, \phi, \omega, V_0)$ has absolutely continuous distribution. If, in addition

$$R_E^0(0, x) = (\Delta + V_0 - \lambda)^{-1}(0, x) \in L^2(\mathbf{Z}^1) \text{ a.e. } E \in I,$$

the operator $H = \Delta + V_0 + V_1(x, \omega)$ has a pure point spectrum μ-a.s. (that is for a.e. $\omega \in \Omega$). Typical example is $V_1(x, \omega) = \epsilon(\omega)\phi(x)$ where $\phi(x)$ has a finite support and an absolutely continuous distribution.. Perturbation of the form $V_1 = \epsilon\phi$, supp $\phi \in [a, b]$ has a finite rank. In many cases it is possible to use perturbation of rank $1 : V_1(x) = \epsilon \cdot \delta_{x_0}(x)$ (Simon-Wolff theorem). We will use system of these similar statements as a **Theorem 2.4'**.

Now I'll describe a few examples which will illustrate some features of the theorems 2.4, 2.4' and typical applications.

Example 1 (A. Gordon). Rare high barriers. Consider the potential $v(x)$ of the form

$$V(x) = \sum_{n-1}^{+\infty} h_n \delta(x - x_n), h_n \to +\infty, L_n = x_n - x_{n-1} \to \infty.$$

It follows from Theorem 2 that condition

$$\lim_{n \to \infty} \sup \sqrt{\frac{L_{n+1}}{L_n}} \cdot \frac{n^{1+\epsilon}}{h_n} = 0 \tag{2.93}$$

is sufficient for the pure point spectrum of H^θ for almost all boundary phases $\theta \in [0, \pi]$.

A. Gordon proved that if

$$\lim_{n \to \infty} \sqrt{\frac{L_{n+1}}{L_n} \cdot \frac{1}{h_n}} > 1, \tag{2.94}$$

then $\Sigma(H^\theta) \cap [-2,2] = \Sigma_{s.c.}(H^\theta) \cap [-2,2]$ for all $\theta \in [o, \pi)$, hence $[-2,2] \subseteq \Sigma_{s.c.}(H^\theta), \theta \in [0, \pi)$. Of course, $[-2,2] = \Sigma(\Delta)$. Outside of $[-2,2]$ the spectrum of $H^\theta = \Delta + V$ is discrete, which follows from oscillatory theorem. We may see that the difference between conditions (2.93) and (2.94) is not very big, especially in the case when $\frac{L_{n+1}}{L_n} \to +\infty$ fast enough, for example, when $x_n = e^{n^\alpha}, \alpha > 1, L_n \sim e^{n^\alpha}, \alpha > 1$.

Gordon's idea is simple. Let $E \in (-2, 2)$, then the amplitude $r(x) = (\psi^2(x-1) + \psi^2(x))^{1/2}$ of any solution ψ of the equation

$$(H - E)\psi = 0$$

is constant between barriers. If $r(\psi(x))$ is the size of the amplitude, and if

$$r(\psi(x)) = A_{n-1}, x_{n-1} < x < x_{n-1}, r(\psi(x)) = A_n, x_n < x < x_{n+1},$$

then it is easy to see that

$$A_{n-1} \cdot \frac{1}{h_n + 1} < A_n < A_{n-1} \cdot (h_n + 1).$$

Then divergence of the series

$$\sum_n \frac{L_{n+1}}{(h_1 + 1)^2 \cdots (h_{n+1} + 1)^2} \tag{2.95}$$

implies that $\psi \notin L^2(\mathbf{Z}_+^1)$. On the other hand, condition

$$\overline{\lim} \ \frac{L_{n+1}}{L_n} \cdot \frac{1}{h_n^2} > 1$$

is sufficient (2.95) to diverge.

It is a very interesting (and, probably, not difficult) problem to find necessary and sufficient conditions on the potentials in example 1 such that H^θ has pure point spectrum for almost all θ.

Example 2. Let $v(x) = \sigma\xi(x), \quad x \in \mathbf{Z}^1$, where $\xi(x, \omega)$ are $i.i.d.r.v$, and coupling const σ can be very small. In addition, suppose (to use Simon-Wolff's theorem) that r.v. $\xi(x, \omega)$ have a common density $p(v)$. We will prove that

$$\Sigma(H) = \Sigma_{pp}(H) \quad \mu - a.s.$$

It's possible to use a special form of the theorem 2.4' (case 4, see above).

Consider the following periodic function

$$v_n(x) = \begin{cases} a, & x = kn - 1 \\ 0, & x \neq kn - 1 \end{cases}$$

with period $T_n = n$.

The simple calculations show that monodromy operator of $H = \Delta + v(x)$ on the interval $[0, n]$ is given by the formula

$$T_{n,E} = \begin{pmatrix} 0 & -1 \\ 1, & -E \end{pmatrix}^{n-1} \begin{pmatrix} 0 & -1 \\ 1 & a - E \end{pmatrix}.$$

If $E = -2\cos\phi$, i.e. $E \in [-2, 2] = \Sigma(\Delta)$, then

$$\begin{pmatrix} 0 & -1 \\ 1 & -E \end{pmatrix}^{n-1} = \frac{1}{\sin\phi} \begin{pmatrix} \sin\phi(n-1) & -\sin\phi n \\ \sin\phi n & \sin\phi(n+1) \end{pmatrix}$$

and

$$Tr(T_{n,E}) = 2\cos n\phi - \frac{a\sin n\phi}{\sin\phi}.$$

Energy E lies in the gaps of periodic Hamiltonian iff $|Tr(T_{n,E})| > 2$. It's easy to see, that for fixed a and arbitrary $E \in (-1, 1)$ (i.e. $\phi \in [0, 2\pi]$) we can find such n that $|Tr(\cdot)| > 2$.

Without loss of generality we can suppose, that $0 \in \text{Supp } p_{\sigma\xi}(v)$. Let $a \neq 0$ be some another point from the same support. If ε is a sufficienty small number then

$$P\{\sigma\xi(x+1) \in (-\varepsilon, \varepsilon), \cdots, \sigma\xi(x+n-1) \in (-\varepsilon, \varepsilon),$$

$$\sigma\xi(x+n) \in (a - \varepsilon, a + \varepsilon)\} \geq \delta^n$$

for suitable $\delta = \delta(\varepsilon) > 0$.

The probability of the N periods (with accurancy ε!) has order δ^{Nn}. For large L, on the interval $[0, L]$ we can meet μ-a.s. the series of $lnL \cdot \gamma(\varepsilon, \delta)$ periods.

Because (even for only one fixed a) the gaps corresponding to some finite number of indexes n cover the interval $[-2, 2]$, we can apply the result mentioned above (Case 4) and use Theorem 2.4'.

If $E \notin [-2, 2]$, we can again use the idea of periodic support (only in the calculations of monodromy operator substituted $sh\,\phi(x)$ instead of $\sin\phi(x)$! or apply the results of the Case 3 (long Bumps).

Now we'll show how to apply our general results to the strongly correlated potentials.

Example 3. Potential on dynamical system.

Let $V(x) = \text{ctg}\,(\pi(\alpha x^2 + \beta))$. This potential is popular in physical literature as one of the models of "quantum chaos". Dynamical system $\varepsilon_x = \alpha x^2 + \beta$ may be included in the skew-shift of the two-dimensional torus S^2, but this fact is not essential for us. (Similarly, as in the previous example, it wasn't essential that $\xi(x)$ are identically distributed). We will consider $V(x)$ as a family of r.v. on the probability space $\Omega = [0,1] \times [0,1]$ with Lebesgue measure μ in the plane.

Introduce family of the events

$$A_{n,x} = \{(\alpha, \beta) : \alpha x^2 + \beta \in [0, 2^{-n} \cdot n^{1+\epsilon}]\}, x \in 2^n, 2^n + 1, \cdots, 2^{n+1} - 1,$$

indicator of corresponding neighborhood

$$\chi_n(t) = \chi_{[0, 2^{-n} n^{1+\epsilon}]}(t),$$

and its Fourier transform

$$\chi_n(t) = \sum_{e=\infty}^{+\infty} C_{\ell,n} e^{i(\ell t 2\pi)}, \quad C_{0,n} = 2^{-n} n^{1+\epsilon}.$$

Then

$$\mu(A_x) = <\chi_n(\alpha x^2 + \beta)> = C_{0,n} = C \cdot 2^{-n} \cdot n^{1+\epsilon}$$

and

$$\mu(A_{n,x} A_{n,y}) = \int_0^1 \int_0^1 d\alpha d\beta \sum_{\ell, \ell_1} C_{\ell,n} \overline{C_{e_1,n}} e^{i\ell((2^n+k)^2 \alpha + \beta)2\pi - i\ell_1((2^n+k_1)\alpha+\beta)2\pi} =$$

$$= \sum_{\ell, \ell_1} \overline{C_{\ell_1,n}} C_{\ell,n} \int_0^1 \int_0^1 d\alpha d\beta (\ell^{i\beta(\ell-\ell_1)2\pi + (\ell x^2 - \ell_1 y^2)2\pi \alpha \cdot 2^n})$$

$$= \sum_\ell \delta(\ell - \ell_1) \delta(\ell x^2 - \ell_1 y^2) C_{\ell_1,n} \cdot \overline{C_{\ell,n}} = \delta(x - y) \sum_\ell |C_{\ell,n}|^2.$$

$$= \delta(x - y) 2^{-n} \cdot n^{1+\epsilon}.$$

This implies that events $A_{n,x}, x \in [2^n, 2^{n+1} - 1]$ are noncorrelated.

Let $V_n = \sum_{x=2^n}^{2^{n+1}-1} I_{A_{n,x}}(\omega)$. Then $\langle V_n \rangle = n^{1+\epsilon}$

and $\text{Var}(V_n) = \sum_x \text{Var}(I_{A_{n,x}}) = 2^n \cdot \mu(A_n, x) \cdot (1 - \mu(A_n, x)) \simeq n^{1+\epsilon}$.

Chebyshev inequality tells us that $\mu\{|V_n - \langle V_n \rangle| > \dfrac{\langle V_n \rangle}{2}\} \leq \dfrac{C}{\langle V_n \rangle} \sim \dfrac{C}{n^{1+\epsilon}}$.

We may apply now the Borel-Cantelli lemma and receive a.e. (α, β) for $n > n_0(\alpha, \beta)$ the existence of at least one peak of the potential $V(x), x \in [2^n, 2^{n+1}-1]$ of the order $|V(x_n)| \geq C \cdot \dfrac{2^n}{n^{1+\epsilon}}$. This means (in virtue of theorem 1) that $\Sigma(H^\theta) = \Sigma_{pp}(H^\theta)$ for almost all θ and almost all $(\alpha, \beta) \in [0,1] \times [0,1]$. Of course, the previous estimation $|V(x_n)| \geq C \cdot \dfrac{2^\mu}{n^{1+\epsilon}}$ is too strong.

In fact, we proved the following result: Let $f(x) = f(x+1), x \in R^1$ be a periodic function and for at least one point $x_0 \in [0,1]$,

$$|f(x)| \geq C \left(\ell n \dfrac{1}{|x - x_0|}\right)^{1+\epsilon}, \quad \epsilon > 0.$$

Then for a potential $V(x) = f(\alpha x^2 + \beta)$ we can apply Theorem 2.4.

The same argument shows that results remain true if we replace dynamical system $\varepsilon_x = \alpha x^2 + \beta$ with any polynomial system $\varepsilon_x = P_N(x)$ containing no less than two free parameters.

Example 4. Gaussian potentials.

Let $V(x), x \geq 0$ be a non stationary Gaussian sequence,

$$\langle V(x) \rangle = 0, \beta(x,y) = \langle V(x)V(y) \rangle, 0 \leq C_1 \leq \beta(x,x) = \text{ Var } V(x) \leq C_2,$$

and $|\beta(x,y)| \leq \dfrac{1}{ln^{1+\varepsilon}|x-y|}, \quad \varepsilon > 0, \quad |x-y| \to \infty.$

As earlier, we need only some hypothesis on non-degeneracy of finite-dimensional distributions, but not stationarity.

Let's remark that the condition of logarithmic decay of correlations is well-known in the theory of Gaussian fields and processes. In this case, the structure of the peaks has bifurcations, inequality

$$|\beta(x,y)| > \frac{1}{ln^{1+\varepsilon}|x-y|}$$

implies that long range behavior of $V(x)$ is approximately the same as of a sequence of independent and identically distributed Gaussian r.v.

It is known that $\max\limits_{x \in [-n,n]} V(x) \sim C\sqrt{ln(n)}$ (under much weaker conditions, see H. Cramer [18]), than we can assume $\max\limits_{2^n \leq x < 2^{n+1}-1} |V(x)|$ has order $\sqrt{n}$. We cannot apply Theorem 2.4 in the case of single peaks. It is necessary to look for blocks of length 3.

Let's divide $[2^n, 2^{n+1})$ in to "triplets":
$Y_1 = (V(2^n)), V(2^n+1), V(2^n+2)), Y_2 = (V(2^n+3), V(2^n+4), V(2^n+5)), \cdots Y_N, N = c \cdot 2^n.$

It's easy to see that $Y_k, k = 1, 2, \cdot N$ have common non-degenerating gaussian distribution with the density

$$\rho(y) = c \exp\left(-(BY, Y)\right), y \in R^3, B = B^* > 0.$$

Then

$$P\{Y_k^{(1)} \geq \delta\sqrt{n}, \quad Y_k^{(2)} \geq \delta\sqrt{n}, \quad Y_k^{(3)} \geq \delta\sqrt{n}\} \quad n \overset{\sim}{\to} \infty$$

$$\sim c_2 \exp(-c\delta^2 n) \cdot n^{-3/2} \quad \text{(Laplace method)}.$$

Let

$$I_k(\omega) = I(\vec{Y}_k \geq \delta\sqrt{n}, k = 1, 2, \cdots N).$$

Then

$$E \sum_{k=1}^{N} I_k = N \cdot P\{\vec{Y}_k \geq \delta \sqrt{n}\} > N \exp^{(-\beta n)}, \beta > 0.$$

Now it is necessary to estimate $\mathrm{Var}\left(\sum\limits_{k=1}^{N} \chi_k\right)$. It's possible to remark that common distribution density of $\vec{Y}_k, \vec{Y}_\ell$ has a form

$$\rho_{k\ell}(y_1, y_2) = c \exp\left(-(By_1, y_1) - (By_2, y_2) + (Cy_1, y_2)\right),$$

where $\|c\| \leq \mathrm{const} \cdot ln^{-1-\epsilon}|k - \ell|$. Simple calculations show that

$$Cov(I_k, I_\ell) \leq \langle I_k \rangle \langle I_\ell \rangle \exp(-cln^{1+\epsilon}|k - \ell|),$$

$$|k - \ell| \geq n_0 > 0.$$

Then

$$\mathrm{Var}\left(\sum_{k=1}^{N} I_k\right) \leq N e^{-\beta n} \sum_{k=1}^{\infty} \exp(-c\, ln^{1+\epsilon}k).$$

Chebyshev inequality, as earlier, gives us that

$$\sum_n \mu \left\{ \left| \sum_{k=1}^{n} I_k - \left\langle \sum_{k=1}^{n} I_k \right\rangle \right| > \frac{1}{2} \left\langle \sum_{k=1}^{N} I_k \right\rangle \right\} < \infty$$

and Borel-Cantelli arguments show that we have enough "triple" high blocks to guarantee due to Case 2 the localization theorem for gaussian potential with the slowly decreasing correlations a.e. $\theta \in [0, \pi]$.

Some additional results related with the theorems 2.4, 2.4′ and their generalizations see in the conclusion of this chapter.

Conclusion to the Chapter II "Localization" (review, remarks, open problems)

The main and still open problem in the localization theory is the problem of "mobility edge" for the Anderson model in high dimensions. According to initial physical publications of Anderson and Mott [6], [60] the situation is as follows.

If $H = \triangle + \sigma\xi(x)$, $x \in \mathbf{Z}^d$, $d \geq 3$ and $\xi(x)$ are i.i.d.r.v with common symmetrical density of distribution $p(v)$ (say, $N(0,1)$), there exists a critical value of coupling const σ_0 such that

1. If $|\sigma| > \sigma_0$ (large disorder) with probability 1

$$\textstyle\sum(H_\sigma) = \sum_{pp}(H_\sigma) \tag{2.95}$$

and more over, corresponding eigenfunctions are decreasing exponentially

2. If $|\sigma| \leq \sigma_0(\sigma_0$ is called "mobility edge) then the spectrum has two components:

2a. For $|E| \geq E_0(\sigma)$ we have, as earlier, an exponential localization μ-a.s.

2b. For $|E| < E_0(\sigma)$ the spectrum is continuous, may be absolutely continuous.

The physical explanations of the possible mechanisms of localization (absence of resonants between different blocks, tunneling and so forth) are very clear and can be translated in the mathematical language (see previous lectures 5, 6, which are using physical ideas in especially understandable form, at least, it is my opinion).

As to explanation of "delocalization" for small σ, this subject is very vague. Probably, the single clear argument is based on percolation theory. If $d \geq 2$ we can find in the potential field "channels" which are going to infinity along the set with "small" values of potential. In this context, the boundary $E(\sigma)$, $|\sigma| \leq \sigma_0$ is similar to the percolation critical level h for the set

$$A_h^+ = \{x \in \mathbf{Z}^d : \quad \sigma\xi(x) \geq h\}.$$

Now we have only one class of models which exhibits the spectral bifurcation of the Anderson's type. This is very popular one-dimensional models with quasi-periodic potential. In the last years hundreds of papers were published in this area and I have no possibility to give exact references. As mentioned above, our subject is the random theory.

Simplest and typical example is "almost-Matie" model on $\mathbf{Z}^1$:

$$H = \triangle + \sigma\cos(n\alpha + \theta). \tag{2.96}$$

Using so-called "Andree-Obrey duality" it's possible to calculate Ljapunov exponent $\gamma = \gamma(E)$ and to prove, that for small σ (in fact, $|\sigma| < 2$) the spectrum is a.c., but for sufficiently large σ (may be, again, $|\sigma| > 2$?), we have the picture of exponential localization.

Technical restrictions include Diophants properties of α (α must be typical so-called Rot's number) and uniform distribution of the parameter θ on $[0, 2\pi]$. In the last case the potential in (2.95) is stationary and ergodic.

Many years ago H. Kumz and B. Souillard had announced the Anderson's phase transition on the homogeneous tree (Bethe lattice), see [48]. Unfortunately, this paper is not correct mathematically, now it is the general opinion of the specialists. Of course, for large σ we have in this case exponential localization (it is the trivial consequence of the general results of [2]), but the problem of delocalization for low disorder in the central part of the spectrum is still open.

I can construct now examples of the graphs (trees with very long one-dimensional branches and increasing index of branching) such that
  1. The growth of the graph is exponential (as for Bethe lattice).
  2. Percolation problem (to say, for i.i.d.r.v $\xi(x)$ with $N(0,1)$ distribution) has nontrivial solution.
  3. Operator $H = \triangle + \sigma\xi(x)$ has μ-a.s. p.p. spectrum for all $\sigma \neq 0$.

This example shows that the problem is a very profound one and percolation ideas cannot be used the trivial direct way.

A few years ago my student L. Bogachev and I published [12] the series of papers where we have developed so-called "mean-field" approach in the localization theory. Unfortunately, the Russian journal "Theoretical and Mathematical Physics", being translated in USA, is not popular in the Western countries. It is the reason of formulating of some result of [12] here.

In the classical statistical mechanics, the mean-field model arises as an approximation of the local potential by the very small, but long range potentials in the increasing system of volumes $Q_n \uparrow \mathbf{Z}^d$. Limiting model has no sense, but we can use the standard method of thermodynamical limiting procedure for studying, for example, magnetization in the ferromagnetic models.

It's known that this model gives phase transition, although the quantity parameters (critical indexes etc.) are not very realistic.

Basing on this idea, we introduced in [12] the following model:

Let $Q_n \uparrow \mathbf{Z}^d$ is an increasing sequence of the subsets of $\mathbf{Z}^d$. Put

$$H_n \psi(x) = \frac{1}{|Q_n|} \sum_{x' \in Q_n} \psi(x') + \sigma \xi(x) \psi(x) \qquad (2.97)$$

"mean-field" Laplacian "Δ" $= \frac{1}{|Q_n|} \sum_{x' \in Q_n}$ has a rank 1, and the spectral problem $H_n \psi = E\psi$ in $L_2(Q_n)$ can be solved explicitly.

For the eigenvalues $E_i, i = 1, 2, \cdots, |Q_n|$ we can obtain the algebraic equation

$$\frac{1}{|Q_n|} \sum_{x' \in Q_n} \frac{1}{E - \sigma \xi(x')} = 1 \qquad (2.98)$$

and corresponding eigenfunctions $\psi_i(x)$ within normalization have a form

$$\psi_i(x) = \frac{1}{E_i - \sigma \xi(x)}. \qquad (2.98')$$

Spectral measure of the given element, say, $\delta_0(x)$ has a form

$$\mu_n^{(d\lambda)} = \sum_{i=1}^{|Q_n|} \frac{1}{||\psi_i||^2} \delta_{E_i}(\lambda) \cdot \frac{1}{|E_i - \sigma \xi(0)|^2}. \qquad (2.99)$$

If the i.i.d.r.v $\xi(x)$ have a bounded continuous density $p(\cdot)$, then we proved [12] that the measures $\mu_n(d\lambda)$ have a p.p. weak limit in probability. It means, that for every $\delta, \varepsilon > 0$ we can find $N = N(\varepsilon, \delta)$ such that uniformly in n

$\mu\{\omega$: we can find not more than N atoms of $\mu_n(d\lambda)$, which have total mass bigger than $1 - \varepsilon\} \geq 1 - \delta.$ $\qquad (2.100)$

Again, for all $\sigma \neq 0$ we have some sort of localization.

It is very interesting to prove the results of this type for the hierarchical Laplacians with finite numbers of scales and with infinite numbers of the "levels" of hierarchy. The most natural definition of this objects is the following one.

Consider the lattice $\mathbf{Z}^d$ (with the elementary scale equal to 1) and the system of sub lattices

$$\mathbf{Z}_2^d = \nu \mathbf{Z}^d, \quad \mathbf{Z}_3^d = \nu^2 \mathbf{Z}_1^d \cdots \mathbf{Z}_k^d = \nu^{k-1} \mathbf{Z}^d \cdots$$

with the scales $\nu, \quad \nu^2 \cdots$. Here ν is a fixed integer number (the ratio of neighboring scales). We can consider the partitions of the $\mathbf{Z}^d$ on the cubes of the different scales. "cubes" of the rank 1 include the single points of $\mathbf{Z}^d$, "cubes" of the rank 2 are defined by inequalities

$$Q_n^i = \{x \in \mathbf{Z}^d : \quad i_1 \leq x_1 < i_1 + \nu, \cdots$$

$$i_d \leq x_d < i_d + \nu\}, \quad i \in \mathbf{Z}_2^d.$$

Every point $x \in \mathbf{Z}^d$ is the element of exactly one cube $Q_r^i(x)$ of every rank $r = 1, 2, \cdots$.

Definition of the hierarchical Laplacian related with this system of partition depends on the weights $P_1, P_2, \cdots, \quad P_i > 0, \quad \sum_{i=1}^{\infty} P_i = 1$ and has a form

$$\bar{\Delta}\psi(x) = P_1\psi(x) + \frac{P_2}{\nu^d} \sum_{x' \in Q_2(x)} \psi(x') + \cdots + \frac{P_n}{\nu^{d(n-1)}} \sum_{x' \in Q_n(x)} \psi(x') + \cdots. \quad (2.101)$$

The operator $\Delta - I$ is the generator of the Markov chain on $\mathbf{Z}^d$ (with continuous time). It is a so-called hierarchical random walk.

Hierarchical Laplacian on $\mathbf{Z}^d$ is essentially more symmetrical than "usual" Laplacian. Its spectrum contains countable many points of the finite multiplicity. Namely, $E_1 = P_1$ corresponds in $L_2(\mathbf{Z}^d)$ to the invariant subspace containing all functions $\psi(x) : \sum_{x' \in Q_2^i} \psi(x') = 0$ for all i.

The eigenvalue $E_2 = P_1 + P_2$ corresponds to the set of eigenfunctions $\psi(x)$, which are constant on the cubes Q_2^i, but

$$\sum_{x' \in Q_3^j} \psi(x') = 0$$

etc. The essential spectrum of Δ contains these points $E_n = \sum_{i=1}^{n} P_i$, $n = 1, 2, \cdots$ and the limiting point $1 = \lim_{n \to \infty} E_n$.

Hierarchical Anderson model is defined by Hamiltonian

$$\bar{H} = H + \sigma\xi(x) \quad (2.102)$$

with i.i.d.r.v. $\xi(\cdot)$, which have a "good" common density $\rho(v)$.

If the weights $P_n, \quad n = 1, 2, \cdots$ in the formula (2.101) are decreasing faster than any exponent, I can prove that μ-a.s the spectrum of $\bar{H}$ is pure point for arbitrary $\sigma > 0$. But the "good caricature" of the original Anderson model on $\mathbf{Z}^d$ is related with the weights which have only power decreasing. Analysis of the localization problem in this case for the operator (2.102) is an open and complicated problem.

Now I am going over to the one-dimensional localization. This subject was developed essentially better and now we have in this area two main directions of the further progress. The first is the quasi one-dimensional disordered systems and the second is one-dimensional random Hamiltonians with strongly correlated potentials.

The simplest and probably most important model of the quasi one-dimensional random Hamiltonian is a Schrödinger operator in a strip with the random potential $\xi(x)$ (i.i.d.r.v.).

If we'll consider a strip with periodic boundary conditions, then the model has a following form. Let Γ_N is an additive group of integer numbers mod N and $\Gamma = \Gamma_N \times \mathbf{Z}^1$ (or $\Gamma_N \times \mathbf{Z}^1_+$). Every point in Γ can be written down as (g, x), $g \in \Gamma_N$, $x \in \mathbf{Z}^1$ and

$$\triangle\psi(g, x) = \psi(g \oplus 1, x) + \psi(g \ominus 1, x) + \psi(g, x + 1) + \psi(g, x - 1). \qquad (2.102)$$

Here $g \oplus 1$, $g \ominus 1$ mean of course addition module N. Anderson operator, as usual, is

$$H = \triangle + \sigma\xi(g, x). \qquad (2.103)$$

The first step in the old approach to the localization theory in the strip is the analysis of the Ljapunov exponents for the product of the independent symplectic $(2N \times 2N)$-matrixes

$$A(x) = \begin{pmatrix} 0 & | & -I \\ -- & | & -- \\ I & | & Q(x) \end{pmatrix} \qquad (2.104)$$

where $Q(x)$ can be expressed in the terms of $\xi(g, x)$, $g \in \Gamma_N$ and energy E. Under minimal conditions on the distribution of $\xi(\cdot)$ the Ljapunov exponents are different and (due to symplectic structure) symmetric

$$\gamma_{2N-k}(E) = -\gamma_{k-1}(E),$$

i.e. for all σ, E

$$\gamma_{2N}(E) > \gamma_{2N-1}(E) > \cdots > \gamma_{N+1}(E) > 0$$

$$-\gamma_1(E) < -\gamma_2(E) < \cdots < \gamma_N(E) < 0$$

$$\gamma_{2N-k}(E) = -\gamma_{k-1}(E).$$

This is a very important result of the I. Goldsheid and G. Margulis [29].

It is sufficient to prove an exponential localization in the case of half-strip for a.e. boundary conditions or for the full strip under some additional restrictions (to say, a.c. of the distribution of $\xi(\cdot)$).

This localization theorem was proved by I. Goldsheid, J. Lacroix and B. Souillard (see details in the monograph [14]).

The cluster expansion method developed in the lecture 7 gives a very elementary approach to the same problem. Suppose, that $[a, b] \in \mathbf{Z}^1$ is an interval. We can consider "block" operator $H^{[a,b]}$ and zero boundary conditions on the boundaries $B_1 = \{(g, a - 1), \ g \in \Gamma_N\}$, $B_2 = \{(g, b + 1), g \in \Gamma_N\}$. Let $R^{[a,b]}_{E,ij}(x, y) = (H^{[a,b]} - E)^{-1}((g_i, x), (g_j, y))$;

$x, y \in [a, b], g_i, g_j \in \Gamma_N$ and $R_E^{[a,b]}(x, y) = \{R_{E,ij}^{[a,b]}(x, y)\}$ be and $N \times N$ resolvent kernel.

For a fixed energy interval $I \in R^1$ the block $\Gamma_N \times [x_n, y_n]$ is not resonant if

$$\|R_{E,\cdot}^{[x_n, y_n]}(x_n, y_n)\| \leq \delta_n, \quad E \in I \tag{2.106}$$

and $\delta_n \to 0$. In the case of the strip the cluster expansion in the theorem 2.4 will have absolutely the same form, only instead of scalar resolvent kernels we have to substitute the matrix expressions (2.106).

When we use the Borel-Cantelly lemma for the measures of the sets $\{E : \|R_{E,\cdot}^{(n)}(\cdot, \cdot)\| > M\}$, we will get the additional factor N^2 (which is equal to the number of the elements in matrix resolvent $\{R_{E,ij}^{(n)}(\cdot, \cdot)\}$). For fixed N it is not essential. In any case, the version of the theorems 2.4 or 2.4' for the case of the strip has practically the same form. The special cases (1, 2, 3, 4) can also be considered without any problem for the "strip localization".

The construction of the barriers for the independent or weakly dependent potentials can be realized according to the same ideas (see examples to the Lecture 7).

The essential and unsolved problem for the Anderson model in the strip is an asymptotical behavior of the smallest positive Ljapunov exponent $\gamma_{N+1}(\sigma, E)$ if $N \to \infty$ and coupling constant σ is small. This problem is not trivial even in the simplest case of i.i.d.r.v $\xi(x)$ with a good distribution density (Gaussian, Cauchy, uniform etc).

Of course, the asymptotical behavior of the "Lajpunov's gap" $\gamma_{N+1}(\sigma, E)$ is closely connected with the problem of Anderson's "mobility edge" in the high dimensions.

For the smallest positive Ljapunov exponent it's possible to find some integral representation in terms of Markov invariant measures on the compact manifolds ($O(N)$ or flag's manifold) of the high dimension. It means, that the problem includes a very interesting and non-trivial algebraic or geometrical point.

I will finish the brief discussion of the quasi one-dimensional theory by the problem of slightly different type.

Let's consider the graphs Γ with the trivial percolation theory (TP-graphs). By the definition graph Γ is an element of this class if for arbitrary small $\delta > 0$ and random Bernulli field

$$\varepsilon(x): \quad \mu(\varepsilon(x) = 1) = \delta, \quad \mu(\varepsilon(x) = 0) = 1 - \delta$$

the random set $\Gamma_\delta = \{x : \varepsilon(x) = 0\}$ has a bounded μ-a.s. connected component.

Class TP includes not only Z^1 or $Z^1 \times A$, $|A| < \infty$, but more complicated graphs. For example, trees with bounded index of the branching and increasing length of the one-dimensional branches (of course, the growth of this tree is lower than exponential), skeletons of the classical fractals sets of the Serpinski gasket type and so forth.

Probably, the Anderson's operators $H = \Delta + \sigma\xi(x)$ for unbounded density and slow tails of distribution have μ-a.s p.p. spectrum for all $\sigma > 0$. But if the r.v. ξ are uniformly bounded, the situation is not clear. What is the analogy of the periodic spectral theory, which we used in such conditions on one-dimention lattice Z^1? What is the interaction between electrons and high fluctuations of the potential (scattering theory)?

I think that all these questions ("random" and "non-random") are very interesting. It is a natural bridge between localization theory and popular now spectral theory of the homogeneous fractals.

In the end of this part I'll formulate a few open mathematical problems for the simplest localization models (on R^1) and give a few remarks about some features of the theorems 2.4, 2.4' and their continuous analogues.

Some of these problems are especially interesting in continuous case, i.e for the operators

$$H^\theta = -\frac{d^2}{dx^2} + V(x), \quad x \in R^1_+$$

$$(2.106)$$

$$\psi(0)\cos\theta + \psi'(0)\sin\theta = 0, \quad \theta \in [0, \pi].$$

Let us formulate the analogue of the main "lattice" localization theorem 2.4 and give some technical details.

Lemma 2.6 on square integrability of the resolvent $R^0_E(a, x)$ for a.e. $E \in I$ implying the localization theorem for H^θ (a.e. $\theta \in [0, \pi]$), is valid in continuous case. In fact, the original Kotani's result [45] was related exactly with spectral problem (2.106).

The single additional requirement is an essential self-ajointness of H. To guarantee this property it's enough to require some weak estimations of the potential $V(x)$ from below: $V(x) \geq \text{const} > -\infty$, $V(x) \geq c_1 - c_2 x^{2-\epsilon}$, $c_2, \epsilon > 0$ and so forth.

If, as earlier, $\{[x_n, y_n], n = 1, 2, \cdot\}$ is a system of "nonresonant" blocks for H, i.e. sequence of the points $0 < x_1 < y_1 < x_2 < y_2 < \cdots$, we can introduce the main "blocks", $(0) = [0, y_1], (1) = [x_1, y_2]$, corresponding resolvents $R^{(n)}_E(x, y)$, $x \in (n)$, graph Γ of the main blocks with neighboring connections, the paths $\tilde\gamma$ on Γ etc.

In Lemma 2.7 (cluster expansion) it is necessary to introduce some modifications. Let $R^{(n-1,n)}_E(x, y)$, $x, y \in (n-1, n) = [x_n, y_n]$ be a resolvent of H on $[x_n, y_n]$ with zero boundary conditions: $\psi(x_n) = \psi(y_n) = 0$ (as for $R_E^{(n)}(x,y)$, $\quad R_E^0(a,x)$). Put

$$K_E^{(n-1,n)}(x_n,y_n) = \frac{\partial^2 R_E^{(n-1,n)}}{\partial x \partial y}(x_n,y_n). \tag{2.107}$$

Then for $a \in (0,x_1) \subset (0), \quad x \in (k)$

$$R_E^{(0)}(a,x) = \sum_{\tilde{\gamma}:(0)\to(k)} R_E^{(\tilde{\gamma})}(0,x) \tag{2.78'}$$

and for $\hat{\gamma} = (0) \to (1) \to (2) \cdots$

$$R_E^{(\hat{\gamma})}(0,x) = R_E^{(0)}(0,x_1) K_E^{(0,1)}(x_1,y_1) R_E^{(1)}(y_1,x_2) \cdots$$

$$K_E^{(1,2)}(x_2,y_2) \cdots.$$

The formula (2.78') has the same structure as (2.78), only instead of $R_E^{(i-1,i)}(x_i,y_i)$ we have to substitute $K_E^{(i-1,i)}(x_i,y_i)$ and so on.

In the complex domain, for sufficiently large ImE this expansion is converging absolutely.

We will call "blocks" $[x_n,y_n], n = 1,2,\cdots$ non-resonant for the energy interval I, if uniformly in $E \in I$

$$|K_E^{(n-1,n)}(x_n,y_n)| \le \delta_n, \quad \delta_n \to 0. \tag{2.108}$$

To prove localization (according to the plan of theorem 2.4) we have to estimate the resolvents of the main blocks $R_E^{(n)}(a,b)$, where for (a,b) there are four variants: (y_n,y_n), (y_n,x_{n+1}), (x_{n+1},x_{n+1}), (x_{n+1},y_n). Kernel $R_E^{(n)}(x,y)$ has a form

$$R_E^{(n)}(x,y) = \sum_{k=1}^{\infty} \frac{\psi_{k,n}(x)\psi_{k,n}(y)}{E - E_{k,n}}, \quad x,y \in (n) \tag{2.109}$$

where $H^{(n)}\psi_{k,n} = E_{k,n}\psi_{k,n}, \psi_{k,n}(x_n) = \psi_{k,n}(y_{n+1} = 0 \quad (\psi_{k,n} \cdot \psi_{m,n}) = \delta(k-m)$

It is known, that

$$e_E^{(n)}(x,y) = \sum_{E_{k,n}<E} \psi_{k,n}(x)\psi_{k,n}(y) \tag{2.110}$$

is a kernel of the projector $\mathcal{E}_E^{(n)}$ in the spectral representation

$$H^{(n)} = \int E d\mathcal{E}_E^{(n)}.$$

If $x = y, \quad y \in (n)$, then $e_E^{(n)}(y,y) = (\mathcal{E}_E^{(n)}\delta_y \cdot \delta_y)$ is a spectral measure (of the element $\delta_y \notin L_2$).

In the lattice theory we had used two essential facts (see Lemma 2.8.):

$$\#\{E_{k,n}\} = L_n = |(n)| + 1$$

$$\text{var } \mathcal{E}. = \sum_{k=1}^{L_n} |\psi_{k,n}(x)| \cdot |\psi_{k,n}(y)| \leq 1.$$

Both these facts are false in the case of $L^2(R^1)$: $\#\{E_{k,n}\} = +\infty$, var $\mathcal{E}. = +\infty$.

Nevertheless, for the fixed bounded energy interval the situation is not bad.

a) If $V(x) \geq 0$, then for every $(n), y \in (n)$ $e_E(y,y) \leq C_0\sqrt{E}$ (2.111')

b) If $E \to \infty$, then $e_E(y,y) \sim c\sqrt{E}$ (2.111")

c) $N^{(n)}(E) = \#\{E_{k,n} < E\} \leq C_0 L_n \sqrt{E}$ (2.111''')

The central statement a) is a consequence of the general Kodaira's theorem for the elliptical nonnegative operators, c) follows from the Schtrum oscillation theorem. Second fact will not be used in the future, its proof is not complicated (see special literature, for example, [67], [3]).

If $V(x)$ is bounded from below on the n-th block: $v(x) \geq -M_n$, then it is necessary only to change E by $E + M_n$ in estimations (2.111).

Gathering together all this information and using the ideology of the theorem 2.4 we can prove the following result.

Theorem 2.5. Let $H^\theta = -\frac{d^2}{dx^2} + V(x)$, $x \geq 0$ be a Schrödinger operator in $L_2(R^1_+)$ with boundary condition $\psi(0)\cos\theta + \psi'(0)\sin\theta = 0$, $\theta \in [0, \pi]$. Assume that for given energy interval $I = [E_0, E_1] \subset R^1$ there exists a partition on the (intersecting) main

blocks $(0), (1), \cdots (n), \cdot$ with boundary blocks $(0,1) = [x_1, y_1], \cdots (n-1, n) = [x_n, y_n], \cdots$ such that

1. Potential $V(x) \in C(R^1_+)$ and

$$V(x) \geq -M_n, \quad x \in (n) = [x_n, y_{n+1}], \quad M_n \uparrow, \quad M_n \geq 0. \quad (2.112)$$

2. For all $E \in I$

$$|K_E^{(n)}(x_n, y_n)| = \left| \frac{\partial^2 R^{(n-1,n)}}{\partial x \partial y}(x_n, y_n) \right| \leq \delta_n \quad (2.113)$$

3. $y_{n+1} \leq c^n A_{1,}, \cdots A_n$ for some const $c > 1$,

$$A_n \geq 1, \quad n = 1, 2, \cdots \quad (2.114)$$

4. $$\sum_n \sqrt{A_n} \delta_n \sqrt{M_n + 1} < \infty \quad (2.115)$$

Then for a.e $\theta \in [0, \pi]$

$$\Sigma(H^\theta) \cap I = \Sigma_{pp}(H^\theta) \cap I. \tag{2.116}$$

If potential $V(x)$ has a suitable low estimation (to guarantee the essential self-ajointness) and instead of condition 4 we have only $\delta_n \to 0$, $n \to \infty$, then $\Sigma(H^\theta) = \Sigma_{sing}(H^\theta)$ in I for all $\theta \in [0, \pi]$. This result actually it's possible to extract form [68].

This theorem contains the majority of all known one-dimensional localization theorems for random potentials (except almost periodic results). To use this theorem for random potential $V(x, \omega)$, $x \in R^1_+$, $\omega \in (\Omega_m, \mathcal{F}_m, \mu)$, we have in the beginning to find (to construct) μ-a.s. the system of non-resonant blocks for given $I \subset R^1$.

These constructions are similar to the corresponding geometrical ideas in the lattice case (cases 1-4, lecture 7).

For example, the case 4 (periodic support, polycrystrals) can be used in continuous case directly. Under conditions (2.90), (2.90'), we can estimate δ_n very well and to prove the localization theorem for many random bounded potentials.

If potential $v(x, \omega)$ is unbounded, we can use the ideology of the high barriers. Corresponding estimation are not so trivial, as in the cases I, II (see lecture 7). The problem is, that, first of all, there are no exact formulas for the resolvents $R_E^{(n-1,n)}(x, y)$; $x, y \in [x_n, y_n]$ and second, we need to estimate not a resolvent kernel, but its derivatives.

The following lemma gives an estimation, sufficient for all applications.

Lemma 2.9. Suppose, that $V(x) \geq h > 0$, $x \in [-\delta, \delta]$. Then for $E \in I$ and sufficiently large h

$$\left| \frac{\partial^2 R_E}{\partial x \partial y}(-\delta, \delta) \right| \leq C(\delta) \sqrt{\int_{-\delta}^{\delta} V^2(x) dx} \, e^{-2\delta\sqrt{h}}. \tag{2.117}$$

Example. Suppose, that $v(x) \geq 0$ and for fixed $\delta > 0$ and infinite sequence $x_n \to \infty$, $x_n \leq c^n$, $c > 1$, $n = 1, 2, \cdots$ we have estimation

$$\sup_{x \in [x_n - \delta, x_n + \delta]} V(x) \geq h_n, \quad h_n \to +\infty.$$

If $\sum_n h_n e^{-2\delta\sqrt{h_n}} < \infty$, then H^θ has a p.p. spectrum for a.e. $\theta \in [0, \pi]$.

A few words about the case of Schrödinger operator in $L_2(R^1)$. If conditions 1-4 of the previous theorem 2.4 are fulfilled separately for $x > 0$ and $x < 0$, we can easily prove that $R_E(x, y) \in L_2(R^1)$ for a.e. $E \in I$. We can not use Simon-Wolff's theorem in this case. Instead, we can use the old "phase approach", which was typical for the initial works on the one-dimensional localization in $L_2(R^1)$.

The following statement goes to B. Souillard at all [19].

Suppose that $H = -\dfrac{d^2}{dx^2} + V(x,\omega)$ is a random Schrödinger operator. If for a.e. $E \in I$ the resolvent kernel is square integrable and for fixed interval $[a, b]$ the conditional distribution of the monodromy matrix $T_E(a, b)$ of the equation $H\psi = E\psi$ under fixed potential $V(x, \omega)$, $x \notin [a, b]$ is absolutely continuous, then, $\sum(H) = \sum_{pp}(H)$ inside I, μ-a.s. This statement (together with Theorem 2.5) gives a possibility to prove the localization theorem for the wide class of potentials of the Markov type. I will not formulate the exact results, you can find corresponding information, details of the proof of Theorem 2.5 and "Simon-Wolff type" lemmas in my lectures [53].

Let us summarize. In both cases, lattice and continuous, if the random potential $V(x, \omega)$ is "sufficiently nondeterministic", the spectrum of $H = \dfrac{d^2}{dx^2} + V(x, \omega)$ is singular μ-a.s and under additional conditions it is pure point. In the stationary situation it means, that Ljapunov exponent $\gamma = \gamma(E) > 0$ for a.e. E.

We can understand the expression "non-deterministic" in the different manner

1. For $x > 0$

$$\langle V(x, \omega) | \mathcal{F} \le 0 \rangle \ne V(x, \omega) \quad \mu - \text{a.s}$$

(It is Kotani's criteria of the singular spectrum [44])

2. Periodic support is rich enough (gaps in the spectrum of periodic Schrödinger operator associated with PS cover R^1). For the stationary potential it's again Kotani's result [46].

3. Potential $V(x, \omega)$ is unbounded from above and corresponding peaks are not too "thin" (Simon-Spencer, [68])

Potentials of the classes 2, 3 can be "deterministic", i.e. Kotanie's theorem [44] does not work.

What are the real boundaries for the class of ergodic homogeneous potentials generating the singular (or, may be, p.p.) spectrum of the corresponding Schrödinger operator? We can try another approach: how to describe the class of potentials, generating a.c. spectrum, or the spectrum with a.c. component. I think, it is a central problem in one-dimensional localization theory. I am sure that the solution of this problem is closely related with the analysis of the KdV equation

$$u_t + 6u_x \cdot u = u_{xxx} \tag{2.118}$$

and especially its statistical solution.

First of all, S. Kotani analyzed recently the class of operators H with a.c. spectrum [47]. As I understand, he is close to the following result (may be under some technical restrictions): if operator $H = -\dfrac{d^2}{dx^2} + V(x,\omega)$ has μ-a..s absolutely continuous spectrum for ergodic $V = V(x,\omega)$, then $v(x,\omega) = u(t_0, x, \omega)$, and $u(t, x, \omega)$ is a solution of (2.118) with periodic initial data $u_0(x,\omega)$.

Such solutions $u(t_0, x, \omega)$ are as a rule almost periodic. It means, that the class of ergodic Schrödinger operators with a.c. spectrum is very narrow. About periodic and almost-periodic solutions of the KdV equation see classical monograph [51].

But the same equation gives a possibility to construct a "very random" potential, which generates an a.c. spectrum. This construction is based on the notion of the "solitons gas", which is interesting by itself.

I'll give now the brief description of our recent paper with B. Simon related with this circle of problems (in preparation):

Suppose that $\phi(x), x \in R^1$ is a fast decreasing function, $|\phi(x)| \leq c \exp(-\delta|x|)$, $c, \gamma > 0$. According to classical scattering theory, operator $H = -\dfrac{d^2}{dx^2} + \phi(x)$ with "elementary" potential ϕ has at most finite number of negative eigenvalues and a.c. spectrum for $E > 0$.

Generalized eigenfunctions, i.e. solutions of the equation

$$-\frac{d^2\psi}{dx^2} + \phi(x)\psi(x) = E\psi(x), \quad E > 0 \tag{2.119}$$

can be constructed in a special form $\psi_I(x)$, which is given by the asymptotics of $\psi_I(x)$, $x \to \pm\infty$:

$$\psi_+(x) = \exp(ikx) + B(k)\exp(-ikx) + \bar{\bar{0}}\,(1), \quad x \to -\infty$$
$$\psi_+(x) = A(k)\exp(ikx) = \bar{\bar{0}}\,(1), \quad x \to +\infty \tag{2.120}$$

and $k = \sqrt{E}, \quad \psi_-(x) = \bar{\psi}_+(x)$. Function $B(k)$ (so-called reflection coefficient) is analytical for real k and we have an alternative:

1. either $B(k) \neq 0$, except may be some discrete set $\{k_i\}$

2. or $B(k) \equiv 0$.

Potentials ϕ with the property 2. are known as "reflectionless" or "solitons". The full analytical description of solitons together with related questions of the scattering theory see [51].

112

Let us consider the "shot noise" ergodic potential of the form

$$V(x,\omega) = \sum_{i=\infty}^{+\infty} \xi_i(\omega)\phi\left(\frac{x - x_i}{\theta_i}\right), \tag{2.121}$$

where (ξ_i, θ_i) are independent and identically distributed for different i and $(x_{i+1} - x_i) = \eta_i$ are again i.i.d.r.v, independent on amplitude and scaling factors $\{\xi_i, \theta_i\}$, $i = 0, \pm 1, \cdots$.

Suppose, that $\mu\{\eta_i \leq L\} = 0$ for some (very large) L and $L_1 = \langle \eta_i \rangle > L$. Using periodic support theorem it's possible to prove, that Ljapunov exponent $\gamma(E)$ of the Schrödinger equation with potential (2.121) is strictly positive (this is the main result of the paper [39], which based on [46]).

But we can now prove more: under standard conditions of a.c. for the distribution of $\{\eta_i\}$ (even for $\xi_i = \text{const}$, $\theta_i = \text{const}$), the spectrum is not only singular but pure point.

The following fact is very important (it was underestimated in [39]). If the potential $\xi\phi\left(\frac{x}{\theta}\right)$ μ-a.s is not a soliton, the Ljapunov exponent $\gamma(E)$, $E > 0$ has an order $\gamma(E) \sim L^{-2}$, $L \to \infty$ (energy E is fixed, $B(\sqrt{E}) \neq 0$. But if $\xi\phi\left(\frac{x}{\theta}\right)$ is a solution μ-a.s, then $\gamma(E) \sim \exp(-cL)$, $c > 0$.

The simplest example, where we have the second possibility is a sum of 1 - solitons with random parameters

$$v(x,\omega) = - \sum_{k=-\infty}^{+\infty} \frac{2\xi_k^2}{ch^2\xi_k(x - x_k)}, \tag{2.122}$$

$\{\xi_k\}$ are i.i.d.r.v.

Now we can try to consider the non-linear "superposition" of these solitons instead of summation (2.122) of the random 1-solitons. Idea is that interaction (due to KdV dynamics) between 1-solitons on the large distance is very small [51], i.e. KdV "superposition" and summation must give the similar results.

To add a sense to this "physical" phrase, we have to construct random ∞-soliton function as a limit of random N-solitons, $N \to \infty$. This program based on explicit formulas for N-solitons and special cluster expansion can be realized and gives the following interesting example:

There exists a homogeneous ergodic random potential $v(x,\omega)$, $x \in R^1$ such that

1. $|v(x,\omega) - v_1(x,\omega)| \leq \exp(-\gamma L)$, $x \in R^1$. Here $v_1(x,\omega)$ is given by formula (2.122).

2. Operator $M = -\dfrac{d^2}{dx^2} + v(x, h)$ has μ-a.s p.p. spectrum for $E < 0$ and a.c. spectrum for $E > 0$.

3. The correlation functions of $v(x, \omega)$ have the property of the "fast splitting". It means that for $h \to \infty$

$$|\langle v(x_1) \cdots v(x_k) v(x_{k+1} + h) \cdots v(x_n + h)\rangle -$$

$$\langle v(x_1) \cdots v(x_k)\rangle\langle v(x_{k+1}) \cdots v(x_n)\rangle| \leq \exp(-\gamma h), \quad h \to \infty.$$

From the physical point of view, the property 3. means a "very good" mixing or an ergodic property. But process $v(x, \omega)$ has analytical realizations and, as a result, is deterministic in Kotani's sense. Periodic support of $v(x, \omega)$ is empty μ-a.s!

Unfortunately, our ∞-soliton process does not depend of t, i.e. can not represent statistical solution of KdV. Construction of such statistical solutions is a very important problem. I think that for "small concentration" of 1-solitons it is a visible problem.

The final remark has an absolutely different nature, however again it is the same problem: relation between p.p. spectrum and continuous (in fact, singular continuous spectrum).

Let's return to the situation of the theorem 2.4 (lattice case) or 2.5 (continuous case): Schrödinger operator on $L_2(\mathbf{Z}_+^1)$, $L_2(R_+^1)$.

If we have a sufficiently rich set of the non-resonant blocks, say, high barriers, the resolvent kernel $R_E(a, x)$ is decreasing very fast a.e. $E \in I$ and, as a result, $R_E(a, \cdot) \in L_2$. According to Kotani's or Simon-Wolff's theorems, we can guarantee p.p. spectrum for almost all boundary conditions (i.e. boundary phases $\theta \in [0, \pi]$. What can we say about "exclusive" set of θ, which has zero Lebesgue measure?

A few years ago I used to think that under some arithmetical conditions on the distribution of "high barriers" and for sufficiently fast increasing of the heights of these barriers, the localization theorem holds for all $\theta \in [0, \pi]$

My student A. Gordon [31] had recently proved a very unexpected result which contradicts my hypothesis: for arbitrary essentially selfadjoint $H = -\dfrac{d^2}{dx^2} + v(x)$, $x \geq 0$ we can find some set $\mathbf{H} \subset [0, \pi]$ of the second category (in the Baire sense) the essential spectrum of H^θ, $\theta \in \mathbf{H}$ is continuous.

In the context of the theorems 2.4, 2.5 the a.c. spectrum is empty μ-a.s and A. Gordon's result means that singular continuous spectrum is generic in the category sense.

This interesting theorem shows that in the space of all potentials $v(x)$, the class of potentials generating p.p. spectrum and class related with singular continuous spectrum are "very mixed".

This is true for many different topologies, as it follows from the paper [37] (B. Simon at all), which contains wide generalization of the A. Gordon's results. Technically the paper [37] is not complicated. It is based on pure "soft" arguments.

Chapter III Intermittency

Lecture 8. Parabolic Anderson model with the stationary random potential.

Notion of intermittency (or intermittent random fields) is very popular in the modern physical literature, especially in the statistical (turbulent) hydrodynamics and magneto hydrodynamics. From the qualitative point of view, the intermittent random fields are distinguished by the formation of strong pronounced spatial structures: sharp peaks, foliations and others giving the main contribution to the physical processes in such media. Ja Zeldovich, one of the famous Russian physicists and astrophysicists, liked to repeat that the Solar magnetic field is intermittent, because more than 99% of the magnetic energy concentrates on the less than 1% of surface area (we know today that this contrast is in fact much more stronger).

Intermittency is a well developed nonuniformity. From this point of view the notion of the intermittency is "orthogonal" to the idea of homogenization.

Mathematical definition of intermittency must be asymptotical, because for the physicist a small number is 0.1, 0.001 at least 10^{-6} , but for the mathematician a small number is the result of the limiting process $\varepsilon \to 0$.

The first "rigorous" definition of the intermittency was proposed and developed in a series of the physical papers, see details in the big review [75]. This definition works well for many interesting evolution problems in the random media (for both cases: stationary and nonstationary RM), including magnetic and temperature fields in the turbulent flow, some schemes of chemical kinetics, biological models ect. Description of the possible applications see again in [75], where it's possible to find additional references.

More flexible definition see in our recent publication with J. Görtner [32]. This lecture will include some results form [32] in the special situation of gaussian potential, but on the other hand asymptotical analysis will be more sound. We'll discuss only the lattice models, although in continuous case it's enough to make just small modifications.

Definition 3.1. let $u(t,x)$, $t \in [0,\infty)$, $x \in \mathbf{Z}^d$ be a family of non-negative, homogeneous and ergodic in space random fields on a join probability space $(\Omega_m, \mathcal{F}_m, \mu)$. Suppose, that all moment functions of order p, $p = 1, 2, \cdots$ are finite for all $t \geq 0$. In particular, functions

$$\Lambda_p(t) = ln\langle u^p(t,x)\rangle, \quad x \in \mathbf{Z}^d, \quad t \geq 0 \tag{3.1}$$

depend only on time t.

Moment Ljapunov exponents (with respect to some monotone scale $A(t) \nearrow \infty$) are the limits

$$\gamma_p = \lim_{t \to \infty} \frac{\Lambda_p(t)}{A(t)} = \lim_{t \to \infty} \frac{ln \langle u^p(t, x) \rangle}{A(t)} \tag{3.2}$$

(of course, if these limits exist).

Family $u(t, \cdot)$ is (asymptotically $t \to \infty$) intermittent, if

$$\gamma_1 < \frac{\gamma_2}{2} < \frac{\gamma_3}{3} < \cdots . \tag{3.3}$$

Definition 3.2. Statistical (or a.s.) Ljapunov exponent with respect to some (my be different) scale $\alpha(t)$, $\alpha(t) \nearrow \infty$ is equal to

$$\hat{\gamma} = \lim_{t \to \infty} \frac{ln\, u(t, x)}{\alpha(t)} (\mu - \text{a.s}). \tag{3.4}$$

Usually (but not always) for the asymptotically intermittent field $u(t, x)$ in stationary RM

$$\lim_{t \to \infty} \frac{\alpha(t)}{A(t)} = 0$$

i.e.

$$\lim_{t \to \infty} \frac{ln\, u(t, x)}{A(t)} = 0\, (\mu - \text{a.s.}). \tag{3.5}$$

Remark 1. From the Hölder inequality it follows, that $\langle u^p(t, x) \rangle^{1/p} \geq \langle u^{p_1}(t, x) \rangle^{1/p_1}$, if $p > p_1$, i.e. **always**

$$\frac{\gamma_p}{p} \geq \frac{\gamma_{p_1}}{p_1}, \quad p > p_1,$$

or

$$\gamma_1 \leq \frac{\gamma_2}{2} \leq \frac{\gamma_3}{3} \leq . \tag{3.3'}$$

Main feature of the intermittency is a "progressive" increase of the moment with its number p: the fourth moment is increasing essentially faster, than the square of the second moment an so on. In different words, inequalities (3.3) must be strict.

According to ergodic theorem,

$$\langle u^p(t, 0) \rangle = \lim_{Q \uparrow \mathbf{Z}^d} \frac{1}{|Q|} \sum_{x \in Q} u^p(t, x). \tag{3.6}$$

We will show now that for $t \to \infty$ a system of "peaks", which for large t are higher and higher and more and more widely spaced gives (under condition (3.3)) the main contribution to the moment $\langle u^p(t, x) \rangle$. Thus, for large t the overwhelming part of the energy (usually it is $\sum_x u^2(t,x)$) concentrates in the "peaks" of the rank 2, and relative area of these "peaks" tends to 0 for $t \to \infty$. "Peaks" related with moments of order p are essentially more "rare" than the "peaks" of the rank p_1, $p > p_1$.

To understand this let us fix some level e, intermediate between two sequential exponents, say,

$$\gamma_1 < e < \frac{\gamma_2}{2},$$

and consider the event

$$E_{t,x} = \{\omega : \quad u(t,x) > \exp(eA(t))\}.$$

On the one hand

$$\mu(E_{t,x}) \leq \exp\{\Lambda_1(t) - eA(t)\} =$$

$$\exp\left\{A(t)\left[\frac{\Lambda_1(t)}{A(t)} - e\right]\right\} \to 0, \quad \text{as } t \to \infty$$

(exponentially in the scale $A(t)$!). It means, that the density of the points x, where $u(t,x) > \exp(eA(t))$, is asymptotically very small. On the other hand,

$$\langle u^2(t,0)\rangle = \langle u^2(t,0)I_{\bar{E}(t,0)}\rangle + \langle u^2(t,0)I_{E(t,0)}\rangle.$$

But the second term can be estimated from above as

$$\exp(2eA(t)) = \exp(\Lambda_2(t) + (2eA(t) - \Lambda_2(t))$$
$$= \langle u^2(t,o)\rangle \cdot \exp\left(2A(t)\left(1 - \frac{\Lambda_2(t)}{2A(t)}\right)\right) = 0(\langle u^2(t,0)\rangle).$$

It means, that

$$\langle u^2(t,0)\rangle \underset{t \to \infty}{\sim} \langle u^2(t,0)I_{E(t,0)}\rangle,$$

in different terms, all the energy asymptotically concentrates on the random set $\{x : u(t,x) > \exp(eA(t))\}$ (for arbitrary $\gamma_1 < e < \frac{\gamma_2}{2}$).

Using convexity arguments, it's easy to prove that in (3.3) only first inequality is essential: if $\gamma_1 < \frac{\gamma_2}{2} \Rightarrow \gamma_1 < \frac{\gamma_2}{2} < \frac{\gamma_3}{3} < \cdots$.

Let's apply these definitions to the special model, so-called Anderson parabolic problem:

$$\frac{\partial u}{\partial t} = k\Delta u + \xi(x)u \tag{3.7}$$

$$u(o,x) \equiv 1.$$

Laplacian will have in this context a slightly different form

$$\Delta \psi(x) = \sum_{|x'-x|=1} (u(t,x') - u(t,x)) \tag{3.8}$$

i.e., $k\triangle$ is a generator of the homogeneous random symmetrical walk x_t on $\mathbf{Z}^d$ with continuous time and the rate of jumps k in all directions $x \to x'$, $|x - x'| = 1$.

Suppose that $\xi(x)$ are i.i.d.r.v. with the standard $N(0,1)$ distribution. In the introduction we had discussed the meaning of the equation (3.7) in the framework of reaction-diffusion equations or branching processes in random media.

The analogy of (3.7) with the localization theory is obvious too: after rescaling of the time, the operator H in the right part of (3.7) can be written down in the form

$$H = \triangle + \frac{1}{k}\xi,$$

i.e. diffusion coefficient k has the same functions as the inverse coupling constant σ.

The only (but very essential) difference is a following one: quantum-mechanical aspects of the localization are related with the Schrödinger equation

$$i\frac{\partial\psi}{\partial t} = H\psi$$

but here we have a parabolic problem

$$\frac{\partial\psi}{\partial t} = H\psi.$$

Asymtotical behaviour of the solution $u(t,x)$ of the problem (3.7) depends only on "upper part" of the spectrum of H. In the case of Schrödinger equation (which is in some sense hyperbolic) the total spectrum is essential.

Equation (3.7) has μ-a.s. the single positive solution, which can be represented in the Kac-Feinmann form

$$u(t,x) = E_x \exp\{\int_0^t \xi(x_s)ds\}. \tag{3.9}$$

I will not be discussing here this rather technical problem. The full analysis of such questions (existence and uniqueness of solutions for (3.7), existence of moments etc.) in maximum generality see [32].

Our goal is a pure analytical one: to study asymptotical behaviour of the solution $u(t,x)$ given by (3.9) and corresponding statistical moments.

Theorem 3.1. For every $p \in N$ and every $t \geq 0$

$$\exp\{\frac{p^2t^2}{2} - 2dkt\} \leq \langle u^p(t,0)\rangle \leq \exp\frac{p^2t^2}{2}, \tag{3.10}$$

i.e. for the scale $A(t) = t^2$ and all $k \geq 0$

$$\gamma_p = \lim_{t\to\infty} \frac{ln\langle u^p(t,0)\rangle}{t^2} = \frac{p^2}{2} \tag{3.11}$$

and

$$\frac{\gamma_p}{p} = \frac{p}{2} \quad \uparrow, \quad p = 1, 2, \cdots.$$

We have intermittency for all diffusion coefficients k in the non-standard scale $A(t) = t^2$, Ljapunov exponents γ_p do not depend from k.

Theorem 3.2. With μ-probability 1

$$\lim_{t \to \infty} \frac{\ln u(t, 0)}{t\sqrt{\ln t}} = \sqrt{2d}. \tag{3.12}$$

We can see that the second scale $\alpha(t) = t\sqrt{\ln t}$ is (asymptotically) essentially smaller than $A(t) = t^2$.

Theorems 3.1, 3.2 contain the statement about asymptotical intermittency for the homogeneous fields $u(t, \cdot), \quad t \to \infty$.

The proof of both theorems in simple. In the paper [32] you can find essentially more general results. In both cases we'll prove separately the upper and lower estimates.

Proof of the theorem 3.1. **Upper estimation.** Applying Hölder's and Jensen's inequalities, we obtain

$$\langle u^p(t, x) \rangle = \langle (E_0 \exp(\int_0^t \xi(x_s) ds))^p \rangle$$

$$\leq \langle E_0 \exp(p \int_0^t \xi(x_s) ds) \rangle = E_0 \langle \exp(p \int_0^t \xi(x_s) ds) \rangle$$

$$\leq \frac{1}{t} \int_0^t ds E_0 \langle \exp(pt\xi(x_s)) \rangle = \exp \frac{p^2 t^2}{2}$$

Lower estimation. If $p = 1$, then $\langle u(t, x) \rangle \geq \langle E_0 \{ \exp(\int_0^t \xi(x_s) ds) \cdot I_{\substack{x(s) \equiv 0 \\ s \in [0,t]}} \} \rangle = \langle \exp \xi(0) t \cdot P_0 \{ x_s \equiv 0, s \in [0, t] \} \rangle = \exp \left(\frac{t^2}{2} \right) \exp(-2dkt)$. We used the obvious facts: the number of the jumps of the random walk $x_s, s \in [0, t]$ is a Poisson process with the rate $2dk$ and $\{ x_s \equiv 0, s \in [0, t] \} = \{ N(t) = 0 \}$.

For $p > 1$, we can consider p independent copies $x^i(t), \quad i = 1, \cdots, p$ of the random walk x_s and to use representation

$$\langle u^p(t, x) \rangle = \left\langle E_0 \left\{ \exp \left(\sum_{i=1}^p \int_0^t \xi(x_s^i) ds \right) \right\} \right\rangle$$

and so on.

Proof of the theorem 3.2. Using Borel-Cantelli's lemmas for the events $A_x^\pm = \{ \xi(x) > (1 \pm \varepsilon)\sqrt{2d \ln |x|} \}, \quad |x| = |x_1| + \cdots + |x_d|$ and standard estimation of the gaussian tails, we can prove (of course, very old and well-known) facts

$$\max_{|x|\leq n}\xi(x) \underset{n\to\infty}{\sim} \sqrt{2d\,\ln n} \quad \mu - \text{a.s.}$$

$$\max_{|x|\leq n}|\xi(x)| \underset{n\to\infty}{\sim} \sqrt{2d\,\ln n} \quad \mu - \text{a.s.}$$

(3.13)

The second elementary fact which will be useful is the nearest future is the following estimation: for the Poisson r.v. Π, $\langle\Pi\rangle = t$ and $n > 2t$

$$P\{\Pi \geq n\} \leq \exp\left\{-n\,\ln n\left(1 - \frac{\ln t}{\ln n} + O\left(\frac{1}{\ln n}\right)\right)\right\}.$$

(3.14)

This estimation works even for fixed t and $n \to \infty$. We had used this fact many times in [32].

At the beginning we will obtain the upper estimation. Let us fix the small $\varepsilon > 0$ and for every large t consider the family of the balls (centered $x_0 = 0$) of the radiuses

$$R_0 = 0, \quad R_1^{(t)} = t^{1+\varepsilon}, \quad R_2^{(t)} = t^{1+2\varepsilon}, \cdots \quad R_n^{(t)} = t^{1+n\varepsilon}, \cdots.$$

According to (3.13), μ-a.s. for all $t > t_0(\varepsilon, \omega)$

$$\max_{|x|\leq R_n}\xi(x) \leq (1+\varepsilon)\sqrt{2d\,\ln R_n}.$$

We have

$$u(t,0) = E_0\{\exp \int_0^t \xi(x_s)ds\} =$$

$$= \sum_{n=0}^{\infty} E_0\{\exp(\int_0^t \xi(x_s)ds)\}I_{R_n\leq N(t)<R_{n+1}}$$

$$\leq \sum_{n=0}^{\infty} \exp\left(t(1+\varepsilon)\sqrt{2d\,\ln R_{n+1}^{(t)}}\right) P\{N(t) \geq R_n^{(t)}\}$$

$$\leq \sum_{n=1}^{\infty} \exp\left(t(1+\varepsilon)\sqrt{2d\,\ln t}\sqrt{1+\varepsilon n}\right) \exp\left(-c(\varepsilon)t^{1+n\varepsilon}\ln t\right) +$$

$$+ \exp\left(t(1+\varepsilon)^{3/2}\sqrt{2d\,\ln t}\right) \leq \exp\left(t(1+\varepsilon)\sqrt{2d\,\ln t}\right)(1+o(1)).$$

(Last term corresponds to $n = 0$, it was important due to (3.14) that

$$1 - \frac{\ln t}{\ln R_n} \geq c(\varepsilon), \quad n = 1, 2, \cdots.$$

Lower estimation. Let us introduce for fixed $\varepsilon > 0$ the ball

$$\{x : \quad |x| \leq t^{1-\varepsilon}\} = B_t^-, \quad R^-(t) = t^{1-\varepsilon}.$$

For all $t \geq t_1(\varepsilon, \omega)$ with probability 1

$$\max_{x\in B_t^-}\xi(x) \geq (1-\varepsilon)\sqrt{2d\,\ln R^-(t)}$$

$$\min_{x\in B_t^-}\xi(x) \geq -(1+\varepsilon)\sqrt{2d\,\ln R^-(t)}.$$

(3.14)

Let $x^*(t)$ be the point of the maximum of the field $\xi(x)$, $x \in B^-(t)$. Of course, $|x^*(t)| \leq R^-(t) = t^{1-\varepsilon}$. Consider one of the shortest paths $\gamma : 0 \to x^*$, $|\gamma| = |x^*(t)| \leq t^{1-\varepsilon}$, and the event $A^*_{\varepsilon,t} = \{x. : \quad x_1 = x^*$, in the interval $s \in [0,1]$ the trajectories x_s are going along $\gamma\}$. It is obvious that

$$P(A^*_{\varepsilon,t}) \geq P\{N_1 = |x^*|\} \cdot \left(\frac{1}{2d}\right)^{|x^*|} \geq$$

$$\geq P\{N_1 \geq t^{1-\varepsilon}\} \left(\frac{1}{2d}\right)^{t^{1-\varepsilon}} \geq$$

$$\geq \exp\{-t^{1-\varepsilon} \ln t\},$$

we have used (3.14). Then

$$u(t,0) \geq E_0 \exp(\int_0^t \xi(x_s)ds) I_{A^*_{\varepsilon,t}} \geq$$

$$\geq \exp(-(1+\varepsilon)\sqrt{2d \ln R^-(t)}) P(A^*_{\varepsilon,t}) \cdot$$

$$\cdot E_{x^*(t)} \exp(\int_1^t \xi(x(s))ds) I_{x(s)\equiv x^*, \, s\in[1,t]}$$

$$\geq \exp(-(1+\varepsilon)\sqrt{2d \ln t})\} \exp\{-t^{1-\varepsilon} \ln t\} \cdot \exp\{\xi(x^*) \cdot$$

$$\cdot (t-1)\} \geq \exp(-2(1+\varepsilon)\sqrt{2d \ln t}) \exp\{-t^{1-\varepsilon} \ln t\} \cdot$$

$$\cdot \exp\{(1-\varepsilon)\sqrt{2d \ln t} \cdot t\} \geq \exp\{(1-2\varepsilon)\sqrt{2d \ln t} \cdot t\}$$

for $t \to \infty$.

Upper and lower estimation together give (3.12).

The results of the theorems 3.1, 3.2 are not satisfactory physically, because the role of diffusivity is not clear. Both lower estimations support the idea of the "strong centers" (which goes to Ja. Zeldowich, see [75]): the main contribution to the solution $u(t,x)$ and its statistical moments is related with the high local maxima of the potential $\xi(x)$ in the ball of the radius of order t.

From the point of view of this idea, the diffusivity k works in two opposite directions. The bigger k, the easier trajectory x_s can reach a "very distant" strong center: this mechanism leads to the increasing of $u(t,x)$ with increasing of k.. From the other hand, the bigger k, the smaller is the probability $\exp(-2dkt)$ to stay long time in the strong center, the smaller is the contribution of this center to the solution. In the gaussian case, where local maximum is increasing very slowly, the second mechanism is essentially stronger and increasing of diffusivity leads to the decreasing of the solution. We will prove now two results, which will give the next terms of asymptotic for $\ln \langle u^p(t,x0) \rangle$, $\ln u(t,x)$.. But let's make several short remarks.

Remark 1. Almost surely Ljapunov exponent in the scale $\alpha(t) = t\sqrt{\ln t}$ is not continuous in k:

$$\tilde{\gamma}(k) = \begin{cases} \sqrt{2d}, & k > 0 \\ \\ 0, & k = 0. \end{cases}$$

Remark 2. The proof of the theorems 3.1, 3.2 works in a very general situation (see [32], where we with J. Görtner have used absolutely the same arguments). The only difference is a lower estimation in the theorem 3.2. We have used the estimation

$$\min_{x \in B_{\bar{t}}} \xi(x) \geq -(1 + \varepsilon)\sqrt{2d \ln R^-(t)},$$

i.e. the symmetry of $\xi(x)$: $\xi(x) \stackrel{\text{dist}}{=} -\xi(x)$. If the negative part of the potential $\xi(x)$ is "stronger' than positive, it is necessary (see [32]) to use more sound percolation arguments.

Remark 3. The following theorems are the small part of the results of our second paper with J. Görtner, which is now in preparation and is a natural continuation of [32].

Theorem 3.3. For $p \in N$ and $t \to \infty$

$$\ln \langle u^p(t, 0) \rangle = \frac{t^2 p^2}{2} - 2d\,kpt + \bar{\bar{0}}\,(t). \tag{3.15}$$

Theorem 3.4. μ-a.s for $t \to \infty$

$$\ln u(t, 0) = t\sqrt{2d \ln t} - 2d\,kt + \bar{\bar{0}}\,(t). \tag{3.16}$$

The proof of both theorems is based on the combination of the probabilistic and spectral ideas. It shows that connections between localization theory and intermittency are very deep. We will begin from the theorem 3.3, which is simpler. For the sake of technical simplicity I consider only the case $p = 1$, general case is similar to this one.

We have already the lower estimation of the form (3.15) (see lower estimation in the theorem 3.1). It's enough to prove (3.15) as an upper estimation.

We will start from the trivial cut-off procedure. Let us consider

$$\langle u(t, 0) \cdot I_{N(t) \geq t^{1+\epsilon}} \rangle$$

$$= \langle \exp(\int_0^t \xi(x_s)ds) \cdot I_{N(t) \geq t^{1+\epsilon}} \rangle \leq \exp\left(\frac{t^2}{2}\right) \cdot$$

$$\cdot P\{N(t) \geq t^{1+\epsilon}\} \leq \exp\left(\frac{t^2}{2} - (1 - \varepsilon)t^{1+\epsilon}\ln t + \bar{\bar{0}}\,(t^{1+\epsilon})\right)$$

We have used the same upper estimation as in theorem 3.1 and (3.14).

Due to lower estimation of the theorem 3.1 from (3.14) we have that

$$\langle u(t,x)\rangle \underset{t\to\infty}{\tilde{\sim}} \langle u(t,x)I_{N(t)<t^{1+\epsilon}}\rangle \leq \tilde{u}(t,x). \tag{3.18}$$

Here $\tilde{u}$ is a solution of the following problem

$$\frac{\partial\tilde{u}}{\partial t} = k\triangle\tilde{u} + \xi(x)\tilde{u}$$

$$\tilde{u}(0,x) = 1, \quad t\geq 0, \quad x\in S_t^d \tag{3.18'}$$

and S_t^d is a cube $|x|\leq [t^{1+\epsilon}]$ with periodic boundary conditions. The number ν_t of points in S_t^d has an order $t^{d(1+\epsilon)}$.

Operator H in the right part of (3.18') (in fact it is a finite matrix) has a discrete spectrum $E_{k,t}$, $k=1,2,\cdots\nu_t$ and orthonormal basis, $\psi_{k,t}(x), x\in S_t^d$. Of course,

$$\tilde{u}(t,x) = \sum_{k=1}^{\nu_t}(1\cdot\psi_{k,t})\psi_{k,t}(x)\exp(E_{k,t}\cdot t) \tag{3.19}$$

and

$$\|\tilde{u}(t,\cdot)\|_{L_2} \leq \exp(\max_k E_{k,t}\cdot t)\|\tilde{u}(0,\cdot)\| =$$

$$\leq ct^{\frac{d}{2}(1+\epsilon)}\exp\{t\cdot\max_k E_{k,t}\}.$$

It means, that

$$\langle u(t,0)\rangle \leq ct^{\frac{d}{2}(1+\epsilon)}\langle\exp\{t\max_k E_{k,t}\}\rangle. \tag{3.20}$$

Factor $t^{\frac{d}{2}(1+\epsilon)}$ is not dangerous, as it contributes to the exponent only a logarithmical term. We have to estimate only $\langle\exp\{t\max_k E_{k,t}\}\rangle$. Consider variational series of the r.v. $\xi(x)$, $x\in S_t^d$:

$$\xi_{(1)} > \xi_{(2)} > \cdots > \xi_{(\nu_t)}.$$

From the minimax principal it follows that

$$\max_k E_{k,t} \leq \max_k \tilde{E}_{k,t}$$

where $\tilde{E}_{k,t}$ are the eigenvalues of the operator

$$\tilde{H} = k\triangle + \xi_{(2)} + \delta_0(x)(\xi_{(1)} - \xi_{(2)}). \tag{3.21}$$

Thus,

$$\max_k E_{k,t} \leq \xi_{(2)} + f_t(\xi_{(1)} - \xi_{(2)}), \tag{3.22}$$

where $f_t(a)$ is a nonnegative eigenvalue of the operator

$$k\Delta + a\delta_0(x), \quad x \in S_t^d, \quad a > 0. \tag{3.23}$$

The expression for $f_t(a)$ has been appearing many times in the intermittency theory (see [14], [53], [75] and so forth), but usually in the case of the full lattice $\mathbf{Z}^d$ instead of S_t^d.

If $S_t^d = [-n, n]^d$, $\nu_t = (2n)^d$, then applying the Fourier transform, we can find that $f_t(a)$ is a maximum root E_0 of the equation

$$\frac{1}{a} = \frac{1}{(2\pi)^d} \sum_{x \in S_t^d} \left(\frac{2\pi}{2n}\right)^d \frac{1}{E + 2kd - 2k\Phi\left(\frac{2\pi x}{2n}\right)}. \tag{3.24}$$

In the limit $t \to \infty$, $n = n(t) \to \infty$ this equation comes to the standard form [53], [75]:

$$\frac{1}{a} = \frac{1}{(2\pi)^d} \int_{[-\pi,\pi]^d} \frac{d\phi}{E + 2dk - 2k\Phi(\phi)} \tag{3.24'}$$

and in both formulas

$$\Phi(\phi) = \sum_{i=1}^{d} \cos \phi_i, \quad i = 1, 2, \cdots d. \tag{3.24''}$$

Because of the convexity of $\frac{1}{x}$, equation (3.24) gives (due to Jensen inequality) that

$$\frac{1}{a} > \frac{1}{E_0 + 2kd - \frac{1}{(2n)^d} \sum_{x \in S_t^d} \Phi\left(\frac{\pi x}{n}\right)} = \frac{1}{E + 2kd},$$

i.e.

$$E_0 = f_t(a) > a - 2kd \tag{3.25}$$

of course, $E_0 = f_t(a) \le a$, because $\Delta \le 0$. Using expansion

$$\frac{1}{E + 2kd - 2k\Phi\left(\frac{\pi x}{n}\right)} = \sum_{s=0}^{\infty} \frac{\left[-2k\Phi\left(\frac{\pi x}{n}\right)\right]^s}{(E + 2kd)^{s+1}}, E > 0$$

and standard formulas of the summation of trigonometrical series we can get

$$\frac{1}{a} = \frac{1}{E + 2kd} \left(1 + \sum_{s=1}^{\infty} \frac{(2k)^{2s} \cdot c_s}{(E + 2kd)^{2s}}\right).$$

It allows to obtain (using (3.25)) the following result:

$$E_0 \le a - 2dk + \sum_{s=1}^{\infty} \frac{(2k)^{2s} c_s}{a^{2s-1}}, \quad a > 2kd.$$

If $a \to \infty$

$$E_0 = f_t(a) = a - 2kd + \frac{k^2 c_2}{a} + \bar{0}\left(\frac{1}{a}\right).$$
(3.26)

Now everything is ready for the final calculations. We have

$$\langle \exp\{t \max_k E_{k,t}\}\rangle \le$$

$$\le \langle \exp\{t(\xi_{(2)} + f_t(\xi_{(1)} - \xi_{(2)}))\}\rangle$$

$$= \iint_{x>y} p_t(x,y) e^{t(y+f_t(x-y))} dx\, dy,$$

where $p_t(x,y)$ is a common distribution density of $\xi_{(1)}, \xi_{(2)}$ and is equal to

$$p_t(x,y) = \nu_t(\nu_t - 1) P^{\nu-2}\{\xi < y\} \phi(x) \phi(y),$$

$$\phi(x) = (2\pi)^{-1/2} \exp\left(-\frac{x^2}{2}\right),$$
(3.27)

i.e.

$$P_t(x,y) \le c\nu_t^2 \exp\left(-\frac{x^2+y^2}{2}\right) \le ct^{2d(1+\varepsilon)} \exp\left(-\frac{x^2+y^2}{2}\right).$$

As earlier, the factor $t^2 d(1+\varepsilon)$ will contribute only const lnt to the exponent of the final answer and we have to evaluate

$$I_t = \iint_{x>y} e^{t(y+f_t(x-y)) - \frac{x^2}{2} - \frac{y^2}{2}} dx\, dy$$
(3.28)

$$= \iint_{\substack{x>y \\ x-y<\delta t}} + \iint_{\substack{x>y \\ x-y>\delta t}} = I_t^{(1)} + I_t^{(2)}.$$

Constant δ soon will be defined but (as $f_t(a) \le a!$)

$$I_t^{(1)} \le \iint_{0<x-y<\delta t} e^{t(y+(x-y)) - \frac{x^2}{2} - \frac{y^2}{2}} dx\, dy$$

$$= \iint_{0<x-y<\delta t} e^{tx - \frac{x^2+y^2}{2}} dx\, dy$$

$$= \iint_{0<v<\delta\sqrt{2}t} e^{\frac{t}{\sqrt{2}}(u+v) - \frac{u^2+v^2}{2}} du\, dv$$

(changing of variables: $u = \frac{x+y}{\sqrt{2}}, v = \frac{x-y}{\sqrt{2}}$)

$$= ce^{t^2/2} \int_{0<v<\delta\sqrt{2}} e^{-\frac{1}{2}\left(v - \frac{t}{\sqrt{2}}\right)^2} dv \le ce^{t^2/2} e^{-\frac{t^2}{2}\left(\delta\sqrt{2} - \frac{1}{\sqrt{2}}\right)^2}$$

for every $\delta < \frac{1}{2}$! It means, that

$$I_t^{(1)} \leq c \exp\left(\frac{t^2}{2} - \gamma\frac{t^2}{2}\right), \quad \gamma > 0$$

i.e. $I_t \sim I_t^{(2)}$ (for arbitrary $\delta < \frac{1}{2}$, to say $\delta = \frac{1}{3}$).

But at the set $x - y > \frac{1}{3}t$ we can use formula (3.26). We obtain

$$I_t^{(2)} = \iint\limits_{x>y} e^{\ t\left(y + x - y - 2dk + \overset{0}{=}\left(\frac{1}{t}\right)\right)}\ p_t(x,y)\ dxdy$$

$$= \iint\limits_{x>y} e^{tx - 2dkt + \overset{0}{=}(1)} p_t(x,y) dxdy \sim$$

$$\sim \text{const } \exp\left(\frac{t^2}{2} - 2dkt\right).$$

We have proved now the theorem 3.3, but in a stronger form

$$ln\langle u(t,0)\rangle = \frac{t^2}{2} - 2kdt + \overset{0}{=}(ln\,t). \tag{3.29}$$

Further analysis shows that

$$\langle u(t,x)\rangle = \frac{t^2}{2} - 2kdt - 2d\,ln\,t + \overset{=}{0}\,(ln\,t). \tag{3.30}$$

The proof of the theorem 3.3 is very instructive: the main contribution to the $\langle u(t,0)\rangle$ gives one extremely high peak of $\xi(x)$ (of order $t/2$), located exactly at the point $x_0 = 0$.

The proof of the theorem 3.4 is based on the same spectral ideology. I will describe only the "sceleton" of the proof, which contains all principal steps, the reader must reconstruct technical details.

About **lower estimation.** We can use the same plan, as in the theorem 3.2, but make more delicate calculations.

First of all, instead of estimations (3.13) for the $\overset{max}{|x|\leq n}\xi(x)$, we may employ the following statement.

For arbitrary $\varepsilon > 0$, the number of the events

$$A_x^+ = \{\xi(x) \geq \sqrt{2d\,ln\,n + (1 + \varepsilon)ln\,ln\,n}\} \tag{3.31}$$

is finite μ-a.s.

But for the $\varepsilon < 0$, we have μ-a.s. infinitely many such events.

It is important that

$$\sqrt{2d \ln n + c \ln \ln \ln n} = \sqrt{2d \ln n} + \overset{0}{=} \left(\frac{\ln \ln n}{\sqrt{\ln n}} \right).$$

Second, we can consider the radius of the ball B_t^-, (which was equal earlier to $R_t^- = t^{1-\varepsilon}$) to be equal to $\dfrac{t}{\ln^2 t}$. The same calculations as earlier will give

$$u(t,x) \geq \exp(t \max_{x \in B_t^-} \xi(x) - 2dkt + \overset{=}{0}(t))$$

and applying (3.31), we will achieve the lower estimation from Theorem 3.4.

Upper estimation. If we chase the system of balls in the corresponding part of the theorem 3.2 more carefully:

$$R_0 = 0, \quad R_1(t) = t \ln^2 t, \cdots R_n(t) = t \ln^{2n} t \cdots$$

and apply (3.31) we'll get, as earlier, the "cut-off" statement:

$$\langle u(t,x) \rangle \sim \langle u(t,x) I_{N(t) \leq t \ln^2 t} \rangle$$

$$u(t,x) \sim \tilde{u}(t,x) (\mu - \text{ a.s.}) \tag{3.18''}$$

For $\tilde{u}(t,x)$ we have the same equation (3.18'), only $S_t^d = \{x : |x| \leq t \ln^2 t\}$.

Spectral theorem gives immediately

$$u(t,x) \leq \exp\{t \max_k E_{k,t} + \overset{0}{=}(\ln t)\} \tag{3.32}$$

and the problem is to find μ-a.s. asymptotical behavior of this upper boundary of the spectrum of operator $H = k\Delta + \xi(x), \quad x \in S_t^d$.

Let us fix small $\varepsilon > 0$ and consider the random set ("high peaks"):

$$\prod_{\varepsilon,t} = \{x : \xi(x) > (1-\varepsilon)\sqrt{2d \ln t}, \quad x \in D_t^d\}.$$

For large t, the number of the points in $\prod_{\varepsilon,t}$ has the order $t^{d(1-(1-r\varepsilon))} > t^{2d\varepsilon}$. I want to prove that the distances between these points are sufficiently large for large t.

Consider (for some $\delta > 0$) the event

$$B_{t,\varepsilon,\delta} = \{\exists(x_1, x_2 \in \prod_{\varepsilon,\delta} : |x_1 - x_2| > t^\delta\}$$

Obviously,

$$\mu(B_{t,\varepsilon,\delta}) \le c(t \ln^2 t)^d t^{6d} P^2 \{\xi(x) > (1-\varepsilon)\sqrt{2d \ln t}\}$$

$$\le c \ln^d t \frac{t^{d+\delta d}}{t^{2d(1-\varepsilon)^2}}.$$

We can consider only integer t (because we are working on $\mathbf{Z}^d$.) It is evident, that for some $\delta = \delta(\varepsilon) > 0$

$$\sum_t \mu(B_{t,\varepsilon,\delta}) < \infty$$

and Borel-Cantelly lemma shows that μ-a.s. for all $t \ge t_0(\omega)$ and all toruses S_t^d we have

$$\min_{x_i,x_j \in \prod_{\varepsilon,t}} |x_i - x_j| > t^\delta.$$

Now,

$$\max_k E_{k,t} \le \max_k \tilde{E}_{k,t},$$

where $\tilde{E}_{k,t}$ are the eigenvalues of the hamiltonian

$$\tilde{H} = k\Delta + \tilde{\xi}(x), \quad x \in S_t^d, \ \tilde{\xi}(x) = \begin{cases} \xi(x), & x \in \Pi_{\varepsilon,t} \\ \\ (1-\varepsilon)\sqrt{2d \ln t}, & x \notin \Pi_{\varepsilon,t}. \end{cases} \tag{3.33}$$

Thus (as in the theorem 3.3),

$$\tilde{E}_{k,t} = (1-\varepsilon)\sqrt{2d \ln t} + \bar{E}_{k,t}$$

and $\bar{E}_{k,t}$ are associated with operator $\bar{H}$, whose potential has "very rare" peaks:

$$\bar{H} = k\Delta + \sum_{x_i \in \prod_{\varepsilon,t}} \eta_i \delta_{x_i}(x),$$

$$\eta_i = (\xi(x_i) - (1-\varepsilon)\sqrt{2d \ln t})_+. \tag{3.34}$$

There are high ones (of the order $\varepsilon\sqrt{2d \ln t}$) among these peaks, and they give the main contribution to the $\max_k \bar{E}_{k,t}$ (or $\tilde{u}(t,x)$).

Interaction between these peaks is extremely small, because the distances are $|x_i - x_j| > t^\delta$, $i \ne j$. As a result, for our estimations we can use different peaks separately. The following general lemma explains the meaning of this "physical statement".

Lemma 3.1. Let us consider the following Schrödinger operator on the lattice $\mathbf{Z}^d$

$$H = k\Delta + \sum_{i=1}^\infty a_i \delta_{x_i}(x)$$

$$a_i \ge 0, \quad \sup a_i = A, \quad |x_i - x_j| \ge B. \tag{3.35}$$

For arbitrary small $\delta > 0$, we can find the (large enough) constants A_0, B_0 such that for $A > A_0$, $B > B_0$ the resolvent $(H - E)^{-1}$ is analytical in the domain

$$E \geq A - 2dk + \delta, \qquad (3.36)$$

i.e.

$$\sup \sum(H) < A - 2dk + \delta.$$

This lemma is working on the torus S_t^d as well (for $t \to \infty$).

Proof of the lemma. For every point x_i we can construct two balls around it:

$$B_i^1 = \{x : |x - x_i| < \frac{B}{4}\}, \quad B_i^2 = \{x : |x - x_i| < \frac{B}{2}\}$$

and introduce the blocks (compare with lecture 7)

$$(i) = B_i^2, \quad i = 1, 2, \cdots$$

$$(0) = \mathbf{Z}^d / \ (\cup_{i=1}^{\infty} B_i^1).$$

Let γ be a graph with the vertexes $(0), (1), \cdots$ and the following transitions $(i) \to (0)$, $i = 1, 2, \cdots$; $(0) \to (i)$ for arbitrary $i = 1, 2, \cdots$.

First entrance point from (0) to (i) locates on the ∂B_i^1, first entrance point from (i) to (0) locates on ∂B_i^2. If A, B are sufficiently large, the positive eigenvalues of the blocks' have operators $H^{(i)}$, $i = 1, 2, \cdots$ has the estimation

$$E^{(i)} > A - 2dk + \delta/2$$

(see consideration of the previous theorem), and for the resolvent kernel

$$R_{x,y}^{(i)} = \left(H^{(i)} - (A - 2dk + \delta)\right)^{-1}(x, y), \quad x \in \partial B_i^1, \quad y \in \partial B_i^2$$

we can prove the estimation

$$|R^{(i)}(x, y)| \leq \left(\frac{2dk}{A + 2dk + \delta}\right)^{|x-y|} \cdot \frac{c}{\delta} \leq$$

$$\leq \left(\frac{c_0}{A}\right)^{B/4} \cdot \frac{c_1}{\delta}. \qquad (3.37)$$

For the resolvent of (0) - block we have similar (and trivial) estimation, which is the consequence of the path expansion (see previous chapter):

$$R^{(0)}(x, y) = \left(H^{(0)} - (A - 2dk + \delta)\right)^{-1}(x, y)$$

$$x \in \partial B_i^{(2)}, \quad u \in \partial B_j^1; \quad i, j = 1, 2, \cdots \qquad (3.38)$$

$$|R^{(0)}(x, y)| \leq C_1 \left(\frac{2dk}{A + 2dk + \delta}\right)^{|x-y|}.$$

Following the idea of the lecture 7, we can "glue together" the resolvents of the different blocks $(i), i = 0, 1, \cdots$ and obtain the "full" resolvent. The corresponding formulas have a form (x_1 is the center of $B_1^{1,2}$, $\quad y \in (k)$)

$$R_E(x_1, y) = (H - E)^{-1}(x_1, y) =$$

$$= \sum_{\tilde{\gamma}:(0) \to (k)} R_E^{(\tilde{\gamma})}(x_1, y). \tag{3.39}$$

If $\tilde{\gamma} : (1) \to (0) \to (i_1) \to (0) \to \cdots$, then

$$R_E^{(\tilde{\gamma})}(x_1, y) = \sum_{Z_1 \in \partial B_1^2, Z_2 \in \partial B_{Z_1}^1, \cdots} R^{(1)}(x_1, Z_1) R_E^{(0)}(Z_1, Z_2) R_E^{(i_1)}(Z_2, Z_3) \cdots. \tag{3.40}$$

Applying the estimations (3.37), (3.38) of the block resolvents, it's easy to prove the convergence of (3.40) for $E > A - 2dk + \delta$.

From the lemma 3.1 it follows that for $t \geq t_0(\omega, \delta)$ and μ-a.s

$$\max_k \tilde{E}_{k,t} \geq \max_{x + S_t^d} \xi(x) - 2dk + \delta$$

and (3.31) gives us immediately the statement of the theorem 3.4 (upper estimation).

Remark 3.1. The lemma 3.1 does not present any information about the structure of the eigen functions $\psi_{k,t}(x)$ on S_t^d. If we will estimate $|\xi_{(k)} - \xi_{(l)}|, k \neq l$ (these differences can not be "too small", i.e. there are no resonants between different high peaks), we can prove that the "upper" eigenfunctions on S_t^d "locate" near the single maxima $\xi_{(k)}, k = 1, 2, \cdots$ of $\xi(x)$, $x \in S_t^d$.

This version of the localization theorem is very important and gives the additional information about the remainder term in (3.30) and the global structure of the "intermitted" solution $u(t, x)$.

Additional and very important information is related with asymptotical properties of the Lifshitz tails for the integrated density of states $N(E)$ (see lecture 5). The logarithmical asymptotics of $N(E)$, $E \to \infty$ is known (see [14], [49]), but for many models, including gaussian i.i.d.r.v. $\xi(x)$, we can find an exact asymptotics for $N(E)$ (details in [32]).

Lecture 9. Non-Stationary Anderson Parabolic Model.

The subject of this lecture is the investigation of asymptotical properties (intermittency) of the solutions of the following stochastic partial differential equation (SPDE):

$$\frac{\partial u(t,x)}{\partial t} = k \triangle u(t,x) + \xi_t(x) u(t,x)$$

$$u(0,x) \equiv 1, \quad t \ge 0, \quad x \in \mathbf{Z}^d \tag{3.41}$$

This equation describes the mean concentration of the particles for the reaction diffusion equation in the non-stationary (turbulent) RM (see introduction). But the real sense of the problem (3.41) is related with the fact, that this equation is a prototype of the more important evolution problems in the turbulent RM, such as heat propagation in the random flow, dynamo-problem etc. (about dynamo-problem see the next lecture).

As earlier, $k > 0$ is a diffusion coefficient, $k\triangle f(d) = k \sum_{|x'-x|=1} (f(x') - f(x))$ is a lattice Laplacian generating on $\mathbf{Z}^d$ a symmetrical random walk with continuous time (and the rate $2dk$ of the jumps).

Potential $\xi_t(x)$ is a generalized random function

$$\xi_t(x) = \dot{W}_t(x), \quad x \in \mathbf{Z}^d, \quad t \in R_+^1 \tag{3.42}$$

where $\{W_t(x), \quad x \in \mathbf{Z}^d\}$ is a family of the standard independent Wiener processes. In different terms, $W_t(x)$ is a gaussian field on $R_+^1 \times \mathbf{Z}^d$ with the properties

$$\langle W_t(x) \rangle = 0, \quad \langle W_{t_1},(x_1)W_{t_2}(x_2) \rangle = (t_1 \wedge t_2)\delta(x_1 - x_2). \tag{3.43}$$

In the spirit of the general SPDE theory (see, for example, [65]) we have to consider this family as a Wiener process in the corresponding functional (Hilbert) space. Of course, it can't be a standard $L_2(\mathbf{Z}^d)$ space, because

$$\{W_t(\cdot)\} \notin L_2(\mathbf{Z}^d), \quad \mathbf{1}(x) \notin L_2(\mathbf{Z}^d).$$

The simplest possibility is to work in the weighted Hilbert space $L_{2,\alpha}(\mathbf{Z}^d)$ with the norm

$$\|f(\cdot)\|_\alpha^2 = \sum_{x \in \mathbf{Z}^d} |f(x)|^2 \alpha^{|x|},$$

$$|x| = |x_1| + \cdots + |x_d|, \quad 0 < \alpha < 1. \tag{3.44}$$

Such special weight $w(x) = \alpha^{|x|}$ is especially convenient, because

$$\alpha\|f(\cdot)\|_\alpha^2 \le \|f(\cdot + h)\|_\alpha^2 \le \frac{1}{\alpha}\|f(\cdot)\|_\alpha^2, \tag{3.45}$$

where $f(\cdot + h)$ is a shift of $f(\cdot)$, $|h| = 1$.

It is easy to check that $\{W_t(\cdot)\} = W_t$ is a Wiener process in $L_{2,\alpha}(\mathbf{Z}^d)$, i.e. covariance operator (3.43) is a nuclear one in $L_{2,\alpha}(\mathbf{Z}^d)$. Of course (by the definition)

$$\xi_t(\cdot) = \dot{W}_t = \dot{W}_t(\cdot) \tag{3.46}$$

is a "white noise" in the Hilbert space $L_{2,\alpha}(\mathbf{Z}^d)$.

Our probabilistic space $(\Omega_m, \mathcal{F}_m, \mu)$ will have now a special structure: $\Omega_m = L_{2,\alpha}(\mathbf{Z}^d)$,
$\mathcal{F}_m = \bigotimes_{x \in \mathbf{Z}^d} \mathcal{F}(x)$, where $\mathcal{F}(x)$ is a σ-algebra of events related with $W_t(x)$, $t \in R_+^L$,
$\mu = \bigotimes_{x \in \mathbf{Z}^d} \mu(x)$. Of course, σ-algebra $\mathcal{F}_m$ contains the filtration $\mathcal{F}_{\leq t}$ of σ-subalgebras of events "before moment $t > 0$".

We have to understand the equation (3.41) as stochastic integral equation

$$u(t,x) = 1 + k \int_0^t \Delta u(s,x) ds + \int_0^t u(s,x) dw_s(x) \tag{3.47}$$

where the solution $u(t, \cdot) \in L_{2,\alpha}(\mathbf{Z}^d)$ and $\mathcal{F}_{\leq t}$ is measurable ($\mathcal{F}_{\leq t}$ adapted).

Last term is the Ito's stochastic integral. Of course, instead of Ito's integral we can use the Strationovich's integral

$$\int_0^t u(s,x) \circ d\, W_s(x). \tag{3.48}$$

If both integrals have a sense, we have a trivial relation between them:

$$\int_0^t u(s,x) \circ d\, w_s(x) = \int_0^t u(s,x) d\, w_s(x) +$$
$$+ \frac{1}{2} [u(\cdot),\, w(\cdot)]_0^t = \int_0^t u(s,x) d\, w_s(x) + \frac{1}{2} \int_0^t u(s,x) ds \tag{3.49}$$

or symbolically

$$u(t,x) \circ d\, w_t(x) = u(t,x) d\, w_s(x) + \frac{1}{2} u(t,x) dt. \tag{3.49'}$$

Formula (3.49') gives the possibility to turn from Ito's SPDE to the Strationvich's SPDE. Namely (in the obvious notations)

$$u^{(S)}(t,x) = u^{(I)}(t,x) \exp(t/2). \tag{3.50}$$

Stratonovich's form of the solution $u^{(S)}(t,x)$ is more natural physically. First of all, it reflects symmetry of the time and second, solution $u^{(S)}(,x)$ is a limit of the solutions with "very short" time correlations. It's possible to find the following result in our (with R. Carmona) memoir [15] containing a full analysis of the equation (3.41).

Theorem. Let us consider equation

$$\frac{\partial u^\epsilon}{\partial t} = k \triangle u^\epsilon + \frac{1}{\sqrt{\epsilon}} \xi(\frac{t}{\epsilon}, x) u^\epsilon$$

$$u^\epsilon(0, x) \equiv 1.$$

(3.41')

Here $\xi(t, x)$ is a regular qaussian field (with continuous realizations in time):

$$\langle \xi(t, x) \rangle = 0, \quad \langle \xi(t_1, x_1) \xi(t_2, x_2) \rangle = \Gamma(t_1 - t_2) \delta(x_1 - x_2).$$

After rescaling $t \to \frac{t}{\epsilon}$, $\xi \to \frac{1}{\sqrt{\epsilon}} \xi$, the field $\frac{1}{\sqrt{\epsilon}} \xi(\frac{t}{\epsilon}, x)$ converges (in the sense of distribution in the Schvartz space) to the "white noise" with a variance $\sigma^2 = 2\pi \hat{\Gamma}(0) = \int_{R^1} \Gamma(\tau) d\tau$.

Solution $u^\epsilon(t, x)$ for every $\epsilon > 0$ exists in the classical sense. If $\epsilon > 0$, then $u^\epsilon(t, x) \xrightarrow{\text{Dist}} u^0(t, x)$ and $u^0(t, x)$ is the solution of the Stratonovich's SPDE:

$$\frac{\partial u^0}{\partial t} = k \triangle u^0 + \sigma u(t, x) o\, d\, w_t(x)$$

$$u^0(0, x) \equiv 1.$$

(3.41")

We had formulated our initial problem in the language of SPDE's, but didn't prove an existence and an uniqueness. This is easy, because Laplacian $\triangle$ is a bounded operator. There is a possibility to use the general theory [65], but direct analysis is simpler than the testing of conditions of the abstract theorems.

Theorem 3.3. Equation (3.47) has an unique solution in $L_{2,\alpha}(\mathbf{Z}^d)$.
Proof: Using the inequality $(a + b + c)^2 \leq 3a^2 + 3b^2 + 3c^2$, Schwartz inequality and trivial estimation (3.45), it's easy to prove, that for $t \leq T$

$$\langle \| \int_0^t \triangle u(s, x) ds \|_\alpha^2 \rangle \leq C(\alpha, T) \int_0^t \langle \| u(s, x) \|_\alpha^2 \rangle ds$$

$$\langle \| \int_0^t u(s, x) d\, w_s(x) \|_\alpha^2 \rangle \leq C(\alpha, T) \int_0^t \langle \| u(s, x) \|_\alpha^2 \rangle ds$$

$$\langle \| 1 + \int_0^t \triangle(s, x) ds + \int_0^t u(s, x) d\, w_s(x) \|_\alpha^2 \rangle \leq$$

$$\leq C(\alpha, T)[1 + \int_0^t \langle \| u(s, \cdot) \|_\alpha^2 \rangle ds].$$

(3.51)

A priori estimation (3.51) together with Bellmann's lemma show (as in the finite dimensional case), that Picard approximations

$$u_0(t, x) \equiv 1$$

$$u_{n+1}(t, x) = 1 + \int_0^t \triangle u_n(s, x) ds + \int_0^t u_n(s, x) dw_s(x)$$

(3.52)

converge to the limit in the Hilbert space $L_{2,\alpha}(\mathbf{Z}^d) \times L_2(\Omega_m, \mu)$.

Uniqueness theorem (due to (3.51)) is also standard. More sophisticated uniqueness theorems (for the more general conditions about space correlations of the "white noise" $\xi_t(x)$) see [15].

The following result allows to obtain the Kac-Feynman representation of the solution

Theorem 3.4. Solution $u^{(S)}(t, x)$ is given by the formula

$$u^{(S)}(t, x) = E_x \exp\{\int_0^t d\, w_s(X_{t-s})\}. \tag{3.53}$$

Here X_{t-s} is a trajectory of the random walk x_s (in inverse time) and Stochastic integral in the exponent has a trivial sense: if $0 < s_1 < s_2 < \cdots < s_\nu < t$ are the moments of the jumps for x_s and $x_s \equiv x_\nu,\quad s \in [0, s_1),\quad x_s = x_{\nu-1},\quad s \in [s_1, s_2), \cdots$ $x_s \equiv x,\quad s \in [s_\nu, t]$, then

$$\int_0^t d\, w_s(x_{t-s}) = (w_{s_1} - w_0)(x_\nu) +$$
$$+ (w_{s_2} - w_{s_1})(x_{\nu-1}) + \cdots + (w_t - w_{s_\nu})(x). \tag{3.53'}$$

The proof is almost trivial and bases on standard Markov construction: $u(t + \Delta t, x) = E_x\{\exp(\int_t^{t+\Delta t} d\, w_s(x_{t+\Delta-s})) u(t, x_\Delta)\}$ etc. See details in [15] for more general case.

It follows from (3.53) that the solution $u(t, x, \omega)$, which is an ergodic and homogeneous field on $\mathbf{Z}^d$ for fixed t, has all the moments:

$$\left\langle \left[u^{(S)}(t, x)\right]^p \right\rangle = \left\langle \left[E_x \exp\left\{\int_0^t d\, w_s(x_{t-s})\right\}\right]^p \right\rangle \leq$$

$$\leq E_x \langle \exp\{p \int_0^t d\, w_s(x_{t-s})\}\rangle = \tag{3.54}$$

$$= E_x \exp\left\{\frac{p^2 t}{2}\right\} = \exp\left\{\frac{p^2 t}{2}\right\}.$$

In the special case $k = 0$, when $x_s \equiv x_0 = x$, the estimation (3.54) is precise, because the solution of the ordinary SDE

$$d\, u_t = u_t \, o \, dw_t$$

has a form

$$u_t = \exp(w_t).$$

In this case, of course,

$$\langle [u_t]^p \rangle = \exp\left\{\frac{p^2 t}{2}\right\}.$$

To prove the asymptotical intermittency $(t \to \infty)$ for the random field $u(t, x)$, we have to study the statistical moments, find scale $A(t)$ (see previous lecture 8) and calculate the corresponding Ljapunov exponents. In the case of the stationary model we used Kac-Feynman representation. In the nonstationary case (and it is the most essential property of the SPDE) we have much more: the exact equations for the correlation functions of the field $u(t, x)$.

Theorem 3.5. For each integer $p \geq 1$, each $t \geq 0$ and $x = (x_1, \cdots, x_p) \in \mathbf{Z}^{pd}$ let us set

$$m_p^{(l)}(t, x) = m_p^{(l)}(t, x_1, \cdots x_p) = \langle u^{(l)}(t, x_1) \cdots u^{(l)}(t, x_p) \rangle.$$

Then these moments (correlation functions) satisfy the following "p-particle" parabolic equation

$$\frac{\partial m_p}{\partial t} = k(\triangle_{x_1} + \cdots + \triangle_{x_p}) m_p + \left(\sum_{i<j} \delta(x_i - x_j) \right) m_p \tag{3.56}$$

$$m_p(0, x) \equiv 1.$$

Equation (3.56) can be written down in the form

$$\frac{\partial m_p}{\partial t} = H_p m_p, \quad H_p = k(\triangle_{x_1} + \cdots + \triangle_{x_p}) + V_p(x) \tag{3.56'}$$

$$V_p(x) = \sum_{i<j} \delta(x_i - x_j), \quad p > 1; V_1(x) \equiv 0$$

and Hamiltonian H_p is a classical "p-particle" Schrödinger operator on the lattice $\mathbf{Z}^{pd}$ with the binary interaction $\delta(x - y)$. We can, of course, discuss this equation in $L_{a,\alpha}(\mathbf{Z}^d)$, but now it is more convenient to work in $L_2(\mathbf{Z}^d)$, where operator H_p is self-ajoint and has the "nice" spectral properties. The spectral analysis of the p-particle Schrödinger operators in $L_2(\mathbf{Z}^d)$ or $L_2(R^d)$ represents a large chapter of the modern mathematical physics (see, for example, the monograph [67], which is mainly dedicated to multiparticle Schrödinger operator).

In the future we will be using mainly Ito's form and $u(t, x)$, $m_p(t, x)$ will be meaning $u^{(l)}(t, x), m_p^{(l)}$ etc.

Proof of the Theorem 3.5. Function $\phi(z_1, \cdots, z_p) = z_1 \cdots z_p$ is smooth and we can apply the Ito's formula to the compound process

$$y_t = u(t, x_1) \cdots u(t, x_p).$$

We know, that

$$dy_t = d\,u(t,x_1)u(t,x_2)\cdots u(t,x_p) + \cdots$$

$$+ u(t,x_1)\cdots u(t,x_{p-1})d\,u(t,x_p) + \tag{3.57}$$

$$+ \frac{1}{2}\sum_{i\leq j}\frac{\partial^2\phi}{\partial x_i \partial x_j}\cdot d\,u(t,x_i)d\,u(t,x_j).$$

But $d\,u(t,x) = \triangle u(t,x)dt + u(t,x)dw_t(x)$ and $\dfrac{\partial^2\phi}{\partial z_i \partial z_j} = z_1\cdots z_{i-1}$

$z_{i+1}\cdots z_{j-1}z_{j+1}\cdots z_p$, $\dfrac{\partial^2\phi}{\partial z_i^2} = 0$. After substitution of these expressions in (3.57) and calculation of expectation we'll obtain (3.56).

The following result reduces the problem of the moments Ljapunov exponents to the spectral theory.

Theorem 3.6. For every $p \geq 1$ and the scale $A(t) \equiv t$ there exists

$$\gamma_p^{(I)}(k) = \lim_{t\to\infty}\frac{ln\langle[u^{(I)}(t,x)]^p\rangle}{t} \tag{3.58}$$

and

$$\gamma_p^{(I)}(k) = \max\sum(H_p) \tag{3.59}$$

where $\sum(H_p)$ is a spectrum of the self-ajoint operator H_p in $L_2(\mathbf{Z}^{pd})$.

Let us remark, that $\gamma_p^{(I)}(k) \geq 0$ (because $\max\sum(\triangle) = 0$ and $V_p(x) \geq 0$).

Theorem 3.7 is a version of the Perron's theorem about asymptotical behavior of the positive semigroups, but in non-compact case. Abstract form of the theorems of this type see [67], for additional details see [15].

I'll give a short draft of the proof containing some additional information.

Let us consider the ball $Q_R = \{x : |x| \leq R\}$ and the "cut-off" problem

$$\frac{\partial u_R}{\partial s} = k\triangle u_R + \xi_t(x)u_R, \quad |x| \leq R, \quad t \geq 0 \tag{3.60}$$

$$u_R(0,x) = I_{|x|<R}; \quad u(t,x) = 0, \quad |x| = R.$$

Of course (Ito's form)

$$u_R(t,x) = E_x \exp\left\{-\frac{t}{2} + \int_0^t dw_s(x_{t-s})\right\}\cdot I_{\max_{s\in[0,t]}|x_s|} < R. \tag{3.61}$$

It is obvious, that

$$u_R(t,x0 \leq u(t,x); \quad |x| \leq R, \quad t > 0.$$

But according to Perron's theorem,

$$\langle u_R^p(t,x)\rangle = \exp(tH_p^{(R)})u_R(0,x)_{t\to\infty}^{\sim}$$

$$\exp(E_0^{(R)}\cdot t)\psi_{R,0}(x)(u_R(0,\cdot)\cdot\psi_{R,0}(x)).$$

(3.62)

Here $H_p^{(R)}$ is a p-particle operator $H_p^{(R)}$ is a ball $\{x: |x| < R\}^p$ with zero boundary condition, $E_0^{(R)}$ is a maximum eigenvalue of $H_p^{(R)}$ and $\psi_{R,0}(x)$ is a corresponding (positive!) eigenfunction.

It follows that from (3.62)

$$\liminf_{t\to\infty} \frac{ln\langle u^p(t,x)\rangle}{t} \geq E_0^{(R)}$$

(3.63)

for arbitrary R. But for $R \to \infty$

$$\lim E_0^{(R)} = \max \sum(H_p)$$

and we have the "half" of the theorem 3.5 (lower estimation).

The main problem with the upper estimation is a consequence from the fact, that $m_p(o,x) \notin L_2$. But we can take $R = R(t) = t^{1+\varepsilon}$. Using (as in the lecture 8) the trivial bound

$$P\{\max_{s\in[0,t]}|x_s|\} \leq P\{\nu_t \leq t^{1+\varepsilon}\} \leq \exp\left(-t^{1+\varepsilon}\right),$$

we can check, that for $t \to \infty$

$$\langle u^p(t,x)\rangle = m_p(t,x,\cdots x) \leq$$

$$\leq u_{t^{1+\varepsilon}}^p(t,x))\cdot(1+\bar{\bar{0}}\,(1)) = e^{tH_p^{(R)}}m_p(0,x).$$

But according to the spectral theorem

$$\|e^{tH_p^{(R)}}m_p(0,\cdot)\| \leq \exp\left(tE_0^{(t^{1+\varepsilon})}\right)\|m_p(0,\cdot)\|$$

$$\leq \exp\left(tE_0^{(t^{1+\varepsilon})}\right)(2t)^{d(1+\varepsilon)}$$

and for $t \to \infty$ it gives us the upper estimation

$$\limsup_{t\to\infty} \frac{ln\langle u^p(t,x)\rangle}{t} \leq E_0 = \max\sum(H_p).$$

(3.64)

Of course, the positivity of the operator $\exp(t\,H_p^{(R)})$ played a central role. Let us remark that semigroup $\exp(t\,H_p^{(R)})$ is positive in both classical senses: in the sense of quadratic forms and in a Markov sense (positivity of the matrix elements).

138

Corollary 3.1. For every $p \geq 1$

$$\gamma_p^{(I)}(k) = \max \sum(H_p) = \sup_{\psi:\, \|\psi\|_2 = 1} (H_p\psi, \psi) =$$

$$= \sup_{\psi:\, \|\psi\|=1} \left(-k \sum_{|h|=1,\, x \in Z^{dp}} (\psi(x+h) - \psi(x))^2 + \sum_{x \in Z^{dp}} V_p(x)\psi^2(x) \right). \tag{3.65}$$

It is a classical variational form for the upper boundary of the spectrum of the self-ajoint operator.

Expression in the brackets is (for fixed ψ) a linear nonincreasing function of k. As a result, we have

Corollary 3.2. For every $p \geq 1$, the moments Ljapunov exponent $\gamma_p^{(I)}(k)$ is a convex nonincreasing function of diffusivity $k \geq 0$.

Let us remember (see remark after (3.54)) that for $k = 0$

$$\gamma_p^{(I)}(0) = \frac{p^2}{2} - \frac{p}{2} = \frac{p(p-1)}{2},$$

i.e.

$$\frac{\gamma_p^{(I)}}{p}(0) = \frac{p-1}{2} \uparrow$$

and we have an intermittency for $k = 0$.

As functions $\gamma_1(\kappa)$ and $\gamma_2(\kappa)$ are continuous, we have (for sufficiently small κ).

$$\gamma_1(\kappa) < \frac{\gamma_2}{2}(\kappa)$$

and it gives the full intermittency:

$$\gamma_1 < \frac{\gamma_2}{2} < \frac{\gamma_3}{3} < \cdots$$

(compare with previous lecture).

But we can obtain essentially better estimations of the moment Ljapunov exponents $\gamma_p(k)$. Corresponding analysis is based on exact formulas for γ_1 and γ_2.

Theorem 3.7. (Ito form)

1. $\gamma_1(k) \equiv 0$

2. $\gamma_2(k)$ is equal to the upper boundary of the spectrum of the two-body Schrödinger operator

$$\tilde{H} = 2\,k\Delta + \delta_0(x), \tag{3.66}$$

which is a result of the removing of the center of mass for the operator

$$H_2 = k(\Delta_{x_1} + \Delta_{x_2}) + \delta(x_1 - x_2).$$

2a. If $d = 1, 2$ then $\gamma_2(k) > 0$ and it is a unique positive solution of the equation

$$1 = \frac{1}{(2\pi)^d} \int_{S^d} \frac{d\phi}{2k\ \Phi(\Phi) + \gamma}, \quad \Phi(\phi) = 2\sum_{i=1}^{d}(1 - \cos\ \phi) \qquad (3.67)$$

2b. If $d \geq 3$ and

$$k \geq k_r = \frac{1}{(2\pi)^d} \int_{S^d} \frac{d\phi}{2\Phi(\phi)}, \qquad (3.68)$$

then $\gamma_2(k) = 0$. If $k < k_r$, then again $\gamma_2(k) > 0$ and is a positive root of the same equation (3.67).

The proof is trivial. Equation for the first moment $m_1(t, x)$ is a pure parabolic one:

$$\frac{\partial m_1}{\partial t} = k \triangle m_1, \quad m_1(0, x) \equiv 1,$$

i.e.

$$m_1(t, x) \equiv 1, \quad t \geq 0, \quad x \in \mathbf{Z}^d$$

and $\gamma_1(k) \equiv 0$. For the second moment we have equation

$$\frac{\partial m_2}{\partial t} = k(\triangle_{x_1} m_2 + \triangle_{x_2} m_2) + \delta(x_1 - x_2)m_2$$

$$m_2(0, x_1, x_2) \equiv 0.$$

It means, of course, that

$$m_2(t, x, x_2) = \psi(t, x_1 - x_2)$$

and for $\psi(t, z)$ we have

$$\frac{\partial \psi}{\partial t} = 2k \triangle \psi + \delta(z)\psi = \tilde{H}\psi, \quad \psi(0, z) \equiv 1.$$

According to theorem 3.7, there are only two possibilities
a) either operator $\tilde{H}$ has no positive eigenvalue $E = E_0(k)$, then $\gamma_2(k) = 0$
b) or for some $E_0(k) > 0$ there exists a positive eigenfunction $\psi_0(x)$:

$$\tilde{H}\psi_0 = (2\ k\ \triangle + \delta_0)\psi_0 = E_0\psi_0 \qquad (3.69)$$

and $\gamma_2(k) = E_0(k)$.

Equation (3.69) can be solved in the Fourier domain. Function $\psi(\phi)$ is, of course, the symbol of Laplacian $(-\triangle)$ (see details in the previous lecture, where we have done the same calculation, or in [15]). Solution of (3.69) produces immediately the statements 2a, 2b of the theorem 3.7.

Corollary 3.3. In the low dimensions $d = 1, 2$ for arbitrary $k > 0$, the family of the fields $u(t, x)$ (i.e. solution of (3.41), which can be understood in both senses: Ito's or

Stratonovich's) has a property of full asymptotical intermittency. In different terms, inequalities

$$\gamma_1(k) < \frac{\gamma_2}{2}(k) < \ \ < \frac{\gamma_p}{p}(k) < \tag{3.70}$$

hold for all $k > 0$.

If $d \geq 3$, then (3.70) has place only for $k < k_{2,r} = \frac{1}{(2\pi)^d} \int_{S^d} \frac{d\phi}{2\Phi(\phi)}$.

If $k > k_{2,r}$, then for $t \to \infty$

$$m_2(t, x) = \langle u^2(t, x) \rangle \to \ \ \text{const}$$

$$m_1(t, x) \equiv 1,$$

i.e. the family of distributions of $u(t, x)$ is tight for $t \to \infty$.

The latter statement of this corollary requires the proof. It is a simple calculation based on the direct Fourier analysis of the equation

$$\frac{\partial m_2}{\partial t} = 2k \, m_2 + \delta_0 \, m_2$$

$$m_2(0, z) \equiv 1.$$

We have no place to discuss such technical details.

In the physical language, the statement of the corollary 3.3 means the following. In dimension $d = 1, 2$ or for $d \geq 3$ with the sufficiently small diffusivity k, the reaction-diffusion particle system related with the Anderson's model (3.41) has a very strong property of aggregation. When $t \to \infty$ "almost all" particles concentrate in a "very high" and "very rare" peaks. It is an analogy of the localization process. If $d \geq 3$ and $k > k_2$, we have a nontrivial mean density of the particles between peaks. Roughly speaking, together with "localization" we have "delocalization", i.e. some analogy of the extended states.

Such "phase transition" with respect to diffusivity k plays the same role in parabolic theory (3.41) as a hypothetical Anderson's "mobility edge' in the quantum disordered physics.

To prove the existence of the "high peaks" generating intermittency of the moments m_p, $p > 2$, we have to estimate the higher Ljapunov exponents $\gamma_p(k)$, $p > 2$. The following result is a part of the general theory [14]. It illustrates the power of the direct probabilistic methods in the pure spectral context. Functional analytical proof of the same result is not trivial.

Theorem 3.8. For any $d \geq 1$ and $p \geq 2$ one has

$$\frac{\gamma_p}{p}(k) \geq \left(\frac{p-1}{2} - 2dk \right)_+ \tag{3.71}$$

and for $p = 2m$

$$\frac{\gamma_p}{p}(k) \le \frac{p-1}{2}\gamma_2\left(\frac{k}{p-1}\right). \tag{3.72}$$

Expression for $\gamma_2(k)$ was obtained in the theorem 3.7.

From (3.71), (3.72) it follows that $\gamma_p(k) > 0$ in dimension $d \ge 3$ iff $k < k_{p,r}$ where $k_{p,r}$ satisfies the estimates:

$$\frac{p-1}{4d} \le k_{p,r} \le \left(2\left[\frac{p}{2}\right] - 1\right)k_{2,r} \tag{3.73}$$

i.e.

$$c_1(d) \le \frac{k_{p,r}}{p} \le c_2(d). \tag{3.73'}$$

Proof. Let $x_t = (x_t^{(1)}, \cdots, x_t^{(p)})$ be a random walk on $\mathbf{Z}^{pd}$ with the generator $k(\Delta_{x_1} + \cdots + \Delta_{x_p})$. Then (3.56) and Kac-Feynman formula give:

$$m_p(t, x) = E_x \exp\{\int_0^t V_p(x_s)ds\}$$

$$V_p(x) = \sum_{i<j} \delta(x_i - x_j), \quad p > 2. \tag{3.74}$$

Lower estimation. It $x = o$, then

$$m_p(t, 0) \ge E_0 \exp\{\int_0^t V_p(x_s)ds\}I_{x_s \equiv 0, s \in [0, t]}$$

$$= \exp\{tV_p(0)\}P\{x_s \equiv 0, s \in [0, t]\} =$$

$$= \exp\left\{\frac{t\,p(p-1)}{2}\right\} \cdot \exp\{-2d\,kpt\}.$$

I.e.

$$\gamma_p(k) \ge \max(0, \frac{p(p-1)}{2} - 2dkp),$$

which is equivalent to (3.71).

Upper estimation. For simplicity, let's consider just the case $p = 4$. We have

$$m_p(t, 0) = E_0 \exp\{\int_0^t[\delta(x_s^{(1)} - x_s^{(2)}) + \delta(x_s^{(3)} - x_s^{(4)})]\}ds$$

$$+ \int_0^t[\delta(x_s^{(1)} - x_s^{(3)}) + \delta(x_s^{(2)} - x_s^{(4)})]ds +$$

$$\int_0^t[\delta(x_s^{(1)} - x_s^{(4)}) + \delta(x_s^{(2)} - x_s^{(3)})]ds\}$$

$$\le E_0\{\exp 3 \int_0^t[\delta(x_s^{(1)} - x_s^{(2)}) + \delta(x_s^{(3)} - x_s^{(4)})]\}$$

$$\left(\text{inequality } abc \le \frac{a^3 + b^3 + c^3}{3}\right)$$

$$\le [E_0\{\exp 3 \int_0^t \delta(x_s^{(1)} - x_s^{(2)})\}]^2$$

(independence of $\left(x_s^{(1)}, x_s^{(2)}\right)$ and $\left(x_s^{(3)}, x_s^{(4)}\right)$).

The expression in the square brackets is the second moment, not with potential $\delta(\cdot)$, but with the potential $3\delta(\cdot)$.

Using definition of $\gamma_2(k)$, after rescaling of the time in the equation

$$\frac{\partial m}{\partial t} = k \triangle m + 3\delta_0 m$$

we'll obtain (for $p = 4$) the upper estimation (3.72).

Now we'll discuss briefly the problem of almost sure Ljapunov exponent for the equation (3.41), which we will understand (as in the biggest part of lecture) in the Ito's form.

First problem is how to define Ljapunov a.s. exponent? The simplest version is to work with the localized initial data, instead of stationary data of the form $u(0, x) \equiv 1$. In the moment analysis it was not essential.

Theorem 3.9. Let $q(s, x, t, y)$, $s \le t$, $x, y \in \mathbf{Z}^d$ be a fundamental solution of the SPDE (3.41), i.e. the solution of the problem

$$\frac{\partial q}{\partial t} = k \triangle_y q(s, x, t, y) + \xi_t(y) q(s, x, t, y) \tag{3.75}$$

$$t \ge s, \quad q(s, x, s, y) = \delta_x(y).$$

Then μ-a.s for all $x, y \in \mathbf{Z}^d$, $s > 0$ there exists

$$\lim_{t \to \infty} \frac{\ln q(s, x, t, y)}{t} = \tilde{\gamma}(k). \tag{3.76}$$

By the definition, $\tilde{\gamma}(k)$ is an a.s. Ljapunov exponent.

Existence and uniqueness of the fundamental solution in $L_{2,\alpha}(\mathbf{Z}^d)$ can be proved exactly by the same method as in the theorem 3.3. It follows from the uniqueness theorem that for $s < u < t$

$$\sum_z q(s, x, u, x) q(u, z, t, y) = q(s, x, t, y) \tag{3.77}$$

and Kac-Feynman representation shows that for $s < t$ and arbitrary $x, y \in \mathbf{Z}^d$

$$q(s, x, t, y) > 0. \tag{3.78}$$

In different terms, "transition operators"

$$Q_{s,t}\psi(x) = \sum_y q(s, x, t, y)\psi(y)$$

make up the homogeneous positive hemygroup (similar to Markov's hemygroups, connected with non-homogeneous Markov chains). Properties (3.77), (3.78) imply that

$$q\,(s,x,u,x)\,q\,(u,x,t,x) < q\,(s,x,t,x)$$

$$s < u < t$$

or

$$ln\,q\,(s,x,u,x) + ln\,q\,(u,x,t,x) < ln\,q\,(s,x,t,x).$$

Notice that the terms factors in the right hand side are independent random variables (because σ-algebras $\mathcal{F}_{[s,u]}$ and $\mathcal{F}_{[u,t]}$ are independent). Their distributions depend only on the lengths of the corresponding time intervals. In such situation we can use (a very special case) of the Kingman's sub additive ergodic theorem (see [22]) and conclude that μ-a.s. there exists

$$\tilde{\gamma}(k) = \lim_{t\to\infty}\ \frac{ln\,q\,(s,x,t,x)}{t}. \tag{376'}$$

Using trivial inequalities of the form

$$q\,(s,x,t,x) \geq g\,(s,x,s_1,y)\,q\,(s_1,y,s_2,y)\cdot$$

$$q\,(s_2,y,t,x), \quad s < s_1 < s_2 < t \tag{3.77'}$$

one can prove that limit in (3.76) does not depend on x and more

$$\tilde{\gamma}(k) = \lim_{t\to\infty}\ \frac{ln\,q(s,x,t,y)}{t}\quad (\mu - \text{a.s.}). \tag{3.76}$$

Additional analysis (which is not trivial) shows, that for the solution $u(t,x)$ of the problem (3.41) we have the same result

$$\lim_{t\to\infty}\ \frac{ln\,u(t,x)}{t} = \tilde{\gamma}(k)\quad (\mu - \text{a.s}).$$

Let us remark, that

$$u(t,x) = \sum_{y\in\mathbf{Z}^d} q(0,y,t,x)$$

i.e.

$$\liminf_{t\to\infty}\frac{ln\,u(t,x)}{t} \geq \tilde{\gamma}(k)\quad (\mu - \text{a.s}).$$

Theorem 3.10. Function $\tilde{\gamma}(k)$ is continuous for $k \geq 0$ (see [11], where corresponding statement is proved in essentially more general situation).

Idea of the proof. Let us consider instead of (3.41) the similar equation:

$$\frac{\partial u}{\partial t} = \triangle u + \sigma\,\xi_t(x)u \tag{3.78}$$

$$u(0,x) \equiv 1.$$

Using the Kac-Feynman representation (as in the theorem 3.4) one can easily prove, that

$$\frac{\ln u(t,x)}{t} = \psi_{t,x}(\sigma)$$

is a convex function of σ. It means, that

$$\lim_{t \to \infty} \frac{\ln u(t,x)}{t} = \psi(\sigma) \quad (\mu - \text{a.s})$$

is a convex (and by the same token) continuous function of σ. More,

$$\lim_{t \to \infty} \frac{\psi(\sigma)}{\sigma} \quad \text{exists}$$

i.e. function $\psi(\sigma)$ is asymptotically linear. But the scaling $t \to \dfrac{t}{k}$ give us possibility to move from the form (3.41) to the form (3.78) of our parabolic problem. It is not difficulty to show, that

$$\tilde{\gamma}(k) = k\psi\left(\frac{1}{k}\right) \tag{3.79}$$

and is equivalent to the statement of the theorem 3.10 (in fact, we had proved more).

If $k = 0$, then as we know

$$u(t,x) = \exp\{w_t(x) - \frac{t}{2}\}$$

i.e.

$$\tilde{\gamma}(0) = \lim_{t \to \infty} \frac{w_t - \frac{t}{2}}{t} = -\frac{1}{2} \quad (\mu - \text{a.s}).$$

For small k we have now $\tilde{\gamma}(k) < 0$, i.e. density of the particles between peaks in the case of full intermittency tends to 0 exponentially if $t \to \infty$. It is an additional essential detail for the "physical picture" of our process (see discussion after corollary 3.3).

There are no explicit formulats or algorithms for the calculation of $\tilde{\gamma}(k)$. Nevertheless, if $k \to 0$ we can give the following asymptotical estimations, which are nontrivial and interesting from the physical point of view.

Theorem 3.11 (see [15]) For suitable const c_1, $c_2 > 0$ and $k \to 0$

$$-\frac{1}{2} + c_2 \frac{1}{\ln \frac{1}{k}} < \tilde{\gamma}(k) < -\frac{1}{2} + c_1 \frac{\ln \ln \frac{1}{k}}{\ln \frac{1}{k}}. \tag{3.90}$$

It means, that for small k a.s. Ljapunov exponent $\tilde{\gamma}(k)$ increases extremely fast. The proof of (3.99) is the most difficult part of our paper with R. Carmona [15].

Lecture 10   Cell-dynamo

Cell-dynamo model introduced in the lecture 1 helps to obtain not only a very good quality picture of the evolution of the real magnetic fields in the turbulent flow of conducting fluid (main examples are magnetic fields of the sun and stars), but allows to get (after choice of the suitable parameters) the excellent quantity agreement with the experimental data. In the most general (non-linear) form, this model is described by the Stochastic parabolic equation of the Anderson type, but in the space of the vector-functions:

$$\frac{\partial \vec{H}}{\partial t} = k \triangle \vec{H}(t,x) + \xi_t(x) o \vec{H} - \varepsilon f(\vec{H})\vec{H}$$

$$\vec{H}(0,x) = \vec{H}_0(x,\omega), \quad t \geq 0, \quad x \in \mathbf{Z}^d.$$

Here $\vec{H}(t,x) \in R^N$ is a vector (of the dimension N) on the product $R_+^1 \times \mathbf{Z}^d$, k is a diffusion coefficient, $f(\cdot) : R^N \to R_+^1$ is a scalar nonconstant function (usually $f(\vec{H}) = |\vec{H}|^\beta$, $\beta > 0$), ε is the parameter of nonlinearity. At last (it is the most important point), $\xi_t(x)$ is a matrix white noise, i.e.

$$\xi_t(x) = \{\dot{W}_{ij}(t), i, j = 1, 2, \cdots N\} \tag{3.77}$$

and $W_{ij}(t)$ are standard independent Wiener processes. In different terms,

$$\langle \dot{W}_{ij}(t)\dot{W}_{i'j'}(t')\rangle = \delta(t-t')\delta_{ii'}\delta_{jj'}. \tag{3.77'}$$

In dynamo applications usually $N = 3$. We will understand (in contrast to previous lecture) equation (3.76) in the Strationovich's sense. We know that this sense is "correct physically", i.e. is the result of the limiting transition from the case of a very "short" time correlations.

The condition (3.77') which guarantees the full isotropy of the cell-dynamo problem, is very essential. In non-isotropic case we can, of course, to write down some (for example, moments) equations, but their analysis is a very hard problem.

The vector case is essentially more complicated than the scalar one. To understand this, let us put $k = \varepsilon = 0$. We have in this case N-dimensional SDE

$$d\vec{H}(t,\cdot) = \{dw_{ij}(t,\cdot)\}o\vec{H}(t,\cdot). \tag{3.78}$$

If $N = 1$ (scalar case), the solution of (3.78) is trivial:

$$H(t,\cdot) = H_0(0,\cdot)\exp\{w(t)\}$$

(see previous lecture). But in general case $N > 1$, because the matrix product is not commutative, we have only the "formula"

$$\vec{H}(t,\cdot) = \prod_{s=0}^{t}(I + dw_{ij}(s))\vec{H}_0(0), \tag{3.78'}$$

where matrix "multiplicative integral" $\prod\limits_{S=0}^{t}(I+dw_{ij}(s))$ is to be understood in common sense:

$$\prod_{s=0}^{t}(I+dw_{ij}(s)) = \lim_{\Delta\to 0}\prod_{k=0}^{[\frac{t}{\Delta}]}\exp(w_{ij}((k+1)\Delta)-w_{ij}(k\Delta)) \tag{3.79}$$

(see details in the excellent Mac Kean's book [50], where it's possible to find not only the construction of the multiplicatie matrix Stochastic integral, but the very interesting applications).

Asymptotical behavior of the product of large number of independent identically distributed random matrixes (usually, elements of $SL(N,R)$ or symplectic group) is a central topic of the Ferstenberg's theory [24]. He introduced in this context the notion of the Ljapunov spectrum and proved (under some conditions) the simplicity of this spectrum. I told already, that Ferstenberg's theory was the main technical moment in the initial version of one-dimensional localization theory [28], [52], [19]. The strongest results related with Ferstenberg theory see in [29].

Unfortunately, expressions for Ljapunov exponents in the general theory [24] (or in [29]) are not constructive. Some analytical (asymptotical) results in this area are related only with the random elements of $SL(2,R)$, more precisely, with a transfer matrix of the one-dimensional Schrödinger operator [7].

It is very interesting, that in continuous (isotropic) case (3.79) Ljapunov exponent (in the different sense) can be calculated explicitly. It was done in the old paper of E. Dynkin [23]. E. Dynkin has used geometrical language (Martin boundaries, diffusion of ellipsoids an so on), but the final results have a clear "Ferstenberg's" form (before publication [24]!). Many authors had repeated E. Dynkin's theorem using different methods [62], [50].

We will start with the finite dimensional problem (3.78), (3.79). Only after full analysis of the Ljapunov spectrum in this special case (where we will be close to [62]), we'll be back to the general problem (3.76). At the beginning, we'll investigate the linear case ($\varepsilon=0$) and then present some results in the non-linear situation.

Let us recall the definition of the a.s. Ljapunov and moment Ljapunov exponents in the form convenient in the future. This definition is based on the general Oseledtz theorem [63].

If $A(t)=\prod\limits_{s=0}^{t}(I+dw(s))$ is a transfer-matrix of our SDE (3.78) and $(h^{(1)},\cdots h^{(k)})$, $k\leq N$ is an initial orthonormal basis, we can introduce the system of the vectors

$$H^{(1)}(t)=A(t)h^{(1)},\cdots,\quad H^{(k)}(t)=A(t)h^{(k)}$$

and corresponding Grahm determinant

$$\triangle_k = [\det\{H^{(i)} \cdot H^{(j)}\}]. \tag{3.80}$$

Then μ-a.s.

a) there exists the limit

$$\lim_{t\to\infty} \frac{\ln \triangle_k}{2t} = \tilde{\gamma}_1 + \cdots + \tilde{\gamma}_k, \quad k = 1, 2, \cdots N \tag{3.81}$$

b) there exists the limit

$$\lim \frac{\ln\langle\triangle_k^p\rangle}{2t} = \gamma_1^{(p)} + \cdots + \gamma_k^{(p)}, \quad \begin{matrix} k = 1, 2, \cdots N \\ p = 1, 2, \cdots \end{matrix}. \tag{3.82}$$

Probably it's better to define the moments Ljapunov exponent by the formulas

$$\lim_{t\to\infty} \frac{\ln \langle\triangle_k^{p/2}\rangle}{t} = \tilde{\gamma}_1^{(p)} + \cdots + \cdots + \tilde{\gamma}_k^{(p)}, \tag{3.82}$$

but we prefer to work only with the analytical functions.

From the formulas (3.81), (3.82) we can find the a.s. Ljapunov exponents $\tilde{\gamma}_i$, $i = 1, 2, \cdots N$ and moments Ljapunov exponents $\gamma_i^{(p)}$, $i = 1, 2, \cdots N$ of the order $p = 1, 2, \cdots$.

Let us remark, that formula (3.79) gives (for $k = N$)

$$\triangle_N = \det \left[\lim_{\triangle\to 0} \prod_{k=0}^{[\frac{t}{\triangle}]} \exp(w(k+1)\triangle) - w(k\triangle)) \right]^2 \tag{3.83}$$

$$= \exp\{2\, tr(w(t) - w(0))\} = \exp\{2\sqrt{N}\tilde{w}(t)\}$$

where $\tilde{w}(t)$ is a standard one-dimensional Wiener process. From the (3.85), it follows that

$$\tilde{\gamma}_1 + \cdots + \tilde{\gamma}_N = 0 \tag{3.84}$$

$$\gamma_1^{(p)} + \cdots + \gamma_N^{(p)} = Np^2, \quad p = 1, 2, \cdots. \tag{3.84'}$$

The equality (3.84) has an interesting sense. In many applications (for example, in magneto hydrodynamics, where $d = N = 3$) the matrix potential has a zero trace (condition of the incompressibility). Of course, then $\det A(t) = \det(\prod_{s=0}^{t}(I + dw(s))) = 1$, i.e. $A(t) \in SL(N, R)$ and $\tilde{\gamma}_1 + \cdots + \tilde{\gamma}_N = 0$. In our situation, of course, $A(t) \notin SL(N, R)$, nevertheless asymptotical result (3.84) is absolutely the same. In some sense, the "flow" generated by matrix potential $\dot{W}_{ij} = \dot{W}$ is "asymptotically incompressible".

In addition, we can remark that general theory [63] gives

$$\tilde{\gamma}_1 \geq \tilde{\gamma}_2 \geq \cdots \geq \tilde{\gamma}_N$$

$$\gamma_1^{(p)} \geq \gamma_2^{(p)} \geq \cdots \geq \gamma_N^{(p)}, \quad p = 1, 2, \cdots. \tag{3.85}$$

We will see in the future, that these inequalities are strict and moreover, that for $i = 1, 2, \cdots N$

$$\gamma_i^{(1)} < \frac{\gamma_i^{(2)}}{2} < \frac{\gamma_i^{(3)}}{3} < \cdots \tag{3.86}$$

(asymptotical intermittency). Both these facts are the results of elementary calculation (similar to the Newman paper [62]).

Theorem 3.11 Ljapunov exponents for the SDI (3.78) is represented by the following explicit formulas

$$\tilde{\gamma}_k = \frac{N - 2k + 1}{2}, \quad k = 1, 2, \cdots N \tag{3.87}$$

$$\gamma_k^{(p)} = p^2 + p\left(\frac{N - 2k + 1}{2}\right) \tag{3.88}$$

$$\frac{\gamma^{(p)}}{p} = p + \frac{N - 2k + 1}{2} = p + \tilde{\gamma}_k. \tag{3.88'}$$

As it was mentioned above, a.s. Ljapunov exponents were calculated by E. Dynkin [23], another approach see [62]. Similar results see in [50]. Moment Ljapunov exponents have not been discussed in these papers, but corresponding calculations are not difficult.

Proof of the theorem. Vectors $h^{(1)}, \cdots h^{(k)}$ and matrix random process $A(t, \omega)$ are generating the system of the imbedded subspaces $L^{(1)}(t) \subset L^{(2)}(t)$ $\subset \cdots \subset L_t^{(k)}$ ("flag"). Here $L^{(i)}(t) = \text{span} \{A(t)h^{(1)},$ $\cdots A(t)h^{(i)}\}, i = 1, 2, \cdots k$. Let $\bar{H}^{(1)}(t) = H^{(1)}(t)$, $\bar{H}^{(2)}(t), \cdots \bar{H}^{(k)}(t)$ is a result of orthogonalization of the vectors $H^{(1)}(t), \cdots, H^{(k)}(t)$:

$$\bar{H}^{(1)}(t) = H^{(1)}(t); \quad \bar{H}^{(2)}(t) = H^{(2)}(t) - \alpha_{11} H^{(1)}(t),$$

$$(\bar{H}^{(2)}(t) \cdot H^{(1)}(t)) = 0; \quad \bar{H}^{(3)}(t) = H^{(3)}(t) - \alpha_{21} H^{(2)} - \alpha_{22} H^{(1)}$$

etc. Of course,

$$\Delta_k^{(t)} = \det\{(H^{(i)} \cdot H^{(j)}), \quad i, j = 1, 2, \cdots k\} =$$

$$= (\bar{H}^{(1)}(t) \cdot \bar{H}^{(1)}(t)) \cdots (\bar{H}^{(k)}(t) \cdot \bar{H}^{(k)}(t)) = \prod_{i=1}^{k} (\bar{H}^{(k)})^2. \tag{3.89}$$

Since $\text{span} \{H^{(1)}(t), \cdots, H^{(i)}(t)\} = \text{span} \{\bar{H}^{(1)}(t), \cdots \bar{H}^{(i)}(t)\} = L^{(i)}(t)$, we have

$$\Delta_k(t + dt) = \det\{[\bar{H}^{(i)}(t) + dA\bar{H}^{(i)}(t)] \cdot [\bar{H}^{(j)}(t) + dA\bar{H}^{(j)}(t)]\}. \tag{3.90}$$

How to understand matrix differential dA? Let us return to the initial equation (3.78) (in the Stratonovich form). As in the scalar case (lecture 9), it can be written down in the Ito form

$$d\vec{H}(t) = dw \cdot \vec{H} + \frac{1}{2}\vec{H}dt. \qquad (3.78')$$

It means, of course (in the obvious notations):

$$(S)\vec{H}(t) = (I)\vec{H}(t)\exp(t/2)$$

$$(S)\tilde{\gamma}_i = (I)\tilde{\gamma}_i + 1/2, \quad i = 1,2,\cdots$$

$$(S)\frac{\gamma_i^{(p)}}{p} = (I)\frac{\gamma^{(p)}}{p} + 1/2.$$

Formulas (3.90) show that we can make all the calculations in the Ito's form and at the last stage we can take into account the additional term $1/2$. This is convenient, because we can use Ito formula.

As $dA(t) = d\{w_{ij}(t)\}$ is a matrix gaussian differential (with independent identically distributed components), the differential

$$dA \cdot H = |\bar{H}^{(i)}| \, dW^{(i)}(t) \qquad (3.91)$$

will be isotropic N-dimensional Wiener vector differential with the variance $|\bar{H}^{(i)}|^2 = \bar{H}^{(i)} \cdot \bar{H}^{(i)}$.

The crucial (but elementary) fact is that for orthoronal $\bar{H}^{(i)}(t)$, $i = 1, 2, \cdots k$, the corresponding Wiener differentials

$$(\vec{H}^{(i)}(t))dw^{(i)}(t), \quad i = 1, r, \cdots$$

are independent. This follows from the relation (notations are tensor!):

$$\langle(dw_{kj} \cdot a_j)(dw_{li} \cdot b_i)\rangle =$$
$$= \delta_{kl} \cdot \delta_{ji}a_jb_i = \delta_{kl}(\vec{a} \cdot \vec{b}). \qquad (3.91)$$

Using the Ito's formula

$$d(\vec{X}(t) \cdot \vec{Y}(t)) = d\vec{X} \cdot \vec{Y} + \vec{X} \cdot d\vec{Y} + d\vec{X} \cdot d\vec{Y}$$

and the obvious fact

$$\vec{a} \cdot \{dw_{ij}\}\vec{b} = |\vec{a}| \, |\vec{b}|dw_{ab}(t),$$

where $dw_{ab}(t)$ is a standard one-dimensional Wiener differential, we can obtain

$$\triangle_k(t + dt) = \begin{vmatrix} (\bar{H}^{(1)} \cdot \bar{H}^{(1)})(t) + 2|H^{(1)}|d\tilde{w}_{(1)} + N|H^{(1)}|^2 dt \cdots \cdots \\ \\ \sqrt{2}|\bar{H}^{(i)}| \, |\bar{H}^{(j)}|d\tilde{w}_{(i,j)} \\ \\ \sqrt{2}|\bar{H}^{(i)}| \, |\bar{H}^{(j)}|d\tilde{w}_{(i,j)} \\ \\ \cdots \cdots (\bar{H}^{(k)} \cdot \bar{H}^{(k)})(t) + 2|\bar{H}^{(k)}|d\tilde{w}_{(k)} + N|H^{(k)}|^2 dt \end{vmatrix}$$

Here $dw_{(i)}$, $i = 1, 2, \cdots k$; $dw_{(ij)}$, $1 \leq i < j \leq k$ are the standard Wiener differentials. Differentials $dw_{(i)}$, $i = 1, 2, \cdot k$ are independent. Expanding this "almost diagonal" determinate, using several times the Ito relation and the obvious fact that $\sum_{i=1}^{k} dw_{(i)} = \sqrt{d}\, dw$, we can receive now the closed Stochastic ODE for the determinant $\triangle_k(t)$:

$$d\triangle_k(t) = 2\sqrt{k}\triangle_k(t)dw(t) + (Nk - k(k - 1))\triangle_k(t)dt. \tag{3.92}$$

Since the equation

$$dx_t = ax_t dw_t + bx_t dt, \quad x_0 = 1$$

has a solution

$$x_t = \exp\left(aw_t + \left(b - \frac{a^2}{2}\right)t\right),$$

we have (from (3.92)) the central formula

$$\triangle_k(t) = e^{2\sqrt{k}w(t) - (Nk - k(k+1))t}. \tag{3.93}$$

According to the definition,

$$(a.s)\lim_{t \to \infty} \frac{\ln \triangle_k(t)}{2t} = \gamma_1 + \cdots + \gamma_k = \frac{Nk - k(k+1)}{2} \text{ (Ito!)} . \tag{3.94}$$

I.e.

$$(I)\tilde{\gamma}_k = \frac{N - 2k}{2}, \quad (S)\tilde{\gamma}_k = \frac{N - 2k + 1}{2}.$$

In addition,

$$\langle \triangle_k^p(t) \rangle = \exp(2kp^2 + p(Nk - k(k + 1))),$$

i.e.

$$\frac{\gamma_1^{(p)} + \cdots + \gamma_k^{(p)}}{p} = kp + \frac{(Nk - k(k + 1))}{2} \quad \text{(Ito)}$$

or

$$(S)\frac{\gamma_k^{(p)}}{p} = p + (S)\tilde{\gamma}_k = p + \frac{N - 2k + 1}{2}$$

$$\frac{\gamma^{(p)} + \cdots + \gamma_N^{(p)}}{p} = pN$$

and the formulas (3.87), (3.88') are proved. Simultaneously we checked on the formulas (3.84), (3.84').

Now we'll discuss the most interesting case, when $k \notin 0$. In the beginning let $\varepsilon = 0$ (linear or kinematic theory). We have to understand equation

$$\frac{\partial H}{\partial t} = k\triangle H + \dot{w}_t(x)oH \tag{3.76'}$$

$$t \geq 0, \quad H = H(t, x) \in R^d$$

as an integral Stochastic equation

$$H(t, x) = H_0(x) + k\int_0^t \triangle H(s, x)ds + \int_0^t dw(s, x)oH \tag{3.76''}$$

in the weighted Hilbert space of the vector functions $L_{2,\alpha}(\mathbf{Z}^d)$ with the norm

$$\|f(x)\|^2 = \sum_{x \in \mathbf{Z}^d} (f(x) \cdot f(x))\alpha^{|x|} \tag{3.95}$$

$$0 < \alpha < 1, \quad f(x) \in R^N, \quad x \in \mathbf{Z}^d.$$

As it was mentioned above, if we prove the existence of the Ito's solution of (3.76''), then (notations are obvious)

$$(S)H(t, x) = (I)H(t, x)\exp\{t/2\}. \tag{3.96}$$

In the probabilistic context, the Ito's calculus is more preferable, as we can use Ito's formulas (compare with the Scalar theory of the lecture 8).

Direct and trivial repetition of the scalar considerations shows, that for the equation (3.76)'' (in both possible forms), the theorem of the existence and uniqueness in $L_{2,\alpha}(\mathbf{Z}^d)$ is valid. Our problem now is to obtain the moments equations and to analyze this equation.

Essential algebraic difference between scalar and vector case is connected with the fact that correlation function of the "magnetic field" $H(t, x)$ is a tensor:

$$M_k = M_k(t, x_1, \cdots x_k) = \{M_{i_1 i_2 \cdots i_k}(t, x_1, \cdots x_k)\} =$$

$$= \{\langle H_{i_1}(t, x_1) \cdot H_{i_2}(t, x_2) \cdots H_{i_k}(t, x_k)\rangle\}.$$

Of course, $i_1, i_2, \cdots i_k \in \{1, 2, \cdots, N\}$. We will study the moment tensor in the Ito's form. Transition to the Stratonovich moment's tensor (according to (3.96)) is trivial:

$$(S)M_k(t, c_1, \cdots x_k) = (I)M_k(t, x_1, \cdots x_k)\exp\left(\frac{kt}{2}\right). \tag{3.97}$$

If the initial "magnetic field" is an isotropic one (we'll discuss only this case, which is natural physically), then

$$M_k(t, \cdot) \equiv 0, \quad k = 1, 3, \cdots.$$

It is the additional reason why we had introduced above only "even" Ljapunov exponents.

Theorem 3.12 Tensor M_p, $p \geq 2$ is a solution of the following "multiparticle" Schrödinger tensor parabolic equation

$$\frac{\partial M_p}{\partial t} = k \sum_{i=1}^{p} \triangle_{x_i} M_p + V_k M_k \tag{3.98}$$

with the initial condition

$$M_p(0, \cdot) = \langle H_{0i_1}, (x_1) \cdots H_{0i_p}(x_p) \rangle.$$

Here

$$V_k \equiv \left\{ V_{i_1 \dots i_k}^{j_1 \dots j_k} \right\} = \sum_{\ell_1 < \ell_2} \delta(x_{\ell_1} - x_{\ell_2}) \delta_{i_1 \dots i_k}^{j_1 \dots j_k} \delta_{i_{\ell_1} i_{\ell_2}} \delta^{j_{\ell_1} j_{\ell_2}}.$$

The proof is simple and almost the same as in the scalar case. We have to calculate $d(H_{i_1}(t, x_1) \cdots H_{i_k}(t, x_k))$, using Ito formula and SDE for $H(t, x)$. The reader can reconstruct the necessary details.

In the most important case $p = 2$ (which is related with magnetic energy) we can construct closed equation for the scalar correlator

$$M_2(t, x_1, x_2) = \langle H(t, x_1) \cdot H(t, x_2) \rangle. \tag{3.99}$$

Theorem 3.13 The scalar correlator $M_2(t, x_1, x_2)$ is a solution of equation

$$\frac{\partial M_2}{\partial t} = k(\triangle_{x_1} + \triangle_{x_2}) M_2 + N \delta(x_1 - x_2) M_2. \tag{3.100}$$

Proof, of course, is the same as earlier, but shorter:

$$d(H(t, x_1) \cdot H(t, x_2)) = (dH(t, x_1) \cdot H(t, x_2)) + (H(t, x_1) \cdot$$

$$\cdot dH(t, x_2)) + (dH(t, x_1) \cdot dH(t, x_2)).$$

But

$$dH(t, x_1) = k \triangle H(t, x) + dw(t, x_1) H(t, x_1),$$

i.e. $(dH(t, x_1) \cdot dH(t, x_2) = dw_{ij}(t, x_1) H_j(t, x_1) dw_{ik}(t, x_2) H_k(t, x_2)$

$$= \sum_{i=1}^{N} dt \cdot \delta(x_1 - x_2) \delta_{jk} H_j(t, x_1) H_k(t, x_2)$$

$$= N dt \delta(x_1 - x_2)(H(t, x) \cdot H(t, x_2)).$$

First two terms, after averaging with respect to μ, will give us $(\triangle_{x_1}+\triangle_{x_2})M_2(t, x_1, x_2)$.

We has solved already the equation (3.100) in the lecture 8. In this case, we earlier decided to use notation $2\gamma_p$ for the even moments of order $2p$.

Upper eigenvalue of the operator $H = 2\triangle + N\delta_0(x_1)$ (the upper boundary of the spectrum of (3.100) for the functions of the form $\psi = \psi(x_1 - x_2)$) is our Ljapunov exponent

$$\gamma_1(k) = N\mu\left(\frac{2k}{N}\right), \tag{3.101}$$

where $\mu(a)$ is the root of the equation

$$a = \frac{1}{(2\pi)^d} \int_{S^d} \frac{d\phi}{4\sum\limits_{i=1}^{d}(1 - \cos\phi_i) + \dfrac{\mu}{a}}. \tag{3.101'}$$

If $d = 1, 2$ we have $\mu = \mu(a) > 0$, $a > 0$. If $d \geq 3$, the equation (3.101') has no solutions for $a > a_r$, i.e. the second moment has no exponential growth for $t \to \infty$ (in fact, this moment is bounded for $k > k_r = \dfrac{Na_r}{2}$).

Let us remark, that for $k = 0$ equation (3.100) gives the obvious solution

$$2\gamma_1(0) = N \Rightarrow \gamma_1(0) = \frac{N}{2} \quad \text{(Ito!)}$$

i.e.

$$\gamma_1(0) = \gamma_1^{(1)}(0) = \frac{N+1}{2} \quad \text{(Stratonovich!)}.$$

This formula is a special case of (3.88'), $p = 1$.

Explicit formulas are not attainable for the higher moments (in general), even in the scalar case. But we can obtain for the Ljapunov a.s exponents and moment exponent (at least for the top exponents, $k = 1$) the minimax representation and estimations, which are similar to the corresponding scalar results (see previous lecture).

Let us start with equation (3.98) for $p = 2\ell$. If initial field $H_0(x, \omega)$ has independent values (for different $x \in \mathbf{Z}^d$) and distribution of $H_0(\cdot, \omega)$ is isotropic, then

$$M_{2\ell}(0, x_1, \cdots, x_{2\ell}) \geq 0.$$

Potential $V_{a\ell}(x_1, \cdots x_{2\ell})$ in (3.98) is also nonnegative. We can rewrite equation (3.98) as an integral one with the kernel $p(s, x_1, y_1) \cdots p(s, x_{2\ell}, y_{1\ell}) = \mathbf{P}(t, x, y)$ generated by operator $k(\triangle_{x_1} + \cdots + \triangle_{x_{2\ell}})$. This "paramatrix" representation gives immediately $M_k(t, x_1, \cdots, x_{2\ell}) \geq 0$ (i.e. all moments of the order 2ℓ are nonnegative).

Now we can repeat the corresponding considerations of the Lecture 9, using Perron theorem and general spectral theorem for symmetrical Hamiltonians. In our case, it will be operator $H_{2\ell}$ in the right part of (2.98). We have to consider this operator in the Hilbert space of the tensor valued function on Z^d.

Theorem 3.14 First moment Ljapunov exponent $\gamma_1^{(2\ell)}(k)$ is given by the formula

$$\gamma_1^{(2\ell)}(k) = \sup \sum (H_{2\ell}) = \sup_{\psi:\|\psi\|=1} -\frac{k}{2} \sum_{|x'-x|=1, x \in Z^d} \|\psi(x') - \psi(x)\|^2 +$$

$$+ (V_{2\ell}\psi, \psi)\}. \quad \text{Here} \tag{3.102}$$
$$\psi(x) = \{\psi_{i_1 \ldots i_{2\ell}}(x), \quad x \in Z^d\} \quad \text{and}$$

$$\|\psi(x)\|^2 = \sum_{x \in Z^d} \psi_{i_1 \ldots i_{2\ell}}(x)\psi^{i_1 \ldots i_{2\ell}}(x).$$

From (3.102) it follows that $\gamma_{2\ell}(k)$ is a decreasing convex (and of course) continuous function of $k \geq 0$. In different terms,

$$\gamma_1^{(2\ell)}(k) = \lim_{t \to \infty} \frac{\ln \langle |H(t, x)|^{2\ell} \rangle}{t}. \tag{3.103}$$

Because $\gamma_2(k) > 0$ for small k (or $d = 1, 2$ and all k, see previous lecture) we have proved the intermittency of $|H(t, x)|$ for small diffusivity (i.e. in the domain, where $\gamma_2 > 0$). Let us remember in addition, that for small k we can use Dynkin's result from the previous theorem 3.11, which corresponds with the case $k = 0$.

Formula (3.102) is not very constructive. It is possible to prove the precise analogy of the corresponding scalar theorem, which gives upper and lower estimations of $\frac{\gamma_{2\ell}(k)}{2\ell}$ for all $k \geq 0$ and

Theorem 3.15 For all $\ell \geq 1$

$$\left(\frac{\gamma_{2\ell}}{2\ell}(0) - 2dk\right)_+ \leq \frac{\gamma_{2\ell}(k)}{2\ell} \leq \frac{2\ell - 1}{2} \gamma_2 \left(\frac{k}{2\ell - 1}\right). \tag{3.104}$$

The proof of the scalar result was based on probabilistic ideas (Kac-Feynman representation and so forth). In vector case, we will use the minimax representation (3.102) and the general functional-analytical concepts.

Lower estimation. Potential tensor $V_{2\ell}(0, \cdots 0)$ has a positive eigenfunction (eigentensor) $\psi_\ell(0) : V_{2\ell}\psi_\ell = \gamma_{2\ell}(0)\psi_\ell$. Corresponding eigenvalue is exactly $\gamma_{2\ell}(0)$ (or, in the old notations $\gamma_1^{(2\ell)}(0)$). Let us consider the test function on $Z^{2\ell d}$:

$$\bar{\psi}(x_1, \cdots x_{2\ell}) = \psi_\ell(0)\delta_0(x_1, \cdots x_{2\ell}).$$

Substitution of this function in the expression (3.102) gives

$$\gamma_{2\ell}(k) \geq -k(\Delta\bar{\psi}, \bar{\psi}) + V_{2\ell}(x)(\bar{\psi}, \bar{\psi}) =$$

$$= \gamma_{2\ell}(0) - 4dk\ell.$$

But, of course, $\gamma_{2\ell}(k) \geq 0$ (because $V_{2\ell} \geq 0$. It allows to obtain the left part in (3.104).

Upper estimation. Again (as in the scalar case), we'll consider the case $2\ell = 4$, $\ell = 2$ for the sake of simplicity.

Equation for the correlation tensor of order 4 has a form

$$\frac{\partial M_4}{\partial t} = k(\triangle_{x_1} + \cdots + \triangle_{x_4}) + N\delta(x_1 - x_2)M_4 +$$

$+$ 5 similar tensor corresponding to pairs (x_1, x_3), (x_1, x_4), (x_2, x_3), (x_2, x_4), (x_3, x_4). Let us introduce the operator of the form

$$\mathcal{H}^{(1,2),(3,4)} = \frac{k}{3}(\triangle_{x_1} + \cdots + \triangle_{x_4}) + \frac{N}{2}(\delta(x_1 - x_2)\langle\cdots\rangle + \delta(x_3 - x_4)\langle\cdots\rangle),$$

which corresponds to all three groups of non-intersecting pairs of indexes 1, 2, 3, 4.

Of course, minimax formula (3.102) means that

$$\gamma_4 \leq 3 \sup_{\|\psi\|=1} (\mathcal{H}^{(1,2),(3,4)}\psi, \psi).$$

But since the operator $\mathcal{H}^{(1,2),(3,4)}$ is a direct sum of two operators $\mathcal{H}^{(1,2)}$, $\mathcal{H}^{(3,4)}$, where, say,

$$\mathcal{H}^{(1,2)} = \frac{k}{3}(\triangle_{x_1} + \triangle_{x_2}) + \frac{N}{2}\delta_0(x_1 - x_2)\langle\cdots\rangle,$$

we have

$$\gamma_4 \leq 6 \sup_{\|\psi\|=1} (\mathcal{H}^{(1,2)}\psi, \psi).$$

Scaling arguments show that

$$\sup_{\|\psi\|} (\mathcal{H}^{(1,2)}\psi, \psi) = \gamma_2\left(\frac{k}{3}\right)$$

and we have proved (for $p = 2\ell = 4$) the upper estimation.

As earlier, the inequalities (3.104) give us for $d \geq 3$ not only the existence of the critical diffusivity $k_{2\ell,r}$, such that

$$\gamma_{2\ell}(k) \equiv 0, \quad k \geq k_{2\ell,r}$$

but the important estimation

$$k_{2\ell,r} \asymp 2\ell, \quad \ell \to \infty.$$

We have no possibility to discuss the (essentially more complicated) problem of the continuity of the a.s. Ljapunov exponent $\tilde{\gamma}_1(k)$, $k \geq 0$.

For $k \to 0$ we can prove (in the special weak form) the statement $\tilde{\gamma}_1(k) \to \tilde{\gamma}_1(0) = \frac{N-1}{2}$ and to give some estimations for the difference $|\tilde{\gamma}_1(k) - \tilde{\gamma}_1(0)|$ (see [58]). Nevertheless, this problem for the vector case is in essence open.

In the conclusion of the lecture we'll discuss briefly the non-linear equation (3.76) for the simplest non-linearity $f(H) = |H|^\beta$, $\beta > 0$.

Theorem 3.16 Uniformly in $k \in [0, k_0]$, where k_0 is a fixed const, and for every $p = 1, 2, \cdots$, the statistical moments of the solution of the cell-dynamo equation (3.76) admit the following asymptotical estimations, if $t \to \infty$, $\varepsilon \to 0$:

$$\left(\frac{c_2}{\varepsilon}\right)^{p/\beta} \leq \langle |H(t,x)|^p \rangle \leq \left(\frac{c_1}{\varepsilon}\right)^{p/\beta}. \tag{3.105}$$

It means that the family of distributions of the vector fields $\varepsilon^{\frac{1}{\beta}} H(t,0)$ (depending on parameters ε, t) is tight. "Physically" it is obvious that there exists a limiting distribution for the field $H(t + \tau, x)$, $t \to \infty$ and the "limiting field" $\bar{H}(\tau, x)$ is ergodic and homogeneous in space and time. Unfortunately, we can not prove this fact. Ergodic theorems for the infinite-dimensional Markov processes is poorly developed branch of the probability theory.

I'll prove theorem 3.16 only for the special case $p = 2$ (which is most important in the physical applications).

Let us consider (as in (3.92)) the scalar correlator

$$M_2(t, x_1 - x_2) = \langle H(t, x_1) \cdot H(t, x_2) \rangle,$$

where (Ito's form)

$$dH(t,x) = k\triangle H(t,x)dt + dw(t,x)H(t,x) - \varepsilon H(t,x) \cdot |H(t,x)|^\beta dt, \quad x \in \mathbf{Z}^d, \quad t \geq 0 \tag{3.106}$$

$$M_2(0, x_1 - x_2) = \sigma_0^2 \delta_0(x_1 - x_2) \tag{3.106'}$$

(Last equality holds, for example, in the case, when $H_0(x, \omega)$ is a field with independent isotropic values).

Applying to the dot product $H(t, x_1) \cdot H(t, x_2)$ the Ito's formula, we'll obtain

$$\frac{\partial M_2^{(t,z)}}{\partial t} = 2k\triangle M_2(t,z) + N\delta_0(z)M_2(t,z) -$$

$$-\varepsilon \langle (H(t, x_1) \cdot H(t, x_2)(|H(t, x_1)|^\beta + H(t, x_2)|^\beta)) \rangle \tag{3.107}$$

$$M_2(0, z) = \sigma_0^2 \delta_0(z).$$

Formula (3.107) is a basis of the moment estimations. Let us start from the upper estimation. Using inequalities

$$|H(t,x_1)|^\beta + |H(t,x_2)|^\beta \geq 2|((H(t,x_1) \cdot H(t,x_2))|^{\beta/2}$$

$$\langle (H(t,x_1) \cdot H(t,x_2))^{1+\beta/2} \rangle \geq \langle (H(t,x_1) \cdot H(t,x_2))) \rangle^{1+\beta/2}$$

and nonegativity of the function $M_2(t,z)$ (it follows, for example, from the Kac-Feynman formula), we can now write down the parabolic non-linear inequality:

$$\frac{\partial M_2(t,x)}{\partial t} \leq 2k \triangle M_2(t,x) + N\delta_0(z)M_2(t,x) -$$

$$-2\varepsilon M_2^{1+\beta/2}(t,z), \tag{3.107'}$$

$$M_2(0,z) = \sigma_0^2 \delta_0(z).$$

For $z = 0$, we can deduce (from (3.107') and using obvious facts: $M_2(t,z) \leq M_2(t,0) \leq \sum_{|z|=1} M_2(t,x)$ etc.) the following relation

$$\frac{\partial M_2(t,0)}{\partial t} \leq 4d\,k\,M_2(t,0) + NM_2(t,0) - 2\varepsilon M_2^{1+\beta/2}(t,0) \tag{3.108}$$

$$M_2(0,0) = \sigma_0^2.$$

Of course, $M_2(t,0) \geq A_2(t)$, where

$$\frac{dA_2}{dt} = (4dk + N)A_2 - 2\varepsilon A_2^{1+\beta/2} \tag{3.109}$$

$$A_2(0) = \sigma_0^2.$$

It is easy to check, that

$$A_2(t) \xrightarrow{t\to\infty} c_0$$

and $(4dk + N)c_0 = 2\varepsilon c_0^{1+\beta/2} \Rightarrow c_0 = \left(\dfrac{4dk + N}{\varepsilon}\right)^{2/\beta}$, and we have proved upper estimation (3.105) in the special case $p = 2$. In the general case we can use the same idea, but calculations are more complicated.

Lower estimation. From equation (3.107) it follows that

$$\frac{\partial M_2(t,0)}{\partial t} \geq (N - 4dk)M_2(t,0) - 2\varepsilon\langle |H(t,x)|^{2+\beta}\rangle \tag{3.110}$$

$$M_2(0,0) = \sigma_0^2.$$

Let $\langle |H(t,x)|^{2+\beta}\rangle = B_\beta(t)$ and $\tilde{A}_2(t)$ is a solution of the equation

$$\frac{d\tilde{A}_2}{dt} = (N - 4dk)\tilde{A}_2 - 2\varepsilon B_\beta(t). \tag{3.111}$$

Upper estimation means that $B_\beta(t) = 0\left(\frac{1}{\varepsilon}\right)^{\frac{2+\beta}{\beta}}$, $t \to \infty$ and, of course, $B_\beta(t) \to$ const. If ε and k are sufficiently small, the solution of (3.111) will be increasing exponentially for small t, because the second term is "very small". It means that the function $\tilde{A}_2(t)$ is uniformly positive, $t \to \infty$, and bounded (upper estimation).

Solution of the problem (3.111) has a form

$$\tilde{A}_2(t) = 2\varepsilon e^{(N-4dk)t} \int_t^\infty e^{-(N-4dk)s} \bar{B}_\beta(s)ds$$

where $\bar{B}_\beta$ is an upper estimation of the moment of order $(2 + \beta)$. For large t

$$\tilde{A}_2(t) \sim 2C\varepsilon \left(\frac{1}{\varepsilon}\right)^{\frac{2+\beta}{\beta}} = c_1 \left(\frac{1}{\varepsilon}\right)^{2/\beta}.$$

But of course, $M_2(t, z) \geq \tilde{A}(t)$, and we have proved the left part in (3.105).

Probably, the estimations (3.105) have place for all k, if $d = 1, 2$. But if $d \geq 3$ and k is large enough, we can prove, that $M_2(t, x_1 - x_2) \to 0$, $t \to \infty$. Non-linearity and large diffusivity can destroy the dynamo process.

Let us emphasize once again, that non-linear cell-dynamo equation has (for small ε and k) two phases of evolution. If initial field $H_0(x, \omega)$ is small enough (or fixed) and $\varepsilon \to 0$, at the first stage we have the exponential increasing of all moments and (probably) the field $H(t, x)$ itself. At the second stage, the "nonlinear potential" in front of ε becomes very large and we have an equilibrium between growth of the solutions due to Ljapunov exponents mechanism and "killing" of the solution by non-linear fraction.

Conclusion to the Chapter III "Intermittency"
(Review of the literature, physical remarks, open problems)

The number of mathematical publications connected with the intermittence theory is not large and I will add only a few references to the recent publications. I few years ago, I started my activity with the goal to prove the mathematical versions of the "physical" results from [53], [75]. Review of some results in this direction see in my talk [54] based mainly on my papers with J. Görtner [32] and with R. Carmona [16]. Recently, A Sznitman has published a series of papers about random walk in the random Poisson environment (see, for example, [72], where one can find additional references).

The model from the paper [72] has (in our terms, see lecture 8) the following form:

$$\frac{\partial u}{\partial t} = k\Delta u - V(x,\omega)u, x \in R^d, t \geq 0$$

$$u(0,x) \equiv 1,$$

$$(3.112)$$

where

$$V(x,\omega) = \sum_i \phi(x - x_i) \qquad (3.112')$$

and $\{x_i\}$ is a Poisson random set with the intemcity $\lambda > 0$, $\phi(x)$ is an "elementary potential" decreasing sufficiently fast (compact supported).

Instead of continuous model (3.112), which was the main subject of the paper [72], we can analyze the lattice model with the similar properties of the potential, say,

$$\frac{\partial u}{\partial t} = k\Delta u + \xi(x,\omega)u$$

$$u(0,x) \equiv 1$$

$$(3.113)$$

where $\xi(x)$ are i.i.d.r.v., uniformly distributed on $[-1, 0]$. In both cases $\sup_x V(x,\omega) = 0$ and $\sum(-k\Delta + V) = [0, \infty)$.

The structure of the edge of the spectrum for both models (3.112), (3.112'), (3.113) is different from the case of gaussian potential (lecture 8). If the potential $\xi(x,\omega)$ is unbounded a.s. and $\max_{|x| \leq R} \xi(x)$ is increasing sufficiently fast, the high local maxima of the $\xi(x,\omega)$ gives the main contribution to the "Lifshitz tails". Corresponding eigenfuctions are similar to δ-functions located at the points, where these "high peaks" are located. For the Poisson type potential (3.112') or potential from the model (3.113) "clearings" of the point field $\{s_i\}$ play the role of "peaks", that is, large areas without points $\{x_i\}$. In the inner part of such area potential $V(x)$ will be "almost zero". The special role of the model (3.112) is caused by the fact that there are deep connections between Kac- Feinman representation of the solution of (3.112) and famous problem on the asymptotical behavior of the "Wiener sausage" volume (see, for example, one of the best known papers on the large deviations theory, M. Donsker, S. Varadhan, [20]).

There exists a very natural program of the investigation of the intermittency for the (say, lattice) Anderson parabolic model (3.113) for different distributions of $\xi(x,\omega)$. The leading terms of asymptotics for the statistical moments or a.s. behavior do not present a very complicated problem (see [32]). Although the answer is not trivial for the bounded r.v. (in the spirit of [72]).

The most important part of the problem is the consideration of the second terms in the asymptotical expansions of the expressions for statistical moments or a.s. representations. These terms must include the diffusion coefficient K. Especially important is the family of the distributions for $\xi(\cdot,\omega)$ with double-exponential tails:

$$\mu\{\omega\ \xi(x,\omega) > x\} \sim \exp\{-c_1 \exp\{c_2 x\}\}.$$

Namely, in this region the "optimal fluctuations" of the potential $\xi(\cdot)$ giving the main contribution to the statistical moments $\langle u^p(t,x)\rangle$, endure bifurcations: instead of "high single peaks" we have "high islands" of the fixed volume (some preliminary information see in [32]).

Another closely related problem is the problem of the asymptotical estimation for the integrated density of states $N^*(\lambda)$ for λ in the neighborhood of the upper boundary of the spectrum. (See lecture 5 for the definition). In the different terms, it is the problem of the "Lifshitz tails". Classical approach to this problem (see [14], [49]) is based on the asymptotical analysis of the first moment for the fundamental solution of the Anderson parabolic problem. First moment allows to obtain the Laplace transform of the density of states $N(\lambda)$. After this step we have to use the Tauberian theorems.

We know presently not only the leading terms for the statistical moments, but also the remainders (see lecture 8) or [72]). This more complicated information allows us to predict the corresponding additional terms for $N(\lambda)$, for instance, not only logarithmical asymptotics, but exact asymptotics or asymptotical expansions). Unfortunately, the Tauberian theorems are "too weak" to make such conclusions (this is a well-known problem, even in the context of the classical spectral analysis for the Laplacians on the compact manyfolds or bounded domains).

However, we can try to use "direct methods", i.e. to classify all "local configurations" of $\xi(x,\omega)$ or (it is the same) "optimal fluctuations". Using claster expansions, we can estimate the interaction between different "optimal fluctuations", for instance, high peaks etc. For the random variables with the distribution having, say, exponential or fractional-exponential tails this program had already been realized (see [54]). Typical result has a following form:

Theorem Suppose that $P\{\xi(\cdot,\omega) > x\} \underset{x \to +\infty}{\sim} \exp\{-cx^\alpha\}$, $\alpha > 2$ (subgaussian tails). Then, if $\lambda \to +\infty$

$$N^*(\lambda) \sim F^*(\lambda) = P\{\xi(\cdot) > \lambda\}. \tag{3.114}$$

Of course, logarithmical asymtotics is trivial $(-\ln N^*(\lambda) \sim -c\lambda^\alpha)$ and has place for all $\alpha > 0$.

It is very important that formula (3.114) is false for $\alpha \leq 2$ (see [54]).

Analysis of the "Lifshitz tails" can be especially interesting if we have the additional parameters in the model (say, coupling const). For instance, one can expect, that for the operator

$$H = \Delta + \sigma\xi(x,\omega)$$

has bifurcations in asymptotics with respect to parameter $a > 0$ if $\sigma \to 0$ and $\lambda = \frac{a}{\sigma}$. It may give us, probably, some weak form of the "delocalization result", if $\sigma \to 0$. This problem in general is still open.

Non-stationary parabolic theory (scalar and vector one) contains also many open problems, even in the linear case.

We have proved (with R. Carmona [15]) an existence of the a.s. Ljapunov exponent $\tilde\gamma(k)$ for the model

$$\frac{\partial u}{\partial t} = k\Delta u + \xi_t(x,\omega)u$$

$$u(0,x) \equiv 1,$$

where $\xi_t(x,\omega) = \dot{w}_t(x,\omega)$ and $w_t(x)$ are i.i.d. Wiener processes for the different points $x \in \mathbf{Z}^d$. Not only is the function $\tilde\gamma(k)$ continuous for $k \geq 0$ but has a special "convexity" property. Nevertheless, the following very natural questions are open:

Problem 1. Prove that for $d = 1, 2$ $\tilde\gamma(k) < 0$ for all $k \geq 0$.

If it is true, then even for very large diffusivity k we have the process of the aggregation of the particles into more and more dense groups getting more and more rare for large t. Physically, it corresponds to the full localization in dimensions $d = 1, 2$ for arbitrary small coupling const $\sigma = 1/k$.

Problem 2. Find for $d \geq 3$ the critical diffusivity $\tilde{k}_r$ such that $\tilde\gamma(k) < 0$, $k < \tilde{k}_r$; $\tilde\gamma(k) \equiv 0$, $k \geq \tilde{k}_r$. What is the meaning of $\tilde{k}_r$ in the language of statistical mechanics? How to prove that for $k > \tilde{k}_r$ there exists a limiting distribution $(t \to \infty)$

for the solution $u(t, x, \omega)$ of non-stationary Anderson model? We can not even prove the corresponding fact for $k > k_{2,r}$ (where we know already about the tightness of the family of the distributions for the field $u(t, \cdot, \omega)$).

Some other problems are more technical.

Problem 3. Prove the analogy of the results of [15] (see lecture 9) for the Anderson parabolic model in R^d. We have to consider the problem

$$\frac{\partial u^\varepsilon}{\partial t} = k \triangle u^\varepsilon + \frac{1}{\sqrt{\varepsilon}} \xi \left(\frac{t}{\varepsilon}, x \right) u^\varepsilon$$

$$\tag{3.115}$$

$$u(0, x) \equiv 1$$

for the gaussian field $\xi(t, x, \omega)$ in space and time, $\langle \xi \rangle = 0$, $\langle \xi(t_1, x_1) \xi(t_2, x_2) \rangle = B(t_1 - t_2, x_1 - x_2)$ with the smooth and fastly decreasing in space and time correlator B.

a.   Prove that $u^\varepsilon(t, x) \xrightarrow{\text{dist}} \bar{u}(t, x)$, where $\bar{u}(t, x)$ is a solution of the Stratonovich form of the stochastic PDE similar to (3.15).

b.   Construct this solution $\bar{u}(t, x)$ in the weighted Sobolev space with the norm

$$\|f\|_{2,\alpha} = \int\limits_{R^d} \exp(-\alpha|x|)(|\nabla f|^2 + f^2) dx.$$

c.   Find the moment Ljapunov exponents $\gamma_p(k, \varepsilon)$ for the problem (3.115) and its SPDE version. Prove the continuity and smoothness of $\gamma_p(\cdot, \cdot)$ with respect to small diffusivity k and time correlation length ε. Calculate the first terms of the corresponding expansions.

d.   The same problem for the a.s. Ljapunov exponent $\tilde{\gamma}(k, \varepsilon)$.

In these directions, something was done in our recent paper [16], but many questions have yet no answers. Especially important and complicated is the problem of a.s. Ljapunov exponents.

Of course, we can try to study not only gaussian potential, but more general fields for both types of models (lattice and continuous ones). For example, it is very natural to analyze the lattice models where $\xi_t(x) = l_t(x, \omega)$ and $l_t(x, \omega)$ is a family of independent Levy processes. Corresponding probabilistic technique is based on the modern version of the Ito calculus for the infinite-divisible processes with the jumps. (See, for instance, the recent paper [1]).

Problem 4. To accomplish the program described above in the problem 3 for the kinematic (linear) dynamo equation

$$\frac{\partial H^\epsilon}{\partial t} = k\triangle H^\epsilon + \text{curl} \left[\frac{1}{\sqrt{\epsilon}} v\left(\frac{t}{\epsilon}, x\right) \times H^\epsilon \right].$$

(3.116)

$$H(0,x) = H_o(x), \quad t \geq 0, quad x \in R^3.$$

Here $H = H(t,x) \in R^3$ is a magnetic field, $v(t,x,\omega)$ is an incompressible (div $v = 0$) random isotropic flow in R^3 with given energy spectrum, $H_0(x,\omega)$ is an initial magnetic field (homogeneous and ergodic in space).

The main technical difficulty of this model is as follows: we can not use such powerful method as Kac-Feynman representation, which played a key role in the scalar theory. Instead of this, we have to work with non-commutative matrix multiplicative integrals.

Continuity of the solutions $u^\epsilon(t,x)$ for $\epsilon \to 0$ (i.e. convergence $u^\epsilon \overset{D}{\to} \bar{u}$) should not be a hard problem. The problem of the moments equations and the ϵ-corrections for such equations (moment equations for $\epsilon = 0$, i.e. for the δ-correlated in time velocity field $v(t,x)$ are well-known) is more interesting and non-trivial. The similar question arises about continuity of the Ljapunov exponents $\gamma_p(k), p = 1, 2$, and $\tilde{\gamma}(k)$, if $k \to 0$. In isotropic δ-correlated case, the equation for the first moment is known in the physical literature: if $M_1(t,x) = \langle H(t,x)\rangle$ then

$$\frac{\partial M_1}{\partial t} = (D + k)\triangle M_1 + \alpha \text{ curl } M_1$$

where D is a turbulent diffusivity and α is so-called helicity (see [75]). Of course, continuity (and even analyticity) of the mean field M_1 with respect to k in this case is obvious. But corresponding problem for the second moment M_2 (especially if $\alpha \neq 0$) and especially for the a.s. exponent $\tilde{\gamma}(k)$ is not trivial. Let us remember (lecture 10), that even in the simplest possible lattice cell-dynamo model, this problem is open by now. From the physical point of view, problem 4 is one of the central problems in the classical turbulent magneto hydro-dynamics.

In conclusion, I'll make a few remarks on the non-linear intermittency theory, more general, about asymptotical behavior of the nonlinear evolution equations in the non-stationary (turbulent RM). It is, more or less, the "terra incognita". Of course, some problems were studied, especially in the context of the reaction-diffusion equation and chemical kinetics. For example, C. Mueller and R. Sowers [61] have investigated the asymptotical behavior of the SPDE equation

$$\frac{\partial u}{\partial t} = k\triangle u + \xi(t,x)u^\gamma,$$

particularly, the phenomenon of the explosion of the solutions for $t \gg 1$ and $\gamma > 3/2$.

F. den Hollander at all [33], [34] have studied the reaction -diffusion equations with the special (non-linear) interaction between particles and so forth.

Probably, the most principal problem in this area is a problem of the convergence to the equilibrium for the non-linear systems in the non-stationary RM (same types of problems are typical for the modern statistical physics). The following problem connected with the scalar Anderson parabolic problem (lecture 9) or its vector version (lecture 10) can be considered as the simplest basic model.

Consider equation

$$\frac{\partial u}{\partial t} = k\triangle u(t,x) + \xi_t(x,\omega)u - \varepsilon u^\gamma \tag{3.117}$$

$$u(0,x) \equiv 1, \quad t \in R^1_+, \quad x \in \mathbf{Z}^d, \quad \gamma = 2,3\cdots.$$

Prove that for $t \to \infty,\quad k \le k_0$

$$u(t+\tau,x) \xrightarrow{\text{dist}} \tilde{u}(\tau,x),$$

where $\tilde{u}(\tau,x)$ is a homogenous in time and space ergodic field. As it was mentioned above, for small k the moments are equal:

$$m_p(t) = \langle u^p(t,x)\rangle \asymp \left(\frac{1}{\varepsilon}\right)^{\frac{p}{\gamma-1}}$$

i.e. the limiting distribution exists (tightness!) but probably it is not unique.

Possible program designed to prove such "ergodic theorem" can have the following structure:

If $k = 0$, we have the family of independent ergodic diffusion processes in every cell x. We can find corresponding moments (if $t \to \infty$) exactly and estimate the time-correlation.

If $k \ne 0$ we can write down the moment equations which are not closed: equation includes $m_\gamma(t,x) = \langle u^\gamma(t,x)\rangle$ (γ is an integer!) etc. In the spirit of statistical mechanics, we may understand these equations as an infinite system for the full collection of correlation functions (similar to Kirkwood-Saltzburg equations in the Ising model). For small k, the operator (linear operator!) of this system must be a contraction in the corresponding functional space) etc.

I hope, that this program (which has many other possible applications) will be realized in the near future.

References

[1] Ahn H., Carmona R., Molchanov S. "Parabolic equations with a Levy potential", Lecture Notes in Control and Information Sciences, 176, Proceed of IFIPWG 7/1 International Conferences, UNCS, 1991, pp. 1-11.

[2] Aizenman M, Molchanov S. "Localization at large disorder and extreme energies: an elementary derivation", 1993, Comm. Math. Phys, 157, pp245-278.

[3] Akhieser N, Glazman A "The theory of linear operators in Hilbert space", Vol. I, II, Ungar, New York, 1961.

[4] Albeverio S., Surgailis D., Molchanov S. "Stratified structure of the Universe and the Burger equation: a probabilistic approach", 1993, to appear in Probability theory and related fields.

[5] Alexander K., Molchanov S. "Percolation of the level sets of the random field with a lattice symmetry", 1994, to appear in Jorn of Stat. Phys.

[6] Anderson P. "Absence of diffusion in certain random lattices", 1958, Phys. Rev, 109, pp. 1492-1501.

[7] Arnold L., Papanicolaon G., Wihstutz V. "Asymptotic analysis of the Ljapunov's exponents and rotation numbers of the random oscillator and applications" 1986, SIAM J. Appl. Math., Vol. 46.

[8] Avellaneda M., Majda A. "Mathematical models with exact renormalization for turbulent transport (1990), Comm. Math. Phys. V. 131, pp. 381-429.

[9] Avellaneda M., Majda A. "Renormalization theory for Eddy Diffusivity in Turbulent Transport", Phys. Rev. Letters, (1992), V. 68, # 20, pp. 3028-3031.

[10] Azbel M., Kaganov M., Lifshitz I. "Electron theory of metals, Consultants Bureau, New York, 1973.

[11] Billingsley "Convergence of Probability measures", (1968), Willey, New York.

[12] Bogachev L., Molchanov S. "Mean-field models in the theory of random media I, II, III; Theor. Math. Phys., I (1989), V. 81, II (1990), V. 82, III (1991), V. 87.

[13] Bulycheva O., Molchanov S. "The necessary conditions for the averaging of one-dimensional random media", Vestnik MGU (Moscow) 1986, #3, pp. 33-38.

[14] Carmona R., Lacroix J. "Spectral theory of Random Schrödinger operator", Birhäuser Verlag, Basel, Boston, Berlin, 1990.

[15] Caromona R., Molchanov S. "Parabolic Anderson model and intermittency", Preprint UCI (1992), to appear in "Memoirs of AMS" (1994).

[16] Carmona R., Molchanov S., Noble J. "Parabolic evolution equation with random gaussian potential", Preprint UCI (1992).

[17] Cycon H., Froese K., Kirsch W., Simon B. "Schrödinger operators with applications to Quantum mechanics and global geometry", (1987), Springer Verlag, Berlin.

[18] Cramer H., Lidbetter H. "Stationary random processes", (1975), Springer-Verlag, Berlin.

[19] Delyon F., Levy Y., Soullard B. "Anderson localization for multidimensional systems at large disorder or low energy", Comm. Math. Phys. (1985), V. 100, pp. 463-470.

[20] Donsker M.D., Varadhan S.R.S. "Asyntotics for the Wiener sausage", Comm. Pure Appl. Math. (1975), V. 28, pp. 525-565.

[21] Von Dreifus H., Klein A. "A new proof of localization in the Anderson tight binding model", Comm. Math. Phys., (1989), V. 124, pp. 285-299.

[22] Durett R. "Probability: theory and examples (1991), Wadsworth and Brooks/Cole Statistics and Probability series".

[23] Dynkin E.B. "Non-negative eigenfunctions of the Laplace Beltrami operator and browinan motion in certain symmetric space", (1961), Soviet Math. (Doklady), V. 2, W 6, pp. 1433-1435.

[24] Ferstenberg H. "Noncommuting random products", (1963), Trans. Amer. Math. Soc., V. 108, pp. 377-428.

[25] Freidlin M. I. "Dirichlet problem for an equations with periodic coefficients", (1964), Probability theory and Appl. V. 9, pp. 133-139.

[26] Freidlin M.I., Wentzell A. D. "Random perturbations of Dynamical Systems", (1984),, Springer-Verlag, Berlin.

[27] Fröhlich J., Martinelli F., Scoppola E., Spencer T. "Constructive proof of localization in the Anderson tight binding model" (1985), Comm. Math. Phys., V. 101, pp. 21-46.

[28] Goldsheid Ya., Molchanov S., Pastur L. "Pure point spectrum of stochastic one dimensional Schrödinger operator", Func. Anal. and Appl. V. 11, #1 (1977).

[29] Goldsheid Ya., Margulis G. "A condition for simplicity of the spectrum of Ljapunov exponents", (1987), Sov. Math. Dokl, V. 35, #2, pp. 309-313.

[30] Gordon A., Jacsič V., Molchanov S., Simon B. "Spectral properties of random Schrödinger operator with unbounded potential", 1990, Caltech, preprint, to appear 1993 in Comm. Math. Phys.

[31] Gordon A. "On exceptional value of the boundary phase for the Schrödinger equation on a half-line", (1992), Russian Mathemat. Surveys, 47, pp. 260-261.

[32] Görtner J., Molchanov S. "Parabolic problems for the Anderson model", (1990) Comm. Math. Phys., V. 132, pp. 613-655.

[33] Greven A., den Hollander F. "Branching random walks in random environment: phase transitions for local and global rates, (1992), Probab. Theory Relation Fields.

[34] Greven A., den Hollander F. "Population growth in random media": I variational formula and phase diagram, II wave front propagation (1991), preprint # 636, University of Heidelberg.

[35] Ibragimov I, Linnik Ju. "Independent and stationary sequences of random variables", 1971, Wolters-Noordhoff publishing Groningern, Holland.

[36] Isichenko M. "Percolation, statistical topography and transport in random media", Rev. Modern Physics, (1992), V. 64, pp. 961-1043.

[37] Jitomirskaya S., Makarov N., del Rio R., Simon B. "Singular continuous spectrum is generic", (1993), Caltech preprint, to appear in Bull. AMS.

[38] Kesten H., "Percolation Theory for Mathematicians" (1982), Bürkhäuser, Boston.

[39] Kirsh W., Kotani S., Simon B. "Absence of absolutely continuous spectrum for one-dimensional random, but deterministic Schrödinger operators", Ann. Inst. H. Poincare, (1985), V. 42, p.383.

[40] Kirsh W., Molchanov S., Pastur L. "One dimensional Schrödinger operator with unbounded potential: pure point spectrum, I, Funct. Anal. and Appl. (1990), #3, p. 24.

[41] Kozlov S. "Averaging of random operators", Mathem. Sbornik, (1979), V. 151, pp. 188-202.

[42] Kozlov S. "The method of averaging and walks in inhomogeneous environments", Russian Math. Surveys, (1985), 40: 2, pp. 73-145.

[43] Kozlov S., Molchanov S. "On conditions under which central limit theorem is applicable to random walk on lattice", Dokl. Acad. Nauk SSSP (1984), V. 278, pp. 531-534, Sov. Math. Dokl 30 (1984), 410-413.

[44] Kotani S. "Ljapunov indices determine absolutely continuous spectra of stationary one-dimensional Schrödinger operator", in Proc. Taneguchi Intern. Symp. on Stochastic Analysis, Katata and Kyoto, (1982), ed. K. Ito, Nort Holland, pp. 225-247.

[45] Kotani S. "Ljapunov exponent and spectra for one-dimensional random Schrödinger operators", (1986) Proc. Conf on Random Matrices and their Applications, Contemporary Math., V. 50, Providence R.I., pp. 277-286.

[46] Kotani S. "Support theorems for random Schrödinger operators", Comm. Math. Phys., (1985), V. 97, pp. 443-452.

[47] Kotani S. "Absolute continuous spectra for one-dimensional ergodic operators", (1993), to appear in Proc. Summer Inst. AMS, Cornell, Ithaka, NY.

[48] Kunz H., Souillard B. "The localization transition on the Bethe lattice" (1983), Journ. Phys. Letters, (Paris), V. 44, pp. 411-414.

[49] Lifshitz I., Gredescul S., Pastur L., "Introduction to the theory of disordered media", (1982), Moscow, Nauka, (1986) Springer-Verlag, Berlin.

[50] McKean H.P., "Stochastic integrals" (1969), Academic press, New York.

[51] Manakov S., Novikov S., Pitaevskii, Zakharov V. "The Theory of Solutions, the method of the inverse problem", Nauka, Moscow, 1980.

[52] Molchanov S. "The structure of eigenfunctions of one-dimensional disordered systems", Izv. Acad. Sci. USSR (1978), V. 2, # 1, pp. 70-101.

[53] Molchanov S. "Lectures on the localization theory", (1990), Preprint, Caltech.

[54] Molchanov S. "Intermittency and localization: new results" (1900), Proc. of the Intern. Congr. Math (Kyoto, Japan), Vol. II, pp. 1091-1103.

[55] Molchanov S. "Ideas in the theory or Random Media", (1991), Acta. Appl. Math., V. 22, pp. 139-282, Kluver Acad. Publish.

[56] Molchanov S., Piterbarg L. "The turbulent diffusion of the temperature gradients", Dokl. Sov. Acad. Sci. (1986), V. 284, #4.

[57] Molchanov S., Piterbarg L. "Localization of the Rossby topographical waves", Dokl. Sov. Acad. Sci. (1990), V. 310, #4.

[58] Molchanov S., Ruzmaikin A "Ljapunov exponents and distribution of the magnetic field in dynamo-model" (1993), to appear in "Proc. conference in Probability. theory and Markov processes", Cornell.

[59] Molchanov S., Piterbarg L. "Heat propagation in random flows" (1992), Russian J. Math. Phys., V. 1, # 1, pp. 18-42.

[60] Mott N., Twose W. "The theory of impurity conduction", (1961), Adv. Phys., V. 10, pp. 107-163.

[61] Mueller C., Tribe R. "A stochastic PDE arising as the limit of a long range contact processes and its phase transition" (1993), Technical Report, Math. Sci. Inst., Cornell.

[62] Newman S. "The distribution of Ljapunov exponents: (1986), Comm. Math. Phys., V. 103, pp. 121-126.

[63] Oseledec V. "A multiplicative ergodic theorem, Ljapunov characteristic numbers in dynamical systems", Trans. Moscow. Math. Soc. (1968), V. 19, pp. 197-231.

[64] Piterbarg L. "Dynamics and prediction of the large scale SST anomalies", (1989) Gidrometeoizdat, Leningrad, Translated in Kluwer (Holland).

[65] Rozovskii B. "Stochastic Differential equations (1991), Kluwer, Holland.

[66] Ruzmaikin A., Liewer P., Fienman J. "Random cell dynamo" (1993), to appear in Gophys. Astroph. Fluid. Dyn.

[67] Reed M., Simon B. "Methods of Modern Mathematical Physics", I-IV (1975-1978), Academic Press, New York.

[68] Simon B., Spencer T. "Trace class perturbation and the absence of absolutely continuous spectrum", (1989) Comm. Math. Phys., v. 125, pp. 113-125.

[69] Simon B., Wolff T. "Singular continuous spectrum under rank one perturbations and localization for random Hamiltonians" (1986), Comm. Pure Appl. Math., V. 39, pp. 75-90.

[70] Sinai Ja. "Limit behavior of one-dimensional random walks in random environment", (1982), Theor. Probab. Appl., 27, pp. 247-258.

[71] Spitzer F. "Principles of random walk" (1976), Springer-Verlag, New York.

[72] Sznitman A.S. "Brownian asymptotics in a Poisson environment", (1991), Preprint ETH-Zentrum (Zürich), to appear in Probab. Theory Relat. Fields.

[73] Papanicolaou G., Varadhan S.R.S. "Boundary value problem with rapidly oscillating random coefficients", (1981), Coll. Math Soc. Janos Bolyai, 27, Random fields, V. 2 North-Holland, Amersterdam - New York, pp. 835-873.

[74] Wegner F. "Bounds of the density of states in disordered systems", (1981) Zeit. Phys. B, Condensed Matter, V. 44, pp. 9-15.

[75] Zeldovich Ya., Molchanov S., Ruzmaikin A., Sokoloff D. "Intermittency, diffusion and generation in a Non-Stationary Random Medium" (1988), Sov. Sci. Rev., Sec C, Vol. 7, pp. 1-110.

Random Walks in Random Environment

Ofer Zeitouni

Department of Electrical Engineering
Department of Mathematics
Technion – Israel Institute of Technology
Haifa 32000, Israel

Originally published in: *École d'Été de Probabilités de Saint-Flour XXXI – 2001*,
Lecture Notes in Mathematics, Vol. **1837**, 191–312, DOI: 10.1007/978-3-540-39874-5,
© Springer-Verlag Berlin Heidelberg 2004, Reprint by Springer-Verlag Berlin Heidelberg 2012

Preface

These notes on random walks in random environments (RWRE) reflect what I hoped to cover in the 15 hours of the St Flour course on this topic, July 9–25, 2001. Of course, this turned out to be over optimistic. Departing even further from the actually delivered lectures, I have taken advantage of the year that elapsed to add some material (especially, related to multi-dimensional walks) and to correct numerous mistakes and omissions.

The manuscript consist roughly of two parts: the first deals with RWRE on $\mathbb{Z}$. The interest in the model began in the early 70's, and with the detailed analysis of RWRE asymptotics in the last decade, has now reached maturity (for an account of the history of the subject and many of the results through the early 90's, see [37]). I have tried to present different tools for the study of such walks, risking some repetition of results in a few cases, and deferring to the bibliographical notes a discussion of refinements and sharpening of the results. It is worthwhile to point out that RWRE's on $\mathbb{Z}$ have already been considered in previous St Flour courses (most notably by Ledrappier [50] and by Molchanov [53]), but the emphasis in this presentation is quite different.

The second part of the notes deals with $\mathbb{Z}^d$. This is currently an active research area, and one hopes that much progress will be made in the next few years. My goal here was to expose the audience to some tools which have proved useful, and to point out several directions where further progress could be made. In several places, I have tried to lay the groundwork for relaxing the often made assumption of i.i.d. environment.

When preparing the notes, and taking into account the time frame of these lectures, it became clear that there were topics that had to be left out. Even the uninitiated will quickly realize that the most glaring omission is the study of RWRE's by renormalization techniques. There are three reasons for this: first, it would take too long to properly expose it. Second, these methods have not yet reached the full scope of their applicability, and in view of very active current research efforts in this direction, any account written now risks being outdated very quickly. And third, an overview of the current status of these techniques can be found in [69] and [70]. Time constraints also did not allow me to discuss random walks on Galton-Watson trees, a topic that has seen much progress in recent years.

Parts of the material presented here is based on joint work, some still unpublished, with F. Comets, A. Dembo, N. Gantert, and Y. Peres. I thank them all, both for the many hours spent together on thinking about RWRE, and for their generosity. I would also like to thank my colleagues in Haifa who suffered through a first draft of these notes in the winter of 2000. In particular, comments from D. Ioffe, H. Kaspi, E. Mayer-Wolf, A. Roitershtein and M. Zerner are gratefully acknowledged. Similarly acknowledged are useful remarks from D. Cheliotis, A. Dembo, N. Gantert, A. Guionnet, H. Kesten, D. Piau, and S.R.S. Varadhan. Comments from participants at the St Flour summer school helped improve the presentation and strengthen numerous results. I am particularly grateful to P. Bougerol who allowed me to incorporate some of his suggestions in the final text of these notes, and D. Ocone and F. Rassoul-Agha for stimulating discussions. Last but not least, I am grateful to J. Picard for the smooth and gentle running of the summer school.

A typographical comment: for aesthetic reasons, I consistently use P_ω^o, E_ω^o, $\mathbb{P}^o$, $\mathbb{E}^o$, etc., when I mean P_ω^0, E_ω^0, $\mathbb{P}^0$, $\mathbb{E}^0$.

1 Introduction

The definition of a RWRE involves two components: first, the *environment*, which is randomly chosen but kept fixed throughout the time evolution, and second, the random walk, which, given the environment, is a time homogeneous Markov chain whose transition probabilities depend on the environment. We do not attempt here a historical review of RWRE's, or in greater generality of motion in homogeneous media, except for stating that we insist on the environment being static, i.e. time independent, and that in general the random walk (conditioned on the environment) is not necessarily reversible.

1.1 Model

We begin with a general setup, that will be specialized later to the cases of interest to us. Let (V, E) denote an (infinite, oriented) graph with countable vertex set V and edges set $E = \{(v, w)\}$ (we allow, but do not require, $(v, v) \in E$). For each $v \in V$, we define its *neighborhood* N_v by

$$N_v = \{w \in V : (v, w) \in E\},$$

throughout assuming that $|N_v| < \infty$, for all $v \in V$.

For each $v \in V$, let $M_1(N_v)$ denote the collection of probability measures on V with support N_v. Formally, an element of $M_1(N_v)$, called a *transition law at v*, is a measurable function $\omega_v : V \to [0, 1]$ satisfying:

$$
\begin{aligned}
&\text{(a)} \quad \omega_v(w) \geq 0 \quad \forall\, w \in V \\
&\text{(b)} \quad \omega_v(w) = 0 \quad \forall\, w \notin N_v \\
&\text{(c)} \quad \sum_{w \in N_v} \omega_v(w) = 1
\end{aligned}
\tag{1.1.1}
$$

Note that if $v \in N_v$ then in (1.1.1c) we allow for $\omega_v(v) > 0$.

We equip $M_1(N_v)$ with the weak topology on probability measures, which makes it into a Polish space. Further, it induces a Polish structure on $\Omega = \prod_{v \in V} M_1(N_v)$. We let $\mathcal{F}$ denote the Borel σ-algebra on Ω (which is the same as the σ-algebra generated by cylinder functions). Given a probability measure P

on $(\Omega, \mathcal{F})$, a *random environment* is an element ω of Ω distributed according to P.

We turn next to define the class of random walks of interest to us. For *each* $\omega \in \Omega$, we define the *random walk in the environment* ω as the time-homogeneous Markov chain $\{X_n\}$ taking values in V with transition probabilities

$$P_\omega(X_{n+1} = w | X_n = v) = \omega_v(w).$$

We use P_ω^v to denote the law induced on $(V^\mathbb{N}, \mathcal{G})$ where $\mathcal{G}$ is the σ-algebra generated by cylinder functions and

$$P_\omega^v(X_0 = v) = 1.$$

In the sequel, we refer to $P_\omega^v(\cdot)$ as the *quenched* law of the random walk $\{X_n\}$. Note that for each $G \in \mathcal{G}$, the map

$$\omega \mapsto P_\omega^v(G)$$

is $\mathcal{F}$-measurable. Hence, we may define the measure $\mathbb{P}^v := P \otimes P_\omega^v$ on $(\Omega \times V^\mathbb{N}, \mathcal{F} \times \mathcal{G})$ from the relation

$$\mathbb{P}^v(F \times G) = \int_F P_\omega^v(G) P(d\omega), \quad F \in \mathcal{F}, G \in \mathcal{G}. \tag{1.1.2}$$

The marginal of $\mathbb{P}^v$ on $V^\mathbb{N}$, denoted also $\mathbb{P}^v$ whenever no confusion occurs, is called the *annealed law* of the random walk $\{X_n\}$; note that under $\mathbb{P}^v$, the random walk in random environment (RWRE) $\{X_n\}$ is *not* a Markov chain!

1.2 Examples

Throughout these notes, we only treat nearest neighbor RWRE's on $\mathbb{Z}^d$:

Nearest neighbor RWRE on $\mathbb{Z}$

Here, we take $V = \mathbb{Z}$ and $E = \cup_{z \in \mathbb{Z}}\{(z, z+1), (z, z)\}$. Then, $N_v = \{v-1, v, v+1\}$ and $M_1(N_v)$ can be identified with the three dimensional simplex; We let $\omega_z^+ := \omega_z(z+1)$, $\omega_z^- := \omega_z(z-1)$, and $\omega_z^0 := \omega_z(z)$. One defines naturally the shift θ on Ω by $(\theta\omega)_z = \omega_{z+1}$. We always make the following assumption: $(\omega, \mathcal{F}, P, \theta)$ is an ergodic system.

It is worthwhile commenting, already at this stage, that for each ω there exists a reversing measure that makes the RWRE reversible. More details are provided in Section 2.1.

Nearest neighbor RWRE on $\mathbb{Z}^d$

Here, $V = \mathbb{Z}^d$ and $E = \cup_{z \in \mathbb{Z}} \{ \cup_{y \sim z} (z, y) \cup (z, z) \}$. For each $v \in V$, N_v contains $2d+1$ vertices, and $M_1(N_v)$ is identified with the $2d+1$-dimensional simplex. One may define the family of shifts $\{\theta^e\}_{|e|_1 = 1}$. As in the case of $d = 1$, we always require P to be ergodic with respect to this family. We write throughout $\omega(x, e) := \omega_x(x + e)$. Unlike the case with $d = 1$, the Markov chain defined by P_ω^v is, in general, not reversible.

Bibliographical notes: the preface section contains relevant bibliography on the RWRE model in $\mathbb{Z}^d$, $d \geq 1$. We mention here some other models of random walks in random media that can be adapted into the general framework presented above, but that will not be considered in these notes:

- *Non nearest neighbor walks: For $\mathbb{Z}^1$, see the recent thesis [7], that includes also a summary of earlier work and in particular of [43]. I am not aware of a systematic study of non nearest neighbor RWRE's on $\mathbb{Z}^d$, see however [79] for some results valid in that generality.*
- *Reversible random walks in random environments in $\mathbb{Z}^d$, $d > 1$: the prime example is the random conductance model, in which bonds on $\mathbb{Z}^d$ carry i.i.d. conductances and modulate the transition mechanism of the walk, see [14]. Other models in the same spirit, and their surprising behavior, are described in [6] and the references therein.*
- *Random walks on Galton-Watson trees: see [15, 51, 52, 59] for recent developments.*

2 RWRE – d=1

This chapter is devoted to the study of the one-dimensional model, where sharp results are available. As a warm-up to the high dimensional case, we sometimes present different proofs of the same statement.

Our exposition progresses from ergodic properties and law of large numbers (Section 2.1), to the study of central limit theorems (Section 2.2), large deviations (Section 2.3), subexponential tail estimates (2.4), and subdiffusive behaviour and aging (Section 2.5). Each section contains a (non-exhaustive!) list pointing to the literature.

2.1 Ergodic theorems

In this section, we are interested in questions concerning transience, recurrence and laws of large numbers, in the most general nearest neighbour one-dimensional setup. Define $\rho_z = \omega_z^- / \omega_z^+$.

Assumption 2.1.1

(A1) *P is stationary and ergodic.*

(A2) $E_P(\log \rho_0)$ *is well defined (with $+\infty$ or $-\infty$ as possible values).*
(A3) $P(\omega_0^+ + \omega_0^- > 0) = 1$.

Theorem 2.1.2 *Assume Assumption 2.1.1. Then,*

(a) $E_P(\log \rho_0) < 0 \quad \Rightarrow \quad \lim_{n\to\infty} X_n = +\infty,$ $\mathbb{P}^o$ *a.s.*

(b) $E_P(\log \rho_0) > 0 \quad \Rightarrow \quad \lim_{n\to\infty} X_n = -\infty,$ $\mathbb{P}^o$ *a.s.*

(c) $E_P(\log \rho_0) = 0 \quad \Rightarrow \quad -\infty = \liminf_{n\to\infty} X_n < \limsup_{n\to\infty} X_n = \infty, \quad \mathbb{P}^o$ *a.s.*

(Note that (a), (b) above imply that X_n is $\mathbb{P}^o$-a.s. transient, whereas (c) implies it is recurrent).

Proof. Due to **(A2)** and **(A3)**, if $P(\omega_0^+ = 0) > 0$ then $P(\omega_0^- = 0) = 0$, and then $E_P(\log \rho_0) = \infty$. To see that (b) holds in this case, let $n_0 = \min\{z > 0 : \omega_z^+ = 0\}$ and $n_i = \max\{z < n_{i-1} : \omega_z^+ = 0\}$. Then, $P(n_0 < \infty) = 1$, hence $\limsup_{n\to\infty} X_n < \infty$, $\mathbb{P}^o$-a.s. Note that by ergodicity, $P(n_i > -\infty) = 1$, and further $P_\omega^{n_i}(X_n \text{ does not hit } n_{i+1}) = 0$, as can be seen either from a coupling with (biased) random walk or from (2.1.4) below. This completes the proof of (b) when $P(\omega_0^+ = 0) > 0$.

A similar argument applies to proving (a) in case $P(\omega_0^- = 0) > 0$. Thus, using **(A3)**, we assume in the sequel that $P(\min(\omega_0^+, \omega_0^-) = 0) = 0$. We begin by deriving some formulae related to one-dimensional random walks in a *fixed* environment. These can be derived concisely by using the link between nearest-neighbor random walks on z and electrical networks, see appendix A.

Fix an environment ω with $|\log \rho_z| < \infty$ for each $z \in \mathbb{Z}$. For $z \in [-m_-, m_+]$, define

$$\mathcal{V}_{m_-,m_+,\omega}(z) := P_\omega^z(\{X_n\} \text{ hits } -m_- \text{ before hitting } m_+).$$

Note that due to the assumption $|\log \rho_z| < \infty$, for each z, it holds that $\mathcal{V}_{m_-,m_+,\omega}(z)$ is well defined as

$$P_\omega^z(\{X_m\} \text{ never hits } [-m_-, m_+]^c) = 0.$$

The Markov property implies that $\mathcal{V}_{m_-,m_+,\omega}(\cdot)$ is harmonic, that is it satisfies

$$\begin{cases} (\omega_z^+ + \omega_z^-)\mathcal{V}_{m_-,m_+,\omega}(z) = \omega_z^- \mathcal{V}_{m_-,m_+,\omega}(z-1) \\ \qquad\qquad\qquad\qquad + \omega_z^+ \mathcal{V}_{m_-,m_+,\omega}(z+1), \quad z \in (-m_-, m_+), \\ \mathcal{V}_{m_-,m_+,\omega}(-m_-) = 1, \quad \mathcal{V}_{m_-,m_+,\omega}(m_+) = 0. \end{cases}$$

$$(2.1.3)$$

Solving (2.1.3), we find

$$\mathcal{V}_{m_-,m_+,\omega}(z) = \frac{\displaystyle\sum_{i=z+1}^{m_+} \prod_{j=z+1}^{i-1} \rho_j}{\displaystyle\sum_{i=z+1}^{m_+} \prod_{j=z+1}^{i-1} \rho_j + \sum_{i=-m_-+1}^{z} \left(\prod_{j=i}^{z} \rho_j^{-1}\right)}. \qquad (2.1.4)$$

(Note that the solution to (2.1.3) is unique due to the maximum principle, hence it is enough to verify that the function in (2.1.4) satisfies (2.1.3).)

Define $S(\omega) = \sum_{n=1}^{\infty} \rho_1 \cdots \rho_n$, $F(\omega) = \sum_{n=0}^{\infty} \rho_0^{-1} \cdots \rho_{-n}^{-1}$. Further, define the events

$$S_+ = \{S(\omega) < \infty\}, \quad \mathcal{F}_+ = \{F(\omega) < \infty\}.$$

Then:

- on $T_+ := \{S_+ \cap \mathcal{F}_+^c\}$, it holds that

$$\lim_{m\to\infty} \left[1 - V_{1,m,\omega}(0)\right] > 0, \lim_{k\to\infty} \lim_{m\to\infty} \left[1 - V_{k,m,\omega}(0)\right] = 1.$$

  Hence, for $\omega \in T_+$,
$$P_\omega^o(\lim_{n\to\infty} X_n = \infty) = 1.$$

- Similarly, for $\omega \in T_- := \{S_+^c \cap \mathcal{F}_+\}$,

$$P_\omega^o(\lim_{n\to\infty} X_n = -\infty) = 1.$$

- Finally, if $\omega \in R := \{S_+^c \cap \mathcal{F}_+^c\}$ then, for any fixed k,

$$1 - \lim_{m\to\infty} V_{k,m,\omega}(0) = \lim_{m\to\infty} V_{m,k,\omega}(0) = 0,$$

  and hence, for $\omega \in R$,

$$P_\omega^o\left(-\infty = \liminf_{n\to\infty} X_n < \limsup_{n\to\infty} X_n = \infty\right) = 1.$$

We observe next that both S_+ and $\mathcal{F}_+$ are invariant events, hence $P(S_+) \in \{0,1\}$, $P(\mathcal{F}_+) \in \{0,1\}$ by the ergodicity of P. Next, $P(S_+) = 1 \Rightarrow P(\mathcal{F}_+) = 0$ by the shift-invariance of P. Thus, it is enough to prove that $P(S_+) = 1$ if and only if $E_P(\log \rho_0) < 0$ and $P(\mathcal{F}_+) = 1$ if and only if $E_P(\log \rho_0) > 0$. We prove the first claim only, the second one possessing a similar proof.

Assume first $c := E_P(\log \rho_0) < 0$. Then, by the ergodic theorem, there exists an $n_0(\omega)$ with $P(n_0(\omega) < \infty) = 1$ such that $\frac{1}{n} \sum_{i=1}^{n} \log \rho_i \le c/2 < 0$ for all $n > n_0(\omega)$. But then, for some $C_1(\omega) < \infty$ P-a.s.,

$$\sum_{n=1}^{\infty} \rho_1 \cdots \rho_n \le C_1(\omega) + \sum_{k=n_0(\omega)+1}^{\infty} e^{kc/2} < \infty, \quad P\text{-a.s}$$

implying $P(S_+) = 1$. Conversely, for $\omega \in S_+$, $\lim_{n\to\infty} \sum_{k=1}^{n} \log \rho_k = -\infty$. But, $\{Y_i = -\log \rho_i\}$ are stationary, and the claim follows from the following well known:

Lemma 2.1.5 (Kesten[40]) *For any real valued, stationary sequence $\{Y_i\}$, fix $Z_n = \sum_{i=1}^{n} Y_i$. Then, one has with probability 1 that the event $\{Z_n \to_{n\to\infty} \infty\}$ implies $\{\liminf_{n\to\infty} Z_n/n > 0\}$.*

Indeed, Kesten's lemma implies that on S_+,

$$E_P(\log \rho_0) = \lim_{n \to \infty} \frac{1}{n} \sum_{k=1}^{n} \log \rho_k < 0, \quad P - \text{a.s.}, \qquad (2.1.6)$$

and $P(S_+) = 1$ thus implies $E_P(\log \rho_0) < 0$ and completes the proof of Theorem 2.1.2. $\qquad \square$

Remarks: 1. P. Bougerol has kindly indicated to me the followig proof of the implication $P(S_+) = 1 \Rightarrow E_P(\log \rho_0) < 0$, which bypasses the use of Kesten's lemma: define the function $f(\omega) = \log S(\omega)$. $P(S_+) = 1$ implies that $f(\omega)$ is well defined. Since $S(\omega) = \rho_1 + \rho_1 S(\theta\omega)$, it holds that $f(\omega) > \log \rho_1 + f(\theta\omega)$, and we conclude by **(A2)** that $(f(\theta\omega) - f(\omega))_+$ is P-integrable and hence $E_P[f(\omega) - f(\theta\omega)] = 0$ (this is Mañe's lemma, apply the ergodic theorem to see it!). Using again $f(\omega) > \log \rho_1 + f(\theta\omega)$, one concludes that $0 > E_P \log \rho_1 = E_P \log \rho_0$, as claimed.

2. If P is i.i.d., Theorem 2.1.2 remains valid when the left hand side of conditions (a), (b), (c) is replaced, respectively, by

$$(\text{a}') : \quad \sum_{n=1}^{\infty} n^{-1} P\left(\prod_{j=1}^{n} \rho_j > 1\right) < \infty.$$

$$(\text{b}') : \quad \sum_{n=1}^{\infty} n^{-1} P\left(\prod_{j=1}^{n} \rho_j < 1\right) < \infty.$$

$$(\text{c}') : \quad \sum_{n=1}^{\infty} n^{-1} P\left(\prod_{j=1}^{n} \rho_j < 1\right) = \sum_{n=1}^{\infty} n^{-1} P\left(\prod_{j=1}^{n} \rho_j > 1\right) = \infty.$$

This is useful in particular when $E_P(\log \rho_0)$ is not well defined. See [67] for details.

Having developed transience and recurrence criteria, we turn to the law of large numbers. We first note that one cannot apply directly ergodic theorems to the sequence X_n/n: The sequence $\{X_n - X_{n-1}\}$ is not even stationary! We will exhibit two approaches to the LLN: The first is based on a hitting times decomposition. The second approach is based on the point of view of the "environment viewed from the particle".

LLN-version I: hitting time decompositions

Introduce the following notations:

$$\overline{S} = \sum_{i=1}^{\infty} \frac{1}{\omega_{(-i)}^+} \prod_{j=0}^{i-1} \rho_{(-j)} + \frac{1}{\omega_0^+} \qquad (2.1.7)$$

$$\overline{F} = \sum_{i=1}^{\infty} \frac{1}{\omega_i^-} \prod_{j=0}^{i-1} \rho_j^{-1} + \frac{1}{\omega_0^-} \qquad (2.1.8)$$

Theorem 2.1.9 *Assume Assumption 2.1.1. Then,*

(a) $\quad E_P(\overline{S}) < \infty \quad \Rightarrow \lim\limits_{n\to\infty} \dfrac{X_n}{n} = \dfrac{1}{E_P(\overline{S})}, \qquad \mathbb{P}^o$ a.s.

(b) $\quad E_P(\overline{F}) < \infty \quad \Rightarrow \lim\limits_{n\to\infty} \dfrac{X_n}{n} = -\dfrac{1}{E_P(\overline{F})}, \qquad \mathbb{P}^o$ a.s.

(c) $\quad E_P(\overline{S}) = \infty \quad and \, E_P(\overline{F}) = \infty \Rightarrow \lim\limits_{n\to\infty} \dfrac{X_n}{n} = 0, \quad \mathbb{P}^o$ a.s.

Remark: In the case that P is i.i.d., (a)–(c) of Theorem 2.1.9 become

(a') $\quad E_P(\rho_0) < 1 \qquad\qquad \Rightarrow \lim\limits_{n\to\infty} \dfrac{X_n}{n} = \dfrac{1 - E_P(\rho_0)}{E_P\left(\frac{1}{w_0^+}\right)}, \qquad \mathbb{P}^o$ a.s.

(b') $\quad E_P(\rho_0^{-1}) < 1 \qquad\qquad \Rightarrow \lim\limits_{n\to\infty} \dfrac{X_n}{n} = -\dfrac{1 - E_P\left(\frac{1}{\rho_0}\right)}{E_P\left(\frac{1}{w_0^-}\right)}, \quad \mathbb{P}^o$ a.s.

(c') $\quad \dfrac{1}{E_P(\rho_0)} \leq 1 \leq E_P(\rho_0^{-1}) \Rightarrow \lim\limits_{n\to\infty} \dfrac{X_n}{n} = 0, \qquad\qquad \mathbb{P}^o$ a.s.

since $E_P \log \rho_0 \leq \log E_P \rho_0$ with a strict inequality whenever P is non-degenerate, it follows that one can find examples where $X_n \to \infty$ $\mathbb{P}^o$-a.s. but $X_n/n \to 0$, $\mathbb{P}^o$-a.s. This does not contradict Kesten's lemma (Lemma 2.1.5) because $\{X_n - X_{n-1}\}$ is not in general a stationary sequence under $\mathbb{P}^o$.

Proof of Theorem 2.1.9

We introduce hitting times which will serve us later too. Let $T_0 = 0$, and

$$T_n = \min\{k : X_k = n\}$$

with the usual convention that the minimum over an empty set is $+\infty$. Set $\tau_0 = 0$ and

$$\tau_n = T_n - T_{n-1}, \quad n \geq 1.$$

Similarly, set

$$T_{-n} = \min\{k : X_k = -n\}$$

and

$$\tau_{-n} = T_{-n} - T_{-n+1}, \quad n \geq 1,$$

the convention being that $\tau_{\pm n} = \infty$ if $T_{\pm n} = \infty$. We have the following lemma:

Lemma 2.1.10 *If* $\limsup\limits_{n\to\infty} X_n = +\infty$, $\mathbb{P}^o$*-a.s., then* $\{\tau_i\}_{i\geq 1}$ *is a stationary and ergodic sequence. If further P is strongly mixing, then* $\{\tau_i\}_{i\geq 1}$ *is also strongly mixing.*

Proof of Lemma 2.1.10

The stationarity of $\{\tau_i\}_{i\geq 1}$ follows from the stationarity of the environment. To see the ergodicity, let $\Xi = [0,1]^{\mathbb{N}}$, let U_Ξ denote the measure on Ξ making all coordinates $\{\xi_i\}$ independent and of uniform law on $[0,1]$, and note that $\{X_n\}$ may be constructed by writing

$$X_{n+1} = X_n + 1_{\{\omega_{X_n}^+ < \xi_{n+1}\}} - 1_{\{\xi_{n+1} \in [\omega_{X_n}^+, \omega_{X_n}^+ + \omega_{X_n}^-)\}} .$$

Suppose $A = A(\omega, \xi) = A(\tau)$ is an event, measurable w.r.t. $\mathcal{G}_n = \sigma\{\tau_i, i \geq 1\}$, which is invariant with respect to the shift $(\theta\tau)_i = \tau_{i+1}$ (we write in the sequel $\theta A = A(\theta\tau)$). We need only show that $P \otimes U_\Xi(A) \in \{0,1\}$. Note however that $\theta^k A$, conditioned on $\sigma\{\omega_i, i \in \mathbb{Z}\}$, is independent of $\xi_1, \ldots, \xi_k$. Thus, since U_Ξ is an i.i.d. law and hence the tail sigma-field of $\{\xi_i\}$ is trival, it follows that $A = \theta^k A$ is, under the above conditioning, independent of $\sigma\{\xi_i, i \geq 1\}$. Thus A depends only on ω. But the shift θ on the sequence $\{\tau_i\}$ induces the usual shift θ on Ω. Thus, $\theta A = A(\theta\omega) = A(\omega)$ and hence $P(A) \in \{0,1\}$.

To prove the strong mixing properties (which we do not actually need in the sequel), consider sets $A_1 \cdots A_k, B_1 \cdots B_j \subset \mathbb{Z}$, and let

$$\mathcal{A} = \bigcap_{i=1}^{k}\{\tau_i \in A_i\}, \qquad \mathcal{B}^m = \bigcap_{i=1}^{j}\{\tau_{m+i} \in B_i\}, .$$

Clearly, $\mathbb{P}^o(\mathcal{B}^m) = \mathbb{P}^o(\mathcal{B}^0)$, and thus we need to prove that whenever $\limsup_{n\to\infty} X_n = \infty$ $\mathbb{P}^o$-a.s., then

$$\lim_{m\to\infty} \mathbb{P}^o(\mathcal{A} \cap \mathcal{B}^m) = \mathbb{P}^o(\mathcal{A})\mathbb{P}^o(\mathcal{B}^0) .$$

Toward this end, let

$$B_i^K = B_i \cap [0, K] \quad \text{and} \quad \mathcal{B}^{m,K} = \bigcap_{i=1}^{j}\{\tau_{m+i} \in B_i^K\} .$$

Fix $\varepsilon > 0$ and then $K = K(\varepsilon)$ large enough such that

$$\mathbb{P}^o(\mathcal{B}^m \setminus \mathcal{B}^{m,K}) = \mathbb{P}^o(\mathcal{B}^0 \setminus \mathcal{B}^{0,K}) \leq \varepsilon$$

which is possible since $\limsup_{n\to\infty} X_n = \infty$, $\mathbb{P}^o$-a.s. Note that, for $m > k$,

$$P_\omega^o(\mathcal{A} \cap \mathcal{B}^{K,m}) = P_\omega^o(\mathcal{A})P_\omega^o(\mathcal{B}^{K,m})$$

and that $P_\omega^o(\mathcal{A})$ is measurable with respect to $\sigma(\omega_i, i \leq k-1)$. On the other hand, since $|X_{n+1} - X_n| = 1$, on the event $\{\tau_{m+i} \leq K, 1 \leq i \leq j\}$, it holds that $X_n \geq m - K$ for $T_m \leq n \leq T_{m+j+1}$. Thus, for $m > K + k$, $P_\omega^o(\mathcal{B}^{K,m})$ is measurable with respect to $\sigma(\omega_i, i \geq m - K)$. It follows from the strong mixing of P that

$$\lim_{m\to\infty} \mathbb{P}^o(\mathcal{A} \cap \mathcal{B}^{K,m}) = \lim_{m\to\infty} E_P(P_\omega^o(\mathcal{A}) P_\omega^o(\mathcal{B}^{K,m}))$$
$$= E_P(P_\omega^o(\mathcal{A})) \cdot E_P(P_\omega^o(\mathcal{B}^{K,0}))$$
$$= \mathbb{P}^o(\mathcal{A}) \mathbb{P}^o(\mathcal{B}^{K,0}). \qquad (2.1.11)$$

On the other hand,

$$\mathbb{P}^o(\mathcal{A} \cap \mathcal{B}^m) - \varepsilon \le \mathbb{P}^o(\mathcal{A} \cap \mathcal{B}^{K,m}) \le \mathbb{P}^o(\mathcal{A} \cap \mathcal{B}^m)$$

while

$$\mathbb{P}^o(\mathcal{B}^m) - \varepsilon \le \mathbb{P}^o(\mathcal{B}^{K,m}) \le \mathbb{P}^o(\mathcal{B}^m)$$

and one concludes from (2.1.11) that

$$\left| \lim_{m\to\infty} \mathbb{P}^o(\mathcal{A} \cap \mathcal{B}^m) - \mathbb{P}^o(\mathcal{A}) \mathbb{P}^o(\mathcal{B}^0) \right| \le \varepsilon$$

which implies the claim since ε is arbitrary. $\square$

Remark: Note that an attempt to mimick this argument in $\mathbb{Z}^d$, $d \ge 1$, with T_i denoting the hitting times of hyperplanes at distance i from the origin, fails because of the extra information contained in the hitting location.

Our strategy consists now of applying the ergodic theorem to the sequence $\{\tau_i\}$. As a first step, we have the

Lemma 2.1.12 *Assume Assumption 2.1.1. Then,*

$$(a) \quad E_{\mathbb{P}^o}(\tau_1) = E_P(\overline{S}),$$
$$(b) \quad E_{\mathbb{P}^o}(\tau_{-1}) = E_P(\overline{F}).$$

Proof. We prove only (a), the proof of (b) being similar. Decompose, with $X_0 = 0$,

$$\tau_1 = \mathbf{1}_{X_1=1} + \mathbf{1}_{X_1=0}(1 + \tau_1') + \mathbf{1}_{X_1=-1}(1 + \tau_0'' + \tau_1''). \qquad (2.1.13)$$

Here, (τ_1') is the first hitting time of 1 after time 1 (possibly infinite), $(1 + \tau_0'')$ is the first hitting time of 0 after time 1, and $1 + \tau_0'' + \tau_1''$ is the first hitting time of 1 after time $1 + \tau_0''$.

Under P_ω^o, the law of τ_1' conditioned on the event $\{X_1 = 0\}$ is identical to the law of τ_1, the law of τ_0'' conditioned on the event $\{X_1 = -1\}$ is $P_{\theta^{-1}\omega}^o(\tau_1 \in \cdot)$, while conditioned on the event $\{X_1 = -1\} \cap \{\tau_0'' < \infty\}$, τ_1'' also has law identical to that of τ_1.

Consider first the case $E_{\mathbb{P}^o}(\tau_1) < \infty$. Then, both $E_\omega^o(\tau_1) < \infty$ and $E_{\theta^{-1}\omega}(\tau_1) < \infty$, P-a.s. Taking expectations in (2.1.13), one gets then

$$E_\omega^o(\tau_1) = 1 + (1 - \omega_0^+) E_\omega^o(\tau_1) + \omega_0^- E_{\theta^{-1}\omega}^o(\tau_1).$$

Hence,

$$E_\omega^o(\tau_1) = \frac{1}{\omega_0^+} + \rho_0 E_{\theta^{-1}\omega}^o(\tau_1).$$

Iterating this equation, we get

$$E_\omega^o(\tau_1) = \frac{1}{\omega_0^+} + \frac{\rho_0}{\omega_{(-1)}^+} + \frac{\rho_0\rho_{(-1)}}{\omega_{(-2)}^+} + \cdots + \frac{\prod_{i=0}^{-(m-1)}\rho_{(-i)}}{\omega_{(-m)}^+}$$
$$+ \left(\prod_{i=0}^{-m+1}\rho_{(-i)}\right)E_{\theta^{-m}\omega}^o(\tau_1). \quad (2.1.14)$$

Omitting the last term, taking expectations on both sides, and then taking $m \to \infty$ using dominated convergence, we get

$$E_{\mathbb{P}^o}(\tau_1) \geq E_P(\overline{S}). \qquad (2.1.15)$$

To see the reverse inequality, note that by (2.1.13), for any $M < \infty$,

$$E_\omega^o(\tau_1\mathbf{1}_{\tau_1<M}) \leq 1 + (1-\omega_0^+)E_\omega^o(\tau_1\mathbf{1}_{\tau_1<M}) + \omega_0^- E_{\theta^{-1}\omega}^o(\tau_1\mathbf{1}_{\tau_1<M}).$$

Iterating, we get that

$$E_\omega^o(\tau_1\mathbf{1}_{\tau_1<M}) \leq \overline{S} + M\prod_{i=0}^{-m+1}\rho_{(-i)}.$$

Taking expectations, we get that

$$E_{\mathbb{P}^o}(\tau_1\mathbf{1}_{\tau_1<M}) \leq E_P(\overline{S}) + ME_P\left(\prod_{i=0}^{-m+1}\rho_{(-i)}\right).$$

Assuming $E_P(\overline{S}) < \infty$ and hence $E_P\left(\prod_{i=0}^{-m+1}\rho_{(-i)}\right) \xrightarrow[m\to\infty]{} 0$, we get that

$$E_{\mathbb{P}^o}(\tau_1\mathbf{1}_{\tau_1<M}) \leq E_P(\overline{S}).$$

Taking $M \to \infty$ and using monotone convergence we conclude, using also (2.1.15), that

$$E_{\mathbb{P}^o}(\tau_1\mathbf{1}_{\tau_1<\infty}) = E_P(\overline{S}),$$

completing the proof that $E_{\mathbb{P}^o}(\tau_1) < \infty \Rightarrow E_{\mathbb{P}^o}(\tau_1) = E_P(\overline{S})$.

It thus remains to show that $E_{\mathbb{P}^o}(\tau_1) = \infty \Rightarrow E_P(\overline{S}) = \infty$. Note next that if $E_P(\log\rho_0) \leq 0$, we have by Theorem 2.1.2 that

$$E_{\mathbb{P}^o}(\tau_1\mathbf{1}_{\tau_1<\infty}) = E_{\mathbb{P}^o}(\tau_1)$$

hence $E_{\mathbb{P}^o}(\tau_1) = \infty$ implies $E_P(\overline{S}) = \infty$. On the other hand, if $E_P(\log\rho_0) > 0$ then $\prod_{i-1}^0\rho_{(-j)} \to_{j\to\infty} \infty$, P-a.s. by the ergodic theorem and hence also $E_P(\overline{S}) = \infty$. This concludes the proof of Lemma 2.1.12.   □

Remark: In fact, a similar proof shows that in the uniformly elliptic case, $E_\omega^o(\tau_1) = \overline{S}$, for *every* environment ω.

An application of Lemmas 2.1.10 and 2.1.12 yields that in case (a)

$$\frac{T_n}{n} = \frac{\sum_{i=1}^{n} \tau_i}{n} \longrightarrow_{n\to\infty} E_{\mathbb{P}^o}(\tau_1) < \infty, \quad \mathbb{P}^o\text{-a.s.}. \tag{2.1.16}$$

On the other hand, we have the following:

Lemma 2.1.17 *Assume $T_n/n \to \alpha$, for some constant $\alpha < \infty$. Then,*

$$\frac{X_n}{n} \underset{n\to\infty}{\longrightarrow} \frac{1}{\alpha}.$$

Proof of Lemma 2.1.17

Let k_n be the unique (random) integers such that

$$T_{k_n} \leq n < T_{k_n+1}.$$

Note that $X_n < k_n + 1$ while $X_n \geq k_n - (n - T_{k_n})$. Hence,

$$\frac{k_n}{n} - \left(1 - \frac{T_{k_n}}{n}\right) \leq \frac{X_n}{n} \leq \frac{k_n+1}{n}.$$

But, $\lim_{n\to\infty} k_n/n = \lim_{n\to\infty} n/T_n$ (due to the existence of the second limit and the definition of k_n). Thus,

$$\frac{1}{\alpha} \geq \limsup_{n\to\infty} \frac{X_n}{n} \geq \liminf_{n\to\infty} \frac{X_n}{n} \geq \frac{1}{\alpha}. \qquad \square$$

Lemma 2.1.17 and (2.1.16) complete the proof of Theorem 2.1.9 in case (a). Case (b) is similar, while case (c) is a minor modification of the above argument and is left out. $\qquad \square$

Bibliographical notes: The proof of Theorem 2.1.2 is essentially from [67], except that the use of Kesten's lemma is borrowed from [1]. See also [50] for an "ergodic" approach. The rest of the section is an adaptation of the argument in [67], which requires a strongly mixing assumption. The proof of ergodicity in Lemma 2.1.10 was suggested to me by P. Bougerol. F. Rassoul-Agha has kindly shown me a different proof of this fact.

Transience and recurrence results for non nearest-neighbour RWRE on $\mathbb{Z}$, in terms of certain Lyapunov exponents of products of random matrices, are developed in [43], see also [50] and [7]. This is further developed in [3], [39], where transience and recurrence criteria for RWRE on graphs of the form $\mathbb{Z} \times G$, G finite, are derived.

LLN-version II: auxiliary Markov chains

We use the evaluation of the LLN as an excuse for introducing the machinery of the "environment viewed from the particle". The first step consists of introducing an auxiliary Markov chain.

Starting from the RWRE X_n, define $\bar{\omega}(n) = \theta^{X_n}\omega$. The sequence $\{\bar{\omega}_n\}$ is a process with paths in $\Omega^{\mathbb{N}}$. What is maybe more useful is that it is in fact a *Markov* process. More precisely:

Lemma 2.1.18 *The process $\{\overline{\omega}(n)\}$ is a Markov process under either P_ω^0 or $\mathbb{P}^0$, with state space Ω and transition kernel*

$$M(\omega, d\omega') = \omega_0^+ \delta_{\theta\omega=\omega'} + \omega_0^- \delta_{\theta^{-1}\omega=\omega'} + \omega_0^0 \delta_{\omega=\omega'}.$$

Proof. For bounded functions $f_i : \Omega \to \mathbb{R}$,

$$E_\omega^o \left(\prod_{i=1}^n f_i(\overline{\omega}(i)) \right) = E_\omega^o \left(\prod_{i=1}^n f_i(\theta^{X_i}\omega) \right)$$

$$= E_\omega^o \left(\prod_{i=1}^{n-1} f_i(\theta^{X_i}\omega) E_\omega^{X_{n-1}}(f_n(\theta^{X_n}\omega)) \right)$$

$$= E_\omega^o \left(\prod_{i=1}^{n-1} f_i(\theta^{X_i}\omega) \left[\omega_{X_{n-1}}^+ f_n(\theta \cdot \theta^{X_{n-1}}\omega) \right. \right.$$

$$\left. \left. + \omega_{X_{n-1}}^- f_n(\theta^{-1} \cdot \theta^{X_{n-1}}\omega) + \omega_{X_{n-1}}^o f_n(\theta^{X_{n-1}}\omega) \right] \right)$$

$$= E_\omega^o \left(\prod_{i=1}^{n-1} f_i(\theta^{X_i}\omega) M f_n(\overline{\omega}(n-1)) \right) \tag{2.1.19}$$

where

$$M f(\omega) = \int f(\omega') M(\omega, d\omega'),$$

which proves the Markov property of $\{\overline{\omega}(n)\}$ under P_ω^o. Integrating both sides of (2.1.19) with respect to P yields the Markov property under $\mathbb{P}^o$.   □

Our next step is to construct an invariant measure for the transition kernel M. In most of this section we will assume that $E_P \log \rho_0 < 0$, implying, by Theorem 2.1.2, that $T_1 < \infty$, $\mathbb{P}^o$-a.s. Whenever $E_{\mathbb{P}^o}(T_1) < \infty$, define the measures

$$Q(B) = E_{\mathbb{P}^o} \left(\sum_{i=0}^{T_1-1} \mathbf{1}_{\{\overline{\omega}(i) \in B\}} \right), \quad \overline{Q}(B) = \frac{Q(B)}{Q(\Omega)} = \frac{Q(B)}{E_{\mathbb{P}^o} T_1}.$$

Using Lemma 2.1.12, one checks that under Assumption 2.1.1 and if $E_P(\bar{S}) < \infty$ then $E_{\mathbb{P}^o} T_1 < \infty$, and $\overline{Q}(\cdot)$ in this case is a probability measure.

Lemma 2.1.20 *Assume Assumption 2.1.1 and $E_P(\bar{S}) < \infty$. Then, $Q(\cdot)$ is invariant under the Markov kernel M, that is*

$$Q(B) = \iint \mathbf{1}_{\omega' \in B} M(\omega, d\omega') Q(d\omega).$$

Proof. We have

$$\iint \mathbf{1}_{\omega' \in B} M(\omega, d\omega') Q(d\omega)$$

$$= \sum_{k=0}^{\infty} E_{\mathbb{P}^o}\left(T_1 > k; \; \mathbf{1}_{\overline{\omega}(k+1) \in B}\right)$$

$$= \sum_{k=0}^{\infty} E_{\mathbb{P}^o}\left(T_1 = k+1; \; \mathbf{1}_{\overline{\omega}(k+1) \in B}\right) + \sum_{k=0}^{\infty} E_{\mathbb{P}^o}\left(T_1 > k+1; \; \mathbf{1}_{\overline{\omega}(k+1) \in B}\right)$$

$$= \mathbb{P}^o(T_1 < \infty; \overline{\omega}(T_1) \in B) + \sum_{k=1}^{\infty} \mathbb{P}^o(T_1 > k; \overline{\omega}(k) \in B).$$

But $\mathbb{P}^o(T_1 < \infty) = 1$ while $\mathbb{P}^o(\overline{\omega}(T_1) \in B) = P(\theta\omega \in B) = P(\omega \in B)$, hence

$$= \sum_{k=0}^{\infty} \mathbb{P}^o(T_1 > k; \overline{\omega}(k) \in B) = Q(B). \qquad \square$$

Define next

$$\Lambda(\omega) = \frac{1}{\omega_0^+}\left[1 + \sum_{i=1}^{\infty} \prod_{j=1}^{i} \rho_j\right].$$

It is not hard to check, by the shift invariance of P, that the condition $E_P(\Lambda(\omega)) < \infty$ is equivalent to $E_P(\overline{S}) < \infty$, c.f. Section 2.1. We next claim the

Lemma 2.1.21 *Under the assumptions of Lemma 2.1.20, it holds that*

$$\frac{dQ}{dP} = \Lambda(\omega).$$

Proof. Note first that by Jensen's inequality, $E_P(\Lambda) < \infty$ implies that $E_P(\log \rho_0) < 0$ and hence $X_n \to_{n \to \infty} \infty$, $\mathbb{P}^o$-a.s., by Theorem 2.1.2. Let $f : \Omega \to \mathbb{R}$ be measurable. Then,

$$\int f dQ = E_{\mathbb{P}^o}\left(\sum_{i=0}^{T_1-1} f(\overline{\omega}_i)\right) = E_{\mathbb{P}^o}\left(\sum_{i \leq 0} f(\theta^i \omega) N_i\right)$$

where $N_i = \{\#k \in [0, T_1) : X_k = i\}$ (note the difference in the role the index i plays in the two sums!). Using the shift invariance of P, we get

$$\int f dQ = \sum_{i \leq 0} E_P\left(f(\theta^i \omega) E_\omega^o N_i\right)$$

$$= \sum_{i \leq 0} E_P\left(f(\omega) E_{\theta^{-i}\omega}^o N_i\right) = E_P\left(f(\omega)\left(\sum_{i \leq 0} E_{\theta^{-i}\omega}^o N_i\right)\right).$$

Hence,

$$\frac{dQ}{dP} = \sum_{i \leq 0} E^o_{\theta^{-i}\omega} N_i, \tag{2.1.22}$$

and the right hand side converges, P-a.s.

In order to prove both the convergence in (2.1.22) and the lemma, we turn to evaluate $E^o_\omega N_i$. Define, for $i \leq 0$,

$$\eta_{i,0} = \min\{k \leq T_1 : X_k = i\}$$
$$\theta_{i,0} = \min\{\eta_{i,0} < k \leq T_1 : X_{k-1} = i, X_k = i - 1\}$$

and, for $j \geq 1$,

$$\eta_{i,j} = \min\{\theta_{i,j-1} < k \leq T_1 : X_k = i\}$$
$$\theta_{i,j} = \min\{\eta_{i,j} < k \leq T_1 : X_{k-1} = i, X_k = i - 1\}$$

(with the usual convention that the minimum over an empty set is $+\infty$). We refer to the time interval $(\theta_{i,j-1}, \eta_{i,j})$ as the j-th excursion from $i - 1$ to i. For any $j \geq 0$, any $i \leq 0$, define

$$U_{i,j} = \{\#\ell \geq 0 : \theta_{i+1,j} < \theta_{i,\ell} < \eta_{i+1,j+1}\}$$
$$Z_{i,j} = \{\#k \geq 0 : X_{k-1} = i, X_k = i, \theta_{i+1,j} < k < \eta_{i+1,j+1}\}.$$

Note that $U_{i,j}$ is the number of steps from i to $i - 1$ during the $j + 1$-th excursion from i to $i + 1$, whereas $Z_{i,j}$ is the number of steps from i to i during the same excursion. The Markov property implies that

$$P^o_\omega\left(U_{i,\ell} = k_\ell, Z_{i,\ell} = m_\ell, \ell = 1, \ldots, L \Big| \{U_{i',j}\}_{i'>i}, \eta_{i+1,L+1} < \infty\right)$$
$$= \prod_{\ell=1}^{L}\left[\left(\frac{\omega_i^-}{\omega_i^- + \omega_i^+}\right)^{k_\ell}\left(\frac{\omega_i^+}{\omega_i^- + \omega_i^+}\right)\left(\frac{\omega_i^0}{\omega_i^0 + \omega_i^+}\right)^{m_\ell}\left(\frac{\omega_i^+}{\omega_i^0 + \omega_i^+}\right)\right]. \tag{2.1.23}$$

Defining $U_i = \sum_j U_{i,j}$, $Z_i = \sum_j Z_{i,j}$, and noting that $\mathbb{P}^o(\{U_i < \infty\} \cap \{Z_i < \infty\}) = 1$ because $X_n \to \infty$, $\mathbb{P}^o$-a.s., (2.1.23) implies that $\{U_i\}$ is under P^o_ω an (inhomogeneous) branching process with geometric offspring distribution of parameter $\frac{\omega_i^-}{\omega_i^- + \omega_i^+}$. Further,

$$E^o_\omega(U_i|U_{i+1}, \cdots, U_0) = \rho_i U_{i+1}$$
$$E^o_\omega(Z_i|U_{i+1}, \cdots, U_0) = \frac{\omega_i^0}{\omega_i^+} U_{i+1} \tag{2.1.24}$$

and using the relation $N_i = U_i + U_{i+1} + Z_i$, $\mathbb{P}^o$-a.s., we get

$$E^o_\omega(N_i|U_{i+1}, \ldots, U_0) = E^o_\omega\left(U_i + U_{i+1} + Z_i|U_{i+1}, \cdots, U_0\right) = \frac{1}{\omega_i^+} E^o_\omega U_{i+1}.$$

Iterating (2.1.24), one gets

$$E_\omega^o N_i = \frac{1}{\omega_i^+} \rho_0 \cdots \rho_{i+1} \,.$$

Hence, using (2.1.22), and the assumption,

$$\frac{dQ}{dP} = \frac{1}{\omega_0^+} \left[1 + \sum_{i=1}^{\infty} \prod_{j=1}^{i} \rho_j \right] < \infty, P\text{-a.s.}$$

which completes the proof of the Lemma.   □

Remark: Note that $dQ/dP > 0$, P-a.s., and hence under the assumption $E_P(\bar{S}) < \infty$ it holds that $Q \sim P$. This fact is true in greater generality, see the discussion in [69] and in Section 3.3 below.

Corollary 2.1.25 *Under the law induced by $\overline{Q} \otimes P_\omega^o$, the sequence $\{\overline{\omega}(n)\}$ is stationary and ergodic.*

Proof. The stationarity follows from the stationarity of $\overline{Q}$. Let $\overline{\theta}$ denote the shift on $\overline{\Omega} = \Omega^{\mathbb{N}}$, that is, for $\overline{\omega} \in \overline{\Omega}, \overline{\theta}\overline{\omega}(n) = \overline{\omega}(n+1)$. Denote by $\overline{P}_\omega$ the law of the sequence $\{\overline{\omega}(n)\}$ with $\overline{\omega}(0) = \omega$, that is, for any measurable sets $B_i \subset \Omega$,

$$\overline{P}_\omega\left(\overline{\omega}(i) \in B_i, i = 1, \ldots, \ell\right)$$
$$= \int_{B_1} \cdots \int_{B_\ell} M(\omega, d\omega^1) M(\omega^1, d\omega^2) \cdots M(\omega^{\ell-1}, d\omega^\ell)$$

and set $\overline{Q} = Q \otimes \overline{P}_\omega$ (as usual, we also use $\overline{Q}$ to denote the corresponding marginal induced on $\overline{\Omega}$).

We need to show that for any invariant A, that is $A \in \overline{\Omega}$ such that $\overline{\theta}A = A$, $\overline{Q}(A) \in \{0, 1\}$. Set $\varphi(\omega) = \overline{P}_\omega(A)$, we claim that $\{\varphi(\overline{\omega}(n))\}$ is a martingale with respect to the filtration $\mathcal{G}_n = \sigma(\overline{\omega}(0), \ldots, \overline{\omega}(n))$: indeed,

$$\varphi(\overline{\omega}(n)) = \overline{P}_{\overline{\omega}(n)}(A) = E_{\overline{Q}}\left(1_{\overline{\theta}^n A}|\mathcal{G}_n\right) = E_{\overline{Q}}\left(1_A|\mathcal{G}_n\right),$$

where the second equality is due to the Markov property and the third due to the invariance of A. Hence, by the martingale convergence theorem,

$$\varphi(\overline{\omega}(n)) \xrightarrow[n \to \infty]{} 1_A, \quad \overline{Q}\text{-a.s.} \tag{2.1.26}$$

Further, $Q(\varphi(\omega) \notin \{0, 1\}) = 0$ because otherwise there exists an interval $[a, b]$ with $\{0\}, \{1\} \notin [a, b]$ and $Q(\varphi(\omega) \in [a, b]) > 0$, while

$$\frac{1}{n} \sum_{0}^{n-1} 1_{\{\varphi(\overline{\omega}(n)) \in [a,b]\}} \to E_{\overline{Q}}\left(1_{\{\varphi(\overline{\omega}(0)) \in [a,b]\}}|\mathcal{J}\right), \tag{2.1.27}$$

where $\mathfrak{I}$ is the invariant σ-field.

Taking expectations in (2.1.27) and using (2.1.26), one concludes that

$$0 = \overline{Q}\Big(\varphi(\overline{w}(0)) \in [a,b]\Big) = \overline{Q}\Big(\varphi(\omega) \in [a,b]\Big),$$

a contradiction. Thus for some measurable $B \subset \Omega$, $\varphi(\omega) = 1_B$, $\overline{Q}$ – a.s.. Further, the Markov property and invariance of A yield that $M1_B = 1_B$, $\overline{Q}$-a.s. and hence P-a.s. But then,

$$1_B = M1_B \geq \omega_0^+ 1_{\theta B}, P\text{-a.s.}.$$

Since $E\Lambda(\omega) < \infty$ implies $P(\omega_0^+ = 0) = 0$, it follows that $1_B \geq 1_{\theta B}$, P-a.s., and then $E_P(1_B) = E_P(1_{\theta B})$ implies that $1_B = 1_{\theta B}$, P-a.s. But then, by ergodicity of P, $P(B) \in \{0,1\}$, and hence $\overline{Q}(B) \in \{0,1\}$. Since $\overline{Q}(A) = E_{\overline{Q}}\varphi(\omega) = \overline{Q}(B)$, the conclusion follows.   □

We are now ready to give the:

Proof of Theorem 2.1.9 - Environment version We begin with case (a), noting that the proof of case (b) is identical by the transformation $\omega_i \mapsto \hat{\omega}_{-i}$, where $\hat{\omega}_i^+ = \omega_i^-, \hat{\omega}_i^- = \omega_i^+$. Set $d(x,\omega) = E_\omega^x(X_1 - x)$. Then

$$X_n = \sum_{i=1}^n (X_i - X_{i-1}) = \sum_{i=1}^n \Big(X_i - X_{i-1} - d(X_{i-1},\omega)\Big) + \sum_{i=1}^n d(X_{i-1},\omega)$$

$$:= M_n + \sum_{i=1}^n d(X_{i-1},\omega). \tag{2.1.28}$$

But, under P_ω^o, M_n is a martingale, with $|M_{n+1} - M_n| \leq 2$; Hence, with $\mathfrak{G}_n = \sigma(M_1,\cdots M_n)$,

$$E_\omega^o(e^{\lambda M_n}) = E_\omega^o\Big(e^{\lambda M_{n-1}} E_\omega^o(e^{\lambda(M_n - M_{n-1})}|\mathfrak{G}_{n-1})\Big)$$

$$\leq E_\omega^o\Big(e^{\lambda M_{n-1}} e^{2\lambda^2}\Big)$$

and hence, iterating, $E_\omega^o(e^{\lambda M_n}) \leq e^{2n\lambda^2}$ (this is a version of Azuma's inequality, see [19, Corollary 2.4.7]). Chebycheff's inequality then implies

$$\frac{M_n}{n} \to 0, \quad \mathbb{P}^o\text{-a.s.}$$

(and even with exponential rate). Next, note that

$$\sum_{i=1}^n d(X_{i-1},\omega) = \sum_{i=1}^n d(0,\overline{w}(i-1)).$$

The ergodicity of $\{\overline{w}(i)\}$ under $\overline{Q} \otimes P_\omega^o$ implies that

$$\frac{1}{n}\sum_{i=1}^{n}d(0,\overline{\omega}(i-1))\longrightarrow E_{\overline{Q}}(d(0,\overline{\omega}(0))),\ \overline{Q}\otimes P_{\omega}^{o}\text{-a.s.}. \tag{2.1.29}$$

But,

$$E_{\overline{Q}}(d(0,\overline{\omega}(0)))$$

$$=\frac{E_{P}\left[\Lambda(\omega)(\omega_{0}^{+}-\omega_{0}^{-})\right]}{E_{P}(\Lambda(\omega))}$$

$$=\frac{1+E_{P}\left(\omega_{1}^{-}\left[\frac{1}{\omega_{1}^{+}}+\sum_{i=2}^{\infty}\prod_{j=2}^{i}\rho_{j}\right]-\omega_{0}^{-}\left[\frac{1}{\omega_{0}^{+}}+\sum_{i=1}^{\infty}\prod_{j=1}^{i}\rho_{j}\right]\right)}{E_{P}(\Lambda(\omega))}$$

$$=\frac{1}{E_{P}(\Lambda(\omega))}=\frac{1}{E_{P}(\overline{S}(\omega))}.$$

Finally, since $E_{P}(\Lambda(\omega))<\infty$, (2.1.29) holds also $\mathbb{P}^{o}$-a.s., completing the proof of the theorem in cases (a),(b).

Case (c) is handled by appealing to Lemma 2.1.12. Suppose $\limsup X_{n}=+\infty,\mathbb{P}^{o}-$ a.s.. Then, $\tau_{1}<\infty$, $\mathbb{P}^{o}$-a.s.. Define $\tau_{i}^{K}=\min(\tau_{i},K)$. Note that under P_{ω}^{o}, the random variables $\{\tau_{i}^{K}\}$ are independent and bounded, and hence, with $G_{n}^{K}=n^{-1}\sum_{i=1}^{n}\tau_{i}^{K}$, we have

$$|G_{n}^{K}-E_{\omega}^{o}G_{n}^{K}|\rightarrow_{n\to\infty}0,\ P_{\omega}^{o}-\text{a.s.}$$

But $f(\omega):=E_{\omega}^{o}\tau_{1}^{K}$ is a bounded, measurable, local function on Ω, and $E_{\omega}^{o}G_{n}^{K}=n^{-1}\sum_{i=1}^{n}f(\theta^{i}\omega)$. Hence, by the ergodic theorem, $E_{\omega}^{o}G_{n}^{K}\rightarrow_{n\to\infty}E_{\mathbb{P}^{o}}\tau_{1}^{K},P-$ a.s.. Since, by Lemma 2.1.12 we have $E_{\mathbb{P}^{o}}\tau_{1}^{K}\rightarrow_{K\to\infty}\infty$, we conclude that

$$\liminf_{n\to\infty}\frac{1}{n}\sum_{i=1}^{n}\tau_{i}\geq\lim_{K\to\infty}E_{\mathbb{P}^{o}}\tau_{1}^{K}=\infty,\mathbb{P}^{o}-\text{a.s.}$$

This immediately implies $\limsup_{n\to\infty}X_{n}/n\leq0,\mathbb{P}^{o}-$ a.s.. The reverse inequality is proved by considering the sequence $\{\tau_{-i}\}$, yielding part (c) of the Theorem. □

Remark: Exactly as in Lemma 2.1.17, it is not hard to check that under Assumption 2.1.1, it holds that

$$\lim_{n\to\infty}\frac{T_{n}}{n}=E_{P}(\overline{S}),\quad\mathbb{P}^{o}-\text{a.s.} \tag{2.1.30}$$

Bibliographical notes: The construction presented here goes back at least to [45]. Our presentation is heavily influenced by [1] and [69].

2.2 CLT for ergodic environments

In this section, we continue to look at the environment from the point of view of the particle. Our main goal is to prove the following:

Theorem 2.2.1 *Assume 2.1.1. Further, assume that for some $\varepsilon > 0$,*

$$E_Q(\overline{S}^{2+\varepsilon}(\omega) + \overline{S}(\theta^{-1}\omega)^{2+\varepsilon}) < \infty, \qquad (2.2.2)$$

and that

$$\sum_{n\geq 1}\sqrt{E_P\left(E_P\left(v_P\overline{S}(\omega) - 1 \,\Big|\, \sigma(\omega_i, i \leq -n)\right)^2\right)} < \infty, \qquad (2.2.3)$$

where $v_P := 1/E_P(\overline{S}(\omega))$. Then, with

$$\sigma_{P,1}^2 := v_P^2 E_Q\left(\omega_0^+(\overline{S}(\omega) - 1)^2 + \omega_0^-(\overline{S}(\theta^{-1}\omega) + 1)^2 + \omega_0^0\right),$$

and

$$\sigma_{P,2}^2 := E_P(v_P\overline{S}(\omega) - 1)^2 + 2\sum_{n=1}^{\infty} E_P\left((v_P\overline{S}(\omega) - 1)(v_P\overline{S}(\theta^n\omega) - 1)\right),$$

we have that

$$\mathbb{P}^o\left(\frac{X_n - nv_P}{\sigma_P\sqrt{n}} > x\right) \longrightarrow_{n\to\infty} \Phi(-x),$$

where

$$\Phi(x) := \frac{1}{\sqrt{2\pi}}\int_{-\infty}^{x} e^{-\frac{\theta^2}{2}}\,d\theta,$$

and $\sigma_P^2 = \sigma_{P,1}^2 + v_P\sigma_{P,2}^2$.

Proof. The basic idea in the proof is to construct an appropriate martingale, and then use the Martingale CLT and the CLT for stationary ergodic sequences. We thus begin with recalling the version of these CLT's most useful to us.

Lemma 2.2.4 ([26], pg. 417) *Suppose $(Z_n, \mathcal{F}_n)_{n\geq 0}$ is a martingale difference sequence, and let $V_n = \sum_{1\leq k\leq n} E(Z_k^2|\mathcal{F}_{k-1})$. Assume that*

(a) $\dfrac{V_n}{n} \to_{n\to\infty} \sigma^2$, *in probability.*

(b) $\dfrac{1}{n}\sum_{m\leq n} E\left(Z_m^2 \mathbf{1}_{\{|Z_m|>\varepsilon\sqrt{n}\}}\right) \to_{n\to\infty} 0$.

Then, $\sum_{i=1}^{n} Z_i/\sigma\sqrt{n}$ converges in distribution to a standard Gaussian random variable.

Lemma 2.2.5 ([26], p. 419) *Suppose $\{Z_n\}_{n\in\mathbb{Z}}$ is a stationary, zero mean, ergodic sequence, and set $\mathcal{F}_n = \sigma(Z_i, i \leq n)$. Assume that*

$$\sum_{n\geq 0}\sqrt{E\left(E(Z_0|\mathcal{F}_{-n})\right)^2} < \infty. \qquad (2.2.6)$$

Then, $\left\{ \sum_{i=1}^{nt} Z_i / \sigma \sqrt{n} \right\}_{t \in [0,1]}$ *converges in distribution to a standard Brownian motion, where*

$$\sigma^2 = EZ_0^2 + 2 \sum_{n=1}^{\infty} E(Z_0 Z_n).$$

We next recall that by Theorem 2.1.9,

$$\frac{X_n}{n} \to v_P, \quad \mathbb{P}^o\text{-a.s.},$$

where $v_P := 1/E_P(\overline{S})$. One is tempted to use the martingale M_n appearing in the environment proof of Theorem 2.1.9 (see (2.1.28)), however this strategy is not so successful because of the difficulties associated with separating the fluctuations in M_n and $\sum_{i=1}^{n} d(X_{i-1}, \omega)$. Instead, write

$$f(x, n, \omega) = x - v_P n + h(x, \omega), \quad x \in \mathbb{Z}.$$

We want to make $f(X_n, n, \omega)$ into a martingale w.r.t. $\mathcal{F}_n := \sigma(X_1, \dots, X_n)$ and the law P_ω^o. This is automatic if we can ensure that

$$E_\omega^{X_n} f(X_{n+1}, n+1, \omega) = f(X_n, n, \omega), \quad P_\omega^o\text{-a.s.} \qquad (2.2.7)$$

Developing this equality and defining $\Delta(x, \omega) = h(x+1, \omega) - h(x, \omega)$, we get that (2.2.7) holds true if a bounded solution to the equation

$$\Delta(x, \omega) = -\left[\frac{\omega_x^+ - \omega_x^- - v_P}{\omega_x^+} \right] + \frac{\omega_x^-}{\omega_x^+} \Delta(x-1, \omega)$$

exists. One may verify that $\Delta(x, \omega) = -1 + v_P \overline{S}(\theta^x \omega)$ is such a solution.

Fixing $h(0, \omega) = 0$, and defining $\overline{M}_0 = 0$ and $\overline{M}_n = f(X_n, n, \omega)$, one concludes that $\overline{M}_n$ is a martingale, and further

$$E_\omega^o \left((\overline{M}_{k+1} - \overline{M}_k)^2 | \mathcal{F}_k \right)$$
$$= \omega_{X_k}^+ v_P^2 (\overline{S}(\theta^{X_k}\omega) - 1)^2 + \omega_{X_k}^- v_P^2 (\overline{S}(\theta^{X_k-1}\omega) + 1)^2 + \omega_{X_k}^0 v_P^2$$
$$= v_P^2 \left[\overline{\omega}(k)_0^+ (\overline{S}(\overline{\omega}_k) - 1)^2 + \overline{\omega}(k)_0^- (\overline{S}(\theta^{-1}\overline{\omega}_k) + 1)^2 + \overline{\omega}(k)_0^0 \right].$$

Hence,

$$\frac{V_n}{n} = \frac{1}{n} \sum_{k=1}^{n} E_\omega^o \left((\overline{M}_{k+1} - \overline{M}_k)^2 | \mathcal{F}_k \right) \xrightarrow[n \to \infty]{} \sigma_{P,1}^2, \quad \mathbb{P}^o\text{-a.s.},$$

using the machinery developed in Section 2.1. The integrability condition (2.2.2) is enough to apply the Martingale CLT (Lemma 2.2.4), and one concludes that for any $\delta > 0$,

$$P\left(\left|P_\omega^o\left(\frac{\overline{M}_n}{\sigma_{P,1}\sqrt{n}} \geq x\right) - \Phi(-x)\right| > \delta\right) \to_{n\to\infty} 0. \qquad (2.2.8)$$

Note that since both $P_\omega^o(\overline{M}_n \geq x\sigma_{P,1}\sqrt{n})$ and $\Phi(x)$ are monotone in x, and that $\Phi(\cdot)$ is continuous, the convergence in (2.2.8) actually is uniform on $\mathbb{R}$. Further, note that

$$h(X_n, \omega) = \sum_{j=1}^{X_n-1} \Delta(j,\omega) = \sum_{j=1}^{nv_P} \Delta(j,\omega) + R_n := Z_n + R_n.$$

Note that, for every $\delta > 0$ and some $\delta_n \to 0$,

$$\mathbb{P}^o\left(\frac{|R_n|}{\sqrt{n}} \geq \delta\right) \leq \mathbb{P}^o\left(|X_n - nv_P| \geq \delta_n n\right)$$

$$+ P\left(\max_{j_-, j_+ \in (-n\delta_n, n\delta_n)} \left|\sum_{i=j_-}^{j_+} \frac{\Delta(i,\omega)}{\sqrt{n}}\right| \geq \delta\right) := P_{1,n}(\delta_n) + P_{2,n}(\delta, \delta_n) \xrightarrow[n\to\infty]{} 0,$$

$$(2.2.9)$$

where the convergence of the first term is due (choosing an appropriate $\delta_n \to_{n\to\infty} 0$ slowly enough) to Theorem 2.1.9 and that of the second one due to $E_P \Delta(i,\omega) = 0$ and the stationary invariance principle (Lemma 2.2.5), which can be applied, for any $\delta_n \to_{n\to\infty} 0$, due to (2.2.3).

Another application of Lemma 2.2.5 yields that

$$\lim_{n\to\infty} P(Z_n \geq z\sqrt{nv_P}\sigma_{P,2}) = \Phi(-z). \qquad (2.2.10)$$

Writing $X_n - nv_P = \overline{M}_n - Z_n - R_n$, and using that $R_n/\sqrt{n} \to_{n\to\infty} 0$ in $\mathbb{P}^o$-probability, one concludes that

$$\lim_{n\to\infty} \mathbb{P}^o\left(\frac{X_n - nv_P}{\sqrt{n}} > x\right) = \lim_{n\to\infty} E_P(P_\omega^o(\overline{M}_n/\sqrt{n} > x + Z_n/\sqrt{n}))$$

$$= \lim_{n\to\infty} E_P\left(\Phi\left(-\frac{x + Z_n/\sqrt{n}}{\sigma_{P,1}}\right)\right), \qquad (2.2.11)$$

where the second equality is due to the uniform convergence in (2.2.8). Combining (2.2.11) with (2.2.10) yields the claim.   □

Remark: The alert reader will have noted that under assumptions (2.1.1) and (2.2.2), and a mild mixing assumption on P which ensures that for any $\delta > 0$ and $\delta_n \to 0$, $P_{2,n}(\delta, \delta_n) \to_{n\to\infty} 0$, c.f. (2.2.9),

$$P_\omega^o\left(\frac{X_n - v_P n - Z_n}{\sqrt{n}\sigma_{P,1}} > x\right) \to_{n\to\infty} \Phi(-x).$$

That is, using a random centering one also has a *quenched* CLT.

Exercise 2.2.12 *Check that the integrability conditions (2.2.2) and (2.2.3) allow for the application of Lemmas 2.2.4 and 2.2.5 in the course of the proof of Theorem 2.2.1.*

Exercise 2.2.13 *Check that in the case of P being a product measure, the assumption (2.2.2) in Theorem 2.2.1 can be dropped.*

Bibliographical notes: The presentation here follows the ideas of [45], as developed in [53]. The latter provides an explicit derivation of the CLT in case $P(\omega_0^0 = 0) = 1$, but it seems that in his derivation only the quenched CLT is derived and the random centering then is missing. A different approach to the CLT is presented in [1], using the hitting times $\{\tau_i\}$; It is well suited to yield the quenched CLT, and under strong assumptions on P which ensure that the random quenched centering vanishes P-a.s., also the annealed CLT. Note however that the case of P being a product measure is not covered in the hypotheses of [1]. See [7] for some further discussion and extensions.

There are situations where limit laws which are not of the CLT type can be exhibited. The proof of such results uses hitting time decompositions, and techniques as discussed in Section 2.4. We refer to Section 2.5 and its bibliographical notes for an example of such a situation and additional information.

2.3 Large deviations

Having settled the issue of the LLN, the next logical step (even if not following the historical development) is the evaluation of the probabilities of large deviations. As already noted in the evaluation of the CLT in Section 2.2, there can be serious differences between quenched and annealed probabilities of deviations. In order to address this, we make the following definitions; throughout this section, $\mathcal{X}$ denotes a completely regular topological space.

Definition 2.3.1 *A function $I : \mathcal{X} \to [0, \infty]$ is a rate function if it is lower semicontinuous. It is a good rate function if its level sets are compact.*

Definition 2.3.2 *A sequence of $\mathcal{X}$ valued random variables $\{Z_n\}$ satisfies the quenched Large Deviations Principle (LDP) with speed n and deterministic rate function I if for any Borel set A,*

$$- I(A^o) \leq \liminf_{n\to\infty} \frac{1}{n} \log P_\omega^o(Z_n \in A) \leq \limsup_{n\to\infty} \frac{1}{n} \log P_\omega^o(Z_n \in A) \leq -I(\overline{A})$$

$$P\text{-a.s.} \quad (2.3.3)$$

where A^o denotes the interior of A, $\overline{A}$ the closure of A, and for any Borel set F,

$$I(F) = \inf_{x\in F} I(x) . \tag{2.3.4}$$

194

214   Ofer Zeitouni

Definition 2.3.5 *A sequence of $\mathcal{X}$ valued random variables $\{Z_n\}$ satisfies the annealed LDP with speed n and rate function I if, for any Borel set A,*

$$-I(A^o) \leq \liminf_{n\to\infty} \frac{1}{n} \log \mathbb{P}^o(Z_n \in A) \leq \limsup_{n\to\infty} \frac{1}{n} \log \mathbb{P}^o(Z_n \in A) \leq -I(\overline{A}) .$$

$$(2.3.6)$$

Finally, we note the

Definition 2.3.7 *A LDP is called weak if the upper bound in (2.3.3) or (2.3.6), holds only with $\overline{A}$ compact.*

For background on the LDP we refer to [19]. It is well known, c.f. [19, Lemma 4.1.4] that if the LDP holds then the rate function is uniquely defined. The following easy lemma is intuitively clear: annealed deviation probabilities allow for atypical fluctuations of the environment and hence are not smaller than corresponding quenched deviation probabilities:

Lemma 2.3.8 *Let $\{A_n\}$ be a sequence of events, subsets of $\Omega \times \mathbb{Z}^N$. Then,*

$$c := \limsup_{n\to\infty} \frac{1}{n} \log \mathbb{P}^o(A_n) \geq \limsup_{n\to\infty} \frac{1}{n} \log P^o_\omega(A_n), \quad P - a.s. \quad (2.3.9)$$

Further,

$$\liminf_{n\to\infty} \frac{1}{n} \log \mathbb{P}^o(A_n) \geq \liminf_{n\to\infty} \frac{1}{n} \log P^o_\omega(A_n), \quad P - a.s. \quad (2.3.10)$$

In particular, if a sequence of $\mathcal{X}$ valued random variables $\{Z_n\}$ satisfies annealed and quenched LDP's with rate functions $I_a(\cdot)$, $I_q(\cdot)$, respectively, then,

$$I_a(x) \leq I_q(x) , \forall x \in \mathcal{X}.$$

Proof. Assume first $c < 0$. Fix $\delta > 0$ and let $B_n^\delta = \{\omega : P^o_\omega(A_n) \geq \exp((c + \delta)n)\}$. Then, by the definition of c, see (2.3.9), and Markov's bound, for n large enough,

$$P(B_n^\delta) \leq e^{-\delta n/2} .$$

Hence, $\omega \in B_n^\delta$ occurs only finitely many times, P-a.s., implying that for P-almost all ω there exists an $n_0(\omega)$ such that for all $n \geq n_0(\omega)$, $P^o_\omega(A_n) < \exp((c + \delta)n)$. Hence,

$$\limsup_{n\to\infty} \frac{1}{n} \log P^o_\omega(A_n) \leq c + \delta , \quad P - \text{a.s.}$$

(2.3.9) follows by the arbitrariness of $\delta > 0$. Next, set $\liminf_{n\to\infty} \frac{1}{n} \log \mathbb{P}^o(A_n)$ $:= c_1 \leq c$. Define $\{n_k\}$ such that

$$\lim_{k\to\infty} \frac{1}{n_k} \log \mathbb{P}^o(A_{n_k}) = c_1 .$$

Apply now the first part of the lemma to conclude that

$$c_1 \geq \limsup_{k \to \infty} \frac{1}{n_k} \log P_\omega^o(A_{n_k}) \geq \liminf_{k \to \infty} \frac{1}{n_k} \log P_\omega^o(A_{n_k}) \geq \liminf_{n \to \infty} \frac{1}{n} \log P_\omega^o(A_n)$$

$$P - \text{a.s.}$$

The case $c = 0$ is the same, except that (2.3.9) is trivial. This completes the proof.   □

Quenched LDP's

The LDP in the quenched setting makes use in its proof of the hitting times $\{\tau_i\}$. Introduce, for any $\lambda \in \mathbb{R}$,

$$\varphi(\lambda, \omega) = E_\omega^o(e^{\lambda \tau_1} 1_{\{\tau_1 < \infty\}}), \quad f(\lambda, \omega) = \log \varphi(\lambda, \omega)$$

$$G(\lambda, P, u) = \lambda u - E_P\Big(f(\lambda, \omega)\Big).$$

We need throughout the following modification of Assumption 2.1.1.

Assumption 2.3.11

(B1) P *is stationary and ergodic,*

(B2) *There exists an* $\varepsilon > 0$ *such that* $P(\omega_0^+ \notin (0, \varepsilon)) P(\omega_0^- \notin (0, \varepsilon)) = 1$,

(B3) $P(\omega_0^+ + \omega_0^- > 0) = 1$, $P(\omega_0^0 > 0, \omega_0^+ \omega_0^- = 0) = 0$, *and* $P(\omega_0^+ = 0) P(\omega_0^- = 0) = 0$.

Note that we allow for the possibility of having one sided transitions (e.g., moves to the right only) of the RWRE. This allows one to deal with the case where "random nodes" are present.

Define

$$\rho_{\min} := \inf[\rho : P(\rho_0 < \rho) > 0],$$

$$\rho_{\max} := \sup[\rho : P(\rho_0 > \rho) > 0],$$

$$\omega_{\max}^0 := \sup[\alpha : P(\omega_0^0 > \alpha) > 0].$$

With P_N denoting the restriction of P to the first N coordinates $\{\omega_i\}_{i=0}^{N-1}$, we say that P is locally equivalent to the product of its marginals if for any N finite, $P_N \sim \otimes^N P_1$.

Finally, we say that a measure P is *extremal* if it is locally equivalent to the product of its marginals and in addition it satisfies the following condition:

(C5) Either $\rho_{\min} \leq 1$ and $\rho_{\max} \geq 1$, or if $\rho_{\min} > 1$ then for all $\delta > 0$, $P(\rho_0 < \rho_{\min} + \delta, \omega_0^0 > \omega_{\max}^0 - \delta) > 0$, or if $\rho_{\max} < 1$ then for all $\delta > 0$, $P(\rho_0 > \rho_{\max} - \delta, \omega_0^0 > \omega_{\max}^0 - \delta) > 0$.

Note that **(C5)**, which is used only in the proof of the annealed LDP, can be read off the support of P_0 and represents an assumption concerning the inclusion of "extremal environments" in the support of P. The introduction of this assumption is not essential and can be avoided at the cost of a slightly more cumbersome proof, see the remarks at the end of this chapter.

For a fixed $\varepsilon > 0$, we denote by $M_1^{e,\varepsilon}$ the set of probability measures satisfying Assumption 2.3.11 with parameter ε in **(B2)**. Define also the maps $F : \Omega \mapsto \Omega$ by $(F\omega)_k^+ = \omega_k^-$, $(F\omega)_k^- = \omega_k^+$, and $(\mathrm{Inv}\,\omega)_k = (F\omega)_{-k}$. We now have:

Theorem 2.3.12 *Assume Assumption 2.3.11.*
a) The random variables $\{T_n/n\}$ satisfy the weak quenched LDP with speed n and convex rate function

$$I_P^{\tau,q}(u) = \sup_{\lambda \in \mathbb{R}} G(\lambda, P, u).$$

b) Assume further that $E_P \log \rho_0 \le 0$. Then, the random variables X_n/n satisfy the quenched LDP with speed n and good convex rate function

$$I_P^q(v) = \begin{cases} v\, I_P^{\tau,q}\left(\frac{1}{v}\right) & , \ 0 < v \le 1 \\ |v|\left(I_P^{\tau,q}\left(\frac{1}{|v|}\right) - E_P(\log \rho_0)\right) & , \ -1 \le v < 0 \end{cases}$$

and

$$I_P^q(0) = \lim_{v \downarrow 0} v I_P^{\tau,q}\left(\frac{1}{v}\right).$$

c) Finally, if $E_P \log \rho_0 > 0$, define $P^{\mathrm{Inv}} := P \circ \mathrm{Inv}^{-1}$. Then, $E_{P^{\mathrm{Inv}}}(\log \rho_0) < 0$, and the LDP for (X_n/n) holds with good convex rate function

$$I_P^q(v) = I_{P^{\mathrm{Inv}}}^q(-v).$$

Proof. It should come as no surprise that we begin with the LDP for T_n/n. We divide the proof of Theorem 2.3.12 into the following steps:

Step I: $E_P \log \rho_0 \le 0$, quenched LDP for T_n/n with convex rate function $I_P^{\tau,q}(\cdot)$:

(I.1)	upper bound, lower tail:	$P_\omega^o(T_n \le nu)$
(I.2)	upper bound, upper tail:	$P_\omega^o(T_n \ge nu)$
(I.3)	lower bound	

Step II: $E_P \log \rho_0 > 0$, quenched LDP for T_n/n with convex rate function $I_P^{\tau,q}(\cdot) + E_P(\log \rho_0)$,
Step III: quenched LDP for X_n/n with convex rate function $I_P^q(\cdot)$.

As a preliminary step we have the following technical lemma, whose proof is deferred:

Lemma 2.3.13 *Assume $P \in M_1^{e,\varepsilon}$ and $E_P(\log \rho_0) \leq 0$; Then*
(a) The convex function $I_P^{\tau,q}(\cdot) : \mathbb{R} \mapsto [0, \infty]$ is nonincreasing on $[1, E_P(\overline{S})]$, nondecreasing on $[E_P(\overline{S}), \infty)$. Further, if $E_P(\overline{S}) < \infty$ then $I_P^{\tau,q}(E_P(\overline{S})) = 0$.
(b) For any $1 < u < E_P(\overline{S})$, there exists a unique $\lambda_0 = \lambda_0(u, P)$ such that $\lambda_0 < 0$ and

$$u = \int \frac{d}{d\lambda} \log \varphi(\lambda, \omega)\Big|_{\lambda=\lambda_0} P(d\omega) . \tag{2.3.14}$$

Further,

$$\inf_{P \in M_1^{e,\varepsilon}} \lambda_0(u, P) > -\infty . \tag{2.3.15}$$

(c) There is a deterministic $\lambda_{\mathrm{crit}} := \lambda_{\mathrm{crit}}(P) \in [0, \infty]$ such that

$$\varphi(\lambda, \omega) \begin{cases} < \infty , & \lambda < \lambda_{\mathrm{crit}} , & P \text{ - a.s.} \\ = \infty , & \lambda > \lambda_{\mathrm{crit}} , & P \text{ - a.s.} \end{cases}$$

with $\lambda_{\mathrm{crit}} < \infty$ if $P(\omega_0^+ \omega_0^- = 0) = 0$. In the latter case, $E_\omega^o(e^{\lambda_{\mathrm{crit}} \tau_1}) < e^{-\lambda_{\mathrm{crit}}}/\varepsilon$, P-a.s., and with

$$u_{\mathrm{crit}} = \begin{cases} \infty , & E_P\left[\frac{E_\omega^o(\tau_1 e^{\lambda_{\mathrm{crit}} \tau_1})}{E_\omega^o(e^{\lambda_{\mathrm{crit}} \tau_1})}\right] = \infty \\ E_P\left(\frac{d}{d\lambda} \log \varphi(\lambda, \omega)\Big|_{\lambda=\lambda_{\mathrm{crit}}}\right) , & E_P\left[\frac{E_\omega^o(\tau_1 e^{\lambda_{\mathrm{crit}} \tau_1})}{E_\omega^o(e^{\lambda_{\mathrm{crit}} \tau_1})}\right] < \infty , \end{cases}$$

and $E_P(\overline{S}) \leq u < u_{\mathrm{crit}}$, there exists a unique $\lambda_0 := \lambda_0(u, P)$ such that $\lambda_0 \geq 0$ and (2.3.14) holds.
(d) Assume P is extremal and further assume that $\rho_{\max} < 1$. Then,

$$\lambda_{\mathrm{crit}} = \overline{\lambda} := -\log\left(\omega_{\max}^0 + \frac{2(1 - \omega_{\max}^0)\sqrt{\rho_{\max}}}{1 + \rho_{\max}}\right) .$$

Further, define

$$\hat{\omega}^+ = (1 - \omega_{\max}^0)/(1 + \rho_{\max}) , \quad \hat{\omega}^- = \rho_{\max} \hat{\omega}^+ , \quad \hat{\omega}^0 = \omega_{\max}^0 ,$$

and let $\tilde{\omega}^{\min}$ denote the deterministic environment with $\tilde{\omega}_k^{\min} = \hat{\omega}$. Then, for any $\lambda \leq \lambda_{\mathrm{crit}}$, and any ω such that $\omega_i^0 \leq \omega_{\max}^0$, $\rho_i \leq \rho_{\max}$,

$$\varphi(\lambda, \omega) \leq \varphi(\lambda, \tilde{\omega}^{\min}) < \infty . \tag{2.3.16}$$

Step I.1: Obviously, it is enough to deal with $u \leq E_P(\overline{S})$. Indeed, for $u > E_P(\overline{S})$ we have by (2.1.30) that $\mathbb{P}^o(T_n \leq nu) \xrightarrow[n \to \infty]{} 1$, and there is nothing to prove. Next, by Chebycheff's inequality, for all $\lambda \leq 0$,

$$P_\omega^o\left(\frac{T_n}{n} \leq u\right) \leq e^{-\lambda n u} E_\omega^o\left(e^{\lambda \sum_{i=1}^n \tau_i}\right) = e^{-\lambda n u} \prod_{i=1}^n E_{\theta^i \omega}^o\left(e^{\lambda \tau_1}\right)$$

$$= e^{-\lambda n u} \prod_{i=1}^n \varphi(\lambda, \theta^i \omega), \quad P \text{ - a.s.}$$

$$(2.3.17)$$

where the first equality is due to the Markov property and the second due to $\tau_i < \infty$, $\mathbb{P}^o$ - a.s. (the null set in (2.3.17) does not depend on λ). An application of the ergodic theorem yields that

$$\frac{1}{n} \log \prod_1^n \varphi(\lambda, \theta^i \omega) \longrightarrow E_P\Big(f(\lambda, \omega)\Big), \quad P \text{ - a.s.}$$

first for all λ rational and then for all λ by monotonicity. Thus,

$$\limsup_{n \to \infty} \frac{1}{n} \log P_\omega^o\left(\frac{T_n}{n} \leq u\right) \leq -\sup_{\lambda \leq 0} G(\lambda, P, u), \quad P \text{ - a.s.}$$

Note that if $E_P(\overline{S}) = \infty$ then clearly $E_P[\log E_\omega^o(e^{\lambda \tau_1})] = \infty$ by Jensen's inequality for $\lambda > 0$, and then $\sup_{\lambda \leq 0} G(\lambda, P, u) = I_P^{\tau,q}(u)$. If $E_P(\overline{S}) < \infty$ then, because $u < E_P(\overline{S})$, it holds that for any $\lambda > 0$,

$$\lambda u - E_P f(\lambda, \omega) \leq \lambda E_P(\overline{S}) - E_P f(\lambda, \omega) \leq 0,$$

where Jensen's inequality was used in the last step. Since $G(0, P, u) = 0$, it follows that also in this case $\sup_{\lambda \leq 0} G(\lambda, P, u) = I_P^{\tau,q}(u)$. Hence,

$$\limsup_{n \to \infty} \frac{1}{n} \log P_\omega^o\left(\frac{T_n}{n} \leq u\right) \leq -I_P^{\tau,q}(u) = -\inf_{w \leq u} I_P^{\tau,q}(w),$$

where the last inequality is due to part a) of Lemma 2.3.13, completing Step I.1.

Step I.2: is similar, using this time $\lambda \geq 0$.

Step I.3: The proof of the lower bound is based on a change of measure argument. We present it here in full detail for $u < u_{\text{crit}}$. Fix $\lambda_0 = \lambda_0(u, P)$ as in Lemma 2.3.13, and set a probability measure $Q_{\omega,n}^o$ such that

$$\frac{dQ_{\omega,n}^o}{dP_\omega^o} = \frac{1}{Z_{n,\omega}} \exp\Big(\lambda_0 T_n\Big), \quad Z_{n,\omega} = \mathbb{E}_\omega^o\Big(\exp\Big(\lambda_0 T_n\Big)\Big),$$

and let $\overline{Q}_{\omega,n}^o$ denote the induced law on $\{\tau_1, \ldots, \tau_n\}$. Due to the Markov property, $\overline{Q}_{\omega,n}^o$ is a product measure, whose first n marginals do not depend on n, hence we will write $\overline{Q}_\omega^o$ instead of $\overline{Q}_{\omega,n}^o$ when integrating over events depending only on $\{\tau_i\}_{i<n}$. But, for any $\delta > 0$,

$$P_\omega^o\left(\frac{T_n}{n} \in (u - \delta, u + \delta)\right)$$

$$\geq \exp\left(-nu\lambda_0 - n\delta|\lambda_0| + \sum_{i=1}^{n} \log\varphi\left(\lambda_0, \theta^i\omega\right)\right)\overline{Q}_\omega^o\left(\left|\frac{T_n}{n} - u\right| \leq \delta\right).$$

$$(2.3.18)$$

By the ergodic theorem and the fact that $u < u_{\text{crit}}$, it holds that

$$E_{\overline{Q}_\omega^o}(T_n/n) \to_{n\to\infty} E_P\left(E_{\overline{Q}_\omega^o}(\tau_1)\right) = u, P\text{ - a.s.}\qquad(2.3.19)$$

where we used again (2.3.14). On the other hand, again because $\lambda_0 < \lambda_{\text{crit}}$ it holds that there exists an $\eta > 0$ such that

$$E_P\left(E_{\overline{Q}_\omega^o}\left(e^{\eta\tau_1}\right)\right) < \infty,$$

implying that

$$\overline{Q}_\omega^o\left(\left|\frac{T_n}{n} - u\right| \geq \delta\right) \xrightarrow[n\to\infty]{} 0, P\text{ - a.s.}\qquad(2.3.20)$$

Combining (2.3.20) with (2.3.18), we get

$$\liminf_{n\to\infty} \frac{1}{n} \log P_\omega^o\left(\frac{T_n}{n} \in (u - \delta, u + \delta)\right)$$

$$\geq -u\lambda_0 - \delta|\lambda_0| + \liminf_{n\to\infty} \frac{1}{n} \sum_{i=1}^{n} \log\varphi(\lambda_0, \theta^i\omega)$$

$$= -u\lambda_0 - \delta|\lambda_0| + E_P(\log\varphi(\lambda_0, \omega))$$

$$= -G(\lambda_0, P, u) - \delta|\lambda_0| = -I_P^{\tau,q}(u) - \delta|\lambda_0|, P\text{ - a.s.}$$

where the first equality is due to the ergodic theorem and the last one to Lemma 2.3.13. This completes Step I.3 when $u < u_{\text{crit}}$, since $\delta > 0$ is arbitrary. For $u > u_{\text{crit}}$, the proof is similar, except that one needs to truncate the variables $\{\tau_i\}$, we refer to [12, Theorem 4] for details. Step I is complete, except for the:

Proof of Lemma 2.3.13

We consider in what follows only the case $P(\omega_0^+ \omega_0^- = 0) = 0$, the modifications in the case where random nodes are allowed are left to the reader.
a) The convexity of $I_P^{\tau,q}(\cdot)$ is immediate from its definition as a supremum of affine functions.

As in the course of the proof of Step I, recall that

$$\sup_{\lambda\in\mathbb{R}} G(\lambda, u, P) = \begin{cases} \sup_{\lambda\leq 0} G(\lambda, u, P), & u < E_P(\overline{S}) \\ \sup_{\lambda\geq 0} G(\lambda, u, P), & u > E_P(\overline{S}) \\ 0, & u = E_P(\overline{S}). \end{cases}$$

The stated monotonicity properties are then immediate.

b)+c) Recall the path decomposition (2.1.13). Exponentiating and taking expectations using $\tau_1 < \infty$, $\mathbb{P}^o$ - a.s., we have that if $\varphi(\lambda, \omega) < \infty$ then

$$\varphi(\lambda, \omega) = \omega_0^+ e^\lambda + \omega_0^0 e^\lambda \varphi(\lambda, \omega) + \omega_0^- e^\lambda \varphi(\lambda, \omega) \varphi(\lambda, \theta^{-1}\omega) . \qquad (2.3.21)$$

Thus $\varphi(\lambda, \omega) < \infty$ implies $\varphi(\lambda, \theta^{-1}\omega) < \infty$, yielding that $1_{\varphi(\lambda,\omega)<\infty}$ is constant P - a.s., and hence for all λ rational, $P(\varphi(\lambda, \omega) < \infty) \in \{0, 1\}$. This, and the monotonicity of $\varphi(\lambda, \omega)$ in λ, immediately yields the existence of a deterministic λ_{crit}. (We note in passing that (2.3.21) gives, by iterating, a representation of $\varphi(\lambda, \omega)$ as a continued fraction, but we do not need this now.) We also conclude from (2.3.21) that for $\lambda < \lambda_{\mathrm{crit}}$ it holds that $\varphi(\lambda, \omega) \leq e^{-\lambda}/\varepsilon$, P-a.s., which implies by monotone convergence that $\varphi(\lambda_{\mathrm{crit}}, \omega) < \infty$, P-a.s.

Next, for $\lambda < 0$ we have that

$$g(\lambda) := \int \frac{E_\omega^o(\tau_1 e^{\lambda\tau_1})}{E_\omega^o(e^{\lambda\tau_1})} P(d\omega) = \int \frac{d}{d\lambda} \log \varphi(\lambda, \omega) P(d\omega) .$$

Further, $g(0) = E_{\mathbb{P}^o}(\tau_1)$, whereas $g(\cdot) \geq 1$ is strictly monotone increasing, satisfying $g(\lambda) \underset{\lambda \to -\infty}{\longrightarrow} 1$. This implies (2.3.14). Finally, to see (2.3.15), note that

$$1 \leq \frac{E_\omega^o(\tau_1 e^{\lambda\tau_1})}{E_\omega^o(e^{\lambda\tau_1})} \leq \frac{\omega_0^+ e^\lambda + E_\omega(\tau_1 e^{\lambda\tau_1} 1_{\tau_1 \geq 2})}{\omega_0^+ e^\lambda}$$

$$\leq 1 + \frac{c e^{3\lambda/2}}{\omega_0^+ e^\lambda} \leq 1 + \frac{c}{\varepsilon} e^{\lambda/2} \underset{\lambda \to -\infty}{\longrightarrow} 1 ,$$

where c is a constant independent of ω or λ. Hence, $g(\lambda) \underset{\lambda \to -\infty}{\longrightarrow} 1$ uniformly in $M_1^{e,\varepsilon}$.

d) Assume that P is extremal. The first inequality in (2.3.16) follows from a simple coupling argument: let $\bar{\varphi}(\lambda) := E_{\hat{\omega}\mathrm{min}}^o[e^{\lambda\tau_1}]$. By the recursions (2.3.21), it holds that if $\bar{\varphi}(\lambda) < \infty$ then as long as $\lambda \leq \bar{\lambda}$ it holds that

$$\bar{\varphi}(\lambda) = \frac{(1 - \omega_{\max}^0 e^\lambda) - \sqrt{(1 - \omega_{\max}^0 e^\lambda)^2 - 4\hat{\omega}^+\hat{\omega}^- e^{2\lambda}}}{2\hat{\omega}^- e^\lambda} .$$

Thus, we have to show that if $\lambda > \lambda_{\mathrm{crit}}$ then $E_\omega^o(e^{\lambda\tau_1}) = \infty$, P-a.s. Since $E_{\hat{\omega}\mathrm{min}}^o(e^{\lambda\tau_1}) = \infty$, we may find an M large enough such that $E_{\hat{\omega}\mathrm{min}}^o(e^{\lambda\tau_1} 1_{\tau_1 < M}) > 1/\epsilon + 1$. Since the last expression is local, i.e. depends only on $\{\omega_i\}_{i=0}^{-M+1}$, it follows (from the assumption of local equivalence to the product of marginals) that with P positive probability, $E_\omega^o(e^{\lambda\tau_1}) > 1/\epsilon$, and hence by part (c) actually $E_\omega^0(e^{\lambda\tau_1}) = \infty$ with P positive probability, and hence with P probability 1. $\qquad \square$

Remark: Before proceeding, we note that a direct consequence of Lemma 2.3.13 is that if $E_P(\log \rho_0) \leq 0$, then for $u < E_P(\bar{S})$,

$$I_P^{\tau,q}(u) = \lambda_0 u - E_P(f(\lambda_0,\omega)) = \sup_{\lambda\in\mathbb{R}} G(\lambda,P,u) > G(0,P,u) = 0$$

since the function $G(\cdot,P,u)$ is strictly concave.

Step II: Recall the transformation Inv : $\Omega \mapsto \Omega$ and the law $P^{\mathrm{Inv}} = P \circ$ Inv^{-1}. Proving the LDP for T_n/n when $E_P(\log\rho_0) > 0$ is the same, by space reversal, as proving the quenched LDP for T_{-n}/n under the law P^{Inv} on the environment. Note that in this case, $E_{P^{\mathrm{Inv}}}(\log\rho_0) < 0$, and further, $P \in M_1^{e,\varepsilon}$ implies that $P^{\mathrm{Inv}} \in M_1^{e,\varepsilon}$. Thus, Step II will be completed if we can prove a quenched LDP for T_{-n}/n for $P \in M_1^{e,\varepsilon}$ satisfying $E_P \log\rho_0 < 0$. We turn to this task now.

Note that if $P(\omega_0^- = 0) > 0$ then $P_\omega^o(T_{-n} < \infty) = 0$ for some $n = n(\omega)$ large enough, and the LDP for T_{-n}/n is trivial. We thus assume throughout that $\omega_0^- \geq \varepsilon$, P-a.s. As a first step in the derivation of the LDP, we compute logarithmic moment generating functions. Define, for any $\lambda \in \mathbb{R}$,

$$\overline{\varphi}(\lambda,\omega) = E_\omega^o(e^{\lambda\tau_{-1}}1_{\{\tau_{-1}<\infty\}}),\quad \overline{f}(\lambda,\omega) = \log\overline{\varphi}(\lambda,\omega).$$

Lemma 2.3.22 *Assume $P \in M_1^{e,\varepsilon}$ and further assume that $\min(\omega_0^+,\omega_0^-) > \varepsilon$, P-a.s. Then,*

$$E_P(\overline{f}(\lambda,\omega)) = E_P(f(\lambda,\omega)) + E_P\log\rho_0 . \tag{2.3.23}$$

Proof of Lemma 2.3.22:

Define the map $I_n : \Omega \mapsto \Omega$ by

$$(I_n\omega)_k = \begin{cases} \omega_k, & k \notin [0,n] \\ (F\omega)_{n-k}, & k \in [0,n]. \end{cases}$$

Introduce

$$\varphi_n(\lambda,\omega) = E_\omega^o(e^{\lambda\tau_1}; \tau_1 < T_{-(n+1)}),\quad \overline{\varphi}_n(\lambda,\omega) = E_\omega^o(e^{\lambda\tau_{-1}}; \tau_{-1} < T_{(n+1)}).$$

We will show below that

$$G_n(\lambda,\omega) := \varphi_n(\lambda,\theta^n\omega)\overline{\varphi}_{n-1}(\lambda,\omega) = \varphi_{n-1}(\lambda,\theta^n\omega)\overline{\varphi}_n(\lambda,\omega) := F_n(\lambda,\omega). \tag{2.3.24}$$

Because $\min(\omega_0^+,\omega_0^-) > \varepsilon$, P-a.s., the function $\log\varphi_n(\lambda,\omega)$ and $\log\overline{\varphi}_n(\lambda,\omega)$ are P-integrable for each n. Taking logarithms in (2.3.24), we find that $E_P(\log\varphi_n(\lambda,\omega))-E_P(\log\overline{\varphi}_n(\lambda,\omega))$ does not depend on n. On the other hand, both terms are monotone in n, hence by monotone convergence either both sides of (2.3.24) are $+\infty$ or both are finite, in which case

$$E_P(\log\varphi(\lambda,\omega)) - E_P(\log\overline{\varphi}(\lambda,\omega)) = E_P\left(\log\left(\frac{\varphi_0(\lambda,\omega)}{\overline{\varphi}_0(\lambda,\omega)}\right)\right) = -E_P(\log\rho_0),$$

yielding (2.3.23).

We thus turn to the proof of (2.3.24). It is straight forward to check, by space inversion, that $F_n(\lambda, I_n\omega) = G_n(\lambda, \omega)$. Thus, the proof of (2.3.24) will be complete once we show that $F_n(\lambda, I_n\omega) = F_n(\lambda, \omega)$. Toward this end, note that by the Markov property,

$$\overline{\varphi}_n(\lambda, \omega) = E_\omega^o(e^{\lambda T_{-1}}; T_{-1} < T_{n+1})$$
$$= E_\omega^o(e^{\lambda T_{-1}}; T_{-1} < T_n)$$
$$+ E_\omega^o(e^{\lambda T_n}; T_n < T_{-1})E_\omega^n(e^{\lambda T_{-1}}; T_{-1} < T_{n+1}).$$

Hence, defining
$$B_n(\lambda, \omega) := E_\omega^o(e^{\lambda T_{-1}}; T_{-1} < T_n),$$
$$C_n(\lambda, \omega) := E_\omega^o(e^{\lambda T_n}; T_n < T_{-1}),$$

one has, using again space reversal and the Markov property in the second equality,

$$F_n(\lambda, \omega) = E_\omega^n(e^{\lambda T_{n+1}}; T_{n+1} < T_0)E_\omega^o(e^{\lambda T_{-1}}; T_{-1} < T_n)$$
$$+ E_\omega^n(e^{\lambda T_{n+1}}; T_{n+1} < T_0)E_\omega^n(e^{\lambda T_{-1}}; T_{-1} < T_{n+1})E_\omega^o(e^{\lambda T_n}; T_n < T_{-1})$$
$$= B_n(\lambda, \omega)B_n(\lambda, I_n\omega)$$
$$+ E_\omega^n(e^{\lambda T_{n+1}}; T_{n+1} < T_0)E_\omega^n(e^{\lambda T_0}; T_0 < T_{n+1})E_\omega^o(e^{\lambda T_{-1}}; T_{-1} < T_{n+1})$$
$$E_\omega^o(e^{\lambda T_n}; T_n < T_{-1})$$
$$= B_n(\lambda, \omega)B_n(\lambda, I_n\omega) + C_n(\lambda, \omega)C_n(\lambda, I_n\omega)F_n(\lambda, \omega),$$

implying the invariance of $F_n(\lambda, \omega)$ under the action of I_n on Ω, except possibly at λ where $C_n(\lambda, \omega)C_n(\lambda, I_n\omega) = 1$. The latter λ is then handled by continuity. This completes the proof of Lemma 2.3.22   □

Step II now is completed by following the same route as in the proof of Step I, using Lemma 2.3.22 to transfer the analytic results of Lemma 2.3.13 to this setup. The details, which are straightforward and are given in [12], are omitted here.   □

Remarks: 1. Note that the conclusion of Lemma 2.3.22 extends immediately, by the ergodic decomposition, to stationary measures $P \in M_1^{s,\varepsilon}$.

2. Lemma 2.3.22 is the key to the large deviations principle, and deserves some discussion. First, by taking $\lambda \uparrow 0$, one sees that if $E_P(\log \rho_0) \le 0$ then $E_P[\log P_\omega^o(\tau_{-1} < \infty)] = E_P(\log \rho_0)$. Next, let $\overline{\tau}_{-1}, \overline{\tau}_{-2}, \overline{\tau}_{-3}, ..., \overline{\tau}_{-N}$ have the distribution of $\tau_{-1}, \tau_{-2}, \tau_{-3}, ...\tau_{-N}$ under P_ω^o conditioned on $T_{-N} < \infty$. In fact the law of $\{\overline{\tau}_{-i}\}_{i=1}^N$ does not depend on N. This can be seen by a discrete h-transform: the distributions of $X_0^{T_{-N}} := (X_0, \ldots, X_{T_{-N}})$ under P_ω^o, conditioned on $T_{-N} < \infty$, $N = 1, 2, \ldots$ form a consistent family whose extension is again a Markov chain. To see this, let $\tilde{P}_{\omega,N}^o := P_\omega^o(\cdot | T_{-N} < \infty)$, restricted to $X_0^{T_{-N}}$. Denoting $x_1^n := (x_1, \ldots, x_n)$, compute (with $x_i > -N$),

$$\tilde{P}^o_{\omega,N}(X_{n+1} = x_n + 1 | X_1^n = x_1^n)$$

$$= \frac{\tilde{P}^o_{\omega,N}(X_{n+1} = x_n + 1, X_1^n = x_1^n)}{\tilde{P}^o_{\omega,N}(X_1^n = x_1^n)}$$

$$= \frac{P^o_{\omega}(X_{n+1} = x_n + 1, X_1^n = x_1^n, T_{-N} < \infty)}{P^o_{\omega}(X_1^n = x_1^n, T_{-N} < \infty)}$$

$$= \frac{P^o_{\omega}(X_{n+1} = x_n + 1, X_1^n = x_1^n) P^o_{\theta^{x_n+1}\omega}(T_{-N-x_n-1} < \infty)}{P^o_{\omega}(X_1^n = x_1^n) P^o_{\theta^{x_n}\omega}(T_{-N-x_n} < \infty)}$$

$$= P^o_{\omega}(X_{n+1} = x_n + 1 | X_1^n = x_1^n) P_{\theta^{x_n+1}\omega}(T_{-1} < \infty)$$

$$= \omega^+_{x_n} P^o_{\theta^{x_n+1}\omega}(T_{-1} < \infty),$$

where we used the Markov property in the third and in the fourth equality. The last term depends neither on N nor on x_1^{n-1}. Therefore, the extension of $(\tilde{P}_{\omega,N})_{N\geq 1}$ is the distribution of the Markov chain with transition probabilities $\tilde{\omega}^+_i = \omega^+_i P_{\theta^{i+1}\omega}(T_{-1} < \infty), \tilde{\omega}^0_i = \omega^0_i, i \in \mathbb{Z}$. In particular, $\bar{\tau}_{-1}, \bar{\tau}_{-2}, \bar{\tau}_{-3}, \ldots$ are independent under P^o_{ω} and, with a slight abuse of notations, form a stationary sequence under $\mathbb{P}^o$. Note now that if we set

$$\bar{\phi}(\lambda, \omega) := E^o_{\omega}(e^{\lambda\bar{\tau}_{-1}}) = \frac{\overline{\varphi}(\lambda, \omega)}{P^o_{\omega}(T_{-1} < \infty)} \tag{2.3.25}$$

then Lemma 2.3.22 tells us that $E_P\bar{\phi}(\lambda, \omega) = E_P\varphi(\lambda, \omega)$. In particular, $\mathbb{E}^o_P(\bar{\tau}_1) = \mathbb{E}^o_P(\tau_1) = E_P(\overline{S})$ if $E_P(\log \rho_0) \leq 0$ and, repeating the arguments leading to the LDP of T_n/n, we find that the sequence of random variables T_{-n}/n, conditioned on $T_{-n} < \infty$, satisfy a quenched LDP under P^o_{ω} with the same rate function as T_n/n!

Step III: By space reversal, it is enough to prove the result for $E_P(\log \rho_0) \leq 0$. Further, as in Step II, it will be enough to consider the case where $\min(\omega^+_0, \omega^-_0) \geq \varepsilon$, P-a.s. Since $I^{\tau,q}_P(\cdot)$ is convex, and since $x \mapsto xf(1/x)$ is convex if $f(\cdot)$ is convex, it follows that $I^q_P(\cdot)$ is convex on $(0,1]$ and on $[-1,0)$ separately. If $\lambda_{\text{crit}}(P) = 0$ then $I^q_P(0) = 0$ and the convexity on $[-1,1]$ follows. In the general case, note that I^q_P is continuous at 0, and $(I^q_P)'(0^-) = -(I^q_P)'(0^+) + E_P(\log \rho_0)$. Note that for $\lambda \leq \lambda_{\text{crit}}$, by the Markov property,

$$E^o_{\omega}(e^{\lambda T_M} \mathbf{1}_{\tau_{-1} < \tau_M}) = E^o_{\omega}(e^{\lambda\tau_{-1}} \mathbf{1}_{\tau_{-1} < \tau_M})\varphi(\lambda, \theta^{-1}\omega)E^o_{\omega}(e^{\lambda T_M}),$$

and hence,

$$1 \geq E^o_{\omega}(e^{\lambda\tau_{-1}} \mathbf{1}_{\tau_{-1} < \tau_M})\varphi(\lambda, \theta^{-1}\omega),$$

leading (by taking $M \to \infty$) to the conclusion that

$$\varphi(\lambda, \omega)\bar{\varphi}(\lambda, \theta^{-1}\omega) \leq 1.$$

Taking logarithms and P-expectations, we conclude that

$$E_P f(\lambda, \omega) + E_P \bar{f}(\lambda, \omega) \leq 0.$$

Combined with Lemma 2.3.22, we deduce that for all $\lambda \leq \lambda_{\text{crit}}$, $2E_P(f(\lambda, \omega)) \leq -E_P(\log \rho_0)$. Hence,

$$(I_P^q)'(0^+) = -E_P\left(\log E_\omega(e^{\lambda_{\text{crit}} T_1})\right) = -E_P\left(f(\lambda_{\text{crit}}, \omega)\right) \geq \frac{1}{2} E_P(\log \rho_0),$$

implying that $(I_P^q)'(0^-) \leq (I_P^q)'(0^+)$, and hence that $I_P^q(\cdot)$ is convex on $[-1, 1]$.

Using the monotonicity of $I_P^{\tau, q}(\cdot)$, c.f. the remark following the proof of Step I, it follows easily that $I_P^q(\cdot)$ is non increasing on $[-1, v_P]$ and non decreasing on $[v_P, 1]$.

Let $v > v_P$. We have

$$P_\omega^o\left(\frac{X_n}{n} \geq v\right) \leq P_\omega^o\left(T_{\lfloor nv \rfloor} \leq n\right) = P_\omega^o\left(\frac{T_{\lfloor nv \rfloor}}{\lfloor nv \rfloor} \leq \frac{n}{\lfloor nv \rfloor}\right).$$

Step I and the monotonicity of $I_P^{\tau, q}(\cdot)$ now imply

$$\limsup_{n \to \infty} \frac{1}{n} \log P_\omega^o\left(\frac{X_n}{n} \geq v\right) \leq -v I_P^{\tau, q}\left(\frac{1}{v}\right),$$

which yields the required upper bound by the monotonicity of $I_P^q(\cdot)$. The same argument applies to yield the desired upper bound on $P_\omega^o\left(\frac{X_n}{n} \leq v\right)$ for $v < 0$, by considering the hitting times $T_{\lceil nv \rceil}$.

In the same way, for any $0 < \eta < \delta/2$,

$$P_\omega^o\left((v+\delta) \geq \frac{X_n}{n} \geq (v-\delta)\right) \geq P_\omega^o\left((1-\eta)n \leq T_{\lceil nv \rceil} \leq n\right),$$

hence, from Step I it follows that for $v \geq 0$,

$$\liminf_{n \to \infty} \frac{1}{n} \log P_\omega^o\left(\frac{X_n}{n} \in (v-\delta, v+\delta)\right) \geq -v I_P^{\tau, q}\left(\frac{1-\eta}{v}\right), \quad P-\text{a.s.},$$

and the lower bound is obtained by letting $\eta \to 0$. The same argument also yields the lower bounds for $v < 0$, using this time the function $I_P^{-\tau, q}(\cdot)$.

Next, we turn to evaluate an upper bound on $P_\omega^o(X_n/n \leq v), 0 \leq v < v_P$, with $v_P \geq 0$. Starting with $v = 0$, let $\eta, \delta > 0$, with $\delta < v_P$. Then,

$$P_\omega^o(X_n \leq 0) \leq P_\omega^o\left(T_{[n\delta]} \geq n\right) + P_\omega^o\left(T_{[n\delta]} < n, \frac{X_n}{n} \leq 0\right) \tag{2.3.26}$$

$$\leq P_\omega^o\left(T_{[n\delta]} \geq n\right) + \sum_{1/\eta \leq k, l; (k+l)\eta \leq 1/\delta} P_\omega^o\left(\frac{T_{[n\delta]}}{n\delta} \in [k\eta, (k+1)\eta)\right) \times$$

$$P_{\theta[n\delta]\omega}^o\left(\frac{T_{-[n\delta]}}{n\delta} \in [l\eta, (l+1)\eta)\right) \sum_{-2n\delta\eta \leq m-n(1-(k+l)\delta\eta) \leq 0} P_\omega^o(X_m \leq 0),$$

by the strong Markov property. Define the random variable

$$a = \limsup_{n \to \infty} \frac{1}{n} \sup_{m:-2n\delta\eta \le m-n \le 0} \log P_\omega^o(X_m \le 0) \ ,$$

and note, using the inequality

$$P_\omega^o(X_n \le 0) \ge P_\omega^o(X_m \le 0) \inf_{i \le 0} P_{\theta^i\omega}^o[X_{n-m} = -(n-m)]$$

with a worst-environment estimate, that

$$a - C\delta\eta \le \limsup_{n \to \infty} \frac{1}{n} \log P_\omega^o[X_n \le 0] \le a \qquad (2.3.27)$$

with $C = -2\log\varepsilon > 0$. The first two probabilities in the right-hand side of (2.3.26) will be estimated using Step I. By convexity, the rate functions $I_P^{\tau,q}$ and $I_P^{-\tau,q} := I_P^{\tau,q} - E_P(\log\rho_0)$ are continuous, so that the oscillation

$$w(\delta;\eta) =$$

$$\max\{|I_P^{\tau,q}(u) - I_P^{\tau,q}(u')| + |I_P^{-\tau,q}(u) - I_P^{-\tau,q}(u')|; u, u' \in [1, 1/\delta], |u - u'| \le \eta\}$$

tends to 0 with η, for all fixed δ. From the proof of Step II, it is not difficult to see that the third term in the right-hand side of (2.3.26) can be estimated similarly (it does not cause problems to consider $P_{\theta[n\delta]\omega}^o$ instead of P_ω^o):

$$\limsup_{n \to \infty} \frac{1}{n} \log P_{\theta[n\delta]\omega}^o \left(\frac{T_{-[n\delta]}}{n\delta} \in [l\eta, (l+1)\eta) \right) \le -\delta\left(I_P^{-\tau,q}(l\eta) - w(\delta;\eta)\right) \ P-\text{a.s.}$$

Finally, we get from (2.3.27) and (2.3.26)

$$a \le C\delta\eta + \max\{-I_P^q(\delta), \max_{1/\eta \le k, l; (k+l)\eta \le 1/\delta}$$

$$[-\delta\eta(kI_P^q(1/k\eta) + lI_P^q(-1/l\eta)) + 2\delta w(\delta;\eta) + (1-(k+l+2)\delta\eta)a]\} \ .$$

By convexity and since $\delta \le v_P$, it holds $kI_P^q(1/k\eta) + lI_P^q(-1/l\eta) \ge (k+l)I_P^q(0) \ge (k+l)I_P^q(\delta)$, and therefore $a' := a + I_P^q(\delta)$ is such that

$$a' \le C\delta\eta + \left(\max_{1/\eta \le k, l; (k+l)\eta \le 1/\delta} [2\delta w(\delta;\eta) + 2\delta\eta I_P^q(\delta) + (1-(k+l+2)\delta\eta)a'] \right)^+ .$$

Computing the maximum for positive a', we derive that $2a' \le C\eta + 2(w(\delta;\eta) + \eta I_P^q(\delta))$. Letting now $\eta \to 0$ and $\delta \to 0$, we conclude that

$$\limsup_{n \to \infty} \frac{1}{n} \log P_\omega^o\left(X_n \le 0\right) \le -I_P^q(0) \ , \quad P-\text{a.s.} \qquad (2.3.28)$$

In fact, the same proof actually shows that

$$\limsup_{n \to \infty} \frac{1}{n} \log P_\omega^o\left(\exists l \ge n : X_l \le 0\right) \le -I_P^q(0) \ , \quad P-\text{a.s.} \qquad (2.3.29)$$

For an arbitrary $v \in [0, v_P)$, we write

$$
P_\omega^o\left(\frac{X_n}{n} \leq v\right) \leq P_\omega^o(\exists \ell \geq n : X_\ell \leq nv) \leq P_\omega^o\left(T_{[nv]} \geq n\right)
$$

$$
+ \sum_{k: v/\eta \leq k < 1/\eta} P_\omega^o\left(\frac{T_{[nv]}}{n} \in [k\epsilon, (k+1)\epsilon)\right) P_{\theta[nv]\omega}^o\left(\exists \ell \geq n - n(k+1)\eta : X_\ell \leq 0\right)
$$

$$(2.3.30)$$

where the two first probabilities in the right-hand side can be estimated using Step I, and concerning the last one we note that from (2.3.29) one has that

$$
\frac{1}{n - n(k+1)\eta} \log P_{\theta[nv]\omega}^o\left(\exists \ell \geq n - n(k+1)\eta : X_\ell \leq 0\right) \to_{n \to \infty} -I_P^q(0),
$$

in probability, and hence a.s. along a random subsequence. Therefore,

$$
\liminf_{n \to \infty} \frac{1}{n} \log P_\omega^o\left(\exists \ell \geq n : X_\ell \leq nv\right)
$$

$$
\leq \limsup_{\eta \to 0}\left(-I_P^q(v) \vee \max_{v/\eta \leq k \leq 1/\eta}[-k\eta I_P^q(v/k\eta) - (1 - k\eta)I_P^q(0)]\right)
$$

$$
= -I_P^q(v) ,
$$

$$(2.3.31)$$

by convexity. But, due to Kingman's sub-additive ergodic theorem, the left hand side of the last expression converges P-a.s., resulting with

$$
\limsup_{n \to \infty} \frac{1}{n} \log P_\omega^o\left(X_n \leq nv\right) \leq -I_P^q(v) , \quad P - a.s..
$$

The upper bound for general subsets of $[0, 1]$ follows by noting the convexity of $I_P^q(\cdot)$. $\qquad\square$

Remarks 1. If P is extremal, a simpler proof of (2.3.28) can be given. Indeed, note that $I_P^q(0) = \lambda_{\mathrm{crit}}(0)$, and, by extremality,

$$
P_\omega^o(X_n \leq 0) \leq P_\omega(X_m \leq 0, \text{ some } m \geq 0) \leq P_{\tilde\omega^{\min}}(X_m \leq 0, \text{ some } m \geq 0)
$$

$$
\leq \sum_{m=n}^\infty P_{\tilde\omega^{\min}}(X_m \leq 0) .
$$

A simple computation reveals that $P_{\tilde\omega^{\min}}(X_m \leq 0) \leq C_\lambda e^{-\lambda m}$ for any $\lambda < \lambda_{\mathrm{crit}}$, yielding that

$$
\limsup_{n \to \infty} \frac{1}{n} \log P_\omega^o(X_n \leq 0) \leq -I_P^q(0) , \quad P - a.s. \qquad (2.3.32)
$$

2. A lot of information is available concerning the shape of the rate function $I_P^q(\cdot)$, and in particular concerning the existence of pieces where the rate function is not strictly convex. We refer to the discussion in [12] for details.

Annealed LDP's

The LDP in the annealed setting also makes use of the hitting times T_n and T_{-n}. For technical reasons, we need to make stronger hypotheses on the environment. To state these, define the empirical process

$$R_n(\omega) := \frac{1}{n} \sum_{i=0}^{n-1} \delta_{\theta^i \omega} \,.$$

R_n takes values in the space $M_1(\Omega)$ of probability measures on Ω, which we equip with the topology of weak convergence. We also need to introduce the specific relative entropy

$$h(\cdot|P) : M_1(\Omega) \mapsto [0,\infty], \ h(Q|P) := \begin{cases} \lim_{N \to \infty} \frac{1}{N} H(Q_N|P_N), \ Q \text{ stationary} \\ \infty, \hspace{3.5cm} \text{otherwise}, \end{cases}$$

where Q_N, P_N denote the restriction of Q, P to the first N coordinates $\{\omega_i\}_{i=0}^{N-1}$ and $H(\cdot|\cdot)$ denotes the relative entropy:

$$H(\mu|\nu) = \begin{cases} \int \log\left(\frac{d\mu}{d\nu}(x)\right) \mu(dx), \ \mu \ll \nu \\ \infty, \hspace{3cm} \text{otherwise} \end{cases} .$$

Assumption 2.3.33

(C1) *P is stationary and ergodic*
(C2) *There exists an $\varepsilon > 0$ such that $\min(\omega_0^+, \omega_0^-) > \varepsilon$, $P - a.s.$,*
(C3) *$\{R_n\}$ satisfies under P the process level LDP in $M_1(\Omega)$ with good rate function $h(\cdot|P)$,*
(C4) *P is locally equivalent to the product of its marginals and, for any stationary measure $\eta \in M_1(\Omega)$ there is a sequence $\{\eta^n\}$ of stationary, ergodic measures with $\eta^n \xrightarrow[n \to \infty]{} \eta$ weakly and $h(\eta^n|P) \to h(\eta|P)$.*
(C5) *P is extremal.*

We note that product measures and Markov processes with bounded transition kernels satisfy **(C1)**–**(C4)** of Assumption 2.3.33, see [27, Lemma 4.8] and [23]. Define now

$$I_P^{\tau,a}(u) = \inf_{\eta \in M_1^{e,\varepsilon}(\Omega)} \left[I_\eta^{\tau,q}(u) + h(\eta|P) \right], \ I_P^a(v) = \inf_{\eta \in M_1^{e,\varepsilon}(\Omega)} \left[I_\eta^q(v) + |v| h(\eta|P) \right] .$$

We now have the annealed analog of Theorem 2.3.12:

Theorem 2.3.34 *Assume Assumption 2.3.33. Then, the random variables $\{T_n/n\}$ satisfy the weak annealed LDP with speed n and rate function $I_P^{\tau,a}(\cdot)$. Further, the random variables X_n/n satisfy the annealed LDP with speed n and good rate function $I_P^a(\cdot)$.*

Proof. Throughout, $M_1^{s,\varepsilon}$ denotes the set of stationary probability measures $\eta \in M_1(\Omega)$ satisfying $\mathrm{supp}\,\eta_0 \subset \mathrm{supp}\,P_0$. If $E_P(\log \rho_0) \leq 0$ then $\lambda_{\mathrm{crit}} = \lambda_{\mathrm{crit}}(P)$ is as in Lemma 2.3.13, whereas if $E_P(\log \rho_0) > 0$ then $\lambda_{\mathrm{crit}} = \lambda_{\mathrm{crit}}(P^{\mathrm{Inv}})$.

Let $M_1^{s,\varepsilon,P} = \{\mu \in M_1^{s,\varepsilon} : \mathrm{supp}\,\mu_0 \subset \mathrm{supp}\,P_0\}$. The following lemma, whose proof is deferred, is key to the transfer of quenched LDP's to annealed LDP's:

Lemma 2.3.35 *Assume P satisfies Assumption 2.3.33. Then, the function $(\mu,\lambda) \mapsto \int f(\lambda,\omega)\mu(d\omega)$ is continuous on $M_1^{s,\varepsilon,P} \times (-\infty,\lambda_{\mathrm{crit}}]$.*

Steps I.1 + I.2: weak annealed LDP upper bound for T_n/n: We have, for $\lambda \leq 0$,

$$\mathbb{P}^o\left(T_n/n \leq u\right) \leq e^{-\lambda n u}\mathbb{E}^o\left(\exp\left(\lambda \sum_{j=1}^n \tau_j\right)\mathbf{1}_{\tau_j<\infty,j=1,\ldots,n}\right)$$

$$= e^{-\lambda n u}E_P\left(\prod_{j=1}^n E_\omega^o\left(e^{\lambda\tau_j}\mathbf{1}_{\tau_j<\infty}\right)\right) = e^{-\lambda n u}E_P\left(\exp\left(\sum_{j=0}^{n-1}f(\lambda,\theta^j\omega)\right)\right)$$

$$= e^{-\lambda n u}E_P\left(\exp\left(n\int f(\lambda,\omega)R_n(d\omega)\right)\right). \quad (2.3.36)$$

By Assumption 2.3.33, $\{R_n\}$ satisfies a LDP with rate function $h(\cdot|P)$. Lemma 2.3.35 ensures that we can apply Varadhan's lemma (see [19, Lemma 4.3.6]) to get

$$\limsup_{n\to\infty}\frac{1}{n}\log E_P\left(\exp\left(n\int f(\lambda,\omega)R_n(d\omega)\right)\right)$$

$$\leq \sup_{\eta\in M_1^{s,\varepsilon}}\left[\int f(\lambda,\omega)\eta(d\omega) - h(\eta|P)\right]. \quad (2.3.37)$$

Going back to (2.3.36), this yields the upper bound

$$\limsup_{n\to\infty}\frac{1}{n}\log \mathbb{P}^o\left(T_n/n \leq u\right) \leq \inf_{\lambda\leq 0}\sup_{\eta\in M_1^{s,\varepsilon}}\left[\int f(\lambda,\omega)\eta(d\omega) - h(\eta|P) - \lambda u\right]$$

$$= -\sup_{\lambda\leq 0}\inf_{\eta\in M_1^{s,\varepsilon}}[G(\lambda,\eta,u) + h(\eta|P)]. \quad (2.3.38)$$

Since $\eta \to -\int f(\lambda,\omega)\eta(d\omega) + h(\eta|P)$ is lower semi-continuous (for $\lambda \leq 0$) and $M_1(\Omega)$ is compact, the infimum in (2.3.38) is achieved for each λ, on measures in $M_1^{s,\varepsilon}$, for otherwise $h(\eta|P) = \infty$. Further, by (2.3.15), the supremum over λ can be taken over a compact set (recall that $\infty > u > 1$!). By the Minimax theorem (see [64, Theorem 4.2] for this version), the min-max is equal to the max-min in (2.3.38). Further, since taking first the supremum in λ in the right hand side of (2.3.38) yields a lower semicontinuous function, an achieving $\bar{\eta}$ exists, and then, due to compactness, there exists actually an achieving pair $\bar{\lambda}, \bar{\eta}$. We will show below that the infimum may be taken over stationary, ergodic measures only, that is

$$\inf_{\eta \in M_1^{s,\varepsilon}} \sup_{\lambda \leq 0} \left(G(\lambda, \eta, u) + h(\eta|P) \right) = \inf_{\eta \in M_1^{e,\varepsilon}} \sup_{\lambda \leq 0} \left(G(\lambda, \eta, u) + h(\eta|P) \right).$$

(2.3.39)

Then,

$$\text{R.H.S. of } (2.3.38) = - \inf_{\eta \in M_1^{e,\varepsilon}} \sup_{\lambda \leq 0} \left(G(\lambda, \eta, u) + h(\eta|P) \right)$$

$$= - \inf_{\eta \in M_1^{e,\varepsilon}} \inf_{w \leq u} \left(I_\eta^{\tau,q}(w) + h(\eta|P) \right). \qquad (2.3.40)$$

The second equality in (2.3.40) is obtained as follows: set $M_u = \{\eta \in M_1^{e,\varepsilon} : E_\eta(E_\omega^o(\tau_1|\tau_1 < \infty)) > u\}$, $M_u^- = \{\eta \in M_1^{e,\varepsilon} : E_\eta(E_\omega^o(\tau_1|\tau_1 < \infty)) \leq u\}$. For $\eta \in M_u$,

$$\inf_{w \leq u} I_\eta^{\tau,q}(w) = I_\eta^{\tau,q}(u) = \sup_{\lambda \in \mathbb{R}} G(\lambda, \eta, u) = \sup_{\lambda \leq 0} G(\lambda, \eta, u).$$

Further, recall that $I_\eta^{\tau,q}(\cdot)$ is convex with minimum value $\max(0, E_\eta(\log \rho_0))$ achieved at $E_\eta(E_\omega^o(\tau_1|\tau_1 < \infty))$. Then, for $\eta \in M_u^-$,

$$\inf_{w \leq u} I_\eta^{\tau,q}(u) = \max(0, E_\eta(\log \rho_0))$$

whereas Jensen's inequality implies that for such η,

$$\sup_{\lambda \leq 0} G(\lambda, \eta, u) = G(0, \eta, u) = \max(0, E_\eta(\log \rho_0)),$$

completing the proof of (2.3.40). Hence,

$$\limsup_{n \to \infty} \frac{1}{n} \log \mathbb{P}^o \left(T_n/n \leq u \right) \leq - \inf_{w \leq u} \inf_{\eta \in M_1^{e,\varepsilon}} \left(I_\eta^{\tau,q}(w) + h(\eta|P) \right)$$

$$= - \inf_{w \leq u} I_P^{\tau,a}(w). \qquad (2.3.41)$$

Turning to the proof of (2.3.39), we have, due to **(C4)** in Assumption 2.3.33, a sequence of stationary, ergodic measures with $\eta^n \to \bar{\eta}$ and $h(\eta^n|P) \to h(\bar{\eta}|P)$. Let λ_n be the maximizers in (2.3.39) corresponding to η^n. We have

$$\inf_{\eta \in M_1^{e,\varepsilon}} \sup_{\lambda \leq 0} \left(\left[\lambda u - \int f(\lambda, \omega) \eta(d\omega) \right] + h(\eta|P) \right) \leq \left[\lambda_n u - \int f(\lambda_n, \omega) \eta^n(d\omega) \right]$$

$$+ h(\eta^n|P). \qquad (2.3.42)$$

W.l.o.g. we can assume, by taking a subsequence, that $\lambda_n \to \lambda^* \leq 0$. Using the joint continuity in Lemma 2.3.35, we have, for $\epsilon' > 0$ and $n \geq N_0(\epsilon')$,

$$\lambda_n u - \int f(\lambda_n, \omega) \eta^n (d\omega) + h(\eta^n | P)$$

$$\leq \left[\lambda^* u - \int f(\lambda^*, \omega) \bar{\eta}(d\omega) \right] + h(\bar{\eta} | P) + \epsilon'$$

$$\leq \inf_{\eta \in M_1^{s,\varepsilon}} \sup_{\lambda \leq 0} \left(\left[\lambda u - \int f(\lambda, \omega) \eta(d\omega) \right] + h(\eta | P) \right) + \epsilon' .$$

But this shows the equality in (2.3.39), since the reverse inequality there is trivial.

The upper bound for the upper tail, that is for $\frac{1}{n} \log P[\infty > \frac{1}{n} \sum_{j=1}^n \tau_j \geq u]$, where $1 < u < \infty$, is achieved similarly. We detail the argument since there is a small gap in the proof presented in [12]. First, exactly as in (2.3.38), one has

$$\limsup_{n \to \infty} \frac{1}{n} \log \mathbb{P}^o (T_n/n \geq u) \leq \inf_{0 \leq \lambda \leq \lambda_{\mathrm{crit}}} \sup_{\eta \in M_1^{s,\varepsilon}} [-G(\lambda, \eta, u) - h(\eta | P)]$$

$$= - \sup_{0 \leq \lambda \leq \lambda_{\mathrm{crit}}} \inf_{\eta \in M_1^{s,\varepsilon}} [G(\lambda, \eta, u) + h(\eta | P)] .$$

$$(2.3.43)$$

One may now apply the min-max theorem to deduce that the right hand side of (2.3.43) equals

$$\inf_{\eta \in M_1^{s,\varepsilon}} \sup_{0 \leq \lambda \leq \lambda_{\mathrm{crit}}} [G(\lambda, \eta, u) + h(\eta | P)] = \inf_{\eta \in M_1^{e,\varepsilon}} \sup_{0 \leq \lambda \leq \lambda_{\mathrm{crit}}} [G(\lambda, \eta, u) + h(\eta | P)] ,$$

where the second equality is proved by the same argument as in (2.3.39). Here a new difficulty arises: the supremum is taken over $\lambda \in [0, \lambda_{\mathrm{crit}}(P)]$, but in general $\lambda_{\mathrm{crit}}(\eta) \geq \lambda_{\mathrm{crit}}(P)$ and hence the identification of the last expression with a variational problem involving $I_\eta^{\tau,q}(\cdot)$ is not immediate. To bypass this obstacle, we note, first by replacing η with $(1 - n^{-1})\eta + n^{-1}P$ and then using again (C4) to approximate with an ergodic measure, that the last expression equals

$$\inf_{\{\eta \in M_1^{e,\varepsilon}, \lambda_{\mathrm{crit}}(\eta) = \lambda_{\mathrm{crit}}(P)\}} \sup_{0 \leq \lambda \leq \lambda_{\mathrm{crit}}} [G(\lambda, \eta, u) + h(\eta | P)] .$$

From here, one proceeds as in the case of the lower tail, concluding that

$$\limsup_{n \to \infty} \frac{1}{n} \log \mathbb{P}^o (T_n/n \geq u)$$

$$\leq - \inf_{\{\eta \in M_1^{e,\varepsilon}, \lambda_{\mathrm{crit}}(\eta) = \lambda_{\mathrm{crit}}(P)\}} \sup_{0 \leq \lambda \leq \lambda_{\mathrm{crit}}} [G(\lambda, \eta, u) + h(\eta | P)]$$

$$= - \inf_{\{\eta \in M_1^{e,\varepsilon}, \lambda_{\mathrm{crit}}(\eta) = \lambda_{\mathrm{crit}}(P)\}} \inf_{w \geq u} I_\eta^{\tau,q}(w) \leq - \inf_{\eta \in M_1^{e,\varepsilon}} \inf_{w \geq u} I_\eta^{\tau,q}(w) .$$

This will then complete the proof of the (weak) upper bound, as soon as we prove the convexity of $I_P^{\tau,a}$. But, the function

$$\sup_{\lambda \in \mathbb{R}} \inf_{\eta \in M_1^{s,\varepsilon}} [G(\lambda, \eta, u) + h(\eta|P)]$$

$$= \sup_{\lambda \in \mathbb{R}} \left[\lambda u + \inf_{\eta \in M_1^{s,\varepsilon}} \left(-\int f(\lambda, \omega) \eta(d\omega) + h(\eta|P) \right) \right], \quad (2.3.44)$$

being a supremum over affine functions in u, is clearly convex in u, while one shows, exactly as in (2.3.39), that

$$\inf_{\eta \in M_1^{s,\varepsilon}} \sup_{\lambda \in \mathbb{R}} [G(\lambda, \eta, u) + h(\eta|P)] = \inf_{\eta \in M_1^{e,\varepsilon}} \sup_{\lambda \in \mathbb{R}} [G(\lambda, \eta, u) + h(\eta|P)] \quad (2.3.45)$$

and therefore

$$\inf_{\eta \in M_1^{s,\varepsilon}} \sup_{\lambda \in \mathbb{R}} [G(\lambda, \eta, u) + h(\eta|P)] = \inf_{\eta \in M_1^{e,\varepsilon}} \left[I_\eta^{\tau,q}(u) + h(\eta|P) \right] = I_P^{\tau,a}(u).$$

Recalling that, as we saw above, supremum and infimum in (2.3.44) can be exchanged, this completes the proof of the upper bounds for the annealed LDP's for T_n/n.

Step I.3: Annealed lower bounds for T_n/n: We will use the following standard argument.

Lemma 2.3.46 *Let P be a probability distribution, $(\mathfrak{F}_n)$ be an increasing sequence of σ-fields and A_n be $\mathfrak{F}_n$-measurable sets, $n = 1, 2, 3, \ldots$. Let (Q_n) be a sequence of probability distributions such that $Q_n[A_n] \to 1$ and*

$$\limsup_{n \to \infty} \frac{1}{n} H(Q_n|P) \Big|_{\mathfrak{F}_n} \le h$$

where $H(\cdot|P)\big|_{\mathfrak{F}_n}$ denotes the relative entropy w.r.t. P on the σ-field $\mathfrak{F}_n$ and h is a positive number. Then we have

$$\liminf_{n \to \infty} \frac{1}{n} \log P[A_n] \ge -h.$$

Proof of Lemma 2.3.46. From the basic entropy inequality ([22], p. 423),

$$Q_n[A_n] \le \frac{\log 2 + H(Q_n|P)\big|_{\mathfrak{F}_n}}{\log(1 + 1/P[A_n])}, \quad A_n \in \mathfrak{F}_n,$$

we have $-Q_n[A_n] \log P[A_n] \le \log 2 + H(Q_n|P)\big|_{\mathfrak{F}_n}$. Dividing by n and taking limits we obtain the desired result. $\quad\square$

We prove the lower bound for the lower tail only, the upper tail being handled by the same truncation as in the quenched case, see [12] for details. For $\eta \in M_1^{e,\varepsilon}$ satisfying $E_\eta(\log \rho_0) \le 0$, define $\overline{Q}_\omega^o$ as in Step I.3 of Theorem 2.3.12, and let $\overline{Q}_\eta^o = \eta(d\omega) \otimes \overline{Q}_\omega^o$. Let $A_n = \{|n^{-1}T_n - u| < \delta\}$. We know already that $\overline{Q}^o_\omega[A^c_n] \xrightarrow[n\to\infty]{} 0$, $\eta -$ a.s., and this implies $\overline{Q}^o_\eta[A^c_n] \xrightarrow[n\to\infty]{} 0$. Let $\mathcal{F}_n :=$ $\sigma(\{\tau_i\}^n_{i=1}, \{\omega_j\}^n_{j=-\infty})$, $\mathcal{F}^\omega_n = \sigma(\{\omega_j\}^n_{j=-\infty})$. Note that

$$\overline{Q}^o_\eta|_{\mathcal{F}_n} = \eta|_{\mathcal{F}^\omega_n}(d\omega) \otimes \overline{Q}^o_\omega|_{\mathcal{F}_n}.$$

Hence,

$$H(\overline{Q}^o_\eta|\mathbb{P}^o)\Big|_{\mathcal{F}_n} = H(\eta|P)\Big|_{\mathcal{F}^\omega_n} + \int H(\overline{Q}^o_\omega|P^o_\omega)\Big|_{\mathcal{F}_n} \eta(d\omega). \qquad (2.3.47)$$

Considering the second term in (2.3.47), we have

$$\frac{1}{n} \int H(\overline{Q}^o_\omega|P^o_\omega)\Big|_{\mathcal{F}_n} \eta(d\omega)$$

$$= -\frac{1}{n} \int \log Z_{n,\omega}\eta(d\omega) + \lambda_0(u,\eta) \int \frac{T_n}{n} d\overline{Q}^o_\omega \eta(d\omega)$$

$$= -\frac{1}{n} \int \sum^n_{j=1} \log \varphi(\lambda_0(u,\eta), \theta^{j-1}\omega)\eta(d\omega) + \lambda_0(u) \int \frac{T_n}{n} d\overline{Q}^o_\omega \eta(d\omega)$$

and we see, as in the proof of the lower bound of Theorem 2.3.12, that

$$\frac{1}{n} \int H(\overline{Q}^o_\omega|P^o_\omega)\Big|_{\mathcal{F}_n} \eta(d\omega) \xrightarrow[n\to\infty]{} \lambda_0(u,\eta)u - E_\eta f(\lambda_0(u,\eta),\omega) \le I^{\tau,q}_\eta(u).$$

Considering the first term in (2.3.47), we know that

$$\limsup_{n\to\infty} \frac{1}{n} H(\eta|P)\Big|_{\mathcal{F}^\omega_n} = h(\eta|P).$$

Hence,

$$\limsup_{n\to\infty} \frac{1}{n} H(\overline{Q}^o_\eta|\mathbb{P}^o)\Big|_{\mathcal{F}_n} \le I^{\tau,q}_\eta(u) + h(\eta|P),$$

and we can now apply Lemma 2.3.46 to conclude that for any $\eta \in M^{e,\varepsilon}_1$ satisfying $E_\eta(\log \rho_0) \le 0$ one has,

$$\liminf_{n\to\infty} E_P(A_n) \ge - \left(I^{\tau,q}_\eta(u) + h(\eta|P)\right).$$

As in the quenched case, one handles $\eta \in M^{e,\varepsilon}_1$ satisfying $E_\eta(\log \rho_0) > 0$ by repeating the above argument with the required (obvious) modifications, replacing $\overline{Q}^o_\omega$ by $\overline{Q}^o_\omega(\cdot|T_n < \infty)$. This completes the proof of Step I. $\square$

Proof of Lemma 2.3.35: For $\kappa > 1$, decompose $\varphi(\lambda,\omega)$ as follows:

$$E^o_\omega(e^{\lambda\tau_1}\mathbf{1}_{\tau_1<\infty}) = E^o_\omega(e^{\lambda\tau_1}; \tau_1 < \kappa) + E^o_\omega(e^{\lambda\tau_1}; \infty > \tau_1 \ge \kappa)$$

$$:= \varphi^\kappa_1(\lambda,\omega) + \varphi^\kappa_2(\lambda,\omega), \qquad (2.3.48)$$

where $(\lambda,\omega) \to \log \varphi^\kappa_1(\lambda,\omega)$ is bounded and continuous. We also have

$$0 \le \log\left(1 + \frac{\varphi_2^\kappa(\lambda, \omega)}{\varphi_1^\kappa(\lambda, \omega)}\right) \le \log\left(1 + \frac{\varphi_2^\kappa(\lambda_{\text{crit}}, \omega)}{\varepsilon e^\lambda}\right).$$

Hence, the required continuity of the function $(\mu, \lambda) \to \int f(\lambda, \omega)\mu(d\omega)$ will follow from (2.3.48) as soon as we show that for any fixed constant $C_1 < 1$,

$$\lim_{\kappa \to \infty} \sup_{\mu \in M_1^{s,\varepsilon,P}} \int \log\left(1 + \frac{\varphi_2^\kappa(\lambda_{\text{crit}}, \omega)}{C_1}\right)\mu(d\omega) = 0. \qquad (2.3.49)$$

If $\rho_{\min} < 1$ and $\rho_{\max} > 1$ then one can easily check, by a coupling argument using **(C4)**, that $\lambda_{\text{crit}} = 0$ (for a detailed proof see [12, Lemma 4]). Then, for each $\epsilon' > 0$ there exists a $\kappa_\mu = \kappa(\epsilon', \mu)$ large enough such that,

$$E_\mu\left(\log\left(1 + \frac{P_\omega^o(\infty > \tau_1 > \kappa_\mu)}{P_\omega^o(\tau_1 < \infty)}\right)\right) < \epsilon'.$$

Further, in this situation, for stationary, ergodic μ,

$$\int f(0, \omega)\mu(d\omega) = \left(-\int \log \rho_0(\omega)\mu(d\omega)\right) \wedge 0. \qquad (2.3.50)$$

In particular, $\mu \mapsto \int f(0, \omega)\mu(d\omega)$, being linear, is uniformly continuous on the compact set $M_1^{s,\varepsilon}$. Therefore, using (2.3.48), one sees that for each such μ one can construct a neighborhood B_μ of μ such that, for each $\nu \in B_\mu \cap M_1^{s,\varepsilon}$,

$$E_\nu\left(\log\left(1 + \frac{P_\omega^o(\infty > \tau_1 > \kappa_\mu + 1)}{P_\omega^o(\tau_1 < \infty)}\right)\right) < \epsilon'.$$

By compactness, it follows that there exists an $\kappa = \kappa(\epsilon')$ large enough such that, for all $\mu \in M_1^{s,\varepsilon}$,

$$E_\mu\left(\log\left(1 + \frac{P_\omega^o(\infty > \tau_1 > \kappa)}{P_\omega^o(\tau_1 < \infty)}\right)\right) < \epsilon'.$$

Using the inequality $\log(1 + cx) \le c\log(1 + x)$, valid for $x \ge 0$, $c \ge 1$, one finds that for κ large enough,

$$\sup_{\mu \in M_1^{s,\varepsilon}} \int \log\left(1 + \frac{\varphi_2^\kappa(0, \omega)}{C_1}\right)\mu(d\omega) \le \epsilon'/C_1,$$

proving (2.3.49) under the condition $\rho_{\min} < 1, \rho_{\max} > 1$.

We next handle the case $\rho_{\max} < 1$. We now complete the proof of Lemma 2.3.35 in the case $\rho_{\min} > 1$. We have $f(\lambda, \omega) \ge \lambda + \log \omega_0^+ \ge \lambda + \log \varepsilon$. We show that $(\lambda, \omega) \mapsto \varphi(\lambda, \omega)$ is continuous as long as $\omega_i \le \omega^{\max}$, $\rho_i \le \rho_{\max}$ and $\lambda \le \lambda_{\text{crit}}$, which is enough to complete the proof. Write, for $\lambda \le \lambda_{\text{crit}}$,

$$E_\omega(e^{\lambda\tau_1}1_{\tau_1<\infty}) = E_\omega(e^{\lambda\tau_1}; \tau_1 < \kappa) + E_\omega(e^{\lambda\tau_1}; \infty > \tau_1 \ge \kappa) \qquad (2.3.51)$$

and observe that the first term in the right hand side of (2.3.51) is continuous as a function of ω and the second term goes to 0 for $\kappa \to \infty$, uniformly in ω. More precisely, due to (2.3.16), for all ω considered here,

$$E_\omega[e^{\lambda \tau_1}; \infty > \tau_1 \geq \kappa] \leq E_{\tilde\omega^{\min}}(e^{\lambda_{\rm crit} \tau_1}; \tau_1 \geq \kappa) \to_{\kappa \to \infty} 0 . \qquad (2.3.52)$$

Finally, in the case $\rho_{\min} > 1$, the conclusion follows from the duality formula (2.3.23) and Remark 1 that follows its proof, by reducing the claim to the case $\rho_{\max} < 1$. $\qquad \square$

Step II: The proof is identical to Step I, and is omitted.

Step III: The proof of all statements, except for the convexity of I_P^a, and the upper bound on $\mathbb{P}^o(X_n \leq nv)$, follow the argument in the quenched case. The latter proofs can be found in [12]. $\qquad \square$

Remarks: 1. We note that under the conditions of Theorem 2.3.34, if $E_P \log \rho_0 \leq 0$ then both $I_P^a(v) \neq 0$ and $I_P^q(v) \neq 0$ for $v \notin [0, v_P]$. Indeed, since $h(\eta|P) \neq 0$ unless $\eta = P$, it holds that $I_P^a(v) = 0$ only if $I_P^q(v) = 0$. If $E_P \log \rho_0 = 0$, $v_P = 0$ and then for any $v \neq 0$, $I_P^a(v) = |v| I_P^{\tau,q}(1/|v|) > 0$ by the remark following the proof of Lemma 2.3.13. On the other hand, if $E_P \log \rho_0 < 0$, the same argument applies for $v > v_P$ while for $v < 0$ we have that $I_P^q(v) \geq -|v| E_P \log \rho_0 > 0$.

2. The condition **(C5)** can be avoided altogether. This is not hard to see if one is interested only in the LDP for T_n/n. Indeed, **(C5)** was used mainly in describing a worst case environment in the course of the proof of Lemma 2.3.35, see also part (d) of Lemma 2.3.13. When it is dropped, the following lemma, whose proof we provide below, replaces Lemma 2.3.35 when deriving the annealed LDP for T_n/n:

Lemma 2.3.53 *Assume P satisfies Assumption 2.3.33 except for* **(C5)**. *Then, $\lambda_{\rm crit}(P)$ depends only on $\mathrm{supp}(P_0)$, and the map $(\mu, \lambda) \mapsto E_\mu(f(\lambda, \omega))$ is continuous on $M_1^{s,\varepsilon,P} \times (-\infty, 0] \cup [0, \lambda_{\rm crit})$.*

Given Lemma 2.3.53, we omit **(C5)** and replace **(C4)** in Assumption 2.3.33 by

(C4') *P is locally equivalent to the product of its marginals and, for any stationary measure $\eta \in M_1(\Omega)$ with $h(\eta|P) < \infty$ there is a sequence $\{\eta^n\}$ of stationary, ergodic measures, locally equivalent to the product of P's marginals, with $\mathrm{supp}((\eta^n)_0) = \mathrm{supp}(P_0)$, $\eta^n \xrightarrow[n \to \infty]{} \eta$ weakly and $h(\eta^n|P) \to h(\eta|P)$.*

One now checks (we omit the details) that all approximations carried out in the proof of the upper bound of the upper tail of T_n/n can still be done, yielding the annealed LDP for T_n/n. To transfer this LDP to an annealed LDP for X_n/n does require a new argument, we refer to [16] for details.

We conclude our discussion of large deviation principles with the:

Proof of Lemma 2.3.53: Set $\Xi = \mathrm{supp}(P_0)$ and define $\bar\lambda := \inf_{\omega \in \Xi^{\mathbb{Z}}} \lambda_{\rm crit}(\omega)$ where

$$\lambda_{\rm crit}(\omega) := \sup\{\lambda \in \mathbb{R} : E_\omega^o(e^{\lambda \tau_1} 1_{\{\tau_1 < \infty\}})\} .$$

By definition, $\lambda_{\text{crit}}(P) \geq \bar{\lambda}$. On the other hand, if $\lambda > \bar{\lambda}$ then there exists a $\bar{\omega} \in \Xi^{\mathbb{Z}}$ with $E^o_{\bar{\omega}}(e^{\lambda \tau_1} \mathbf{1}_{\{\tau_1 < \infty\}}) = \infty$. Fix $K = e^{-\lambda}/\varepsilon$, and using monotone convergence, choose an M large enough such that

$$\varphi_{M,\bar{\omega}}(\lambda) := E^o_{\bar{\omega}}(e^{\lambda \tau_1} \mathbf{1}_{\{\tau_1 < M\}}) > K + 1.$$

Since $\varphi_{M,\bar{\omega}}$ depends only on $\{\omega_i, i \in (-M, 0)\}$, it holds that with positive P-probability, $E^o_\omega(e^{\lambda \tau_1} \mathbf{1}_{\{\tau_1 < M\}}) \geq K + 1$. But, if $\lambda \leq \lambda_{\text{crit}}(P)$ it follows from the recursions (2.3.21) that $\varphi(\lambda, \omega) < K$, P-a.s., a contradiction unless $\bar{\lambda} = \lambda_{\text{crit}}(P)$. In particular, $\lambda_{\text{crit}}(P)$ depends only on $\text{supp}(P_0)$. Note that the characterization of $\lambda_{\text{crit}}(P)$ as $\bar{\lambda}$ implies that for any $\mu \in M_1^{s,\varepsilon,P}$ it holds that $\lambda_{\text{crit}}(\mu) \geq \lambda_{\text{crit}}(P)$.

Next, as in the course of the proof of Lemma 2.3.35, see (2.3.49), it is enough to show that for any $\lambda < \lambda_{\text{crit}}(P)$,

$$\lim_{\kappa \to \infty} \sup_{\mu \in M_1^{s,\varepsilon,P}} \int \varphi_2^\kappa(\lambda, \omega) \mu(d\omega) = 0. \qquad (2.3.54)$$

But, since $\varphi(\lambda, \omega) \leq e^{-\lambda}/\epsilon$ μ-a.s. for all $\mu \in M_1^{s,\varepsilon,P}$ (use again the recursions (2.3.21) and that $\lambda_{\text{crit}}(\mu) \geq \lambda_{\text{crit}}(P)$), it holds that

$$\int \varphi_2^\kappa(\lambda, \omega) \mu(d\omega) \leq e^{(\lambda - \lambda_{\text{crit}})M} \frac{e^{-\lambda}}{\varepsilon},$$

yielding immediately (2.3.54). □

Bibliographical notes: The first quenched LDP result is due to Greven and Den Hollander, [34], who proved it for i.i.d. environments using the method of the environment viewed from the particle. Our derivation here follows the hitting times approach developed in [12], except that the proof of Lemma 2.3.22 follows the article [58]. Extensions of the LDP's in this chapter to more general models allowing for (non geometric) holding times is presented in [16], where the derivation avoids completely coupling arguments and thus bypasses altogether the need for (C5) in deriving the annealed LDP for X_n/n.

The "process level LDP" for R_n was first proved in [23] in the context of Markov chains with law P satisfying appropriate regularity conditions. It was extended to various ergodic situation in [55] and [56], see also [11]. We refer to [27] and [19, Chapter 6] for further information. Our presentation of the annealed LDP follows here [12], where additional information on the shape of the rate functions etc. can be found. Note that [12] treats the case $P(\omega_0^0 = 0) = 1$. In the exposition here, we corrected and simplified some of the arguments in [12], following [16], where a RWRE with general (i.e., not necessarily geometric) holding times is considered. Finally, a completely different approach to the derivation of the LDP, both annealed and quenched, is described in [79].

216

2.4 The subexponential regime

We saw in Section 2.3 that, at least when P satisfies Assumption 2.3.33 and $E_P \log \rho_0 \leq 0$, we have that for any δ small enough, any $v \in [0, v_P]$,

$$\lim_{n \to \infty} \frac{1}{n} \log P_\omega^o \left(\frac{X_n}{n} \in (v - \delta, v + \delta) \right)$$

$$= \lim_{n \to \infty} \frac{1}{n} \log \mathbb{P}^o \left(\frac{X_n}{n} \in (v - \delta, v + \delta) \right) = 0. \quad (2.4.1)$$

Our goal in this section is to obtain more precise information on the rate of convergence in (2.4.1). Surprisingly, it turns out that it is better to consider first the annealed case.

Throughout this section, we impose the following assumption on the law P. Together with **(C4)** there, it implies Assumption 2.3.33.

Assumption 2.4.2

(D1) *There exists an $\varepsilon > 0$ such that $\min(\omega_0^+, \omega_0^-) > \varepsilon$, P-a.s.*
(D2) $\rho_{\min} < 1$, $\rho_{\max} > 1$, *and $E_P \log \rho_0 \leq 0$.*
(D3) *P is α-mixing with $\alpha(n) = \exp(-n(\log n)^{1+\eta})$ for some $\eta > 0$; that is, for any ℓ-separated measurable bounded by 1 functions f_1, f_2,*

$$E_P \left(f_1(\omega)(f_2(\omega) - E_P f_2(\omega)) \right) \leq \alpha(\ell).$$

(functions f_i are ℓ separated if f_i is measurable on $\sigma(\omega_j, j \in I_i)$ with I_i intervals satisfying $\mathrm{dist}(I_i, I_k) > \ell$ for any $i \neq k$).

It is known that **(D3)** implies **(C1)** and **(C3)** of Assumption 2.3.33, see [10]. In particular, letting $\overline{R}_k := k^{-1} \sum_{i=0}^{k-1} \log \rho_i$, it implies that $\overline{R}_k$ satisfies the LDP with good rate function $J(\cdot)$. We add the following assumption on $J(\cdot)$:

(D4) $J(0) > 0$.

Condition **(D4)** implies that $E_P(\log \rho_0) < 0$. Define next $s := \min_{y \geq 0} \frac{1}{y} J(y)$. Note that the condition $E_P(\overline{S}) < \infty$ and the existence of a LDP for $\overline{R}_k$ with good rate function $J(\cdot)$ are enough to imply, by Varadhan's lemma, that $0 \geq \sup_y (y - J(y))$, and in particular that $s \geq 1$. (In the case where P is a product measure, we can identify s as satisfying $E_P(\rho_0^s) = 1$, and then $E_P(\overline{S}) < \infty$, which is equivalent to $E_P(\rho_0) < 1$, implies that $s > 1$.)

Annealed subexponential estimates

Theorem 2.4.3 *Assume P satisfies Assumption 2.4.2, and $v_P > 0$. Then, for any $v \in (0, v_P)$ and any $\delta > 0$ small enough,*

$$\lim_{n \to \infty} \frac{\log \mathbb{P}^o \left(\frac{X_n}{n} \in (v - \delta, v + \delta) \right)}{\log n} = 1 - s.$$

Proof. We begin by proving the lower bound. Fix $0 < v-\delta < v-4\eta < v < v_P$; let

$$L_k = \max\left\{n \geq T_k : (k - X_n)\right\}$$

denote the largest excursion of $\{X.\}$ to the left of k after hitting it. Observe that the event $\{n^{-1}X_n \in (v - \delta, v + \delta)\}$ contains the event

$$A := \left\{\frac{(v - 4\eta)n}{v_P} < T_{(v-2\eta)n} < n,\, L_{(v-2\eta)n} < \frac{\eta n}{2},\, T_{vn} > n\right\}, \qquad (2.4.4)$$

namely, the RWRE hits $(v - 2\eta)n$ at about the expected time, from which point its longest excursion to the left is less than $\eta n/2$, but the RWRE does not arrive at position vn by time n.

Next, note that by (2.1.4),

$$P_\omega^o\left(L_{(v-2\eta)n} \geq \eta n/2\right) \leq \sum_{i=0}^{\infty} \prod_{j=-(\eta n/2-1)}^{i} \rho_{(v-2\eta)n+j} \cdot \qquad (2.4.5)$$

Hence, using the LDP for $\overline{R}_k$, we have for all n large enough

$$\mathbb{P}^o\left(L_{(v-2\eta)n} \geq \eta n/2\right) \leq \sum_{i=0}^{\infty} E\left(e^{(\eta n/2+i)\overline{R}_{\eta n/2+i}}\right)$$

$$\leq e^{-\eta n \sup_y (y-J(y))/4} \leq e^{-\delta_1 n} \qquad (2.4.6)$$

for some $\delta_1 > 0$. Thus, for all n large enough,

$$\mathbb{P}^o(A) \geq \mathbb{P}^o\left(\frac{(v - 4\eta)n}{v_P} < T_{(v-2\eta)n} < n,\, T_{vn} > n\right) - e^{-\delta_1 n}$$

$$\geq \mathbb{E}^o\left(P_\omega^o\left(\frac{(v - 4\eta)n}{v_P} < T_{(v-2\eta)n} < n\right)\right.$$

$$\left. P_\omega^{(v-\eta)n}\left(T_{vn} > \frac{4\eta n}{v_P},\, L_{(v-\eta)n<\eta n/2}\right)\right) - e^{-\delta_1 n}$$

$$\geq B \cdot C - \alpha(\eta n/2) - 2e^{-\delta_1 n},$$

where

$$B = \mathbb{P}^o\left(\frac{(v - 4\eta)n}{v_P} < T_{(v-2\eta)n} < n\right)$$

$$C = \mathbb{P}^o\left(T_{\eta n} > \frac{4\eta n}{v_P}\right).$$

and $\alpha(\cdot)$ is as in **(D3)**.

Next, note that $B \to_{n\to\infty} 1$ by (2.1.16). We will prove below that for any $\delta' > 0$,

$$C \geq n^{1-s-2\delta'} \qquad (2.4.7)$$

and this implies that for all n large, $\mathbb{P}^o(A) \geq n^{1-s-4\delta'}$, which yields the required lower bound (recall δ' is arbitrary!) as soon as we prove (2.4.7).

Turning to the proof of (2.4.7), fix y such that $\frac{J(y)}{y} \leq s + \frac{\delta'}{4}, K = [n^{\frac{\delta'}{4}}]$, $k = [\frac{1}{y}\log n]$, and set $\overline{m}_K = [\eta n/K]$. Now, using **(D3)**,

$$P\left(\bigcap_{j=1}^{\overline{m}_K}\{\overline{R}_k(\theta^{jK}\omega) \leq y\}\right)$$

$$\leq \left(P(\overline{R}_k(\omega) \leq y)\right)^{\overline{m}_K} + \overline{m}_K\alpha(K-k)$$

$$= \left(1 - P(\overline{R}_k(\omega) > y)\right)^{\overline{m}_K} + \overline{m}_K\alpha(K-k)$$

$$\leq \left(1 - e^{-k(J(y)+\frac{\delta'y}{4})}\right)^{\overline{m}_K} + \overline{m}_K\alpha(K-k) \leq 1 - n^{1-s-\delta'},$$

for all n large enough. Hence,

$$P\left(\exists j \in \{1, \cdots, \overline{m}_K\} : \overline{R}_k(\theta^{jK}\omega) > y\right) \geq n^{1-s-\delta'}. \tag{2.4.8}$$

On the other hand, let ω and $j \leq \overline{m}_K$ be such that $\overline{R}_k(\theta^{jK}\omega) > y$. Then, using (2.1.6) in the second inequality, for such ω,

$$P_\omega^o\left(T_{\eta n} > \frac{4\eta n}{v_P}\right) \geq P_\omega^{jK}\left(T_k > \frac{4\eta n}{v_P}\right) \geq (1 - e^{-ky})^{\frac{4\eta n}{v_P}}$$

$$\geq \left(1 - \frac{1}{n}\right)^{\frac{6\eta n}{v_P}} \geq e^{-\frac{8\eta}{v_P}}. \tag{2.4.9}$$

Combining (2.4.8) and (2.4.9), we conclude that

$$C \geq n^{1-s-\delta'} \cdot e^{-\frac{8\eta}{v_P}},$$

as claimed.

We next turn to the proof of the upper bounds. We may and will assume that $s > 1$, for otherwise there is nothing to prove. We first note that, for some $\delta'' := \delta''(\delta) > 0$,

$$\mathbb{P}^o\left(\frac{X_n}{n} \in (v - \delta, v + \delta)\right) \leq \mathbb{P}^o\left(\frac{X_n}{n} < v + \delta\right)$$

$$\leq \mathbb{P}^o(T_{n(v+2\delta)} > n) + \mathbb{P}^o(L_0 > n\delta)$$

$$\leq \mathbb{P}^o(T_{n(v+2\delta)} > n) + e^{-\delta''n} \tag{2.4.10}$$

where the stationarity of P was used in the second inequality, and (2.4.6) in the third. Thus, the required upper bound follows once we show that for any $v < v_P$, any $\delta' > 0$,

$$\mathbb{P}^o(T_{nv} > n) \leq n^{1-s+\delta'} \tag{2.4.11}$$

for all n large enough.

Set $a := \sup_y (y - J(y))$. Because $s > 1$ and $J(0) > 0$, it holds that $a < 0$. Fix $A > -s/a$, and set $k = k(n) = A \log n$. Next, define the process $\{Y_n\}$ in $\mathbb{Z}^{\mathbb{N}}$ and the hitting times $\tilde{T}_{ik} = \min(n \geq 0 : Y_n = ik), i = 0, 1, \cdots$ such that the only change between the processes $\{X_n\}$ and $\{Y_n\}$ is that the process $\{Y_n\}_{n \geq \tilde{T}_{ik}}$ is reflected at position $(i-1)k$ (with a slight abuse of notations, we continue to use P_ω^o, $\mathbb{P}^o$ to denote the law of $\{Y_n\}$ as well as that of $\{X_n\}$). Set $m_k = [vn/k] + 1$, and $\tilde{\tau}_k^{(i)} = \tilde{T}_{ik} - \tilde{T}_{(i-1)k}, i = 1, \cdots, m_k$. Note that the $\tilde{\tau}_k^{(i)}$ are identically distributed, each stochastically dominated by T_k. Hence, $\mathbb{E}^o \tilde{T}_{ik} \leq \mathbb{E}^o T_{ik}$. On the other hand, fixing $\lambda \in (1/s, 1)$, we will see below (cf. Lemma 2.4.16) that $\mathbb{E}^o(T_k^{1/\lambda}) \leq ck^{1/\lambda}$ for some $c := c(\lambda)$, yielding, by Hölder's inequality, that

$$\mathbb{E}^o T_k \leq \mathbb{E}^o(\tilde{T}_k) + \mathbb{P}^o(L_0 \geq k)^{1-\lambda} \mathbb{E}^o(T_k^{1/\lambda})^\lambda \leq \mathbb{E}^o(\tilde{T}_k) + ck\mathbb{P}^o(L_0 \geq k)^{1-\lambda}.$$

Thus, using (2.4.6) and the fact that $\mathbb{E}^o(T_k)/k = v_P$, we conclude that $\lim_{k \to \infty} \mathbb{E}^o T_k / \mathbb{E}^o \tilde{T}_k = 1$, implying that $\mathbb{E}^o \tilde{T}_k / k \to_{k \to \infty} 1/v_P$.

Next, note that on the event $\{L_{ik} < k$ for $i = 0, \cdots, m_k\}$, the processes $\{X_n\}$ and $\{Y_n\}$ coincide for $n < T_{m_k k}$. Hence

$$\mathbb{P}^o(T_{nv} > n) \leq \mathbb{P}^o(\tilde{T}_{m_k k} > n) + m_k \mathbb{P}^o(L_0 > k). \tag{2.4.12}$$

But, as in (2.4.6), for k large enough

$$\mathbb{P}^o(L_0 > k) \leq E_P(e^{k(\overline{R}_k + \delta)}) \leq e^{\log n(Aa + \delta')} \leq n^{-s + \delta''},$$

where $\delta'' := \delta''(\delta) \to_{\delta \to 0} 0$. Since $m_k < n$, the second term in (2.4.12) is of the right order, and the upper bound follows as soon as we prove that, for n large enough

$$\mathbb{P}^o(\tilde{T}_{m_k k} > n) \leq n^{1 - s + \delta'}. \tag{2.4.13}$$

To see (2.4.13), note that $\tilde{T}_{m_k k} = \sum_{i=1}^{m_k} \tilde{\tau}_k$, with $\mathbb{E}^o(\tilde{\tau}_k^{(i)})/k = \mathbb{E}^o(\tilde{T}_k)/k \to 1/v_P$. Hence, for some $\eta > 0$, using that $km_k \leq v < v_P$,

$$\mathbb{P}^o(\tilde{T}_{m_k k} > n) \leq \mathbb{P}^o\left(\sum_{i=1}^{m_k} \left(\tilde{\tau}_k^{(i)} - \mathbb{E}^o(\tilde{\tau}_k^{(i)})\right) > 4\eta n\right)$$

$$\leq 4\mathbb{P}^o\left(\sum_{i=1}^{\lceil m_k/4 \rceil} \left(\tilde{\tau}_k^{(4i)} - \mathbb{E}^o(\tilde{T}_k)\right) > \eta n\right).$$

Note that the quenched law of $\tilde{\tau}_k^{(4i)}$ depends on $\{\omega_j, j \in I_i\}$ where $I_i = \{4i - k, 4i - k + 1, \cdots, 4i + k\}$. Let $\{\overline{\tau}_k^{(i)}\}$ be i.i.d. random variables such that for any Borel set G, $P(\overline{\tau}_k^{(i)} \in G) = \mathbb{P}^o(\tilde{\tau}_k^{(i)} \in G)$. Then, by iterating the definition of $\alpha(\cdot)$, one has that

$$\mathbb{P}^o \left(\sum_{i=1}^{\lceil m_k/4 \rceil} \left(\tilde{\tau}_k^{(4i)} - \mathbb{E}^o(\tilde{T}_k) \right) > \eta n \right) \le$$

$$P \left(\sum_{i=1}^{\lceil m_k/4 \rceil} \left(\overline{\tau}_k^{(4i)} - \mathbb{E}^o(\tilde{T}_k) \right) > \eta n \right) + \frac{m_k \alpha(2k)}{4}. \quad (2.4.14)$$

We recall that

$$\frac{m_k \alpha(2k)}{4} \le o(n^{1-s}). \quad (2.4.15)$$

The following estimate, whose proof is deferred, is crucial to the proof of (2.4.13):

Lemma 2.4.16 *For each $\kappa < s$, there exists a constant $c(\kappa) < \infty$ such that*

$$\mathbb{E}^o(T_k)^\kappa \le c(\kappa) k^\kappa. \quad (2.4.17)$$

By Markov's inequality, for any $\kappa < \kappa' < s$,

$$P(\overline{\tau}_k^{(4i)} - E\overline{\tau}_k^{(4i)} > \eta n) \le \frac{1}{(\eta n)^{\kappa'}} E |\overline{\tau}_k^{(4i)} - E\overline{\tau}_k^{(4i)}|^{\kappa'} \le n^{-\kappa}$$

where n is large enough and we used Lemma 2.4.16 and the fact that $E((\overline{\tau}_k^{(4i)})^{\kappa'}) = \mathbb{E}^o((\tilde{\tau}_k^{(4i)})^{\kappa'}) \le \mathbb{E}^o(T_k^{\kappa'})$. Hence, (see [54, (1.3),(1.7a)]),

$$P \left(\sum_{i=1}^{\lceil m_k/4 \rceil} \left(\overline{\tau}_k^{(4i)} - E\overline{\tau}_k^{(4i)} \right) > \eta n \right)$$

$$\le \lceil \frac{m_k}{4} \rceil P(\overline{\tau}_k^{(4)} - E\overline{\tau}_k^{(4)} > \eta n) + \frac{1}{2} n^{1-\kappa} \le n^{1-\kappa}.$$

Since $\kappa < s$ is arbitrary, this completes the proof, modulo the

Proof of Lemma 2.4.16

Note first that by Minkowski's inequality, for any $k \ge 1$,

$$\mathbb{E}^o(T_k^\kappa) = \mathbb{E}^o \left(\sum_{i=1}^{k} \tau_i \right)^\kappa \le k^\kappa \mathbb{E}^o \tau_1^\kappa.$$

Hence, it will be enough to prove that

$$\mathbb{E}^o(\tau_1^\kappa) < \infty. \quad (2.4.18)$$

To prove (2.4.18), we build upon the techniques developed in the course of proving Lemma 2.1.21. Indeed, recall the random variables $U_{i,j}, Z_{i,j}$ and N_i defined there, and note that since $\tau_1 = \sum_{i=-\infty}^{o} N_i$, it is enough to estimate

$$\mathbb{E}^o \left(\sum_{i=-\infty}^0 N_i \right)^\kappa = \mathbb{E}^o \left(\sum_{i=-\infty}^0 U_i + U_{i+1} + Z_i \right)^\kappa \leq C_\varepsilon \mathbb{E}^o \left(\sum_{i=-\infty}^0 U_i \right)^\kappa .$$

$$(2.4.19)$$

An important step in the evaluation of the RHS in (2.4.19) involves the computation of moments of U_i. To present the idea, consider first the case $\kappa > 2$, and write

$$U_i = \sum_{j=1}^{U_{i+1}} G_j$$

where, under P_ω^o, the G_j are i.i.d. geometric random variables, independent of $\{U_{i+1}, \cdots U_0\}$, of parameter $\frac{\omega_i^-}{\omega_i^- + \omega_i^+}$. Hence,

$$E_\omega^o(U_i^2) = E_\omega^o \left(\sum_{j=1}^{U_{i+1}} (G_j - E_\omega^o G_j) + \sum_{j=1}^{U_{i+1}} E_\omega^o G_j \right)^2 \qquad (2.4.20)$$

$$\leq c_\delta E_\omega^o \left(\sum_{j=1}^{U_{i+1}} (G_j - E_\omega^o G_j) \right)^2 + (1+\delta)(E_\omega^o G_j)^2 \cdot E_\omega^o(U_{i+1}^2)$$

$$\leq c_\delta' E_\omega^o(U_{i+1}) \cdot E_\omega^o(G_j^2) + (1+\delta)\rho_i^2 E_\omega^o(U_{i+1}^2) .$$

Here, c_δ, c_δ' are constants which depend on δ only. Since $E_\omega^o(G_j^2)$ is uniformly (in ω) bounded, and $E_\omega^o(U_{i+1}) = \rho_i$, we get

$$E_\omega^o(U_i^2) \leq c_\delta'' \rho_i E_\omega^o U_{i+1} + (1+\delta)\rho_i^2 E_\omega^o(U_{i+1}^2) .$$

Iterating and using (cf. (2.1.24)) that $E_\omega^o U_{i+1} = \prod_{j=i+1}^0 \rho_j$, we conclude the existence of a constant c_δ''' such that

$$E_\omega^o(U_i^2) \leq c_\delta''' \left(\sum_{j=0}^{|i|} \left(\prod_{k=-j}^0 \rho_k + \prod_{k=-j}^0 (\rho_k^2(1+\delta)) \right) \right) ,$$

and hence

$$\mathbb{E}^o(U_i^2) \leq c_\delta''' \sum_{j=0}^{|i|} \left(E_P \prod_{k=-j}^0 \rho_k + E_P \prod_{k=-j}^0 \left(\rho_k^2(1+\delta) \right) \right) . \qquad (2.4.21)$$

Note that, by Varadhan's lemma (see [19, Theorem 4.3.1]), for any constant β,

$$\lim_{n \to \infty} \frac{1}{n} \log \mathbb{E}^o \left(\prod_{k=-n}^0 \rho_k^\beta \right) = \sup_y \left(\beta y - J(y) \right) = \sup_y y \left(\beta - \frac{J(y)}{y} \right) := \beta'(\beta) ,$$

$$(2.4.22)$$

and $\beta'(\beta) < 0$ as soon as $\beta < s$. Hence, substituting in (2.4.21), and choosing δ such that $\log(1 + \delta) < \beta'(\beta)/4$, we obtain that for some constant c_δ''',

$$\mathbb{E}^o(U_i^2) \leq c_\delta''''e^{-i\beta'(2)/2},$$

implying that

$$\sqrt{\mathbb{E}^o\left(\sum_{i=-\infty}^{0} N_i\right)^2} \leq C_\varepsilon^{\frac{1}{2}}\sqrt{\sum_{i=-\infty}^{0}\mathbb{E}^o(U_i^2)} < \infty.$$

A similar argument holds for any integer $\kappa < s$: mimicking the steps leading to (2.4.21), we get that

$$E_\omega^o(U_i^\kappa) \leq c_\delta'''\left(\sum_{j=0}^{|i|}\left(\prod_{k=-j}^{0}\rho_k^{\kappa/2} + \prod_{k=-j}^{0}\rho_k^\kappa\right)\right),$$

and using (2.4.22) and an induction on lower (integer) moments, we get that $\mathbb{E}^o\left(\sum_{i=-\infty}^{0} N_i\right)^\kappa < \infty$ for all $\kappa < s$ integer. Finally, to handle $\lfloor s \rfloor < \kappa < s$, we replace (2.4.20) by

$$E_\omega^o(U_i^\kappa) \leq c_\delta'E_\omega^o(U_{i+1}^{\kappa/2\vee 1})E_\omega^o(G_j^\kappa) + (1 + \delta)\rho_i^\kappa E_\omega^o(U_{i+1}^\kappa)$$
$$\leq c_\delta''(E_\omega^o U_i^{\lceil\kappa/2\rceil})^{\frac{\kappa/2}{\lceil\kappa/2\rceil}} + (1 + \delta)\rho_i^\kappa E_\omega^o(U_{i+1}^\kappa),$$

and one proceeds as before. $\qquad\square$

Quenched subexponential estimates

Theorem 2.4.23 *Assume P satisfies Assumption 2.4.2, and $v_P > 0$. Then, for any $v \in (0, v_P)$, any $\eta > 0$, and any $\delta > 0$ small enough,*

$$\liminf_{n\to\infty}\frac{1}{n^{1-1/s+\eta}}\log P_\omega^o\left(\frac{X_n}{n} \in (v - \delta, v + \delta)\right) = 0, \quad P - a.s. \quad (2.4.24)$$

Further,

$$\limsup_{n\to\infty}\frac{1}{n^{1-1/s-\eta}}\log P_\omega^o\left(\frac{X_n}{n} < v\right) = -\infty, \quad P - a.s.. \quad (2.4.25)$$

Proof. Starting with the lower bound, we have, using (2.4.4) and (2.4.5), that for some $\delta_1(\omega) > 0$,

$$P_\omega^o(\frac{X_n}{n} \in (v - \delta, v + \delta)) \geq P_\omega^o\left(\frac{(v - 4\eta)n}{v_P} < T_{(v-2\eta)n} < n\right)$$
$$P_\omega^{(v-\eta)n}\left(T_{vn} > \frac{4\eta}{v_P}n, L_{(v-\eta)n<\eta n}\right) - e^{-\delta_1(\omega)n}.$$

By (2.1.16), $P_\omega^o(\frac{X_n}{n} \in (v - \delta, v + \delta)) \to_{n\to\infty} 1$, P-a.s. On the other hand, as in the proof of (2.4.8), fix y such that $J(y)/y \leq s + \delta'/4$, $k = \lfloor (1 - \delta')/ys \rfloor$, and $K = n^{\delta'/4}$. Then, one checks as in the annealed case that

$$P\left(\forall j \in \{1, \cdots, \overline{m}_K\} : \overline{R}_k(\theta^{jK}\omega) \leq y\right) \leq \frac{1}{n^2},$$

and one concludes by the Borel-Cantelli lemma that there exists an $n_0(\omega)$ such that for all $n_0(\omega)$, there exists a $j \in \{1, \ldots, \overline{m}_K\}$ such that $\overline{R}_k(\theta^{jK}\omega) > y$. The lower bound (2.4.24) now follows as in the proof of (2.4.9).

Turning to the proof of the upper bound (2.4.25), as in the annealed setup it is straightforward to reduce the proof to proving

$$\lim_{n\to\infty} \frac{1}{n^{1-1/s-\delta}} \log P_\omega^o(T_n > n/v) = -\infty. \qquad (2.4.26)$$

We provide now a short sketch of the proof of (2.4.26) before getting our hands dirty in the actual computations. Divide the interval $[0, nv]$ into blocks of size roughly $k = k_n := n^{1/s+\delta}$. By using the annealed bounds of Theorem 2.4.3, one knows that $P(T_k > k/v) \sim k^{1-s}$. Hence, taking appropriate subsequences, one applies a Borel-Cantelli argument to control uniformly the probability $P_\omega^{ik}(T_{(i+1)k} > k/v)$, c.f. Lemma 2.4.28.

The next step involves a decoupling argument. Define

$$\overline{T}_{(i+1)k} = \inf \{t > T_{ik} : X_t = (i + 1)k \text{ or } X_t = (i - 1)k\}. \qquad (2.4.27)$$

Then one shows that for all relevant blocks, that is $i = \pm 1, \pm 2, \ldots, \pm n/k$, $P_\omega^{ik}(\overline{T}_{(i+1)k} \neq T_{(i+1)k})$ is small enough. Therefore, we can consider the random variables $\overline{T}_{(i+1)k} - T_{ik}$ instead of $T_{(i+1)k} - T_{ik}$, which have the advantage that their dependence on the environment is well localized. This allows us to obtain a uniform bound on the tails of $\overline{T}_{(i+1)k} - T_{ik}$, for all relevant i, see (2.4.30).

The final step involves estimating how many of the k-blocks will be traversed from right to left before the RWRE hits the point nv. This is done by constructing a simple random walk (SRW) S_t whose probability of jump to the left dominates $P_\omega^{ik}(T_{(i+1)} \neq \overline{T}_{(i+1)k})$ for all relevant i. The analysis of this SRW will allow us to claim (c.f. Lemma 2.1.17) that the number of visits to a k-block after entering its right neighbor is negligible. Thus, the original question on the tail of T_n is replaced by a question on the sum of (dominated by i.i.d.) random variables, which is resolved by means of the tail estimates obtained in the second step.

A slight complication is presented by the need to work with subsequences in order to apply the Borel-Cantelli lemma at various places. Going from subsequences to the original n sequence is achieved by means of monotonicity arguments. Indeed, by monotonicity, note that it is enough to prove the result when, for arbitrary δ small enough, n is replaced by the subsequence $n_j = \lfloor j^{2/\delta} \rfloor$, since $n_{j+1}/n_j \to_{j\to\infty} 1$.

Turning to the actual proof, fix $C_n = n^\delta$, $k = k_j = \dfrac{C_{n_j} n_j^{1/s}}{1 - \varepsilon}$ for some $1 > \varepsilon > 0$, $b_n = C_n^{-\delta}$, and $I_j = \left\{ -\left[\frac{n_j}{k_j}\right] - 1, \cdots, \left[\frac{n_j}{k_j}\right] + 1 \right\}$. Finally, fix $v' < v$ and $\overline{T}_{(i+1)k}$ as in (2.4.27). (We will always use $\overline{T}_{(i+1)k}$ in conjunction with the RWRE started at ik!). We now claim the:

Lemma 2.4.28 *For P – a.e. ω, there exists a $J_0(\omega)$ such that for all $j > J_0(\omega)$, and all $i \in I_j$,*

$$P_\omega^{ik}\left(\frac{T_{(i+1)k_j}}{k_j} > \frac{1}{v'}\right) \le b_{n_j} . \tag{2.4.29}$$

Further, for all $j > J_0(\omega)$, and each $i \in I_j$, and for $x \ge 1$,

$$P_\omega^{(ik)}\left(\frac{\overline{T}_{(i+1)k_j}}{k_j} > \frac{x}{v'}\right) \le (2b_{n_j})^{[x/2] \vee 1} . \tag{2.4.30}$$

Proof of Lemma 2.4.28. By Chebycheff's bound,

$$P\left(P_\omega^{ik}\left(\frac{T_{(i+1)k_j}}{k_j} > \frac{1}{v'}\right) > b_{n_j}\right) \le \frac{1}{b_{n_j}} \mathbb{P}^{ik}\left(\frac{T_{(i+1)k_j}}{k_j} > \frac{1}{v'}\right)$$

$$\le \frac{1}{b_{n_j}} k_j^{1 - s + o(1)} ,$$

where the last inequality follows from Theorem 2.4.3. Hence,

$$P\left(P_\omega^{ik}\left(\frac{T_{(i+1)k_j}}{k_j} > \frac{1}{v'}\right) > b_{n_j} \text{ for some } i \in I_j\right) \le 3\left[\frac{n_j}{k_j}\right] \cdot \frac{1}{b_{n_j}} \cdot k_j^{1 - s + o(1)}$$

$$\le \frac{3}{n_j^{\delta(s - o(1) - \delta)}} \le \frac{4}{j^{2(s - o(1) - \delta)}}$$

and (2.4.29) follows from the Borel-Cantelli lemma. (2.4.30) follows by iterating this inequality and using the Markov property. $\qquad\square$

Recall that $a = \sup_y (y - J(y)) < 0$ and let $0 < \theta < -\frac{a}{1 - \varepsilon/4}$, $d_n^\theta = e^{-\theta n^{1/s} C_n}$. We now have:

Lemma 2.4.31 *For P – a.e. ω, there is a $J_1(\omega)$ s.t. for all $j \ge J_1(\omega)$, all $i \in I_j$,*

$$P_\omega^{ik}\left(\overline{T}_{(i+1)k_j} \ne T_{(i+1)k_j}\right) \le d_{n_j}^\theta .$$

Proof of Lemma 2.4.31. Again, we use the Chebycheff bound:

$$P\left(P_\omega^{ik}\left(\overline{T}_{(i+1)k_j} \neq T_{(i+1)k_j}\right) > d_{n_j}^\theta, \text{ some } i \in I_j\right)$$

$$\leq \frac{1}{d_{n_j}^\theta} \cdot \frac{3n_j}{k_j} \, \mathbb{P}^o\left(\overline{T}_{k_j} \neq T_{k_j}\right)$$

$$\leq \frac{1}{d_{n_j}^\theta} \cdot \frac{3n_j}{k_j} \cdot \exp\left(-k_j a(1 - \varepsilon/2)\right)$$

$$\leq 3 \, n_j^{1-\frac{1}{s}-\delta} \exp\left(n_j^{\frac{1}{s}+\delta}\left(\frac{a}{1 - \varepsilon/4} + \theta\right)\right),$$

where the second inequality follows again from (2.1.4) and the LDP for $\overline{R}_k$. The conclusion follows from the Borel-Cantelli lemma. $\qquad\square$

We need one more preliminary computation related to the bounds in (2.4.30). Let $\{Z_{k_j}^{(i)}\}$, $i = 1, 2, \ldots$ denote a sequence of i.i.d. positive random variables, with

$$P\left(\frac{Z_{k_j}^{(i)}}{k_j} < \mu'\right) = 0, \quad P\left(\frac{Z_{k_j}^{(i)}}{k_j} > \mu'x\right) = \left(2b_{n_j}\right)^{[x/2]\vee 1}, \quad x \geq 1.$$

Note now that for any $\lambda > 0$, and any $\varepsilon > 0$,

$$E\left(\exp\left(\lambda \frac{Z_{k_j}^{(i)}}{k_j}\right)\right) = \int_0^\infty P\left(\frac{Z_{k_j}^{(i)}}{k_j} > \frac{\log u}{\lambda}\right) du$$

$$\leq e^{\lambda\mu'(1+\varepsilon)} + \int_{e^{\lambda\mu'(1+\varepsilon)}}^\infty \left(2b_{n_j}\right)^{\left[\frac{\log u}{2\lambda\mu'(1+\varepsilon)}\right]\vee 1} du$$

$$= e^{\lambda\mu'(1+\varepsilon)} + g_j. \tag{2.4.32}$$

where $g_j \to_{j\to\infty} 0$.

In order to control the number of repetitions of visits to k_j–blocks, we introduce an auxiliary random walk. Let S_t, $t = 0, 1, \ldots$, denote a simple random walk with $S_0 = 0$ and

$$P\left(S_{t+1} = S_t + 1 \big| S_t\right) = 1 - P\left(S_{t+1} = S_t - 1 \big| S_t\right) = 1 - d_n^\theta.$$

Set $M_{n_j} = \frac{1}{C_{n_j}} n_j^{1-\frac{1}{s}}$.

Lemma 2.4.33 *For θ as in Lemma 2.4.31, and n large enough,*

$$P\left(\inf\{t : S_t = \left[\frac{n_j}{k_j}\right]\} > M_{n_j}\right) \leq \exp\left(-\frac{\theta\varepsilon}{2} n_j\right).$$

Proof of Lemma 2.4.33.

$$P\left(\inf\left\{t: S_t = \left[\frac{n_j}{k_j}\right]\right\} > M_{n_j}\right) \leq P\left(\frac{S_{[M_{n_j}]}}{M_{n_j}} < \frac{n_j}{k_j M_{n_j}}\right)$$

$$= P\left(\frac{S_{[M_{n_j}]}}{M_{n_j}} < 1 - \varepsilon\right) \leq 2\,e^{-M_{n_j} h_{n_j}(1-\varepsilon)},$$

where the last inequality is a consequence of Chebycheff's inequality and the fact that $d_n^\theta < \varepsilon$. Here,

$$h_n(1-x) = (1-x)\log\left(\frac{1-x}{1-d_n^\theta}\right) + x\log\frac{x}{d_n^\theta}.$$

Using $h_n(1-x) \geq -\frac{2}{e} - x\log d_n^\theta$, we get

$$P\left(\frac{S_{[M_{n_j}]}}{M_{n_j}} < 1 - \varepsilon\right) \leq 2\,e^{2M_{n_j}/e}\,e^{+\varepsilon M_{n_j}\log d_{n_j}^\theta} \leq e^{-\frac{\varepsilon}{2}\theta n_j}. \qquad \square$$

We are now ready to prove (2.4.26). Note that, for all $j > J_0(\omega)$, and all $i \in I_j$, we may, due to (2.4.30), construct $\{Z_{k_j}^{(i)}\}$ and $\{\overline{T}_{(i+1)k_j}\}$ on the same probability space such that for all $i \in I_j$, $P_\omega^{ik}(Z_{k_j}^{(i)} \geq \overline{T}_{(i+1)k_j}) = 1$. Fix $1/v_P > 1/v' > 1/v$ and $\varepsilon > 0$ small enough. Recalling that under the law P_ω^o, the random variables $\overline{T}_{k_j}^{(i)} := \overline{T}_{(i+1)k_j} - T_{ik_j}$ are independent, we obtain, with $\{S_t\}$ defined before Lemma 2.4.33, and j large enough,

$$P_\omega^o(T_{n_j} > n_j/v) \leq P\left(\inf\left\{t: S_t = \left[\frac{n_j}{k_j}\right]\right\} > M_{n_j}\right) + P\left(\sum_{i=1}^{M_{n_j}} Z_{k_j}^{(i)} > n_j/v\right)$$

$$\leq e^{-\theta\varepsilon n_j/2} + P\left(\frac{1}{M_{n_j}}\sum_{i=1}^{M_{n_j}}\frac{Z_{k_j}^{(i)}}{k_j} > 1/v(1-\varepsilon)\right)$$

$$\leq e^{-\theta\varepsilon n_j/2} + \left[E\left(\exp\left(\lambda\frac{Z_{k_j}^{(i)}}{k_j^{(i)}}\right)\right)\cdot e^{-\lambda(1-\varepsilon)/v}\right]^{M_{n_j}}$$

$$\leq e^{-\theta\varepsilon n_j/2} + \left(e^{\lambda(1/v'+2\varepsilon/v-1/v)} + g_j e^{-\lambda(1-\varepsilon)/v}\right)^{M_{n_j}}$$

$$\leq e^{-\theta\varepsilon n_j/2} + \left(e^{-\lambda\varepsilon/v}\right)^{M_{n_j}},$$

where Lemma 2.4.33 was used in the second inequality and (2.4.32) in the fourth. Since $\lambda > 0$ is arbitrary, (2.4.26) follows. $\qquad \square$

Remarks: 1. A study of the proof of the annealed estimates shows that the strong mixing condition (**D3**) can be replaced by the sightly milder one that $\alpha(n) = \exp(-Cn)$ for some C large enough such that (2.4.15) holds, *if* one also assumes the existence of a LDP for $\overline{R}_k$. In this form, the assumption is satisfied for many Markov chains satisfying a Doeblin condition.

2. It is worthwhile noting that the transfer of the annealed estimates to the quenched setting required very few assumptions on the environment, besides the existence of a LDP for $\overline{R}_k$. This technique, as we will see, is not limited to the one-dimensional setup, and works well in situations where a drift is present.

3. One may study by similar techniques also the case where $E_P(\overline{S}) < \infty$ but $\rho_{max} = 1$ with $\alpha := P(\rho_{max} = 1) > 0$. The rate of decay is then quite different: at least when the environment is i.i.d., the annealed rate of decay in Theorem 2.4.3 is exponential with exponent $n^{1/3}$, see [18], whereas the quenched one has exponent $n/(\log n)^2$, see [30], and it seems both proofs extend to the mixing setup. By adapting the method of enlargement of obstacles to this setup, one actually can show more in the i.i.d. environment case: it holds then that,

$$\lim_{\delta \to 0} \lim_{n \to \infty} \frac{1}{n^{1/3}} \log \mathbb{P}^o \left(\frac{X_n}{n} \in (v - \delta, v + \delta) \right) = -\frac{3}{2} \left| \frac{\pi \log \alpha}{2} \right|^{2/3}, \qquad (2.4.34)$$

and

$$\lim_{\delta \to 0} \lim_{n \to \infty} \frac{(\log n)^2}{n} \log P^o \left(\frac{X_n}{n} \in (v - \delta, v + \delta) \right) = -\frac{|\pi \log \alpha|^2}{8} (1 - \frac{v}{v_P}), \qquad (2.4.35)$$

see [60] and [61]. (Note that the lower bounds in (2.4.34) and (2.4.35) are not hard to obtain, by constructing "neutral" traps. The difficulty lies in matching the constants in the upper bound to the ones in the lower bound.) The technique of enlargement of obstacles in this context is based on considerably refining the classification of blocks used above when going from annealed to quenched estimates, by introducing the notion of "good" and "bad" blocks (and double blocks...)

4. One can check, at least in the i.i.d. environment case, that when $\rho_{max} = 1$ with $\alpha = 0$ then intermediate decay rates, between Theorems 2.4.3, 2.4.23 and (2.4.34), (2.4.35) can be achieved. We do not elaborate further here.

5. Again in the case of i.i.d. environment and the setup of Theorem 2.4.23, one can show, c.f. [30], that

$$\limsup_{n \to \infty} \frac{1}{n^{1-1/s}} \log P^o_\omega \left(\frac{X_n}{n} \in (v - \delta, v + \delta) \right) = 0, \quad P - a.s. \qquad (2.4.36)$$

This is due to fluctuations in the length of the "significant" trap where the walk may stay for large time. Based on the study of these fluctuations, it is reasonable to conjecture that

$$\liminf_{n \to \infty} \frac{1}{n^{1-1/s}} \log P^o_\omega \left(\frac{X_n}{n} \in (v - \delta, v + \delta) \right) = -\infty, \quad P - a.s.,$$

explaining the need for δ in the statement of Theorem 2.4.23. This conjecture has been verified only in the case where $P(\rho_{min} = 0) > 0$, i.e. in the presence of "reflecting nodes", c.f. [29, 28].

Bibliographical notes: The derivation in this section is based on [18] and [30]. Other relevant references, giving additional information not described here, are described in the remarks at the end of the section, so we only mention them here without repeating the description given there: [29, 60, 61].

2.5 Sinai's model: non standard limit laws and aging properties

Throughout this section, define $\overline{R}_k = k^{-1} \sum_{i=1}^{k-1} \log \rho_i(\text{sign } i)$. We assume the following

Assumption 2.5.1

(E1) *Assumption 2.1.1 holds.*

(E2) $E_P \log \rho_0 = 0$, *and there exists an* $\varepsilon > 0$ *such that* $E_P |\log \rho_0|^{2+\varepsilon} < \infty$.

(E3) P *is strongly mixing, and the functional invariance principle holds for* $\sqrt{k}\,\overline{R}_k/\sigma_P$; *that is,* $\{\sqrt{k}\,\overline{R}_{[kt]}/\sigma_P\}_{t \in \mathbb{R}}$ *converges weakly to a Brownian motion for some* $\sigma_P > 0$ *(sufficient conditions for such convergence are as in Lemma 2.2.4).*

(In the i.i.d. case, note that $\sigma_P^2 = E_P(\log \rho_0)^2$). Define

$$W^n(t) = \frac{1}{\log n} \sum_{i=0}^{\lfloor (\log n)^2 t \rfloor} \log \rho_i \cdot (\text{sign } t)$$

with $t \in \mathbb{R}$. By Assumption 2.5.1, $\{W^n(t)\}_{t \in \mathbb{R}}$ converges weakly to $\{\sigma_P B_t\}$, where $\{B_t\}$ is a two sided Brownian motion.

Next, we call a triple (a, b, c) with $a < b < c$ a valley of the path $\{W^n(\cdot)\}$ if

$$W^n(b) = \min_{a \leq t \leq c} W^n(t),$$
$$W^n(a) = \max_{a \leq t \leq b} W^n(t),$$
$$W^n(c) = \max_{b \leq t \leq c} W^n(t).$$

The *depth* of the valley is defined as

$$d_{(a,b,c)} = \min(W^n(a) - W^n(b), W^n(c) - W^n(b)).$$

If (a, b, c) is a valley, and $a < d < e < b$ are such that

$$W^n(e) - W^n(d) = \max_{a \leq x < y \leq b} W^n(y) - W^n(x)$$

then (a, d, e) and (e, b, c) are again valleys, which are obtained from (a, b, c) by a *left refinement*. One defines similarly a *right refinement*. Define

$$c_0^n = \min\{t \geq 0: \quad W^n(t) \geq 1\}$$
$$a_0^n = \max\{t \leq 0: \quad W^n(t) \geq 1\}$$
$$W^n(b_0^n) = \min_{a_0^n \leq t \leq c_0^n} W^n(t).$$

(b_0^n is not uniquely defined, however, due to Assumption 2.5.1, with P-probability approaching 1 as $n \to \infty$, all candidates for b_0^n are within distance converging to 0 as $n \to \infty$; we define b_0^n then as the smallest one in absolute value.)

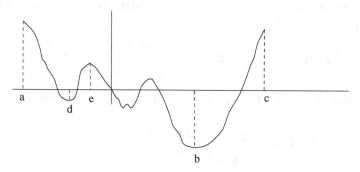

Fig. 2.5.1. Left refinement of (a, b, c)

One may now apply a (finite) sequence of refinements to find the *smallest* valley $(\overline{a}^n, \overline{b}^n, \overline{c}^n)$ with $\overline{a}^n < 0 < \overline{c}^n$, while $d_{(\overline{a}^n, \overline{b}^n, \overline{c}^n)} \geq 1$. We define similarly the smallest valley $(\overline{a}_\delta^n, \overline{b}_\delta^n, \overline{c}_\delta^n)$ such that $d_{(\overline{a}_\delta^n, \overline{b}_\delta^n, \overline{c}_\delta^n)} \geq 1 + \delta$. Let

$$A_n^{J,\delta} = \left\{ \begin{array}{l} \omega \in \Omega : \overline{b}^n = \overline{b}_\delta^n, \text{ any refinement } (a,b,c) \text{ of } (\overline{a}_\delta^n, \overline{b}_\delta^n, \overline{c}_\delta^n) \text{ with} \\ b \neq \overline{b}^n \text{ has depth } < 1 - \delta, |\overline{a}_\delta^n| + |\overline{c}_\delta^n| \leq J, \\ \min_{t \in [\overline{a}^n, \overline{c}^n] \backslash [\overline{b}^n - \delta, \overline{b}^n + \delta]} W^n(t) - W^n(\overline{b}^n) > \delta^3 \end{array} \right\}$$

then it is easy to check by the properties of Brownian motion that

$$\lim_{\delta \to 0} \lim_{J \to \infty} \lim_{n \to \infty} P(A_n^{J,\delta}) = 1. \tag{2.5.2}$$

The following is the main result of this section:

Theorem 2.5.3 *Assume* $P(\min(\omega_0^-, \omega_0^+) < \varepsilon) = 0$ *and Assumption 2.5.1. For any* $\eta > 0$,

$$\mathbb{P}^o\left(\left|\frac{X_n}{(\log n)^2} - \overline{b}^n\right| > \eta\right) \underset{n \to \infty}{\to} 0.$$

Proof. Fix $\delta < \eta/2, J$ and n large enough with $\omega \in A_n^{J,\delta}$. For simplicity of notations, assume in the sequel that ω is such that $\overline{b}^n > 0$. Write

$a^n = \bar{a}^n(\log n)^2, b^n = \bar{b}^n(\log n)^2, c^n = \bar{c}^n(\log n)^2$, with similar notations for $a_\delta^n, b_\delta^n, c_\delta^n$. Define

$$\overline{T}_{b,n} = \min\{t \geq 0 : X_t = b^n \text{ or } X_t = a_\delta^n\}.$$

By (2.1.4),

$$P_\omega^o\left(X_{\overline{T}_{b,n}} = a_\delta^n\right) \leq \frac{1}{1 + \frac{\exp\{(\log n)(W^n(\bar{a}_\delta^n) - W^n(\bar{b}^n))\}}{Jn(\log n)^2}} \leq \frac{J(\log n)^2}{n^\delta}. \quad (2.5.4)$$

On the other hand, let $\tilde{T}_{b,n}$ have the law of $\overline{T}_{b,n}$ except that the walk $\{X.\}$ is reflected at a_δ^n, and define similarly $\tilde{\tau}_1$. Using the same recursions as in (2.1.14), we have that

$$E_\omega^o(\tilde{\tau}_1) = \frac{1}{\omega_0^+} + \frac{\rho_0}{\omega_{(-1)}^+} + \cdots + \frac{\prod_{i=0}^{a_\delta^n+2} \rho_{-i}}{\omega_{a_\delta^n-1}^+} + \prod_{i=0}^{a_\delta^n+1} \rho_{-i}.$$

Hence, with $\tilde{\omega}_i = \omega_i$ for $i \neq a_\delta^n$ and $\tilde{\omega}_{a_\delta^n}^+ = 1$, for all n large enough,

$$E_\omega^o(\overline{T}_{b,n}) \leq E_\omega^o(\tilde{T}_{b,n}) = \sum_{i=1}^{b^n} \sum_{j=0}^{i-1-a_\delta^n} \frac{\prod_{k=1}^j \rho_{i-k}}{\omega_{(i-j-1)}^+}$$

$$\leq \frac{1}{\varepsilon} \sum_{i=1}^{b^n} \sum_{j=0}^{i-1-a_\delta^n} e^{(\log n)(W^n(i) - W^n(i-j))} \leq \frac{2J^2}{\varepsilon} e^{\log n(1-\delta)} \leq n^{1-\frac{\delta}{2}}.$$

We thus conclude that

$$P_\omega^o\left(\overline{T}_{b,n} < n, \quad X_{\overline{T}_{b,n}} = b^n\right) \underset{n\to\infty}{\longrightarrow} 1$$

implying that

$$P_\omega^o\left(T_{b^n} < n\right) \underset{n\to\infty}{\longrightarrow} 1. \quad (2.5.5)$$

Next note that another application of (2.1.4) yields

$$P_\omega^{b^n-1}(X. \text{ hits } b^n \text{ before } a_\delta^n) \geq 1 - n^{-(1+\frac{\delta}{2})}$$

$$P_\omega^{b^n+1}(X. \text{ hits } b^n \text{ before } c_\delta^n) \geq 1 - n^{-(1+\frac{\delta}{2})}. \quad (2.5.6)$$

On the same probability space, construct a RWRE $\{\tilde{X}_t\}$ with the same transition mechanism as $\{X_t\}$ except that it is reflected at a_δ^n, i.e. replace ω by $\tilde{\omega}$. Then, using (2.5.6),

$$P_\omega^o\left(\left|\frac{X_n}{(\log n)^2} - \bar{b}^n\right| > \delta\right) \leq P_\omega^o\left(T_{b^n} > n\right) + \max_{t \leq n} P_\omega^{b^n}\left(\left|\frac{X_t}{(\log n)^2} - \bar{b}^n\right| > \delta\right)$$

$$\leq P_\omega^o\left(T_{b^n} > n\right) + \left[1 - (1 - n^{-(1+\frac{\delta}{2})})^n\right]$$

$$+ \max_{t \leq n} P_\omega^{b^n}\left(\left|\frac{\tilde{X}_t}{(\log n)^2} - \bar{b}^n\right| > \delta\right).$$

Hence, in view of (2.5.2) and (2.5.5), the theorem holds as soon as we show that

$$\sup_{\omega \in A_n^{J,\delta}} \max_{t \leq n} P_\omega^{\bar{b}^n} \left(\left| \frac{\tilde{X}_t}{(\log n)^2} - \bar{b}^n \right| > \delta \right) \xrightarrow[n \to \infty]{} 0. \tag{2.5.7}$$

To see (2.5.7), define

$$f(z) = \frac{\prod_{a_\delta^n + 1 \leq i < z} \omega_i^+}{\prod_{a_\delta^n + 1 \leq i < z} \omega_{i+1}^-}, \quad \bar{f}(z) = \frac{f(z)}{f(b^n)},$$

(as usual, the product over an empty set of indices is taken as 1. $\bar{f}(\cdot)$ corresponds to the invariant measure for the resistor network corresponding to $\tilde{X}_\cdot$). Next, define the operator

$$(Ag)(z) = \bar{\omega}_{z-1}^+ g(z-1) + \bar{\omega}_{z+1}^- g(z+1) + \bar{\omega}_z^0 g(z) \tag{2.5.8}$$

where $\bar{\omega}_z = \omega_z$ for $z > a_\delta^n$, $\bar{\omega}_{a_\delta^n}^+ = 1, \bar{\omega}_{a_\delta^n - 1}^+ = 0$. Note that $A\bar{f} = \bar{f}$, and further that

$$P_\omega^{b^n}(\tilde{X}_t = z) = A^t \mathbf{1}_{b^n}(z).$$

Since $\bar{f}(z) \geq \mathbf{1}_{b^n}(z)$ and A is a positive operator, we conclude that

$$P_\omega^{b^n}(\tilde{X}_t = z) \leq \bar{f}(z).$$

But, for z with $|z/(\log n)^2 - \bar{b}^n| > \delta$, it holds that $\bar{f}(z) \leq e^{-\delta^3 \log n}$, and hence

$$P_\omega^{b^n}(\tilde{X}_t = z) \leq n^{-\delta^3}.$$

Thus, for $\omega \in A_n^{J,\delta}$, using the fact that the second inequality in (2.5.6) still applies for $\tilde{X}$,

$$\max_{t \leq n} P_\omega^{\bar{b}^n} \left(\left| \frac{\tilde{X}_t}{(\log n)^2} - \bar{b}^n \right| > \delta \right) \leq (\bar{b}^n + \delta)(\log n)^2 n^{-\delta^3} + 1 - \left(1 - n^{-(1+\delta/2)}\right)^n,$$

yielding (2.5.7) and completing the proof of the theorem.   □

We next turn to a somewhat more detailed study of the random variable $\bar{b}^n$. By replacing 1 with t in the definition of $\bar{b}^n$, one obtains a process $\{\bar{b}^n(t)\}_{t \geq 0}$. Further, due to Assumption 2.5.1, the process $\{\bar{b}^n(t/\sigma_P)\}_{t \geq 0}$ converges weakly to a process $\{\bar{b}(t)\}_{t \geq 0}$, defined in terms of the Brownian motion $\{B_t\}_{t \geq 0}$; Indeed, $\bar{b}(t)$ is the location of the bottom of the smallest valley of $\{B_t\}_{t \geq 0}$, which surrounds 0 and has depth t. Throughout this section we denote by $\mathcal{Q}$ the law of the Brownian motion $B_\cdot$. Our next goal is to characterize the process $\{\bar{b}(t)\}_{t \geq 0}$. Toward this end, define

$$m_+(t) = \min\{B_s : 0 \leq s \leq t\}, m_-(t) = \min\{B_{-s} : 0 \leq s \leq t\}$$
$$T_+(a) = \inf\{s \geq 0 : B_s - m_+(s) = a\},$$
$$T_-(a) = \inf\{s \geq 0 : B_{-s} - m_-(s) = a\}$$
$$s_\pm(a) = \inf\{s \geq 0 : m_\pm(T_\pm(a)) = B_{\pm s}\},$$
$$M_\pm(a) = \sup\{B_{\pm\eta} : \quad 0 \leq \eta \leq s_\pm(a)\}.$$

Next, define $W_\pm(a) = B_{s_\pm(a)}$. It is not hard to check that the pairs $(M_+(\cdot), W_+(\cdot))$ and $(M_-(\cdot), W_-(\cdot))$ form independent Markov processes. Define finally

$$H_\pm(a) = (W_\pm(a) + a) \vee M_\pm(a).$$

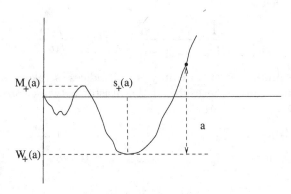

Fig. 2.5.2. The random variables $(M_+(a), W_+(a), s_+(a))$

We now have the

Theorem 2.5.9 *For each $a > 0$, $\mathcal{Q}(\bar{b}(a) \in \{s_+(a), -s_-(a)\}) = 1$. Further, $\bar{b}(a) = s_+(a)$ iff $H_+(a) < H_-(a)$.*

Proof. Note that $\mathcal{Q}(H_+(a) = H_-(a)) = 0$. That $\bar{b}(a) \in \{s_+(a), -s_-(a)\}$ is a direct consequence of the definitions, i.e. assuming $\bar{b}(a) > 0$ and $\bar{b}(a) \neq s_+(a)$ it is easy to show that one may refine from the right the valley defining $\bar{b}(a)$, contradicting minimality. We begin by showing, after Kesten [41], that $\bar{b}(a) = s_+(a)$ iff either

$$W_-(a) > W_+(a), \quad M_+(a) < (W_-(a) + a) \vee M_-(a) \tag{2.5.10}$$

or

$$W_-(a) < W_+(a), \quad M_-(a) > (W_+(a) + a) \vee M_+(a). \tag{2.5.11}$$

Indeed, assume $\bar{b}(a) = s_+(a)$, and $W_-(a) > W_+(a)$. Let $(\alpha, \bar{b}(a), \gamma)$ denote the minimal valley defining $\bar{b}(a)$. If $-s_-(a) \leq \alpha$, then

$$M_-(a) = \max\{B_{-s} : s \in (0, s_-(a))\} \geq B_{-\alpha}$$
$$= \max\{B_s : -\alpha \leq s \leq \bar{b}(a)\} \geq M_+(a) \tag{2.5.12}$$

implying (2.5.10). On the other hand, if $-s_-(a) > \alpha$, refine $(\alpha, \bar{b}(a), \gamma)$ on the left (find α', β' with $\alpha < \alpha' < \beta' < \bar{b}(a)$), such that

$$B_{\beta'} - B_{\alpha'} = \max_{\alpha < x < y < \bar{b}(\alpha)} (B_y - B_x) \geq M_+(a) - W_-(a)$$

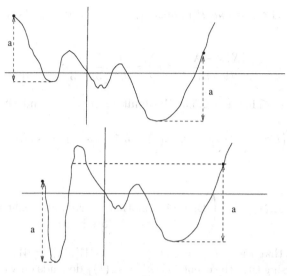

Fig. 2.5.3. $\bar{b}(a) = s_+(a)$

and thus minimality of $(\alpha, \bar{b}(a), \gamma)$ implies that $M_+(a) - W_-(a) < a$, implying (2.5.10).

We thus showed that if $\bar{b}(a) = s_+(a)$ and $W_-(a) > W_+(a)$ then (2.5.10) holds. On the other hand, if (2.5.10) holds, we show that $\bar{b}(a) = s_+(a)$ by considering the cases $\alpha \le -s_-(a)$ and $-s_-(a) < \alpha$ separately. In the former case, necessarily $\gamma > s_+$, for otherwise $M_-(\alpha) \le B_\gamma \le M_+(a) \le W_-(a) + a$ which together with $\bar{b}(a) = -s_-(a)$ would imply that the depth of $(\alpha, \bar{b}(a), \gamma)$ is smaller than a. Thus, under (2.5.10) if $\alpha \le -s_-(a)$ then $\gamma > s_+$, and in this case $\bar{b}(a) = s_+(a)$ since $B_{s_+(a)} < B_{-s_-(a)}$. Finally, if $\alpha > -s_-(a)$ then $\bar{b}(a) \ne -s_-(a)$ and hence $\bar{b}(a) = s_+(a)$.

Hence, we showed that if $W_-(a) > W_+(a)$ then (2.5.10) is equivalent to $\bar{b}(a) = s_+(a)$. Interchanging the positive and negative axis, we conclude that if $W_-(a) < W_+(a)$, then $\bar{b}(a) = -s_-(a)$ iff $M_+(a) < (W_+(a)+1) \vee M_+(a)$. This completes the proof that $\bar{b}(a) = s_+(a)$ is equivalent to (2.5.10) or (2.5.11).

To complete the proof of the theorem, assume first $W_-(a) > W_+(a)$. Then, $\bar{b}(a) = s_+(a)$ iff (2.5.10) holds, i.e. $M_+(a) < (W_-(a) + a) \vee M_-(a) = H_-(a)$. But $H_-(a) \ge W_-(a) + a \ge W_+(a) + a$, and hence $M_+(a) < H_-(a)$ is equivalent to $M_+(a) \vee (W_+(a) + a) < H_-(a)$, i.e. $H_+(a) < H_-(a)$. The case $W_+(a) < W_-(a)$ is handled similarly by using (2.5.11). □

One may use the representation in Theorem 2.5.9 in order to evaluate explicitly the law of $\bar{b}(a)$ (note that $\bar{b}(a) \overset{\mathcal{L}}{=} a^2 \bar{b}(1)$ by Brownian scaling). This is done in [41], and we do not repeat the construction here. Our goal is to use Theorem 2.5.9 to show that Sinai's model exhibits *aging* properties. More precisely, we claim that

Theorem 2.5.13 *Assume $P(\min(\omega_0, \omega_0^+) < \varepsilon) = 0$ and Assumption 2.5.1. Then, for $h > 1$,*

$$\lim_{\eta \to 0} \lim_{n \to \infty} \mathbb{P}^o \left(\frac{|X_{n^h} - X_n|}{(\log n)^2} < \eta \right) = \frac{1}{h^2} \left[\frac{5}{3} - \frac{2}{3} e^{-(h-1)} \right]. \qquad (2.5.14)$$

Proof. Applying Theorem 2.5.3, the limit in the left hand side of (2.5.14) equals

$$\mathcal{Q}\left(\overline{b}(h) = \overline{b}(1) \right) = 2\mathcal{Q}\left(\overline{b}(h) = \overline{b}(1) = s_+(1) = s_+(h) \right).$$

Note that

$$\mathcal{Q}(s_+(h) = s_+(1)) = \mathcal{Q}\left(\begin{array}{c} \text{Brownian motion, started at height 1,} \\ \text{hits } h \text{ before hitting } 0 \end{array} \right) = \frac{1}{h}.$$

Hence, using that on $s_+(1) = s_+(h)$ one has $W_+(1) = W_+(h), M_+(1) = M_+(h)$, and using that the event $\{s_+(h) = s_+(1)\}$ depends only on increments of the path of the Brownian motion after time $T_+(1)$, one gets

$$\mathcal{Q}\left(\overline{b}(h) = \overline{b}(1) \right) = \frac{2}{h} \mathcal{Q}\left(H_+(1) < H_-(1), (W_+(1) + h) \vee M_+(1) < H_-(h) \right). \qquad (2.5.15)$$

Next, let

$$\tau_0 = \min\{t > s_-(1) : B_{-t} = W_-(1) + 1\}$$
$$\tau_h = \min\{t > \tau_0 : B_{-t} = W_-(1) + h \text{ or } B_t = W_-(1)\}.$$

Note that $\tau_h - \tau_0$ has the same law as that of the hitting time of $\{0, h\}$ by a Brownian motion Z_t with $Z_0 = 1$. (Here, $Z_t = B_{-(\tau_0 + t)} - W_-(1)$!). Further, letting $I_h = \mathbf{1}_{\{B_{\tau_h} = W_-(1)\}} (= \mathbf{1}_{\{Z_{\tau_h - \tau_0} = 0\}})$, it holds that

$$W_-(h) = W_-(1) + I_h \tilde{W}_-(h)$$

$$M_-(h) = \begin{cases} M_-(1), & I_h = 0 \\ M_-(1) \vee (\overline{M}_-(h) + W_-(1) + 1) \vee (\tilde{M}_-(h) + W_-(1)), & I_h = 1 \end{cases}$$

where $(\tilde{W}_-(h), \tilde{M}_-(h))$ are independent of $(W_-(1), M_-(1))$ and possess the same law as $(W_-(h), M_-(h))$, while $\overline{M}_-(h)$ is independent of both $(W_-(1), M_-(1))$ and $(\tilde{W}_-(h), \tilde{M}_-(h))$ and has the law of the maximum of a Brownian motion, started at 0, killed at hitting -1 and conditioned not to hit $h - 1$. (See figure 2.5.4 for a graphical description of these random variables.)

Set now

$$\hat{M}_-(h) = \begin{cases} h, & I_h = 0 \\ 1 + \overline{M}_-(h), & I_h = 1, \end{cases}$$

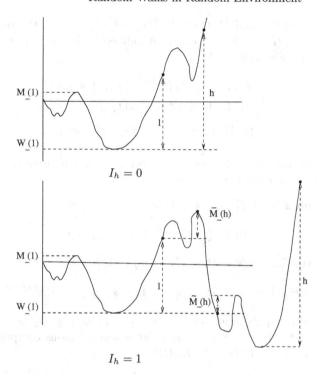

$M_-(1)$

$W_-(1)$

$I_h = 0$

$\tilde{M}_-(h)$

$M_-(1)$

$W_-(1)$

$\hat{M}_-(h)$

$I_h = 1$

Fig. 2.5.4. Definition of auxiliary variables

$\tilde{H}_-(h) = (\tilde{W}_-(h) + h) \vee \tilde{M}_-(h)$ and $\Gamma(h) = \max(\tilde{H}_-(h), \hat{M}_-(h))$. Note that $\tilde{H}_-(h)$ has the same law as $H_-(h)$ but is independent of $\overline{M}_-(h)$. Further, it is easy to check that $(W_-(h) + h) \vee M_-(h) = (W_-(1) + \Gamma_h) \vee M_-(1)$ (note that either $M_-(h) = M_-(1)$ or $M_-(h) > M_-(1)$ but in the latter case, $M_-(h) \leq W_-(1) + \Gamma(h)$.) We have the following lemma, whose proof is deferred:

Lemma 2.5.16 *The law of $\Gamma(h)$ is $\frac{1}{h}\delta_h + \frac{h-1}{h}U[1, h]$, where $U[1, h]$ denotes the uniform law on $[1, h]$.*

Substituting in (2.5.15), we get that

$$\mathcal{Q}(\overline{b}(h) = \overline{b}(1)) = \mathcal{Q}\left(E_{\mathcal{Q}}(\overline{b}(h) = \overline{b}(1)|\Gamma(h))\right) = \frac{2}{h^2}\left[\int_1^h Q(t)dt + Q(h)\right]$$
(2.5.17)

where

$$Q(t) = \mathcal{Q}\left(H_+(1) < H_-(1), H_+(h) < H_-(t) \,|\, s_+(h) = s_+(1), s_-(1) = s_-(t)\right).$$

In order to evaluate the integral in (2.5.17), we need to evaluate the joint law of $(H_+(1), H_+(t))$ (the joint law of $(H_-(1), H_-(t))$ being identical). Since

$0 \leq H_+(1) \leq 1$ and $H_+(1) \leq H_+(t) \leq H_+(1) + t - 1$, the support of the law of $(H_+(1), H_+(t))$ is the domain A defined by $0 \leq x \leq 1$, $x \leq y \leq x + t - 1$. Note that for $(z, w) \in A$,

$$\mathfrak{Q}(H_+(1) \leq z, H_+(t) \leq w \mid s_+(1) = s_+(h))$$
$$= \mathfrak{Q}(M_+(1) \leq z \wedge w, W_+(1) \leq -[(1 - z) \vee (t - w)]$$
$$= \mathfrak{Q}\left(M_+(1) \leq z, W_+(1) \leq -(t - w)\right).$$

We now have the following well known lemma. For completeness, the proof is given at the end of this section:

Lemma 2.5.18 For $z + y \geq 1$, $0 \leq z \leq 1$, $y \geq 0$,

$$\mathfrak{Q}(M_+(1) \leq z, W_+(1) \leq -y) = ze^{-(z+y-1)}.$$

Lemma 2.5.18 implies that, for $(z, w) \in A$, $t > 1$,

$$\mathfrak{Q}(H_+(1) \leq z, H_+(t) \leq w \mid s_+(1) = s_+(h)) = ze^{-(z+t-w-1)}. \qquad (2.5.19)$$

Denote by B_1 the segment $\{0 \leq x = y \leq 1\}$ and by B_2 the segment $\{t - 1 \leq y = x + t - 1 \leq t\}$. We conclude, after some tedious computations, that the conditional law of $(H_+(1), H_+(t))$:

- possesses the density $f(z, w) = (1 - z)e^{-z}e^{-w-(t-1)}$, $\quad (z, w) \in A \backslash (B_1 \cup B_2)$
- possesses the density $\tilde{f}(z, y) = (1 - z)e^{-(t-1)}$, $\quad z = w \in B_1$
- possesses the density $\overline{f}(z, z + t - 1) = z$, $\quad w = z + t - 1 \in B_2$.

Substituting in the expression for $Q(t)$, we find that

$$Q(t) = \frac{5}{12}e^{-(h-t)} + \frac{1}{12}e^{-(h+t-2)}.$$

Substituting in (2.5.19), the theorem follows. $\qquad \square$

Proof of Lemma 2.5.16: Note that $\mathfrak{Q}(I_h = 0) = 1/h$, and in this case $\Gamma_h = h$. Thus, we only need to consider the case where $I_h = 1$ and show that under this conditioning, $\max(H_-(h), 1 + \overline{M}_-(h))$ possesses the law $U[1, h]$. Note that by standard properties of Brownian motion,

$$\mathfrak{Q}(\hat{M}_-(h) \leq \xi \mid I_h = 1) = \frac{\frac{\xi-1}{\xi}}{\frac{h-1}{h}}.$$

We show below that the law of $\tilde{H}_-(h)$, which is identical to the law of $H_-(h)$, is uniform on $[0, h]$. Thus, using independence, for $\xi \in [1, h]$,

$$\mathfrak{Q}(\Gamma_h < \xi \mid I_h = 1) = \frac{h(\xi - 1)\xi}{\xi(h - 1)h} = \frac{\xi - 1}{h - 1},$$

i.e. the law of Γ_h conditioned on $I_h = 1$ is indeed $U[1, h]$.

It thus only remains to evaluate the law of $H_-(h)$. By Brownian scaling, the law of $H_-(h)$ is identical to the law of $hH_+(1)$, so we only need show that the law of $H_+(1)$ is uniform on $[0, 1]$. This in fact is a direct consequence of Lemma 2.5.18. $\qquad\square$

Proof of Lemma 2.5.18:   Let Q^x denote the law of a Brownian motion $\{Z_t\}$ starting at time 0 at x. The Markov property now yields

$$Q(M_+(1) \leq z, W_+(1) \leq -y) = Q^o(\{Z.\} \text{ hits } z - 1 \text{ before hitting } z)$$
$$Q^{z-1}(M_+(1) \leq z, W_+(1) \leq -y)$$
$$= zQ^o(M_+(1) \leq 1, W_+(1) \leq -y - z + 1)$$
$$= zQ^o(W_+(1) \leq -(y + z - 1)). \qquad (2.5.20)$$

For $x \geq 0$, let $f(x) := Q(W_+(1) \leq -x)$. The Markov property now implies

$$f(x + \epsilon) = f(x)Q^{-x}(W_+(1) \leq -(x + \epsilon)) = f(x)f(\epsilon).$$

Since $f(0) = 1$ and $f(\epsilon) = 1 - \epsilon + o(\epsilon)$, it follows that $f(x) = e^{-x}$. Substituting in (2.5.20), the lemma follows. $\qquad\square$

Bibliographical notes: Theorem 2.5.3 is due to [66]. The proof here follows the approach of Golosov [31], who dealt with a RWRE reflected at 0, i.e. with state space $\mathbb{Z}_+$. In the same paper, Golosov evaluates the analogue of Theorem 2.5.9 in this reflected setup, and in [32] he provides sharp (pathwise) localization results. These are extended to the case of a walk on $\mathbb{Z}$ in [33]. The statement of Theorem 2.5.9 and the proof here follow the article [41], where an explicit characterization of the law of $\bar{b}(1)$ is provided. The same characterization appears also in [33]. The aging properties of RWRE (Theorem 2.5.13) were first derived heuristically in [24], to which we refer for additional aging properties and discussion. The derivation here is based on [17]. The right hand side of formula (2.5.14) appears also in [33], in a slightly different context. We mention that results of iterated logarithm types, and results concerning most visited sites for Sinai's RWRE, can be found in [35], [36]. See [65] for a recent review. Finally, extensions of the results in this section and a theorem concerning the dichotomy between Sinai's regime and the classical CLT for ergodic environments can be found in [7].

Limit laws for transient RWRE in an i.i.d. environment appear in [42]. One distinguishes between CLT limit laws and stable laws: recall the parameter s introduced in Section 2.4. The main result of [42] is that if $s > 2$, a CLT holds true (see Section 2.2 for other approaches), whereas for $s \in (0, 2)$ a Stable(s) limit law holds true. Note that this is valid even when $s < 1$, i.e. when $v_P = 0$! It is an interesting open problem to extend the results concerning stable limit laws to non i.i.d. environments. Some results in this direction are forthcoming in the Technion thesis of A. Roitershtein.

3 RWRE – $d > 1$

3.1 Ergodic Theorems

In this section we present some of the general results known concerning $0 - 1$ laws and laws of large numbers for nearest neighbour RWRE in $\mathbb{Z}^d$. Even if considerable progress was achieved in recent years, the situation here is, unfortunately, much less satisfying than for $d = 1$.

A standing assumption throughout this section is the following:

Assumption 3.1.1

(A1) *P is stationary and ergodic, and satisfies a ϕ-mixing condition: there exists a function $\phi(l) \underset{l\to\infty}{\to} 0$ such that any two l-separated events A, B with $P(A) > 0$,*

$$\left| \frac{P(A \cap B)}{P(A)} - P(B) \right| \leq \phi(l).$$

(A2) *P is uniformly elliptic: there exists an $\varepsilon > 0$ such that*

$$P(\omega(0, e) \geq \varepsilon) = 1, \qquad \forall e \in \{\pm e_i\}_{i=1}^d.$$

(Events A, B are *l-separated* if the shortest lattice path connecting A and B is of length l or more.)

Remark: I have recently learnt that Assumption **(A1)** implies, in fact, that P is finitely dependent, c.f. [5]. On the other hand, the basic structure of what appears in the rest of this section remains unchanged if P is *mixing on cones*, see [13], and thus I have kept the proof in its original form.

Fix $\ell \in \mathbb{R}^d \setminus \{0\}$, and consider the events

$$A_{\pm \ell} = \{ \lim_{n \to \infty} X_n \cdot \ell = \pm \infty \}.$$

We have the

Theorem 3.1.2 *Assume Assumption 3.1.1. Then*

$$\mathbb{P}^o(A_\ell \cup A_{-\ell}) \in \{0, 1\}.$$

Proof. We begin by constructing an extension of our probability space: recall that the RWRE was defined by means of the law $\mathbb{P}^o = P \otimes P_\omega^o$ on $(\Omega \times (\mathbb{Z}^d)^{\mathbb{N}}, \mathcal{F} \times \mathcal{G})$. Set $W = \{0\} \cup \{\pm e_i\}_{i=1}^d$ and $\mathcal{W}$ the cylinder σ-algebra on $W^{\mathbb{N}}$. We now define the measure

$$\overline{\mathbb{P}}^o = P \otimes Q_\varepsilon \otimes \overline{P}_{\omega,\varepsilon}^o$$

on

$$\left(\Omega \times W^{\mathbb{N}} \times (\mathbb{Z}^d)^{\mathbb{N}}, \quad \mathcal{F} \times \mathcal{W} \times \mathcal{G} \right)$$

in the following way: Q_ε is a product measure, such that with $\varepsilon = (\varepsilon_1, \varepsilon_2, \ldots)$ denoting an element of $W^{\mathbb{N}}$, $Q_\varepsilon(\varepsilon_1 = \pm e_i) = \varepsilon$, $i = 1, \cdots, d$, $Q_\varepsilon(\varepsilon_1 = 0) = 1 - 2\varepsilon d$. For each fixed ω, ε, $\overline{P}^o_{\omega,\varepsilon}$ is the law of the Markov chain $\{X_n\}$ with state space $\mathbb{Z}^d$, such that $X_0 = 0$ and, for each $e \in W$, $e \neq 0$,

$$\overline{P}^o_{\omega,\varepsilon}(X_{n+1} = z + e | X_n = z) = \mathbf{1}_{\{\varepsilon_{n+1}=e\}} + \frac{\mathbf{1}_{\{\varepsilon_{n+1}=0\}}}{1 - 2d\varepsilon}[\omega(z, z + e) - \varepsilon].$$

It is not hard to check that the law of $\{X_n\}$ under $\overline{\mathbb{P}}^o$ coincides with its law under $\mathbb{P}^o$, while its law under $Q_\varepsilon \otimes \overline{P}^o_{\omega,\varepsilon}$ coincides with its law under P^o_ω.

We will prove the theorem for $\ell = (1, 0 \ldots 0)$, the general case being similar but requiring more cumbersome notations. Note that for any $u < v$, the walk cannot visit infinitely often the strip $u \leq z \cdot \ell \leq v$ without crossing the line $z \cdot \ell = v$. More precisely, with

$$T_v = \inf\{n \geq 0 : \quad X_n \cdot \ell \geq v\}, \tag{3.1.3}$$

we have

$$\mathbb{P}^o\Big(\#\{n > 0 : X_n \cdot \ell \geq u\} = \infty, T_v = \infty\Big) = 0. \tag{3.1.4}$$

Indeed, note that for any z with $u \leq z \cdot \ell \leq v$, and any ω,

$$P^z_\omega(X_{v-u} \cdot \ell \geq v) = Q_\varepsilon \otimes P^z_{\omega,\varepsilon}(X_{v-u} \cdot \ell \geq v) \geq \varepsilon^{v-u},$$

yielding (3.1.4) by the strong Markov property.

Assume next that $\mathbb{P}^o(A_\ell) > 0$. Set $D = \inf\{n \geq 0 : X_n \cdot \ell < X_0 \cdot \ell\}$. Clearly, $\mathbb{P}^o(D = \infty) > 0$, because if $\mathbb{P}^o(D = \infty) = 0$ then $\mathbb{P}^z(D < \infty) = 1$ $\forall z \in \mathbb{Z}^d$, and thus P-a.s., for all $z \in \mathbb{Z}^d$, $P^z_\omega(D < \infty) = 1$. This implies by the Markov property that

$$\liminf_{n \to \infty} X_n \cdot \ell \leq 0, \quad \mathbb{P}^o\text{-a.s.},$$

contradicting $\mathbb{P}^o(A_\ell) > 0$.

Define $\mathcal{O}_\ell$ to be the event that $X_n \cdot \ell$ changes its sign infinitely often. We next show that whenever $\mathbb{P}^o(A_\ell) > 0$, then $\mathbb{P}^o(\mathcal{O}_\ell) = 0$. Set $M = \sup_n X_n \cdot \ell$, fix $v > 0$ and note by (3.1.4) that

$$\mathbb{P}^o(\mathcal{O}_\ell \cap \{M < v\}) = 0. \tag{3.1.5}$$

We next prove that if $\mathbb{P}^o(A_\ell) > 0$ then $\mathbb{P}^o(\mathcal{O}_\ell \cap \{M = \infty\}) = 0$, by first noting that

$$\mathbb{P}^o(\mathcal{O}_\ell \cap \{M = \infty\}) = \overline{\mathbb{P}}^o(\mathcal{O}_\ell \cap \{M = \infty\}).$$

Then, set $\mathcal{G}_n = \sigma((\varepsilon_i, X_i), i \leq n)$, fix $L > 0$ and, setting $S_0 = 0$, define recursively $\mathcal{G}_n$ stopping times as follows:

$$R_k = \inf\Big\{n \geq S_k : X_n \cdot \ell < 0\Big\},$$

$$S_{k+1} = \inf\{n \geq R_k : X_{n-L} \cdot \ell$$

$$\geq \max\{X_m \cdot \ell : m \leq n - L\}, \varepsilon_{n-1} = \varepsilon_{n-2} = \ldots = \varepsilon_{n-L} = e_1\Big\}.$$

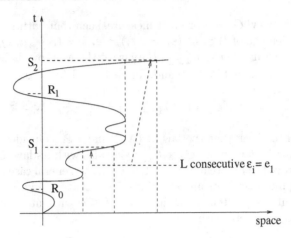

Fig. 3.1.1. Definition of the hitting times (S_k, R_k)

On $\mathcal{O}_\ell \cap \{M = \infty\}$, all these stopping times are finite. Now, at each time $S_k - L$ the walk enters a half space it never visited before, and then *due to the action of the ε sequence alone*, it proceeds L steps in the direction e_1. Formally, "events in the σ-algebra $\mathcal{G}_{S_k}$ are L-separated from $\sigma(\omega_z : z \cdot \ell \geq X_{S_k} \cdot \ell)$". Note that, using $\mathbb{P}^o(A_\ell) > 0$ in the second inequality,

$$\overline{\mathbb{P}}^o(R_0 < \infty) = \overline{\mathbb{P}}^o(D < \infty) < 1,$$

whereas, using θ to denote both time and space shifts as needed from the context,

$$\overline{\mathbb{P}}^o(R_1 < \infty) \leq \overline{\mathbb{P}}^o(R_0 < \infty, R_0 \circ \theta_{X_{S_1}} < \infty)$$

$$= \sum_{z \in \mathbb{Z}^d} \overline{\mathbb{P}}^o(R_0 < \infty, R_0 \circ \theta_z < \infty, X_{S_1} = z)$$

$$= \sum_{z \in \mathbb{Z}^d} \sum_{n \in \mathbb{N}} E_{P \otimes Q_\varepsilon}\left(\overline{P}^o_{\omega,\varepsilon}(R_0 < \infty, X_{S_1} = z, S_1 = n) \cdot \overline{P}^o_{\theta^z \omega, \theta^n \varepsilon}(R_0 < \infty)\right).$$

Note that $\overline{P}^o_{\theta^z \omega, \theta^n \varepsilon}(R_0 < \infty)$ is measurable on $\sigma(\omega_x : x \cdot \ell \geq z \cdot \ell) \times \sigma(\varepsilon_i, i > n)$, whereas $\overline{P}^o_{\omega,\varepsilon}(R_0 < \infty, X_{S_1} = z, S_1 = n)$ is measurable on $\sigma(\omega_x : x \cdot \ell \leq z \cdot \ell - L) \times \sigma(\varepsilon_i, i \leq n)$. Hence, by the ϕ-mixing property of P and the product structure of Q_ε,

$$\overline{P}^o(R_1 < \infty) \leq \sum_{z \in \mathbb{Z}^d} \sum_{n \in \mathbb{N}} \left[E_{P \otimes Q_\varepsilon}\left(\overline{P}^o_{\omega,\varepsilon}(R_0 < \infty, X_{S_1} = z, S_1 = n) \right) \right.$$

$$\left. \cdot E_{P \otimes Q_\varepsilon}\left(\overline{P}^o_{\omega,\varepsilon}(R_0 < \infty) \right) \right]$$

$$+ \phi(L) \sum_{z \in \mathbb{Z}^d} \sum_{n \in \mathbb{N}} E_{P \otimes Q_\varepsilon}\left(\overline{P}^o_{\omega,\varepsilon}(R_0 < \infty, X_{S_1} = z, S_1 = n) \right)$$

$$\leq (\overline{P}^o(R_0 < \infty))^2 + \phi(L)\overline{P}^o(R_0 < \infty) \leq (\overline{P}^o(D < \infty) + \phi(L))^2.$$

Repeating this procedure, we conclude that $\overline{P}^o(\mathcal{O}_\ell \cap \{M = \infty\}) \leq \overline{P}^o(R_k < \infty) \leq (\mathbb{P}^o(D < \infty) + \phi(L))^{k+1}$. Since k is arbitrary and $\phi(L) \underset{L \to \infty}{\to} 0$, we conclude that $\overline{P}^o(\mathcal{O}_\ell \cap \{M = \infty\}) = 0$, yielding with the above that $\mathbb{P}^o(\mathcal{O}_\ell) = 0$ as soon as $\mathbb{P}^o(A_\ell) > 0$. In a similar manner one proves that $\mathbb{P}^o(A_{-\ell}) > 0$ also implies $\mathbb{P}^o(\mathcal{O}_\ell) = 0$.

Assume now $1 > \mathbb{P}^o(A_\ell \cup A_{-\ell})$. Then one can find a v such that $\mathbb{P}^o(X_n \cdot \ell \in [-v, v]$ infinitely often) > 0. Therefore, $\mathbb{P}^o(\mathcal{O}_\ell) > 0$, implying by the above $\mathbb{P}^o(A_\ell) = \mathbb{P}^o(A_{-\ell}) = 0$.   □

Remark: It should be obvious that one does not need the full strength of **(A1)** in Assumption 3.1.1, and weaker forms of mixing suffice. For an example of how this can be relaxed, see [13].

Bibliographical notes: The 0-1 law described in this section is due to Kalikow [38], who handled the i.i.d. setup. Our proof borrows from [82], which, still in the i.i.d. case, relaxes the uniform ellipticity assumption A2. In that paper, they show that a stronger 0-1 law holds if P is a product measure and $d = 2$, namely they show that $\mathbb{P}^o(A_\ell) \in \{0,1\}$, while that last conclusion is false for certain mixing environments with elliptic, but not uniformly elliptic, environments.

3.2 A Law of Large Numbers in $\mathbb{Z}^d$

Our next goal is to prove a law of large numbers. Unfortunately, at this point we are not able to deal with general non i.i.d. environments (see however Remark 2 following the proof of Theorem 3.2.2), and further the case of i.i.d. environments does offer some simplifications. Thus, throughout this section we make the following assumptions:

Assumption 3.2.1 *P is a uniformly elliptic, i.i.d. law on Ω.*

The main result of this section is the following:

Theorem 3.2.2 *Assume Assumption 3.2.1 and that $\mathbb{P}^o(A_\ell \cup A_{-\ell}) = 1$. Then, there exist deterministic $v_\ell, v_{-\ell}$ (possibly zero) such that*

$$\lim_{n \to \infty} \frac{X_n \cdot \ell}{n} = v_\ell 1_{A_\ell} + v_{-\ell} 1_{A_{-\ell}}, \quad \mathbb{P}^o\text{-a.s.}$$

(See (3.2.8) for an expression for v_ℓ. When $v_\ell \neq 0$ for some ℓ, we say that the walk is *ballistic*).

Proof. As in Section 3.1 we will take here $\ell = (1, 0, \cdots, 0)$. Further, we assume throughout that $\mathbb{P}^o(A_\ell) > 0$. The proof is based on introducing a renewal structure, as follows: Define $\overline{S}_0 = 0, M_0 = \ell \cdot X_0$,

$$\overline{S}_1 = T_{M_0+1} \leq \infty, \quad \overline{R}_1 = D \circ \theta_{\overline{S}_1} + \overline{S}_1 \leq \infty,$$
$$M_1 = \sup\{\ell \cdot X_m, \quad 0 \leq m \leq \overline{R}_1\} \leq \infty$$

and by induction, for $k \geq 1$,

$$\overline{S}_{k+1} = T_{M_k+1} \leq \infty, \quad \overline{R}_{k+1} = D \circ \theta_{\overline{S}_{k+1}} + \overline{S}_{k+1} \leq \infty,$$
$$M_{k+1} = \sup\{\ell \cdot X_m, \quad 0 \leq m \leq \overline{R}_{k+1}\} \leq \infty.$$

The times $\overline{S}_1, \overline{S}_2, \ldots$, are called "fresh times", and the locations $X_{\overline{S}_1}, X_{\overline{S}_2}$, $\cdots$, are "fresh points": at the time $\overline{S}_k$, the path X. visits for the first time after $\overline{S}_{k-1}$ and after hitting again the hyperplane $X_{\overline{S}_{k-1}} \cdot \ell - 1$, a fresh part of the environment. Note that $(\overline{S}_i, \overline{R}_i)$ are related to, but differ slightly from, (S_i, R_i) introduced in Section 3.1. Clearly,

$$0 = \overline{S}_0 \leq \overline{S}_1 \leq \overline{R}_1 \leq \overline{S}_2 \leq \cdots \leq \infty$$

and the inequalities are strict if the left member is finite. Define:

$$K = \inf\{k \geq 1 : \overline{S}_k < \infty, \overline{R}_k = \infty\} \leq \infty,$$
$$\tau_1 = \overline{S}_K \leq \infty.$$

τ_1 is called a "regeneration time", because after τ_1, $X \cdot \ell$ never falls behind $X_{\tau_1} \cdot \ell$.

By the same argument as in the proof of Theorem 3.1.2, $\mathbb{P}^o(\overline{R}_k < \infty) \leq \mathbb{P}^o(D < \infty)^k \xrightarrow[k \to \infty]{} 0$ because $\mathbb{P}^o(A_\ell) > 0$ implies $\mathbb{P}^o(D < \infty) < 1$. On the other hand, on A_ℓ, $\overline{R}_k < \infty \Rightarrow S_{k+1} < \infty$, $\mathbb{P}^o$-a.s., and hence

$$\mathbb{P}^o(A_\ell \cap \{K = \infty\}) = \mathbb{P}^o(A_\ell \cap \{\tau_1 = \infty\}) = 0.$$

Define now the measure

$$\mathbb{Q}^o(\cdot) = \mathbb{P}^o(\cdot \mid \{\tau_1 < \infty\}) = \mathbb{P}^o(\cdot \mid A_\ell)$$

and set

$$\mathcal{G}_1 = \sigma\left(\tau_1, X_0, \cdots, X_{\tau_1}, \{\omega(y, \cdot)\}_{\ell \cdot y < \ell \cdot X_{\tau_1}}\right).$$

Note that since $\{D = \infty\} \subset \{\tau_1 < \infty\}$, we have that $\{D = \infty\} \in \mathcal{G}_1$. We have the following crucial lemma, whose proof is a simple exercise in the application of the Markov property, is omitted. It is here that the i.i.d. assumption on the environment plays a crucial role:

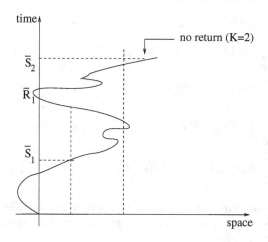

Fig. 3.2.1. Regeneration structure

Lemma 3.2.3 *For any measurable sets A, B,*

$$\mathbb{Q}^o\Big(\{X_{\tau_1+n} - X_{\tau_1}\}_{n\geq 0} \in A, \{\omega(X_{\tau_1} + y, \cdot)\}_{y\cdot\ell\geq 0} \in B\Big)$$
$$=\mathbb{P}^o\Big(\{X_n\}_{n\geq 0} \in A, \{\omega(y, \cdot)\}_{y\cdot\ell\geq 0} \in B|\{D = \infty\}\Big).$$

In fact,

$$\mathbb{Q}^o\Big(\{X_{\tau_1+n} - X_{\tau_1}\}_{n\geq 0} \in A, \{\omega(X_{\tau_1} + y, \cdot)\}_{y\cdot\ell\geq 0} \in B|\mathcal{G}_1\Big)$$
$$=\mathbb{P}^o\Big(\{X_n\}_{n\geq 0} \in A, \{\omega(y, \cdot)\}_{y\cdot\ell\geq 0} \in B|\{D = \infty\}\Big). \qquad (3.2.4)$$

Proof of Lemma 3.2.3 Clearly, it suffices to prove (3.2.4). Let h denote a $\mathcal{G}_1$ measurable random variable. Set $1_A := 1_{\{X_n-X_0\}_{n\geq 0}\in A}$, $1_B := 1_{\{\omega(y,\cdot)\}_{y\cdot\ell\geq 0}}$. Further, note that for each $k \in \mathbb{N}$, $x \in \mathbb{Z}^d$, there exists a random variable $h_{x,k}$, measurable with respect to $\sigma(\{\omega(y, \cdot)\}_{\ell\cdot y<x\cdot\ell}, \{X_i\}_{i\leq\overline{S}_k}, \overline{S}_k)$, such that on the event $\{\tau_1 = \overline{S}_k, X_{\overline{S}_k} = x\}$, $h = h_{x,k}$ (this follows from the $\mathcal{G}_1$ measurability of h. Then, using $\bar\theta$ to denote spatial shift and θ to denote temporal shift,

$$E_{\mathbb{P}^o}\left(1_A \circ \theta^{\tau_1} \cdot 1_B \circ \bar\theta^{X_{\tau_1}} \cdot h \cdot 1_{\tau_1 < \infty}\right)$$

$$= \sum_{k \geq 1} \sum_{x \in \mathbb{Z}^d} E_P\left(E_\omega^o\left(1_{\overline{S}_k < \infty} 1_{\overline{R}_k = \infty} 1_{X_{\overline{S}_k} = x} 1_A \circ \theta^{\overline{S}_k} \cdot 1_B \circ \bar\theta^x \cdot h_{x,k}\right)\right)$$

$$= \sum_{k \geq 1} \sum_{x \in \mathbb{Z}^d} E_P\left(1_B \circ \bar\theta^x E_\omega^o\left(1_{\overline{S}_k < \infty} 1_{D \circ \bar\theta_{\overline{S}_k} = \infty} 1_{X_{\overline{S}_k} = x} 1_A \circ \theta^{\overline{S}_k} \cdot h_{x,k}\right)\right)$$

$$= \sum_{k \geq 1} \sum_{x \in \mathbb{Z}^d} E_P\left(1_B \circ \bar\theta^x E_\omega^x\left(1_{D = \infty} 1_A\right) E_\omega^o\left(h_{x,k} 1_{\overline{S}_k < \infty} 1_{X_{\overline{S}_k} = x}\right)\right)$$

$$= \sum_{k \geq 1} \sum_{x \in \mathbb{Z}^d} E_P\left(1_B 1_{D = \infty} 1_A\right) E_P\left(h_{x,k} 1_{\overline{S}_k < \infty} 1_{X_{\overline{S}_k} = x}\right),$$

where we used the Markov property in the next to last equality and the i.i.d. structure of the environment in the last one. Substituting in the above trivial A, B, one concludes that

$$E_{\mathbb{P}^o}\left(h \cdot 1_{\tau_1 < \infty}\right) = P(\{D = \infty\}) \sum_{k \geq 1} \sum_{x \in \mathbb{Z}^d} E_P\left(h_{x,k} 1_{\overline{S}_k < \infty} 1_{X_{\overline{S}_k} = x}\right).$$

Hence,

$$E_{\mathbb{Q}^o}\left(1_A \circ \theta^{\tau_1} \cdot 1_B \circ \bar\theta^{X_{\tau_1}} \cdot h\right) = E_{\mathbb{Q}^p}(h) E_{\mathbb{P}^o}\left(1_A 1_B | \{D = \infty\}\right),$$

concluding the proof of the lemma. $\square$

Consider now τ_1 as a function of the path $(X_n)_{n \geq 0}$ and set

$$\tau_{k+1} = \tau_k(X.) + \tau_1(X_{\tau_k +} - X_{\tau_k}),$$

with $\tau_{k+1} = \infty$ on $\{\tau_k = \infty\}$ (the sequence $\{\tau_k\}$ enumerates times such that for all $k < m < n$, $X_k \cdot \ell < X_m \cdot \ell \leq X_n \cdot \ell$). By the definition and the fact that $\mathbb{P}^o(A_\ell \cap \{\tau_1 = \infty\}) = 0$, we have that $\mathbb{P}^o(A_\ell \cap \{\tau_k = \infty\}) = 0$. Setting

$$\mathcal{G}_k = \sigma\left(\tau_1, \ldots, \tau_k, \quad X_0, \cdots, X_{\tau_k}, \quad \{\omega(y, \cdot)\}_{\ell \cdot y < \ell \cdot X_{\tau_k}}\right),$$

an obvious rerun of the proof of Lemma 3.2.3 yields that

$$\mathbb{Q}^o\left(\{X_{\tau_k + n} - X_{\tau_k}\}_{n \geq 0} \in A, \{\omega(X_{\tau_k + y}, \cdot)\}_{\ell \cdot y \geq 0} \in B | \mathcal{G}_k\right)$$

$$= \mathbb{P}^o\left(\{X_n\}_{n \geq 0} \in A, \{\omega(y, \cdot)\}_{\ell \cdot y \geq 0} \in B | \{D = \infty\}\right).$$

We thus conclude that under $\mathbb{Q}^o$,

$$\left(X_{\tau_2} - X_{\tau_1}, \tau_2 - \tau_1\right), \cdots, \left(X_{\tau_{k+1}} - X_{\tau_k}, \tau_{k+1} - \tau_k\right)$$

are i.i.d. pairs of random variables, independent of (X_{τ_1}, τ_1), such that

$$\mathbb{Q}^o(X_{\tau_2} - X_{\tau_1} \in C_1, \tau_2 - \tau_1 \in C_2) = \mathbb{P}^o(X_{\tau_1} \in C_1, \tau_1 \in C_2 | \{D = \infty\}).$$

Next, we have the following lemma, whose proof is deferred:

Lemma 3.2.5 (Zerner)

$$\mathbb{E}^o(X_{\tau_1} \cdot \ell | \{D = \infty\}) = \frac{1}{\mathbb{P}^o(D = \infty)}.$$

We are now ready to complete the proof of Theorem 3.2.2. Assume first that $\mathbb{E}^o(\tau_1 | \{D = \infty\}) < \infty$. Then, by the law of large numbers, and the renewal structure,

$$\frac{\tau_k}{k} \rightarrow_{k \to \infty} \mathbb{E}^o(\tau_1 | \{D = \infty\}), \qquad \mathbb{Q}^o\text{-a.s.} \qquad (3.2.6)$$

$$\frac{X_{\tau_k} \cdot \ell}{k} \rightarrow_{k \to \infty} \mathbb{E}^o(X_{\tau_1} \cdot \ell | \{D = \infty\}), \quad \mathbb{Q}^o\text{-a.s.} \qquad (3.2.7)$$

(note that the finiteness of the expression in the right hand side of (3.2.7) is trivial if the right hand side of (3.2.6) is finite, and Lemma 3.2.5 is not needed in this case).

Hence,

$$\frac{X_{\tau_k} \cdot \ell}{k} \xrightarrow{k \to \infty} \frac{\mathbb{E}^o(X_{\tau_1} \cdot \ell | \{D = \infty\})}{\mathbb{E}^o(\tau_1 | \{D = \infty\})} =: v_\ell, \quad \mathbb{Q}\text{-a.s.} \qquad (3.2.8)$$

Mimicking now the argument at the end of the proof of Lemma 2.1.5, we conclude that $\frac{X_n \cdot \ell}{n} \rightarrow_{n \to \infty} v_\ell, \mathbb{Q}^o$-a.s., in the case $\mathbb{E}^0(\tau_1 | \{D = \infty\}) < \infty$.

On the other hand, Lemma 3.2.5 implies that (3.2.7) holds true even when $\mathbb{E}^o(\tau_1 | \{D = \infty\}) = \infty$. But then, $\tau_k / k \rightarrow_{k \to \infty} \infty$, $\mathbb{Q}^o$-a.s. With $v_\ell = 0$ in this case, we conclude that

$$\frac{X_{\tau_k} \cdot \ell}{\tau_k} \xrightarrow{k \to \infty} v_\ell = 0, \quad \mathbb{Q}^o\text{-a.s.}$$

Finally, setting k_n such that $\tau_{k_n} \leq n < \tau_{k_n+1}$, we have that $k_n \rightarrow_{n \to \infty} \infty$ and $k_n / n \xrightarrow[n \to \infty]{} 0$, $\mathbb{Q}^o$-a.s. because $n / k_n \geq \tau_{k_n} / k_n$. Thus,

$$\frac{X_n \cdot \ell}{n} \leq \frac{X_{\tau_{k_n+1}} \cdot \ell}{k_n + 1} \cdot \frac{k_n + 1}{n} \xrightarrow[n \to \infty]{} 0, \ \mathbb{Q}^o\text{-a.s.}$$

Since $\liminf_{n \to \infty} \frac{X_n \cdot \ell}{n} \geq 0$, $\mathbb{Q}^o$-a.s., we conclude

$$\frac{X_n \cdot \ell}{n} \xrightarrow[n \to \infty]{} 0, \quad \mathbb{Q}^o\text{-a.s.} \qquad \square$$

Remarks:

1. Note that on A_ℓ, $v_\ell > 0$ if $\mathbb{E}^o(\tau_1 | \{D = \infty\}) < \infty$ and $v_\ell = 0$ otherwise.
2. It is clear from the proof that in fact, if $\mathbb{E}^o(\tau_1 | \{D = \infty\}) < \infty$, then the result of Theorem 3.2.2 can be strengthened to

$$\frac{X_n}{n} \xrightarrow{n \to \infty} \frac{\mathbb{E}^o(X_{\tau_1} | \{D = \infty\})}{\mathbb{E}^o(\tau_1 | \{D = \infty\})}, \ \mathbb{Q}^o\text{-.a.s.}$$

3. In the stationary, ϕ-mixing case, one can prove that the times $\{\tau_i\}$ are well defined, and form a mixing sequence. What I have not been able to show is that they are identically distributed under $\mathbb{Q}^o$ (they seem not!). Modifying slightly the definition of $(\overline{R}_k, \overline{S}_k)$ by adding an L-safeguard as in Section 3.1, the results of this section extend immediately to the case where the environment is K-dependent (i.e., $\{\omega(x,\cdot)\}_{x\cdot\ell\leq 0}$ and $\{\omega(x,\cdot)\}_{x\cdot\ell > K}$ are independent). This applies, e.g., in the setup considered in [63]. The extension to a mixing setup is more complicated, and some results applicable there can be found in [13].

4. Still discussing mixing environements, some progress has been made using the approach of the environment viewed from the particle. We mention here [44] and in particular the recent preprint [62]. The latter preprint uses a-priori estimates concerning regeneration times in the ballistic case to construct an invariant measure for the environment viewed from the particle which is absolutely continuous with respect to P on certain half-spaces, and deduces a LLN using that measure.

Proof of Lemma 3.2.5

Recall that we consider $\ell = (1, 0, \dots, 0)$. Then,

$$\mathbb{Q}^o\Big(\{\exists k : X_{\tau_k} \cdot \ell = i\}\Big)$$

$$= \frac{\sum_{y\in\mathbb{Z}^{d-1}} E\Big(P_\omega^o(\{\exists k : X_{\tau_k} = (i,y)\}, A_\ell)\Big)}{\mathbb{P}^o(A_\ell)}$$

$$= \frac{\sum_{y\in\mathbb{Z}^{d-1}} E\Big(P_\omega^o(T_i < \infty, X_{T_i} = (i,y), D \circ \theta_{T_i} = \infty)\Big)}{\mathbb{P}^o(A_\ell)}$$

$$= \frac{\sum_{y\in\mathbb{Z}^{d-1}} E\Big(P_\omega^o(T_i < \infty, X_{T_i} = (i,y))P_\omega^{(i,y)}(D = \infty)\Big)}{\mathbb{P}^o(A_\ell)}$$

$$= \frac{\mathbb{P}^o(T_i < \infty)\mathbb{P}^o(D = \infty)}{\mathbb{P}^o(A_\ell)} \xrightarrow[i\to\infty]{} \mathbb{P}^o(D = \infty) \tag{3.2.9}$$

(since $\mathbb{P}^o(A_\ell \cup A_{-\ell}) = 1$ and $\lim_{i\to\infty} \mathbb{P}^o(\{T_i < \infty\} \cap A_{-\ell}) = 0$). On the other hand,

$$\lim_{i\to\infty} \mathbb{Q}^o\Big(\{\exists k : X_{\tau_k} \cdot \ell = i\}\Big) = \lim_{i\to\infty} \mathbb{Q}^o\Big(\{\exists k \geq 2 : X_{\tau_k} \cdot \ell = i\}\Big)$$

(because $\mathbb{Q}^o(\tau_k > i) \xrightarrow[i\to\infty]{} 0$)

$$= \lim_{i \to \infty} \sum_{n \geq 1} \mathbb{Q}^o \left(\{ \exists k \geq 2 : X_{\tau_k} \cdot \ell = i, X_{\tau_1} \cdot \ell = n \} \right)$$

$$= \lim_{i \to \infty} \sum_{n \geq 1} \mathbb{Q}^o \left(\{ \exists k \geq 2 : (X_{\tau_k} - X_{\tau_1}) \cdot \ell = i - n, X_{\tau_1} \cdot \ell = n \} \right)$$

$$= \lim_{i \to \infty} \sum_{n \geq 1} \mathbb{Q}^o (X_{\tau_1} \cdot \ell = n) \cdot \mathbb{Q}^o \left(\{ \exists k \geq 2 : (X_{\tau_k} - X_{\tau_1}) \cdot \ell = i - n \} \right).$$

But, recall that by the renewal theorem,

$$\mathbb{Q}^o \left(\exists k \geq 2 : (X_{\tau_k} - X_{\tau_1}) \cdot \ell = i - n \right) \xrightarrow[i \to \infty]{} \frac{1}{E_{\mathbb{Q}^o}((X_{\tau_2} - X_{\tau_1}) \cdot \ell)}$$

and hence, by dominated convergence,

$$\lim_{i \to \infty} \mathbb{Q}^o \left(\{ \exists k : X_{\tau_k} \cdot \ell = i \} \right) = \frac{\sum_{n \geq 1} \mathbb{Q}^o (X_{\tau_1} \cdot \ell = n)}{E_{\mathbb{Q}^o}((X_{\tau_2} - X_{\tau_1}) \cdot \ell)} = \frac{1}{E_{\mathbb{Q}^o}((X_{\tau_2} - X_{\tau_1}) \cdot \ell)}.$$
$$(3.2.10)$$

Comparing (3.2.9) and (3.2.10), we conclude that

$$E_{\mathbb{Q}^o} \left((X_{\tau_2} - X_{\tau_1}) \cdot \ell \right) = \frac{1}{\mathbb{P}^o(D = \infty)} < \infty. \qquad \square$$

Theorem 3.2.2 assumes that $\mathbb{P}^o(A_\ell \cup A_{-\ell}) = 1$, and in that situation provided a LLN if $\mathbb{P}^o(A_\ell) \in \{0, 1\}$. A recent improvement to Theorem 3.2.2, due to Zerner [83], actually shows that if a 0-1 law holds true, a LLN holds, at least for i.i.d. environments. More precisely, one has the following:

Theorem 3.2.11 *There exist deterministic* $v_\ell, v_{-\ell}$ *(possibly zero) such that*

$$\lim_{n \to \infty} \frac{X_n \cdot \ell}{n} = v_\ell 1_{A_\ell} + v_{-\ell} 1_{A_{-\ell}}, \quad \mathbb{P}^o\text{-}a.s. \qquad (3.2.12)$$

An immediate corollary, obtained by applying Theorem 3.2.11 d times with respect to the basis $\ell = e_i$, $i = 1, \ldots, d$, is the following:

Corollary 3.2.13 *Assume that* $\mathbb{P}^o(A_\ell) \in \{0, 1\}$ *for every* ℓ. *Then there exists a deterministic* v *(possibly zero) such that*

$$\lim_{n \to \infty} \frac{X_n}{n} = v, \quad \mathbb{P}^o\text{-}a.s.$$

Proof of Theorem 3.2.11: (sketch) In view of the 0-1 law Theorem 3.1.2 and of Theorem 3.2.2, all that remains to prove is that if $\mathbb{P}^o(A_\ell \cup A_{-\ell}) = 0$ then $X_n \cdot \ell/n \to 0$, $\mathbb{P}^o$-a.s. The complete proof for that is given in [83], and we provide next a brief description.

Consider the set of visits to the hyperplane $\mathcal{H}_m := \{ z : z \cdot \ell = m \}$, defining $\tau_m^0 = T_m$ and $\tau_m^i = \min\{ n > \tau_m^{i-1} : X_n \cdot \ell = m \}$. Fixing an integer L, let

$$h_{m,L} = \sup_{i \geq 0} \{\tau_m^i - \tau_m^0 : \tau_m^i < T_{m+L}\}$$

be the diameter of the set of visits to $\mathcal{H}_m$ before T_{m+L}. For any constant $c > 0$, let

$$F_{M,L}(c) = \frac{\#\{0 \leq m \leq M : h_{m,L} \leq c\}}{M+1}$$

denote the fraction of m's smaller than M such that the time between the first and last visit to $\mathcal{H}_m$ before T_{m+L} is smaller than c. The first observation, which is a deterministic (combinatorial) computation that we skip, is that for any path with $\liminf_{n \to \infty} X_n \cdot \ell/n > 0$ there exists a constant c such that

$$\inf_{L \geq 1} \limsup_{M \to \infty} F_{M,L}(c) > 0,$$

that is, roughly, there is a fraction of m's for which the time between first and last visits of $\mathcal{H}_m$ (before hitting $\mathcal{H}_{m+L}$) is not too large.

Assume now that $\mathbb{P}^o(\limsup X_n \cdot \ell/n > 0) > 0$. Then, by the above observation, there is some $c > 0$ such that

$$\mathbb{P}^o(\limsup_{L \to \infty} \limsup_{M \to \infty} F_{M,L}(c) > 0) > 0. \tag{3.2.14}$$

But on the event $\{h_{m,L} \leq c\}$, the last point visited in $\mathcal{H}_m$ before hitting $\mathcal{H}_{m+L}$ is at most at distance c from X_{T_m} and has been visited at most c times before T_{m+L}. Thus, there is a $z \in \mathcal{H}_0$ with $|z|_1 \leq c$, and an $1 \leq r \leq c$ such that the r-th visit to $X_{T_m} + z$ occurs before T_{m+L} and the walk does not backtrack from $\mathcal{H}_m$ after this r-th visit. Denoting the last event by $B_{m,L}^1(z,r)$, it follows that

$$F_{M,L}(c) \leq \frac{1}{M+1} \sum_{z \in \mathcal{H}_0, |z|_1 \leq c} \sum_{r=1}^{c} \sum_{m=0}^{M} 1_{B_{m,L}^1(z,r)}.$$

Noting that the summation over r and z is over a finite set, and combining the last inequality with (3.2.14), it follows that for some z and r,

$$\mathbb{P}^o(\limsup_{L \to \infty} \limsup_{M \to \infty} \frac{1}{M+1} \sum_{m=0}^{M} 1_{B_{m,L}^1(z,r)} > 0) > 0. \tag{3.2.15}$$

While the events $\{B_{m,L}^1(z,r)\}_m$ are not independent, some independence can be restored in the following way: construct independent (given the environment) copies $Y^{.y}$ of the RWRE, starting at y. Define the event $B_{m,L}(z,r)$ as the union of $B_{m,L}^1(z,r)$ with the event that $X.$ does not hit $X_{T_m} + z$ for the r-th time before T_{m+L}, but $Y^{.X_{T_m}+z}$ does not backtrack from $\mathcal{H}_m$ before it hits $\mathcal{H}_{m+L}$. An easy computation involving the Markov property shows that for each fixed $i = 0, 1, \ldots, L-1$, the events $\{B_{jL+i,L}(z,r)\}_j$ are independent, with

$$\mathbb{P}^o(B_{jL+i,L}(z,r)) = \mathbb{P}^o(D \geq T_L).$$

(Here and in the sequel, we abuse notations by still using $\mathbb{P}^o$ to denote the annealed law on the enlarged probability space that supports the extra Y^y walks). Hence, since we have from (3.2.15) that

$$\mathbb{P}^o(\limsup_{L \to \infty} \limsup_{M \to \infty} \frac{1}{M+1} \sum_{i=0}^{L-1} \sum_{j=0}^{[M/L]} \mathbf{1}_{B_{jL+i,L}^1(z,r)} > 0) > 0, \qquad (3.2.16)$$

it follows, by the standard law of large numbers, that

$$\mathbb{P}^o(D = \infty) = \limsup_{L \to \infty} \mathbb{P}^o(D \geq T_L) > 0.$$

But from (3.1.4), we have that $\mathbb{P}^o(A_\ell) \geq \mathbb{P}^o(D = \infty) > 0$. In particular, this shows that $\mathbb{P}^o(A_\ell) = 0$ implies that $\limsup X_n \cdot \ell/n \leq 0$, $\mathbb{P}^o$-a.s. Repeating this argument with $-\ell$ instead of ℓ completes the proof of the theorem. □

Bibliographical notes:

The proof here follows closely [76], except that Lemma 3.2.5 is due to private communication with Martin Zerner. The improvement Theorem 3.2.11 is based on [83].

The ballistic LLN has been proved for certain non iid environments in [13]. Alternative approaches to ballistic LLN's using the environment viewed from the particle were developped in [44] and in great generality in [62].

There are only a few LLN results in the non-ballistic case, see the bibliographical notes of Section 3.3.

3.3 CLT for walks in balanced environments

The setup in this section is the following:

Assumption 3.3.1

(B1) P *is stationary and ergodic.*
(B2) P *is balanced: for* $i = 1, \cdots, d$, $P(\omega(x, x + e_i) = \omega(x, x - e_i)) = 1$.
(B3) P *is uniformly elliptic: there exists an* $\varepsilon > 0$ *such that for* $i = 1, \cdots, d$,

$$P(\omega(x, x + e_i) > \varepsilon) = 1.$$

Unlike the situation in Section 2.1, we do not have an explicit construction of invariant measures at our disposal. The approach toward the LLN and CLT uses however (B2) in an essential way: indeed, note that in the notations of (2.1.28),

$$d(x, \omega) = \sum_{i=1}^{d} e_i \Big[\omega(x, x + e_i) - \omega(x, x - e_i)\Big] = 0.$$

Hence, the processes $(X_n(i))_{n \geq 0}, i = 1, \cdots, d$, are martingales, with, denoting $\mathcal{F}_n = \sigma(X_0, \cdots X_n)$,

$$E_\omega^o((X_n(i) - X_{n-1}(i))(X_n(j) - X_{n-1}(j))|\mathcal{F}_{n-1}) = 2\delta_{ij}\omega(X_{n-1}, X_{n-1} + e_i).$$

Since $|\omega(\cdot,\cdot)| \le 1$ P-a.s., it immediately follows that $X_n/n \xrightarrow[n\to\infty]{} 0$, $\mathbb{P}^o$-a.s. Further, the multi-dimensional CLT (compare with Lemma 2.2.4) yields that if there exists a deterministic vector $\mathbf{a} = (a_1, \cdots, a_d)$ such that

$$\frac{1}{n}\sum_{k=1}^n \omega(X_{k-1}, X_{k-1} + e_i) \xrightarrow[n\to\infty]{} \frac{a_i}{2} > 0, \quad \mathbb{P}^o\text{-a.s.}, \tag{3.3.2}$$

then, for any bounded continuous function $f : \mathbb{R}^d \to \mathbb{R}$, and any $y \in \mathbb{R}$,

$$\lim_{n\to\infty} P_\omega^o\left(f\left(\frac{X_n}{\sqrt{n}}\right) \le y\right) \tag{3.3.3}$$

$$= \frac{1}{(2\pi)^{d/2}\prod_{i=1}^d \sqrt{a_i}}\int_{\mathbb{R}^d} 1_{\{f(\mathbf{x})\le y\}} \exp\left(-\sum_{i=1}^d \frac{x_i^2}{2a_i}\right)\prod_{i=1}^d dx_i, \quad P\text{-a.s.}$$

Our goal in this section is to demonstrate such a CLT, and to study transience and recurrent questions for the RWRE.

Central limit theorems

Theorem 3.3.4 *Assume Assumption 3.3.1. Then, there exists a deterministic vector $\mathbf{a}$ such that (3.3.2) holds true. Consequently, the quenched CLT (3.3.3) holds true.*

Remark 3.3.5 *In fact, the above observations yield not only a CLT in the form of (3.3.3) but also a trajectorial CLT for the process $\{X_{[nt]}/\sqrt{n},\ t \in [0,1]\}$.*

Proof of Theorem 3.3.4

As in Section 2.1, the key to the proof of (3.3.2) is to consider the environment viewed from the particle. Define $\bar\omega(n) = \theta^{X_n}\omega$, and the Markov transition kernel

$$M(\omega, d\omega') = \sum_{e_i}\left[\omega(0, e_i)\delta_{\theta^{e_i}\omega=\omega'} + \omega(0, -e_i)\delta_{\theta^{-e_i}\omega=\omega'}\right]. \tag{3.3.6}$$

As in Lemma 2.1.18, the process $\bar\omega(n)$ is Markov under either P_ω^o or $\mathbb{P}^o$. Mimicking the proof of Corollary 2.1.25, if we can construct a measure Q on Ω which is absolutely continuous with respect to P and such that it is invariant under the Markov transition M, we will conclude, as in Corollary 2.1.25, that $\bar\omega(n)$ is stationary and ergodic and hence

$$\frac{1}{n}\sum_{i=1}^n \omega(X_{n-1}, X_{n-1} + e_i) = \frac{1}{n}\sum_{i=1}^n \bar\omega(n)(0, e_i) \xrightarrow[n\to\infty]{} \frac{a_i}{2}$$

$$:= E_Q\bar\omega(0, e_i) \ge \varepsilon, \mathbb{P}^o\text{-a.s.}, \tag{3.3.7}$$

yielding (3.3.2). Our effort therefore is directed towards the construction of such a measure. Naturally, such measures will be constructed from periodic modifications of the RWRE, and require certain a-priori estimates on harmonic functions. We state these now, and defer their proof to the end of the section. The estimates we state are slightly more general than needed, but will be useful also in the study of transience and recurrence.

We let $|x|_\infty := \max_{i=1}^d |x_i|$ and define $D = D_R(x_0) = \{x \in \mathbb{Z}^d : |x-x_0|_\infty < R\}$. The generator of the RWRE, under P_ω, is the operator

$$(L_\omega f)(x) = \sum_{i=1}^d \omega(x, x + e_i)\Big[f(x + e_i) + f(x - e_i) - 2f(x)\Big].$$

For any bounded $E \subset \mathbb{Z}^d$ of cardinality $|E|$, set $\partial E = \{y \in E^c : \exists x \in$

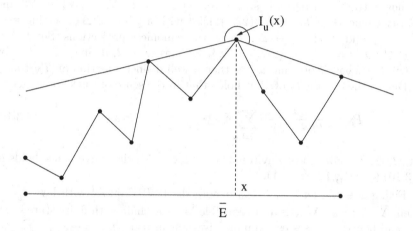

Fig. 3.3.1. The normal set at $x \in E$

$E, |x - y|_\infty = 1\}$, $\overline{E} = E \cup \partial E$, and $\mathrm{diam}(E) = \max\{|x - y|_\infty : x, y \in \overline{E}\}$. For any function $u : \mathbb{Z}^d \to \mathbb{R}$, we define the *normal set* at a point $x \in E$ as

$$I_u(x) = \{s \in \mathbb{R}^d : u(z) \le u(x) + s \cdot (z - x), \forall z \in \overline{E}\}.$$

Finally, for any $q > 0$, E and u as above, define

$$\|g\|_{E,q,u} := \Big(\frac{1}{|E|} \sum_{x \in E} 1_{\{I_u(x) \ne \emptyset\}} |g(x)|^q\Big)^{1/q}, \|g\|_{E,q} := \Big(\frac{1}{|E|} \sum_{x \in E} |g(x)|^q\Big)^{1/q}.$$

Then we have the following:

Lemma 3.3.8 *There exists a constant $C = C(\varepsilon, d)$ such that*

(a) *(maximum principle) For any $E \subset \mathbb{Z}^d$ bounded, any functions u and g such that*

$$L_\omega u(x) \geq -g(x), \quad x \in E$$

satisfy

$$\max_{x \in E} u(x) \leq C \mathrm{diam}(E) |E|^{1/d} \|g\|_{E,d,u} + \max_{x \in \partial E} u^+(x).$$

(b) *(Harnack inequality) Any function $u \geq 0$ such that*

$$L_\omega u(x) = 0, \qquad x \in D_R(x_0), \tag{3.3.9}$$

satisfies

$$\frac{1}{C} u(x_0) \leq u(x) \leq C u(x_0), \quad x \in D_{R/2}(x_0).$$

We now introduce a periodic structure. Set $\Delta_N = \{-N, \cdots, N\}^d \subset \mathbb{Z}^d$ and identify elements of $T_N = \mathbb{Z}^d/(2N+1)\mathbb{Z}^d$ with a point of Δ_N, setting $\pi_N : \mathbb{Z}^d \to T_N$ and $\hat{\pi}_N : \mathbb{Z}^d \to \Delta_N$ to be the canonical projections. Set $\Omega^N = \{\omega \in \Omega : \theta^x \omega = \omega, \forall x \in (2N+1)\mathbb{Z}^d\}$. For any $\omega \in \Omega$, define $\omega^N \in \Omega^N$ by $\omega^N(x) = \omega(\hat{\pi}_N x)$. Note that ω^N is then a well defined function on T_N too.

Due to the ergodicity of P, it holds that in the sense of weak convergence,

$$P_N := \frac{1}{(2N+1)^d} \sum_{x \in \Delta_N} \delta_{\theta^x \omega^N} \xrightarrow[N \to \infty]{} P, \quad P\text{-a.s.} \tag{3.3.10}$$

Let $\Omega_0 \subset \Omega$ denote those environments ω for which the convergence holds in (3.3.10) (clearly, $P(\Omega_0) = 1$).

Fixing $\omega \in \Omega_0$, let $(X_{n,N})_{n \geq 0}$ denote the RWRE on $\mathbb{Z}^d$ with law $P_{\omega^N}^{X_{0,N}}$. Then, $\overline{X}_{n,N} := \pi_N X_{0,N}$ is an irreducible Markov chain with finite state space T_N, and hence it possesses a unique invariant measure $\mu_N = \frac{1}{(2N+1)^d} \sum_{x \in T_N} \phi_N(x)\delta_x$. Setting $\overline{\omega}^N(n) := \theta^{X_{n,N}} \omega^N$, it follows that $\overline{\omega}^N(n)$ is an irreducible Markov chain with finite state space $S_N := \{\theta^x \omega^N\}_{x \in \Delta_N}$ and transition kernel M. Its unique invariant measure, supported on Ω^N, is then easily checked to be of the form

$$Q_N = \frac{1}{(2N+1)^d} \sum_{x \in \Delta_N} \phi_N(\pi_N x)\delta_{\theta^x \omega^N}.$$

Partitioning the state space S_N into finitely many *disjoint* states $\{\omega_\alpha^N\}_{\alpha=1}^K$, set $C_N(\alpha) = \{x \in \Delta_N : \theta^x \omega^N = \omega_\alpha^N\}$. Then,

$$f_N := \frac{dQ_N}{dP_N} = \sum_{\alpha=1}^K \mathbf{1}_{\{\omega = \omega_\alpha^N\}} \frac{1}{|C_N(\alpha)|} \sum_{x \in C_N(\alpha)} \phi_N(\pi_N x).$$

We show below, as a consequence of part (a) of Lemma 3.3.8, that there exists a constant $C_2 = C_2(\varepsilon, d)$, independent of N, such that

$$\|\phi_N(\pi_N \cdot)\|_{D_{N+1}(0), d/d-1} \leq C_2. \tag{3.3.11}$$

Thus, using Jensen's inequality in the first inequality and (3.3.11) in the second,

$$\int f_N^{d/d-1} dP_N = \sum_{\alpha=1}^{K} \left[\frac{1}{|C_N(\alpha)|} \sum_{x \in C_N(\alpha)} \phi_N(\pi_N x) \right]^{d/d-1} \frac{|C_N(\alpha)|}{(2N+1)^d}$$

$$\leq \sum_{\alpha=1}^{K} \sum_{x \in C_N(\alpha)} \phi_N(\pi_N(x))^{d/d-1} \frac{1}{(2N+1)^d}$$

$$= \frac{1}{(2N+1)^d} \sum_{x \in \Delta_N} \phi_N(\pi_N(x))^{d/d-1} \leq C_2^{(d-1)/d}. \tag{3.3.12}$$

Note that f_N extends to a measurable function on Ω, and the latter is, due to (3.3.12), uniformly integrable with respect to P_N. Thus, any weak limit of Q_N is absolutely continuous with respect to P, and further it is invariant with respect to the Markov kernel M.

Let $E = \{\omega : \frac{dQ}{dP} = 0\}$. By invariance, $E_Q M 1_E = E_Q 1_E = 0$, and hence $M 1_E \leq 1_E$, P-a.s. But, $M 1_E \geq \varepsilon \sum_{i=1}^{d} (1_E \circ \theta^{e_i} + 1_E \circ \theta^{-e_i})$. Hence, $1_E \geq 1_E \circ \theta^{\pm e_i}$, P-a.s. Since P is stationary, $1_E = 1_E \circ \theta^{\pm e_i}$, P-a.s., and hence by ergodicity (considering the invariant event $\cap_{x \in \mathbb{Z}^d} (\theta^x)^{-1} E$) $P(E) \in \{0, 1\}$. But $Q \ll P$ implies $P(E) = 0$. Hence, $Q \sim P$, as claimed (further, by (3.3.7), Q is then uniquely defined).

It thus only remains to prove (3.3.11). Fix a function g on T_N, and define the resolvent

$$R^{\omega^N} g(x) := \sum_{j=0}^{\infty} \left(1 - \frac{1}{N^2} \right)^j E_{\omega^N}^x g(\overline{X}_{j,N})$$

$$= \sum_{j=0}^{\infty} \left(1 - \frac{1}{N^2} \right)^j E_{\omega^N}^x g \circ \pi_N(X_{j,N}), \quad x \in T_N$$

and the stopping times $\tau_0 = 0, \tau_1 = \tau := \min\{k \geq 1 : |X_{k,N} - X_{0,N}| \geq N\}$ and $\tau_{k+1} = \tau \circ \theta^k + \tau_k$. Since for $x \in \mathbb{Z}^d$ with $|x - X_{0,N}| < N$ it holds that $L_{\omega^N} E_{\omega^N}^x \left(\sum_{j=0}^{\tau-1} g \circ \pi_N(X_{j,N}) \right) = -g(x)$, we have by Lemma 3.3.8(a) that for some constant $C = C(\varepsilon, d)$,

$$\sup_{|x - X_{0,N}| < N} \left| E_{\omega^N}^x \left(\sum_{j=0}^{\tau-1} g \circ \pi_N(X_{j,N}) \right) \right| \leq CN^2 \|g\|_{D_{N+1}(0), d}. \tag{3.3.13}$$

Since $(X_{n,N})_{n \geq 0}$ is a martingale, it follows from Doob's inequality that, for any $K \geq 1$,

$$P^o_{\theta^x\omega^N}[\tau \leq K] \leq 2 \sum_{i=1}^{d} P^o_{\theta^x\omega^N}\left[\sup_{n\leq K} X_n(i) \geq N\right]$$

$$\leq \frac{2}{N} \sum_{i=1}^{d} E^o_{\theta^x\omega^N}\left((X_K(i))_+\right) \leq \frac{2d}{N}\sqrt{K}.$$

Hence, using $K = N^2/8d^2$,

$$E^o_{\theta^x\omega^N}\left(\left(1 - \frac{1}{N^2}\right)^\tau\right) \leq \frac{2d}{N}\sqrt{K} + \left(1 - \frac{1}{N^2}\right)^K \leq C_3 \qquad (3.3.14)$$

where $C_3 = C_3(d) < 1$ is independent of N. Thus, using the strong Markov property, (3.3.13) and (3.3.14),

$$|R^{\omega^N} g(x)| = \sum_{m\geq 0} E^x_{\omega^N}\left(\sum_{\tau_m \leq j < \tau_{m+1}} \left(1 - \frac{1}{N^2}\right)^j g \circ \pi_N(X_{j,N})\right)$$

$$\leq \sum_{m\geq 0} E^x_{\omega^N}\left(\left(1 - \frac{1}{N^2}\right)^{\tau_m} E^{X_{\tau_m,N}}_{\omega^N} \sum_{j=0}^{\tau-1} g \circ \pi_N(X_{j,N})\right)$$

$$\leq \sum_{m\geq 0} \left(\sup_{x\in\mathbb{Z}^d} E^x_{\omega^N}\left(\left(1 - \frac{1}{N^2}\right)^\tau\right)\right)^m \cdot \sup_{x\in\mathbb{Z}^d} E^x_{\omega^N}\left(\sum_{j=0}^{\tau-1} g \circ \pi_N(X_{j,N})\right)$$

$$\leq C_4 N^2 \|g\|_{D_{N+1}(0),d}$$

where $C_4 = C_4(d,\varepsilon)$. Using the invariance of ϕ_N, we now get

$$\|\phi_N(\pi_N\cdot)\|_{D_{N+1}(x_0),d/d-1} = \|\phi_N(\pi_N\cdot)\|_{D_{N+1}(0),d/d-1}$$

$$= \sup_{g:\|g\|_{D_{N+1}(0),d}\leq 1} \frac{1}{|D_{N+1}(0)|} \sum_{y\in D_{N+1}(0)} \phi_N(\pi_N y)g(y)$$

$$= \frac{1}{N^2} \sup_{g:\|g\|_{D_{N+1}(0),d}\leq 1} \sum_{k\geq 0} \left(1 - \frac{1}{N^2}\right)^k \frac{1}{(2N+1)^d} \sum_{x\in\Delta_N} \phi_N(x) E^x_{\omega^N}(g \circ \pi_N(X_{k,N}))$$

$$\leq C_2$$

with $C_2 = C_2(d,\varepsilon)$, proving (3.3.11). $\square$

Proof of Lemma 3.3.8

(a) We may assume without loss of generality that $\max_{x\in\partial E} u(x) \leq 0$, $g \geq 0$, $g \neq 0$ and that $u \geq 0$ is not identically 0. Let $\bar{u} = \max_{x\in\overline{E}} u = u(x_0)$, some $x_0 \in E$. Then, for s satisfying $|s|_\infty < \bar{u}/\text{diam}(\overline{E})$, it holds that

$$u(x_0) + s\cdot(x - x_0) > 0, \quad \forall x \in \overline{E}.$$

Hence, with $t = \inf\{\rho \geq 0 : u(x_0) + s{\cdot}(x - x_0) + \rho > u(x), \forall x \in \overline{E}\}$, we have that $u(x) = u(x_0) + s{\cdot}(x - x_0) + t$, some $x \in \overline{E}$, and hence $u(x) + s{\cdot}(z - x) = u(x_0) + s{\cdot}(z - x_0) + t \geq u(z)$, $\forall z \in \overline{E}$. Hence,

$$s \in I_u(x) \subset \bigcup_{x \in E} I_u(x), \text{ for all } s \text{ with } |s|_\infty < \frac{\overline{u}}{\text{diam}(\overline{E})}. \tag{3.3.15}$$

Assume $s \in I_u(x)$. Then, with $e \in \{\pm e_i\}$, and $v(y) = u(x) + s{\cdot}(y - x)$,

$$0 = \omega(x, x + e)\Big(2v(x) - v(x + e) - v(x - e)\Big)$$
$$\leq \omega(x, x + e)(2u(x) - u(x + e) - u(x - e)),$$

and hence,

$$0 \leq \omega(x, x + e)(2u(x) - u(x + e) - u(x - e))$$
$$\leq \sum_{i=1}^{d} \omega(x, x + e_i)(2u(x) - u(x + e_i) - u(x - e_i)) = -L_\omega u(x) \leq g(x).$$

Hence,

$$\Big(u(x) - u(x - e)\Big) - \Big(u(x + e) - u(x)\Big) \leq \frac{g(x)}{\omega(x, x + e)} \leq \frac{g(x)}{\varepsilon}.$$

Because $s \in I_u(x)$, it holds that

$$u(x + e) - u(x) \leq s \cdot e \leq u(x) - u(x - e)$$

and hence

$$u(x) - u(x - e) - \frac{g(x)}{\varepsilon} \leq s \cdot e \leq u(x) - u(x - e), \quad \forall s \in I_u(x). \tag{3.3.16}$$

Using (3.3.15) in the first inequality and (3.3.16) in the second, we have that

$$\left(\frac{2\overline{u}}{\text{diam}(\overline{E})}\right)^d \leq \left|\bigcup_{x \in E} I_u(x)\right| \leq \sum_{x \in E}\left(\frac{g(x)}{\varepsilon}\right)^d \mathbf{1}_{\{I_u(x) \neq \emptyset\}}.$$

Hence,

$$\overline{u} \leq C_0(d, \varepsilon)\text{diam}(\overline{E})|E|^{1/d}\left(\frac{1}{|E|}\sum_{x \in E}|g(x)|^d \mathbf{1}_{\{I_u(x) \neq \emptyset\}}\right)^{\frac{1}{d}},$$

completing the proof of part (a).

(b) It is enough to consider $x_0 = 0$. We begin with some estimates. For parts of the proof, it is easier to work with L_2 (instead of L_∞) balls. Set

256

$B_R = \{x \in \mathbb{Z}^d : |x|_2 < R\}$. We first deduce from part (a) that for any $p \le d$ and $\sigma < 1$, there exists a constant $C_1 = C_1(p, \sigma, d)$ such that

$$\max_{x \in B_{\sigma R}} u(x) \le C_1 \left(\frac{1}{|B_R|} \sum_{x \in B_R} |u^+(x)|^p \right)^{\frac{1}{p}}. \tag{3.3.17}$$

Indeed, define $\eta(x) = \left(1 - \frac{|x|^2}{R^2}\right)^{2d/p}$. A Taylor expansion reveals that for some $C_2 = C_2(p, d)$, it holds that

$$|\eta(x \pm e_i) - \eta(x)| < \frac{C_2}{R}, \ |\eta(x + e_1) + \eta(x - e_1) - 2\eta(x)| \le \frac{C_2}{R^2}. \tag{3.3.18}$$

Fix $\kappa_i = \kappa_i(x) \in [0, 1]$, $i = 1, \ldots, d$, set $\nu(x) = \eta(x)u(x)$, $x \in B_R$, and

$$\hat{L}_\omega \nu(x) = \sum_{i=1}^d \hat{\omega}(x, x + e_i)(\nu(x + e_i) + \nu(x - e_i) - 2\nu(x))$$

where

$$\hat{\omega}(x, x + e_i) = \begin{cases} \omega(x, x + e_i)\left[\frac{\kappa_i}{\eta(x - e_i)} + \frac{1 - \kappa_i}{\eta(x + e_i)}\right], & |x|^2 \le R^2 - 4R \\ \omega(x, x + e_i), & R^2 \ge |x|^2 > R^2 - 4R \end{cases}.$$

Then, a tedious computation reveals that, on the set $|x|^2 \le R^2 - 4R$,

$$-\hat{L}_\omega \nu(x) = -L_\omega u(x)$$
$$-2\sum_i \frac{\kappa_i(\nu(x + e_i) - \nu(x)) + (1 - \kappa_i)(\nu(x) - \nu(x - e_i))}{\eta(x + e_i)\eta(x - e_i)}[\eta(x + e_i) - \eta(x - e_i)]$$
$$+\sum_i \frac{u(x)}{\eta(x + e_i)\eta(x - e_i)}$$
$$\left[2(\eta(x + e_i) - \eta(x))(\eta(x) - \eta(x - e_i)) - \eta(x)(\eta(x + e_i) + \eta(x - e_i) - 2\eta(x))\right]$$
$$\le C_3(d, p)\left[\sum_i \frac{|\kappa_i(\nu(x + e_i) - \nu(x)) + (1 - \kappa_i)(\nu(x) - \nu(x - e_i))|}{R} + \frac{u(x)}{R^2}\right]$$

where we used (3.3.9) in the first equality.

If for such x, $I_\nu(x) \ne \phi$, then by the proof in part (a), there exists a vector $q \in I_\nu(x)$ with $|q| \le \frac{\nu(x)}{R - |x|_\infty}$, and one may find a $\kappa_i \in [0, 1]$ such that

$$\kappa_i(\nu(x + e_i) - \nu(x)) + (1 - \kappa_i)(\nu(x) - \nu(x - e_i)) = q_i.$$

Thus, on $\{I_\nu(x) \ne \phi\} \cap \{x : |x|^2 \le R^2 - 4R\}$, it holds that $-\hat{L}_\omega \nu(x) \le C_4(d, p)\frac{u(x)}{R^2}$. On the other hand, when $|x|^2 \ge R^2 - 4R$, recalling that $u \ge 0$, it holds that

$$-\Big(\nu(x+e_i)+\nu(x-e_i)-2\nu(x)\Big) \le 2\eta(x)u(x) \le C_5(d,p)\frac{u(x)}{R^2}\eta(x)^{\frac{d-p}{d}}$$

and in conclusion,

$$-\hat{L}_\omega\nu(x) \le g(x), \quad x \in B_R$$

where

$$\left|g(x)\mathbf{1}_{I_\nu(x)\neq\phi}\right| \le \frac{C_6(d,p)u(x)}{R^2}.$$

Applying part (a) of the lemma, we get (3.3.17).

Next, let $\sigma < \tau < 1$, and set

$$\mathbf{u}_\sigma = \min_{x \in B_{\sigma R}} u(x), \quad \mathbf{u}_\tau = \min_{x \in B_{\tau R}} u(x).$$

We claim that (3.3.17) implies the existence of a constant $\gamma = \gamma(d,\sigma,\tau,\varepsilon)$ such that

$$\mathbf{u}_\tau \ge \gamma\mathbf{u}_\sigma. \tag{3.3.19}$$

Indeed, set $\bar{\eta}(x) = (R^2 - |x|^2)^\beta$ with $\beta > 2 \vee 1/\sigma$ and $w(x) = \mathbf{u}_\sigma R^{-2\beta}\bar{\eta}(x) - u(x)$. Then, $w(x) \le 0$ on $B_{\sigma R} \cup B_R^c$, and $L_\omega w = \mathbf{u}_\sigma R^{-2\beta}L_\omega\bar{\eta}$ on B_R. But, there is an $R_1(\beta)$ such that on $B_R \backslash B_{\sigma R}, R > R_1$,

$$L_\omega\bar{\eta}(x) \ge \begin{cases} 0, & |x| < R \\ -C(\beta,d,\varepsilon)R^{2(\beta-1)}, & |x| = R \end{cases},$$

implying by part (a) that on $B_R \backslash B_{\sigma R}, R > R_1(\beta)$,

$$w(x) \le C(\beta,d,\varepsilon)\mathbf{u}_\sigma R^2 R^{-2\beta}\left(\frac{1}{R^d}\sum_{|x|=R}R^{2(\beta-1)d}\right)^{\frac{1}{d}} \le \frac{C(\beta,d,\varepsilon)}{R^{1/d}}\mathbf{u}_\sigma.$$

Thus,

$$\mathbf{u}_\tau \ge \mathbf{u}_\sigma\left[(1-\tau^2)^\beta - \frac{C(\beta,d,\varepsilon)}{R^{1/d}}\right].$$

We conclude that there exists an $R_0 = R_0(\sigma,\tau,d,\varepsilon)$ and $\gamma = \gamma(d,\sigma,\tau,\varepsilon)$ such that for all $R > R_0$, (3.3.19) holds. On the other hand, for $R < R_0$ (but $(1-\tau)R > 1!$), (3.3.19) is trivial by finitely many applications of the equality $L_\omega u = 0$. Thus, (3.3.19) is always satisfied.

A conclusion of (3.3.19) is that if $L_\omega u = 0$ on B_R, $\sigma < 1$, and $\Gamma \subset B_{\sigma R} \subset B_{\tau R} \subset B_R$, letting $\mathbf{u}_\Gamma = \min_{x \in \Gamma} u(x)$, we have that for some $\delta = \delta(\varepsilon,d)$,

$$|\Gamma| \ge \delta|B_{\sigma R}| \Longrightarrow \mathbf{u}_\tau \ge \gamma\mathbf{u}_\Gamma. \tag{3.3.20}$$

Indeed, define $\nu = \mathbf{u}_\Gamma - u$ and conclude from (3.3.17) that

$$\max_{x \in B_{\sigma R/2}}\nu(x) \le C_1\left(\frac{1}{|B_{\sigma R}|}\sum_{x \in B_{\sigma R}}\nu^+(x)\right) \le C_1(1-\delta)\max_{x \in B_{\sigma R}}\nu(x)$$

and hence, taking $\delta < 1$ such that $C_1(1-\delta) < 1/2$,

$$\mathbf{u}_\Gamma - \min_{x \in B_{\sigma R/2}} u(x) \le C_1(1-\delta)(\mathbf{u}_\Gamma - \mathbf{u}_\sigma) \le \frac{1}{2}(\mathbf{u}_\Gamma - \mathbf{u}_\sigma),$$

from which one concludes that $\mathbf{u}_\Gamma \le \mathbf{u}_{\sigma/2}$. (3.3.20) follows from combining this and (3.3.19).

We finally use the following covering argument. Fix a cube $Q \subset \mathbb{Z}^d$. For $t > 0$, set

$$\Gamma_t = \{x \in Q : u(x) > t\}.$$

Note that if $Q' = Q'(z,r)$ is any cube in $\mathbb{Z}^d$, centered at z and of side r, (3.3.20) implies that

$$|\Gamma_t \cap Q'| \ge \delta|Q'| \Rightarrow u(x) \ge \gamma t, \text{ some } \gamma = \gamma(\delta, d, \varepsilon). \qquad (3.3.21)$$

Define, for any $A \subset Q$,

$$A_\delta = \bigcup_{\substack{\{r,z\} \\ z \in (\frac{1}{2}\mathbb{Z})^d}} \{Q'(z,3r) \cap Q : |A \cap Q'(z,r)| \ge |Q'(z,r)|\}.$$

Then, cf. [78, Lemma 3] for a proof, either $A_\delta = Q$ or $|A_\delta| \ge |\Gamma|/\delta$. Thus, if $|\Gamma_t| \ge \delta^s|Q|$, then iterating (3.3.21) and the above, $\inf_{x \in Q} u(x) \ge \gamma^s t$. Choosing s such that $\delta^s \le \frac{|\Gamma_t|}{|Q|} \le \delta^{s-1}$, we conclude that $\inf_{x \in D_R} u(x) \ge \gamma t \left(\frac{|\Gamma_t|}{|D_R|}\right)^{\log \gamma/\log \delta}$. Hence, with $p < \log \delta/\log \gamma := p'$, and $\mathbf{u} = \min_{D_R} u$, we have

$$\frac{1}{|D_R|}\sum_{x \in D_R}|u(x)|^p = p\int_{\mathbf{u}}^\infty t^{p-1}\left(\frac{1}{|D_R|}\sum_{x \in D_r}1_{u(x)\ge t}\right)dt$$

$$= p\int_{\mathbf{u}}^\infty t^{p-1}\left(\frac{|\Gamma_t|}{|D_R|}\right)dt$$

$$\le c(p)\,\mathbf{u}^{p'}\int_{\mathbf{u}}^\infty \frac{t^{p-1}}{t^{p'}}dt = c(p,p')\mathbf{u}^p,$$

for some constants $c(p), c(p,p')$, since $p' + 1 - p > 1$. Combining this and (3.3.17) yields the lemma.   □

Transience and recurrence of balanced walks

The main result in this section is the following:

Theorem 3.3.22 *Assume Assumption 3.3.1. Then the RWRE* $(X_n)_{n\ge 0}$ *is transient if $d \ge 3$ and recurrent if $d = 2$.*

Proof. We begin with the transience statement. Fix $d \geq 3$, K large, and define $r_i = K^i$, with $B_i = \{x : |x|_\infty \leq r_i\}$. Set $\tau_0 = 1$ and

$$\tau_i = \min\{n > \tau_{i-1} : X_n \in \partial B_i\}.$$

We use the following uniform estimate on exit probabilities, that actually is stronger than needed: there exists some constant $C = C(\delta, \varepsilon, d) > 0$ such that, if $\Omega_0 = \{\omega : \omega(z, z + e_i) = \omega(z, z - e_i) > \varepsilon, i = 1, \ldots, d, \forall z \in \mathbb{Z}^d\}$,

$$\sup_{\omega \in \Omega_0} P_\omega^o(|X_n| < L, n = 1, \cdots, L^{2(1+\delta)}) \leq Ce^{-CL^{2\delta}}. \tag{3.3.23}$$

There are many ways to prove (3.3.23), including a coupling argument. We use here an optimal control trick. Let $\{B_n\}_{n \geq 0}$ denote a sequence of i.i.d. Bernoulli(1/2) random variables, independent of the environment, of law Q. Then, X_n can be constructed as follows:

$$P_{\omega,B}^o\left(X_{n+1} = X_n + e_i | X_n = x, X_{n-1}, \cdots X_0\right) = 2\omega(x, e_i)\mathbf{1}_{2B_{n+1}-1=\pm 1}.$$

(As in Section 3.1, $Q \times P_{\omega,B}^o$, when restricted to $(\mathbb{Z}^d)^\mathbb{N}$, equals P_ω^o.) Set $\mathcal{G}_n = \sigma(B_0, B_1, \cdots, B_n, X_0, \cdots, X_n, (\omega_z)_{z \in \mathbb{Z}^d})$. An admissible control $\alpha = (\alpha_n)_{n \geq 0}$ is a sequence of $\mathcal{G}_n$ measurable function taking values in $\mathcal{A} := [2\varepsilon, \frac{1}{2} - \varepsilon(d-1)]$. Then define the $\mathbb{Z}$-valued controlled process (Y_n^α) by $Y_0 = 0$ and

$$P\left(Y_{n+1}^\alpha = Y_n^\alpha \pm 1 | \mathcal{G}_n, Y_0^\alpha, \cdots, Y_n^\alpha\right) = \alpha_n \mathbf{1}_{2B_{n+1}-1=\pm 1}.$$

Note that, by taking $\hat{\alpha}_n = 2\omega(X_n, e_1)$, we may construct $(Y_n^{\hat\alpha})$ and X_n on the same probability space such that $Y_n^{\hat\alpha} = X_n$, $Q \times P_{\omega,B}^o$-a.s. Thus,

$$\sup_{\omega \in \Omega_0} P_\omega^o\left(|X_n|_\infty < L, n = 1, \ldots, L^{2(1+\delta)}\right)$$

$$\leq \sup_{\omega \in \Omega_0} \sup_\alpha Q \times P_{\omega,B}^o\left(|Y_n^\alpha| < L, n = 1, \cdots, L^{2(1+\delta)}\right). \tag{3.3.24}$$

Let $g_{n,\omega}(x) = \sup_\alpha Q \times P_{\omega,B}(|Y_i^\alpha| < L, i = 1, \cdots, n|Y_0 = x)$ (it turns out eventually that $g_{n,\omega}$ does not depend on ω!) Then, due to the Markov property, $g_{n,\omega}(\cdot)$ must satisfy the dynamic programming equation

$$g_{n,\omega}(x) = \begin{cases} \max_{\alpha \in \mathcal{A}}\left(\frac{\alpha}{2}\left(g_{n+1,\omega}(x+1) + g_{n-1,\omega}(x-1)\right)\right) + (1-\alpha)g_{n-1,\omega}(x)\right), & |x| < L \\ 0, & |x| \geq L \end{cases}$$

and $g_{0,\omega}(x) = \mathbf{1}_{|x|<L}$. Next, we note that $g_{n,\omega}(\cdot)$ satisfies

$$g_{n,\omega}(x+1) + g_{n,\omega}(x-1) - 2g_{n,\omega}(x) \leq 0. \tag{3.3.25}$$

For $n = 0$ this is immediate, and hence

$$g_{1,\omega}(x) = g_{0,\omega}(x) + \varepsilon\Big(g_{0,\omega}(x+1) + g_{0,\omega}(x-1) - 2g_{0,\omega(x)}\Big).$$

We then have that $g_{1,\omega}(x)$ satisfies (3.3.25), and the argument can be iterated. We further conclude that

$$g_{n,\omega}(x) = g_{n-1,\omega}(x) + \varepsilon\Big(g_{n-1,\omega}(x+1) + g_{n-1,\omega}(x-1) - 2g_{n-1,\omega}(x)\Big). \quad (3.3.26)$$

Thus, $g_{n,\omega}(x)$ is nothing but the probability that a simple random walk on $\mathbb{Z}$ with geometric $(1 - 2\varepsilon)$ holding times, stays confined in a strip of size L for $L^{2(1+\delta)}$ units of time (note that (3.3.26) possesses a unique solution, which does not depend on $\omega \in \Omega_0$!). The conclusion (3.3.23) follows from solving (3.3.26) and combining it with (3.3.24).

From (3.3.23), we conclude that $E_\omega^o(\tau_{i+2}) \le Cr_{i+2}^{2(1+\delta)}$, for all i large enough, all $\omega \in \Omega_0$, where $\mathbb{P}(\Omega_0) = 1$. Thus,

$$Cr_{i+2}^{2(1+\delta)} \ge E_\omega^o\Big(E_\omega^o(\# \text{ visits of } X_n \text{ at } B_{i-1} \text{ for } n \in (\tau_i + 1, \cdots, \tau_{i+2})|X_{\tau_i})\Big)$$

$$= E_\omega^o\Bigg(\sum_{y \in B_{i-1}} E_\omega^{X_{\tau_i}}(\# \text{ visits at } y \text{ before } \tau_{i+2})\Bigg)$$

$$\ge \sum_{y \in B_{i-1}} E_\omega^o\Big(E_{\theta - y\omega}^{X_{\tau_i} - y}(\# \text{ visits at } 0 \text{ before } \tau_{i+1})\Big)$$

$$\ge C \sum_{y \in B_{i-1}} \max_{z \in E_i}\Big(E_{\theta - y\omega}^z(\# \text{ of visits at } 0 \text{ before } \tau_{i+1})\Big)$$

where $E_i = \{x : \frac{r_i}{2} < |x|_\infty < \frac{3r_i}{2}\}$, and Harnack's inequality (Lemma 3.3.8) was used in the last step. Taking P-expectations, we conclude that

$$Cr_{i+1}^{2(1+\delta)} \ge C \sum_{y \in B_{i-1}} \mathbb{E}^o\Big(\max_{z \in E_i} E_{\theta - y\omega}^z(\# \text{ of visits at } 0 \text{ before } \tau_{i+1})\Big)$$

$$\ge C \sum_{y \in B_{i-1}} \mathbb{E}^o\Big(E_{\theta - y\omega}^{X_{\tau_i}}(\# \text{ of visits at } 0 \text{ before } \tau_{i+1})\Big)$$

$$= C \sum_{y \in B_{i-1}} \mathbb{E}^o\Big(E_\omega^{X_{\tau_i}}(\# \text{ of visits at } 0 \text{ before } \tau_{i+1})\Big)$$

$$= C'(r_{i-1})^d \mathbb{E}^o\Big(E_\omega^{X_{\tau_i}}(\# \text{ of visits at } 0 \text{ before } \tau_{i+1})\Big),$$

where the shift invariance of P was used in the next to last equality. Therefore,

$$\mathbb{E}^o(\# \text{ of visits at } 0 \text{ between } \tau_i + 1 \text{ and } \tau_{i+1}) \le C'' r_i^{2+\delta-d}.$$

Hence, for $d \ge 3$,

$$\mathbb{E}^o(\# \text{ of visits at } 0) \leq C'' \sum_{i=1}^{\infty} r_i^{2+\delta-d} < \infty,$$

implying that P-a.s., $E_\omega^o(\# \text{ of visits at } 0) < \infty$, i.e. (X_n) is transient if $d \geq 3$.
Turning to $d = 2$, we recall the following lemma:

Lemma 3.3.27 (Derrienic[20]) *Let (Y_i) be a stationary and ergodic lattice valued sequence, and set $S_n = \sum_{i=1}^{n} Y_i$. Define*

$$R_n = \{\# \text{ of sites visited up to time } n\}.$$

Then,

$$\frac{R_n}{n} \xrightarrow[n \to \infty]{} \text{Prob}(S_i \neq 0, i \geq 1).$$

Proof. The sequence R_n is sub-additive and hence, by Kingman's ergodic sub-additive theorem, $R_n/n \to_{n \to \infty} a$, a.s. and in L^1, for some constant a. Noting that $R_{n+1} = R_n \circ \theta + 1_{Y_1 \notin \{\cup_{i=2}^{n+1} S_i\}}$, it holds that $(R_{n+1} - R_n \circ \theta) \xrightarrow[n \to \infty]{} 1_A \circ \theta$, where $A = \{S_i \neq 0, i \geq 1\}$. Thus, $ER_n/n \xrightarrow[n \to \infty]{} E1_A =: a$. $\qquad\square$

Under the measure on the environment Q introduced in this section, the increments $\{X_{n+1} - X_n\}$ are stationary and ergodic. Letting R_n denote the range of the RWRE up to time n, we have that

$$\frac{R_n}{n} \longrightarrow_{n \to \infty} Q \times P_\omega^o(\text{no return to } 0), \quad Q\text{-a.s.}$$

But, due to the CLT (Theorem 3.3.4 and Remark 3.3.5), for any $\delta > 0$,

$$\liminf_{n \to \infty} P_\omega^o\left(\frac{R_n}{n} < \delta\right) > 0, \quad Q\text{-a.s.}$$

Hence, for any $\delta > 0$,

$$P_\omega^o(\text{no return to } 0) < \delta, \quad Q\text{-a.s.}$$

and hence also P-a.s. This concludes the recurrence proof. $\qquad\square$
Remark: It is interesting to note that the transience (for $d \geq 3$) and recurrence (for $d = 2$) results are *false* for certain balanced, elliptic environments in Ω_0 (however, the P-probability of these environments is, of course, null). A simple example that exhibits the failure of recurrence for $d = 2$ was suggested by N. Gantert: fix $0.25 < p < 0.5$ and $q = 0.5 - p$. With $x = (x_1, x_2) \in \mathbb{Z}^2$, define

$$\omega(x, e) = \begin{cases} \frac{1}{4}, & x_1 = x_2, |e| = 1 \\ p, & \begin{cases} e = \pm e_2, |x_1| > |x_2| \\ \text{or} \\ e = \pm e_1, |x_1| < |x_2| \end{cases} \\ q, & \begin{cases} e = \pm e_1, |x_1| > |x_2| \\ \text{or} \\ e = \pm e_2, |x_1| < |x_2| \end{cases} \end{cases}$$

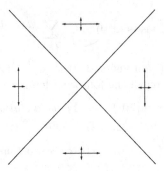

Fig. 3.3.2. Transient balanced environment, $d = 2$

Define
$$\nu(x) = \begin{cases} 1, & x \neq 0 \\ 4q, & x = 0. \end{cases}$$

Then, $\nu(\cdot)$ is an excessive measure, i.e.

$$(L_\omega^* \nu)(x) := \sum_{e:|e|=1} \omega(x-e, e)\nu(x-e) \leq \nu(x), \quad x \in \mathbb{Z}^2.$$

If $\{X_n\}$ was recurrent, then every excessive measure needs to equal the (unique) invariant measure. But, with

$$\nu((1,0)) = 1 > (\nu L_\omega)((1,0)) = 2q + 0.5.$$

Thus, $\nu(\cdot)$ is not invariant, contradicting the recurrence of the chain.

The intuitive idea behind the example above is that for points far from the origin, the "radial component" of the walk behaves roughly like a Bessel process of dimension $2 + \delta$, some $\delta > 0$, implying the transience. A similar argument, only more complicated, allows one to construct environments in $d \geq 3$ where the radial component behaves like a Bessel process of dimension $2 - \delta$, some $\delta > 0$. It is not hard to prove, using Lyapunov functions techniques, that there exists a $\kappa(d) < 1/2d$ such that if $d \geq 3$ and the balanced environment is such that $\min_{e:|e|=1} \omega(x, e) > \kappa(d)$ then the walk is transient.

Bibliographical notes: The basic CLT under Assumption 3.3.1 is due to Lawler [47], who transferred to the discrete setting some results of Papanicolau and Varadhan. An extension to the case of non nearest neighbour walks appears in [48]. The Harnack principle (Lemma 3.3.8) was provided in [49], and in greater generality in [46], whose approach we follow, after a suggestion by Sznitman (see also [69]).

The proof of the transience part in Theorem 3.3.22 was suggested by G. Lawler in private communication. The proof of the recurrence part is due to H. Kesten, also in private communication. A recent independent proof appears in [8]. Finally, the examples mentioned at the end of the section go back to Krylov (in the context of diffusions), with this version based on discussions with Comets and Gantert.

We comment that there are very few results on LLN's and CLT's for non balanced, non ballistic walks. One exception is the result in [9], where renormalization techniques are used to prove a (quenched) CLT in symmetric (not-balanced!) environments with small disorder. Another case, in which some of the RWRE coordinates perform a simple random walk, is analysed in details in [4], using cut-times of the random walk instead of the regeneration times used in Section 3.5.

3.4 Large deviations for nestling walks

In this section, we derive an LDP for a class of nearest neighbour random walks in random environment, in $\mathbb{Z}^d$. For reasons that will become clearer below, we need to restrict attention to environments which satisfy a condition on the support of P, which we call, after M. Zerner, "nestling environments". For technical reasons, we also need to make an independence assumption (see however the remark at the end of this section).

Define $d(\omega) := \sum_{e:|e|=1} \omega(0, e)e$, and let $P_d := P \circ d^{-1}$ denote the law of $d(\omega)$ under P.

Assumption 3.4.1

(C1) *P is i.i.d.*
(C2) *P is elliptic: there exists an $\varepsilon > 0$ such that $P(\omega(z, z + e_i) \geq \varepsilon) = P(\omega(z, z - e_i) \geq \varepsilon) = 1$, $i = 1, \cdots, d$.*
(C3) *(Nestling property): $0 \in conv\ (supp(P_d))$.*

We elaborate below on the nestling assumption. Clearly, balanced walks are nestling, but one may construct examples, as in $d = 1$, of nestling environments with ballistic behaviour.

For any $y \in \mathbb{R}^d$, we denote by $[y]$ the point in $\mathbb{Z}^d$ with $1 > y_i - [y]_i \geq 0$. For $z \in \mathbb{Z}^d$, we let $T_z = \inf\{n \geq 0 : X_n = z\}$. As in Section 2.3, the key to our approach to large deviation results for (X_n) is a large deviation principle for $T_{[nz]}$, $z \in \mathbb{R}^d$, stated next.

Theorem 3.4.2 *(a) Assume P is ergodic and elliptic. For any $z \in \mathbb{R}^d$, $|z|_1 = 1$, any $\lambda \leq 0$, the following deterministic limit exists P-a.s.*

$$a(\lambda, z) := \lim_{n \to \infty} \frac{1}{n} \log E^o_\omega (e^{\lambda T_{[nz]}} \mathbf{1}_{T_{[nz]} < \infty}).$$

(b) Further assume Assumption 3.4.1, and define

$$I_{T,z}(s) = \sup_{\lambda < 0} (\lambda s - a(\lambda, z)).$$

Then, $T_{[nz]}/n$ satisfies, P-a.s., under P_ω^o, a (weak) LDP, with rate function $I_{T,z}(s)$. That is,

$$\lim_{\delta \to 0} \limsup_{n \to \infty} \frac{1}{n} \log P_\omega^o(T_{[nz]}/n \in (s - \delta, s + \delta))$$

$$= \lim_{\delta \to 0} \liminf_{n \to \infty} \frac{1}{n} \log P_\omega^o(T_{[nz]}/n \in (s - \delta, s + \delta)) = -I_{T,z}(s), \quad P - a.s.$$

$$(3.4.3)$$

(Note that $I_{T,z}(s) = \infty$ for $s < 1$).

With Theorem 3.4.2 at hand, we may state the LDP for X_n/n. Define, for $x \in \mathbb{R}^d$,

$$I(x) = \begin{cases} |x|_1 I_{T, x/|x|_1}(1/|x|_1), & |x|_1 \le 1 \\ \infty, & \text{otherwise} \end{cases}.$$

Obviously, $a(\lambda, z)$ is defined for any $z \in \mathbb{R}^d \setminus \{0\}$, and is by definition homogeneous in $|z|_1$. An easy computation then reveals that $I(x) = \sup_{\lambda < 0}(\lambda - a(\lambda, x))$. We have the

Theorem 3.4.4 *Assume Assumption 3.4.1. Then, P-a.s., the random variables X_n/n satisfy the LDP in $\mathbb{R}^d$ with good, convex rate function $I(\cdot)$. That is,*

$$\lim_{\delta \to 0} \limsup_{n \to \infty} \frac{1}{n} \log P_\omega^o\left(\frac{X_n}{n} \in B_x(\delta)\right) = \lim_{\delta \to \infty} \liminf_{n \to \infty} \frac{1}{n} \log P_\omega^o\left(\frac{X_n}{n} \in B_x(\delta)\right)$$

$$= -I(x), \quad P - a.s.$$

Proof of Theorem 3.4.2

The idea behind the proof is relatively simple, and is related to our proof of large deviations for $d = 1$. However, there are certain complications in the proof of the lower bound, which can be overcome at present only under the nestling assumption.

a) We begin by defining, for $\lambda \le 0$,

$$a_{n,m}(\lambda, z) := \log E_\omega^{[mz]}\left(e^{\lambda T_{[nz]}} \mathbf{1}_{T_{[nz]} < \infty}\right).$$

We then have (since the time to reach $[nz]$ is not larger than the time to reach $[nz]$, when one is forced also to first visit $[mz]$), that

$$a_{n,0}(\lambda, z) \ge a_{m,0}(\lambda, z) + a_{n,m}(\lambda, z).$$

Further, we note that due to **C2**, there exists a constant $C(\lambda, \varepsilon)$ such that $n^{-1}|a_{n,0}(\lambda, z)| \le C(\lambda, \varepsilon)$, for all ω with $\omega(x, x + e) \ge \epsilon$, all $x \in \mathbb{Z}^d$ and e such that $|e| = 1$. Thus, by Kingman's subadditive ergodic theorem,

$$\frac{a_{n,0}(\lambda, z)}{n} \xrightarrow[n \to \infty]{} a(\lambda, z), \quad P\text{-a.s.} \tag{3.4.5}$$

b) By Chebycheff's inequality, (3.4.5) immediately implies the upper bound in (3.4.3). Thus, all our effort is now concentrated in proving the lower bound.

Toward this end, note that by Jensen's inequality, the deterministic function $a(\cdot, z)$ is convex, and thus differentiable a.e. We denote by $\mathcal{D}$ the set of $\lambda < 0$ such that $a(\cdot, z)$ is differentiable at λ. Recall that a point $s \in \mathbb{R}_+$ is an exposed point of $I_{T,z}(\cdot)$ if for some $\lambda < 0$ ("the exposing plane") and all $t \neq s$,

$$\lambda t - I_{T,z}(t) > \lambda s - I_{T,z}(s).$$

It is straightforward to check, see e.g., [19, Lemma 2.3.9(b)] that if $y = a'(\lambda, z)$ for some $\lambda \in \mathcal{D}$, then $I_{T,z}(y) = \lambda y - a(\lambda, z)$, and further y is an exposed point of $I_{T,z}(\cdot)$, with exposing plane λ.

As we already saw, it is then standard (see, e.g., [19, Theorem 2.3.6(b)]) that the lower bound in (3.4.3) holds for any exposed point. Thus, it only remains to handle points which are not exposed. Toward this end, define (using the monotonicity to ensure the existence of the limit!)

$$s_+ := \lim_{\lambda \to 0, \lambda \in \mathcal{D}} a'(\lambda, z) \leq \infty.$$

Note that, for any $s \geq s_+$, $I_{T,z}(s) = -\lim_{\lambda \to 0} a(\lambda, z)$.

The approach toward the lower bound is different when $s \geq s_+$ (case a) and $s < s_+$ (case b): in case a, a strategy which will achieve a lower bound consists of spending first some time in a "trap" at the neighborhood of the origin, returning to the origin and then getting to $[nz]$ within time roughly ns_-^η, where $s_-^\eta < s_+$ is an exposed point with $|I_{T,z}(s_-^\eta) - I_{T,z}(s_+)| \leq \eta$. The nestling assumption is crucial to create the trap. In case b, the achieving strategy consists of finding an intermediate point, progressing faster than needed toward the intermediate point, and then progressing slower than expected toward $[nz]$. To control the behavior of the walk starting at intermediate points, the independence assumption comes in handy.

Turning to case a, the role of the nestling assumption is evident in the following lemma:

Lemma 3.4.6 *Assume Assumption 3.4.1. Then, there exists an $\Omega_0 \subset \Omega$ with $P(\Omega_0) = 1$ with the following property: for each $\delta > 0$ and each $\omega \in \Omega_0$, there exists an $R(\delta, \omega)$ and an $n_0 = n_0(\delta, \omega)$ such that, for any $n > n_0$ even,*

$$P_\omega^o(|X_m|_2 \leq R(\delta, \omega), \quad m = 1, \cdots, n-1, X_n = 0) \geq e^{-\delta n}.$$

Proof of Lemma 3.4.6

We begin by constructing a "trap". As a preliminary, with $x \in \mathbb{Z}^d$, and $\bar{\imath}$ such that $|x_{\bar{\imath}}| \geq |x_j|$, $j = 1, \cdots, d$ (and hence $|x|_2 \leq \sqrt{d}|x_{\bar{\imath}}|$) we have, defining $y_x = x - \text{sign}(x_i)e_i$, that

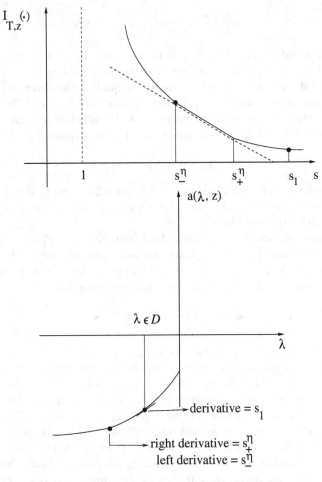

Fig. 3.4.1. exposed points and differentiability

$$|x|_2 - |y_x|_2 = \frac{|x_{\bar{i}}|^2 - (|x_{\bar{i}}| - 1)^2}{|x|_2 + |y_x|_2} \geq \frac{1}{2\sqrt{d}} \, .$$

Fix $\kappa = \varepsilon\delta/32\sqrt{d}$ and $F(x) = (1 - \kappa^2|x|_2^2) \vee 0$. Call a site $x \in \mathbb{Z}^d$ "successful" if

$$x \cdot \sum_{i=1}^{d} \Big(\omega(x, x + e_i)e_i - \omega(x, x - e_i)e_i \Big) \leq 1 \, .$$

Due to **(C3)**,

$$P(x \text{ is successful}) > 0 \, .$$

and hence, by the independence assumption **(C1)**,

$$P(\text{all sites } x \in B_{1/\kappa}(0) \text{ are successful}) > 0. \tag{3.4.7}$$

Fix now $\omega \in \Omega$ such that all sites $x \in B_{1/\kappa}(0)$ are successful. We next claim that for such ω,

$$\sum_i \omega(x, x \pm e_i) F(x \pm e_i) \geq e^{-\delta/3} F(x). \tag{3.4.8}$$

Indeed, for $|x|_2 \geq 1/\kappa$ this is obvious, for $1/\kappa - \varepsilon/4\sqrt{d} < |x|_2 < 1/\kappa$ this follows from the ellipticity assumption (C2) while for $|x|_2 < 1/\kappa - \varepsilon/4\sqrt{d}$ this follows from a Taylor expansion. Thus, $e^{\delta n/3} F(X_n)$ is, for such ω, a submartingale under P_ω^o, and we have that for all $n \geq 1$,

$$e^{-\delta n/3} = e^{-\delta n/3} E_\omega^o F(X_0) \leq e^{-\delta n/3} E_\omega^o \left(e^{\delta n/3} F(X_n) \right) \leq P_\omega^o \left(|X_n|_2 < \frac{1}{\kappa} \right). \tag{3.4.9}$$

Fixing n_1 even large enough such that $e^{-\delta n_1/3} \varepsilon^{\sqrt{d}/\kappa} \geq e^{-2\delta n_1/3}$, we conclude that for such ω,

$$P_\omega^o \left(|X_m|_\infty \leq n_1, m = 1, \ldots, n_1 - 1, \quad X_{n_1} = 0 \right) \geq e^{-2\delta n_1/3}.$$

Due to (3.4.7) and (C1), there exists (P-a.s.) an $x_0 = x_0(\omega, \delta)$ such that all sites in $B_{1/\kappa}(x_0)$ are successful. Set $m_0 = m_0(\omega, \delta) := \sum_{i=1}^d |x_0(\omega, \delta)(i)|$. Due to the ellipticity assumption (C2), we have

$$P_\omega^o \left(X_{m_0} = x_0(\omega, \delta) \right) \geq \varepsilon^{m_0}, \quad P_\omega^{x_0} (X_{m_0} = 0) \geq \varepsilon^{m_0}.$$

Next set $R(\delta, \omega) := n_1 + 2m_0 + 1$. Define $K = \lfloor (n - 2m_0)/n_1 \rfloor$. We then have, using the Markov property, that

$$P_\omega^o \left(|X_m|_2 \leq R(\delta, \omega), \quad m = 1, \cdots, n, \quad X_n = 0 \right)$$
$$\geq P_\omega^o \left(X_{m_0} = x_0(\omega, \delta) \right) P_\omega^{x_0} \left(|X_m - x_0|_\infty \leq n_1, X_{n_1} = 0 \right)^K$$
$$P_\omega^{x_0} (X_{m_0} = 0) P_\omega^o (X_{n - Kn_1 - 2m_0} = 0)$$
$$\geq \varepsilon^{2m_0} \varepsilon^{n_1} \cdot e^{-\frac{2\delta}{3} K n_1} \geq e^{-\delta n},$$

for all $n > n_0(\delta, \varepsilon, \omega)$. $\qquad\square$

Equipped with Lemma 3.4.6 we may complete the proof in case a. Indeed, all we need to prove is that for any $\delta > 0$,

$$\liminf_{n \to \infty} \frac{1}{n} \log P_\omega^o \left(T_{[nz]}/n \in (s - \delta, s + \delta) \right) = -I_{T,z}(s_+), \quad P - a.s.$$

Fix $\eta > 0$ and an exposed point s_-^η with $|I_{T,z}(s_-^\eta) - I_{T,z}(s_+)| \leq \eta$. Due to the Markov property, for all n such that $|nz|_\infty > R(\delta, \omega)$,

$$P_\omega^o\left(T_{[nz]}/n \in (s - \delta, s + \delta)\right)$$

$$\geq P_\omega^o\left(|X_m|_2 \leq R(\delta, \omega),\, m = 1, \cdots, \lceil n(s - s_-^\eta)\rceil,\, X_{\lceil n(s-s_-^\eta)\rceil} = 0\right)$$

$$P_\omega^o\left(T_{[nz]}/n \in (s_-^\eta - \delta',\, s_-^\eta + \delta')\right)$$

where $\delta' = \delta \cdot s_-^\eta/2s$, and hence,

$$\liminf_{n\to\infty} \frac{1}{n} \log P_\omega^o\left(T_{[nz]}/n \in (s - \delta, s + \delta)\right) \geq -\delta - I_{T,z}(s_-^\eta).$$

Since δ is arbitrary and $I_{T,z}(s_-^\eta) \underset{\eta\to 0}{\to} I_{T,z}(s_+)$, the proof is concluded for $s \geq s_+$.

Turning to case b, recall that our plan is to consider intermediate points. This requires a slight strengthening of the convergence of $a(\lambda, z)$. We state this in the following Lemma, whose proof is deferred.

Lemma 3.4.10 *Assume Assumption 3.4.1 and set* $\nu \in (0,1)$. *Then, for* $z \in \mathbb{R}^d$, $|z|_1 = 1$, *and any* $\lambda < 0$, *we have*

$$\lim_{n\to\infty} \frac{1}{n} \log E_\omega^{[\nu nz]}\left(e^{\lambda T_{[nz]}}\, \mathbf{1}_{T_{[nz]}<\infty}\right) = (1 - \nu)a(\lambda, z), \quad P - a.s.$$

Assuming Lemma 3.4.10, we complete the proof of part b. Note the existence, for any $\eta \geq 0$, of $s_-^\eta < s < s_+^\eta$ such that s_-^η, s_+^η are exposed, and further

$$\left| I_{T,z}(s) - \left(\frac{s - s_-^\eta}{s_+^\eta - s_-^\eta}\right) I_{T,z}(s_-^\eta) - \left(\frac{s_+^\eta - s}{s_+^\eta - s_-^\eta}\right) I_{T,z}(s_+^\eta) \right| < \eta. \qquad (3.4.11)$$

Set $\nu := (s_+^\eta - s)/(s_+^\eta - s_-^\eta)$. By the Markov property,

$$P_\omega^o\left(T_{[nz]}/n \in (s - \delta, s + \delta)\right)$$

$$\geq P_\omega^o\left(T_{[\nu nz]}/n \in (s_-^\eta - \delta',\, s_-^\eta + \delta')\right) P_\omega^{[\nu nz]}\left(T_{[nz]}/n \in (s_+^\eta - \delta',\, s_+^\eta + \delta')\right)$$

where $\delta' = \frac{\min(\nu, 1-\nu)\delta}{2}$. Due to Lemma 3.4.10, and the fact that s_+^η, s_-^η are exposed points of $I_{T,z}(\cdot)$, one concludes that

$$\lim_{\delta\to\infty} \liminf_{n\to\infty} \frac{1}{n} \log P_\omega^o\left(T_{[nz]}/n \in (s - \delta, s + \delta)\right)$$

$$\geq -\left[\left(\frac{s - s_-^\eta}{s_+^\eta - s_-^\eta}\right) I_{T,z}(s_-^\eta) + \left(\frac{s_+^\eta - s}{s_+^\eta - s_-^\eta}\right) I_{T,z}(s_+^\eta)\right].$$

Using (3.4.11), this completes the proof of the theorem. $\square$

Proof of Theorem 3.4.4

Fix x and δ as in the statement of the theorem. Then, using the ellipticity assumption **(C2)**, for any n large enough,

$$P_\omega^o\left(\frac{X_n}{n} \in B_x(\delta)\right) \geq P_\omega^o\left(T_{[nx]} \in n\left(1 - \frac{\delta}{2}, 1 + \frac{\delta}{2}\right)\right) \varepsilon^{n\delta/2}$$

and the lower bound follows from Theorem 3.4.2.

Turning to the upper bound, note that $|nB_x(\delta) \cap \mathbb{Z}^d| \leq C_\delta n^d$, and that

$$P_\omega^o\left(\frac{X_n}{n} \in B_x(\delta)\right) = \sum_{y \in nB_x(\delta) \cap \mathbb{Z}^d} P_\omega^o(X_n = y).$$

Further, note that $P_\omega^o(X_n = y) \leq P_\omega^o(T_{[y]} \leq n)$, and that due to the ellipticity **(C2)**,

$$\sup_{y \in nB_x(\delta)} P_\omega^o(X_n = y) \leq \varepsilon^{-n\sqrt{d}\delta} P_\omega^o\left(T_{[nx]} \leq n(1 + \delta)\right)$$

and hence,

$$\lim_{\delta \to \infty} \limsup_{n \to \infty} \frac{1}{n} \log P_\omega^o\left(\frac{X_n}{n} \in B_x(\delta)\right)$$

$$\leq \lim_{\delta \to \infty} \limsup_{n \to \infty} \frac{1}{n} \log P_\omega^o\left(T_{[nx]} \leq n(1 + \delta)\right)$$

$$\leq -\inf_{0 \leq \eta \leq 1} I(\eta x), \quad P - a.s.$$

The monotonicity of $I(\eta \cdot)$ in η, which is induced from that of $I_{T,z}(\cdot)$, completes the proof.   □

Proof of Lemma 3.4.10

By the ellipticity assumption **(C2)**, $\frac{1}{n} \log E_\omega^{[\nu n z]}$ $(e^{\lambda T_{[nz]}} \mathbf{1}_{T_{[nz]} < \infty})$ is uniformly bounded. Further, it possesses the same law as $\frac{1}{n} \log E_\omega^o(e^{\lambda T_{[n(1-\eta)z]}} \mathbf{1}_{T_{[n(1-\eta)z]} < \infty})$. Thus,

$$\frac{1}{n} \log E_\omega^{[\nu n z]}\left(e^{\lambda T_{[nz]}} \mathbf{1}_{T_{[nz]} < \infty}\right) \xrightarrow[n \to \infty]{P} (1 - \nu)a(\lambda, z). \tag{3.4.12}$$

Our goal is thus to prove that the convergence in (3.4.12) is in fact a.s.

Toward this end, as a first step we truncate appropriately the expectation. Set

$$N_x = \#\{ \text{ visits at } x \text{ before } T_{[nz]} \},$$

and $N = \sup_{x \in \mathbb{Z}^d} N_x$. We show that, for some $\delta < 1$,

$$\limsup_{n \to \infty} \frac{1}{n} \log \frac{E_\omega^{[\nu n z]}(e^{\lambda T_{[nz]}} \mathbf{1}_{T_{[nz]} < \infty})}{E_\omega^{[\nu n z]}(e^{\lambda T_{[nz]}} \mathbf{1}_{T_{[nz]} < \infty} \mathbf{1}_{N < n^\delta})} = 0, \quad P - a.s. \tag{3.4.13}$$

Indeed, note first that

$$E_\omega^{[\nu n z]}\left(e^{\lambda T_{[nz]}}\, \mathbf{1}_{T_{[nz]}<\infty}\right) \le E_\omega^{[\nu n z]}\left(e^{\lambda T_{[nz]}}\, \mathbf{1}_{T_{[nz]}<\infty}\, \mathbf{1}_{N<n^\delta}\right)$$
$$+ \sum_{x\in\mathbb{Z}^d} E_\omega^{[\nu n z]}\left(e^{\lambda T_{[nz]}}\, \mathbf{1}_{T_{[nz]}<\infty}\, \mathbf{1}_{N_x>n^\delta}\right).$$

But, due to the Markov property,

$$E_\omega^{[\nu n z]}\left(e^{\lambda T_{[nz]}}\, \mathbf{1}_{T_{[nz]}<\infty}\, \mathbf{1}_{N_x>n^\delta}\right)$$
$$\le \sum_{k=[n^\delta]+1}^{\infty} E_\omega(e^{\lambda T_x}\mathbf{1}_{T_x<T_{[nz]}})^k E_\omega^{[\nu n z]}\left(e^{\lambda T_{[nz]}}\, \mathbf{1}_{T_x<T_{[nz]}<\infty}\right)$$
$$\le \frac{e^{\lambda n^\delta}}{1-e^\lambda} E_\omega^{[\nu n z]}\left(e^{\lambda T_{[nz]}}\, \mathbf{1}_{T_{[nz]}<\infty}\right),$$

and hence,

$$E_\omega^{[\nu n z]}\left(e^{\lambda T_{[nz]}}\, \mathbf{1}_{T_{[nz]}<\infty}\right) \le \frac{E_\omega^{[\nu n z]}\left(e^{\lambda T_{[nz]}}\, \mathbf{1}_{T_{[nz]}<\infty}\, \mathbf{1}_{N<n^\delta}\right)}{1-n^d e^{\lambda n^\delta}/(1-e^\lambda)},$$

yielding (3.4.13). Further, due to the ellipticity assumption **(C2)**, it holds that for some constant $K=K(\lambda)$ large enough,

$$E_\omega^{[\nu n z]}(e^{\lambda T_{[nz]}}\, \mathbf{1}_{T_{[nz]}<Kn}\, \mathbf{1}_{N<n^\delta}) \ge E_\omega^{[\nu n z]}(e^{\lambda T_{[nz]}}\, \mathbf{1}_{T_{[nz]}<\infty}\, \mathbf{1}_{N<n^\delta})/2 .$$

Thus, it suffices to consider

$$g_\omega^\delta = \log E_\omega^{[\nu n z]}(e^{\lambda T_{[nz]}}\, \mathbf{1}_{T_{[nz]}<Kn}\, \mathbf{1}_{N<n^\delta}) .$$

Denote by $\mathcal{P}_{k,\delta}$ the set of nearest neighbour paths (γ_n) on $\mathbb{Z}^d$ with $\gamma_0=[\nu n z]$, $\gamma_k=[nz]$, and $N(\gamma)\le n^\delta$. For $e\in\{\pm e_i\}_{i=1}^d =: \mathcal{E}$, set

$$N_{x,e}(\gamma) = \#\{\text{ steps from } x \text{ to } x+e \text{ of } \gamma \text{ before } T_{[nz]}(\gamma)\}.$$

Then, with $\beta(x,x+e)=\log\omega(x,x+e)$ and $D_{Kn}=\{x\in\mathbb{Z}^d : |x|_\infty \le Kn\}$,

$$g_\omega^\delta = \log \sum_{k\le Kn} e^{\lambda k} \sum_{\gamma\in\mathcal{P}_{k,\delta}} \prod_{x\in D_{Kn}} \prod_{e\in\mathcal{E}} e^{\beta(x,x+e)N_{x,e}(\gamma)} .$$

We use the following concentration inequality, which is a slight variant of [77, Theorem 6.6].

Lemma 3.4.14 (Talagrand) *Let $\mathcal{K}\subset\mathbb{R}^{d_1}$ be compact and convex. Let μ be a law supported on $\mathcal{K}$, and let $f:\mathcal{K}^N\to\mathbb{R}$ be convex and of Lipschitz constant L. Finally, let M_N denote the median of f with respect to $\mu^{\otimes N}$, i.e. M_N is the smallest number such that*

$$\mu^{\otimes N}(f \leq M_N) \geq \frac{1}{2}, \ \mu(f \geq M_N) \geq \frac{1}{2}.$$

Then, there exists a constant $C = C(\mathcal{K})$, independent of f, μ, such that for all $t > 0$,

$$\mu^{\otimes N}(|f - M_N| \geq t) \leq C \exp(-Ct^2/L^2).$$

To apply Lemma 3.4.14, note that

$$\left| \frac{\partial g_\omega^\delta}{\partial \beta(x, x+e)} \right|$$

$$\leq \frac{1}{\varepsilon} \frac{\sum_{k \leq Kn} e^{\lambda k} \sum_{\gamma \in \mathcal{P}_{k,\delta}} N_{x,e}(\gamma) \prod_{x' \in D_{Kn}} \prod_{e \in \mathcal{E}} e^{\beta(x', x'+e)N_{x',e}(\gamma)}}{\sum_{k \leq Kn} e^{\lambda k} \sum_{\gamma \in \mathcal{P}_{k,\delta}} \prod_{x' \in D_{Kn}} \prod_{e \in \mathcal{E}} e^{\beta(x', x'+e)N_{x',e}(\gamma)}}.$$

$$(3.4.15)$$

Thus, using Jensen's inequality in the first inequality,

$$\sum_{x \in D_{Kn}} \left| \frac{\partial g_\omega^\delta}{\partial \beta(x, x+e)} \right|^2 \leq \frac{1}{\varepsilon} \sum_{x \in D_{Kn}}$$

$$\frac{\sum_{k \leq Kn} e^{\lambda k} \sum_{\gamma \in \mathcal{P}_{k,\delta}} N_{x,e}(\gamma)^2 \prod_{x' \in D_{Kn}} \prod_{e \in \mathcal{E}} e^{\beta(x', x'+e)N_{x',e}(\gamma)}}{\sum_{k \leq Kn} e^{\lambda k} \sum_{\gamma \in \mathcal{P}_{k,\delta}} \prod_{x' \in D_{Kn}} \prod_{e \in \mathcal{E}} e^{\beta(x', x'+e)N_{x,e}(\gamma)}}$$

$$\leq \frac{Kn^{1+\delta}}{\varepsilon}.$$

It is immediate to see that on the other hand g_ω^δ is a convex function of $\{\beta(x, x+e)\}$. Hence, by Lemma 3.4.14 and the above,

$$P(|g_\omega^\delta - Eg_\omega^\delta| > tn) \leq C_1 e^{-C_1 n^{1-\delta}},$$

where $C_1 = C_1(\epsilon, \delta)$. The Borel-Cantelli lemma then completes the proof of Lemma 3.4.10.   □

Remarks: 1. In the proof above, the independence assumption **(C1)** was used in two places. The first is the construction of traps (Lemma 3.4.6), where the independence assumption may be replaced by the requirement that P, when restricted to finite subsets, be equivalent to a product measure. More seriously, the product structure was used in the application of Talagrand's Lemma 3.4.14. It is plausible that this can be bypassed, e.g. using the techniques in [68].

2. S. R. S. Varadhan has kindly indicated to me a direct argument which gives the quenched LDP for the position, for ergodic environments, without passing through hitting times. Fix $\epsilon > 0$, and define X_n^ϵ to be the RWRE with geometric holding times of parameter $1/\epsilon$. Fix a deterministic v, with $|v|_1 < 1$, and define

$$g(m, n) = P_{\theta[mv]\omega}^0 (X_{m-n}^\epsilon - X_0^\epsilon = [(n-m)v]).$$

Then, $g(0, n + m) \geq g(0, m)g(m, n + m) > 0$ for all $n, m \geq 1$. Consequently, by Kingman's ergodic sub-additive theorem,

$$\frac{1}{n} \log g(0, n) \to_{n \to \infty} -I^\epsilon(v), P - a.s.,$$

for some deterministic $I^\epsilon(v)$. From this it follows (see e.g. [19, Theorem 4.1.11]) that X_n^ϵ/n satisfies the (quenched) LDP with convex, good rate function $I^\epsilon(\cdot)$. Finally, it is easy to check that $I^\epsilon(\cdot) \to_{\epsilon \to 0} I(\cdot)$ (even uniformly on compacts) and that

$$\limsup_{\epsilon \to 0} \limsup_{n \to \infty} \frac{1}{n} \log P_\omega^o(|X_n - X_n^\epsilon| > \delta n) = -\infty, P - a.s.,$$

from which it follows that X_n/n satisfies the quenched LDP with deterministic, convex, good rate function $I(\cdot)$.

3. Returning to the i.i.d. nestling setup, a natural question is whether one may prove an annealed large deviations principle for the position. A partial answer is given by the following. Fix a direction ℓ and recall the time $D = D(\ell)$ introduced in Section 3.2. Define $T_k^\ell = \min\{n : (X_n - X_0) \cdot \ell \geq k\}$. Then, for any $\lambda \in \mathbb{R}$,

$$\mathbb{E}^o(e^{\lambda T_{k+m}^\ell} \mathbf{1}_{\{D(\ell)=\infty\}}) \geq E_P\left(E_\omega^o\left(e^{\lambda T_k^\ell} \mathbf{1}_{\{D(\ell)>T_k^\ell\}}\right) E_\omega^{X_{T_k^\ell}}\left(e^{\lambda T_m^\ell} \mathbf{1}_{\{D(\ell)=\infty\}}\right)\right)$$

$$\geq \mathbb{E}^o(e^{\lambda T_k^\ell} \mathbf{1}_{\{D(\ell)=\infty\}}) \mathbb{E}^o(e^{\lambda T_m^\ell} \mathbf{1}_{\{D(\ell)=\infty\}}),$$

and hence, by sub-additivity, the following limit exists:

$$\lim_{k \to \infty} \frac{1}{k} \log \mathbb{E}^o(e^{\lambda T_k^\ell} \mathbf{1}_{\{D(\ell)=\infty\}}) =: g(\ell, \lambda).$$

One can check that if the conclusions of Lemma 3.5.11 hold then also, for $-\lambda > 0$ small enough,

$$\limsup_{k \to \infty} \frac{1}{k} \log \mathbb{E}^o(e^{\lambda T_k^\ell}) = \lim_{k \to \infty} \frac{1}{k} \log \mathbb{E}^o(e^{\lambda T_k^\ell} \mathbf{1}_{\{D'=\infty\}}),$$

and hence for such λ,

$$g(\ell, \lambda) = \limsup_{k \to \infty} \frac{1}{k} \log \mathbb{E}^o(e^{\lambda T_k^\ell}).$$

An interesting open question is to use this argument, in the nestling setup, to deduce a LDP and to relate the annealed and quenched rate functions.

Bibliographical notes: Large deviations for the position X_n of nestling RWRE in $\mathbb{Z}^d, d > 1$ were first derived in Zerner's thesis [80]. Zerner uses a martingale differences argument instead of Lemma 3.4.14. With the same technique, he also derives a more general version of Lemma 3.4.10, under the name *"uniform shape theorem". The large deviations for the hitting times $T_{[nz]}$ are implicit in his approach.*

A recent paper of Varadhan [79] develops the quenched large deviations alluded to in remark 2 above, and a corresponding annealed LDP. He also obtains information on the zero set of the annealed and quenched rate functions, and in particular proves in a great generality that they coincide. The techniques are quite different from those presented here.

3.5 Kalikow's condition

We introduce in this section a condition on the environment, due to Kalikow, which ensures that the RWRE is "ballistic". Suppose P is elliptic, and let U be a strict subset of $\mathbb{Z}^d$, with $0 \in U$, and define on $U \cup \partial U$ an auxiliary Markov chain with transition probabilities

$$\hat{P}_U(x, x+e) = \begin{cases} \dfrac{\mathbb{E}^o\left[\sum_{n=0}^{\tau_{U^c}} \mathbf{1}_{\{X_n=x\}}\omega(x,x+e)\right]}{\mathbb{E}^o\left[\sum_{n=0}^{\tau_{U^c}} \mathbf{1}_{\{X_n=x\}}\right]}, & x \in U, |e| = 1 \\ 1 & x \in \partial U, e = 0 \end{cases} \quad (3.5.1)$$

where $\tau_{U^c} = \min\{n \geq 0 : X_n \in \partial U\}$ (note that the expectations in (3.5.1) are finite due the Markov property and ellipticity). The transition kernel $\hat{P}_U$ weights the transitions $x \mapsto x + e$ according to the occupation time of the vertex x before exiting U. We denote by $\hat{E}_U$ expectations with respect to the measure $\hat{P}_U$.

The following is a basic consequence of the definition of $\hat{P}_U(\cdot, \cdot)$:

Lemma 3.5.2 (Kalikow) *Assume $\hat{P}_U(\tau_{U^c} < \infty) = 1$. Then, $\hat{P}_U(X_{\tau_{U^c}} = v) = \mathbb{P}^o(X_{\tau_{U^c}} = v), v \in \partial U$. In particular, $\mathbb{P}^o(\tau_{U^c} < \infty) = 1$.*

Proof of Lemma 3.5.2:

Set $g_\omega(x) = E^o_\omega\left(\sum_{n=0}^{\tau_{U^c}} \mathbf{1}_{\{X_n=x\}}\right)$. Then

$$\hat{P}_U(x, y) = \frac{E(g_\omega(x)\omega(x, y))}{E(g_\omega(x))}, \quad x \in U, \ y \in U \cup \partial U. \quad (3.5.3)$$

But, due to the Markov property,

$$g_\omega(x) = \mathbf{1}_{\{x=0\}} + \sum_{z \in U} \omega(z, x) g_\omega(z),$$

and hence, using (3.5.3),

$$\sum_{x \in U} \Big(E(g_\omega(x))\Big)\hat{P}_U(x, y) + \mathbf{1}_{\{y=0\}} = E(g_\omega(y)).$$

Set $\hat{\pi}_n(y) = \hat{E}_U\left(\sum_{j=0}^{\tau_{U^c} \wedge n} \mathbf{1}_{\{X_j=y\}}\right)$. Then, $\hat{\pi}_0(y) = \mathbf{1}_{\{y=0\}}$ and

$$\hat{\pi}_{n+1}(y) = 1_{\{y=0\}} + \sum_{x \in U} \hat{P}_U(x,y)\hat{\pi}_n(x).$$

Then, for $y \in U \cup \partial U$,

$$E\Big(g_\omega(y)\Big) - \hat{\pi}_{n+1}(y) = \sum_{x \in U} \hat{P}_U(x,y)\Big(E(g_\omega(x)) - \hat{\pi}_n(x)\Big).$$

Since $E(g_\omega(y)) - \hat{\pi}_0(y) \geq 0$, it follows by the positivity of $\hat{P}_U(x,y)$ that for $y \in U \cup \partial U$,

$$\hat{E}_U\left(\sum_{n=0}^{\tau_{U^c}} 1_{\{X_n=y\}}\right) = \lim_{n\to\infty} \hat{\pi}_n(y) \leq E(g_\omega(y)).$$

Taking $y \in \partial U$ yields

$$\hat{P}_U(X_{\tau_{U^c}} = y) \leq \mathbb{P}^o(X_{\tau_{U^c}} = y), \quad y \in \partial U.$$

On the other hand, $\sum_{y \in \partial U} \hat{P}_U(X_{\tau_{U^c}} = y) = 1$ because $\hat{P}_U(\tau_{U^c} < \infty) = 1$ by assumption. Hence

$$\mathbb{P}^o(X_{\tau_{U^c}} = y) = \hat{P}_U(X_{\tau_{U^c}} = y), \quad \forall y \in \partial U. \qquad \square$$

We are now ready to introduce Kalikow's condition. Fix a hyperplane by picking a point $\ell \in \mathbb{R}^d \backslash \{0\}$, $|\ell|_1 \leq 1$. Define

$$\varepsilon_\ell := \inf_{U, x \in U} \sum_{|e|=1} (\ell \cdot e)\hat{P}_U(x, x+e)$$

where the infimum is over all connected strict subsets of $\mathbb{Z}^d$ containing 0. We say that *Kalikow's condition with respect to ℓ holds* if $\varepsilon_\ell > 0$. Note that ε_ℓ acts as a drift in the direction ℓ for the Markov chain $\hat{P}_U$.

A consequence of Lemma 3.5.2 is the following:

Theorem 3.5.4 *Assume that P satisfies Assumption 3.1.1. If Kalikow's condition with respect to ℓ holds, then $\mathbb{P}^o(A_\ell) = 1$. If further P is an i.i.d. measure then $v_\ell > 0$.*

Proof. Fix $U_L = \{z \in \mathbb{Z}^d : |z \cdot \ell| \leq L\}$. Let $\hat{X}_{n,L}$ denote the Markov chain with $\hat{X}_{0,L} = 0$ and transition law $\hat{P}_{U_L}(x, x+e)$. Set the local drift at x, $\hat{d}(x) = \sum_{|e|=1} e\hat{P}_{U_L}(x, x+e)$, and recall that $\hat{X}_{n,L} - \sum_{i=0}^{n-1} \hat{d}(\hat{X}_{i,L})$ is a martingale, with bounded increments. It follows that for some constant C,

$$\hat{P}_{U_L}\left(\sup_{0 \leq n \leq N} |\hat{X}_{n,L} - \sum_{i=0}^{n-1} \hat{d}(\hat{X}_{i,L})| > \delta N\right) \leq Ce^{-C\delta^2 N}. \tag{3.5.5}$$

On the other hand, $\sum_{i=0}^{n-1} \hat{d}(\hat{X}_{i,L}) \cdot \ell \geq \varepsilon_\ell(n \wedge \tau_{U_L^c})$ while $|\hat{X}_{n,L} \cdot \ell| \leq L+1$. We thus conclude from (3.5.5) that

$$\hat{P}_{U_L}\left(\tau_{U_L^c} > \frac{L+1}{\varepsilon_\ell} + \frac{\delta}{\varepsilon_\ell}N\right) \leq Ce^{-C\delta^2 N/\varepsilon_\ell^2},$$

and hence, for some C_1 independent of L, and all L large,

$$\hat{P}_{U_L}\left(|\hat{X}_{\tau_{U_L^c}} \cdot \ell - L| > 1\right) \leq C_1 e^{-C_1 L}.$$

It follows from Lemma 3.5.2 that

$$\mathbb{P}^o\left(|X_{\tau_{U_L^c}} \cdot \ell - L| > 1\right) \leq C_1 e^{-C_1 L}.$$

A similar argument shows that $\mathbb{P}^o(D = \infty) > 0$: indeed, take now $U_{L,+} = \{z \in \mathbb{Z}^d : 0 \leq z \cdot \ell \leq L\}$. Arguing as above, one finds that for some $C_2 > 0$ independent of L,

$$\hat{P}_{U_{L,+}}\left(|\hat{X}_{\tau_{U_{L,+}^c}} - L| \leq 1\right) > C_2,$$

implying that

$$\mathbb{P}^o\left(|X_{\tau_{U_{L,+}^c}} - L| \leq 1\right) > C_2.$$

Thus, $\mathbb{P}^o(D = \infty) > 0$, and then, by an argument as in the proof of Theorem 3.1.2, $\mathbb{P}^o(A_\ell) > 0$. By Theorem 3.1.2, it follows that $\mathbb{P}^o(A_\ell \cup A_{-\ell}) = 1$. Due to (3.5.5), it holds that $\mathbb{P}^o(\limsup_{n\to\infty} X_n \cdot \ell = \infty) = 1$. We thus conclude that $\mathbb{P}^o(A_\ell) = 1$.

To see that if P is i.i.d. then $v_\ell > 0$, recall the regeneration times $\{\tau_i\}_{i\geq 1}$ introduced in Section 3.2. By Lemma 3.2.5, it suffices to prove that $\mathbb{E}^o(\tau_1 | D = \infty) < \infty$. Let $U_{m,k,-} = \{z \in \mathbb{Z}^d : |z| < k, z \cdot \ell < m\}$, and set $T_{m,k} := T_{U_{m,k,-}}$ with $T_m = \lim_{k\to\infty} T_{m,k} = \min\{n : X_n \cdot \ell \geq m\}$, $m \geq 1$. By Kalikow's condition,

$$\mathbb{E}^o\left(\sum_{n=0}^{T_{m,k}} 1_{\{X_n=x\}} \sum_{e:|e|=1} w(x, x+e)\ell \cdot e\right) \geq \varepsilon_\ell \mathbb{E}^o\left(\sum_{n=0}^{T_{m,k}} 1_{\{X_n=x\}}\right),$$

and hence, summing over $x \in U_{m,k,-}$ and recalling that $X_{i+1} - X_i - d(\theta^{X_i}\omega)$ is a martingale difference sequence, one gets

$$1 + m \geq \mathbb{E}^o(X_{T_{m,k}} \cdot \ell) \geq \varepsilon_\ell \mathbb{E}^o(T_{m,k}),$$

and taking $k \to \infty$ one concludes that $m+1 \geq \varepsilon_\ell \mathbb{E}^o(T_m)$. In particular,

$$\mathbb{E}^o(\liminf_{m\to\infty} T_m/m) \leq \liminf_{m\to\infty} \mathbb{E}^o(T_m/m) \leq 1/\varepsilon_\ell. \qquad (3.5.6)$$

Since $\tau_i \to_{i\to\infty} \infty$, $\mathbb{P}^o$-a.s., one may find a (random) sequence k_m such that $\tau_{k_m} \leq T_m < \tau_{k_m+1}$. By definition,

$$\ell \cdot X_{\tau_{k_m}} \leq \ell \cdot X_{T_m} \leq \ell \cdot X_{\tau_{k_m+1}},$$

and further,

$$\ell \cdot X_{\tau_k}/k \to_{k\to\infty} \mathbb{E}^o(\ell \cdot X_{\tau_1}|\{D = \infty\}) = \frac{1}{\mathbb{P}^o(D = \infty)} < \infty, \mathbb{P}^o - a.s.,$$

due to Lemma 3.2.5 and (3.2.7). Thus, $k_m/m \to_{m\to\infty} \mathbb{E}^o(X_{\tau_1}\cdot\ell|\{D = \infty\})^{-1}$, $\mathbb{P}^o$-a.s. But, since $\tau_{k_m}/k_m \to_{m\to\infty} \mathbb{E}^o(\tau_1|\{D = \infty\}) \in [1, \infty]$, $\mathbb{P}^o$-a.s., it follows that

$$\liminf_{m\to\infty} \frac{T_m}{m} \geq \liminf_{m\to\infty} \frac{\tau_{k_m}}{k_m}\frac{k_m}{m} = \lim_{m\to\infty} \frac{\tau_{k_m}}{k_m}\frac{k_m}{m} = \frac{\mathbb{E}^o(\tau_1|\{D = \infty\})}{\mathbb{E}^o(X_{\tau_1}\cdot\ell|\{D = \infty\})}$$

$$= \mathbb{E}^o(\tau_1|\{D = \infty\})\mathbb{P}^o(D = \infty).$$

Since (3.5.6) implies that $\liminf_{m\to\infty} T_m/m < \infty$, we conclude that $\mathbb{E}^o(\tau_1|\{D = \infty\}) < \infty$, and hence $v_\ell > 0$. □

By noting that if Kalikow's condition holds for some ℓ_0 then it holds for all ℓ in a neighborhood of ℓ_0, one gets immediately the

Corollary 3.5.7 *Assume that P satisfies assumption 3.2.1. If Kalikow's condition with respect to some ℓ holds, then there exists a deterministic v such that*

$$\frac{X_n}{n} \to_{n\to\infty} v, \quad P - a.s.$$

The following is a sufficient condition for Kalikow's condition to hold true:

Lemma 3.5.8 *Assume P is i.i.d. and elliptic. Then Kalikow's condition with respect to ℓ holds if*

$$\inf_{f\in\mathcal{F}} \frac{E\left(\frac{\sum_{e:|e|=1} \omega(0,e)\ell\cdot e}{\sum_{e:|e|=1} \omega(0,e)f(e)}\right)}{E\left(\frac{1}{\sum_{e:|e|=1} \omega(0,e)f(e)}\right)} > 0, \tag{3.5.9}$$

where $\mathcal{F}$ denotes the collection of nonzero functions on $\{e : |e| = 1\}$ taking values in $[0, 1]$.

Proof. Fix U a strict subset of $\mathbb{Z}^d$, $x \in U$, and let $\tau_x = \min\{n \geq 0 : X_n = x\}$. Define $g(x, y, \omega) := E_\omega^y(1_{\{\tau_x < \tau_{U^c}\}})$. Note that $g(x, y, \omega)$ is independent of ω_x. Next,

$$\sum_{|e|=1} (\ell \cdot e)\hat{P}_U(x, x + e)$$

$$= \frac{E\left(E_\omega^o\left(\sum_{n=0}^{\tau_{U^c}} 1_{\{X_n=x\}}\sum_{e:|e|=1} \omega(x, x+e)e\cdot\ell\right)\right)}{E\left(E_\omega^o\left(\sum_{n=0}^{\tau_{U^c}} 1_{\{X_n=x\}}\right)\right)}$$

$$= \frac{E\left(g(x, 0, \omega)E_\omega^x\left(\sum_{n=0}^{\tau_{U^c}} 1_{\{X_n=x\}}\sum_{e:|e|=1} \omega(x, x+e)e\cdot\ell\right)\right)}{E\left(g(x, 0, \omega)E_\omega^x\left(\sum_{n=0}^{\tau_{U^c}} 1_{\{X_n=x\}}\right)\right)}.$$

Under P_ω^x, the process X_n is a Markov chain with Geometric($\sum_{e:|e|=1} \omega(x, x+e)g(x, x+e, \omega)$) number of visits at x. The last equality and the Markov property then imply

$$\sum_{|e|=1} (\ell \cdot e)\hat{P}_U(x, x+e)$$

$$= \frac{E\left(g(x, 0, \omega)\frac{\sum_{e:|e|=1} \omega(x, x+e)e\cdot\ell}{\sum_{e:|e|=1} \omega(x, x+e)g(x, x+e, \omega)}\right)}{E\left(g(x, 0, \omega)\frac{1}{\sum_{e:|e|=1} \omega(x, x+e)g(x, x+e, \omega)}\right)}$$

$$= \frac{E\left(\frac{\sum_{e:|e|=1} \omega(x, e)\ell\cdot e}{\sum_{e:|e|=1} \omega(x, e)g(x, x+e, \omega)/g(x, 0, \omega)}\right)}{E\left(\frac{1}{\sum_{e:|e|=1} \omega(x, e)g(x, x+e, \omega)/g(x, 0, \omega)}\right)}$$

$$\geq \inf_{f\in\mathcal{F}} \frac{E\left(\frac{\sum_{e:|e|=1} \omega(x, e)\ell\cdot e}{\sum_{e:|e|=1} \omega(x, e)f(e)}\right)}{E\left(\frac{1}{\sum_{e:|e|=1} \omega(x, e)f(e)}\right)} = \inf_{f\in\mathcal{F}} \frac{E\left(\frac{\sum_{e:|e|=1} \omega(0, e)\ell\cdot e}{\sum_{e:|e|=1} \omega(0, e)f(e)}\right)}{E\left(\frac{1}{\sum_{e:|e|=1} \omega(0, e)f(e)}\right)} > 0,$$

where the first inequality is due to the independence of $g(x, x+e, \omega)/g(x, 0, \omega)$ in ω_x. $\qquad\square$

An easy corollary is the following

Corollary 3.5.10 *Assume P satisfies Assumption 3.2.1. If either*
(a) $\mathrm{supp}(P_d) \subset \{z \in \mathbb{R}^d : \ell \cdot z \geq 0\}$ *but* $\mathrm{supp}(P_d) \not\subset \{z \in \mathbb{R}^d : \ell \cdot z = 0\}$,
or
(b) $E((\sum_{e:|e|=1} \omega(0, e)e \cdot \ell)_+) > \frac{1}{\varepsilon}E((\sum_{e:|e|=1} \omega(0, e)e \cdot \ell)_-)$,
then Kalikow's condition with respect to ℓ holds.

In particular, non-nestling walks or walks with drift "either neutral or pointing to the right" satisfy Kalikow's condition with respect to an appropriate hyperplane ℓ. Further there exist truly nestling walks which do satisfy Kalikow's condition.

Remark: It is interesting to note that when P is elliptic and i.i.d. and $d = 1$, Kalikow's condition is equivalent to $v \neq 0$, i.e. to the walk being "ballistic". For $d > 1$, it is not clear yet whether there exist walks with $\mathbb{P}^o(A_\ell) > 0$ but with zero speed. Such walks, if they exist, necessarily cannot satisfy Kalikow's condition.

Our next goal is to provide tail estimates on $X_{\tau_1} \cdot \ell$ and on τ_1. Our emphasis here is in providing a (relatively) simple proof and not the sharpest possible result. For the latter we refer to [71]. In this spirit we throughout assume $\ell = e_1$.

Lemma 3.5.11 *Assume P is i.i.d. and satisfies Kalikow's condition. Then, there exists a constant c such that*

$$\mathbb{E}^o(\exp c X_{\tau_1} \cdot \ell) < \infty.$$

Proof. Recall the notations of Section 3.2 and write

$$\mathbb{E}^o(\exp(c\,X_{\tau_1}\cdot\ell)) = \sum_{k\geq 1}\mathbb{E}^o\Big(\exp(c\,X_{\tau_1}\cdot\ell)\mathbf{1}_{\{K=k\}}\Big)$$

$$= \sum_{k\geq 1}\sum_{x\in\mathbb{Z}^d} e^{c\,x\cdot\ell}E\Big(E^o_\omega\Big(\mathbf{1}_{\{X_{\overline{S}_k}=x\}}\mathbf{1}_{\{\overline{S}_k<\infty\}}\Big)\cdot P^x_\omega(D=\infty)\Big)$$

$$= \mathbb{P}^o(D=\infty)E\left(\sum_{k\geq 1}\sum_{x\in\mathbb{Z}^d} e^{c\,x\cdot\ell}E^o_\omega\Big(\mathbf{1}_{\{X_{\overline{S}_k}=x\}}\mathbf{1}_{\{\overline{S}_k<\infty\}}\Big)\right)$$

$$= \mathbb{P}^o(D=\infty)\sum_{k\geq 1}\mathbb{E}^o\Big(\exp(c\,X_{\overline{S}_k}\cdot\ell)\mathbf{1}_{\{\overline{S}_k<\infty\}}\Big). \qquad (3.5.12)$$

But, using the Markov property,

$$\mathbb{E}^o\Big(\exp(c\,X_{\overline{S}_k}\cdot\ell)\,\mathbf{1}_{\{\overline{S}_k<\infty\}}\Big)$$

$$\leq \mathbb{E}^o\Big(\exp(c\,X_{\overline{S}_{k-l}}\cdot\ell)\,\mathbf{1}_{\{\overline{S}_{k-1}<\infty\}}\Big)\mathbb{E}^o\Big(\exp(c\,M_0\cdot\ell)\,\mathbf{1}_{\{D<\infty\}}\Big).$$

Hence,

$$\mathbb{E}^o\Big(\exp(c\,X_{\overline{S}_k}\cdot\ell)\mathbf{1}_{\{\overline{S}_k<\infty\}}\Big)\leq\mathbb{P}^o(D=\infty)\sum_{k\geq 0}\Big(\mathbb{E}^o(\exp(c\,M_0\cdot\ell)\mathbf{1}_{\{D<\infty\}}\Big)^k.$$

Note, using $\ell=e_1$, that

$$\mathbb{E}^o\Big(e^{cM_1\cdot\ell}\mathbf{1}_{\{D<\infty\}}\Big)$$

$$= \sum_{k=1}^\infty e^{ck}\mathbb{E}^o\Big(\mathbf{1}_{\{M_0\cdot\ell=k\}}\mathbf{1}_{\{D<\infty\}}\Big)$$

$$= \sum_{k=1}^\infty e^{ck}\sum_{y\in\mathbb{Z}^{d-1}}\mathbb{E}^o\Big(E^o_\omega\Big(\mathbf{1}_{\{X_{T_k}=(k,y)\}}E^{(k,y)}_\omega(T_0<T_{k+1})\Big). \qquad (3.5.13)$$

Using Kalikow's condition and a computation as in Theorem 3.5.4, one has that for some $c_1>0$, $\mathbb{P}^o\Big(\sum_{|y|>\frac{2k}{\varepsilon\ell}}\mathbf{1}_{\{X_{T_k}=(k,y)\}}\Big)\leq e^{-c_1 k}$, while $\mathbb{P}^o(T_{-k}<T_1)\leq e^{-c_1 k}$. Hence, substituting in (3.5.13), one has

$$\mathbb{E}^o\Big(e^{cM_1\cdot\ell}\mathbf{1}_{\{D<\infty\}}\Big)\leq\sum_{k=1}^\infty e^{ck}\left(\left(\frac{4k}{\varepsilon\ell}\right)^{d-1}+1\right)e^{-c_1 k}.$$

Taking $c<c_1$, the lemma follows.　　　□

A direct consequence of Lemma 3.5.11 is that

$$\limsup_{n\to\infty}\frac{1}{n}\log\mathbb{P}^o(X_{\tau_1}\cdot\ell>vn)\leq-\beta(v) \qquad (3.5.14)$$

where $\beta(v)>0$ for $v>0$.

With a proof very similar to that of Lemma 3.5.11, we have in fact the

Lemma 3.5.15 *Assume P is i.i.d. and satisfies Kalikow's condition. Set $X^* = \sup_{0 \leq n \leq \tau_1} |X_n|$. Then, there exists a constant c' such that*

$$\mathbb{E}^o(\exp c' X^*) < \infty.$$

We next turn to obtaining tail estimates on τ_1. Here, due to the presence of "traps", one cannot in general expect exponential decay as in (3.5.14). We aim at proving the following result.

Theorem 3.5.16 *Assume P satisfies Assumption 3.2.1. and Kalikow's condition. Then, with $d \geq 2$, there exists an $\alpha > 1$ such that for all u large,*

$$\mathbb{P}^o(\tau_1 > u) \leq e^{-(\log u)^\alpha}.$$

In particular, τ_1 possesses all moments.

Proof. Recall that we take $\ell = e_1$ and, for $L > 0$, set $c_L = (-L, L) \times \left(-\frac{2L}{\varepsilon \ell}, \frac{2L}{\varepsilon \ell}\right)^{d-1}$. Note that, with c as in Lemma 3.5.11,

$$\mathbb{P}^o(\tau_1 \geq u) \leq \mathbb{P}^o\left(\tau_1 \geq u, X_{\tau_1} \cdot \ell \leq L\right) + \mathbb{P}^o\left(X_{\tau_1} \cdot \ell \geq L\right)$$
$$\leq e^{-cL/2} + \mathbb{P}^o(\tau_L > \tau_{c_L}) + \mathbb{P}^o(\tau_{c_L} = \tau_L \geq u)$$

where

$$\tau_L = \inf\{t : X_t \cdot \ell \geq L\} \text{ and}$$
$$\tau_{c_L} = \inf\{t : X_t \notin c_L\}.$$

Hence, by Kalikow's condition,

$$\mathbb{P}^o(\tau_1 \geq u) \leq e^{-\bar{c}L/2} + \mathbb{P}^o\left(\tau_{c_L} = \tau_L \geq u\right) \qquad (3.5.17)$$

for some constant $\bar{c} > 0$.

The heart of the proof of Theorem 3.5.16 lies in the following lemma, whose proof is deferred.

Lemma 3.5.18 *There exist a $\beta < 1$ and $\xi > 1$ such that for any $c > 0$,*

$$\limsup_{L \to \infty} \frac{1}{L^\xi} \log \mathbb{P}\left(P_\omega^o(X_{\tau_{U_L}} \cdot \ell \geq L) \leq e^{-cL^\beta}\right) < 0,$$

where $U_L = \{z \in \mathbb{Z}^d : |z \cdot \ell| \leq L\}$.

Accepting Lemma 3.5.18, let us complete the proof of Theorem 3.5.16. Toward this end, set $\Delta(u) = a \log u$ with a small enough such that $\varepsilon^{\Delta(u)} > \frac{1}{u^{1/6}}$, and set $L = L(u) = (\log u)^{\bar{\alpha}}$, with $N = L/\Delta = \frac{1}{a}(\log u)^{\bar{\alpha}-1}$ and $2 - \bar{\alpha} = \beta$ where β is as in Lemma 3.5.18. Observe that

$$\mathbb{P}^0(\tau_{c_L} \geq u) \leq \left(\frac{1}{2}\right)^{(\log u)^{\overline{\alpha}}} + P^0\left(\exists\, x_1 \in c_L, P_\omega^{x_1}\left(\tau_{c_L} > \frac{u}{(\log u)^{\overline{\alpha}}}\right) \geq \frac{1}{2}\right)$$

$$=: \left(\frac{1}{2}\right)^{(\log u)^{\overline{\alpha}}} + P(\mathcal{R}). \tag{3.5.19}$$

Note that

$$\frac{u}{(\log u)^\alpha} \cdot P_\omega^{x_1}\left(\tau_{c_L} > \frac{u}{(\log u)^{\overline{\alpha}}}\right)$$

$$\leq E_\omega^{x_1}(\tau_{c_L}) = E_\omega^{x_1}\left(\sum_{i=1}^{\tau_{c_L}} 1\right) = E_\omega^x\left(\sum_{y \in c_L} \sum_{i=1}^{\tau_{c_L}} 1_{\{X_n=y\}}\right)$$

$$= \sum_{y \in c_L} \frac{E_\omega^x\left(1_{\{\tau_y < \tau_{c_L}\}}\right)}{E_\omega^y\left(1_{\{\tau_y > \tau_{c_L}\}}\right)} \leq |c_L| \frac{1}{\inf\limits_{y \in c_L} P_\omega^y\left(1_{\{\tau_y\}} > \tau_{c_L}\right)},$$

with $\tau_y = \inf\{t : X_t = y\}$. Hence,

$$\inf_{y \in c_L} P_\omega^y(\tau_y > \tau_{c_L}) \leq \frac{|c_L|(\log u)^{\overline{\alpha}}}{u \inf\limits_{x_1 \in c_L} P_\omega^{x_1}\left(\tau_{c_L} > \frac{u}{(\log u)^\alpha}\right)}.$$

Hence, on $\mathcal{R}$ there exists a y with $\mathbb{P}_\omega^y(\tau_y > \tau_{c_L}) \leq \frac{1}{u^{4/5}}$, for all u large enough.

Set $A_i = \{z \in \mathbb{Z}^d : z \cdot \ell = i\Delta\}$. By ellipticity (recall $\varepsilon^{\Delta(u)} > 1/u^{1/6}$), it follows that on the event $\mathcal{R}$, there exists an $i_0 \in [-N+2, N-1]$ and an $x \in A_{i_0}$ such that

$$P_\omega^{x_0}\left(\tau_{(i_0-1)\Delta} > \tau_{c_L}\right) \leq \frac{1}{\sqrt{u}} \tag{3.5.20}$$

where $\tau_{(i_0-1)\Delta} = \inf\{t : X_t \cdot \ell = (i_0-1)\Delta\}$. Set

$$X_i = -\log \min_{z \in A_i \cap c_L} P_\omega^z\left(\tau_{(i-1)\Delta} > \tau_{(i+1)\Delta}\right).$$

Then, the Markov property implies that

$$P_\omega^x\left(\tau_{(i-1)\Delta} > \tau_{c_L}\right) \geq \exp\left(-\sum_{j=i}^{N-1} X_j\right) \varepsilon^{\Delta(u)}$$

$$\geq \exp\left(-\sum_{j=i}^{N-1} X_j\right) \frac{1}{u^{1/6}}.$$

Thus, using (3.5.20)

$$P(\mathcal{R}) \leq P\left(\sum_{i=-N+1}^{N-1} X_j \geq \frac{\log u}{12}\right) \leq 2N \sup_{-N+1 \leq i \leq N-1} P\left(X_i \geq \frac{\log u}{24N}\right). \tag{3.5.21}$$

But, using the shift invariance of P and the definition of $\{X_i\}$,

$$2N \sup_{N+1 \leq i \leq N-1} P\left(X_i \geq \frac{\log u}{24N}\right) \leq |c_L| P\left(P_\omega^o(\tau_{-\Delta} > \tau_\Delta) \leq e^{-(\log u)^{2-\overline{\alpha}}b}\right)$$

for $b = \frac{a}{24}$. The theorem follows by an application of Lemma 3.5.18.   □

Proof of Lemma 3.5.18

The interesting aspect in proving the Lemma is the fact that one constructs *lower* bounds on $P_\omega^o(X_{\tau_{u_L}} \cdot \ell \geq L)$ for many configurations. Toward this end, fix $1 > \beta > \overline{\beta}$, $\gamma \in (\frac{1}{2}, 1)$, $\chi = \frac{1-\overline{\beta}}{1-\gamma} < \overline{\beta} < 1$ such that $d(\overline{\beta} - \chi) > 1$ (for $\overline{\beta}$ close enough to 1, one may always find a γ close to 1 such that this condition is satisfied, if $d \geq 2$). Set next $L_0 = L^\chi$, $L_1 = L^{\overline{\beta}}$, $N_0 = L_1/L_0$ (for simplicity, assume that L_0, L_1, L_0^γ and N_0 are all integer).

Let R be a rotation of $\mathbb{Z}^d$ such that $\tilde{R}(\ell) = \tilde{R}(e_1) = \frac{v}{|v|}$, and define

$$B_1(z) = \tilde{R}\left(z + [0, L_0]^d\right) \cap \mathbb{Z}^d$$

$$B_2(z) = \tilde{R}\left(z + [-L_0^\gamma, L_0 + L_0^\gamma]^d\right) \cap \mathbb{Z}^d$$

and $\partial_+ B_2(z) = \partial B_2(z) \cap \left\{x : x \cdot \frac{v}{|v|} \geq L_0 + L_0^\gamma\right\}$.

We say that $z \in L_0 \mathbb{Z}^d$ is *good* if $\sup_{x \in B_1(z)} P_\omega^x(X_{\tau_{B_2(z)}} \notin \partial_+ B_2(z)) \leq \frac{1}{2}$ and say that it is *bad* otherwise. The following estimate is a direct consequence of Kalikow's condition and Lemma 3.5.15.

Lemma 3.5.22 *For $\gamma \in (1/2, 1)$,*

$$\limsup_{L_0 \to \infty} L_0^{1-2\gamma} \log P(0 \text{ is bad }) < 0.$$

Proof of Lemma 3.5.22

Set $u = \max\{\overline{u} : \sup_{x \in B_2(0)} x \cdot \ell \geq \overline{u}\}$ and set $L_u = \sup\{n \geq 0 : X_n \cdot \ell \leq u\}$. Define $\pi(z) = z - \frac{z \cdot v}{|v|^2} v$. Setting $K_n = \sup\{k \geq 0 : \tau_k < n\}$, it holds that $n \leq L_u \Rightarrow K_n \leq u$. Setting $w \in \mathbb{R}^d$ with $w \cdot v = 0$, and $|w|_1 = 1$ one has

$$X_n \cdot w = X_{\tau_{K_n}} \cdot w + (X_n - X_{\tau_{K_n}}) \cdot w$$
$$\leq X_{\tau_{K_n}} \cdot w + X^* \circ \theta_{\tau_{K_n}}.$$

Hence,

$$\mathbb{P}^o\left(\sup_{0\leq n\leq L_u} X_n\cdot w \geq u^\gamma\right) \leq \sum_{0\leq k\leq u} \mathbb{P}^o\left(X_{\tau_k}\cdot w + X^*\circ\theta_{\tau_k} > u^\gamma\right)$$

$$\leq \sum_{0\leq k\leq u}\mathbb{P}^o\left(\left(X_{\tau_k}-X_{\tau_1}\right)\cdot w > \frac{u^\gamma}{3}\right) + u\,\mathbb{P}^o\left(X_{\tau_1}\cdot w > \frac{u^\gamma}{3}\right)$$

$$+ u\,\mathbb{P}^o\left(X^* > \frac{u^\gamma}{3}\Big| D=\infty\right)$$

$$\leq \sum_{0\leq k\leq u}\mathbb{P}^o\left(\left(X_{\tau_k}-X_{\tau_1}\right)\cdot w > \frac{u^\gamma}{3}\right) + \frac{2u}{\mathbb{P}^o(D=\infty)}\mathbb{P}^o\left(X^* > \frac{u^\gamma}{3}\right).$$

Note that by Lemma 3.5.15, $\mathbb{P}^o\left(X^* > \frac{u^\gamma}{3}\right) \leq e^{-c_0 u^\gamma}$ while the random variables $(X_{\tau_{i+1}}-X_{\tau_i})\cdot w$ are i.i.d., of zero mean and finite exponential moments. In particular,

$$\mathbb{P}^o\left(\left|\frac{(X_{\tau_k}-X_{\tau_1})\cdot w}{k}\right| > \frac{u^\gamma}{3k}\right) \leq e^{-c_0 k \frac{u^{2\gamma}}{9k^2}} \leq e^{-c_0 u^{2\gamma-1}}$$

by moderate deviations (see e.g., [19, Section 3.7]).

Since $\gamma > 2\gamma-1$, we conclude that

$$\limsup_{u\to\infty}\frac{1}{u^{2\gamma-1}}\log\mathbb{P}^o\left(\sup_{0\leq n\leq L_u} X_n\cdot w \geq u^\gamma\right) < 0$$

and hence

$$\limsup_{u\to\infty}\frac{1}{u^{2\gamma-1}}\log\mathbb{P}^o\left(\sup_{0\leq n\leq L_u} |\pi(X_n)| \geq u^\gamma\right) < 0. \qquad (3.5.23)$$

Fix now $x \in B_1(0)$. Then, for some $c=c(d)$,

$$\mathbb{P}^x\left(X_{\tau_{B_2(0)}} \notin \partial_+ B_2(0)\right) \leq \mathbb{P}^o\left(\sup_{0\leq n\leq L_{cu}} \pi(X_n\cdot w) \geq u^\gamma\right)$$

$$+ \mathbb{P}^o\left(X_{\tau_{V_u}}\cdot\ell < 0\right)$$

where $V_u = \left\{z : \frac{-L_0^\gamma}{c} \leq z\cdot\ell \leq cL_0^\gamma\right\}$ and the conclusion follows from (3.5.23) and Kalikow's condition. $\qquad \square$

Construct now the following subsets of U_L:

Set $M = \left\{z \in L_0\mathbb{Z}^d, z=(0,\bar z), \bar z \in \left\{-\frac{L_1}{L_0},\cdots,0,\frac{L_1}{L_0}\right\}^{d-1} L_0\right\}$. For $z \in M$,

set $\mathrm{Row}(z) = \cup_{j=N_-(z)}^{N_+(z)} B_1(z+jL_0\ell)$ where

$$N_-(z) = \min\{j : B_1(z+jL_0\ell)\cap\{x : x\cdot\ell\geq 0\}\neq\emptyset\} \geq -c\frac{L}{L_0}$$

$$N_+(z) = \max\{j : B_1(z+jL_0\ell)\cap\{x : x\cdot\ell\geq 0\}\neq\emptyset\} \leq c\frac{L}{L_0}$$

Random Walks in Random Environment   303

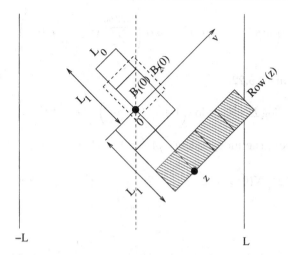

Fig. 3.5.1.

for some constant c depending on v and uniformly bounded, and set $T = \bigcup_{j \in \{-\frac{L_1}{L_0}, \cdots, 0, \cdots, \frac{L_1}{L_0}\}^{d-1}} \{jL_0\}$.

The idea behind the proof is that if one of the rows $\{R(z)\}_{z \in M}$, say $R(z_0)$, contains mostly good blocks, a good strategy for the event $(X_{\tau_{u_L}} \cdot \ell) \geq L$ is to force the walker started at x to first move to z, then move to the right successively without leaving $\bigcup_{z \in \mathrm{Row}(z_0)} B_2(z)$ until exiting from u_L. More precisely, let $N(z_0)$ denote the number of bad blocks in $\bigcup_{z \in \mathrm{Row}(z_0)} B_1(z)$. Then, for some constants c_i, using ellipticity and the definition of good boxes,

$$P_\omega^o(X_{\tau_{u_L}} \cdot \ell \geq L) \geq \varepsilon^{L_1} \left(\frac{1}{2} \varepsilon^{2(d-1)} L_0^\gamma\right)^{cL/L_0} (\varepsilon^{L_0})^{N(z_0)}$$

$$= e^{-c_4(L^{\overline{\beta}} + L^\chi N(z_0))}.$$

Hence, for an arbitrary constant c_6 and all L large enough ($L > g(c_6, c)$ for some fixed function $g(\cdot)$),

$$P\left(P_\omega^o\left(X_{\tau_{u_L}} \cdot \ell \geq L\right) \leq e^{-c\overset{*}{L}^\beta}\right) \leq P\left(\{\not\exists z_0 \in M : N(z_0) \leq c_6 L^{\overline{\beta}-\chi}\}\right)$$

$$\leq \left[P\left(N(0) \geq c_6 L^{\overline{\beta}-\chi}\right)\right]^{\left(\frac{cL_1}{L_0}\right)^{(d-1)}}.$$

using the independence between even rows in Figure 3.5.1. But note that $N(0) = \sum_{i=1}^{L/2L_0} (Y_i + Z_i)$ where $\{Y_i\}$ are i.i.d., $\{Z_i\}$ are i.i.d., $\{0, 1\}$ valued, $P(Y_i = 1) = P(B_1(0))$ is a bad block (the division to (Y_i, Z_i) reflects the division to even and odd blocks, which creates independence). Hence,

$$P\left(N(0) \geq c_6 L^{\bar{\beta}-x}\right) \leq 2P\left(\sum_{i=1}^{L/2L_0} Y_i \geq \frac{c_6}{2} L^{\bar{\beta}-x}\right).$$

Since, by Lemma 3.5.22,

$$\log E(e^{Y_i}) \leq \log\left(1 + e^{-c_7 L_0^{2\gamma-1}}\right) \leq e^{-c_7 L_0^{2\gamma-1}},$$

we conclude from the independence of the Y_i's that

$$P\left(N(0) \geq c_6 L^{\bar{\beta}-x}\right) \leq 2e^{-\frac{c_6}{2} L^{\bar{\beta}-x}} e^{e^{-c_7 L_0^{2\gamma-1}} \cdot \frac{L}{2L_0}}$$

$$\leq e^{-c_8 L^{\bar{\beta}-x}}.$$

Hence,

$$P\left(P_\omega^o\left(X_{\tau_{u_L} \cdot \ell} \geq L\right) \leq e^{-cL^\beta}\right) \leq e^{-c_9 L^{(\bar{\beta}-x)d}} \leq e^{-L^\xi}$$

for some $\xi > 1$, as claimed. □

Remark The restriction to $d \geq 2$ in Theorem 3.5.16 is essential: as we have seen in the case $d = 1$, one may have ballistic walks (and hence, in $d = 1$, satisfying Kalikow's condition) with moments $m_r := E^o(\tau_1^r)$ of the regeneration time τ_1 being finite only for small enough $r > 1$.

We conclude this section by showing that estimates of the form of Theorem 3.5.16 lead immediately to a CLT. The statement is slightly more general than needed, and does not assume Kalikow's condition but rather some of its consequences.

Theorem 3.5.24 *Assume Assumption 3.2.1, and further assume that $\mathbb{P}^o(A_\ell)$ $= 1$ and that the regeneration time τ_1 satisfies $E^o(\tau_1^{(2+\delta)}) < \infty$ for some $\delta > 0$. Then, under the annealed measure $\mathbb{P}^o$,*

$$X_n/n \to_{n\to\infty} v := \frac{E^o(X_{\tau_2} - X_{\tau_1})}{E^o(\tau_2 - \tau_2)} \neq 0, \quad \mathbb{P}^o - a.s., \qquad (3.5.25)$$

and $(X_n - nv)/\sqrt{n}$ converges in law to a centered Gaussian vector.

Proof. The LLN (3.5.25) is a consequence of Theorem 3.2.2 and its proof. To see the CLT, set

$$\xi_i = X_{\tau_{i+1}} - X_{\tau_i} - (\tau_{i+1} - \tau_i)v, \quad S_n := \sum_{i=1}^{n} \xi_i,$$

and $\Xi = E^o(\xi_1 \xi_1^T)$. It is not hard to check that Ξ is non-degenerate, simply because $\mathbb{P}^o(|\xi_1| > K) > 0$ for each $K > 0$. Then S_n is under $\mathbb{P}^o$ a sum of i.i.d. random variables possessing finite $2 + \delta$-th moments, and thus $S_{[nt]}/\sqrt{n}$ satisfies the invariance principle, with covariance matrix Ξ. Define

$$\nu_n = \min\left\{ j : \sum_{i=1}^{j} (\tau_{i+1} - \tau_i) > n \right\}.$$

Note that in $\mathbb{P}^o$ probability, $n/\nu_n \to \mathbb{E}^o(\tau_2 - \tau_1) < \infty$. Hence, by time changing the invariance principle, see e.g. [2, Theorem 14.4], $S_{\nu_n}/\sqrt{\nu_n}$ converges in $\mathbb{P}^o$ probability to a centered Gaussian variable of covariance Ξ. On the other hand, for any positive η,

$$\mathbb{P}^o(|S_{\nu_n} - (X_n - nv)| > \eta\sqrt{n}) \leq \mathbb{P}^o(\exists i \leq n : (\tau_{i+1} - \tau_i) > \eta\sqrt{n}/2)$$
$$+ \mathbb{P}^o(\tau_1 > \eta\sqrt{n}/2)$$
$$\leq \frac{(n+1)\mathbb{P}^o(\tau_1 > \eta\sqrt{n}/2)}{\mathbb{P}^o(D')} \to_{n\to\infty} 0,$$

where we used the moment bounds on $\mathbb{E}^o(\tau_2 - \tau_1)^{2+\delta}$ and the fact that $\mathbb{P}^o(D') > 0$ in the last limit. This yields the conclusion. Further, one observes that the limiting covariance of $X_n/\sqrt{n}$ is $\Xi/(\mathbb{E}^o(\tau_2 - \tau_1))$. □
A direct conclusion of Theorem 3.5.24 is that under Kalikow's condition, $X_n/\sqrt{n}$ satisfies an annealed CLT.

Bibliographical notes: Lemma 3.5.2, Kalikow's condition, the fact that it implies $\mathbb{P}^o(A_\ell) = 1$, and Lemma 3.5.8 appeared in [38]. The argument for $v_\ell > 0$ under Kalikow's condition is due to Sznitman and Zerner [76], who also observed Corollary 3.5.10. [71] proves that in the i.i.d. environment case, $a(0, z) = 0$ if and only if $z = tv$, some $t > 0$. The estimates in Theorem 3.5.16 are a weak form of estimates contained in [71]. Finally, [81] characterizes, under Kalikow's condition, the speed v as a function of Lyapunov exponents closely related to the functions $a(\lambda, z)$.

In a recent series of papers, Sznitman has shown that many of the conclusions of this section remain valid under a weaker condition, Sznitman's (T) or (T') conditions, see [74, 73, 75]

Appendix
Markov chains and electrical networks: a quick reminder

With (V, E) as in Section 1.1, let $C_e \geq 0$ be a *conductance* associated to each edge $e \in E$. Assume that we can write

$$\omega_v(w) = \frac{C_{vw}}{\sum_{w \in N_v} C_{vw}} := \frac{C_{vw}}{C_v}.$$

To each such graph we can associate an electrical network: edges are replaced by conductors with conductance C_{vw}. The relation between the electrical network and the random walk on the graph is described in a variety of texts, see e.g. [25] for an accessible summary or [57] for a crash course. This relation is based on the uniqueness of harmonic functions on the network, and is best described as follows: fix two vertices $v, w \in V$, and apply a unit voltage between v and w. Let $V(z)$ denote the resulting voltage at vertex z. Then,

$$P_\omega^z(\{X_n\} \text{ hits } v \text{ before hitting } w) = V(z).$$

Recall that for any two vertices v, w, the *effective conductance* $C^{\text{eff}}(v \leftrightarrow w)$ is defined by applying a unit voltage between v and w and measuring the outflow of current at v. In formula, this is equivalent to

$$C^{\text{eff}}(v \leftrightarrow w) = \sum_{v' \in N_v} [1 - V(v')]C_{vv'} = \sum_{w' \in N_w} V(w')C_{ww'}.$$

For any integer r, the effective conductance $C_{v,r}$ between v and the horocycle of distance r from v is the effective conductance between v and the vertex r' in a modified graph where all vertices in the horocycle have been identified. We set then $C_{v,\infty} := \lim_{r \to \infty} C_{v,r}$. The effective conductance obeys the following rule:

Combination rule: Edges in parallel can be combined by summing their conductances. Futher, the effective conductance between vertices v, w is not affected if, at any vertex $w' \notin \{v, w\}$ with $N_{w'} = \{v', z'\}$, one removes the edges (v', w') and (z', w') and replaces the conductance $C_{v',z'}$ by

$$\overline{C}_{v',z'} = C_{v',z'} + \left(\frac{1}{C_{v',w'}} + \frac{1}{C_{z',w'}}\right)^{-1}.$$

(This formula applies even if an edge $C_{v'.w'}$ is not present, by taking $C_{v',z'} = 0$.)

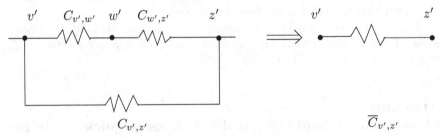

Exercise A.1 *Prove formulae (2.1.3) and (2.1.4).*

Markov chains of the type discussed here possess an easy criterion for recurrence: a vertex v is recurrent if and only if the effective conductance $C_{v,\infty}$ between v and ∞ is 0. A sufficient condition for recurrence is given by means of the Nash-Williams criterion (see [57, Corollary 9.2]). Recall that an edge-cutset Π separating v from ∞ is a set of edges such that any path starting at v which includes vertices of arbitrarily large distance from v must include some edge in Π.

Lemma A.2 (Nash-Williams) *If Π_n are disjoint edge-cutsets which sepa-rate v from ∞, then*

$$C_{v,\infty} \leq \left(\sum_n \left(\sum_{e \in \Pi_n} C_e\right)^{-1}\right)^{-1}.$$

As an application of the Nash-Williams criterion, we prove that a product of independent Sinai's walks is recurrent. Recall that a Sinai walk (in dimension 1) is a RWRE satisfying Assumption 2.5.1. For simplicity, we concentrate here on Sinai's walk without holding times and define a product of Sinai's walk in dimension d as the RWRE on $\mathbb{Z}^d$ constructed as follows: for each $v \in \mathbb{Z}^d$, set $N_v = \times_{i=1}^d (v_i - 1, v_i + 1)$ and let $\Omega = \times_{i=1}^d (M_1(N_v))^{\mathbb{Z}}$. For $z \in \mathbb{Z}^d$, we set $\omega_{i,z}^+ = \omega_i(z_i, z_i+1)$, $\omega_{i,z}^- = \omega_i(z_i, z_i-1)$ and $\rho_i(z) = \omega_{i,z}^-/\omega_{i,z}^+$. We equip Ω with a product of measures $P = \times_{i=1}^d P_i$, such that each P_i is a product measure which also satisfies Assumption 2.5.1. For a fixed $\omega \in \Omega$, define the RWRE in environment ω as the Markov chain (of law P_ω^o) such that $P_\omega^o(X_0 = 0) = 1$ and, for $v \in \{-1,1\}^d$, $P_\omega^o(X_{n+1} = x + v|X_n = x) = \prod_{i=1}^d \omega_i(x_i, x_i + v_i)$. Define

$$C(x,v) = \prod_{i=1}^d \left(\left(\prod_{j_i=x_i}^{-1} \rho_i(j_i)\right)\left(\prod_{j_i=0}^{x_i-1} \rho_i(j_i)^{-1}\right)(\rho_i(x_i)^{-1})^{(v_i+1)/2}\right),$$

where by definition a product over an empty set of indices equals 1. Then, the resistor network with conductances $C(x,v)$ is a model for the product of Sinai's RWRE. Define

$$B_i^n(t) = -\frac{1}{\sqrt{n}}\sum_{j=0}^{\lfloor nt\rfloor} \log \rho_i(j) \cdot (\text{sign } t).$$

Then,

$$C(x,v) \leq \varepsilon^{-d} \prod_{i=1}^d e^{\sqrt{n}B_i^n(x_i/n)}.$$

Taking as cutsets Π_n the set of edges $(x, x+v)$ with $|x|_\infty = n$, $v_i \in -1, 1$ and $|x+v|_\infty = n+1$, we thus conclude that

$$\left(\sum_{e \in \Pi_n} C_e\right) \leq \varepsilon^{-d}\sum_{i=1}^d (e^{\sqrt{n}B_i^n(1)} + e^{\sqrt{n}B_i^n(-1)})$$

$$\prod_{j=1,j\neq i}^d \left(\sum_{k=-n}^n e^{\sqrt{n}B_j^n(k/n)}\right) =: D_n.$$

Since P_i are product measures, we have by Kolmogorov's 0-1 law that $P(\liminf_{n\to\infty} D_n = 0) \in \{0,1\}$. On the other hand, for all n large enough, we have by the CLT that

288

$$P(D_n < e^{-n^{1/4}}) \geq P(B_i^n(1) \leq -1, B_i^n(-1)$$
$$\leq -1, \sup_{-1 \leq t \leq 1} B_i^n \leq 1/2d, i = 1, \ldots, d) \geq c,$$

for some constant $c > 0$ independent of n. Thus, by Fatou's lemma, $P(\liminf_{n \to \infty} D_n = 0) > 0$, and hence $= 1$ by the above mentioned 0-1 law. We conclude from Nash's criterion (Lemma A.2) that $C_{0,\infty} = 0$, establishing the recurrence as claimed.

Exercise A.3 *Extend the above considerations to Sinai's walk with holding times and non product measures P_i.*

Bibliographical notes: The classical reference for the link between electrical networks and Markov chain is the lovely book [25]. The application to the proof of recurrence of products of Sinai's walks was prompted by a question of N. Gantert and Z. Shi.

References

1. S. Alili, Asymptotic behaviour for random walks in random environments, *J. Appl. Prob.* **36** (1999) pp. 334–349.
2. P. Billingsley, *Convergence of probability measures*, 2-nd. edition, Wiley (1999).
3. E. Bolthausen and I. Goldsheid, Recurrence and transience of random walks in random environments on a strip, *Comm. Math. Phys* **214** (2000), pp. 429–447.
4. E. Bolthausen, A. S. Sznitman and O. Zeitouni, Cut points and diffusive random walks in random environments, *Ann. Inst. H. Poincaré - Prob. Stat.* **39** (2003), pp. 527–555.
5. R. C. Bradley, A caution on mixing conditions for random fields, *Stat. & Prob. Letters* **1989**, pp. 489–491.
6. M. Bramson and R. Durrett, Random walk in random environment: a counterexample?, *Comm. Math. Phys.* **119** (1988), pp. 199–211.
7. J. Brémont, Marches aléatoire en milieu aléatoire sur $\mathbb{Z}$; Dynamique d'applications localement contractantes sur le cercle. *Thesis*, Université de Rennes I, 2002.
8. J. Brémont, Récurrence d'une marche aléatoire symétrique dans $\mathbb{Z}^2$ en milieu aléatoire, *preprint* (2000).
9. J. Bricmont and A. Kupiainen, Random walks in asymmetric random environments, *Comm. Math. Phys* **142** (1991), pp. 345–420.
10. W. Bryc and A. Dembo, Large deviations and strong mixing, *Ann. Inst. H. Poincaré - Prob. Stat.* **32** (1996) pp. 549–569.
11. F. Comets. Large deviations estimates for a conditional probability distribution. Applications to random interacting Gibbs measures. *Prob. Th. Rel. Fields* **80** (1989) pp. 407–432.
12. F. Comets, N. Gantert and O. Zeitouni, Quenched, annealed and functional large deviations for one dimensional random walk in random environment, *Prob. Th. Rel. Fields* **118** (2000) pp. 65–114.
13. F. Comets and O. Zeitouni, A law of large numbers for random walks in random mixing environments, to appear, *Annals Probab.* (2003).

14. A. De Masi, P. A. Ferrari, S. Goldstein and W. D. Wick, An invariance principle for reversible Markov processes. Applications to random motions in random environments, *J. Stat. Phys.* **55** (1989), pp. 787—855.

15. A. Dembo, N. Gantert, Y. Peres and O. Zeitouni, Large deviations for random walks on Galton-Watson trees: averaging and uncertainty, *Prob. Th. Rel. Fields* **122** (2002), pp. 241–288.

16. A. Dembo, N. Gantert and O. Zeitouni, Large deviations for random walk in random environment with holding times, to appear, *Annals Probab.* (2003).

17. A. Dembo, A. Guionnet and O. Zeitouni, Aging properties of Sinai's random walk in random environment, *XXX preprint archive*, math.PR/0105215 (2001).

18. A. Dembo, Y. Peres and O. Zeitouni, Tail estimates for one-dimensional random walk in random environment, *Comm. Math. Physics* **181** (1996) pp. 667–684.

19. A. Dembo and O. Zeitouni, *Large deviation techniques and applications*, 2nd edition, Springer, New-York (1998).

20. Y. Derriennic, Quelques application du théorème ergodique sous-additif, *Asterisque* **74** (1980), pp. 183–201.

21. J. D. Deuschel and D. W. Stroock, *Large deviations*, Academic Press, Boston (1989).

22. M. D. Donsker and S. R. S. Varadhan, Asymptotic evaluation of certain Markov process expectations for large time III, *Comm. Pure Appl. Math.* **29** (1976) pp. 389-461.

23. M. D. Donsker and S. R. S. Varadhan. Asymptotic evaluation of certain Markov process expectations for large time, IV. *Comm. Pure Appl. Math.* **36** (1983) pp. 183-212.

24. P. Le Doussal, C. Monthus and D. S. Fisher, Random walkers in one-dimensional random environments: exact renormalization group analysis. *Phys. Rev. E* **59** (1999) pp. 4795–4840.

25. P. G. Doyle and L. Snell, *Random walks and electric networks*, Carus Math. Monographs **22**, MAA, Washington (1984).

26. R. Durrett, *Probability: theory and examples*, 2nd ed., Duxbury Press, Belmont (1996).

27. H. Föllmer, *Random fields and diffusion processes*, Lecture Notes in Mathematics **1362** (1988), pp. 101–203.

28. N. Gantert, Subexponential tail asymptotics for a random walk with randomly placed one-way nodes, *Ann. Inst. Henri Poincaré - Prob. Stat.* **38** (2002) pp. 1–16.

29. N. Gantert and O. Zeitouni, Large deviations for one dimensional random walk in a random environment - a survey, *Bolyai Society Math Studies* **9** (1999) pp. 127–165.

30. N. Gantert and O. Zeitouni, Quenched sub-exponential tail estimates for one-dimensional random walk in random environment, *Comm. Math. Physics* **194** (1998) pp. 177–190.

31. A. O. Golosov, Limit distributions for random walks in random environments, *Soviet Math. Dokl.* **28** (1983) pp. 18–22.

32. A. O. Golosov, Localization of random walks in one-dimensional random environments, *Comm. Math. Phys.* **92** (1984) pp. 491–506.

33. A. O. Golosov, On limiting distributions for a random walk in a critical one dimensional random environment, *Comm. Moscow Math. Soc.* **199** (1985) pp. 199–200.

34. A. Greven and F. den Hollander, Large deviations for a random walk in random environment, *Annals Probab.* **22** (1994) pp. 1381–1428.

35. Y. Hu and Z. Shi, The limits of Sinai's simple random walk in random environment, *Annals Probab.* **26** (1998), pp. 1477–1521.

36. Y. Hu and Z. Shi, The problem of the most visited site in random environments, *Prob. Th. Rel. Fields* **116** (2000), pp. 273–302.

37. B. D. Hughes, *Random walks and random environments*, Oxford University Press (1996).

38. S. A. Kalikow, Generalized random walks in random environment, *Annals Probab.* **9** (1981), pp. 753–768.

39. M. S. Keane and S. W. W. Rolles, Tubular recurrence, *Acta Math. Hung.* **97** (2002), pp. 207–221.

40. H. Kesten, Sums of stationary sequences cannot grow slower than linearly, *Proc. AMS* **49** (1975) pp. 205–211.

41. H. Kesten, The limit distribution of Sinai's random walk in random environment, *Physica* **138A** (1986) pp. 299–309.

42. H. Kesten, M. V. Kozlov and F. Spitzer, A limit law for random walk in a random environment, *Comp. Math.* **30** (1975) pp. 145–168.

43. E. S. Key, Recurrence and transience criteria for random walk in a random environment, *Annals Probab.* **12** (1984), pp. 529–560.

44. T. Komorowski and G. Krupa, The law of large numbers for ballistic, multi-dimensional random walks on random lattices with correlated sites, *Ann. Inst. H. Poincaré - Prob. Stat.* **39** (2003), pp. 263–285.

45. S. M. Kozlov, The method of averaging and walks in inhomogeneous environments, *Russian Math. Surveys* **40** (1985) pp. 73–145.

46. H.-J. Kuo and N. S. Trudinger, Linear elliptic difference inequalities with random coefficients, *Math. of Computation* **55** (1990) pp. 37–53.

47. G.F. Lawler, Weak convergence of a random walk in a random environment, *Comm. Math. Phys.* **87** (1982) pp. 81–87.

48. G.F. Lawler, A discrete stochastic integral inequality and balanced random walk in a random environment, *Duke Mathematical Journal* **50** (1983) pp. 1261–1274.

49. G.F. Lawler, Estimates for differences and Harnack inequality for difference operators coming from random walks with symmetric, spatially inhomogeneous, increments, *Proc. London Math. Soc.* **63** (1991) pp. 552–568.

50. F. Ledrappier, *Quelques propriétés des exposants charactéristiques*, Lecture Notes in Mathematics **1097**, Springer, New York (1984).

51. R. Lyons, R. Pemantle and Y. Peres, Ergodic theory on Galton–Watson trees: speed of random walk and dimension of harmonic measure, *Ergodic Theory Dyn. Systems* **15** (1995), pp. 593–619.

52. R. Lyons, R. Pemantle and Y. Peres, Biased random walk on Galton–Watson trees, *Probab. Theory Relat. Fields* **106** (1996), pp. 249–264.

53. S. A. Molchanov, *Lectures on random media*, Lecture Notes in Mathematics **1581**, Springer, New York (1994).

54. S. V. Nagaev, Large deviations of sums of independent random variables, *Annals Probab.* **7** (1979) pp. 745–789.

55. S. Olla, Large deviations for Gibbs random fields. *Prob. Th. Rel. Fields* **77** (1988) pp. 343–357.

56. S. Orey and S. Pelikan, Large deviation principles for stationary processes. *Annals Probab.* **16** (1988), pp. 1481–1495.

57. Y. Peres, *Probability on trees: an introductory climb*, Lecture notes in Mathematics **1717** (1999) P. Bernard (Ed.), pp. 195–280.

58. D. Piau, Sur deux propriétés de dualité de la marche au hasard en environnement aléatoire sur $\mathbb{Z}$, *preprint* (2000).

59. D. Piau, Théorème central limite fonctionnel pour une marche au hasard en environment aléatoire, *Ann. Probab.* **26** (1998), pp. 1016–1040.

60. A. Pisztora and T. Povel, Large deviation principle for random walk in a quenched random environment in the low speed regime, *Annals Probab.* **27** (1999) pp. 1389–1413.

61. A. Pisztora, T. Povel and O. Zeitouni, Precise large deviation estimates for one-dimensional random walk in random environment, *Prob. Th. Rel. Fields* **113** (1999) pp. 135–170.

62. F. Rassoul-Agha, A law of large numbers for random walks in mixing random environment, to appear, *Annals Probab.* (2003).

63. L. Shen, Asymptotic properties of certain anisotropic walks in random media, *Annals Applied Probab.* **12** (2002), 477–510.

64. M. Sion, On general minimax theorems, *Pacific J. Math* **8** (1958), pp. 171–176.

65. Z. Shi, Sinai's walk via stochastic calculus, in *Milieux Aléatoires*, F. Comets and E. Pardoux, eds., Panoramas et Synthèses **12**, Société Mathématique de France (2001).

66. Ya. G. Sinai, The limiting behavior of a one-dimensional random walk in random environment, *Theor. Prob. and Appl.* **27** (1982) pp. 256–268.

67. F. Solomon, Random walks in random environments, *Annals Probab.* **3** (1975) pp. 1–31.

68. A. S. Sznitman, *Brownian motion, obstacles and random media*, Springer-Verlag, Berlin (1998).

69. A. S. Sznitman, *Lectures on random motions in random media*, In DMV seminar **32**, Birkhauser, Basel (2002).

70. A. S. Sznitman, Milieux aléatoires et petites valeurs propres, in *Milieux Aléatoires*, F. Comets and E. Pardoux, eds., Panoramas et Synthèses **12**, Société Mathématique de France (2001).

71. A. S. Sznitman, Slowdown estimates and central limit theorem for random walks in random environment, *JEMS* **2** (2000), pp. 93–143.

72. A. S. Sznitman, Slowdown and neutral pockets for a random walk in random environment, *Probab. Th. Rel. Fields* **115** (1999), pp. 287–323.

73. A. S. Sznitman, On a class of transient random walks in random environment, *Annals Probab.* **29** (2001), pp. 724–765.

74. A. S. Sznitman, An effective criterion for ballistic behavior of random walks in random environment, *Probab. Theory Relat. Fields* **122** (2002), pp. 509–544.

75. A. S. Sznitman, On new examples of ballistic random walks in random environment, *Annals Probab.* **31** (2003), pp. 285–322.

76. A. S. Sznitman and M. Zerner, A law of large numbers for random walks in random environment, *Annals Probab.* **27** (1999) pp. 1851–1869.

77. M. Talagrand, A new look at independence, *Annals Probab.* **24** (1996), pp. 1–34.

78. N. S. Trudinger, Local estimates for subsolutions and supersolutions of general second order elliptic quasilinear equations, *Invent. Math.* **61** (1980), pp. 67–79.

79. S. R. S. Varadhan, Large deviations for random walks in a random environment, *preprint* (2002).

312   Ofer Zeitouni

80. M. P. W. Zerner, Lyapounov exponents and quenched large deviations for multidimensional random walk in random environment, *Annals Probab.* **26** (1998), pp. 1446–1476.

81. M. P. W. Zerner, Velocity and Lyapounov exponents of some random walks in random environments, *Ann. Inst. Henri Poincaré - Prob. Stat.* **36** (2000), pp. 737–748.

82. M. P. W. Zerner and F. Merkl, A zero-one law for planar random walks in random environment, *Annals Probab.* **29** (2001), pp. 1716–1732.

83. M. P. W. Zerner, A non-ballistic law of large numbers for random walks in i.i.d. random environment, *Elect. Comm. in Probab.* **7** (2002), pp. 181–187.

CORRECTIONS - Lecture notes on RWRE

Page numbers refer to the published version, Springer Lecture Notes in Mathematics, Volume 1837, pp. 191–312.

1. Page 200, line 6, reverse $<$ to $>$ in first indicator.

2. Page 202, lines 2,3,10,12,13, replace $\rho_{(-i)}$ by ρ_i.

3. Page 206: the right side of equation (2.1.23) should be replaced by

$$\prod_{\ell=1}^{L} \binom{m+k}{k} (\omega_i^0)^{m_\ell} (\omega_i^-)^{k_\ell} \omega_i^+ .$$

4. Page 212, line -2 (Remark): As F. Rassoul-Agha pointed out to me, the argument given only shows that

$$P\left(\left| P_\omega^o\left(\frac{X_n - v_P n - Z_n}{\sqrt{n}\sigma_{P,1}} > x \right) - \Phi(-x) \right| > \delta \right) \to_{n\to\infty} 0,$$

which gives less than a full-blown quenched CLT; To give a full quenched CLT requires an additional estimate. *Update: Jon Peterson, in his thesis, has completed the details of this argument, by using hitting times. See arXiv:0810.0257v1 [math.PR], and also Ilya Goldsheid's article "Simple transient random walks in one-dimensional random environment: the central limit theorem", Probab. Theory Related Fields 139 (2007), pp. 41–64.*

5. Page 213, section 2.3: A better version of this argument, that avoids some of the coupling arguments and hence gives better conditions, is available at
   Dembo, A., Gantert, N., and Zeitouni, O., " Large Deviations for Random Walk in Random Environment with holding times", *Annals Probab.* 32 (2004), pp. 996–1029.

6. Page 219, line 2, replace $\leq \delta$ by $< \delta$.

7. Page 228: line 1, write $M_1^{s,\epsilon} = M_1^{s,\epsilon,P}$ and erase in line 5 the sentence "Let .....}."

8. Page 230, last display, last line: add (twice) $h(\eta|P)$.

9. Page 232, display below (2.3.47), last line: replace $\lambda_0(u)$ by $\lambda_0(u,\eta)$.

10. Page 239, line below (2.4.13), replace $\tilde{\tau}_k$ by $\tilde{\tau}_k^{(i)}$.

11. Page 240, (2.4.14) and (2.4.15): (2.4.14) does not follow from the α mixing condition $(D3)$. It does follow if one assumes β mixing instead. Alternatively, (2.4.14) holds true if instead of the last summand in the right hand side one writes $n^2 m_k \alpha(2k)/4 + m_k \mathbb{P}^o(\tilde{\tau}^{(1)} > n^2)/4 =: B(n)$. Using Lemma 2.4.16 and the definition of $\alpha(2k)$, one then replaces (2.4.15) by the estimate $B(n) \leq o(n^{1-s})$.

12. Page 247, (2.4.34), a factor $(1-v/v_P)^{1/3}$ is missing on the right hand side.

13. Page 251, lines 3 and 19, replace $P_\omega^{\bar{b}^n}$ by $P_\omega^{b_n}$. Line 19, last display in proof, replace in right side $(\bar{b}^n + \delta)$ by J.

14. Page 252, equation (2.5.12), replace $B_{-\alpha}$ by B_α (recall $\alpha < 0$!).

15. Page 255, display (2.5.17): replace $\mathcal{Q}(E_\mathcal{Q}(\bar{b}(h) = \bar{b}(1)|\Gamma(h)))$ by $E_\mathcal{Q}(\mathcal{Q}(\bar{b}(h) = \bar{b}(1)|\Gamma(h)))$.

16. Page 256, line 4 and (2.5.19), condition on $s_+(1) = s_+(t)$. Line 5, add) at end of line. Line 16, replace $f(z,\omega)$ by $f(z,w)$ and replace $e^{-w-(t-1)}$ by $e^{w-(t-1)}$.

17. Page 258: all of the multi-dimensional chapter 3 actually assumes that $P(\omega(0,0) > 0) = 0$, that is no holding times. This should have been stated explicitly as part of (**A2**).

18. Page 264: in lines 5 and 9, P should be replaced by $\bar{\mathbb{P}}^o$ (twice in each line). In line 11, $\mathbb{Q}^p$ should be $\mathbb{Q}^o$. Finally, in line -7, X_{τ_k+y} should be replaced by $X_{\tau_k} + y$.

19. Page 265, in the left hand of (3.2.8), one should divide by τ_k, not by k.

20. Page 268, line 8, replace $\liminf$ by $\limsup$.

21. Page 270, change the index of summation in both sums in (3.3.7) from i to k, replace X_{i-1} by X_{k-1}, and $\overline{\omega}(i)$ by $\overline{\omega}(k)$.

22. Page 279, line 15, should have $\mathcal{A} := [2\varepsilon, 1-2\varepsilon(d-1)]$. Line 19, $Y_n^{\hat{\alpha}} = X_n \cdot e_1$. Display below (3.3.24), $g_{n+1,\omega}(x+1)$ should be $g_{n-1,\omega}(x+1)$.

23. Page 300, lines 5,6, replace x by x_1 (x_1 is as in (3.5.19)). Line 8, replace inf in the right hand side by sup, and α by $\bar{\alpha}$. In (3.5.20), replace x_0 by x, and in lines -4 and -3, replace i by i_0. It is a good idea to replace X_i by ξ_i in the second half of the page.

Thanks to Antar Bandyopadhyay, Nina Gantert, Ghaith Hiary, Achim Klenke, Jon Peterson, Mike Weimerskirch, and Firas Rassoul-Agha for their comments.

Frank den Hollander

Random Polymers

École d'Été de Probabilités
de Saint-Flour XXXVII – 2007

Originally published in: *École d'Été de Probabilités de Saint-Flour XXXVII – 2007*,
Lecture Notes in Mathematics, Vol. **1974**, III–XIII, 1–258, DOI: 10.1007/978-3-642-00333-2,
© Springer-Verlag Berlin Heidelberg 2009, Reprint by Springer-Verlag Berlin Heidelberg 2012

Preface

This monograph contains the worked out and expanded notes of the lecture series I presented at the 37–TH PROBABILITY SUMMER SCHOOL IN SAINT-FLOUR, 8–21 JULY 2007. The goal I had set myself for these lectures was to provide an up-to-date account of some key developments in the *mathematical* theory of polymer chains, focusing on a number of models that are at the heart of the subject. In order to achieve this goal, I decided to limit myself to single polymers living on a lattice, to consider only models for which a transparent picture has emerged, and from the latter select those that lead to challenging open questions capable of attracting future research. Needless to say, my choice was influenced by my personal taste and involvement.

Polymers are studied intensively in mathematics, physics, chemistry and biology. Our focus will lie at the interface between *probability theory* and *equilibrium statistical physics*. To fully appreciate the results to be described, the reader needs a basic knowledge of both these areas. No other background is required. We will look at a number of *paradigm* models that exhibit interesting phenomena. The key objects of interest will be free energies, phase transitions as a function of underlying parameters and associated critical behavior, scaling properties of path measures in the different phases and associated invariance principles, as well as effects of randomness in the interactions. The emphasis will be on techniques coming from large deviation theory, combinatorics, ergodic theory and variational calculus.

We start with TWO BASIC MODELS of polymer chains: *simple random walk* and *self-avoiding walk*. After having collected a few key properties of these models, which serve to set the stage, we turn to the main body of the monograph, which is divided into two parts.

In PART A, we look at four models of POLYMERS WITH SELF-INTERACTION: (1) *soft polymers*, where self-intersections are not forbidden but are penalized, resulting in a repulsive interaction modeling the effect of "steric hindrance"; (2) *elastic polymers*, where self-intersections are penalized in a way that depends on their distance along the chain, in such a way that long loops are less penalized than short loops; (3) *polymer collapse*, where due to attractive interactions the polymer may roll itself up to form a ball; (4) *polymer adsorption*, where the polymer interacts with a linear substrate to which it may be attracted, either moving on both sides of the substrate (pinning at an interface) or staying on one side of the substrate (wetting of a surface).

In PART B, we look at five models of POLYMERS IN RANDOM ENVIRONMENT: (1) *charged polymers*, where positive and negative charges are arranged randomly along the chain, resulting in a mixture of repulsive and attractive interactions; (2) *copolymers near a linear selective interface*, where the polymer consists of a random concatenation of two types of monomers that interact differently with two solvents separated by a linear interface; (3) *copolymers near a random selective interface*, where the linear interface is replaced by a percolation-type interface; (4) *random pinning and wetting of polymers*, where the polymer interacts with a linear interface or surface consisting of different types of atoms or molecules arranged randomly; (5) *polymers in a random potential*, where the polymer interacts with different types of atoms or molecules arranged randomly in space.

All the results that are presented come with a complete mathematical proof. Nonetheless, there are a few places where proofs are a bit sketchy and the reader is referred to the literature for further details. Not doing so would have meant lengthening the exposition considerably. Still, even where proofs are tight I have taken care that the reader can always hold on to the main line of the argument.

All chapters can be read *essentially independently*. Each chapter tells a story that is *self-contained*, both in terms of content and of notation. Each chapter ends with a brief description of a number of important *extensions* (added to further enlarge the panorama) and with a number of *challenges* for the future (ranging from "doable in principle" via "very tough indeed" to "almost beyond hope"). An index with key words is added after the references, to help the reader connect the terminology that is used in the different chapters. For the topics covered in Parts A and B, I believe to have caught most of the relevant *mathematical* literature. There is a huge literature in physics and chemistry, of which only a few snapshots are being offered.

The choice I made of what material to cover was not driven by content alone. I also wanted to exhibit a number of key techniques that are currently available in the area and are being developed further. Thus, the reader will encounter the *method of local times* (Chapters 3, 6 and 8), *large deviations* and *variational calculus* (Chapters 3, 6, 9 and 10), the *lace expansion* in combination with the *induction approach* (Chapters 4 and 5), *generating functions* (Chapters 6 and 7), the *method of excursions* (Chapters 7, 9 and 11), the *subadditive ergodic theorem* (Chapters 9, 10 and 11), *partial annealing estimates* (Chapters 9, 10 and 11), *coarse-graining* (Chapter 10), and *martingales* (Chapter 12).

I greatly benefited from reading overview works that address various *mathematical* aspects of polymers, in particular, the monographs by Barber and Ninham [12], Madras and Slade [230], Hughes [175], Vanderzande [300], van der Hofstad [154], Sznitman [288], Janse van Rensburg [188], Slade [280], and Giacomin [116], the review papers by van der Hofstad and König [165], Bolthausen [28], and Soteros and Whittington [283], as well as the PhD theses of Caravenna [51], Pétrélis [263], and Vargas [302]. If the present monograph contributes towards making the area more accessible to a broad mathematical readership, as the above works do, then I will consider my goal reached.

While planning this monograph, I decided not to touch upon combinatorial counting techniques, exact enumeration methods and power series analysis, which provide invaluable insight for models that are too hard to handle analytically. Nor will the reader find a description of knotted polymers, which have many fascinating properties and a broad range of applications, nor of branched polymers, which are related to superprocesses arising as scaling limit. These are rapidly growing subjects, which the reader is invited to explore. For overviews, see Dušek [93], Guttmann [137], Orlandini and Whittington [257], and Guttmann [138]. Similarly, there is no discussion of models of two or more polymers interacting with each other, like in a polymer melt, nor of models dealing with dynamical aspects of polymers, such as reptation in a polymer melt. For overviews, see de Gennes [114], and Doi and Edwards [92]. The literature offers plenty of possibilities for the latter two topics as well, but so far the mathematics is rather thin.

I am grateful to Marek Biskup, Erwin Bolthausen, Andreas Greven, Remco van der Hofstad, Wolfgang König, Nicolas Pétrélis, Gordon Slade, Stu Whittington and Mario Wüthrich for co-authoring the joint papers we wrote on random polymers and for the many interesting and enjoyable discussions we have shared over the years. I am further grateful to Anton Bovier, Matthias Birkner, Thierry Bodineau, Francesco Caravenna, Francis Comets, Giambattista Giacomin, Tony Guttmann, Neil O'Connell, Andrew Rechnitzer, Chris Soteros, Alain-Sol Sznitman, Fabio Toninelli, Ivan Velenik and Lorenzo Zambotti for fruitful exchange on various occasions.

Matthias Birkner, Giambattista Giacomin, Remco van der Hofstad and Gordon Slade read parts of the prefinal draft and offered a number of useful remarks. Stu Whittington commented on three drafts in various stages of development, patiently answered a long list of questions and provided many references. He generously offered his guidance, which has been both stimulating and reassuring. Nicolas Pétrélis helped me to prepare my lectures in Saint-Flour and assisted me afterwards to finish the present monograph. Not only did we spend many hours together discussing the content, Nicolas carefully went through the full text and drew many of the figures. He was an indispensable companion in bringing the whole enterprise to a good end.

It is a pleasure to thank the staff of EURANDOM in Eindhoven for providing so many opportunities to do quiet research in a stimulating environment. It continues to be an honor and a pleasure to be affiliated with the institute, where most of the above colleagues are at home. Over the years, my research has been amply supported by NWO (Netherlands Organization for Scientific Research), which I gratefully acknowledge as well.

VIII   Preface

Finally, I thank Jean Picard for the invitation to lecture at the Saint-Flour summer school and for the pleasant exchange we have had before, during and after the event.

Mathematical Institute
Leiden University
P.O. Box 9512
2300 RA Leiden
The Netherlands
www.math.leidenuniv.nl/~denholla/
December 2008

Frank den Hollander

Contents

X   Contents

305

1

Introduction

1.1 What is a Polymer?

A *polymer* is a large molecule consisting of *monomers* that are tied together by chemical bonds. The monomers can be either small units (such as CH_2 in polyethylene; see Fig. 1.1) or larger units with an internal structure (such as the adenine-thymine and cytosine-guanine base pairs in the DNA double helix; see Fig. 1.2). Polymers abound in nature because of the *multivalency* of atoms like carbon, silicon, oxygen, nitrogen, sulfur and phosphorus, which are capable of forming long concatenated structures.

Polymer Classification. Polymers come in two varieties: (1) HOMOPOLY-MERS, with all their monomers identical (such as polyethylene); (2) COPOLY-MERS, with two or more different types of monomer (such as DNA). The order of the monomer types in copolymers can be either periodic (e.g. in agar) or random (e.g. in carrageenan).

An important classification of polymers is into SYNTHETIC POLYMERS (such as nylon, polyethylene and polystyrene) and NATURAL POLYMERS (also called biopolymers). Major subclasses of the latter are: (a) *proteins* (strings of amino-acids), the chief constituents of all living objects, carrying out a multitude of tasks; (b) *nucleic acids* (DNA, RNA), the building blocks of genes that are the very core of life processes; (c) *polysaccharides* (e.g. agar, amylose, carrageenan, cellulose), which form part of the structure of animals and plants and provide an energy source; (d) *lignin* (plant cement), which fills up the space between cellulose fibres; (e) *rubber*, occurring in the fluid of latex cells in certain trees and shrubs. Apart from (a)–(e), which are *organic* materials, clays and minerals are *inorganic* examples of natural polymers. Synthetic polymers typically are homopolymers, natural polymers typically are copolymers (with notable exceptions). Bacterial polysaccharides tend to be periodic, plant polysaccharides tend to be random.

Yet another classification of polymers is into LINEAR and BRANCHED. In the former, the monomers have one reactive group (such as CH_2), leading

F. den Hollander, *Random Polymers*,
Lecture Notes in Mathematics 1974, DOI: 10.1007/978-3-642-00333-2_1,
© Springer-Verlag Berlin Heidelberg 2009, Reprint by Springer-Verlag Berlin Heidelberg 2012

Fig. 1.1. Polyethylene.

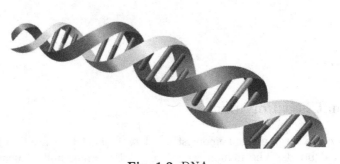

Fig. 1.2. DNA.

to a linear organization as a result of the polymerization process. In the latter, the monomers have two or more reactive groups (such as hydroxy acid), leading to an intricate network with multiple cross connections. Most natural polymers are linear, like DNA, RNA, proteins and the polysaccharides agar, alginate, amylose, carrageenan and cellulose. Some polysaccharides are branched, like amylopectin. Many synthetic polymers are linear, and many are branched. An example of a branched polymer is rubber, both natural and synthetic.

Polymerization. The chemical process of building a polymer from monomers is called *polymerization*. The size of a polymer, i.e., the number of constituent monomers (also called the degree of polymerization) may vary from 10^3 up to 10^{10} (smaller molecules do not qualify to be called polymers, larger molecules have not been recorded). Human DNA has $10^9 - 10^{10}$ base pairs, lignin consists of $10^6 - 10^7$ phenyl-propanes, while polysaccharides carry $10^3 - 10^4$ sugar units. Both in synthetic and in natural polymers, the size distribution may either be broad, with numbers varying significantly from polymer to polymer (e.g. in nylons and in polysaccharides) or be narrow (e.g. in proteins and in DNA). In synthetic polymers the size distribution can be made narrow through specific polymerization methods. The size of the monomer units varies from $1.5\,\text{Å}$ (for CH_2 in polyethylene) to $20\,\text{Å}$ (for the base pairs in DNA), with $1\,\text{Å} = 10^{-10}\,\text{m}$. For more background on the structure of polymers, the reader is referred to Green and Milne [129].

The chemical bonds in a polymer are flexible, so that the polymer can arrange itself in many different *spatial configurations*. The longer the chain, the more involved these configurations tend to be. For instance, the polymer can wind around itself to form knots, can be extended due to repulsive forces between the monomers as a result of excluded-volume, or can collapse to a ball due to attractive van der Waals forces between the monomers or repulsive forces between the monomers and a poor solvent. It can also interact with a surface on which it may or may not be adsorbed, or it can live in a slit between two confining surfaces.

As mentioned above, the polymer can be either homogeneous (homopolymer) or inhomogeneous (copolymer). A typical example of the latter is a copolymer whose monomers carry positive and negative charges, randomly arranged along the chain. Another example is a copolymer consisting of hydrophobic and hydrophilic monomers. If such a copolymer is placed near an interface separating oil and water, then it may try to wiggle around the interface in order to match the monomers as much as possible, and in doing so stay closely tied to the interface.

Targets. In the present monograph we will consider various different models of *linear* polymers, aimed at describing a variety of different physical settings, of the type alluded to above. Key quantities of interest will be the number of different spatial configurations, the typical end-to-end distance (subdiffusive, diffusive or superdiffusive), the space-time scaling limit, the fraction of monomers at an interface or adsorbed onto a surface, the average length and height of excursions away from an interface or a surface, all typically in the limit as the polymer gets long. We will pay special attention to the free energy of the polymer in this limit, and to the presence of *phase transitions* as a function of underlying model parameters, signalling drastic changes in behavior when these parameters cross critical values. We will also study the effect of randomness in the interactions.

Classical monographs on polymers with a physical and chemical orientation are Flory [102] and de Gennes [114]. (It is worthwhile to read their Nobel lectures: Flory [103] and de Gennes [115].) Monographs with a mathematical orientation are Barber and Ninham [12], Madras and Slade [230], Hughes [175], Vanderzande [300], Janse van Rensburg [188] and Giacomin [116].

1.2 What is the Model Setting?

Our polymers will live on the d-dimensional Euclidean lattice $\mathbb{Z}^d$, $d \geq 1$. They will be modelled as *random paths* on this lattice, where the monomers are the vertices in the path, and the chemical bonds connecting the monomers are the edges in the path.

The choice of model will depend on two key objects. Namely, for each $n \in \mathbb{N}_0 = \mathbb{N} \cup \{0\}$ we need to specify

$\mathcal{W}_n$ = a collection of allowed n-step *paths* on $\mathbb{Z}^d$,

H_n = a Hamiltonian that associates an *energy* to each path in $\mathcal{W}_n$. $\qquad$ (1.1)

4   1 Introduction

As we will see, there is flexibility in the choice of $\mathcal{W}_n$, depending on the particular application we have in mind. We will be considering both undirected and directed paths (see Figs. 1.3 and 1.4). The choice of H_n will be driven by the underlying physics, and captures the interaction of the polymer with itself and/or its environment. Typically, H_n depends on one or two parameters, including temperature.

For each $n \in \mathbb{N}_0$, the *random polymer* of length n is defined by assigning to each $w \in \mathcal{W}_n$ a probability given by

$$P_n(w) = \frac{1}{Z_n} e^{-H_n(w)}, \qquad w \in \mathcal{W}_n, \tag{1.2}$$

where Z_n is the normalizing partition sum. This is called the *Gibbs measure* associated with the pair $(\mathcal{W}_n, H_n)$, and it describes the polymer in *equilibrium* with itself and/or its environment (at fixed temperature and fixed polymer

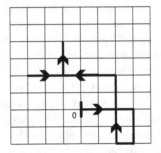

Fig. 1.3. A 19-step path on $\mathbb{Z}^2$, modeling a polymer in the plane with 20 monomers tied together by 19 chemical bonds. The steps in the path are: 3 east, 2 south, 1 west, 4 north, 5 west, 2 east and 2 north.

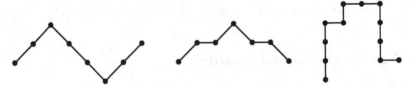

Fig. 1.4. Three examples of directed paths on $\mathbb{Z}^2$, in combinatorics commonly referred to as ballot paths (with allowed steps $\nearrow$ or $\searrow$), generalized ballot paths (with allowed steps $\nearrow$, $\rightarrow$ and $\searrow$), and partially directed self-avoiding walks (with allowed steps $\uparrow$, $\downarrow$ or $\rightarrow$, subject to no self-intersections). Ballot paths and generalized ballot paths that begin and end at the same horizontal line and lie entirely on or above this line are called Dyck paths, repectively, Motzkin paths. Without the one-sidedness restriction they are called bilateral Dyck paths, respectively, bilateral Motzkin paths.

length). Under this Gibbs measure, paths with a low energy have a high probability, while paths with a high energy have a low probability.[1]

Note that $(P_n)_{n \in \mathbb{N}}$ in general is *not* a consistent family of probability distributions, i.e., P_n is not the projection of P_{n+1} obtained by summing out the position of the $(n+1)$-st monomer. Rather, for each n we have a different distribution, modeling a polymer chain of a fixed length.

The polymer measure in (1.2) will be our main focus in PART A. In PART B we will consider models in which the Hamiltonian also depends on a *random environment* (e.g. randomly ordered charges or monomer types). We will generically denote this random environment by ω and its probability distribution by $\mathbb{P}$. We will write H_n^ω to exhibit the ω-dependence of the Hamiltonian. Three types of Gibbs measures will make their appearance:

(1) The *quenched* Gibbs measure

$$P_n^\omega(w) = \frac{1}{Z_n^\omega} e^{-H_n^\omega(w)}, \qquad w \in \mathcal{W}_n. \tag{1.3}$$

(2) The *average quenched* Gibbs measure

$$\mathbb{E}(P_n^\omega(w)) = \int P_n^\omega(w) \, \mathbb{P}(d\omega), \qquad w \in \mathcal{W}_n. \tag{1.4}$$

(3) The *annealed* Gibbs measure

$$\mathbb{P}_n(w) = \frac{1}{\mathbb{Z}_n} \int e^{-H_n^\omega(w)} \, \mathbb{P}(d\omega), \qquad w \in \mathcal{W}_n. \tag{1.5}$$

The latter is used to model a polymer whose random environment is not frozen but takes part in the equilibration. We will mostly be interested in the behavior of the polymer in the limit as $n \to \infty$, typically after some appropriate scaling.[2]

We will not (!) consider models where the length or the configuration of the polymer changes with time (e.g. due to growing or shrinking, or to a Metropolis dynamics associated with the Hamiltonian for an appropriate choice of allowed transitions). These *non-equilibrium* situations are very interesting and challenging indeed, but so far the available mathematics is very thin.

---

[1] W. Kuhn, in the 1930's, seems to have been the first to put forward the "ensemble description" of polymers given by (1.2) (see Flory [100], Chapter I). The Gibbs measure in (1.2) maximizes the *entropy* $\sum_{w \in \mathcal{W}_n} [-P_n(w) \log P_n(w)]$ subject to the constraint of constant *energy* $\sum_{w \in \mathcal{W}_n} P_n(w) H_n(w)$, as required by equilibrium statistical physics.

[2] The reader must carefully distinguish between the upper index ω, labeling the random environment, and the argument w, labeling the path, even though these symbols look very much alike. In Part B they will appear side by side in many formulas.

1.3 The Central Role of Free Energy

The *free energy* of the polymer is defined as

$$f = \lim_{n \to \infty} \frac{1}{n} \log Z_n, \qquad (1.6)$$

or, in the presence of a random environment,

$$f = \lim_{n \to \infty} \frac{1}{n} \log Z_n^\omega \qquad \omega - a.s. \qquad (1.7)$$

The limit in (1.7) typically is constant ω-a.s. (i.e., constant on a set of ω's with $\mathbb{P}$-measure 1), a property referred to as "self-averaging w.r.t. ω".

Existence. It can frequently be shown that the partition sum Z_n satisfies the inequality

$$Z_n \leq Z_m \, Z_{n-m} \qquad \forall 0 \leq m \leq n. \qquad (1.8)$$

For instance, this typically occurs when H_n assigns a repulsive self-interaction to the polymer, in which case the inequality follows by viewing the n-polymer as a concatenation of an m-polymer and an $(n - m)$-polymer. We will see examples in Part A. It follows from (1.8) that $n \mapsto \log Z_n$ is a *subadditive sequence*, i.e.,

$$\log Z_n \leq \log Z_m + \log Z_{n-m} \qquad \forall 0 \leq m \leq n, \qquad (1.9)$$

and, with $f_n = \frac{1}{n} \log Z_n$, the latter implies that

$$f = \lim_{n \to \infty} f_n = \inf_{n \in \mathbb{N}} f_n \quad \text{exists in } [-\infty, \infty) \qquad (1.10)$$

(for a proof see e.g. Madras and Slade [230], Lemma 1.2.2). If, moreover, $\inf_{w \in \mathcal{W}_n} H_n \leq Cn$ for some $C < \infty$, then $-\infty$ can be excluded as limiting value. The same is true when H_n assigns an attractive self-interaction to the polymer, in which case the inequalities in (1.8–1.9) are reversed, inf in (1.10) is replaced by sup, and ∞ is excluded as limiting value as soon as $|\mathcal{W}_n| \leq e^{Cn}$ and $\inf_{w \in \mathcal{W}_n} H_n \geq -Cn$ for some $C < \infty$. When H_n assigns both repulsive and attractive interactions to the polymer (as e.g. in a self-avoiding walk with self-attraction), then the above argument is generally not available, and the existence of the free energy either remains open or has to be established by other means.

In the presence of a random environment ω, the subadditivity property in (1.9) must be replaced by a random form of subadditivity. When applicable, $n \mapsto \log Z_n^\omega$ is called a *subadditive random process* (rather than a subadditive sequence), and Kingman's subadditive ergodic theorem [214] implies the ω-a.s. existence of the free energy in (1.7). We will see examples in Part B.

Convexity. Suppose that the Hamiltonian depends linearly on a single parameter $\beta \in \mathbb{R}$, and write $\beta H_n(w)$, $Z_n(\beta)$, P_n^β, $f_n(\beta)$ and $f(\beta)$ to exhibit this dependence. Then $\beta \mapsto f_n(\beta)$ is convex. Indeed, using Hölder's inequality, we have

$$
\begin{aligned}
Z_n\left(\tfrac{1}{2}\beta_1 + \tfrac{1}{2}\beta_2\right) &= E_n\left(e^{\left(\frac{1}{2}\beta_1 + \frac{1}{2}\beta_2\right)H_n}\right) \\
&\leq E_n\left(e^{\beta_1 H_n}\right)^{\frac{1}{2}} E_n\left(e^{\beta_2 H_n}\right)^{\frac{1}{2}} = Z_n(\beta_1)^{\frac{1}{2}} Z_n(\beta_2)^{\frac{1}{2}},
\end{aligned}
\tag{1.11}
$$

which yields

$$
f_n(\tfrac{1}{2}(\beta_1 + \beta_2)) \leq \tfrac{1}{2}f_n(\beta_1) + \tfrac{1}{2}f_n(\beta_2),
\tag{1.12}
$$

i.e., $\beta \mapsto f_n(\beta)$ is convex. Passing to the limit $n \to \infty$, we get that also $\beta \mapsto f(\beta)$ is convex.

Differentiability. Convexity implies continuity and, when combined with a sign, also monotonicity. Moreover, at those values of β where $f(\beta)$ is differentiable, convexity implies that

$$
f'(\beta) = \lim_{n \to \infty} f_n'(\beta).
\tag{1.13}
$$

The latter observation is important, because it tells us that

$$
\begin{aligned}
f'(\beta) &= \lim_{n \to \infty} \frac{1}{n} \log Z_n'(\beta) \\
&= \lim_{n \to \infty} \frac{1}{n} \sum_{w \in \mathcal{W}_n} [-H_n(w)] \, P_n^\beta(w).
\end{aligned}
\tag{1.14}
$$

What this says is that $-\beta f'(\beta)$ is the limiting energy per monomer under the Gibbs measure. Thus, at those values of β where the free energy fails to be differentiable this quantity is discontinuous, signalling the occurrence of a *phase transition*, i.e., a drastic change in the behavior of the typical path.

A similar argument applies when other parameters are in play. Derivatives of the free energy w.r.t. these parameters indicate how quantities related to these parameters change at a phase transition. We will see plenty of examples as we go along.

2

Two Basic Models

In this chapter we consider two basic models for a polymer chain: *simple random walk* (SRW), describing a polymer with no self-interaction, and *self-avoiding walk* (SAW), describing a polymer with a self-interaction given by the "excluded-volume effect", i.e., no site can be occupied by more than one monomer. Our goal is to give a *quick summary* of what is known for these models in order to set the stage for the physically more realistic models treated in Parts A and B. We will focus on those results that are relevant to later chapters and *omit proofs*, for which we refer the reader to the literature. A standard reference to SRW is the monograph by Spitzer [284]. A standard reference to SAW is the monograph by Madras and Slade [230]. Figs. 2.1 and 2.2 below are borrowed from Bill Casselman and Gordon Slade.

2.1 Simple Random Walk

Simple random walk (SRW) on $\mathbb{Z}^d$ is the random process $(S_n)_{n \in \mathbb{N}_0}$ defined by

$$S_0 = 0, \qquad S_n = \sum_{i=1}^{n} X_i, \quad n \in \mathbb{N}, \tag{2.1}$$

where

$$X = (X_i)_{i \in \mathbb{N}} \tag{2.2}$$

is an i.i.d. sequence of random variables taking values in $\mathbb{Z}^d$ with marginal law

$$P(X_1 = x) = \begin{cases} \frac{1}{2d}, & x \in \mathbb{Z}^d \text{ with } \|x\| = 1, \\ 0, & \text{otherwise,} \end{cases} \tag{2.3}$$

with $\| \cdot \|$ the Euclidean norm. If we think of the path of this process as modeling a polymer chain, then X_i (the step at time i) corresponds to the orientation of the chemical bond between the $(i-1)$-st and i-th monomer, while S_n (the position at time n) corresponds to the location of the n-th

F. den Hollander, *Random Polymers*,
Lecture Notes in Mathematics 1974, DOI: 10.1007/978-3-642-00333-2_2,
© Springer-Verlag Berlin Heidelberg 2009, Reprint by Springer-Verlag Berlin Heidelberg 2012

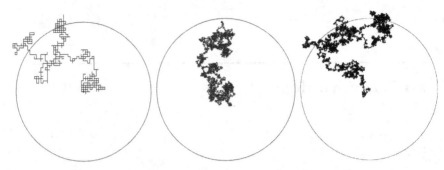

Fig. 2.1. Simulation of a SRW on $\mathbb{Z}^2$ taking $n = 10^3$, 10^4 and 10^5 steps. The circles have radius $\sqrt{n}$, in units of the step size.

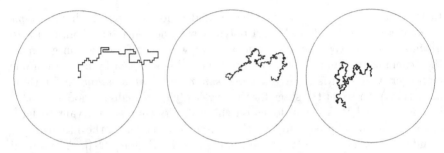

Fig. 2.2. Simulation of a SAW on $\mathbb{Z}^2$ with $n = 10^2$, 10^3 and 10^4 steps. The circles have radius $n^{3/4}$, in units of the step size.

monomer, the end-point of the polymer. We write P, E to denote probability and expectation w.r.t. SRW.

The SRW defined in (2.1–2.3) corresponds to choosing the set of paths and the Hamiltonian in (1.1) as

$$\mathcal{W}_n = \left\{ w = (w_i)_{i=0}^n \in (\mathbb{Z}^d)^{n+1} : w_0 = 0, \|w_{i+1} - w_i\| = 1 \ \forall \, 0 \le i < n \right\},$$
$$H_n \equiv 0,$$

$$(2.4)$$

and the path measure P_n in (1.2) as the uniform distribution on $\mathcal{W}_n$. This describes a *random polymer without interaction*, with $(S_i)_{i=0}^n$ taking values in $\mathcal{W}_n$. Clearly, $|\mathcal{W}_n| = (2d)^n$, and P_n is the projection of P onto $\mathcal{W}_n$. Consequently, $(P_n)_{n \in \mathbb{N}_0}$ is a consistent family. Below we collect a few key properties of SRW.

1. Transition probabilities. Let

$$\widehat{p}(k) = E\left(e^{i(k \cdot X_1)}\right) = \frac{1}{2d} \sum_{\substack{x \in \mathbb{Z}^d \\ \|x\|=1}} e^{i(k \cdot x)} = \frac{1}{d} \sum_{j=1}^d \cos(k_j),$$

$$(2.5)$$

$$k = (k_1, \ldots, k_d) \in [-\pi, \pi)^d,$$

denote the characteristic function of X_1, with $(\cdot, \cdot)$ the inner product on $\mathbb{R}^d$. Then

$$\widehat{p}_n(k) = E\left(e^{i(k \cdot S_n)}\right) = [\widehat{p}(k)]^n, \qquad k \in [-\pi, \pi)^d, \, n \in \mathbb{N}_0. \qquad (2.6)$$

Inversion gives

$$P(S_n = x) = \frac{1}{(2\pi)^d} \int_{[-\pi,\pi)^d} e^{-i(k \cdot x)} \, [\widehat{p}(k)]^n \, dk, \qquad x \in \mathbb{Z}^d, \, n \in \mathbb{N}_0, \quad (2.7)$$

which is the Fourier representation of the two-point function. The $(r+1)$-point functions follow from the two-point function via the Markov property (in accordance with (2.6)):

$$P\left(S_{n_1} = x_1, \ldots, S_{n_r} = x_r\right) = \prod_{s=1}^{r} P\left(S_{n_s - n_{s-1}} = x_s - x_{s-1}\right) \qquad (2.8)$$

(with $n_0 = x_0 = 0$). Note that SRW exhibits *diffusive behavior*:

$$E(S_n) = 0 \text{ and } E(\|S_n\|^2) = n \text{ for all } n \in \mathbb{N}_0. \qquad (2.9)$$

2. Local limit theorem. Using that $\widehat{p}(k) = 1$ if and only if $k = 0$ and $\widehat{p}(k) = 1 - \frac{1}{2d}\|k\|^2 + o(\|k\|^2)$ as $k \to 0$, we deduce from (2.7) that

$$\begin{aligned} P(S_{2n} = 0) &= \frac{1}{(2\pi)^d} \int_{[-\pi,\pi)^d} [\widehat{p}(k)]^{2n} \, dk \\[2mm] &\sim \frac{1}{(2\pi)^d} \int_{\mathbb{R}^d} e^{-\frac{1}{d}\|k\|^2 n} \, dk \qquad (2.10) \\[2mm] &= 2 \left(\frac{d}{4\pi n}\right)^{\frac{d}{2}} \qquad \text{as } n \to \infty. \end{aligned}$$

A similar calculation gives

$$P(S_n = x) \sim 2 \left(\frac{d}{2\pi n}\right)^{\frac{d}{2}} e^{-\frac{d}{2n}\|x\|^2} \qquad \text{as } n \to \infty \text{ and } \|x\|/n \to 0, \quad (2.11)$$

provided x and n have the same parity.

3. Green function. Let

$$G(x) = \sum_{n \in \mathbb{N}_0} P(S_n = x), \qquad x \in \mathbb{Z}^d, \qquad (2.12)$$

denote the average number of visits to x by SRW. Then $G(x) = \infty$ for all $x \in \mathbb{Z}^d$ when $d = 1, 2$ (recurrence) and $G(x) < \infty$ for all $x \in \mathbb{Z}^d$ when $d \geq 3$ (transience). Inserting (2.7) into (2.12), we get

$$G(x) = \frac{1}{(2\pi)^d} \int_{[-\pi,\pi)^d} e^{-i(k \cdot x)} \left[1 - \widehat{p}(k)\right]^{-1} dk, \qquad x \in \mathbb{Z}^d. \tag{2.13}$$

For $\|x\|$ large, the integral is dominated by the small k-values and so, similarly as in (2.10), a little calculation gives

$$G(x) \sim \frac{1}{(2\pi)^d} \int_{\mathbb{R}^d} e^{-i(k \cdot x)} \, 2d\|k\|^{-2} \, dk$$

$$= C_d \, \|x\|^{-d+2} \qquad \text{as } \|x\| \to \infty \text{ when } d \geq 3, \tag{2.14}$$

where $C_d = \frac{d}{2}\Gamma(\frac{d}{2} - 1)\pi^{-d/2} = 2/(d-2)\omega_d$ with Γ the Gamma-function and ω_d the volume of the unit ball in $\mathbb{R}^d$.

4. Return probability. The probability that SRW returns to the origin, written F_d, equals $F_d = 1 - [G(0)]^{-1}$ (Spitzer [284], Section 1). For $d = 1, 2$, $F_d = 1$ (recurrence), while for $d \geq 3$, $0 < F_d < 1$ (transience), e.g. $F_3 \approx 0.340$, $F_4 \approx 0.193$ and $F_5 \approx 0.135$ (see Griffin [131]).

5. Functional central limit theorem. The scaling limit of SRW is *Brownian motion* on $\mathbb{R}^d$ (see Fig. 2.1), i.e., under the law P_n of SRW,

$$\left(\frac{1}{\sqrt{n}} S_{\lfloor nt \rfloor}\right)_{0 \leq t \leq 1} \implies (B_t)_{0 \leq t \leq 1} \qquad \text{as } n \to \infty, \tag{2.15}$$

with $\implies$ denoting convergence in distribution on the space of càdlàg paths endowed with the Skorohod topology (Billingsley [14], Section 14). In Fourier language, the central limit theorem, i.e., (2.15) at $t = 1$, reads

$$\widehat{p}_n\left(\frac{k}{\sqrt{n}}\right) = e^{-\frac{1}{2d}\|k\|^2 \, [1 + o(1)]} \qquad \text{as } n \to \infty \text{ and } \|k\|^2/n \to 0. \tag{2.16}$$

2.2 Self-avoiding Walk

Self-avoiding walk (SAW) on $\mathbb{Z}^d$ corresponds to choosing the set of paths and the Hamiltonian in (1.1) as

$$\mathcal{W}_n = \left\{w = (w_i)_{i=0}^n \in (\mathbb{Z}^d)^{n+1} \colon w_0 = 0, \, \|w_{i+1} - w_i\| = 1 \; \forall 0 \leq i < n, \right.$$

$$\left. w_i \neq w_j \; \forall 0 \leq i < j \leq n\right\},$$

$$H_n \equiv 0, \tag{2.17}$$

and the path measure P_n in (1.2) as the uniform distribution on $\mathcal{W}_n$. This describes a *random polymer with self-exclusion*. Because of the self-avoidance constraint, $(P_n)_{n \in \mathbb{N}_0}$ is not (!) a consistent family.

Let $c_n = |\mathcal{W}_n|$. For $d = 1$ we trivially have $c_n = 2$ for all $n \in \mathbb{N}$. For $d \geq 2$ no closed form expression is known for c_n, but for small and moderate n it can be computed by exact enumeration methods (see Guttmann [137]). The current record is $n = 71$ for $d = 2$ (Jensen [200], [201]), $n = 30$ for $d = 3$ and $n = 24$ for $d \geq 4$ (Clisby, Liang and Slade [68]). Larger n can be handled either via numerical simulation (presently up to $n = 2^{25} \approx 3.3 \times 10^7$ in $d = 3$) or with the help of extrapolation techniques. Some rigorous information on the asymptotic behavior of c_n as $n \to \infty$ is available. We collect the main results below.

1. Concatenation and subadditivity. Pick $m, n \in \mathbb{N}$. Any two SAWs of length m and n can be concatenated. This may or may not produce a SAW of length $m+n$, depending on whether the concatenation is or is not self-avoiding (see Fig. 2.3). Consequently,

$$c_{m+n} \leq c_m c_n, \qquad m, n \in \mathbb{N}, \tag{2.18}$$

i.e., $n \mapsto \log c_n$ is subadditive. From this in turn it follows that (see Madras and Slade [230], Lemma 1.2.2)

$$\lim_{n \to \infty} (c_n)^{1/n} = \inf_{n \in \mathbb{N}} (c_n)^{1/n} = \mu \in [1, \infty) \quad \text{exists.} \tag{2.19}$$

The logarithm of the limit, $\log \mu$, is referred to as the *connective constant*. For SRW we have $\mu = 2d$.

Since $d^n \leq c_n \leq 2d(2d-1)^{n-1}$, it follows that $d \leq \mu \leq 2d-1$. The precise value of $\mu = \mu(d)$ is not known. Approximate values are $\mu(2) \approx 2.638$ and $\mu(3) \approx 4.684$, with many more decimal places known: see Jensen [200], [201], and Clisby, Liang and Slade [68]. In the latter paper, an asymptotic expansion of $\mu(d)$ in powers of $1/2d$ is derived containing 13 terms. The value of μ is *not* universal, i.e., it changes when we take a different lattice in the same dimension (e.g. the triangular lattice in $d = 2$ or the body-centered cubic lattice in $d = 3$).

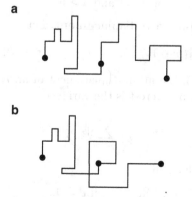

Fig. 2.3. Concatenation of two self-avoiding walks. The first concatenation is self-avoiding, the second is not.

14   2 Two Basic Models

2. Scaling for c_n. Exact enumeration and numerical simulation suggest that

$$c_n = \begin{cases} A\,\mu^n\,n^{\gamma-1}\,[1+o(1)], & \text{if } d \neq 4, \\ A\,\mu^n\,(\log n)^{\frac{1}{4}}\,[1+o(1)], & \text{if } d = 4, \end{cases} \qquad \text{as } n \to \infty, \qquad (2.20)$$

with A a non-universal amplitude and γ a universal *critical exponent*. The values of γ are

$$\gamma = 1 \ (d=1), \quad \tfrac{43}{32} \ (d=2), \quad 1.157\ldots \ (d=3), \quad 1 \ (d \geq 5). \qquad (2.21)$$

For $d = 1$, the scaling behavior in (2.20) is trivial: $A = 2$, $\mu = 1$, $\gamma = 1$. For $d \geq 5$, a proof based on the *lace expansion* was given by Hara and Slade [147], [148]; this proof is *computer-assisted*, and provides error bounds replacing $o(1)$. For $d = 2$ it is known that if (!) SAW has a conformally invariant scaling limit, then (!) this scaling limit is $\text{SLE}_{8/3}$, the *Schramm-Loewner Evolution* with parameter 8/3. The value $\gamma = 43/32$, which was first predicted by Nienhuis [248] based on non-rigorous arguments, would be implied by this scaling. Theoretical support for scaling to $\text{SLE}_{8/3}$ is provided in Lawler, Schramm and Werner [227], numerical support in Kennedy [209] (see Fig. 2.2 for a simulation). An overview on SLE is given in the Saint-Flour lectures by Werner [309]. For $d = 4$ work in progress by Brydges and Slade (private communication) based on *renormalization group theory* gives hope for a proof of the logarithmic correction. The case $d = 3$ remains open, but there are accurate estimates in Guida and Zinn-Justin [134] and in Clisby, Liang and Slade [68].

By noting that (2.20) implies

$$c_{2n}/(c_n)^2 \sim \begin{cases} A^{-1}(2/n)^{\gamma-1}, & \text{if } d \neq 4, \\ A^{-1}(\log n)^{-\frac{1}{4}}, & \text{if } d = 4, \end{cases} \qquad \text{as } n \to \infty, \qquad (2.22)$$

we see that γ has an interpretation as an *intersection* exponent, describing the rate of decay of the probability that two n-step SAWs starting at 0 have no site in common (apart from 0 itself). Note that this probability tends to 0 for $2 \leq d \leq 4$ and to A^{-1} for $d = 1$ and $d \geq 5$.

3. Scaling for mean-square displacement. Let

$$c_n(x) = |\{w \in \mathcal{W}_n \colon w_n = x\}|, \qquad x \in \mathbb{Z}^d, \ n \in \mathbb{N}_0. \qquad (2.23)$$

Then $\frac{1}{c_n} \sum_{x \in \mathbb{Z}^d} x c_n(x)$, the mean displacement of an n-step SAW, is zero by symmetry. A quantity of interest is the variance

$$v_n = \frac{1}{c_n} \sum_{x \in \mathbb{Z}^d} \|x\|^2 c_n(x). \qquad (2.24)$$

This is believed to behave like

$$v_n = \begin{cases} D\,n^{2\nu}\,[1+o(1)], & \text{if } d \neq 4, \\ D\,n(\log n)^{\frac{1}{4}}\,[1+o(1)], & \text{if } d = 4, \end{cases} \qquad \text{as } n \to \infty, \qquad (2.25)$$

with D a non-universal diffusion constant and ν a universal *critical exponent*. (Compare (2.25) with (2.9).) The values of ν are

$$\nu = 1 \ (d=1), \quad \tfrac{3}{4} \ (d=2), \quad 0.588\ldots \ (d=3), \quad \tfrac{1}{2} \ (d \geq 5). \qquad (2.26)$$

As with (2.20), for $d = 1$ the scaling behavior in (2.25) is trivial, while for $d \geq 5$ a proof was given by Hara and Slade [147], [148] (see also Slade [277], [278], [279]). These two cases correspond to *ballistic*, respectively, *diffusive* behavior. The situation for $d = 2, 3, 4$ is the same as for (2.20). For a random walk on a hierarchical lattice that is effectively four-dimensional, the scaling in the second line of (2.25) was proved by Brydges and Imbrie [41] with the help of renormalization group theory. There is promise for these techniques to carry over to $\mathbb{Z}^4$.

4. Functional central limit theorem. It was proved in Hara and Slade [147], [148] (see also Slade [277], [278], [279]) that, for $d \geq 5$ under the law of P_n of the n-step SAW,

$$\left(\frac{1}{D\sqrt{n}} S_{\lfloor nt \rfloor} \right)_{0 \leq t \leq 1} \implies (B_t)_{0 \leq t \leq 1} \qquad \text{as } n \to \infty, \qquad (2.27)$$

which is the analogue of (2.15). Let, as in (2.6),

$$\widehat{c}_n(k) = \frac{1}{c_n} \sum_{w \in \mathcal{W}_n} e^{i(k \cdot w_n)}, \qquad k \in [-\pi, \pi]^d, \, n \in \mathbb{N}_0. \qquad (2.28)$$

Then (2.27) at $t = 1$ reads

$$\widehat{c}_n \left(\frac{k}{D\sqrt{n}} \right) = e^{-\frac{1}{2d} \|k\|^2 \, [1+o(1)]} \qquad \text{as } n \to \infty \text{ and } \|k\|^2 \leq C < \infty, \qquad (2.29)$$

which is the analogue of (2.16).

5. Green function. It was proved in Hara, van der Hofstad and Slade [146], and Hara [145] that there exists a $C > 0$ such that

$$G_\mu(x) = \sum_{n \in \mathbb{N}_0} c_n(x) \, \mu^{-n} \sim C \|x\|^{-d+2} \qquad \text{as } \|x\| \to \infty \text{ for } d \geq 5, \qquad (2.30)$$

which is the analogue of (2.14). Thus, the diffusive scaling manifests itself also at the level of the Green function. For SRW we have $\mu = 2d$ and $c_n(x)\mu^{-n} = P(S_n = x)$ (compare with (2.12)).

6. Flory argument. The following heuristic argument in support of (2.26) is due to Flory [101]. Suppose that SAW of length n has radius L. Then its density is $\asymp n/L^d$ monomers per site (where $\asymp$ means that the ratio of the two sides is bounded above and below by strictly positive and finite constants). Therefore its "repulsive energy" per site is $\asymp (n/L^d)^2$, and so its total repulsive energy is $\asymp n^2/L^d$. Now, the probability that SRW has radius L is roughly $\exp[-L^2/n]$. Hence, the probability that SAW has radius L is roughly $\exp[-\{n^2/L^d + L^2/n\}]$. Put $L = n^\nu$. Then the term between braces equals $n^{2-d\nu} + n^{2\nu-1}$, which is minimal when $2 - d\nu = 2\nu - 1$, or

$$\nu = \frac{3}{d+2}. \tag{2.31}$$

For this choice of ν the term between braces takes the value $n^{(4-d)/(d+2)}$, and so we need $1 \le d \le 4$ for the total repulsive energy not to vanish. For $d \ge 5$, the total repulsive energy vanishes and we are in the diffusive regime with $\nu = \frac{1}{2}$. (Madras and Slade [230], Section 2.2, gives a more refined version of the above argument.)

The value in (2.31) for the critical exponent in (2.25) when $1 \le d \le 4$ fits with (2.26), except for $d = 3$, where it is off by 2 percent. Thus, Flory's simple heuristics is remarkably accurate. Still, nobody has yet succeeded in turning it into a rigorous argument. For $d = 2, 3, 4$ it has not even been proved that v_n grows faster than $n^{1/2}$ and slower than n.

The fact that SAW has the same scaling behavior as SRW for $d \ge 5$ is expressed by saying that "SAW and SRW are in the same universality class". Correspondingly, $d = 4$ is called the *upper critical dimension*. In physical jargon, for $d \ge 5$ SAW has "mean-field behavior", meaning that its critical exponents cease to depend on dimension and are the same as for SRW (which has no interaction). The intuitive reason for the crossover is that in low dimension long loops of SRW are dominant, causing the excluded-volume effect in SAW to be long-ranged, whereas in high dimension short loops are dominant, causing it to be short-ranged. The behavior in (2.24–2.26) is called *ballistic* in $d = 1$, *subballistic* and *superdiffusive* in $d = 2, 3, 4$, and *diffusive* in $d \ge 5$.

Over the years, SAW has been a breeding ground for the development of methods that have been successfully applied to a variety of polymer models. Key examples are *concatenation arguments* (Hammersley [141]), *folding arguments* (Hammersley and Welsh [143], Hammersley and Whittington [144]), *pattern theorems* (Kesten [211]), and the *lace expansion* (Brydges and Spencer [44]). Also, SAW has led to the development of the so-called *pivot algorithm* to simulate self-avoiding lattice paths (Madras and Sokal [231]) and to advanced *exact numerations methods* (Guttman [137]), both of which have had considerable spin-off in other areas of probability theory and statistical physics as well. Guttmann [138] contains an extensive overview of rigorous, approximate and numerical results for lattice polygons, i.e., self-avoiding loops, including results obtained by exact enumeration methods.

Part A

Polymers with Self-interaction

OUTLINE:

In Part A we look at a four variations on the two "plain vanilla" models described in Chapter 2.

- In Chapters 3 and 4, we consider a polymer for which self-intersections are not forbidden (as in SAW) but are penalized, resulting in a repulsive interaction modeling the effect of "steric hindrance". This model, which we call *soft polymer*, may be viewed as an "interpolation" between SRW and SAW. Our main result will be that the soft polymer is ballistic in $d = 1$ and diffusive in $d \geq 5$, like SAW. We will obtain the full scaling behavior.
- In Chapter 5, we consider a polymer for which self-intersections are penalized in a way that depends on their distance along the chain, in such a way that long loops are less penalized than short loops. This model, which we call *elastic polymer*, provides a crude way of incorporating the effect of "stiffness" of a polymer (i.e., its resistance against making sharp bends). We will show that the smaller the stiffness, the lower the critical dimension for diffusive behavior.
- In Chapter 6, we add a reward for self-touchings (pairs of monomers occupying neighboring sites), which results in a mix of repulsive and attractive interactions. This is a model for a polymer in a poor solvent. If the attraction is strong enough to compete with the repulsion, then the polymer *collapses* from a "random coil" to a "compact ball". We will show that there are three phases – extended, collapsed and localized – with different scaling properties.
- In Chapter 7, finally, we consider a polymer that interacts with a linear substrate: monomers at the substrate receive a reward. If the attraction is strong enough, then the polymer *adsorbs* onto the substrate. We consider both the case where the substrate is penetrable ("pinning by an interface") and where it is impenetrable ("wetting of a surface"). The latter is a model for paint on a wall: at low temperature the paint sticks to the wall, at high

temperature it does not. We also look at what happens when a *force* is applied to the right endpoint of an adsorbed polymer, pulling it away from and off the substrate with the help of optical tweezers. In addition, we look at a polymer in a slit between two impenetrable substrates and compute the effective force the polymer exerts on the substrates, which can be either attractive or repulsive. This is a model for colloidal dispersions of particles with polymers attached to them.

What is challenging about the models to be described below is that the polymers appear as random objects with a *long-range interaction*: monomers that are far apart along the chain can meet and can interact with each other. This places polymers in a league of their own, with specific questions driven by specific applications. As such, polymers are different from more classical probabilistic objects like Markov chains, percolation, the contact process or interacting diffusions. Consequently, their study requires the development of proper ideas and techniques.

3

Soft Polymers in Low Dimension

In Chapters 3 and 4 we consider a variation of the SAW in which self-intersections are not forbidden but are penalized. We refer to this as the *soft polymer*. In Chapter 3 will show that the soft polymer has *ballistic* behavior in $d = 1$. The proof uses a Markovian representation of the local times of one-dimensional SRW (a powerful technique that is useful also for other models), in combination with large deviation theory, variational calculus and spectral calculus. In Chapter 4 we will show that the soft polymer has *diffusive* behavior in $d \geq 5$. The proof there uses a combinatorial expansion technique called the lace expansion, and is based on the idea that in high dimension SAW can be viewed as a "perturbation" of SRW.

The above scaling says that in $d = 1$ and $d \geq 5$ the soft polymer is in the *same universality class* as SAW. This is expected to be true also for $2 \leq d \leq 4$, but a proof is missing.

In Section 3.1 we define the model, in Section 3.2 we state the main result, a large deviation principle (LDP) for the location of the right endpoint. In Section 3.3 we outline a five-step program to prove the LDP for bridge polymers, i.e., polymers confined between their endpoints. This program is carried out in Section 3.4. In Section 3.5 we remove the bridge condition and prove the full LDP. It will turn out that the rate function has an interesting *critical value strictly below the typical speed*. The main technique that is used is the *method of local times*.

3.1 A Polymer with Self-repellence

The soft polymer on $\mathbb{Z}^d$ treated in Chapters 3 and 4 is defined by choosing the set of paths and the Hamiltonian in (1.1) as

$$\mathcal{W}_n = \left\{ w = (w_i)_{i=0}^n \in (\mathbb{Z}^d)^{n+1} \colon w_0 = 0, \|w_{i+1} - w_i\| = 1 \ \forall \, 0 \leq i < n \right\},$$
$$H_n(w) = \beta I_n(w),$$

$$(3.1)$$

F. den Hollander, *Random Polymers*,
Lecture Notes in Mathematics 1974, DOI: 10.1007/978-3-642-00333-2_3,
© Springer-Verlag Berlin Heidelberg 2009, Reprint by Springer-Verlag Berlin Heidelberg 2012

where $\beta \in [0, \infty)$ and

$$I_n(w) = \sum_{\substack{i,j=0 \\ i<j}}^{n} 1_{\{w_i = w_j\}} \tag{3.2}$$

is the *intersection local time* of w. This model goes under the name of *weakly self-avoiding random walk*: every self-intersection contributes an energy β to the Hamiltonian and is therefore penalized by a factor $e^{-\beta}$. Another name used in the literature is Domb-Joyce model. Think of β as a *strength of self-repellence* parameter.

We write P_n^β to denote the law of the soft polymer of length n with parameter β, as in (1.2). We add a factor $(1/2d)^n$ to P_n^β in order to be able to compare it with the law P_n of SRW, i.e., we put

$$P_n^\beta(w) = \frac{1}{Z_n^\beta} e^{-\beta I_n(w)} P_n(w), \qquad w \in \mathcal{W}_n, \tag{3.3}$$

so that we may think of P_n^β as an exponential tilting of P_n. Thus, P_n^β is the law of a *random process* $(S_i)_{i=0}^n$ with weak self-repellence, taking values in $\mathcal{W}_n$. Note that, like for SAW in Section 1.2, $(P_n^\beta)_{n \in \mathbb{N}_0}$ is not (!) a consistent family when $\beta \in (0, \infty)$. The case $\beta = 0$ corresponds to SRW, the case $\beta = \infty$ to SAW.

In what follows we focus on the case $d = 1$. In Chapter 4 we deal with the case $d \geq 5$.

3.2 Weakly Self-avoiding Walk in Dimension One

Intuitively, we expect that typical paths under the measure P_n^β hang around the origin for a while and then wander off to infinity at a strictly positive speed because of the self-repellence (there is a trivial symmetry between moving to the left and moving to the right). Ballistic behavior was first shown by Bolthausen [25], without existence and identification of the speed. Theorems 3.1 and 3.2 below, which are taken from Greven and den Hollander [130], establish existence and identify the speed in terms of a variational problem. See also den Hollander [168], Chapter IX.

Theorem 3.1. *For every $\beta \in (0, \infty)$ there exists a $\theta^*(\beta) \in (0, 1)$ such that*

$$\lim_{n \to \infty} P_n^\beta\left(\left|\frac{1}{n}S_n - \theta^*(\beta)\right| \leq \epsilon \;\middle|\; S_n \geq 0\right) = 1 \text{ for all } \epsilon > 0. \tag{3.4}$$

Theorem 3.2. *The function $\beta \mapsto \theta^*(\beta)$ can be computed in terms of a variational problem. It follows from the solution of this variational problem that*

$$\beta \mapsto \theta^*(\beta) \text{ is analytic on } (0, \infty),$$

$$\lim_{\beta \downarrow 0} \theta^*(\beta) = 0, \qquad \lim_{\beta \to \infty} \theta^*(\beta) = 1. \tag{3.5}$$

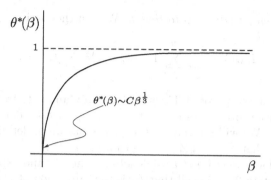

Fig. 3.1. The linear speed of the soft polymer.

The quantity $\theta^*(\beta)$ is *the speed of the soft polymer with strength of repellence* β. In Section 3.6 we will see that $\beta \mapsto \theta^*(\beta)$ looks like the curve in Fig. 3.1.

Theorems 3.1 and 3.2 will follow from the following *large deviation principle*, which is the main result of the present chapter.

Theorem 3.3. *For every* $\beta \in (0,\infty)$ *the family* $(P_n^{+,\beta})_{n\in\mathbb{N}_0}$ *defined by*

$$P_n^{+,\beta}(\cdot) = P_n^\beta\left(\frac{1}{n}S_n \in \cdot \,\bigg|\, S_n \geq 0\right) \tag{3.6}$$

satisfies a large deviation principle (LDP) on $[0,1]$ *with rate* n *and with rate function* I_β, *identified in (3.66) below, having* $\theta^*(\beta)$ *as its unique zero.*

The full proof of this LDP will have to wait until Section 3.5 (see Theorem 3.14 and Fig. 3.4). For the definition of LDP, we refer the reader to Dembo and Zeitouni [82], Chapter 1, and den Hollander [168], Chapter III. In essence, Theorem 3.3 says that

$$\lim_{\delta\downarrow 0} \lim_{n\to\infty} \frac{1}{n}\log P_n^{+,\beta}\big([\theta-\delta,\theta+\delta]\big) = -I_\beta(\theta). \tag{3.7}$$

Before we get going on the proof of Theorem 3.3, we first rewrite the definition of P_n^β in (3.3) in a way that is more convenient. Let

$$\widehat{I}_n(w) = \sum_{i,j=0}^{n} 1_{\{w_i=w_j\}}. \tag{3.8}$$

Then $\widehat{I}_n(w) = 2I_n(w) + (n+1)$. Hence, we may as well put $\widehat{I}_n(w)$ in the exponential weight factor (which only changes β to 2β). Henceforth we write $I_n(w)$ again, suppressing the overscript.

The following object is of paramount importance for the argument given below. Define

$$\ell_n(x) = \sum_{i=0}^{n} 1_{\{w_i=x\}}, \qquad x \in \mathbb{Z}, n \in \mathbb{N}_0, \tag{3.9}$$

i.e., the *local time at site x up to time n*. We can then write

$$I_n(w) = \sum_{x \in \mathbb{Z}} \sum_{i,j=0}^{n} 1_{\{w_i = w_j = x\}} = \sum_{x \in \mathbb{Z}} \ell_n(x)^2. \tag{3.10}$$

We thus see that the proof of Theorem 3.3 really amounts to understanding the large deviation properties of the random sequence $\{\ell_n(x)\}_{x \in \mathbb{Z}}$ under the law P_n of SRW. We will see that this sequence has an underlying *Markovian structure*. Note that $\sum_{x \in \mathbb{Z}} \ell_n(x) = n + 1$ for all $n \in \mathbb{N}$.

In what follows we write P, E to denote probability and expectation w.r.t. SRW (as in Section 2.1). Recall that P_n is the projection of P onto $\mathcal{W}_n$.

3.3 The Large Deviation Principle for Bridges

In order to obtain the desired LDP for $P_n^{+,\beta}(\frac{1}{n} S_n \in \cdot)$, we begin by deriving an LDP under the restriction that the path be a *bridge*, i.e., that it lies between its endpoints. This restriction will be crucial for the proof in Section 3.4, and will only be removed in Section 3.5.

Folding a path into a bridge. Our first lemma shows that the bridge condition does not change the normalizing constant.

Lemma 3.4. *For $n \to \infty$,*

$$E\left(e^{-\beta I_n} 1_{\{S_n \geq 0\}}\right) = e^{o(n)} E\left(e^{-\beta I_n} 1_{\circledast_n}\right), \tag{3.11}$$

with $I_n = I_n((S_i)_{i=0}^n)$ and

$$\circledast_n = \{S_0 \leq S_i \leq S_n \ \forall 0 \leq i \leq n\}. \tag{3.12}$$

Proof. The proof uses a folding argument due to Hammersley and Welsh [143]. Fix n.

First, suppose that the path is a *half-bridge* to the right, i.e., $S_i > S_0$ $\forall 0 < i \leq n$. We can then do a reflection procedure starting from the left endpoint of the path, as follows. Put $i_0 = 0$ and, for $j = 1, 2, \ldots$, define (R_j, i_j) recursively as

$$R_j = \max_{i_{j-1} < i \leq n} (-1)^j (S_{i_{j-1}} - S_i),$$
$$i_j = \text{the largest } i \text{ where the maximum is attained.}$$

The recursion is stopped at the smallest integer k such that $i_k = n$. What this definition says is that R_j is the span of the subwalk $(S_{i_{j-1}}, \ldots, S_n)$. Each subwalk $(S_{i_{j-1}}, \ldots, S_{i_j})$ lies strictly on one side of the point $S_{i_{j-1}}$, and

$$R_1 + \cdots + R_k \leq n \quad \text{and} \quad R_1 > R_2 > \cdots > R_k \geq 1. \tag{3.13}$$

If, for $j = 1, 2, \ldots, k - 1$, we reflect $(S_{i_j}, \ldots, S_n)$ around the point S_{i_j}, then we end up with a bridge, i.e., a path satisfying $S_0 < S_i \leq S_n \; \forall 0 < i \leq n$. Moreover, this bridge is less penalized than the original path because it has less self-intersections.

Next, we drop the assumption that the path be a half-bridge and only suppose that $S_n \geq 0$. Let

$$i_- = \min\left\{0 \leq i \leq n: S_i = \min_{0 \leq j \leq n} S_j\right\},$$

$$i_+ = \max\left\{0 \leq i \leq n: S_i = \max_{0 \leq j \leq n} S_j\right\}. \tag{3.14}$$

Then, when $i_- > 0$ and $i_+ < n$, both $(S_{i_-}, \ldots, S_0)$ and $(S_n, \ldots, S_{i_+})$ are half-bridges, and the above reflection procedure applies. If we fold both pieces outwards after the reflection procedure is through, then we end up with a bridge. (The cases $i_- = 0$ and $i_+ = n$ need no reflection.) Hence, we conclude that

$$E\left(e^{-\beta I_n} 1_{\{S_n \geq 0\}}\right) \leq N_n^2 \, E\left(e^{-\beta I_n} 1_{\circledast_n}\right), \tag{3.15}$$

where N_n is the number of solutions of (3.13) summed over k (which is the number of ordered partitions of $\{1, \ldots, n\}$). However, it is known that $N_n = \exp[O(\sqrt{n})]$ (see Madras and Slade [230], Theorem 3.1.4), so this factor is harmless and the claim follows. $\square$

The LDP for bridges. Our main result, whose proof will be given in Section 3.4, is the following LDP for the speed of the bridge soft polymer.

Theorem 3.5. *For every $\beta \in (0, \infty)$ the family $(P_n^{\beta,\text{bridge}})$, $n \in \mathbb{N}_0$, defined by*

$$P_n^{\beta,\text{bridge}}(\cdot) = P_n^\beta\left(\frac{1}{n} S_n \in \cdot \;\middle|\; \circledast_n\right) \tag{3.16}$$

satisfies the LDP on $(0, 1]$ with rate n and with rate function J_β identified in (3.21) and Lemma 3.12 below (see Fig. 3.3 below). The unique zero of J_β is $\theta^*(\beta)$ in Theorem 3.1.

To prove Theorem 3.5 we will carry out the following *program*:

(I) Pick $\theta \in (0, 1]$ and consider the quantity

$$P_n^\beta\left(S_n = \lceil \theta n \rceil \mid \circledast_n\right) = \frac{\widehat{K}_n(\theta)}{\int_{\theta \in (0,1]} \mathrm{d}(\theta n) \widehat{K}_n(\theta)}, \tag{3.17}$$

where

$$\widehat{K}_n(\theta) = E\left(e^{-\beta I_n} 1_{\{S_n = \lceil \theta n \rceil\}} 1_{\circledast_n}\right) \tag{3.18}$$

($\lceil \theta n \rceil$ and n must have the same parity). The value $\theta = 0$ is not relevant for bridges.

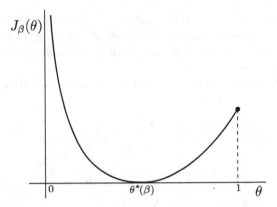

Fig. 3.2. The rate function J_β for bridge soft polymers.

(II) Show that there exists a function $\widehat{J}_\beta\colon (0,1] \to (0,\infty)$ such that

$$\lim_{n\to\infty} \frac{1}{n} \log \widehat{K}_n(\theta) = -\widehat{J}_\beta(\theta), \tag{3.19}$$

with the property that $\theta \mapsto \widehat{J}_\beta(\theta)$ is continuous, strictly convex and minimal at $\theta^*(\beta)$. Identify $\widehat{J}_\beta$ in terms of a variational problem.

(III) Combine (I) and (II), to obtain

$$\lim_{n\to\infty} \frac{1}{n} \log P_n^\beta \big(S_n = \lceil \theta n \rceil \,|\, \circledast_n\big) = -J_\beta(\theta), \tag{3.20}$$

with

$$J_\beta(\theta) = \widehat{J}_\beta(\theta) - \inf_{\theta \in (0,1]} \widehat{J}_\beta(\theta). \tag{3.21}$$

Evidently, $\theta \mapsto J_\beta(\theta)$ is also continuous, strictly convex and minimal at $\theta^*(\beta)$, which is its unique zero (see Fig. 3.2).

The argument in Section 3.4 will show that the same results as in (3.19–3.20) apply when θ is replaced by $\theta_n \to \theta$ as $n \to \infty$, which is why we get Theorem 3.5.

The above *program* will be carried out in Section 3.4, in five steps organized as Sections 3.4.1–3.4.5. The first two steps are a preparation that is needed to get the key quantities in the right format for applying large deviation theory. The actual application of large deviation theory and the analysis of the ensuing variational problem are carried out in the last three steps. The computation is technical but powerful, and can be carried over to other one-dimensional models as well.

After we have completed the proof of Theorem 3.5 we will show how to remove the bridge condition. This is done in Section 3.5 and leads to Theorem 3.3, the LDP we are actually after. It will turn out that the associated rate function is different from J_β but still has $\theta^*(\beta)$ as its unique zero (see Fig. 3.4 below), which is why Theorems 3.1–3.2 will follow as corollaries.

3.4 Program of Five Steps

3.4.1 Step 1: Adding Drift

We begin by going through a number of rewrites of the quantity $K_n(\theta)$ defined in (3.18).

Fix $\theta \in (0,1)$. (The case $\theta = 0$ is degenerate, the case $\theta = 1$ is trivial.) Let P_θ, E_θ denote probability and expectation for the random walk with drift θ (i.e., with probabilities $\frac{1}{2}(1 + \theta)$ and $\frac{1}{2}(1 - \theta)$ to step to the right and to the left, respectively). Then we can write (3.18) as

$$\widehat{K}_n(\theta) = (1 - \theta)^{-\frac{n-\lceil\theta n\rceil}{2}} (1 + \theta)^{-\frac{n+\lceil\theta n\rceil}{2}} \widetilde{K}_n(\theta), \tag{3.22}$$

with

$$\widetilde{K}_n(\theta) = E_\theta\left(e^{-\beta I_n} 1_{\{S_n = \lceil\theta n\rceil\}} 1_{\circledast_n}\right). \tag{3.23}$$

Indeed, every path from 0 to $\lceil\theta n\rceil$ makes the same number of steps to the left and to the right, so we pick up a simple Radon-Nikodym factor. Thus it suffices to study the asymptotics of $\widetilde{K}_n(\theta)$, i.e., *our task now is to relate the soft polymer with drift θ to the random walk with drift θ.*

The advantage of the reformulation in (3.22–3.23) is that the path does not care to return to $[0, S_n]$ after time n.

Lemma 3.6. *For every $\theta \in (0,1)$ and $n \in \mathbb{N}_0$,*

$$E_\theta\left(e^{-\beta I_n} 1_{\{S_n = \lceil\theta n\rceil\}} 1_{\circledast_n}\right) = \frac{1}{\theta} E_\theta\left(e^{-\beta I_n} 1_{\{S_n = \lceil\theta n\rceil\}} 1_{\circledast_n \cap \circledcirc_n}\right), \tag{3.24}$$

with

$$\circledcirc_n = \{S_i > S_n \ \forall i > n\}. \tag{3.25}$$

Proof. Simply use that I_n does not depend on S_i for $i > n$, and that $P_\theta(\circledcirc_n) = \theta$ for all n (Spitzer [284], Section 1). □

An important consequence of Lemma 3.6 is that on the event $\{S_n = \lceil\theta n\rceil\} \cap \circledast_n \cap \circledcirc_n$ we may write

$$I_n = \sum_{x=0}^{\lceil\theta n\rceil} \ell(x)^2, \tag{3.26}$$

where

$$\ell(x) = \sum_{i\in\mathbb{N}_0} 1_{\{S_i = x\}}, \qquad x \in \mathbb{Z}, \tag{3.27}$$

is the *total local time at site x.* Indeed, this follows from the observation that on the event $\{S_n = \lceil\theta n\rceil\} \cap \circledast_n \cap \circledcirc_n$ we have

$$\ell_n(x) = \begin{cases} \ell(x), & \text{if } 0 \le x \le \lceil\theta n\rceil, \\ 0, & \text{otherwise.} \end{cases}$$

Therefore we may rewrite (3.23) as

$$\widetilde{K}_n(\theta) = \frac{1}{\theta} E_\theta\left(e^{-\beta \sum_{x=0}^{\lceil \theta n\rceil} \ell^2(x)} 1_{\{S_n=\lceil \theta n\rceil\}} 1_{\circledast_n \cap \odot_n}\right). \tag{3.28}$$

Note that time n has been replaced by space $\lceil \theta n\rceil$ in (3.28). The total local times turn out to have a nice structure, as we show next.

3.4.2 Step 2: Markovian Nature of the Total Local Times

In this section we show that $\{\ell(x)\}_{x\in\mathbb{N}_0}$ admits a nice Markovian description. This will allow us to deduce the asymptotics of $\widetilde{K}_n(\theta)$ from an LDP for Markov chains. Let

$$m(x) = \sum_{i\in\mathbb{N}_0} 1_{\{S_i=x,\, S_{i+1}=x+1\}}, \qquad x\in\mathbb{Z}, \tag{3.29}$$

be the *total number of jumps from x to $x+1$*. Then, on the event $\{S_n = \lceil \theta n\rceil\}\cap \circledast_n \cap \odot_n$, the total number of jumps from $x+1$ to x equals

$$\sum_{i\in\mathbb{N}_0} 1_{\{S_i=x+1,\, S_{i+1}=x\}} = m(x) - 1, \qquad 0\le x\le\lceil \theta n\rceil, \tag{3.30}$$

because the net number of jumps along the edge between x and $x+1$ must be $+1$. Since $\ell(x)$ is the sum of the number of jumps to x coming from the left and from the right, we have

$$\ell(x) = m(x-1) + m(x) - 1_{\{x>0\}}, \qquad 0\le x\le\lceil \theta n\rceil. \tag{3.31}$$

Moreover, on the event $\{S_n = \lceil \theta n\rceil\}\cap \circledast_n \cap \odot_n$, the total time spent between 0 and $\lceil \theta n\rceil$ is $n+1$. Therefore

$$\{S_n = \lceil \theta n\rceil\}\cap \circledast_n \cap \odot_n$$
$$= \left\{\sum_{x=0}^{\lceil \theta n\rceil} [m(x-1) + m(x) - 1_{\{x>0\}}] = n+1,\, m(-1) = 0,\, m(\lceil \theta n\rceil) = 1\right\}. \tag{3.32}$$

Therefore we may rewrite (3.28) as

$$\widetilde{K}_n(\theta) = \frac{1}{\theta} E_\theta\left(e^{-\beta \sum_{x=0}^{\lceil \theta n\rceil} [m(x-1)+m(x)-1_{\{x>0\}}]^2}\right.$$
$$\left. \times 1_{\left\{\sum_{x=0}^{\lceil \theta n\rceil} [m(x-1)+m(x)-1_{\{x>0\}}]=n+1\right\}} 1_{\{m(-1)=0,\, m(\lceil \theta n\rceil)=1\}}\right). \tag{3.33}$$

The indicator $1_{\{x>0\}}$ and the restrictions $m(-1) = 0$ and $m(\lceil \theta n\rceil) = 1$ are to be thought of as harmless boundary terms.

The main reason for the reformulation in (3.33) is the following fact, which goes back to Knight [215].

Lemma 3.7. *For every $\theta \in (0,1)$ under the law P_θ, $\{m(x)\}_{x \in \mathbb{N}_0}$ is a Markov chain on state space $\mathbb{N}$ with transition kernel*

$$P_\theta(i,j) = \binom{i+j-2}{i-1}\left(\frac{1+\theta}{2}\right)^i\left(\frac{1-\theta}{2}\right)^{j-1}, \qquad i,j \in \mathbb{N}. \tag{3.34}$$

Proof. Fix x. If $m(x) = i$, then the edge $(x, x+1)$ receives i upcrossings and $i-1$ downcrossings. For $s = 1, \ldots, i-1$, let Z_s denote the number of upcrossings of $(x+1, x+2)$ in between the s-th upcrossing and the s-th downcrossing of $(x, x+1)$. Let Z denote the number of upcrossings of $(x+1, x+2)$ after the i-th upcrossing of $(x, x+1)$, which is different from the others because no further downcrossing of $(x, x+1)$ is allowed. Since the random walk has drift θ, the probability that it makes a loop excursion to the right of $x+1$ is $\frac{1-\theta}{2}$. Hence, we have

$$P_\theta(Z_s = k \mid m(x) = i) = \left(\frac{1+\theta}{1-\theta}\right)\left(\frac{1-\theta}{2}\right)^{k+1}, \qquad k \in \mathbb{N}_0, \ s = 1, \ldots, i-1,$$

$$P_\theta(Z = k \mid m(x) = i) = \frac{1+\theta}{1-\theta}\left(\frac{1-\theta}{2}\right)^k, \qquad k \in \mathbb{N}.$$

$$\tag{3.35}$$

Since $m(x+1) = j$ means that $Z_1 + \cdots + Z_{i-1} + Z = j$, we see that our process is Markov: it is irrelevant for the outcome of $m(x+1)$ what the random walk does to the left of x, only the value of $m(x)$ matters. Moreover, $Z_s + 1$, $s = 1, \ldots, i-1$, have the same law as Z. Since $(Z_1+1) + \cdots + (Z_{i-1}+1) + Z = i+j-1$ and since the number of ways $i+j-1$ can be divided into i pieces of length ≥ 1 equals the binomial factor in (3.34), we obtain the formula for $P_\theta(i,j)$ in (3.34). $\square$

The proof of Lemma 3.7 shows that $\{m(x)\}_{x \in \mathbb{N}_0}$ is a branching process with one immigrant and with an offspring distribution that has mean smaller than 1. Therefore it is a positive recurrent Markov chain.

3.4.3 Step 3: Key Variational Problem

We have now completed our rewrite of $K_n(\theta)$ in (3.18) and are ready to apply large deviation theory. In this section we derive the key variational problem underlying the LDP for bridges in Theorem 3.5. This can be done along fairly standard lines. However, in order not to get lost in too many technicalities, the reader is asked to make a few small "leaps of faith".

The nice fact about the representation in (3.33) is that $\widetilde{K}_n(\theta)$ can be expressed in terms of the *pair empirical measure* associated with $\{m(x)\}_{x \in \mathbb{N}_0}$. To that end, define

$$L_N^2 = \frac{1}{N}\sum_{x=0}^{N-1} \delta_{(m(x-1),m(x))}, \qquad N \in \mathbb{N}, \tag{3.36}$$

with periodic boundary conditions $(m(-1) = m(N))$, and let

$$F_\beta(\nu) = -\beta \sum_{i,j\in\mathbb{N}} (i+j-1)^2 \nu(i,j),$$

$$A_\theta = \Big\{\nu \in \widetilde{\mathcal{M}}_1(\mathbb{N}\times\mathbb{N}): \sum_{i,j\in\mathbb{N}} (i+j-1)\nu(i,j) = \frac{1}{\theta}\Big\},$$

(3.37)

where

$$\widetilde{\mathcal{M}}_1(\mathbb{N}\times\mathbb{N}) = \text{the set of probability measures}$$
$$\text{on } \mathbb{N}\times\mathbb{N} \text{ with identical marginals.}$$

(3.38)

Then (3.33) becomes

$$\widetilde{K}_n(\theta) = e^{o(n)} E_\theta\Big(e^{NF_\beta(L_N^2)} 1_{\{L_N^2 \in A_\theta\}}\Big) \text{ with } N = \lceil\theta n\rceil + 1. \qquad (3.39)$$

Indeed,

1. The exponential factor in (3.33) equals the one in (3.39), with a negligible error arising from forcing the periodic boundary condition in the definition of L_N^2.
2. The first constraint in (3.33) is asymptotically the same as the constraint in (3.39), because we replaced $(n+1)/(\lceil\theta n\rceil+1)$ by $1/\theta$, which will *a posteriori* be justified by the continuity of the function $\theta \mapsto \widetilde{J}_\beta(\theta)$ appearing in Lemma 3.8 below (see Sections 3.4.4–3.4.5).
3. The second constraint in (3.33) is negligible as $n \to \infty$.

The reason for introducing the representation in (3.39) is that it allows us to use the LDP for the empirical pair measure L_N^2, based on the Markov property established in Lemma 3.7.

Lemma 3.8. *For every* $\theta \in (0,1)$,

$$\lim_{n\to\infty} \frac{1}{n} \log \widetilde{K}_n(\theta) = -\widetilde{J}_\beta(\theta), \qquad (3.40)$$

with

$$\widetilde{J}_\beta(\theta) = \theta \inf_{\nu\in A_\theta} \big[-F_\beta(\nu) + I_{P_\theta}^2(\nu)\big], \qquad (3.41)$$

where

$$I_{P_\theta}^2(\nu) = \sum_{i,j\in\mathbb{N}} \nu(i,j) \log\Big(\frac{\nu(i,j)}{\bar\nu(i)P_\theta(i,j)}\Big). \qquad (3.42)$$

Proof. The formula in (3.42) is the *weak* rate function in the *weak* LDP (see den Hollander [168], Section III.6) for $(L_N^2)_{N\in\mathbb{N}}$ under the law of the Markov chain $\{m(x)\}_{x\in\mathbb{N}_0}$ with transition kernel P_θ. In order to apply Varadhan's Lemma (see den Hollander [168], Section III.3) we need the LDP, i.e., we need to overcome the technical difficulty that the state space $\mathbb{N}$ is infinite. This can be handled via a truncation argument because, as was observed at the end of Section 3.4.2, the Markov chain $\{m(x)\}_{x \in \mathbb{N}_0}$ has strong recurrence properties (see Greven and den Hollander [130] for more details).

Next we apply Varadhan's Lemma to (3.39), which is an exponential integral restricted to the set A_θ. Here another technical difficulty arises: the weak LDP "needs to be transferred from $\widetilde{\mathcal{M}}_1(\mathbb{N} \times \mathbb{N})$ to A_θ" (see Dembo and Zeitouni [82], Lemma 4.1.5). The result of the usual manipulations reads, somewhat informally,

$$
\begin{aligned}
\widetilde{K}_n(\theta) &= e^{o(n)} \, E_\theta \Big(e^{N F_\beta(L_N^2)} 1_{\{L_N^2 \in A_\theta\}} \Big) \\
&= e^{o(n)} \int_{A_\theta} e^{N F_\beta(L_N^2)} P_\theta(L_N^2 \in d\nu) \\
&= e^{o(n)} \, e^{N \sup_{\nu \in A_\theta} [F_\beta(\nu) - I_{P_\theta}^2(\nu)]}, \qquad N \to \infty,
\end{aligned}
\tag{3.43}
$$

which proves the claim because $N = \lceil \theta n \rceil + 1$. $\square$

At this point we recall (3.22) and (3.23), and rewrite Lemma 3.8 as follows:

Lemma 3.9. *For every $\theta \in (0,1)$,*

$$
\lim_{n \to \infty} \frac{1}{n} \log \widehat{K}_n(\theta) = -\widehat{J}_\beta(\theta),
\tag{3.44}
$$

with

$$
\widehat{J}_\beta(\theta) = \theta \inf_{\nu \in A_\theta} \big[- F_\beta(\nu) + I_{P_0}^2(\nu) \big],
\tag{3.45}
$$

i.e., the same variational formula as in Lemma 3.8 but with P_θ replaced by P_0, given by (recall Lemma 3.7)

$$
P_0(i,j) = \binom{i+j-2}{i-1} \Big(\frac{1}{2} \Big)^{i+j-1}, \qquad i,j \in \mathbb{N}.
\tag{3.46}
$$

Proof. Simply note that $\nu \in A_\theta$ implies $\sum_{i \in \mathbb{N}} i\bar{\nu}(i) = \frac{1+\theta}{2\theta}$, so that

$$
\begin{aligned}
I_{P_0}^2(\nu) - I_{P_\theta}^2(\nu) &= \sum_{i,j \in \mathbb{N}} \nu(i,j) \log\big[(1+\theta)^i (1-\theta)^{j-1} \big] \\
&= \frac{1+\theta}{2\theta} \log(1+\theta) + \frac{1-\theta}{2\theta} \log(1-\theta), \qquad \nu \in A_\theta,
\end{aligned}
\tag{3.47}
$$

which makes the prefactor in (3.22) cancel out. $\square$

Thus we have identified $\widehat{J}_\beta(\theta)$ for $\theta \in (0,1)$, which is the function we were after in Section 3.3. The same formulas as in (3.44–3.45) apply for the degenerate case $\theta = 1$, as is easily checked by direct computation. Finally, (3.21) gives us J_β, the rate function in the LDP for bridge soft polymers in Theorem 3.5.

3.4.4 Step 4: Solution of the Variational Problem in Terms of an Eigenvalue Problem

We next proceed to give the solution of the variational problem in (3.45), leading to the qualitative shape of the function $\theta \mapsto J_\beta(\theta)$ anticipated in Fig. 3.2. The variational problem requires us to minimize a non-linear functional under a linear constraint. It is possible to find the solution in terms of a certain eigenvalue problem that is well-behaved, and we will see that the outcome is relatively simple.

Fix $\beta \in (0, \infty)$ and $r \in \mathbb{R}$, and let $A_{r,\beta}$ be the $\mathbb{N} \times \mathbb{N}$ matrix with components

$$A_{r,\beta}(i,j) = e^{r(i+j-1)-\beta(i+j-1)^2} P_0(i,j), \qquad i,j \in \mathbb{N}. \tag{3.48}$$

The parameter r will be seen to play the role of a Lagrange multiplier needed to handle the constraint in (3.45).

Lemma 3.10. *Fix $\beta \in (0, \infty)$. For every $r \in \mathbb{R}$, $A_{r,\beta}$ is a self-adjoint operator on $l^2(\mathbb{N})$ having a unique largest eigenvalue $\lambda_{r,\beta}$ and corresponding eigenvector $\tau_{r,\beta}$ (normalized as $\|\tau_{r,\beta}\|_2 = 1$).*

Proof. Since $A_{r,\beta}$ is strictly positive and has rapidly decaying tails, the assertion follows from standard Perron-Frobenius theory. In fact, $A_{r,\beta}$ has the so-called Hilbert-Schmidt property $\sum_{i,j \in \mathbb{N}} A_{r,\beta}(i,j)^2 < \infty$ and, consequently, is a compact operator (see Dunford and Schwartz [94], Section XI.6). $\square$

The eigenvalue $\lambda_{r,\beta}$ has the following properties:

Lemma 3.11. (i) $(r, \beta) \mapsto \lambda_{r,\beta}$ *is analytic on* $\mathbb{R} \times (0, \infty)$.
(ii) $\lim_{r \to -\infty} \frac{\partial}{\partial r} \log \lambda_{r,\beta} = 1$ *and* $\lim_{r \to \infty} \frac{\partial}{\partial r} \log \lambda_{r,\beta} = \infty$ *for all* $\beta \in (0, \infty)$.
(iii) $r \mapsto \log \lambda_{r,\beta}$ *is strictly convex for all* $\beta \in (0, \infty)$.

Proof. Here is a quick sketch. For details we refer to Greven and den Hollander [130].
(i) Analyticity holds because $\lambda_{r,\beta}$ has multiplicity 1 and all elements of the matrix $A_{r,\beta}$ are analytic.
(ii) This follows from straightforward estimates on the eigenvector $\tau_{r,\beta}$ for $r \to -\infty$ and $r \to \infty$, respectively.
(iii) Convexity follows from the observations:

1. $\lambda_{r,\beta} = \sup_{x \in l^2(\mathbb{N}): \ x>0, \|x\|_2=1} \sum_{i,j \in \mathbb{N}} x(i) A_{r,\beta}(i,j) x(j)$;
2. $r \mapsto \log A_{r,\beta}(i,j)$ is linear for all $i, j \in \mathbb{N}$ and $\beta \in (0, \infty)$;
3. log-convexity is preserved under taking sums and suprema.

Strict convexity follows from convexity in combination with (i) and (ii). $\square$

With the help of Lemma 3.11 we can express $\widehat{J}_\beta(\theta)$ in terms of the eigenvalue $\lambda_{r,\beta}$ for some r depending on θ.

Lemma 3.12. *Fix $\beta \in (0, \infty)$. Then, for every $\theta \in (0,1)$,*

$$\widehat{J}_\beta(\theta) = r - \theta \log \lambda_{r,\beta}\Big|_{r=r_\beta(\theta)}, \tag{3.49}$$

where $r_\beta(\theta) \in \mathbb{R}$ is the unique solution of the equation

$$\frac{1}{\theta} = \frac{\partial}{\partial r} \log \lambda_{r,\beta}. \tag{3.50}$$

Proof. The fact that (3.50) has a solution for all $\theta \in (0,1)$ and that this solution is unique follows from Lemma 3.11 (see Fig. 3.3).

Consider the following family of pair probability measures:

$$\nu_{r,\beta}(i,j) = \frac{1}{\lambda_{r,\beta}} \tau_{r,\beta}(i) A_{r,\beta}(i,j) \tau_{r,\beta}(j), \qquad i,j \in \mathbb{N}. \tag{3.51}$$

One easily checks that $\nu_{r,\beta} \in \widetilde{\mathcal{M}}_1(\mathbb{N} \times \mathbb{N})$. Compute

$$
\begin{aligned}
I_{P_0}^2(\nu_{r,\beta}) &= \sum_{i,j} \nu_{r,\beta}(i,j) \log \left(\frac{\nu_{r,\beta}(i,j)}{\bar{\nu}_{r,\beta}(i) P_0(i,j)} \right) \\
&= \sum_{i,j} \nu_{r,\beta}(i,j) \left[r(i+j-1) - \beta(i+j-1)^2 - \log \lambda_{r,\beta} + \log \frac{\tau_{r,\beta}(j)}{\tau_{r,\beta}(i)} \right],
\end{aligned}
\tag{3.52}
$$

where we use that $\bar{\nu}_{r,\beta}(i) = \tau_{r,\beta}^2(i)$. Since $\nu_{r,\beta}$ has identical marginals, the last term vanishes and we end up with the simple expression

$$I_{P_0}^2(\nu_{r,\beta}) = \frac{r}{\theta} + F_\beta(\nu_{r,\beta}) - \log \lambda_{r,\beta}, \tag{3.53}$$

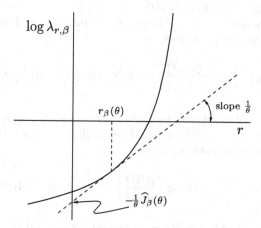

Fig. 3.3. Identification of $\widehat{J}_\beta(\theta)$.

provided (!) $\nu_{r,\beta} \in A_\theta$. This expression says that

$$-\theta\big[F_\beta(\nu_{r,\beta}) - I_{P_0}^2(\nu_{r,\beta})\big] = r - \theta \log \lambda_{r,\beta}. \tag{3.54}$$

We thus see from Lemma 3.9 that the claim in Lemma 3.12 is correct provided (!) we can prove the following two properties:

(i) $r = r_\beta(\theta)$ implies $\nu_{r,\beta} \in A_\theta$;
(ii) $\nu_{r_\beta(\theta),\beta}$ is a minimizer of the variational problem in (3.45).

Property (i): Compute

$$\sum_{i,j}(i+j-1)\nu_{r,\beta}(i,j) = \frac{1}{\lambda_{r,\beta}}\sum_{i,j}\tau_{r,\beta}(i)\Big[\frac{\partial}{\partial r}A_{r,\beta}(i,j)\Big]\tau_{r,\beta}(j)$$

$$= \frac{1}{\lambda_{r,\beta}}\frac{\partial}{\partial r}\Big[\sum_{i,j}\tau_{r,\beta}(i)A_{r,\beta}(i,j)\tau_{r,\beta}(j)\Big] = \frac{1}{\lambda_{r,\beta}}\frac{\partial}{\partial r}\lambda_{r,\beta} = \frac{\partial}{\partial r}\log\lambda_{r,\beta}. \tag{3.55}$$

The second equality uses that $A_{r,\beta}\tau_{r,\beta} = \lambda_{r,\beta}\tau_{r,\beta}$ and $\|\tau_{r,\beta}\|_2 = 1$ for all r,β. Hence (3.50) indeed guarantees (i).

Property (ii): If $\nu \in A_\theta$, then we can write

$$-\theta\big[F_\beta(\nu) - I_{P_0}^2(\nu)\big]$$

$$= [r - \theta\log\lambda_{r,\beta}] + \theta\sum_{i,j}\nu(i,j)\log\left(\frac{\nu(i,j)}{\bar{\nu}(i)\frac{1}{\lambda_{r,\beta}}A_{r,\beta}(i,j)}\right) \tag{3.56}$$

$$= [r - \theta\log\lambda_{r,\beta}] + \theta\sum_{i,j}\nu(i,j)\log\left(\frac{\nu(i,j)}{\bar{\nu}(i)}\frac{\sqrt{\bar{\nu}_{r,\beta}(i)\bar{\nu}_{r,\beta}(j)}}{\nu_{r,\beta}(i,j)}\right),$$

where we once again use that $\bar{\nu}_{r,\beta}(i) = \tau_{r,\beta}^2(i)$. The first term is precisely the value we found when $\nu = \nu_{r,\beta}$. Moreover, because ν has identical marginals the second term simplifies further to

$$\theta\sum_{i,j}\nu(i,j)\log\left(\frac{\nu(i,j)}{\bar{\nu}(i)}\frac{\bar{\nu}_{r,\beta}(i)}{\nu_{r,\beta}(i,j)}\right) = \theta\sum_i\bar{\nu}(i)H\Big([\nu(i)]\,\big|\,[\nu_{r,\beta}(i)]\Big), \tag{3.57}$$

where $[\nu(i)], [\nu_{r,\beta}(i)] \in \mathcal{M}_1(\mathbb{N})$ (= the set of probability measures of $\mathbb{N}$) are defined by $[\nu(i)](j) = \nu(i,j)/\nu(i)$ and $[\nu_{r,\beta}(i)](j) = \nu_{r,\beta}(i,j)/\nu_{r,\beta}(i)$, while $H(\cdot\,|\,\cdot)$ denotes relative entropy, defined as

$$H(\mu_1\,|\,\mu_2) = \sum_i\mu_1(i)\log\left(\frac{\mu_1(i)}{\mu_2(i)}\right), \qquad \mu_1,\mu_2 \in \mathcal{M}_1(\mathbb{N}). \tag{3.58}$$

Clearly, the r.h.s. of (3.57) is ≥ 0 with equality if and only if $\nu = \nu_{r,\beta}$. $\square$

3.4.5 Step 5: Identification of the Speed

Lemma 3.12 gives us a nice representation of $\widehat{J}_\beta(\theta)$, $\theta \in (0,1)$, in terms of the family of eigenvalues $\lambda_{r,\beta}$, $r \in \mathbb{R}$. As we saw in Fig. 3.2, $\theta^*(\beta)$ is to be identified as the unique minimum of $\theta \mapsto \widehat{J}_\beta(\theta)$.

Lemma 3.13. *Fix $\beta \in (0,\infty)$. Then*

$$\frac{1}{\theta^*(\beta)} = \frac{\partial}{\partial r} \log \lambda_{r,\beta}\Big|_{r=r^*(\beta)}, \tag{3.59}$$

with $r^*(\beta) \in (0,\infty)$ the unique solution of the equation

$$\lambda_{r,\beta} = 1. \tag{3.60}$$

Proof. The fact that (3.60) has a solution for all $\beta \in (0,\infty)$ and that this solution is unique follows from Lemma 3.11 (see Fig. 3.3).

Differentiate $\widehat{J}_\beta(\theta)$ with respect to θ to obtain

$$\frac{\partial}{\partial \theta} \widehat{J}_\beta(\theta) = \frac{\partial}{\partial \theta} r_\beta(\theta) - \log \lambda_{r_\beta(\theta),\beta} - \theta \Big[\frac{\partial}{\partial \theta} r_\beta(\theta)\Big]\Big[\frac{\partial}{\partial r}\log \lambda_{r,\beta}\Big]_{r=r_\beta(\theta)}. \tag{3.61}$$

However, the first and the third term cancel out because of (3.50), so we get

$$\frac{\partial}{\partial \theta} \widehat{J}_\beta(\theta) = -\log \lambda_{r_\beta(\theta),\beta}. \tag{3.62}$$

This is zero if and only if θ is such that $\lambda_{r_\beta(\theta),\beta} = 1$, i.e., the minimum $\theta^*(\beta)$ of $\theta \mapsto \widehat{J}_\beta(\theta)$ is found by solving (3.60). After that we put

$$r_\beta(\theta^*(\beta)) = r^*(\beta) \tag{3.63}$$

and use (3.50). Note that, by Lemma 3.12,

$$\frac{\partial^2}{\partial \theta^2} \widehat{J}_\beta(\theta) = -\frac{1}{\theta} \frac{\partial}{\partial \theta} r_\beta(\theta) > 0 \tag{3.64}$$

(see Fig. 3.3 and note that $\theta \mapsto r_\beta(\theta)$ has a negative slope), so that $r_\beta(\theta^*(\beta))$ is indeed the unique minimizer of $\theta \mapsto \widehat{J}_\beta(\theta)$. $\square$

Lemmas 3.11–3.13 yield Fig. 3.3. This finishes our analysis of the rate function J_β for bridge soft polymers, and the proof of Theorem 3.5 is now complete.

Note that $\theta^*(\beta)$ is the unique minimum of $\widehat{J}_\beta$ and the unique minimum and zero of J_β (recall (3.21) and Fig. 3.2).

3.5 The Large Deviation Principle without the Bridge Condition

In Sections 3.4.1–3.4.5 we have proved Theorem 3.5, the LDP for bridge soft polymers. In this section we give a quick sketch of how to obtain the LDP without the bridge condition, i.e., Theorem 3.3, which implies Theorems 3.1 and 3.2. Remarkably, it turns out that the rate function in this LDP has a linear piece between 0 and a *critical speed* $\theta^{**}(\beta)$ that is strictly smaller than $\theta^*(\beta)$ (see Fig. 3.4). What is written below developed out of discussions with W. König.

Theorem 3.14. *For every $\beta \in (0, \infty)$ the family $(P_n^{+,\beta})_{n\in\mathbb{N}}$ defined by*

$$P_n^{+,\beta}(\cdot) = P_n^\beta \left(\frac{1}{n} S_n \in \cdot \ \Big| \ S_n \geq 0 \right) \tag{3.65}$$

satisfies the LDP on $[0,1]$ with rate n and with rate function I_β given by

$$I_\beta(\theta) = \begin{cases} J_\beta(\theta), & \text{if } \theta \geq \theta^{**}(\beta), \\ I_\beta(0) + \frac{\theta}{\theta^{**}(\beta)}[J_\beta(\theta^{**}(\beta)) - I_\beta(0)], & \text{if } \theta \leq \theta^{**}(\beta), \end{cases} \tag{3.66}$$

where $\theta^{**}(\beta)$ is the unique solution of the equation

$$J_\beta(\theta) - I_\beta(0) = \theta \frac{\partial}{\partial\theta} J_\beta(\theta), \tag{3.67}$$

and $I_\beta(0)$ is identified in Lemma 3.15 below. Moreover, $\theta^{**}(\beta) \in (0, \theta^*(\beta))$.

The linear piece in (3.66) can be understood as follows. If the soft polymer is required to move at a speed $\theta < \theta^{**}(\beta)$, then it prefers to violate the bridge condition by moving at speed $\theta^{**}(\beta)$ between 0 and $\lceil \theta n \rceil$, and making two loops, one below 0 and one above $\lceil \theta n \rceil$. The penalty for making these loops

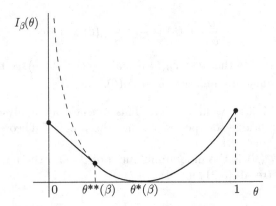

Fig. 3.4. The rate function I_β (compare with Fig. 3.2).

is less than the penalty for staying locked up like a bridge. The total length of these loops is proportional to $\theta^{**}(\beta) - \theta$, i.e., the penalty for not behaving like a bridge grows linearly with $\theta^{**}(\beta) - \theta$.

The analogue of Lemmas 3.12 and 3.13 reads as follows (compare (3.46) and (3.48) with (3.69) and (3.68)):

Lemma 3.15. *Fix* $\beta \in (0, \infty)$. *For* $r \in \mathbb{R}$, *let* $A_{r,\beta}^{\circlearrowleft}$ *be the* $\mathbb{N} \times \mathbb{N}$ *–matrix with components*

$$A_{r,\beta}^{\circlearrowleft}(i, j) = e^{r(i+j) - \beta(i+j)^2} P_0^{\circlearrowleft}(i, j), \qquad i, j \in \mathbb{N}, \tag{3.68}$$

with

$$P_0^{\circlearrowleft}(i, j) = 1_{\{i \neq 0\}} P_0(i, j + 1) + 1_{\{i=j=0\}}. \tag{3.69}$$

Let $\lambda_{r,\beta}^{\circlearrowleft}$ *be the unique largest eigenvalue of* $A_{r,\beta}^{\circlearrowleft}$ *acting as an operator on* $l^2(\mathbb{N})$. *Then*

$$I_\beta(0) = r^{**}(\beta) - r^*(\beta), \tag{3.70}$$

with $r^{**}(\beta) \in (0, \infty)$ *the unique solution of the equation*

$$\lambda_{r,\beta}^{\circlearrowleft} = 1. \tag{3.71}$$

Proof. See den Hollander [168], Chapter IX. $\square$

It is easy to show that $r^{**}(\beta) > r^*(\beta)$ for all $\beta \in (0, \infty)$. Since (3.67) says that $r^{**}(\beta) = r_\beta(\theta^{**}(\beta))$, with $\theta \mapsto r_\beta(\theta)$ defined in Lemma 3.12, it follows that $\theta^{**}(\beta) < \theta^*(\beta)$ for all $\beta \in (0, \infty)$.

In conclusion, we have proved Theorem 3.3 and identified the rate function I_β in terms of the families of principal eigenvalues $(r, \beta) \mapsto \lambda_{r,\beta}$ and $(r, \beta) \mapsto \lambda_{r,\beta}^{\circlearrowleft}$ of the operators $A_{r,\beta}$ and $A_{r,\beta}^{\circlearrowleft}$ defined in (3.48) and (3.68). These families are analytically well-behaved and can also be easily computed numerically (see Greven and den Hollander [130]).

3.6 Extensions

(1) Theorem 3.1 has been extended to a central limit theorem by König [218]. The standard deviation, denoted by $\sigma^*(\beta)$, turns out to be given by the formula

$$\frac{1}{\sigma^{*2}(\beta)} = \frac{\partial^2}{\partial \theta^2} J_\beta(\theta) \Big|_{\theta = \theta^*(\beta)} = \frac{\partial^2}{\partial \theta^2} I_\beta(\theta) \Big|_{\theta = \theta^*(\beta)}. \tag{3.72}$$

Numerical computation gives the picture in Fig. 3.5.

(2) Van der Hofstad and den Hollander [157] prove that

$$\lim_{\beta \downarrow 0} \beta^{-\frac{1}{3}} \theta^*(\beta) = C \text{ for some } C > 0 \tag{3.73}$$

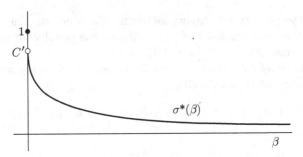

Fig. 3.5. The spread of the soft polymer.

(recall Fig. 3.1), while van der Hofstad, den Hollander and König [158] prove that

$$\lim_{\beta \downarrow 0} \sigma^*(\beta) = C' \text{ for some } C' \neq 1. \tag{3.74}$$

These asymptotic formulas show that, even in the limit of weak self-repellence, the behavior of the soft polymer cannot be understood via a perturbation argument around the non-repellent SRW. Van der Hofstad [153] derives rigorous bounds on C and C'. Numerically, $C \approx 1.1$ and $C' \approx 0.63$. These constants are related to a Brownian version of the polymer model – called the Edwards model – defined in (3.78) below. We refer to van der Hofstad, den Hollander and König [159] for the analogous law of large numbers (first proved by Westwater [312]) and central limit theorem.

The heuristics behind the scaling in (3.73) is as follows. Suppose that $S_n \approx \theta n$ and that $0 \leq S_i \leq S_n$ for all $0 \leq i \leq n$. The Hamiltonian $H_n = \beta \sum_{x \in \mathbb{Z}} \ell_n(x)^2$ is minimal when the local times are constant, i.e., when $\ell_n(x) \approx n/S_n \approx 1/\theta$ for $0 \leq x \leq S_n$, in which case $H_n \approx \beta(\theta n)(1/\theta)^2 = (\beta/\theta)n$. The probability under P, the law of SRW, that $S_n \approx \theta n$ is roughly $\exp[-(\theta n)^2/2n]$. Consequently, the contribution to the partition sum coming from paths with $S_n \approx \theta n$ is roughly $\exp[-\{(\beta/\theta) + \frac{1}{2}\theta^2\}n]$. The term between braces in the exponent is minimal when $\theta = \beta^{\frac{1}{3}}$.

(3) Van der Hofstad, den Hollander and König [158] prove that if β is replaced by β_n satisfying

$$\lim_{n \to \infty} \beta_n = 0 \quad \text{and} \quad \lim_{n \to \infty} n^{\frac{3}{2}} \beta_n = \infty, \tag{3.75}$$

then the law of large numbers and central limit theorem apply with $\theta^*(\beta)$ and $\sigma^*(\beta)$ replaced by

$$\theta^*(\beta_n) \sim C(\beta_n)^{\frac{1}{3}} \quad \text{and} \quad \sigma^*(\beta_n) \sim C'. \tag{3.76}$$

In comparison with (3.73–3.74), this shows that the weak interaction limit has a degree of *universality*. Van der Hofstad, den Hollander and König [161] offer a coarse-graining argument showing that, in one dimension, self-repellent random walks scale to self-repellent Brownian motions. The proof is based on cutting the path into pieces, controlling the interaction between the different pieces, and applying the invariance principle to the single pieces. The scaling properties are shown to be stable against adding self-attraction, provided the self-repellence remains dominant. We will return to polymers with self-repellence and self-attraction in Chapter 6.

(4) König [216], [217] (extending earlier work by Alm and Janson [9]) considers the case where the random walk is "spread out", e.g., it draws its step uniformly from the set $\{-L, \ldots, L\}$ for some $L \in \mathbb{N}$. For this case, the SAW problem is interesting. It is shown that for every $\beta \in (0, \infty]$ and $L \in \mathbb{N}$ the self-avoiding polymer has a speed $\theta^*(\beta, L) \in (0, L)$. Aldous [3] – assuming that the speed existed – had earlier conjectured that

$$\lim_{L \to \infty} L^{-\frac{2}{3}} \theta^*(\infty, L) = C'' \text{ for some } C'' \in (0, \infty). \qquad (3.77)$$

This conjecture was based on a scaling result for the self-intersection local time of the spread-out random walk in the limit as $L \to \infty$. The conjecture in (3.77) was subsequently proved in van der Hofstad, den Hollander and König [161], where it was shown that $C'' = C3^{-\frac{1}{3}}$.

(5) A continuum version of the Domb-Joyce model – called the Edwards model – is analyzed in Kusuoka [224] and van der Hofstad, den Hollander and König [159]. The Hamiltonian for this model is

$$H\big((B_t)_{t \in [0,T]}\big)$$
$$= -\beta \int_0^T \mathrm{d}s \int_0^T \mathrm{d}t \, \delta(B_s - B_t) = -\beta \int_{\mathbb{R}} L(T, x)^2 \, \mathrm{d}x, \qquad T \geq 0,$$
$$(3.78)$$

where $\delta(\cdot)$ is the Dirac delta-function, $(B_t)_{t \geq 0}$ is standard Brownian motion and $L(T, x)$ is its local time at position x up to time T. The behavior is ballistic, and both a law of large numbers and a central limit theorem apply, with speed $C\beta^{\frac{1}{3}}$ and spread C' (which provide the link with the weak interaction limit of the Domb-Joyce model). The corresponding LDP is proved in van der Hofstad, den Hollander and König [160].

3.7 Challenges

(1) Prove that $\beta \mapsto \theta^*(\beta)$ is non-decreasing (see Fig. 3.1). Even though this property seems intuitively plausible, it is actually deep (see Greven and den Hollander [130]) and remains open. Similarly, prove that $\beta \mapsto \sigma^*(\beta)$ is non-increasing (see Fig. 3.5). It is not hard to compute $\lambda_{r,\beta}$ numerically and get support for the monotonicity of both quantities. Coupling arguments do not work: P_n^β's for different values of β are hard to compare, because the normalizing partition sum depends on β (recall (3.3)).

(2) Prove the functional central limit theorem, i.e., show that under the law $P_n^{+,\beta}$ we have

$$\left(\frac{1}{\sigma^*(\beta)\sqrt{n}}\left[S_{\lfloor tn\rfloor} - \theta^*(\beta)\,\lfloor tn\rfloor\right]\right)_{0\leq t\leq 1} \Longrightarrow (B_t)_{0\leq t\leq 1} \quad \text{as } n \to \infty, \quad (3.79)$$

with standard Brownian motion as limit. The proof of (3.79) should not be hard: the method of local times employed in Sections 3.4.1–3.4.3 is very powerful and should allow us to get the multivariate analogues of the law of large numbers and the central limit theorem, together with the appropriate form of tightness.

(3) Derive a functional LDP extending Theorem 3.5.

(4) Van der Hofstad, den Hollander and König [161] have extended the LDP for the speed of the endpoint to the weak interaction limit in (3.75), but only for speeds that are not too small. Extend the proof to all speeds.

(5) Try to put some rigor into the following heuristic observation (put forward in van der Hofstad, den Hollander and König [161]), which argues in favor of the critical exponent $\nu = \frac{3}{4}$ for the two-dimensional soft polymer (recall (2.26)) based on the result in Section 3.6, Extension (3). Consider simple random walk on the slit $\{-L,\ldots,L\} \times \mathbb{Z}$ with periodic boundary conditions (see Fig. 3.6). Write

$$S = (S_i)_{i=0}^n = (S^{(1)}, S^{(2)}) = (S_i^{(1)}, S_i^{(2)})_{i=0}^n \qquad (3.80)$$

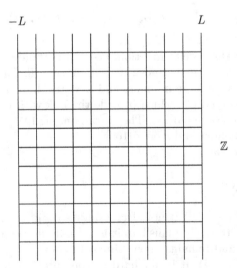

Fig. 3.6. The slit $\{-L,\ldots,L\} \times \mathbb{Z}$ with periodic boundary conditions.

for its two components, and note that S makes a self-intersection if and only if both $S^{(1)}$ and $S^{(2)}$ make a self-intersection. Under the soft polymer measure P_n^β, we have

$$|S_n^{(1)}| \asymp L \quad \text{and} \quad |S_n^{(2)}| \asymp L^{-\frac{1}{3}}n \quad \text{as } n \to \infty. \tag{3.81}$$

The first claim is trivial. The second claim comes from the fact that $S^{(1)}$ self-intersects one out of $(2L+1)$ times. Hence, the motion of $S^{(2)}$ is comparable to that of the one-dimensional soft polymer with self-repellence parameter $\beta_n = \beta/(2L+1)$. Therefore, according to (3.76), $S^{(2)}$ moves a distance of order

$$(\beta_n)^{\frac{1}{3}}n \asymp L^{-\frac{1}{3}}n \tag{3.82}$$

in time n. Now, the two scales in (3.81) coincide when $L = n^{\frac{3}{4}}$. Then, the two components run on the same scale, and consequently the slit is wide enough for the motion of S to be comparable to that of the two-dimensional soft polymer. For $L = n^{\frac{3}{4}}$, both $S_n^{(1)}$ and $S_n^{(2)}$ run on scale $n^{\frac{3}{4}}$, and hence so does S_n. (Note that $\beta_n \asymp n^{-\frac{3}{4}}$ when $L = n^{\frac{3}{4}}$, which indeed satisfies (3.75).)

4

Soft Polymers in High Dimension

In this chapter we consider the same model as in Chapter 3, but for $d \geq 5$ instead of $d = 1$. Our goal is to prove *diffusive behavior*. The tool to achieve this is the so-called *lace expansion*, a combinatorial technique that hinges on the idea that in high dimensions the soft polymer can be viewed as a "perturbation" of SRW. For the exposition below, we borrow from van der Hofstad [156], Section 2, and Slade [280], Chapter 3. Figs. 4.2–4.4 are borrowed from Gordon Slade, Fig. 4.5 from Bill Casselman and Gordon Slade.

4.1 Weakly Self-avoiding Walk in Dimension Five or Higher

The lace expansion was introduced by Brydges and Spencer [44] to prove diffusive behavior of weakly self-avoiding walk in $d \geq 5$. To obtain convergence of the lace expansion, a "small parameter" is needed. For that purpose, in [44] the parameter β was taken to be small. It is more interesting, however, to replace SRW by a random walk that is "spread out", i.e., to allow the walk to choose its steps randomly from the set

$$\Omega = \{x \in \mathbb{Z}^d \colon 0 < \|x\|_\infty \leq L\} \tag{4.1}$$

with $\| \cdot \|_\infty$ the supremum norm and $L \in \mathbb{N}$ (see Fig. 4.1). If L is large, then self-intersections are so few that the spread-out soft polymer can be viewed as a "perturbation" of the spread-out random walk with no interaction. The small parameter is $1/|\Omega|$.

Instead of (3.1–3.2), we pick as our set of n-step paths

$$\mathcal{W}_n = \left\{ w = (w_i)_{i=0}^n \in (\mathbb{Z}^d)^{n+1} \colon w_0 = 0, \ w_{i+1} - w_i \in \Omega \ \forall 0 \leq i < n \right\} \tag{4.2}$$

and we keep as our Hamiltonian

$$H_n(w) = \beta I_n(w). \tag{4.3}$$

F. den Hollander, *Random Polymers*,
Lecture Notes in Mathematics 1974, DOI: 10.1007/978-3-642-00333-2_4,
© Springer-Verlag Berlin Heidelberg 2009, Reprint by Springer-Verlag Berlin Heidelberg 2012

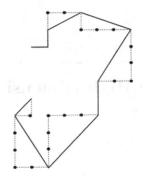

Fig. 4.1. A path drawing steps from Ω with $L = 3$.

Our soft polymer measure is

$$P_n^\beta(w) = \frac{1}{Z_n^\beta} e^{-\beta I_n(w)}, \qquad w \in \mathcal{W}_n, \tag{4.4}$$

which differs from (3.3) in that we drop the reference measure P_n. This is harmless because P_n is the uniform distribution. We do this because most of the present chapter is about combinatorics rather than about probability.

For SRW (corresponding to $L = 1$) we may take $1/2d$ as the small parameter for large d. However, choosing L large has the advantage that we will be able to prove diffusive behavior for *all $d \geq 5$* and *all $\beta \in (0, \infty)$*. The threshold value for L will depend on d and β, but will be finite.

The goal of the present chapter is to derive Theorem 4.1 below, which is the analogue of (2.20–2.21), (2.25–2.26) and (2.29) for SAW in $d \geq 5$ stated in Section 2.2. To formulate this theorem, we introduce some notation. Let $\mathcal{W}_n(x) = \{w \in \mathcal{W}_n \colon w_n = x\}$, and

$$Z_n(x) = \sum_{w \in \mathcal{W}_n(x)} e^{-\beta I_n(w)}, \ x \in \mathbb{Z}^d,$$

$$Z_n = \sum_{x \in \mathbb{Z}^d} Z_n(x), \tag{4.5}$$

$$\widehat{Z}_n(k) = \sum_{x \in \mathbb{Z}^d} e^{i(k \cdot x)} Z_n(x), \ k \in [-\pi, \pi)^d.$$

where we suppress the dependence on β. In Section 2.2, where we were dealing with $\beta = \infty$ and $L = 1$, these objects were denoted by $c_n(x)$, c_n and $\widehat{c}_n(k)$, respectively. Note that

$$Z_n^{-1} Z_n(x) = P_n^\beta(S_n = x),$$

$$-Z_n^{-1} \nabla^2 \widehat{Z}_n(0) = E_n^\beta(\|S_n\|^2), \tag{4.6}$$

$$Z_n^{-1} \widehat{Z}_n(k) = E_n^\beta\left(e^{i(k \cdot S_n)}\right),$$

with $\nabla^2 \widehat{Z}_n = \nabla \cdot (\nabla \widehat{Z}_n)$. Let

$$D(x) = |\Omega|^{-1} \, 1\{x \in \Omega\} \tag{4.7}$$

denote the uniform distribution on Ω, and $\sigma^2 = \sum_{x \in \mathbb{Z}^d} \|x\|^2 D(x)$.

Theorem 4.1. *Fix $d \geq 5$ and $\beta \in (0, \infty)$. Then there exists an L_0 (depending on d, β) such that for all $L \geq L_0$ and as $n \to \infty$,*

$$\begin{aligned}
Z_n &= A\mu^n \, [1 + o(1)], \\
-Z_n^{-1} \nabla^2 \widehat{Z}_n(0) &= \sigma^2 v n \, [1 + o(1)], \\
Z_n^{-1} \widehat{Z}_n \left(k / \sqrt{\sigma^2 v n} \right) &= e^{-\frac{1}{2d} \|k\|^2} [1 + o(1)], \qquad k \in \mathbb{R}^d,
\end{aligned} \tag{4.8}$$

where $\mu, A, v > 0$ are constants (depending on d, L, β) and the error term in the last line is uniform in k provided $\|k\|^2 / \log n$ is sufficiently small.

In view of (4.6), the first line of (4.8) gives the scaling of the partition sum, with μ the analogue of the connective constant for SAW, the second line gives the scaling of the mean-square displacement, with $\sigma^2 v$ playing the role of the "renormalized" diffusion constant, while the third line is the central limit theorem in Fourier language.

Note that the self-repellence "renormalizes" the diffusion constant, which equals $\sigma^2 v$ instead of σ^2. We will see in Section 4.6 that $v > 1$, with v tending to 1 as $L \to \infty$.

The proof of Theorem 4.1 is given in Sections 4.2–4.6. In Sections 4.2–4.3 we derive the *lace expansion* for $Z_n(\cdot)$, which leads to a *recursion relation* for $Z_n(\cdot)$ in n. In Section 4.4 we describe *diagrammatic estimates* that bound the coefficients in this recursion relation. In Section 4.5 we state *induction hypotheses* that exploit these bounds to control the asymptotics of $Z_n(\cdot)$ as $n \to \infty$. Finally, in Section 4.6 the results are collected to complete the proof.

What follows is an *elegant but difficult computation*. The core idea is due to Brydges and Spencer [44]. Over the years, this idea has been developed and refined into a powerful tool capable of describing critical behavior in a variety of probabilistic models. For an overview we refer the reader to the Saint-Flour lectures by Slade [280].

4.2 Expansion

4.2.1 Graphs and Connected Graphs

For $w \in \mathcal{W}_n$ and $s, t \in \mathbb{N}_0$ with $s < t$, define

$$U_{st}(w) = \begin{cases} -\lambda(\beta), & \text{if } w_s = w_t, \\ 0, & \text{if } w_s \neq w_t, \end{cases} \tag{4.9}$$

with

$$\lambda(\beta) = 1 - e^{-\beta} \in (0,1). \tag{4.10}$$

Then the normalizing partition sum in (1.2) equals

$$Z_n = \sum_{x \in \mathbb{Z}^d} Z_n(x) \quad \text{with} \quad Z_n(x) = \sum_{w \in \mathcal{W}_n(x)} \prod_{0 \le s < t \le n} (1 + U_{st}(w)). \tag{4.11}$$

We take a closer look at the product.

For $w \in \mathcal{W}_n$ and $a, b \in \mathbb{N}_0$ with $a < b$, define

$$K[a,a](w) = 1 \quad \text{and} \quad K[a,b](w) = \prod_{a \le s < t \le b} (1 + U_{st}(w)), \tag{4.12}$$

and abbreviate $(a,b) \subset \mathbb{N}_0$ to denote the interval of integers strictly between a and b, and $[a,b] \subset \mathbb{N}_0$ to denote the interval including a and b.

Definition 4.2. (i) *Given an interval I, a pair $\{s,t\}$ of elements of I with $s < t$ is called an edge; st is short-hand notation for $\{s,t\}$. A set of edges is called a graph. The set of all graphs on $[a,b]$ is denoted by $\mathcal{G}[a,b]$.*
(ii) *A graph $\Gamma \in \mathcal{G}[a,b]$ is said to be connected if both a and b are endpoints of edges in Γ and, in addition, for any $c \in (a,b)$ there are $s, t \in [a,b]$ such that $s < c < t$ and $st \in \Gamma$, i.e., $\cup_{st \in \Gamma}(s,t) = (a,b)$ as intervals. The set of all connected graphs on $[a,b]$ is denoted by $\mathcal{C}[a,b]$.*

In words, a graph Γ on $[a,b]$ is connected if all the elements of (a,b) lie strictly under the "arc" of some edge in Γ (see Fig. 4.2). The edges symbolize the *self-intersections* of the path. Indeed, with the help of Definition 4.2(i) we may write

$$K[a,b](w) = \sum_{\Gamma \in \mathcal{G}[a,b]} \prod_{st \in \Gamma} U_{st}(w). \tag{4.13}$$

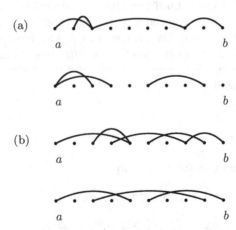

Fig. 4.2. Graphs in which an edge st is represented by an arc joining s and t. The graphs in (a) are not connected, whereas the graphs in (b) are connected.

Next, with the help of Definition 4.2(ii), define

$$J[a, a](w) = 1 \quad \text{or} \quad J[a, b](w) = \sum_{\Gamma \in \mathcal{C}[a,b]} \prod_{st \in \Gamma} U_{st}(w). \tag{4.14}$$

The following lemma, which is a kind of renewal equation, links $K[a, b](w)$ to $J[a, b](w)$.

Lemma 4.3. *For all $a, b \in \mathbb{N}_0$ with $a < b$,*

$$K[a, b](w) = K[a + 1, b](w) + \sum_{m=a+1}^{b} J[a, m](w)K[m, b](w). \tag{4.15}$$

Proof. Note that $K[a + 1, b](w)$ is the contribution to the sum in the right-hand side of (4.13) coming from all graphs Γ for which a is not in an edge of Γ. To resum the contribution due to the remaining graphs, we argue as follows. For Γ containing an edge that contains a, let $m = m(\Gamma)$ be the largest value of m such that the set of edges in Γ with both ends in the interval $[a, m]$ forms a connected graph on $[a, m]$. Then the sum over Γ factorizes into sums over connected graphs on $[a, m]$ and arbitrary graphs on $[m, b]$. Hence, we have

$$K[a, b](w) = K[a + 1, b](w) + \sum_{m=a+1}^{b} \left(\sum_{\Gamma \in \mathcal{C}[a,m]} \prod_{st \in \Gamma} U_{st}(w) \right) K[m, b](w), \tag{4.16}$$

which proves the claim. □

4.2.2 Recursion Relation

The importance of Lemma 4.3 lies in the following. For $m \in \mathbb{N}_0$ and $x \in \mathbb{Z}^d$, define

$$\pi_m(x) = \sum_{w \in \mathcal{W}_n(x)} J[0, m](w), \tag{4.17}$$

and note that $\pi_1 \equiv 0$ because $0 \notin \Omega$ defined in (4.1).

Lemma 4.4. *For $n \in \mathbb{N}$ and $x \in \mathbb{Z}^d$,*

$$Z_n(x) = (|\Omega|D * Z_{n-1})(x) + \sum_{m=2}^{n} (\pi_m * Z_{n-m})(x), \tag{4.18}$$

with D defined in (4.7), $*$ denoting convolution in space, and $Z_0 \equiv 1$.

Proof. Recall from (4.11–4.12) that $Z_n(x) = \sum_{w \in \mathcal{W}_n(x)} K[0, n](w)$. Therefore, when we sum (4.15) with $a = 0$ and $b = n$ over $w \in \mathcal{W}_n$, note that the sums over disjoint intervals factorize, and use (4.17), we get the claim. □

As we will see, (4.17) will be the central object of our analysis. The key feature of (4.18) is that it is a *recursion relation*: $Z_n(\cdot)$ is expressed in terms of $Z_m(\cdot)$, $0 \leq m \leq n - 1$, and certain *unknown* $\pi_m(\cdot)$, $1 \leq m \leq n$. The idea is that the latter are small when L is large and fall off rapidly with m. Hence we may view (4.18) as a *perturbation* of the trivial recursion where the second term is absent, whose solution is

$$Z_n(x) = |\Omega|^n \, D^{*n}(x), \qquad n \in \mathbb{N}_0. \tag{4.19}$$

The second factor in the right-hand side of (4.19) is the distribution at time n of the random walk whose steps are drawn from D (the uniform distribution on Ω).

In order to make the perturbation argument work, we need to find a way to *bound* $\pi_m(\cdot)$, $1 \leq m \leq n$, in terms of $Z_m(\cdot)$, $0 \leq m \leq n - 1$. Indeed, this will open up the possibility of *induction* on n. Deriving this bound requires finding a more tractable representation of $\pi_m(\cdot)$, obtained by performing a resummation of (4.17). It is here that the notion of *lace* enters the stage.

4.3 Laces

4.3.1 Laces and Compatible Edges

Definition 4.5. *A lace is a minimally connected graph, i.e., a connected graph for which the removal of any edge results in a disconnected graph. The set of laces on $[a, b]$ is denoted by $\mathcal{L}[a, b]$. The set of laces on $[a, b]$ consisting of exactly N edges is denoted by $\mathcal{L}^{(N)}[a, b]$.*

We write $L \in \mathcal{L}^{(N)}[a, b]$ as $L = \{s_1 t_1, \ldots, s_N t_N\}$, with $s_l < t_l$ for each $l = 1, \ldots, N$. The fact that L is a lace is equivalent to a certain ordering of the s_l and t_l. For $N = 1$ we simply have $a = s_1 < t_1 = b$. For $N \geq 2$, we have

$$L \in \mathcal{L}^{(N)}[a, b] \Longleftrightarrow$$
$$a = s_1 < s_2, \ s_{l+1} < t_l \leq s_{l+2} \ (l = 1, \ldots, N - 2), \ s_N < t_{N-1} < t_N = b. \tag{4.20}$$

Thus, L divides $[a, b]$ into $2N - 1$ subintervals

$$[s_1, s_2], [s_2, t_1], [t_1, s_3], [s_3, t_2], \ldots, [s_N, t_{N-1}], [t_{N-1}, t_N]. \tag{4.21}$$

The intervals numbered $3, 5, \ldots, 2N - 3$ may have zero length when $N \geq 3$, while all other intervals have length at least 1 (see Fig. 4.3).

Given a connected graph $\Gamma \in \mathcal{C}[a, b]$, the following prescription *associates* to Γ a *unique* lace $L_\Gamma \subset \Gamma$: The lace L_Γ consists of edges $s_1 t_1, s_2 t_2, \ldots$ with $t_1, s_1, t_2, s_2, \ldots$ determined (in this order) by

$$t_1 = \max\{t \colon at \in \Gamma\},$$
$$s_1 = a, \tag{4.22}$$

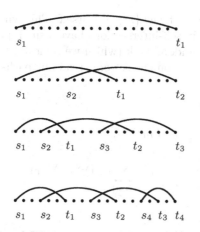

Fig. 4.3. Laces in $\mathcal{L}^{(N)}[a,b]$ for $N = 1, 2, 3, 4$, with $s_1 = a$ and $t_N = b$.

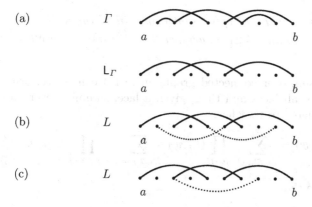

Fig. 4.4. (a) A connected graph Γ and its associated lace $L = L_\Gamma$. (b) The dotted edges are compatible with the lace L. (c) The dotted edge is not compatible with the lace L.

and for $i = 1, 2, \ldots$ by

$$
\begin{aligned}
t_{i+1} &= \max\{t \colon \exists s < t_i \text{ such that } st \in \Gamma\}, \\
s_{i+1} &= \min\{s \colon st_{i+1} \in \Gamma\}.
\end{aligned}
\tag{4.23}
$$

The prescription terminates when $t_{i+1} = b$.

Definition 4.6. *Given a lace L, the set of all edges $st \notin L$ such that $L_{L \cup \{st\}} = L$ is denoted by $\mathcal{C}(L)$. Edges in $\mathcal{C}(L)$ are said to be compatible with L (see Fig. 4.4).*

Note that the construction in (4.22–4.23) looks for the longest edges possible. What Definition 4.6 says is that $st \in \mathcal{C}(L)$ if and only if the largest lace that can be found as a subset of $L \cup \{st\}$ is L itself. This will be important later on, because long loops in self-intersections have a small probability under the law of the spread-out random walk (which we removed from (4.4) because it is the uniform distribution on the set of spread-out paths defined in (4.2)).

4.3.2 Resummation

Lemma 4.7. *For $x \in \mathbb{Z}^d$ and $m \in \mathbb{N}$,*

$$\pi_m(x) = \sum_{N=1}^{\infty} (-1)^N \, \pi_m^{(N)}(x) \tag{4.24}$$

with

$$\pi_m^{(N)}(x) = \sum_{w \in \mathcal{W}_m(x)} \sum_{L \in \mathcal{L}^{(N)}[0,m]} \prod_{st \in L} (-U_{st}(w)) \prod_{s't' \in \mathcal{C}(L)} (1 + U_{s't'}(w)), \tag{4.25}$$

where $\mathcal{L}^{(N)}[0, m]$ is the set of laces on $[0, m]$ with N edges. Note that the factor $(-1)^N$ is inserted into (4.24) to arrange that $\pi_m^{(N)}(x) \geq 0$ for all N, m, x (recall (4.9)).

Proof. The sum over connected graphs in (4.14) can be performed by first summing over all laces and then, given a lace, summing over all connected graphs associated to that lace:

$$J[a, b](w) = \sum_{L \in \mathcal{L}[a,b]} \prod_{st \in L} U_{st}(w) \sum_{\Gamma: \; L_\Gamma = L} \prod_{s't' \in \Gamma \setminus L} U_{s't'}(w). \tag{4.26}$$

Next we note that

$$L_\Gamma = L \text{ if and only if } L \text{ is a lace}, \quad L \subset \Gamma, \quad \Gamma \setminus L \subset \mathcal{C}(L). \tag{4.27}$$

This is due to the fact that the lace L_Γ is obtained from Γ by looking for maxima and minima. Indeed, $L_\Gamma = L$ is equivalent to the property that an edge not in L is never chosen in (4.22–4.23). Using (4.27), we get that the second sum in (4.26) equals $\prod_{s't' \in \mathcal{C}(L)}(1 + U_{s't'}(w))$, which gives (recall (4.17))

$$\pi_m(x) = \sum_{w \in \mathcal{W}_m(x)} \sum_{L \in \mathcal{L}[0,m]} \prod_{st \in L} U_{st}(w) \prod_{s't' \in \mathcal{C}(L)} (1 + U_{s't'}(w)). \tag{4.28}$$

Splitting this out according to the number of edges in the laces, we get the claim. □

The factor $\prod_{st \in L}(-U_{st}(w))$ in (4.25) forces w to self-intersect at the times that are the endpoints of the edges in L (recall the second line of (4.9)). Hence, walks contributing to (4.25) are constrained to have the topology indicated

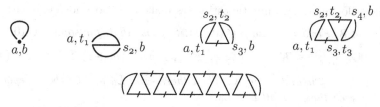

Fig. 4.5. Self-intersections required for a walk w with $\prod_{st} U_{st}(w) \neq 0$, for the laces with $N = 1, 2, 3, 4$ edges depicted in Fig. 4.3. The picture for $N = 11$ is added too.

by the diagrams in Fig. 4.5. The N-th diagram has $2N - 1$ subwalks. The factor $\prod_{s't' \in \mathcal{C}(L)} (1 + U_{s't'}(w))$ carries the self-repellence along the compatible edges. Hence, each subwalk is self-repellent (because all edges under a lace are compatible with that lace), while different subwalks are only partially (!) mutually self-repellent (because some of the compatible edges link the subwalks while others do not). The number of loops in a diagram is equal to the number of edges in the corresponding lace. The lines that are slashed correspond to subwalks that may have zero steps; the other lines correspond to subwalks that have at least one step. Note that the sum in (4.24) terminates at $N = m$.

A key feature of (4.25) is that $\mathcal{L}^{(N)}[0, m]$ is a *highly restrictive set of graphs*. Indeed, the N-th diagram requires the path to make N returns to sites visited earlier. If L is large, then such returns are "unlikely" under the law of the spread-out random walk, and so $\pi_m^{(N)}$ is expected to fall off rapidly with m and N.

Note that $|\mathcal{L}^{(N)}[0, n]|$ is bounded from above by the number of possible choices of $0 = s_0 < \cdots < s_N < n$ times the number of possible choices of $0 < t_1 < \cdots < t_N = n$, which is $[n^{N-1}/(N-1)!]^2$. Hence, summing on N, we see that

$$|\mathcal{L}[0, n]| \leq e^{2n} \ll 2^{\binom{n}{2}} = |\mathcal{G}[0, n]|. \tag{4.29}$$

4.4 Diagrammatic Estimates

The key idea to exploit Lemmas 4.4 and 4.7 is the following:

- Because the $2N - 1$ subwalks in the N-th diagram have a *repulsive* interaction, we get an upper bound on $\pi_m^{(N)}(x) \geq 0$ by ignoring the self-repellence *between* the different subwalks, retaining only the self-repellence *within* each subwalk.

After having done so, each subwalk becomes an independent self-repellent walk of a certain length, say l, starting and ending at certain sites, say u and v. The weight of this subwalk is precisely $Z_l(v - u)$. Thus, what we need to do to control the expansion is to keep track of the various summations encoded in the diagrams.

The following diagrammatic estimates make the above idea explicit.

Lemma 4.8. *Fix $d \geq 5$ and $\beta \in (0, \infty)$. Suppose that, for some $z \in (0, \infty)$,*

$$\|Z_m\|_1 \, z^m \leq K, \qquad \|Z_m\|_\infty \, z^m \leq K |\Omega|^{-1} (m+1)^{-d/2} \qquad (4.30)$$

for all $1 \leq m \leq n$ and some K. Then there exist L_0 and C (both depending on d, K) such that, for all $L \geq L_0$,

$$z^m \sum_{x \in \mathbb{Z}^d} \|x\|^q \, |\pi_m(x)| \leq C |\Omega|^{-1} (m+1)^{-(d-q)/2} \qquad (4.31)$$

for all $2 \leq m \leq n+1$ and $q = 0, 2, 4$.

Proof. We consider the case $q = 0$, and show how to estimate the terms with $N = 1$ and $N = 2$ (depicted by the first two figures in Fig. 4.5) for $2 \leq m \leq n+1$. A similar argument applies for $q = 2, 4$ and $N \geq 3$.

$\underline{N = 1}$: Return to (4.25). There is only one lace, namely, $L = \{0, m\}$, and all other edges are compatible. Hence we get (recall the first diagram in Fig. 4.5)

$$\sum_{x \in \mathbb{Z}^d} \pi_m^{(1)}(x) = \pi_m^{(1)}(0) \leq \sum_{y \in \mathbb{Z}^d} D(y) Z_{m-1}(y) \leq \|Z_{m-1}\|_\infty. \qquad (4.32)$$

Inserting (4.30) into (4.32), we get

$$z^m \sum_{x \in \mathbb{Z}^d} \pi_m^{(1)}(x) \leq CK |\Omega|^{-1} (m+1)^{-d/2}. \qquad (4.33)$$

$\underline{N = 2}$: Neglecting the self-repellence between the three subwalks in the second diagram in Fig. 4.5, we get

$$\begin{aligned}
\sum_{x \in \mathbb{Z}^d} \pi_m^{(2)}(x) &\leq \sum_{x \in \mathbb{Z}^d} \sum_{\substack{m_1, m_2, m_3 \geq 1 \\ m_1 + m_2 + m_3 = m}} Z_{m_1}(x) Z_{m_2}(x) Z_{m_3}(x) \\
&\leq 3! \sum_{\substack{m_1 \geq m_2 \geq m_3 \geq 1 \\ m_1 + m_2 + m_3 = m}} \|Z_{m_1}\|_\infty \|Z_{m_2}\|_\infty \|Z_{m_3}\|_1.
\end{aligned} \qquad (4.34)$$

Inserting (4.30) into (4.34), we get

$$\begin{aligned}
z^m \sum_{x \in \mathbb{Z}^d} \pi_m^{(2)}(x) &\leq 3! \sum_{\substack{m_1 \geq m_2 \geq m_3 \geq 1 \\ m_1 + m_2 + m_3 = m}} \|Z_{m_1}\|_\infty z^{m_1} \|Z_{m_2}\|_\infty z^{m_2} \|Z_{m_3}\|_1 z^{m_3} \\
&\leq 3! K^3 |\Omega|^{-2} \sum_{\substack{m_1 \geq m_2 \geq m_3 \geq 1 \\ m_1 + m_2 + m_3 = m}} (m_1 + 1)^{-d/2} (m_2 + 1)^{-d/2} \\
&\leq 3! K^3 |\Omega|^{-2} \left(\frac{m}{3} + 1 \right)^{-d/2} \sum_{m_2 \geq m_3 \geq 1} (m_2 + 1)^{-d/2} \\
&\leq C K^3 |\Omega|^{-2} (m+1)^{-d/2},
\end{aligned} \qquad (4.35)$$

where the last inequality uses $d > 4$ to make the sum in the third line finite.

For $N \geq 3$ similar estimates hold, which can be derived by induction on N. Each time N increases by 1, an extra factor $K^2|\Omega|^{-1}$ appears, coming from the two extra subwalks in the diagram making an extra loop. The key point is that $K^2|\Omega|^{-1} < 1$ when L is large enough, so that the prefactors are summable (!) in N. For details we refer to van der Hofstad and Slade [166]. □

The estimate in (4.31) for $q = 0, 2$ is needed to obtain the scaling in (4.8), the estimate for $q = 4$ to control error terms.

4.5 Induction

The key fact about Lemma 4.8 is that the bounds assumed in (4.30) for $1 \leq m \leq n$ are turned into a bound in (4.31) valid for $2 \leq m \leq n + 1$. This opens up the possibility of *induction* on n. Indeed, with (4.31) we have control over the coefficients in the recursion relation in Lemma 4.4.

4.5.1 Notation

The induction will be done in Fourier language. For $k \in [-\pi, \pi)^d$, let $\widehat{D}(k)$, $\widehat{\pi}_m(k)$ and $\widehat{\pi}_m^{(N)}(k)$ denote the Fourier transforms of $D(x)$, $\pi_m(x)$ and $\pi_m^{(N)}(x)$. It terms of the latter, (4.18) reads

$$\widehat{Z}_n(k) = |\Omega|\widehat{D}(k)\widehat{Z}_{n-1}(k) + \sum_{m=1}^{n} \widehat{\pi}_m(k)\widehat{Z}_{n-m}(k). \qquad (4.36)$$

In what follows we will need four *induction hypotheses*. We will state these in Section 4.5.3, but first we introduce some notation and in Section 4.5.2 provide its background.

First, because

$$\|Z_n\|_1 = Z_n = \widehat{Z}_n(0), \qquad \|Z_n\|_\infty \leq \|\widehat{Z}_n\|_1, \qquad (4.37)$$

Lemma 4.8 can be reformulated as saying that

- bounds on $\widehat{Z}_m(0)$ and $\|\widehat{Z}_m\|_1$ for $1 \leq m \leq n$ imply bounds on $\widehat{\pi}_m(0)$, $\nabla^2\widehat{\pi}_m(0)$ for $2 \leq m \leq n + 1$.

Second, to formulate the induction hypotheses we need some more notation. Put $z_0 = z_1 = 1$, and define z_n recursively by

$$z_{n+1} = \frac{1}{|\Omega|}\left[1 - \sum_{m=2}^{n+1} \widehat{\pi}_m(0)\,[z_n]^m\right], \qquad n \in \mathbb{N}. \qquad (4.38)$$

Also put

$$v_n(z) = \frac{B_n(z)}{1 + C_n(z)}, \qquad n \in \mathbb{N}, \qquad (4.39)$$

with

$$B_n(z) = |\Omega|z - \sigma^{-2} \sum_{m=2}^{n} \nabla^2 \widehat{\pi}_m(0)\, z^m, \quad C_n(z) = \sum_{m=2}^{n} (m-1)\, \widehat{\pi}_m(0)\, z^m.$$
(4.40)

Fix $\delta_1, \delta_2, \delta_3 > 0$ such that

$$0 < \frac{d-4}{2} - \delta_1 < \delta_2 < \delta_2 + \delta_3 < 1 \wedge \frac{d-4}{2}.$$
(4.41)

We further need five constants satisfying

$$K_3 \gg K_1 \gg K_4 \gg 1, \quad K_2 \gg K_1, K_4, \quad K_5 \gg K_4.$$
(4.42)

Define the intervals

$$I_n = \left[z_n - K_1 |\Omega|^{-1} n^{-(d-2)/2},\, z_n + K_1 |\Omega|^{-1} n^{-(d-2)/2} \right], \quad n \in \mathbb{N}, \quad (4.43)$$

and abbreviate

$$\widehat{Z}_m(k; z) = \widehat{Z}_m(k)\, z^m, \qquad \widehat{\pi}_m(k; z) = \widehat{\pi}_m(k)\, z^m.$$
(4.44)

4.5.2 Heuristics

Here is the heuristics behind the above notation. In anticipation of the first line of (4.8), substitute $\widehat{Z}_n(0; \mu^{-1}) \to A$ as $n \to \infty$ into the recursion relation (4.36), with A and μ yet to be determined. This gives the relation

$$1 = |\Omega|\, \mu^{-1} + \sum_{m=2}^{\infty} \widehat{\pi}_m(0; \mu^{-1}).$$
(4.45)

The sum has not (!) yet been proved to converge. The recursion for $(z_n)_{n \in \mathbb{N}}$ defined in (4.38) approximates the relation in (4.45) by discarding the terms in the sum with $m > n$, which cannot be handled at the n-th stage of the induction argument. This is done in anticipation of $z_n \to \mu^{-1}$ as $n \to \infty$.

Differentiate the recursion relation (4.36) twice w.r.t. k, set $k = 0$, and use that odd derivatives vanish because of symmetry, to get

$$\nabla^2 \widehat{Z}_n(0; z) = |\Omega|\, z \left[\nabla^2 \widehat{Z}_{n-1}(0; z) - \sigma^2 \widehat{Z}_{n-1}(0; z) \right]$$
$$+ \sum_{m=2}^{n} \left[\widehat{\pi}_m(0; z) \nabla^2 \widehat{Z}_{n-m}(0; z) + \nabla^2 \widehat{\pi}_m(0; z) \widehat{Z}_{n-m}(0; z) \right],$$
(4.46)

using that $\widehat{D}(0) = 1$ and $\nabla^2 \widehat{D}(0) = -\sigma^2$. Approximate, for z close to z_n,

$$|\Omega|\, z \approx 1 - \sum_{m=2}^{n} \widehat{\pi}_m(0; z), \qquad \widehat{Z}_{n-m}(0; z) \approx \widehat{Z}_{n-1}(0; z).$$
(4.47)

Substitute this approximation into (4.46), and use the first half of (4.40), to get

$$\nabla^2 \widehat{Z}_n(0;z) - \nabla^2 \widehat{Z}_{n-1}(0;z)$$

$$\approx -\sigma^2 B_n(z)\widehat{Z}_{n-1}(0;z) + \sum_{m=2}^{n} \widehat{\pi}_m(0;z)\left[\nabla^2 \widehat{Z}_{n-m}(0;z) - \nabla^2 \widehat{Z}_{n-1}(0;z)\right].$$

$$(4.48)$$

Next, in anticipation of the second line of (4.8) and $v_n(z_n) \to v$ as $n \to \infty$, approximate, for z close to z_n,

$$\nabla^2 \widehat{Z}_{n-m}(0;z) - \nabla^2 \widehat{Z}_{n-1}(0;z) \approx (m-1)\,\sigma^2\, v_n(z)\, \widehat{Z}_{n-1}(0;z). \qquad (4.49)$$

Substitute this approximation into (4.48), and use the second half of (4.40), to get

$$\nabla^2 \widehat{Z}_n(0;z) - \nabla^2 \widehat{Z}_{n-1}(0;z) \approx -\sigma^2 B_n(z)\widehat{Z}_{n-1}(0;z) + \sigma^2 v_n(z) C_n(z) \widehat{Z}_{n-1}(0;z). \qquad (4.50)$$

The left-hand side of (4.50) is expected to be close to $-\sigma^2 v_n(z)\widehat{Z}_{n-1}(0;z)$. Equate the latter with the right-hand side of (4.50), to arrive at the definition of $v_n(z)$ in (4.39).

4.5.3 Induction Hypotheses

Our induction hypotheses read as follows.

- For all $z \in I_n$ and all $1 \leq m \leq n$:
  (IH1) $|z_m - z_{m-1}| \leq K_1|\Omega|^{-1}m^{-d/2}$.
  (IH2) $|v_m(z) - v_{m-1}(z)| \leq K_2|\Omega|^{-1}m^{-(d-2)/2}$.
  (IH3) For k such that $1 - \widehat{D}(k) \leq \delta_2 m^{-1}\log m$,

$$\widehat{Z}_m(k;z) = \prod_{l=1}^{m}\left\{1 - v_l[1 - \widehat{D}(k)] + r_l(k;z)\right\} \qquad (4.51)$$

  with

$$|r_l(0;z)| \leq K_3|\Omega|^{-1}l^{-(d-2)/2},$$
$$|r_l(k;z) - r_l(0;z)| \leq K_3|\Omega|^{-1}[1 - \widehat{D}(k)]l^{-\delta_3}. \qquad (4.52)$$

  (IH4) For k such that $1 - \widehat{D}(k) > \delta_2 m^{-1}\log m$,

$$|\widehat{Z}_m(k;z)| \leq K_4[1 - \widehat{D}(k)]^{-2-\delta_1}m^{-d/2},$$
$$|\widehat{Z}_m(k;z) - \widehat{Z}_{m-1}(k;z)| \leq K_5[1 - \widehat{D}(k)]^{-1-\delta_1}m^{-d/2}. \qquad (4.53)$$

The role of (IH1–IH2) is to control the rate at which z_m and $v_m(z_m)$ tend to μ^{-1} and v, respectively, as $m \to \infty$. The role of (IH3–IH4) is to control the behavior of $\widehat{Z}_n(k;z)$ for small k and large k, respectively, the first coming from

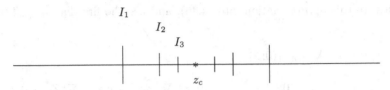

Fig. 4.6. Nested sequence of intervals $(I_n)_{n \in \mathbb{N}}$ converging to $z_c = 1/\mu$.

an expansion of the recursion relation in (4.36) around $k = 0$, the second being an error term that is needed to bound $\|\widehat{Z}_m\|_1$ (recall (4.37) and the remark following it). The factor $[1 - \widehat{D}(k)]^{-2-\delta_1}$ in (4.53) diverges as $k \to 0$, but is integrable for $d > 4$ when $\delta_1 > 0$ is taken small enough. It is *vital* that the induction hypotheses are taken for z on the *nested sequence* $(I_n)_{n \in \mathbb{N}}$ (see Fig. 4.6).

The key fact now is the following.

Lemma 4.9. (IH1)–(IH4) *can be advanced subject to* (4.41–4.42).

Proof. The advancement of (IH1)–(IH4) is technically involved. The proof starts from (4.36),

$$\widehat{Z}_{n+1}(k; z) = z\widehat{D}(k)\,\widehat{Z}_n(k; z) + \sum_{m=2}^{n+1} \widehat{\pi}_m(k; z)\,\widehat{Z}_{n+1-m}(k; z), \qquad (4.54)$$

and uses the bounds in the Fourier version of Lemma 4.8 (recall (4.37) and the remark following it). We refer to van der Hofstad [156], Section 2.4, for details. Note that *initialization* of the induction is trivial: the inequalities in (IH1–IH4) are easily seen to be true for $m = 1$, when the walk consists of a single step only. $\square$

4.6 Proof of Diffusive Behavior

With Lemma 4.9, we know that the inequalities in (IH1–IH4) are valid for *all* $n \in \mathbb{N}$. Finally, we show how this implies Theorem 4.1.

Proof. For $z \in I_n$, (IH1) gives

$$\begin{aligned}
|z_{n-1} - z| &\le |z_n - z| + |z_n - z_{n-1}| \\
&\le K_1 |\Omega|^{-1} n^{-(d-2)/2} + K_1 |\Omega|^{-1} n^{-d/2} \\
&\le K_1 |\Omega|^{-1} (n-1)^{-(d-2)/2},
\end{aligned} \qquad (4.55)$$

and so $I_n \subset I_{n-1}$. This monotonicity is crucial, because it allows us to control z and shows that $\lim_{n \to \infty} z_n = z_c$ with $\{z_c\} = \cap_{n \in \mathbb{N}} I_n$ (see Fig. 4.6). Thus (IH1–IH4) can be applied at z_c.

By passing to the limit $n \to \infty$ in (4.38), we see that z_c solves the equation

$$z_c = \frac{1}{|\Omega|} \left[1 - \sum_{m=2}^{\infty} \widehat{\pi}_m(0; z_c) \right], \tag{4.56}$$

where the sum converges because after the induction is completed we know that $|\widehat{\pi}_m(0; z_c)|$ falls off like $|\Omega|^{-1} m^{-d/2}$ for large m. Thus, μ in the first line of (4.8) equals (recall (4.45))

$$\mu = 1/z_c. \tag{4.57}$$

For $k = 0$, (IH3) reduces to $\widehat{Z}_m(0; z) = \prod_{l=1}^{m}[1 + r_l(0; z)]$. Therefore the bound on $r_l(0)$ in (4.52) implies that

$$\lim_{n \to \infty} \widehat{Z}_n(0) \mu^{-n} = A = \prod_{l=1}^{\infty}[1 + r_l(0; \mu^{-1})]. \tag{4.58}$$

This proves the first line in (4.8).

Picking $k/\sqrt{\sigma^2 vn}$ instead of k in (4.51), we get from (IH3) that

$$\widehat{Z}_n\left(\frac{k}{\sqrt{\sigma^2 vn}}\right) \mu^{-n}$$

$$= \prod_{l=1}^{n} \left\{ 1 - v_l(\mu^{-1}) \left[1 - \widehat{D}\left(\frac{k}{\sqrt{\sigma^2 vn}}\right) \right] + r_l\left(\frac{k}{\sqrt{\sigma^2 vn}}; \mu^{-1}\right) \right\}$$

$$= [1 + o(1)] A \prod_{l=1}^{n} \left\{ 1 - v_l(\mu^{-1}) \left[1 - \widehat{D}\left(\frac{k}{\sqrt{\sigma^2 vn}}\right) \right] \right\} \tag{4.59}$$

$$= [1 + o(1)] A \, e^{-\frac{1}{2d}\|k\|^2},$$

where we use that for $l \to \infty$ (recall (4.7))

$$1 - v_l(\mu^{-1}) \left[1 - \widehat{D}\left(\frac{k}{\sqrt{\sigma^2 vn}}\right) \right] = e^{-\frac{1}{vn} v_l(\mu^{-1}) \frac{1}{2d}\|k\|^2 [1+o(1)]}, \tag{4.60}$$

$$v_l(\mu^{-1}) = v[1 + o(1)],$$

with v given by (recall (4.40))

$$v = \frac{B_\infty(\mu^{-1})}{1 + C_\infty(\mu^{-1})}. \tag{4.61}$$

This proves the third line in (4.8). The second part of (4.60) follows from (4.39) and (IH2), and needs $d > 4$ to guarantee that $B_\infty(\mu^{-1})$ and $C_\infty(\mu^{-1})$ are finite.

The second line in (4.8) follows by a Taylor expansion of $\widehat{Z}_n(k; \mu^{-1})$ around $k = 0$ together with the bounds in (IH3).

Note that (IH4) is not used in the above. However, this inequality is needed to control the error terms. □

Thus, we have finally completed the proof of Theorem 4.1. The constants in Theorem 4.1 are expressed in terms of $\widehat{\pi}_m$. The bounds on $\widehat{\pi}_m$, together with exact enumeration for small and moderate m, allow for an estimate of these constants (see e.g. Clisby, Liang and Slade [68] for SAW with $L = 1$). For instance, it can be deduced from (4.61) that $v > 1$ for L large enough, tending to 1 as $L \to \infty$.

4.7 Extensions

(1) The induction method in Section 4.5 was introduced in van der Hofstad, den Hollander and Slade [162] and was subsequently refined in van der Hofstad and Slade [166]. In the latter paper it is shown that, in some sense, (IH1–IH4) are universal induction hypotheses for diffusive scaling. The advantage of the induction method is that it avoids taking the Laplace transform in time, which is non-trivial to invert. Theorem 4.1 extends to spread-out random walk whose steps are not drawn uniformly from the set Ω defined in (4.1). See [166] for details.

(2) There are other ways to manipulate the recursion relation in Lemma 4.4 and the diagrammatic estimates in Lemmas 4.8. One is via generating functions (Hara and Slade [147]), an approach explained in detail in Madras and Slade [230]. An alternative approach, due to Bolthausen and Ritzmann [32], is via Banach fixed-point theorems. Here, (4.54) with $k = 0$ and $z = z_c$ is viewed as a fixed-point equation for the sequence $(\widehat{Z}_m[z_c]^m)_{m\in\mathbb{N}}$. The operator acting on this sequence is shown to be a contraction when β is sufficiently small. Control of the π_m is done via an induction scheme, reminiscent of that described in Section 4.5. The advantage of this approach is that it avoids taking the Fourier transform altogether, working in real space and time.

(3) Van der Hofstad, den Hollander and Slade [162] and van der Hofstad and Slade [166] extend the third line of (4.8) to a local central limit theorem. Rather than considering the probability that the right endpoint of the soft polymer is at a single site at a distance $O(\sqrt{n})$ from the origin, it is necessary to consider the probability that it lies in a box whose size tends to infinity as $n \to \infty$. To appreciate why, note that for SAW ($\beta = \infty$) the path cannot return to the origin.

(4) The third line of (4.8) is a statement about the asymptotic behavior of the two-point function $Z_n(x)$ as $n, \|x\| \to \infty$. Similar scaling results can be derived for the higher-point functions, defined by

$$Z_{n_1,\ldots,n_r}(x_1,\ldots,x_r) = \sum_{w\in\mathcal{W}_n} e^{-\beta I_n(w)} 1_{\{w_{n_1}=x_1,\ldots,w_{n_r}=x_r\}}, \qquad (4.62)$$

for $r = 2, 3, 4 \ldots$. Indeed, even the functional central limit can be proved, with Brownian motion as the scaling limit. We refer to van der Hofstad and Slade [167] for details.

(5) The estimates in Section 4.4–4.5 are valid uniformly in $\beta \in (0, \infty)$ for L sufficiently large. Indeed, the estimates only use that $\lambda(\beta)$ in (4.10) satisfies $\lambda(\beta) \leq 1$. Therefore, Theorem 4.1 equally well applies to the spread-out SAW ($\beta = \infty$). For $L = 1$, the first and second line of (4.8) were proved for $\beta = \infty$ in Hara and Slade [147], [148]. The proof is computer-assisted: for small and moderate m the diagrams are computed via exact enumeration, while for large m bounds on the diagrams are used. This is needed to get convergence without a small parameter.

(6) Heydenreich, van der Hofstad and Sakai [152] perform the lace expansion for SAW with long-range steps, i.e., the reference measure P_n is not that of SRW (which we actually dropped from (4.4) because it is uniform) but of a random walk whose step distribution has a polynomial tail. The critical dimension above which diffusive behavior occurs is computed as a function of the tail exponent, and turns out to drop from 4 and 0 as the tail becomes thicker. This is because self-intersections are less likely for long-range random walk. (A similar drop will be encountered for the elastic polymer described in Chapter 5, for a related but different reason.) Heydenreich [151] shows that the path measure converges weakly as $n \to \infty$ to that of an α-stable process, with α depending on the tail exponent.

(7) Ueltschi [299] looks at a spread-out SAW with self-attraction, i.e., (4.9) is replaced by

$$U_{st}(w) = \begin{cases} 1, & \text{if } \|w_s - w_t\| = 0, \\ -(1 - e^{\gamma}), & \text{if } \|w_s - w_t\| = 1, \\ 0, & \text{otherwise,} \end{cases} \tag{4.63}$$

where $\gamma \in (0, \infty)$ is a binding energy between neighboring monomers. It is shown that Theorem 4.1 carries over in the following sense: the first line of (4.8) is true when $d \geq 5$ and $\gamma \leq \gamma_0$, while the second line is true when $d \geq 5$, $\gamma \leq \gamma_0$ and $L \geq L_0$ (with γ_0 and L_0 depending on d). The step distribution of the SAW is assumed to have a particular scaling form and to be sufficiently "smooth", which makes the model somewhat restrictive. The proof relies on the lace expansion. The convergence of the lace expansion is the hard issue, because of the presence of the self-attraction (recall the remarks made at the beginning of Section 4.4). Convergence is achieved by playing out the self-attraction against the self-avoidance, showing that the effective interaction is repulsive. In particular, Z_n is shown to be submultiplicative ($Z_{m+n} \leq Z_m Z_n$ for all $m, n \in \mathbb{N}$), implying that the free energy exists. We will return to polymers with self-repellence and self-attraction in Chapter 6.

(8) Networks of mutually-avoiding self-avoiding walks have been considered in van der Hofstad and Slade [167] and in Holmes, Járai, Sakai and Slade [174]. Gaussian scaling behavior is proved for the spread-out model in $d > 4$. The diffusion constant depends on the topology of the network: each node of the network contributes to this constant in a way that depends on how the self-avoiding walks are tied together, i.e., on the topology of the network (for an

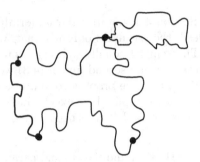

Fig. 4.7. An example of a network of SAWs.

example, see Fig. 4.7). This problem belongs to the topic of branched polymers, which is not addressed in the present monograph.

4.8 Challenges

(1) Prove that the constant v in Theorem 4.1 is a non-decreasing function of β. In view of the expression found in (4.61), this requires a strong control on the coefficients in the lace expansion.

(2) Extend Theorem 4.1 to SAW ($\beta = \infty$) in $d \geq 5$ with $L = 1$ without the help of the computer. This seems a very hard challenge.

(3) Derive an LDP for the endpoint of the soft polymer in $d \geq 5$, either with L large or β small. Determine how the rate function in this LDP depends on β. For $d = 1$, the LDP was derived in Chapter 3. The results in Bolthausen and Ritzmann [32] yield an upper bound on the rate function for $d \geq 5$ and β small.

(4) For the continuum version of the Domb-Joyce model – the Edwards model based on Brownian mentioned in Section 3.6, Extension (5) – the existence of the soft polymer measure is non-trivial when $d \geq 2$, unlike the situation in $d = 1$ where the Hamiltonian is given by (3.78). Indeed, for $d \geq 2$ the self-repellence via the Dirac delta-function is not properly defined and some regularization is needed. Varadhan [301] gives a construction of the soft polymer measure for $d = 2$, Bolthausen [26] for $d = 3$ (simplifying and improving earlier work by Westwater [310], [311] and Kusuoka [223]). The construction uses a renormalization procedure that is necessary to deal with the accumulation of short self-intersections in Brownian motion. For $d = 2$ the soft polymer measure is absolutely continuous w.r.t. the law of Brownian motion, for $d = 3$ it is not. For further details we refer to the Saint-Flour lectures by Bolthausen [28], Chapter 1. The challenge is to understand the behavior of this measure as a function of time and of the self-repellence parameter.

(5) What can be done for the spread-out random walk with self-attraction beyond the restrictive model studied in Ueltschi [299]?

5

Elastic Polymers

In this chapter we take a brief look at a version of the soft polymer where the penalty of the self-intersections decays with the loop length, i.e., the difference between the times at which the self-intersection occurs. This model is called the *elastic polymer*. Interestingly, it will turn out that this model has diffusive behavior in any $d \geq 1$ as soon as the decay is sufficiently fast, namely, the critical dimension for diffusive behavior is *lower* than $d = 4$ and depends on the parameter controlling the *rate of decay* of the penalty as a function of the loop length. We will see that the scaling behavior of the elastic polymer is in fact highly sensitive to the value of this parameter. The lace expansion that was used in Chapter 4 can be carried over with minor modifications to arrive at the main scaling result.

Section 5.1 defines the model, while Section 5.2 describes the main result, which is taken from van der Hofstad, den Hollander and Slade [162].

5.1 A Polymer with Decaying Self-repellence

We consider the model in which the set of paths and the Hamiltonian in (3.1–3.2) are replaced by

$$\mathcal{W}_n = \left\{ w = (w_i)_{i=0}^n \in (\mathbb{Z}^d)^{n+1} \colon w_0 = 0, \|w_{i+1} - w_i\| = 1 \ \forall \, 0 \leq i < n \right\},$$
$$H_n(w) = \beta I_n^\kappa(w),$$

$$\text{(5.1)}$$

where $\beta, \kappa \in (0, \infty)$ and

$$I_n^\kappa(w) = \sum_{\substack{i,j=0 \\ i<j}}^n (j - i)^{-\kappa} 1_{\{w_i = w_j\}}, \tag{5.2}$$

and (3.3) is replaced by

$$P_n^{\beta,\kappa}(w) = \frac{1}{Z_n^{\beta,\kappa}} \, e^{-\beta I_n^\kappa(w)}, \qquad w \in \mathcal{W}_n, \tag{5.3}$$

F. den Hollander, *Random Polymers*,
Lecture Notes in Mathematics 1974, DOI: 10.1007/978-3-642-00333-2_5,
© Springer-Verlag Berlin Heidelberg 2009, Reprint by Springer-Verlag Berlin Heidelberg 2012

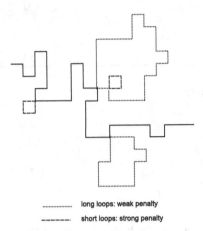

·················· long loops: weak penalty

------ short loops: strong penalty

Fig. 5.1. A path with long loops (thin dash) and short loops (thick dash). Long loops are less penalized than short loops.

where, as in (4.4), we drop the uniform reference measure P_n. Thus, self-intersections occurring after a short time lapse receive a larger penalty than self-intersections occurring after a long time lapse. In other words, short loops are more costly than long loops, and so we may think of the extra parameter κ as measuring the degree of *elasticity* of the polymer (see Fig. 5.1).

Our main theorem is the following analogue of Theorem 4.1, which holds under the restriction

$$\kappa + \frac{d-4}{2} > 0. \tag{5.4}$$

In what follows, β is the small parameter, rather than $|\Omega|^{-1}$ in Chapter 4.

Theorem 5.1. Fix $d \geq 1$. If (5.4) holds, then there exists a constant $\beta_0 > 0$ (depending on d, κ) such that for all $\beta \in (0, \beta_0)$ and as $n \to \infty$,

$$Z_n = A\mu^n [1 + o(1)],$$
$$-Z_n^{-1} \nabla^2 \widehat{Z}_n(0) = Dn [1 + o(1)], \tag{5.5}$$
$$Z_n^{-1} \widehat{Z}_n \left(k/\sqrt{Dn} \right) = e^{-\frac{1}{2d} \|k\|^2} [1 + o(1)], \qquad k \in \mathbb{R}^d,$$

where $\mu, A, D > 0$ are constants (depending on d, β, κ) and the error term in the last line is uniform in k provided $\|k\|^2 / \log n$ is sufficiently small.

Thus, the critical dimension is lowered from $d = 4$ to $d = 0 \vee (4 - 2\kappa)$. For instance, in $d = 1$ there is diffusive behavior as soon as $\kappa > \frac{3}{2}$.

5.2 The Lace Expansion Carries over

The proof of Theorem 5.1 can be built on the lace expansion approach in Chapter 4. Indeed, all we have to do is adapt the definition of $U_{st}(w)$ in (4.9) to

$$U_{st}(w) = \begin{cases} -\lambda_{t-s}(\beta, \kappa), & \text{if } w_s = w_t, \\ 0, & \text{if } w_s \neq w_t, \end{cases} \tag{5.6}$$

with

$$\lambda_m(\beta, \kappa) = 1 - e^{-\beta m^{-\kappa}}, \qquad m \in \mathbb{N}. \tag{5.7}$$

In particular, all the expansion formulas leading up to Lemma 4.4 and (4.24–4.25) carry over verbatim. The key difference is that the weights of the subwalks in the successive diagrams (recall Fig. 4.5) decay faster with their length as when $\kappa = 0$. This is why we get better convergence. More precisely, because

$$\lambda_m(\beta, \kappa) \leq \beta m^{-\kappa}, \tag{5.8}$$

Lemma 4.8 carries over with (4.31) replaced by

$$\sum_{x \in \mathbb{Z}^d} \|x\|^q |\pi_m(x)| z^m \leq C\beta(m+1)^{-(d+2\kappa-q)/2} \tag{5.9}$$

for $2 \leq m \leq n$ and $q = 0, 2, 4$, subject to (4.30) and β being small enough. Indeed, this is easily checked by repeating the estimates in (4.32–4.35), taking advantage of (5.8). In particular, for $q = 0, 2$ the bound in (5.9) yields (recall the notation introduced in (4.44))

$$|\hat{\pi}_m(0; z_c)| \leq C\beta(m+1)^{-(\frac{d}{2}+\kappa)}, \quad |\nabla^2 \hat{\pi}_m(0; z_c)| \leq C\beta(m+1)^{-(\frac{d}{2}+\kappa-1)}, \tag{5.10}$$

which shows that $\frac{d}{2} + \kappa > 2$ is needed to get convergence, for instance, for the quantity v_n in (4.61), based on the definitions of B_n and C_n in (4.39–4.40). This explains the restriction in (5.4).

For details of the full argument, we refer to van der Hofstad, den Hollander and Slade [162]. The induction hypotheses (IH1–IH4) formulated in Section 4.5 made their first appearance in that paper, dealing with the elastic polymer subject to (5.4). The soft polymer is the special case with $\kappa = 0$.

As in Sections 4.4–4.6, β may be replaced by $|\Omega|^{-1}$, in which case the lace expansion converges for L large enough.

It is important that the interaction in (5.2) is *repulsive*, resulting in $U_{st}(w) \leq 0$ in (5.6). Indeed, this played a key role in Section 4.4, where we estimated the quantity $\pi_m^{(N)}$ defined in (4.25). Removing the interaction *between* the subwalks in the diagrams in Fig. 4.3 results in an upper bound on $\pi_m^{(N)}$ for each N and m (recall the proof of Lemma 4.8)).

5.3 Extensions

(1) It is shown in van der Hofstad, den Hollander and Slade [162] that also the local central limit theorem holds for the elastic polymer subject to (5.4), provided a box of growing size is considered (recall Section 4.7, Extension (3)).

(2) Van der Hofstad [155], extending an earlier result by Kennedy [208], proves that in $d = 1$ the elastic polymer is ballistic for $\kappa \in [0, 1]$ and β sufficiently large. The proof uses the lace expansion, perturbing around the one-sided deterministic walk instead of SRW and employing induction hypotheses that are adapted to proving ballistic behavior. It has not yet been proved that the same result is true for all $\beta \in (0, \infty)$.

(3) The soft polymer with $\beta \in (-\infty, 0)$ (i.e., self-intersections are rewarded rather than penalized) is localized: with a probability tending to 1 as $n \to \infty$ the path eventually jumps back and forth between two sites (Oono [251], [252]). Brydges and Slade [42], [43] consider the soft polymer with β replaced by $\beta_n = \beta n^{-\kappa}$, $\kappa \in [0, \infty)$, allowing $\beta \in \mathbb{R}$. This model is attractive for $\beta < 0$ and repulsive for $\beta > 0$. They show the following:

(a) For $d \geq 1$ and $\kappa \in [0, 1)$ the polymer is *collapsed* for all $\beta \in (-\infty, 0)$, i.e., there is a finite box (whose size depends on d, κ and β) such that the polymer of length n lies entirely inside this box with a probability that tends to 1 exponentially fast as $n \to \infty$.

(b) For $d \geq 1$ and $\kappa = 1$ there is a critical β_c^* such that the polymer is *diffusive* for all $\beta \in (\beta_c^*, \infty)$, with $\beta_c^* = 0$ in $d = 1$ and $\beta_c^* < 0$ for $d \geq 2$. For $d \geq 3$ the diffusion constant equals 1 and the scaling limit is Brownian motion, while for $d = 2$ the diffusion constant is "renormalized" and the scaling limit is the two-dimensional (attractive) Edwards model constructed in Varadhan [301] (recall Challenge (3) in Section 4.8).

(c) For $d = 1$ and $\kappa = \frac{3}{2}$ the polymer is *diffusive* for all $\beta \in \mathbb{R}$. The diffusion constant is "renormalized" and the scaling limit is the one-dimensional (attractive) Edwards model with Hamiltonian (3.78).

Bolthausen and Schmock [33], [34] show that

(d) For $d \geq 1$ and $\kappa = 1$ there is a critical β_c such that the polymer is *collapsed* for all $\beta \in (-\infty, \beta_c)$, with $\beta_c = 0$ in $d = 1$ and $\beta_c < 0$ for $d \geq 2$.

Clearly, $\beta_c \leq \beta_c^*$. It is not known whether $\beta_c = \beta_c^*$ for $d \geq 2$ (see Fig. 5.2). Formulas for these two critical thresholds are

$$\beta_c = \sup \left\{ \beta \in \mathbb{R} : \limsup_{n \to \infty} Z_n^{1/n} > 1 \right\},$$

$$\beta_c^* = \inf \left\{ \beta \in \mathbb{R} : \sup_{n \in \mathbb{N}} Z_n^* < \infty \right\}, \tag{5.11}$$

Fig. 5.2. Phase diagram for the weakly attracting soft polymer with $d \geq 2$ and $\kappa = 1$, with the collapsed phase for $\beta \in (-\infty, \beta_c)$ and the diffusive phase for $\beta \in (\beta_c^*, \infty)$.

where

$$Z_n = E\left(\exp\left[-\beta n^{-\kappa} I_n\right]\right),$$
$$Z_n^* = E\left(\exp\left[-\beta n^{-\kappa}(I_n - E(I_n))\right]\right), \qquad (5.12)$$

with I_n the intersection local time in (3.2). Note that if $\beta_c = \beta_c^*$, then the phase transition between collapsed and diffusive is discontinuous in $d \geq 3$ (and possibly continuous in $d = 2$).

For further details we refer to the Saint-Flour lectures by Bolthausen [28], Chapter 2. Collapse transitions are well known in statistical physics, but mathematically they turn out to be hard. We return to this topic in Chapter 6.

5.4 Challenges

Very little information is available when $\kappa \in (0, \frac{4-d}{2}]$. Neither the large deviation approach in Chapter 3 nor the lace expansion approach in Chapter 4 are able to handle this regime.

(1) It was conjectured by Caracciolo, Parisi and Pelissetto [50] that for $1 \leq d \leq 4$ the critical exponent $\nu(\kappa)$ for the end-to-end distance, defined by

$$v_n = \sum_{w \in \mathcal{W}_n} \|w_n\|^2 P_n^{\beta,\kappa}(w) = D n^{2\nu(\kappa)}[1 + o(1)] \qquad \text{as } n \to \infty, \qquad (5.13)$$

equals (see Fig. 5.3)

$$\nu(\kappa) = \nu_{\mathrm{SAW}} \wedge \nu_{\mathrm{MF}}(\kappa) \qquad (5.14)$$

with ν_{SAW} the critical exponent for the self-avoiding walk found in (2.26), and $\nu_{\mathrm{MF}}(\kappa)$ the exponent for the "mean-field version" of the elastic polymer given by

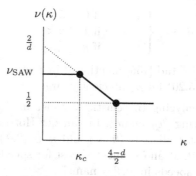

Fig. 5.3. Conjectured critical exponent of the elastic polymer for $1 \leq d \leq 4$. Here, $\kappa_c = \frac{4-d}{d+2}$ when $\nu_{\mathrm{SAW}} = \frac{3}{d+2}$, the Flory value in (2.31). The slope of the linear piece between κ_c and $\frac{4-d}{2}$ is $-\frac{1}{d}$. Proofs are known only for $\kappa > \frac{4-d}{2}$ and for $\kappa \leq \kappa_c = 1$ when $d = 1$ and β is sufficiently large. Note that the behavior is superdiffusive and subballistic for $\kappa \in (\kappa_c, \frac{4-d}{2})$.

$$\nu_{\mathrm{MF}}(\kappa) = \begin{cases} \frac{1}{2}, & \text{if } \kappa \geq \frac{4-d}{2}, \\ \frac{2-\kappa}{d}, & \text{if } \kappa < \frac{4-d}{2}. \end{cases} \tag{5.15}$$

The challenge is to provide a proof of this conjecture. In [50], the scaling in (5.13–5.15) is amply supported by simulations.

The formula for $\nu_{\mathrm{MF}}(\kappa)$ in (5.15) comes from a *Gaussian approximation* of the path measure in (5.3). Namely, (5.2) is replaced by

$$I_n^G(w) = \sum_{\substack{i,j=0 \\ i<j}}^n (G^{-1})(i,j)\,(w_i, w_j), \qquad w \in \mathcal{W}_n, \tag{5.16}$$

for a given covariance matrix $G = (G(i,j))_{i,j\in\mathbb{N}_0}$, satisfying $G(i,j) = G(0, j-i)$, $i,j \in \mathbb{N}_0$, with $(\cdot,\cdot)$ the inner product on $\mathbb{Z}^d$. Subsequently, the functional

$$G \mapsto \lim_{n\to\infty} \frac{1}{n} \left\{ \sum_{w\in\mathcal{W}_n} P_n^{\beta,G}(w)\,\beta \left[I_n^\kappa(w) - I_n^G(w) \right] - \log Z_n^G \right\} \tag{5.17}$$

is minimized w.r.t. to G, where $P_n^{\beta,G}$ is the path measure associated with I_n^G and Z_n^G is its normalizing partition sum. This variational computation leads to a minimizer $G_{\min}$ whose Fourier transform

$$\widehat{G}_{\min}(k) = \sum_{l\in\mathbb{N}_0} G_{\min}(0,l)\,e^{i\,kl}, \qquad k \in [0, 2\pi), \tag{5.18}$$

scales like

$$\widehat{G}_{\min}(k) \asymp k^{-2\nu_{\mathrm{MF}}(\kappa)-1}, \qquad k \downarrow 0, \tag{5.19}$$

with $\nu_{\mathrm{MF}}(\kappa)$ given by (5.15). The rationale behind (5.14) is that the Gaussian approximation overestimates the critical exponent $\nu(\kappa)$, because (5.16) carries long-range interactions whereas (5.2) carries only local interactions.

For $d = 1$, (5.14–5.15) amount to

$$\nu(\kappa) = \begin{cases} 1, & \text{if } 0 \leq \kappa \leq 1, \\ 2 - \kappa, & \text{if } 1 < \kappa < \frac{3}{2}, \\ \frac{1}{2}, & \text{if } \kappa \geq \frac{3}{2}. \end{cases} \tag{5.20}$$

The result by van der Hofstad [155] mentioned in Section 5.3, Extension (3), proves the first line of (5.20) for β sufficiently large.

(2) The difficulty in analyzing the elastic polymer in $d = 1$ in the regime $\kappa \in (0,1]$ is an underlying *degeneracy*. In van der Hofstad and König [165] it is conjectured that, for all $\beta \in (0,\infty)$ and $\kappa \in (0,1)$, the speed of the polymer satisfies an LDP, at an exponential cost for speeds in $(\theta^*, 1]$ but at a subexponential cost for speeds in $[0, \theta^*]$, namely,

$$\lim_{n\to\infty} n^{-(1-\kappa)} \log P_n^{\beta,\kappa}(S_n = \lceil \theta n \rceil) = -\frac{\beta}{2(1-\kappa)}\,\frac{(\theta^* - \theta)^{1-\kappa}}{(\theta^*)^{2-\kappa}}, \qquad \theta \in [0, \theta^*], \tag{5.21}$$

with $\theta^* = \theta^*(\beta, \kappa) \in (0,1)$ the speed. The challenge is to prove this conjecture.

A heuristic explanation of (5.21), similar in the spirit to what we found in Section 3.5 for the one-dimensional soft polymer, runs as follows. For $\theta \in [0, \theta^*)$, the best strategy for the polymer is to first move at speed θ^* for a time $\frac{(\theta^* + \theta)}{2\theta^*} n$ and afterwards move at speed $-\theta^*$ for a time $\frac{(\theta^* - \theta)}{2\theta^*} n$, so as to arrive at site θn at time n. In each of the two stretches, the polymer visits a site $1/\theta^*$ times (on average), so that the interaction it accumulates between the two stretches equals $n^{1-\kappa}$ times the right-hand side of (5.21). This is the cost compared to the strategy in which the polymer moves at speed θ^* for a time n.

(3) Find out whether for $d = 1$ and $\kappa = 1$ there is a critical value $\beta_c \in (0, \infty)$ such that the elastic polymer is ballistic for $\beta > \beta_c$ and subdiffusive for $\beta < \beta_c$. Or is $\beta_c = 0$? (Recall Fig. 5.3.)

(4) For $d = 1$, try to derive a functional central limit theorem and an LDP. These are the analogues of Challenges (2) and (3) in Section 3.7. The method of local times used in Chapter 3 is a powerful tool that can potentially be adapted to deal with the time lapses between the successive visits to a site.

(5) For the soft polymer with $d \geq 2$ and $\beta_n = \beta n^{-1}$, find out whether or not $\beta_c = \beta_c^*$. (See Extension (3) in Section 5.3.)

(6) For the soft polymer with $d = 1$ and $\beta_n = \beta n^{-\kappa}$, prove that the variance of the end-to-end distance of the polymer grows like $n^{\nu(\kappa)}$ as $n \to \infty$ with

$$\nu(\kappa) = \kappa - 1 \text{ for } \beta < 0 \text{ and } \kappa \in (1, \tfrac{3}{2}). \tag{5.22}$$

If this were true, then as κ is varied (rather than β) the polymer makes a gradual crossover from collapsed to diffusive behavior. This is in contrast to $d \geq 2$, where we know from the observations made in Section 5.2, Extension (3), that the polymer makes a sharp crossover at $\kappa = 1$ when $\beta > \beta_c^*$. We already know from (3.75–3.76) that $\nu(\kappa) = 1 - \tfrac{1}{3}\kappa$ for $\beta > 0$ and $\kappa \in (0, \tfrac{3}{2})$.

6

Polymer Collapse

In this chapter we look at the soft polymer (studied in Chapters 3–4) and add to the penalty for self-intersections a reward for self-touchings, i.e., a negative energy is associated with contacts between any two monomers that are not connected to each other within the polymer chain. This is a model of a *polymer in a poor solvent*: when the polymer does not like to make contact with a solvent it is immersed in, it tries to make contact with itself in order to push out the solvent. An example is polystyrene dissolved in cyclohexane. At temperatures above 35 degrees Celsius the cyclohexane is a good solvent, at temperatures below 30 it is a poor solvent. When cooling down, the polystyrene undergoes a transition from a "random coil" to a "compact ball" (see e.g. S.-T. Sun, I. Nishio, G. Swislow and T. Tanaka [287] and S.F. Sun, C.-C. Chou and R.A. Nash [286]).

In Section 6.1 we consider an undirected model, in Section 6.2 a directed model. We will see that there are three phases: *extended, collapsed* and *localized*. The transition between the last two phases is relatively well understood, the transition between the first two phases is not. The localized phase is not present in the case of SAW (infinite self-repellence).

In Section 4.7, Extension (7), and Section 5.3, Extension (3), we already encountered a polymer with an attractive interaction. In the present chapter we are facing a model with mixed interactions, both repulsive and attractive. Due to the presence of the latter, the lace expansion method does not work without modification. We will use *local times* and *generating functions* instead.

6.1 An Undirected Polymer in a Poor Solvent

Our choice for the set of paths and for the Hamiltonian is

$$\mathcal{W}_n = \left\{ w = (w_i)_{i=0}^n \in (\mathbb{Z}^d)^{n+1} \colon w_0 = 0, \ \|w_{i+1} - w_i\| = 1 \ \forall \, 0 \le i < n \right\},$$
$$H_n(w) = \beta I_n(w) - \gamma J_n(w),$$

$$(6.1)$$

F. den Hollander, *Random Polymers*,
Lecture Notes in Mathematics 1974, DOI: 10.1007/978-3-642-00333-2_6,
© Springer-Verlag Berlin Heidelberg 2009, Reprint by Springer-Verlag Berlin Heidelberg 2012

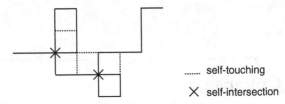

Fig. 6.1. A polymer with self-intersections and self-touchings.

where $\beta, \gamma \in (0, \infty)$ and

$$
\begin{aligned}
I_n(w) &= \sum_{\substack{i,j=0 \\ i<j}}^{n} 1_{\{\|w_i - w_j\| = 0\}}, \\
J_n(w) &= \frac{1}{2d} \sum_{\substack{i,j=0 \\ i<j-1}}^{n} 1_{\{\|w_i - w_j\| = 1\}},
\end{aligned}
\tag{6.2}
$$

are the number of *self-intersections*, respectively, the number of *self-touchings* (divided by $2d$ to account for the $2d$ possible directions). In this model, the polymer suffers on-site repulsion and off-site attraction: each self-intersection contributes an energy β and is penalized by a factor $e^{-\beta}$, while each self-touching contributes an energy $-\gamma$ and is rewarded by a factor e^{γ} (see Fig. 6.1).

Our choice of the path measure in (1.2) is

$$
P_n^{\beta,\gamma}(w) = \frac{1}{Z_n^{\beta,\gamma}} e^{-\beta I_n(w) + \gamma J_n(w)} P_n(w), \qquad w \in \mathcal{W}_n, \tag{6.3}
$$

where P_n is the law of SRW. In what follows we will drop the orderings $i < j$ and $i < j - 1$ from (6.2), as in (3.8). This amounts to replacing β, γ by $2\beta, 2\gamma$ in the Hamiltonian and adding to it a constant term $\beta(n+1) - \gamma n$. The advantage of dropping the ordering is that we can write

$$
\begin{aligned}
H_n(w) &= \beta \widehat{I}_n(w) - \gamma \widehat{J}_n(w), \\
\widehat{I}_n(w) &= \sum_{x \in \mathbb{Z}^d} \ell_n(x)^2, \\
\widehat{J}_n(w) &= \frac{1}{2d} \sum_{\substack{x \in \mathbb{Z}^d \\ \|e\|=1}} \ell_n(x)\ell_n(x+e),
\end{aligned}
\tag{6.4}
$$

which brings us back to the *local times* that played such an important role in Chapter 3. We henceforth write $I_n(w), J_n(w)$ again, dropping the overscript.

We will prove the following two theorems, both of which are due to van der Hofstad and Klenke [163]. For $L \in \mathbb{N}$, abbreviate $\Lambda_L = [-L, L]^d \cap \mathbb{Z}^d$.

Theorem 6.1. *If $\gamma < \beta$, then the polymer is minimally extended, i.e., there exists an $\epsilon_0 = \epsilon_0(\beta, \gamma) > 0$ such that for all $0 < \epsilon \leq \epsilon_0$ there exists a $c = c(\beta, \gamma, \epsilon) > 0$ such that*

$$P_n^{\beta,\gamma}\left(S_i \in \Lambda_{\lceil \epsilon n^{1/d} \rceil} \ \forall 0 \leq i \leq n\right) \leq e^{-cn} \qquad \forall n \in \mathbb{N}. \tag{6.5}$$

Theorem 6.2. *If $\gamma > \beta$, then the polymer is exponentially localized, i.e., there exist $c = c(\beta, \gamma) > 0$ and $L_0 = L_0(\beta, \gamma) \in \mathbb{N}$ such that*

$$P_n^{\beta,\gamma}\left(S_i \in \Lambda_L \ \forall 0 \leq i \leq n\right) \geq 1 - e^{-cLn} \qquad \forall n \in \mathbb{N}, \ L \geq L_0. \tag{6.6}$$

Thus, at $\gamma = \beta$ a transition takes place from a phase in which the polymer is spread out over a box of volume at least order n to a phase in which it is confined to a finite box.

The proofs of Theorems 6.1 and 6.2 are given in Sections 6.1.1 and 6.1.2. We will see that the key ingredients in the proofs are the following properties for the partition sum:

$$\begin{aligned}
\gamma < \beta: \quad & \limsup_{n \to \infty} \frac{1}{n} \log Z_n^{\beta,\gamma} \in (-\infty, 0), \\
\gamma > \beta: \quad & \liminf_{n \to \infty} \frac{1}{n^2} \log Z_n^{\beta,\gamma} \in (0, \infty).
\end{aligned} \tag{6.7}$$

Note that existence of the limits does not follow from subadditivity arguments, because of the mixture of repellent and attractive interactions (recall Section 1.3).

6.1.1 The Minimally Extended Phase

Proof. Fix $n \in \mathbb{N}$ and $\gamma < \beta$. From (6.4), we have

$$\begin{aligned}
H_n(w) &= \beta I_n(w) - \gamma J_n(w) \\
&= (\beta - \gamma) \sum_{x \in \mathbb{Z}^d} \ell_n(x)^2 + \frac{\gamma}{4d} \sum_{\substack{x \in \mathbb{Z}^d \\ \|e\| = 1}} [\ell_n(x) - \ell_n(x + e)]^2 \\
&\geq (\beta - \gamma) \sum_{x \in \mathbb{Z}^d} \ell_n(x)^2.
\end{aligned} \tag{6.8}$$

For $L \in \mathbb{N}$, let

$$\mathcal{C}_{n,L} = \{w \in \mathcal{W}_n : w_i \in \Lambda_L \ \forall 0 \leq i \leq n\} \tag{6.9}$$

be the event that the path stays confined to Λ_L up to time n. Since $\sum_{x \in \mathbb{Z}^d} \ell_n(x) = n + 1$, it follows from (6.8) that

$$H_n(w) \geq (\beta - \gamma)(n + 1)^2 |\Lambda_L|^{-1} \qquad \forall w \in \mathcal{C}_{n,L}. \tag{6.10}$$

On the other hand, writing w^* for the path that makes n steps in the 1-st direction, we have from (6.1) and (6.3–6.4) that

$$Z_n^{\beta,\gamma} \geq \frac{1}{(2d)^n} e^{-H_n(w^*)}$$

$$= \frac{1}{(2d)^n} e^{-\beta(n+1)+(\gamma/2d)[2+2(n-1)]} \tag{6.11}$$

$$= e^{-\beta} \left[\frac{1}{2d} e^{-\beta+(\gamma/d)} \right]^n.$$

Combining (6.3) and (6.10–6.11) with the bound $|\mathcal{C}_{n,L}| \leq (2d)^n$, we obtain

$$P_n^{\beta,\gamma}(\mathcal{C}_{n,L}) \leq e^{-(\beta-\gamma)(n+1)^2 |\Lambda_L|^{-1}} e^{\beta} \left[2d\, e^{\beta-(\gamma/d)} \right]^n \leq e^{-cn}, \tag{6.12}$$

with the last inequality being true for all $L \geq L_n = \lceil \epsilon n^{1/d} \rceil$ and $0 < \epsilon \leq \epsilon_0$, for some $\epsilon_0 = \epsilon_0(\beta,\gamma) > 0$ and $c = c(\beta,\gamma,\epsilon) > 0$. This proves (6.5). □

Note that $Z_n^{\beta,\gamma} \leq Z_n^{\beta-\gamma,0}$ by (6.8), while subadditivity gives

$$\lim_{n\to\infty} \frac{1}{n} \log Z_n^{\beta-\gamma,0} = \inf_{n\in\mathbb{N}} \frac{1}{n} \log Z_n^{\beta-\gamma,0} \leq \frac{1}{2} \log Z_2^{\beta-\gamma,0}$$

$$= \frac{1}{2} \log\left[\left(1 - \frac{1}{2d}\right) + \frac{1}{2d} e^{-(\beta-\gamma)} \right] < 0. \tag{6.13}$$

This explains the first line of (6.7).

6.1.2 The Localized Phase

Proof. The proof uses two lemmas. The first is an estimate of the minimum of the Hamiltonian, the second says that an energetic penalty of order Ln is associated with leaving the box Λ_L prior to time n, provided L is large enough to contain a minimizing path for the Hamiltonian.

For $n, L \in \mathbb{N}$, define

$$M_n = M_n(\beta,\gamma) = \min_{w\in\mathcal{W}_n} H_n(w),$$

$$\mathcal{W}_n \setminus \mathcal{C}_{n,L} = \{w \in \mathcal{W}_n : w_i \notin \Lambda_L \text{ for some } 0 \leq i \leq n\}. \tag{6.14}$$

Lemma 6.3. *If $\gamma > \beta$, then there exist $a = a(\beta,\gamma) > 0$ and $b = b(\beta,\gamma) > 0$ such that $-bn \leq M_n + an^2 \leq bn$ for all $n \in \mathbb{N}$.*

Lemma 6.4. *If $\gamma > \beta$, then $H_n(w) \geq M_n + \frac{1}{6}aLn$ for all $w \in \mathcal{W}_n \setminus \mathcal{C}_{n,L}$ when L is large enough.*

Before giving the proof of these lemmas, we complete the proof of Theorem 6.2. Estimate

$$P_n^{\beta,\gamma}(\mathcal{W}_n \setminus \mathcal{C}_{n,L}) = \frac{1}{Z_n^{\beta,\gamma}} \sum_{w \in \mathcal{W}_n \setminus \mathcal{C}_{n,L}} e^{-H_n(w)} \frac{1}{(2d)^n}$$

$$\leq \left[\frac{1}{(2d)^n} e^{-M_n}\right]^{-1} e^{-M_n - \frac{1}{6}aLn} |\mathcal{W}_n \setminus \mathcal{C}_{n,L}| \frac{1}{(2d)^n} \quad (6.15)$$

$$\leq (2d)^{2n} e^{-\frac{1}{6}aLn}$$

$$\leq e^{-cLn},$$

where the first inequality uses Lemma 6.4 and the bound $Z_n^{\beta,\gamma} \geq (2d)^{-n}e^{-M_n}$, the second inequality uses the bound $|\mathcal{W}_n \setminus \mathcal{C}_{n,L}| \leq (2d)^n$, while the third inequality is true for some $c > 0$ provided $L \geq L_0$ with $L_0 = \frac{12}{a}\log(2d)$. $\square$

Note that $(2d)^{-n}e^{-M_n} \leq Z_n^{\beta,\gamma} \leq e^{-M_n}$, and so Lemma 6.3 gives

$$\lim_{n\to\infty} \frac{1}{n^2} \log Z_n^{\beta,\gamma} = a. \quad (6.16)$$

This explains the second line of (6.7).

Proof of Lemma 6.3

Proof. The idea is that the minimum in (6.14) is achieved by a path that stays inside Λ_L for some $L \in \mathbb{N}$ large enough, making a minimal number of self-intersections and a maximal number of self-touchings. To get a crude idea, consider a path w^* that fills Λ_L approximately uniformly. Then $[\ell_n(x)](w^*) \approx n/|\Lambda_L|$ for $x \in L$ and hence, by (6.4),

$$H_n(w^*) \approx \beta|\Lambda_L|\left(\frac{n}{|\Lambda_L|}\right)^2 - \frac{\gamma}{2d}\left[|\Lambda_L|2d + O(|\partial\Lambda_L|)\right]\left(\frac{n}{|\Lambda_L|}\right)^2$$

$$= -(\gamma - \beta)\frac{n^2}{|\Lambda_L|}\left[1 + O(|\partial\Lambda_L|/|\Lambda_L|)\right]. \quad (6.17)$$

By picking L large enough, we get $M_n \leq H_n(w^*) \leq -a_1n^2$ for some $a_1 > 0$. Conversely, since $\sum_{x \in \mathbb{Z}^d}[\ell_n(x)](w) = n + 1$ for all $w \in \mathcal{W}_n$, we see from (6.8) that $H_n(w) \geq -(\gamma-\beta)(n+1)^2$ for all $w \in \mathcal{W}_n$, and so we also get $M_n \geq -a_2n^2$ for some $a_2 > 0$. Thus, M_n indeed is of order $-n^2$.

It turns out that in a minimizing path the local times are not close to uniform on Λ_L, but rather have a certain profile that is given by a variational problem. To understand why, let $F = F^{\beta,\gamma}$ be the $\mathbb{Z}^d \times \mathbb{Z}^d$-matrix defined by

$$F(x,y) = \begin{cases} -\beta, & \text{if } \|x - y\| = 0, \\ \gamma/2d, & \text{if } \|x - y\| = 1, \\ 0 & \text{otherwise.} \end{cases} \quad (6.18)$$

Then the Hamiltonian in (6.4) can be written as the symmetric bilinear form $-H_n = \langle \ell_n, F\ell_n \rangle$ (i.e., the inner product of ℓ_n with $F\ell_n$). Define

$$[\tilde{\ell}_n](w) = [\ell_n](w) - \tfrac{1}{2}1_{w_0} - \tfrac{1}{2}1_{w_n}, \qquad (6.19)$$

i.e., we take away $\tfrac{1}{2}$ from the local times at the endpoints. Since $|F(x,y)| \leq \gamma$ for all $x, y \in \mathbb{Z}^d$ and $\sum_{x \in \mathbb{Z}^d} \tilde{\ell}_n(x) = n$, we have

$$-\gamma(2n+1) \leq H_n(w) + \langle \tilde{\ell}_n, F\tilde{\ell}_n \rangle \leq \gamma(2n+1). \qquad (6.20)$$

Now, let $\mathcal{F}$ denote the set of all probability measures on $\mathbb{Z}^d$ that may arise as limit points of the sequence $(\tfrac{1}{n}\tilde{\ell}_n)_{n \in \mathbb{N}}$. Then, clearly,

$$\frac{1}{n^2} \langle \tilde{\ell}_n, F\tilde{\ell}_n \rangle \leq a \quad \text{with } a = \max_{f \in \mathcal{F}} \langle f, Ff \rangle, \qquad (6.21)$$

and so (6.20) gives $-H_n(w) \leq an^2 + \gamma(2n+1)$ for all $w \in \mathcal{W}_n$. This proves the lower bound on M_n with $b = 3\gamma$. The upper bound on M_n follows by noting that for every $f \in \mathcal{F}$ and $n \in \mathbb{N}$ there exists a $w \in \mathcal{W}_n$ such that $\sum_{x \in \mathbb{Z}^d} |\tilde{\ell}_n(x) - nf(x)| \leq c$ for some $c > 0$. For details we refer to van der Hofstad and Klenke [163]. $\square$

It is easy to show that the variational formula for a in (6.21) has a maximizer with finite support. Uniqueness, however, is not obvious. In $d = 1$ the maximizer is known to be unique modulo translations and can be computed explicitly (van der Hofstad, Klenke and König [164]).

Proof of Lemma 6.4

Proof. The proof is based on a cutting argument. If $w \in \mathcal{W}_n \setminus \mathcal{C}_{n,L}$, then w leaves Λ_L. Cutting Λ_L by some hyperplane, we can look at the local times on the two sides of this hyperplane and estimate the loss associated with these local times not contributing to the self-touchings. Below we only give the main steps of the argument. For details we again refer to van der Hofstad and Klenke [163].

Fix $L \in 3\mathbb{N}$ and $w \in \mathcal{W}_n \setminus \mathcal{C}_{n,L}$. Without loss of generality we may assume that $w_i^1 > L$ for some $0 \leq i \leq n$. Consider the hyperplanes

$$\mathrm{HP}(y) = \{x \in \mathbb{Z}^d \colon x^1 = y\}, \qquad y \in \mathbb{Z}. \qquad (6.22)$$

Since $\sum_{x \in \mathbb{Z}^d} \tilde{\ell}_n(x) = n$, there exists a $y_0 \in (\tfrac{1}{3}L, \tfrac{2}{3}L) \cap \mathbb{N}$ such that

$$\sum_{x \in \mathrm{HP}(y_0-1) \cup \mathrm{HP}(y_0) \cup \mathrm{HP}(y_0+1)} \tilde{\ell}_n(x) \leq 3\,\frac{n}{\tfrac{1}{3}L-1} = \frac{9n}{L-3}. \qquad (6.23)$$

Decompose the path into pieces left of $\mathrm{HP}(y_0)$ and right of $\mathrm{HP}(y_0)$. These pieces give rise to their own local times (with the endpoint correction), which are denoted by $\tilde{\ell}_n^+$ and $\tilde{\ell}_n^-$, satisfying $\tilde{\ell}_n = \tilde{\ell}_n^+ + \tilde{\ell}_n^-$. By (6.20), we have

$$-H_n - \gamma(2n+1) \leq \langle \tilde{\ell}_n, F\tilde{\ell}_n \rangle = \langle \tilde{\ell}_n^+, F\tilde{\ell}_n^+ \rangle + \langle \tilde{\ell}_n^-, F\tilde{\ell}_n^- \rangle + 2\langle \tilde{\ell}_n^-, F\tilde{\ell}_n^+ \rangle, \quad (6.24)$$

where, by (6.21),

$$\langle \tilde{\ell}_n^+, F\tilde{\ell}_n^+ \rangle \le a(m^+)^2,$$
$$\langle \tilde{\ell}_n^-, F\tilde{\ell}_n^- \rangle \le a(m^-)^2, \tag{6.25}$$

with $m^+ = \sum_{x \in \mathbb{Z}^d} \tilde{\ell}_n^+(x)$ and $m^- = \sum_{x \in \mathbb{Z}^d} \tilde{\ell}_n^-(x)$. Moreover, we may estimate

$$2\langle \tilde{\ell}_n^-, F\tilde{\ell}_n^+ \rangle$$
$$= \frac{\gamma}{d} \sum_{x \in \mathrm{HP}(y_0)} [\tilde{\ell}_n^-(x - e^1)\tilde{\ell}_n^+(x) + \tilde{\ell}_n^-(x)\tilde{\ell}_n^+(x + e^1)]$$

$$\le \frac{\gamma}{d}\left[\sum_{x \in \mathrm{HP}(y_0-1)} \tilde{\ell}_n^-(x) \sum_{x \in \mathrm{HP}(y_0)} \tilde{\ell}_n^+(x) + \sum_{x \in \mathrm{HP}(y_0)} \tilde{\ell}_n^-(x) \sum_{x \in \mathrm{HP}(y_0+1)} \tilde{\ell}_n^+(x) \right]$$

$$\le \frac{\gamma}{d} \frac{20n}{L} (m^- \wedge m^+), \tag{6.26}$$

where in the last line we use (6.23) and note that $9n/(L - 3) \le 10n/L$ when $L \ge 30$. Now, $m^- \wedge m^+ \ge \frac{1}{3}L$ by construction and, since $m^- + m^+ = n$, we therefore have $m^- m^+ \ge (m^- \wedge m^+) \frac{1}{2}n \ge \frac{1}{6}Ln$. Consequently, combining Lemma 6.3 and (6.24–6.26), we get

$$-H_n - \gamma(2n + 1) \le a(m^+)^2 + a(m^-)^2 + \frac{20\gamma n}{dL}(m^- \wedge m^+)$$
$$= a(m^+ + m^-)^2 + \left[\frac{20\gamma n}{dL}(m^- \wedge m^+) - 2am^-m^+ \right]$$
$$\le -M_n + bn + \left(\frac{20\gamma}{dL} - a \right) n(m^- \wedge m^+). \tag{6.27}$$

The term between brackets in the last line is negative for L large enough, in which case we get

$$-H_n \le -M_n + bn + \left(\frac{20\gamma}{dL} - a \right) \frac{1}{3}Ln + \gamma(2n + 1). \tag{6.28}$$

The sum of the last three terms is $\le -\frac{1}{6}aLn$ for L large enough, and so we get the claim. $\square$

6.1.3 Conjectured Phase Diagram

It is conjectured in van der Hofstad and Klenke [163] that on the critical line $\gamma = \beta$ the polymer lives on scale $n^{1/(d+1)}$, more precisely,

$$\sum_{w \in \mathcal{W}_n} \|w_n\|^2 P_n^{\beta,\gamma}(w) \sim D n^{2/(d+1)} (\log n)^{-1/(d+1)} \tag{6.29}$$

for some $D > 0$. For $d = 1$, this conjecture is proved in van der Hofstad, Klenke and König [164].

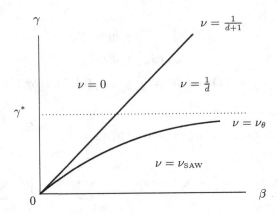

Fig. 6.2. Conjectured phase diagram and critical exponents for the polymer with self-repellence and self-attraction in $d \geq 2$. Only the existence of the critical line $\gamma = \beta$ and the critical exponent $\nu = 0$ in the region above the critical line have been proved. Below the critical line it is known that $\nu \geq 1/d$.

Theorem 6.1 says nothing about the scale of the polymer in the minimally extended phase other than that it is at least of order $n^{1/d}$. As explained in Brak, Owczarek and Prellberg [38], it is believed that there is a second phase transition at a critical value $\gamma_c = \gamma_c(\beta) < \beta$ at which the polymer moves from scale $n^{1/d}$ ("collapsed phase") to scale $n^{\nu_{\mathrm{SAW}}}$ ("extended phase"), with ν_{SAW} the critical exponent for SAW in (2.26). This second critical curve is believed to have a finite asymptote $\lim_{\beta \to \infty} \gamma_c(\beta) = \gamma^*$, the latter being the critical value for the SAW-version of the model with $\beta = \infty$ and $\gamma \in (0, \infty)$. At this critical curve, the critical exponent is believed to be (see B. Duplantier and H. Saleur [95], F. Seno and A. Stella [273])

$$\nu_\theta = \begin{cases} \frac{4}{7}, & \text{if } d = 2, \\ \frac{1}{2}, & \text{if } d \geq 3. \end{cases} \tag{6.30}$$

(The index θ is used here because in the physics literature the phase transition point for the SAW-version of the model is called the θ-point.)

Summarizing the above remarks, we get that the phase diagram for $d \geq 2$ is believed to look like Fig. 6.2. For $d = 1$ there is no collapse transition. Indeed, we know from Theorem 6.5 that below the critical line $\gamma = \beta$ the polymer moves past the point ϵn with high probability for ϵ small enough, so that its motion is ballistic like SAW in $d = 1$.

6.2 A Directed Polymer in a Poor Solvent

In this section we turn to a *directed* version of the model in order to get our hands on the collapse transition, which for the undirected model is conjectured to occur at the second critical curve in Fig. 6.2. Rather than looking at the mean-square displacement, which for the directed model is of less interest, we will prove the existence of a critical value $\gamma_c > 0$ for the free energy at $\beta = \infty$. The results described below are taken from Brak, Guttmann and Whittington [36].

Our choice for the set of paths and the Hamiltonian in (1.1) is

$$\mathcal{W}_n = \big\{ w = (w_i)_{i=0}^n \in (\mathbb{N}_0 \times \mathbb{Z})^{n+1} \colon w_0 = 0, \, w_1 - w_0 = \rightarrow,$$
$$w_{i+1} - w_i \in \{\uparrow, \downarrow, \rightarrow\} \, \forall 0 < i < n, \, w_i \neq w_j \, \forall 0 \leq i < j \leq n \big\},$$
$$H_n(w) = -\gamma J_n(w),$$

$$(6.31)$$

where $\uparrow$, $\downarrow$ and $\rightarrow$ denote steps between neighboring sites in the north, south and east direction, respectively, the first step is required to be east for later convenience, $\gamma \in \mathbb{R}$ and

$$J_n(w) = \sum_{\substack{i,j=0 \\ i < j-1}}^n 1_{\{\|w_i - w_j\| = 1\}}. \tag{6.32}$$

An example of a path in $\mathcal{W}_n$ is drawn in Fig. 6.3 (recall also Fig. 1.4). Our choice for the path measure in (1.2) is

$$P_n^\gamma(w) = \frac{1}{Z_n^\gamma} \, e^{\gamma J_n(w)}, \qquad w \in \mathcal{W}_n, \tag{6.33}$$

where we again drop the uniform reference measure P_n. Thus, each self-touching is weighted by a factor e^γ, and we allow for both positive (=attractive) and negative (=repulsive) values of γ. The case $\gamma = 0$ corresponds to the non-interacting directed random walk. Note that because the path is self-avoiding the law in (6.33) corresponds to the law in (6.3) with $\beta = \infty$.

In Section 6.2.1 we prove that there is a collapse transition at a critical value $\gamma_c > 0$. In Section 6.2.2 we look at the limiting number of self-touchings per monomer in the collapsed phase.

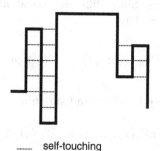

········ self-touching

Fig. 6.3. A directed SAW with self-touchings.

6.2.1 The Collapse Transition

The free energy of the directed polymer defined in (6.31–6.33) is given by (recall (1.6))

$$f(\gamma) = \lim_{n \to \infty} \frac{1}{n} \log Z_n^\gamma, \tag{6.34}$$

whenever the limit exists. The following theorem establishes existence and shows that there are two phases: a *collapsed* phase and an *extended* phase.

Theorem 6.5. *The free energy exists, is finite, and has a collapse transition at $\gamma_c = \log x_c$, with $x_c \approx 3.382975$ the unique positive solution of the cubic equation $x^3 - 3x^2 - x - 1 = 0$. The collapsed phase corresponds to $\gamma > \gamma_c$, the extended phase to $\gamma < \gamma_c$ (see Fig. 6.5).*

Proof. We begin by noting that the partition sum $Z_n^\gamma = \sum_{w \in \mathcal{W}_n} e^{-\gamma J_n(w)}$ can be written as the power series

$$Z_n(x) = \sum_{m=0}^{\infty} c_n(m) x^m, \quad x \in [0, \infty), \ n \in \mathbb{N}_0, \tag{6.35}$$

with $x = e^\gamma$ and

$$\begin{aligned} c_n(m) &= |\{w \in \mathcal{W}_n \colon J_n(w) = m\}| \\ &= \text{the number of } n\text{-step paths with } m \text{ self-touchings} \end{aligned} \tag{6.36}$$

$(c_0(0) = 1)$. For each n the sum in (6.35) is finite, because $c_n(m) = 0$ when $m \geq n$.

1. The existence of the free energy can be proved with the help of a standard subadditivity argument, as explained in Section 1.3. If we *concatenate* two paths, one with k steps and l self-touchings and one with $n - k$ steps and $m - l$ self-touchings – putting a single step to the east in between to guarantee that no self-touchings arise between the two paths – then we get a path with $n + 1$ steps and m self-touchings. Since different concatenated paths are obtained for different choices of l, we have

$$c_{n+1}(m) \geq \sum_{l=0}^{m} c_k(l) c_{n-k}(m - l) \qquad \forall\, k, m, n \in \mathbb{N}_0, \ k \leq n. \tag{6.37}$$

Multiplying this inequality by x^{m+1} and summing over m, we get

$$Z_{n+1}(x) \geq x Z_k(x) Z_{n-k}(x) \qquad \forall\, k, n \in \mathbb{N}_0, \ k \leq n. \tag{6.38}$$

Putting $\widetilde{Z}_{n+1}(x) = x Z_n(x)$, $n \in \mathbb{N}_0$, we get the inequality

$$\widetilde{Z}_{n+2}(x) \geq \widetilde{Z}_{k+1}(x) \widetilde{Z}_{n-k+1}(x) \qquad \forall\, k, n \in \mathbb{N}_0, \ k \leq n, \tag{6.39}$$

i.e., $n \mapsto \log \widetilde{Z}_n(x)$ is superadditive on $\mathbb{N}$. Consequently,

$$\widetilde{f}(x) = \lim_{n\to\infty} \frac{1}{n} \log \widetilde{Z}_n(x) \tag{6.40}$$

exists and equals

$$\widetilde{f}(x) = \sup_{n\in\mathbb{N}} \frac{1}{n} \log \widetilde{Z}_n(x) \in (-\infty, \infty]. \tag{6.41}$$

The free energy in (6.34) is $f(\gamma) = \widetilde{f}(e^{\gamma})$.

2. The finiteness of the free energy follows from the observation that $c_n(m) = 0$ for $m \geq n$ and $\sum_{m=0}^{\infty} c_n(m) \leq 3^n$, which give $\widetilde{Z}_{n+1}(x) = x Z_n(x) \leq [3(x \vee 1)]^n$ and $\widetilde{f}(x) \leq \log[3(x \vee 1)]$.

3. Lemma 6.6 below gives a closed form expression for the generating function

$$G = G(x, y) = \sum_{n=0}^{\infty} Z_n^{\gamma} y^n = \sum_{n=0}^{\infty} \sum_{m=0}^{n} c_n(m) x^m y^n, \quad x, y \in [0, \infty). \tag{6.42}$$

By analyzing the singularity structure of $G(x, y)$ we will be able to compute $f(\gamma)$. Indeed, our task will be to identify the critical curve $x \mapsto y_c(x)$ in the (x, y)-plane below which $G(x, y)$ has no singularities and on or above which it does, because this identifies the free energy as

$$f(\gamma) = -\log y_c(e^{\gamma}), \quad \gamma \in \mathbb{R}. \tag{6.43}$$

We will see that this curve has the shape given in Fig. 6.4.

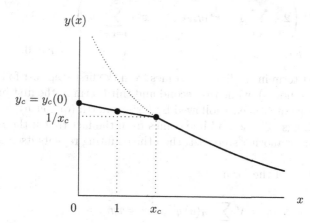

Fig. 6.4. The domain of convergence of the generating function $G(x, y)$ for the directed SAW with self-touchings lies below the solid curve. The dotted line and the piece of the solid curve on $[x_c, \infty)$ are the hyperbola $xy = 1$. The point x_c is identified with the collapse transition, because this is where the free energy $f(\gamma) = f(\log x)$ is non-analytic.

Lemma 6.6. *For $x, y \in [0, \infty)$ the generating function is given by the formal power series*

$$G(x, y) = -\frac{aH(x, y) - 2y^2}{bH(x, y) - 2y^2}, \tag{6.44}$$

where

$$a = y^2(2 + y - xy), \quad b = y^2(1 + x + y - xy), \quad H(x, y) = y\frac{g_{0,1}(x, y)}{g_{1,1}(x, y)}, \tag{6.45}$$

with

$$g_{r,1}(x, y) = y^r \left(1 + \sum_{k=1}^{\infty} \frac{(y - q)^k y^{2k} q^{\frac{1}{2}k(k+1)}}{\prod_{l=1}^{k}(yq^l - y)(yq^l - q)} q^{kr}\right), \quad q = xy, \; r = 1, 2. \tag{6.46}$$

Proof. For $r \in \mathbb{N}_0$, let $c_{n,r}(m)$ denote the number of n-step paths with m self-touchings that make precisely r steps either north or south immediately after the first step east. Let

$$g_r = g_r(x, y) = \sum_{n=0}^{\infty} \sum_{m=0}^{\infty} c_{n,r}(m) \, x^m y^n. \tag{6.47}$$

Then

$$G = G(x, y) = \sum_{r=0}^{\infty} g_r(x, y). \tag{6.48}$$

The generating functions g_r satisfy the recursion relation

$$g_r = y^{r+1}\left(2 + \sum_{s=0}^{r}(1 + x^s)g_s + (1 + x^r)\sum_{s=r+1}^{\infty} g_s\right), \quad r \in \mathbb{N}, \tag{6.49}$$

$$g_0 = y + yG, \qquad\qquad\qquad\qquad\qquad\qquad\qquad r = 0.$$

Indeed, the first term in the first line comes from the first step east followed by r steps south (or north), while the second and third term in the first line come from the subsequent step east followed by k steps north (or south), leading to $k \wedge r$ self-touchings. The second line comes from the fact that if the first step east is followed by another step east, then the counting repeats itself after the first step.

Try a solution of the form

$$g_r = \lambda^r \sum_{l=0}^{\infty} p_l(q) \, q^{lr}, \quad q = xy, \, r \in \mathbb{N}, \tag{6.50}$$

with $p_0(q) = 1$. Substitution of (6.50) into the first line of (6.49) gives

$$p_l(q) = \frac{\lambda \, (y - q) \, yq^l}{(\lambda q^l - y)(\lambda q^l - q)} p_{l-1}(q), \quad l \in \mathbb{N}, \tag{6.51}$$

provided λ solves $\lambda^2 - (y+q)\lambda + yq = 0$, i.e.,

$$\lambda \in \{\lambda_1, \lambda_2\} = \{y, q\}. \tag{6.52}$$

Thus, the general solution of (6.49) is of the form

$$g_r = C_1 g_{r,1} + C_2 g_{r,2}, \quad r \in \mathbb{N}, \tag{6.53}$$

with C_1, C_2 arbitrary functions of y, q, to be determined by the boundary conditions, and

$$g_{r,i} = (\lambda_i)^r \left(1 + \sum_{k=1}^{\infty} \frac{(\lambda_i)^k (y-q)^k y^k q^{\frac{1}{2}k(k+1)}}{\prod_{l=1}^{k}(\lambda_i q^l - y)(\lambda_i q^l - q)} q^{kr} \right), \quad i = 1, 2, r \in \mathbb{N}_0. \tag{6.54}$$

Next, for $x > 1$ and $0 < y < 1$ such that $xy < 1/(1+\sqrt{2})$, we have

$$\lim_{r \to \infty} q^{-r} g_{r,1} = 0, \quad \lim_{r \to \infty} q^{-r} g_{r,2} = 1, \quad \lim_{r \to \infty} q^{-r} g_r = 0. \tag{6.55}$$

Indeed, the first two limits follow straight from (6.54), while the last limit follows from monotonicity in x, y and the observation that $f(0) = \log(1+\sqrt{2})$. Hence $C_2 = 0$, and we are left with

$$g_r = C_1 g_{r,1}, \quad r \in \mathbb{N}. \tag{6.56}$$

To determine C_1, we note that

$$g_0 = y + yG = \tfrac{1}{2} C_1 g_{0,1}, \qquad g_1 = a + bG = C_1 g_{1,1}, \tag{6.57}$$

with a and b given by (6.45). The first identity follows from (6.49), because for $r = 0$ the right-hand side of the first line of (6.49) reduces to $2(y + yG)$ and equals $C_1 g_{0,1}$ by the above argument (note that (6.54) includes $r = 0$). The second identity follows from the first line of (6.49) for $r = 1$, giving $g_1 = y^2(2 + 2g_0 + (1+x)(G - g_0)$. Solving (6.57) for G, we get the expression in (6.44). $\square$

4. The function $G(x, y)$ in (6.44) can be analyzed via a closer study of the function $H(x, y)$ in (6.45-6.46). The latter is a quotient of two q-hypergeometric functions that – as shown in Brak, Guttmann and Whittington [36] – can be expressed in the form of a *continued fraction*. The advantage is that this continued fraction is easy to treat numerically. The analysis of the continued fraction representation was subsequently refined in Owczarek, Prellberg and Brak [259]. It turns out that $G(x, y)$ has no zeroes in the denominator when $x > x_c$ and $0 \leq y \leq 1/x$, with x_c the unique positive solution of the cubic equation $x^3 - 3x^2 - x - 1 = 0$, i.e., $x_c \approx 3.382975$. The denominator has three zeroes:

- $(x, y) = (0, y_c)$, with y_c the solution of the cubic equation $1 - 2y - y^3 = 0$, i.e., $y_c \approx 0.453397$.
- $(x, y) = (1, -1 \pm \sqrt{2})$.

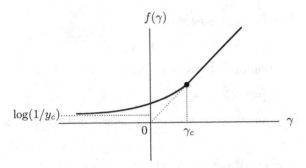

Fig. 6.5. Plot of the free energy per monomer.

Two of these zeroes are indicated in Fig. 6.4. In this way, three points are identified on the piece of the curve in Fig. 6.4 on $[0, x_c]$. The rest of this piece follows from a numerical analysis of the continued fraction representation of $H(x, y)$. We refer to [36] and [259] for details.

Fig. 6.4 yields the free energy via (6.43). This completes the proof of Theorem 6.5. $\square$

By the convexity of the free energy as a function of $\log x$, it follows that the curve in Fig. 6.4 is continuous on $(0, \infty)$. This curve has not been shown to be analytic on $(0, x_c)$, although this is expected to be true.

The regime $x > 1$ corresponds to attraction, the regime $0 \leq x < 1$ to repulsion. The collapse transition occurs at x_c, corresponding to $\gamma_c = \log x_c$. Note that (see Fig. 6.5)

$$f(\gamma) \begin{cases} = \gamma, & \text{if } \gamma \geq \gamma_c, \\ > \gamma, & \text{if } \gamma < \gamma_c. \end{cases} \tag{6.58}$$

6.2.2 Properties of the Two Phases

As in (1.14), the derivative of the free energy, whenever it exists, is the limiting number of self-touchings per monomer:

$$f'(\gamma) = \lim_{n \to \infty} \frac{1}{n} \sum_{w \in \mathcal{W}_n} J_n(w) \, P_n^\gamma(w). \tag{6.59}$$

Theorem 6.7. *The limiting number of self-touchings per monomer is 1 in the collapsed phase and < 1 in the extended phase. The phase transition is second order.*

Proof. A closer analysis of the curve in Fig. 6.4 based on Lemma 6.6, i.e., on the identification of $G(x, y)$ in (6.44) and the continued fraction representation of $H(x, y)$ in (6.45), reveals that

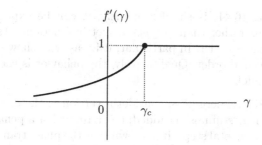

Fig. 6.6. Plot of the number of self-touchings per monomer.

$$f'(\gamma) \begin{cases} = 1, & \text{if } \gamma \geq \gamma_c, \\ < 1, & \text{if } \gamma < \gamma_c, \end{cases} \tag{6.60}$$

with $\gamma \mapsto f'(\gamma)$ continuous at γ_c (see Fig. 6.6). We again refer to [36] and [259] for details. □

The case $\gamma = -\infty$ ($x = 0$) corresponds to the directed polymer avoiding self-touchings altogether, and $\log(1/y_c)$ is its entropy per step. In the collapsed phase the entropy per step is zero, in the extended phase it is strictly positive.

6.3 Extensions

(1) Van der Hofstad and Klenke [163] show that for the model in Section 6.1 a central limit holds when $d = 1$ and $\gamma \leq \beta - \frac{1}{2}\log 2$, i.e., under $P_n^{\beta,\gamma}$ as $n \to \infty$,

$$\frac{|S_n| - \theta^* n}{\sigma^* \sqrt{n}} \Longrightarrow Z, \tag{6.61}$$

where Z is a standard normal random variable, and $\theta^* = \theta^*(\beta, \gamma) \in (0, 1]$ and $\sigma^* = \sigma^*(\beta, \gamma) \in (0, \infty)$ are the speed, respectively, the spread of the polymer. This is the analogue of Extension (1) in Section 3.6 for the soft polymer.

(2) Van der Hofstad, den Hollander and König [161] study the weak interaction limit in $d = 1$ of the model in Section 6.1, i.e., β, γ in (6.1–6.3) are replaced by β_n, γ_n tending to zero as $n \to \infty$. Provided the self-repellence dominates, the behavior is shown to be similar as that of the soft polymer treated in Chapter 3. Compare with Section 3.6, Extension (3).

(3) Owczarek, Prellberg and Brak [259] consider a version of the directed model in Section 6.2 in which the horizontal steps take discrete values, as before, but the vertical steps take continuous values. The number of self-touchings is replaced by the total length where vertical pieces lie next to each other (compare with Fig. 6.3). It turns out that the corresponding generating

function replacing (6.44) is simpler, namely, it can be expressed in terms of Bessel functions rather than q-hypergeometric functions. As a result, the singularity analysis is easier. In particular, it is easier to show that the phase transition is of second order. Qualitatively, the behavior is the same as that of the original model.

(4) Owczarek and Prellberg [258] consider a version of the directed model in Section 6.2 to which "stiffness" is added, i.e., a reward or a penalty is given to two consecutive horizontal steps. It is shown that the phase transition becomes first order in the case of positive stiffness (=reward), but remains second order in the case of negative stiffness (=penalty).

(5) Tesi, Janse van Rensburg, Orlandini and Whittington [289], [290] perform a Monte Carlo study of SAWs in $d = 3$ with self-attraction. They find that $\gamma_c \in [0.274, 0.282]$ and $\nu_\theta \in [0.48, 0.50]$. In [290] it is shown that for SAWs the free energy exists for $\gamma \leq 0$, while this remains open for $\gamma > 0$. On the other hand, it is shown that for self-avoiding polygons the free energy exists for all $\gamma \in \mathbb{R}$. For $\gamma \leq 0$ the two free energies are shown to coincide. This is expected to be the case also for $\gamma > 0$, as the numerical estimates indicate.

Janse van Rensburg, Orlandini, Tesi and Whittington [193] consider self-avoiding polygons consisting of a random concatenation of different types of monomers, allowing for different interaction strengths between different pairs of monomers. The free energy is shown to exist and to be self-averaging, i.e., to be sample-independent. The proof requires a delicate concatenation technique, in which the randomness of the polymer is changed along. Polymers with a random interaction will be treated in Part B of the present monograph.

(6) Kumar, Jensen, Jacobsen and Guttmann [222] look at a two-dimensional version of the polymer with self-attraction in which the path is a SAW constrained to be a bridge, i.e., constrained to lie between the vertical lines through its endpoints, and a force $\phi \in (0, \infty)$ is applied to the right endpoint of the path in the positive vertical direction. This force is modelled by adding a term $-\phi w_n$ to the Hamiltonian: ϕw_n is the work exerted by the force to move the right endpoint a distance w_n away from the left endpoint. With the help of exact enumeration (up to $n = 55$), combined with extrapolation techniques (Guttmann [137]), the "force-temperature diagram" is computed, showing for what values of ϕ and γ the polymer is collapsed, extended or stretched, with the mean-square displacement being of order $n^{1/d}$, n^ν and n, respectively, with ν the critical exponent of SAW. The model is a caricature for folding-unfolding transitions in proteins. We will return to polymers and forces in Section 7.3: see, in particular, (7.62) and Figs. 7.8–7.11. See Haupt, Senden and Sevick [150], Gunari, Balasz and Walker [135], and Gunari and Walker [136] for experiments on a force applied to a polymer in a poor solvent.

(7) Ioffe and Velenik [184] consider a version of the model in Section 6.1 in which the Hamiltonian takes the form (compare with (6.4))

$$H_n^{\psi,\phi}(w) = \sum_{x \in \mathbb{Z}^d} \psi(\ell_n(x)) - (\phi, w_n), \tag{6.62}$$

where $\psi \colon \mathbb{N}_0 \to [0, \infty)$ is non-decreasing with $\psi(0) = 0$, and $\phi \in \mathbb{R}^d$ is a force acting on the endpoint of the polymer. The reference measure P_n in the path measure $P_n^{\psi,\phi} = (Z_n^{\psi,\phi})^{-1} \exp[-H_n^{\psi,\phi}] P_n$ can be either SRW or a random walk with drift. Two cases are considered: (1) ψ is superlinear (repulsive interaction); (2) ψ is sublinear (attractive interaction), with the additional assumption that $\lim_{\ell \to \infty} \psi(\ell)/\ell = 0$. It is shown that in the attractive case there is a convex set $K = K(\psi) \subset \mathbb{R}^d$, with $\mathrm{int}(K) \ni 0$, such that the polymer is in a collapsed phase when $\phi \in \mathrm{int}(K)$ and in an extended phase when $\phi \notin K$. It is further shown that in the repulsive case the polymer is in an extended phase for all $\phi \in \mathbb{R}^d$. With the help of a coarse-graining technique it is shown that in the extended phase the transversal fluctuations of the polymer (in the direction perpendicular to ϕ or to the drift of the reference random walk) are diffusive.

6.4 Challenges

(1) Extend the central limit theorem in (6.61) for $d = 1$ to the full minimally extended phase, i.e., $\gamma < \beta$. Subsequently, extend the central limit theorem to an LDP, analogous to the LDP derived in Chapter 3 for the soft polymer.

(2) Show that $\theta^*(\beta, \gamma)$ in (6.61) is increasing in β and deceasing in γ, with $\lim_{\gamma \uparrow \beta} \theta^*(\beta, \gamma) = 0$ for all $\beta \in (0, \infty)$. This is the analogue of Challenge (1) in Section 3.7 for the soft polymer.

(3) As indicated in Fig. 6.2, show that $\nu = \frac{1}{d+1}$ at the critical curve $\gamma = \beta$ for $d \geq 2$.

(4) Prove that the second critical curve in Fig. 6.2 exists and has the qualitative shape that is drawn, and show that the critical exponents on either side take the values as indicated. For $2 \leq d \leq 4$ this is a very serious challenge indeed, because little is known rigorously in these dimensions even for the soft polymer. But for $d \geq 5$ there is some hope, because the critical exponents are known for the soft polymer (see Chapter 4). The lace expansion does not work without modification because of the presence of attractive interactions. Recall the remark made at the beginning of Section 4.4. Recall also Section 4.7, Extension (7).

(5) Write down a continuum version of the model defined in (6.1–6.3) and investigate its behavior. For $d = 1$ this is in the spirit of Extension (5) in Section 3.6. What about $d = 2, 3$, in the spirit of Challenge (3) in Section 4.8?

84   6 Polymer Collapse

(6) Study the version of the directed model in Section 6.2 in which the vertical steps take values in $\mathbb{Z}^d$ rather than in $\mathbb{Z}$, i.e., a $(1+d)$-dimensional directed SAW. Does the model have the same qualitative behavior in $d \geq 2$ as in $d = 1$?

(7) What is the order of the phase transition between the collapsed phase and the extended phase in the model mentioned in Section 6.3, Extension (7)? Does the order depend on d?

7

Polymer Adsorption

In this chapter we study a polymer in the vicinity of a *linear substrate*. Each monomer on the substrate feels a *binding energy*, resulting in an attractive interaction between the polymer and the substrate. We will consider the two situations where the substrate is:

(1) *penetrable* ('pinning at an interface"),
(2) *impenetrable* ("wetting of a surface").

Our focus will be on the free energy and on the occurrence of a phase transition between a *localized phase*, where the polymer stays close to the substrate, and a *delocalized phase*, where it wanders away from the substrate (see Fig. 7.1). The main technical tools that we will use are *generating functions* and the *method of excursions*. In Sections 7.1–7.4 we focus on directed SAW's, in Section 7.5 on undirected SAW's.

Early references on pinning and wetting of polymers (both for directed and undirected random walk models) are Rubin [272], Burkhardt [47], van Leeuwen and Hilhorst [228], Fisher [98], Privman, Forgacs and Frisch [269], Carvalho and Privman [63], Forgacs, Privman and Frisch [105], and an overview by De'Bell and Lookman [81]. Early results on scaling properties were obtained by Brak, Essam and Owczarek [35] (for ballot paths; recall Fig. 1.4), Whittington [315], Isozaki and Yoshida [186] (for generalized ballot paths), and Roynette, Vallois and Yor [271] (for Brownian paths). More recent references will be mentioned as we go along.

Pinning and wetting may occur not only when the polymer is attracted by the substrate, but also when the polymer is repelled by the solvent(s) around the substrate. Alternatively, the polymer may sit at an interface between two incompatible solvents to reduce the interfacial tension. The monograph by Fleer, Cohen Stuart, Cosgrove, Scheutjens and Vincent [99] describes a wealth of experiments with polymers at interfaces.

In Section 7.1 we look at a directed model for the "pinned polymer", compute the free energy, derive path properties in the two phases, and identify the order of the phase transition. In Section 7.2 we do the same for the "wetted

F. den Hollander, *Random Polymers*,
Lecture Notes in Mathematics 1974, DOI: 10.1007/978-3-642-00333-2_7,
© Springer-Verlag Berlin Heidelberg 2009, Reprint by Springer-Verlag Berlin Heidelberg 2012

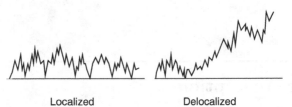

Localized   Delocalized

Fig. 7.1. The localized and delocalized phase for the wetted polymer.

polymer", exploiting the fact that the mathematics is essentially the same. In Section 7.3 we look at the pinned and the wetted polymer when a force is applied to its right endpoint perpendicular to and away from the substrate (e.g. with the help of optical tweezers or atomic force microscopy), we compute the free energy and compute the force that is needed to pull the polymer off the substrate when the interaction parameters are in the localized phase. In Section 7.4 we look at a polymer in a slit between two impenetrable substrates. We show that such a polymer exerts an effective force on the two substrates, which is repulsive or attractive depending on the relative strengths of the adsorption energies at the two walls. In Section 7.5, finally, we describe what is known about the pinned and the wetted polymer when it is undirected, for which the analysis is much harder.

We refer to Giacomin [116], Chapter 2, for an account of the main mathematical ingredients of directed polymer pinning and wetting via the *method of excursions*. Much of what is written in Sections 7.1–7.3 follows the arguments put forward there.

7.1 A Polymer Near a Linear Penetrable Substrate: Pinning

In this section we consider the model where the set of paths and the Hamiltonian are

$$\mathcal{W}_n = \left\{ w = (i, w_i)_{i=0}^n \colon w_0 = 0, \, w_i \in \mathbb{Z} \, \forall \, 0 \le i \le n \right\},$$
$$H_n(w) = -\zeta K_n(w),$$
(7.1)

with $\zeta \in \mathbb{R}$ and

$$K_n(w) = \sum_{i=1}^n 1_{\{w_i = 0\}}, \qquad w \in \mathcal{W}_n.$$
(7.2)

In words, $\mathcal{W}_n$ denotes the set of all n-step directed paths in $\mathbb{N}_0 \times \mathbb{Z}$ starting from the origin (see Fig. 1.4 for the special cases of ballot paths and generalized ballot paths), and $H_n(w)$ assigns an energy $-\zeta$ to each visit of w to $\mathbb{N} \times \{0\}$. The path measure is

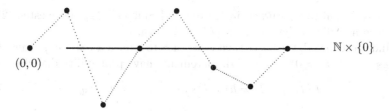

Fig. 7.2. A 7-step path in $\mathbb{N}_0 \times \mathbb{Z}$ that makes 2 visits to $\mathbb{N} \times \{0\}$.

$$P_n^\zeta(w) = \frac{1}{Z_n^\zeta} e^{\zeta K_n(w)} P_n(w), \qquad w \in \mathcal{W}_n, \tag{7.3}$$

where

- P_n is the projection onto $\mathcal{W}_n$ of the path measure P of a directed irreducible *random walk*, i.e., under P the increments $w_{i+1} - w_i$, $i \in \mathbb{N}_0$, are i.i.d. $\mathbb{Z}$-valued with a common law whose support generates $\mathbb{Z}$.

The choice in (7.1–7.3) models a two-dimensional directed polymer in $\mathbb{N}_0 \times \mathbb{Z}$ where each visit to the substrate $\mathbb{N} \times \{0\}$ contributes a weight factor e^ζ, which is a reward when $\zeta > 0$ and a penalty when $\zeta < 0$ (see Fig. 7.2).

Let $S = (S_i)_{i \in \mathbb{N}_0}$ denote the random walk with law P starting from $S_0 = 0$. For $k \in \mathbb{N}$, define

$$\begin{aligned} a(k) &= P(S_i \neq 0 \;\forall\, 1 \leq i \leq k), \\ b(k) &= P(S_i \neq 0 \;\forall\, 1 \leq i < k,\, S_k = 0). \end{aligned} \tag{7.4}$$

Throughout the sequel we will assume that $\sum_{k \in \mathbb{N}} b(k) = 1$, i.e., S is a recurrent random walk under the law P. In addition, we will assume that $b(\cdot)$ is *regularly varying at infinity*, i.e.,

$$b(k) \sim k^{-1-a} L(k) \quad \text{as } k \to \infty \text{ through } \Lambda \tag{7.5}$$

for some $a \in (0, \infty)$ and some slowly varying function L, where $\Lambda = \{k \in \mathbb{N}: b(k) > 0\}$. Since $a(k) = \sum_{l=k+1}^\infty b(l)$, $k \in \mathbb{N}$, it follows from (7.5) that

$$a(k) \sim \frac{1}{a} k^{-a} L(k) \quad \text{as } k \to \infty. \tag{7.6}$$

Note that (7.5–7.6) imply that $a(k)/b(k) \leq Ck$, $k \in \Lambda$, for some $C < \infty$.

For SRW we have $\Lambda = 2\mathbb{N}$, $a = \frac{1}{2}$ and L constant (Spitzer [284], Section 1). To exhibit another example, take the random walk whose increments have distribution $P(S_1 = x) = |x|^{-1-\chi}/2\zeta(1+\chi)$, $x \in \mathbb{Z} \setminus \{0\}$, with ζ the Riemann zeta-function and $\chi \in (0, \infty)$. This random walk has $\Lambda = \mathbb{N} \setminus \{1\}$ and is recurrent when $\chi \in [1, \infty)$ (Spitzer [284], Section 8). Moreover, if $\chi \in (1, \infty)$, then $a = \frac{1}{2} \wedge [1 - (1/\chi)]$ and L is constant, except when $\chi = 2$, in which case L is logarithmic. This is an example of a much larger class of random walks with polynomial tails, studied in Kesten and Spitzer [212] and Kesten [210], for which a scaling behavior as in (7.5) is known.

Under the law P, the successive visits of S to 0 form a *renewal process* $T = (T_l)_{l \in \mathbb{N}_0}$ (with $T_0 = 0$) whose i.i.d. increments have probability distribution

$$P(T_{l+1} - T_l = k) = b(k), \qquad k \in \mathbb{N}, \, l \in \mathbb{N}_0. \tag{7.7}$$

This fact will drive the computations. Indeed, we will see that the results below are valid for general renewals not necessarily arising from random walk, as long as (7.5) is in force. For convenience, in the proofs of the theorems to be stated below we will assume that $\Lambda = \mathbb{Z}$. The proofs are easily adapted to the situation where the support of $b(\cdot)$ is smaller.

7.1.1 Free Energy

Our starting point is the existence of the free energy.

Theorem 7.1. $f(\zeta) = \lim_{n \to \infty} \frac{1}{n} \log Z_n^\zeta$ *exist for all* $\zeta \in \mathbb{R}$ *and is non-negative.*

Proof. Let

$$Z_n^{\zeta,0} = E \left(\exp \left[\zeta \sum_{i=1}^n 1_{\{S_i = 0\}} \right] 1_{\{S_n = 0\}} \right) \tag{7.8}$$

be the partition sum for the polymer constrained to end at the substrate. We begin by showing that there exists a $C < \infty$ such that

$$Z_n^{\zeta,0} \leq Z_n^\zeta \leq (1 + Cn) Z_n^{\zeta,0} \qquad \forall n \in \mathbb{N}. \tag{7.9}$$

The lower bound is obvious. The upper bound is proved as follows. By conditioning on the last hitting time of 0 prior to time n, we may write

$$Z_n^\zeta = Z_n^{\zeta,0} + \sum_{k=1}^n Z_{n-k}^{\zeta,0} \, a(k) = Z_n^{\zeta,0} + \sum_{k=1}^n Z_{n-k}^{\zeta,0} \, b(k) \, \frac{a(k)}{b(k)}. \tag{7.10}$$

As noted below (7.6), we have $a(k)/b(k) \leq Ck$, $k \in \mathbb{N}$, for some $C < \infty$. Without the factor $a(k)/b(k)$, the last sum in (7.10) is precisely $Z_n^{\zeta,0}$, and so we get the upper bound in (7.9).

The proof of the existence of the free energy is completed by noting that

$$Z_n^{\zeta,0} \geq Z_m^{\zeta,0} \, Z_{n-m}^{\zeta,0} \qquad \forall m, n \in \mathbb{N}, \, m \leq n, \tag{7.11}$$

which follows by inserting an extra indicator $1_{\{S_m = 0\}}$ into (7.8) and using the Markov property of S at time m. This inequality says that $n \mapsto \log Z_n^{\zeta,0}$ is superadditive, which implies that (recall Section 1.3)

$$f(\zeta) = \lim_{n \to \infty} \frac{1}{n} \log Z_n^{\zeta,0} \quad \text{exists.} \tag{7.12}$$

The limit equals the free energy because of (7.9).

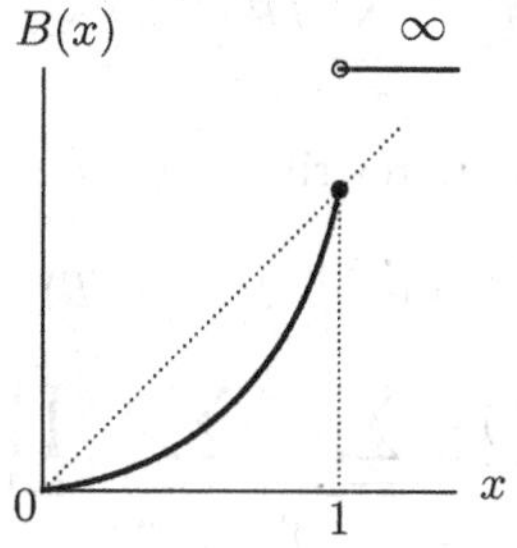

Fig. 7.3. Qualitative picture of $x \mapsto B(x)$.

Finally, the trivial bound $Z_n^\zeta \geq a(n)$, obtained by inserting into (7.8) the indicator of the event that the path never returns to the interface, implies that $f(\zeta) \geq 0$ because of (7.6). □

Our next theorem relates the free energy to the generating function for the length of the *excursions away from the interface*. Let (see Fig. 7.3)

$$B(x) = \sum_{k=1}^{\infty} x^k\, b(k), \qquad x \in [0, \infty). \tag{7.13}$$

Theorem 7.2. *The free energy is given by*

$$f(\zeta) = \begin{cases} 0, & \text{if } \zeta \leq 0, \\ r(\zeta), & \text{if } \zeta > 0, \end{cases} \tag{7.14}$$

where $r(\zeta)$ is the unique solution of the equation

$$B(e^{-r}) = e^{-\zeta}, \qquad \zeta > 0. \tag{7.15}$$

Proof. For $\zeta \leq 0$, we have the trivial bounds

$$a(n) \leq Z_n^\zeta \leq 1 \qquad \forall n \in \mathbb{N}, \tag{7.16}$$

which by (7.6) imply that $f(\zeta) = 0$.

For $\zeta > 0$, we look at the constrained partition sum $Z_n^{\zeta,0}$ and write

$$Z_n^{\zeta,0} = \sum_{m=1}^{n} \sum_{\substack{j_1,\ldots,j_m \in \mathbb{N} \\ j_1 + \cdots + j_m = n}} \prod_{i=1}^{m} e^\zeta\, b(j_i). \tag{7.17}$$

Let

$$b^\zeta(k) = e^{\zeta - r(\zeta)k}\, b(k), \qquad k \in \mathbb{N}. \tag{7.18}$$

By (7.15), this is a probability distribution on $\mathbb{N}$. Moreover, because $r(\zeta) > 0$, we have

$$M_{b\zeta} = \sum_{k\in\mathbb{N}} k b^\zeta(k) < \infty. \tag{7.19}$$

With the help of (7.18), we may rewrite (7.17) as

$$Z_n^{\zeta,0} = e^{r(\zeta)n} P^\zeta(n \in T),$$

$$P^\zeta(n \in T) = \sum_{m=1}^{n} \sum_{\substack{j_1,\dots,j_m \in \mathbb{N} \\ j_1+\cdots+j_m=n}} \prod_{i=1}^{m} b^\zeta(j_i), \tag{7.20}$$

where $T = (T_l)_{l\in\mathbb{N}_0}$ (with $T_0 = 0$) is the renewal process whose i.i.d. increments have probability distribution

$$P^\zeta(T_{l+1} - T_l = k) = b^\zeta(k), \qquad k \in \mathbb{N}, \, l \in \mathbb{N}_0. \tag{7.21}$$

By the renewal theorem (Asmussen [10], Theorem I.2.2), we have

$$\lim_{n\to\infty} P^\zeta(n \in T) = \frac{1}{M_{b\zeta}}. \tag{7.22}$$

Combining (7.9), (7.20) and (7.22), we find that $f(\zeta) = \lim_{n\to\infty} \frac{1}{n} \log Z_n^{\zeta,0} = r(\zeta)$. $\square$

Theorem 7.2 shows that $\zeta \mapsto f(\zeta)$ is non-analytic at $\zeta_c = 0$. Since $x \mapsto B(x)$ is strictly increasing and analytic on $(0,1)$, it follows from (7.15) that $\zeta \mapsto f(\zeta)$ is strictly increasing and analytic on $(0,\infty)$. Consequently, $\zeta_c = 0$ is the only point of non-analyticity of f.

For SRW we have (Spitzer [284], Section 1)

$$B(x) = 1 - \sqrt{1 - x^2}, \qquad x \in [0,1], \tag{7.23}$$

and so (see Fig. 7.4)

$$r(\zeta) = \tfrac{1}{2}\zeta - \tfrac{1}{2}\log(2 - e^{-\zeta}), \qquad \zeta > 0. \tag{7.24}$$

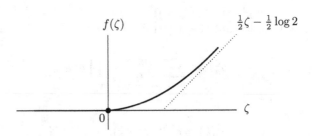

Fig. 7.4. Plot of the free energy for pinned SRW as found in (7.14) and (7.24).

7.1.2 Path Properties

Our next theorem identifies the path behavior in the two phases, and confirms the naive view of localized and delocalized behavior based on the free energy.

Note that the proof of Theorem 7.2 shows that P_∞^ζ, the weak limit as $n \to \infty$ of P_n^ζ defined in (7.3), is the path measure whose excursions are i.i.d. with length distribution

$$P^\zeta(T_1 - T_0 = k) = e^{\zeta - r(\zeta)k} b(k), \qquad k \in \mathbb{N}, \qquad (7.25)$$

and with distribution at length k equal to that of the random walk conditioned on returning to the origin after k steps. For $\zeta > 0$ this process is positive recurrent $(r(\zeta) > 0)$, for $\zeta = 0$ it is positive recurrent when $M_b = \sum_{k \in \mathbb{N}} kb(k) < \infty$ and null recurrent $(r(\zeta) = 0)$ when $M_b = \infty$, while for $\zeta < 0$ it is transient $(r(\zeta) = 0)$.

Theorem 7.3. *Under P_n^ζ as $n \to \infty$:*
(a) *If $\zeta > 0$, then the path hits $\mathbb{N} \times \{0\}$ with a strictly positive density, while the length and the height of the largest excursion away from $\mathbb{N} \times \{0\}$ up to time n is of order $\log n$.*
(b) *If $\zeta < 0$, then the path hits $\mathbb{N} \times \{0\}$ finitely many times.*

Proof. (a) The first claim follows from the observation that P_∞^ζ is positive recurrent when $\zeta > 0$. Note also that, as argued in Section 1.3, the derivative of the free energy equals the limiting average fraction of monomers on the substrate, i.e.,

$$f'(\zeta) = \lim_{n \to \infty} \frac{1}{n} \sum_{w \in \mathcal{W}_n} K_n(w) P_n^\zeta(w), \qquad \zeta \neq 0, \qquad (7.26)$$

where we recall (7.2–7.3). Theorem 7.2 shows that this derivative is strictly positive for all $\zeta > 0$.

To prove the second claim, we define, for $k \in \mathbb{N}_0$,

$$\text{gap}_k(T) = [T_{l(k)+1} - T_{l(k)}] 1_{\{k \notin T\}}, \qquad l(k) = \max\{l \in \mathbb{N}: T_l \leq k\}. \quad (7.27)$$

and, for $n \in \mathbb{N}$,

$$\text{maxgap}_n(T) = \max_{0 \leq k < n} \text{gap}_k(T). \qquad (7.28)$$

Then

$$P_n^{\zeta,0}\left(\text{maxgap}_n(T) > C \log n\right) \leq \sum_{k=0}^{n-1} P_n^{\zeta,0}\left(\text{gap}_k(T) > C \log n\right). \qquad (7.29)$$

To estimate the right-hand side we observe that, for every $0 \leq k_1 < k_2 \leq n$,

$$P_n^{\zeta,0}\big(k_1 \in T, k_2 \in T, (k_1, k_2) \cap T = \emptyset\big) = \frac{1}{Z_n^{\zeta,0}} Z_{k_1}^{\zeta,0} b(k_2 - k_1) e^{\zeta} Z_{n-k_2}^{\zeta,0}$$

$$\leq \frac{1}{Z_{k_2-k_1}^{\zeta,0}} b(k_2 - k_1) e^{\zeta},$$

$$(7.30)$$

where in the last line we use the inequality $Z_n^{\zeta,0} \geq Z_{k_1}^{\zeta,0} Z_{k_2-k_1}^{\zeta,0} Z_{n-k_2}^{\zeta,0}$. It follows from (7.12) and (7.30) that there exists a $C_1 < \infty$ such that

$$P_n^{\zeta,0}\big(\mathrm{gap}_k(T) > N\big) \leq C_1 e^{-\frac{1}{2}Nf(\zeta)} \qquad \forall N, n \in \mathbb{N}, \, 0 \leq k < n. \qquad (7.31)$$

Substituting this bound into (7.29), we get

$$\lim_{n\to\infty} P_n^{\zeta,0}\big(\mathrm{maxgap}_n(T) > C_2 \log n\big) = 0 \qquad (7.32)$$

as soon as $C_2 > 2/f(\zeta)$, which is finite for all $\zeta > 0$. Hence, the maximal gap is of order $\log n$ as $n \to \infty$.

(b) The claim follows from the observation that P_∞^ζ is transient when $\zeta < 0$. Note also that Theorem 7.2 shows that $f'(\zeta) = 0$ for all $\zeta < 0$, i.e., the density at which the path hits $\mathbb{N} \times \{0\}$ is zero. $\square$

The fact that in the localized phase the largest excursion away from the substrate up to time n is of order $\log n$ has the following intuitive explanation. The free energy f is strictly positive, and the probability for the polymer to be away from the substrate during a time j decays like e^{-fj} as $j \to \infty$. This is because the polymer contributes an amount fj to the free energy when it stays near the substrate, but contributes 0 when it moves away from the substrate.

The average fraction of adsorbed monomers is linear near $\zeta_c = 0$ when the phase transition is second order (see Fig. 7.5), and discontinuous at $\zeta_c = 0$ when the phase transition is first order.

It follows from Theorem 7.2 that

$$\lim_{\zeta\to\infty} f'(\zeta) = \frac{1}{k^*} \quad \text{with } k^* = \min \Lambda = \min\{k \in \mathbb{N}: b(k) > 0\}. \qquad (7.33)$$

This is the maximal fraction of monomers that can be adsorbed when the reward for adsorption tends to infinity.

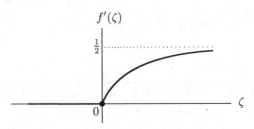

Fig. 7.5. Plot of the average fraction of adsorbed monomers for SRW as found in (7.14) and (7.24).

7.1.3 Order of the Phase Transition

Theorem 7.4. *Recall (7.5). There exists a slowly varying function L^* such that*

$$f(\zeta) \sim \zeta^{1/(1 \wedge a)} L^*(1/\zeta), \qquad \zeta \downarrow 0. \tag{7.34}$$

Moreover, L^* is constant in the following three cases: (1) $a \in (1, \infty)$; (2) $a = 1$ and $M_b = \sum_{k=1}^{\infty} kb(k) < \infty$; (3) $a \in (0, 1)$ and L is constant.

Proof. It follows from (7.5) and a standard Abelian theorem (see Bingham, Goldie and Teugels [15], Corollary 8.1.7) that, in the limit as $r \downarrow 0$, the generating function defined in (7.13) scales like

$$1 - B(e^{-r}) \sim \begin{cases} \frac{1}{a} \Gamma(1 - a) r^a L(1/r), & \text{if } a \in (0, 1), \\[2mm] \left[\int_1^{\infty} \frac{1}{s} L(s) e^{-rs} \, ds \right] r, & \text{if } a = 1, M_b = \infty, \\[2mm] M_b r, & \text{if } a = 1, M_b < \infty \text{ or } a \in (1, \infty), \end{cases} \tag{7.35}$$

where Γ is the Gamma-function. Inserting this into (7.15) and using (7.14), we get the claim. For details, see Giacomin [116], Section 2.1. $\square$

Theorem 7.4 shows that the order of the phase transition depends on a and L. Indeed, the phase transition is first order when $a \in (1, \infty)$ or when $a = 1$ and $M_b < \infty$, while for $m \in \mathbb{N} \setminus \{1\}$ it is m-th order when $a \in [\frac{1}{m}, \frac{1}{m-1})$ and L is constant.

For SRW, we have $a = \frac{1}{2}$ and L constant, and so the phase transition is second order (see Fig. 7.4).

7.2 A Polymer Near a Linear Impenetrable Substrate: Wetting

In this section we investigate in what way the results in Section 7.1 are to be modified when the substrate is impenetrable, i.e., when the set of paths $\mathcal{W}_n$ in (7.1) is replaced by (see Fig. 7.6)

$$\mathcal{W}_n^+ = \left\{ w = (i, w_i)_{i=0}^n : w_0 = 0, w_i \in \mathbb{N}_0 \ \forall 0 \leq i \leq n \right\}. \tag{7.36}$$

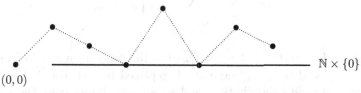

Fig. 7.6. A 7-step path in $\mathbb{N}_0 \times \mathbb{N}_0$ that makes 2 visits to $\mathbb{N} \times \{0\}$.

Accordingly, we write $P_n^{\zeta,+}(w)$, $Z_n^{\zeta,+}$ and $f^+(\zeta)$ for the path measure, partition sum and free energy in this version of the model.

The free energy can be computed using the same excursion approach as in Section 7.1. Theorem 7.1 carries over immediately. The analogue of Theorem 7.2 reads as follows. For $k \in \mathbb{N}$, let

$$b^+(k) = P(S_i > 0 \; \forall 1 \le i < k, \; S_k = 0). \tag{7.37}$$

This is a *defective* probability distribution, i.e., $\sum_{k \in \mathbb{N}} b^+(k) < 1$. Define

$$B^+(x) = \sum_{k=1}^{\infty} x^k \, b^+(k), \qquad x \in [0, \infty), \tag{7.38}$$

and put

$$\widetilde{B}(x) = \frac{B^+(x)}{B^+(1)}, \qquad \zeta_c^+ = \log \frac{1}{B^+(1)} > 0. \tag{7.39}$$

Theorem 7.5. *The free energy is given by*

$$f^+(\zeta) = \begin{cases} 0, & \text{if } \zeta \le \zeta_c^+, \\ r^+(\zeta), & \text{if } \zeta > \zeta_c^+, \end{cases} \tag{7.40}$$

where $r^+(\zeta)$ is the unique solution of the equation

$$\widetilde{B}(e^{-r}) = e^{-(\zeta - \zeta_c^+)}, \qquad \zeta > \zeta_c^+. \tag{7.41}$$

Proof. The same bounds as in (7.9) hold, so that

$$f^+(\zeta) = \lim_{n \to \infty} \frac{1}{n} \log Z_n^{\zeta,0,+}, \tag{7.42}$$

where, as before, the upper index 0 refers to the polymer being constrained to end at the substrate.

Write, similarly as in (7.17),

$$Z_n^{\zeta,0,+} = \sum_{m=1}^{n} \sum_{\substack{j_1,\ldots,j_m \in \mathbb{N} \\ j_1 + \cdots + j_m = n}} \prod_{i=1}^{m} e^{\zeta} \, b^+(j_i). \tag{7.43}$$

Shifting ζ, we may rewrite this as

$$Z_n^{\zeta,0,+} = \sum_{m=1}^{n} \sum_{\substack{j_1,\ldots,j_m \in \mathbb{N} \\ j_1 + \cdots + j_m = n}} \prod_{i=1}^{m} e^{\zeta - \zeta_c^+} \, \widetilde{b}(j_i), \tag{7.44}$$

with $\widetilde{b}(k) = b^+(k)/B^+(1)$, $k \in \mathbb{N}$. This has the same form as (7.17), except that ζ is replaced by $\zeta - \zeta_c^+$ and $b(\cdot)$ is replaced by $\widetilde{b}(\cdot)$. But $\widetilde{b}(\cdot)$ is a *non-defective* probability distribution, and so we can simply copy the proof of Theorem 7.2 to get the claim. $\square$

Localization on an impenetrable substrate is harder than on a penetrable substrate, because the polymer suffers a larger loss of entropy. This is the reason why $\zeta_c^+ > 0$ in Theorem 7.5.

To get the analogue of Theorem 7.4, we need to assume, as in (7.5), that

$$b^+(k) \sim k^{-1-a} L(k) \quad \text{as } k \to \infty \text{ through } \Lambda^+, \tag{7.45}$$

for some $a \in (0, \infty)$ and some slowly varying function L, where $\Lambda^+ = \{k \in \mathbb{N}: b^+(k) > 0\} \subseteq \Lambda$. Under this assumption, the theorem and its proof carry over verbatim, after we shift ζ by ζ_c^+. Note that $\widetilde{b}(\cdot)$ inherits the scaling of $b^+(\cdot)$ in (7.45).

For SRW we have $b^+(k) = \frac{1}{2}b(k)$, $k \in \mathbb{N}$. Hence

$$\zeta_c^+ = \log 2, \quad \widetilde{b}(\cdot) = b(\cdot), \quad \widetilde{B}(\cdot) = B(\cdot), \tag{7.46}$$

implying that $f^+(\zeta) = f(\zeta - \zeta_c^+)$, $\zeta \in \mathbb{R}$. Thus, for SRW the free energy *simply suffers a shift*. For more general random walks, however, it is *not* the case that $\widetilde{b}(\cdot) = b(\cdot)$. Even their tail exponents may be different. The latter may be seen from the relation

$$1 - \sum_{k=1}^{\infty} x^k b^+(k) = \exp\left[-\sum_{k=1}^{\infty} \frac{1}{k} x^k P(S_k = 0)\right], \quad x \in [0, \infty), \tag{7.47}$$

for which we refer the reader to Spitzer [284], Section 17. Indeed, this relation shows that $b^+(k) \asymp P(S_k = 0)/k$ as $k \to \infty$ (subject to a regulary varying tail condition on $P(S_k = 0)$ implied by (7.45)). For instance, for the Riemann random walk mentioned at the end of Section 7.1, if $\chi \in (1, 2)$, then $P(S_k = 0) \asymp k^{-1/\chi}$ as $k \to \infty$ (where $\asymp$ means that the ratio of the two sides is bounded above and below by strictly positive and finite constants), and so the tail exponent of $\widetilde{b}(\cdot)$ is $1 + (1/\chi)$, which is different from the tail exponent $2 - (1/\chi)$ we found for $b(\cdot)$. Only for $\chi \in [2, \infty)$ are both exponents the same, and equal to $\frac{3}{2}$ (with a logarithmic correction for $\chi = 2$).

7.3 Pulling a Polymer off a Substrate by a Force

A polymer can be pulled off a substrate by a force. This can be done, for instance, with the help of *optical tweezers*. Here, a focussed laser beam is used, containing a narrow region – called the beam waist – in which there is a strong electric field gradient. When a dielectric particle, a few nanometers in diameter, is placed in the waist, it feels a strong attraction towards the center of the waist. One can chemically attach such a particle to the end of the polymer and then pull on the particle with the laser beam, thereby effectively exerting a force on the polymer itself. Current experiments allow for forces in the range of 10^{-12} Newton. With such microscopically small forces the structural, mechanical and elastic properties of polymers can be probed.

In Section 7.3.1 we look at the pinned polymer, in Section 7.3.2 at the wetted polymer.

7.3.1 Force and Pinning

To model the effect of the force on a pinned polymer we return to (7.1–7.3) and replace the Hamiltonian by

$$H_n(w) = -\zeta K_n(w) - \phi w_n, \tag{7.48}$$

where $\phi \in \mathbb{R}$ is a force in the upward direction acting on the right endpoint of the polymer (ϕw_n is the work exerted by the force to move the endpoint a distance w_n away from the interface). We write $Z_n^{\zeta, \phi}$ to denote the partition sum and

$$f(\zeta, \phi) = \lim_{n \to \infty} \frac{1}{n} \log Z_n^{\zeta, \phi} \tag{7.49}$$

to denote the free energy. Without loss of generality, we will assume that $\phi > 0$.

For convenience of exposition, we restrict to the case where the reference random walk can only make steps of size ≤ 1, namely,

$$P(S_1 = -1) = P(S_1 = +1) = \tfrac{1}{2}p, \quad P(S_1 = 0) = 1 - p, \quad p \in (0, 1]. \tag{7.50}$$

Theorem 7.6. *For every $\zeta \in \mathbb{R}$ and $\phi > 0$, the free energy exists and is given by*

$$f(\zeta, \phi) = f(\zeta) \vee g(\phi), \tag{7.51}$$

with $f(\zeta)$ the free energy of the pinned polymer without force given by Theorem 7.2, and

$$g(\phi) = \log \big[p \cosh(\phi) + (1 - p) \big]. \tag{7.52}$$

Proof. We have

$$Z_n^{\zeta, \phi} = Z_n^{\zeta, 0} + \sum_{k=1}^{n} Z_{n-k}^{\zeta, 0} \, Z_k^{+, \phi} \tag{7.53}$$

with $Z_n^{\zeta, 0}$ the partition sum of the pinned polymer without force constrained to end at the substrate, and

$$Z_k^{+, \phi} = \sum_{x \in \mathbb{Z}} e^{\phi x} \, b^+(k, x), \quad b^+(k, x) = P(S_i > 0 \; \forall 1 \leq i < k, \; S_k = x). \tag{7.54}$$

We will show that

$$g(\phi) = \lim_{k \to \infty} \frac{1}{k} \log Z_k^{+, \phi} \tag{7.55}$$

exists and is given by (7.52). Combining (7.12), (7.53) and (7.55), we get the claim.

The contribution to the sum in (7.54) coming from $x \in \mathbb{Z} \backslash \mathbb{N}$ is bounded above by $1/(1 - e^{-\phi}) < \infty$. For $x \in \mathbb{N}$, on the other hand, we can use the *reflection principle*, which gives

$$b^+(k, x) = \tfrac{p}{2}\left[P(S_{k-1} = x - 1) - P(S_{k-1} = x + 1)\right], \qquad k \in \mathbb{N}. \qquad (7.56)$$

Hence, for $k \in \mathbb{N}$,

$$
\begin{aligned}
Z_k^{+,\phi} &= O(1) + \tfrac{p}{2} \sum_{x \in \mathbb{N}} e^{\phi x}\left[P(S_{k-1} = x - 1) - P(S_{k-1} = x + 1)\right] \\
&= O(1) + p\sinh(\phi) \sum_{x \in \mathbb{N}} e^{\phi x}\, P(S_{k-1} = x) \\
&= O(1) + p\sinh(\phi) \sum_{x \in \mathbb{Z}} e^{\phi x}\, P(S_{k-1} = x) \qquad\qquad (7.57) \\
&= O(1) + p\sinh(\phi)\, E(e^{\phi S_{k-1}}) \\
&= O(1) + p\sinh(\phi)\left[p\cosh(\phi) + (1 - p)\right]^{k-1},
\end{aligned}
$$

where the error term changes from line to line. This proves that, indeed, the limit in (7.55) exists and is given by (7.52). □

Formula (7.51) tells us that the force either leaves most of the polymer adsorbed (when $f(\zeta, \phi) = f(\zeta) > g(\phi)$) or pulls most of the polymer off (when $f(\zeta, \phi) = g(\phi) > f(\zeta)$). It further tells us that a phase transition occurs at those values of ζ and ϕ where $f(\zeta) = g(\phi)$, i.e., the critical value of the force is given by

$$\phi_c(\zeta) = g^{-1}(f(\zeta)), \qquad \zeta \in \mathbb{R}, \qquad (7.58)$$

with g^{-1} the inverse of g. Note that $\phi_c(\zeta) = 0$ for $\zeta \leq 0$ and $\phi_c(\zeta) > 0$ for $\zeta > 0$. In the latter regime, the phase transition is first order because $\partial f(\zeta, \phi)/\partial\phi$ is discontinuous at $\phi = \phi_c(\lambda)$. Also note that the phase transition is first order in ζ when $\phi > 0$ and second order in ζ when $\phi = 0$.

For our choice of random walk the analogue of (7.23) reads

$$B_p(x) = 1 - \sqrt{[1 - (1 - p)x]^2 - [px]^2}, \qquad x \in [0, 1]. \qquad (7.59)$$

Consequently, $f(\zeta) = -\log B_p^{-1}(e^{-\zeta})$, with B_p^{-1} the inverse of B_p, and so

$$\phi_c(\zeta) = \cosh^{-1}\left(\frac{1}{p}\left[\frac{1}{B_p^{-1}(e^{-\zeta})} - (1 - p)\right]\right), \qquad \zeta > 0. \qquad (7.60)$$

In what follows we look at $p = 1$ and $p \in (0, 1)$ separately.

● Case $p = 1$

For $p = 1$ (SRW), the formula in (7.60) reduces to

$$\phi_c(\zeta) = \tanh^{-1}(1 - e^{-\zeta}), \qquad \zeta > 0, \qquad (7.61)$$

which we plot in Fig. 7.7. It is easily checked that $\zeta \mapsto \phi_c(\zeta)$ is strictly concave, with $\phi_c(\zeta) \sim \zeta$, $\zeta \downarrow 0$, and $\phi_c(\zeta) \sim \tfrac{1}{2}\zeta$, $\zeta \to \infty$.

An alternative way of exhibiting Fig. 7.7 is to put

$$\zeta = 1/T, \quad \phi = F/T, \quad F_c(T) = \phi_c(1/T)/T, \tag{7.62}$$

with $T > 0$ the physical temperature and F the physical force. This leads to the picture in Fig. 7.8, which shows that the *force-temperature diagram* is strictly increasing.

- **Case $p \in (0, 1)$**

From (7.59) we have

$$B_p(x) \sim \begin{cases} (1-p)x, & \text{as } x \downarrow 0, \\ 1 - \sqrt{2p(1-x)}, & \text{as } x \uparrow 1. \end{cases} \tag{7.63}$$

It therefore follows from (7.60) and (7.62) that

$$F_c(T) = \begin{cases} \frac{1}{p}, & \text{as } T \to \infty, \\ 1 + T[\log(1-p) - \log(\frac{1}{2}p)] + o(T), & \text{as } T \downarrow 0, \end{cases} \tag{7.64}$$

which is plotted in Fig. 7.9. Note that $T \mapsto F_c(T)$ goes through a *minimum* when $p \in (\frac{2}{3}, 1)$. This is remarkable, since it says that at a fixed force F slightly below 1 the polymer is adsorbed both for small and for large temperatures, but is desorbed for moderate temperatures. This type of behavior is referred to as a *re-entrant phase transition*.

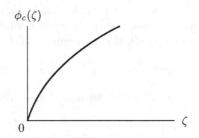

Fig. 7.7. Plot of $\zeta \mapsto \phi_c(\zeta)$ for $p = 1$.

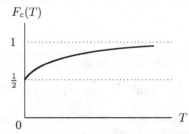

Fig. 7.8. Plot of $T \mapsto F_c(T)$ for $p = 1$.

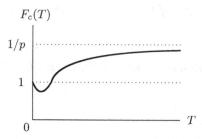

Fig. 7.9. Plot of $T \mapsto F_c(T)$ for $p \in (\frac{2}{3}, 1)$.

Since $\zeta_c = 0$, at zero force the polymer is adsorbed (pinned) for every $0 \leq T < \infty$, and becomes desorbed (depinned) only at $T = \infty$, implying that $F_c(T) > 0$ for all $0 \leq T < \infty$. This explains why the critical curves in Figs. 7.8–7.9 do not hit the horizontal axis.

Here is a heuristic explanation of Fig. 7.9. First consider the case $p = \frac{2}{3}$, where the steps in the east, north-east and south-east direction are equally probable, so that all paths have the same probability. For every $0 < T < \infty$, the polymer makes excursions away from the substrate and therefore has a strictly positive entropy. Part of this entropy is lost when a force is applied and part of the polymer (near its endpoint) is pulled away from the substrate and is forced to move upwards steeply. As T increases, the effect of this entropy loss on the free energy increases (because "free energy = energy − [temperature × entropy]"). This in turn must be counterbalanced by a larger force to achieve desorption and, consequently, $\partial F_c(T)/\partial T > 0$ for $0 < T < \infty$. The case $T = 0$ is special, because the polymer is fully pinned and has no entropy, implying that $\lim_{T \downarrow 0} \partial F_c(T)/\partial T = 0$. Next consider the case $p \in (\frac{2}{3}, 1)$. Then the steps in the east direction are disfavored over the steps in the north-east and south-east direction, and this tends to move the polymer farther away from the substrate for every $0 < T < \infty$. Consequently, $F_c(T) < F_c(0)$ for small T, i.e., $T \mapsto F_c(T)$ decreases for small T, an effect that becomes more pronouced as p increases. At larger T, entropy again dominates in the free energy and hence $T \mapsto F_c(T)$ increases when T is large enough. Finally, for the case $p \in (0, \frac{2}{3})$ the effect is precisely the opposite for small T.

7.3.2 Force and Wetting

A similar calculation can be done for the wetting version of the model. The main difference is that the critical curve shifts over ζ_c^+. The analogue of Theorem 7.6 reads as follows.

Theorem 7.7. *For every $\zeta \in \mathbb{R}$ and $\phi > 0$, the free energy exists and is given by*

$$f^+(\zeta, \phi) = f^+(\zeta) \vee g(\phi), \tag{7.65}$$

with $f^+(\zeta)$ the free energy of the wetted polymer without force given by Theorem 7.5, and

$$g(\phi) = \log\left[p\cosh(\phi) + 1 - p\right]. \tag{7.66}$$

Proof. The proof is the same as that of Theorem 7.6. □

● **Case $p = 1$**

The analogue of (7.61), with $\zeta_c^+ = \log 2$ for $p = 1$, reads

$$\phi_c^+(\zeta) = \tanh^{-1}\left(1 - e^{-(\zeta - \zeta_c^+)}\right), \qquad \zeta > \zeta_c^+, \tag{7.67}$$

which is Fig. 7.7 shifted to the right by ζ_c^+, and gives a force-temperature diagram as plotted in Fig. 7.10.

● **Case $p \in (0, 1)$**

The analogue of (7.64) reads

$$F_c^+(T) = \begin{cases} 0, & \text{as } T \to \infty, \\ 1 + T[\log(1 - p) - \log(\frac{1}{2}p)] + o(T), & \text{as } T \downarrow 0, \end{cases} \tag{7.68}$$

which is plotted in Fig. 7.11. Note that $T \mapsto F_c(T)$ goes through a *maximum* when $p \in (0, \frac{2}{3})$. At a fixed force F slightly above 1 the polymer is desorbed

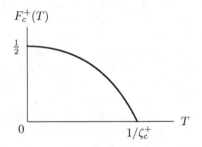

Fig. 7.10. Plot of $T \mapsto F_c^+(T)$ for $p = 1$.

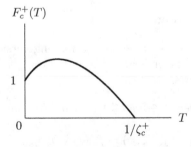

Fig. 7.11. Plot of $T \mapsto F_c^+(T)$ for $p \in (0, \frac{2}{3})$.

both for small and for large temperatures, but is adsorbed for moderate temperatures. This is precisely the *reverse* of what we had for pinning.

Since $\zeta_c^+ > 0$, at zero force the polymer is adsorbed (wetted) for small temperature and becomes desorbed (dewetted) at large temperature, which is why the critical curves in Figs. 7.10–7.11 hit the horizontal axis at $T = 1/\zeta_c^+$.

A heuristic explanation of Fig. 7.11 runs parallel to that of Fig. 7.9 given at the end of Section 7.3.1. When a force is applied, part of the polymer (near its endpoint) is pulled away from the substrate. For the case $p = \frac{2}{3}$, where all paths have equal probability, this comes with a gain of entropy (unlike the loss of entropy we had for pinning), because the polymer feels less of the hard-wall constraint. As T increases, the effect of this entropy gain on the free energy increases. This is counterbalanced by a smaller force to achieve desorption and, consequently, $\partial F_c(T)/\partial T < 0$. For the case $p \in (\frac{2}{3}, 1)$ (where steps in the east direction are disfavored), the polymer is farther away from the substrate, resulting in $F_c(T) < F_c(0)$, while for the case $p \in (0, \frac{2}{3})$ (where steps in the east direction are favored), the effect is precisely the opposite. For small T this explains the sign of the slope of the critical curve. At larger T, entropy dominates in the free energy and $T \mapsto F_c(T)$ decreases until it hits the horizontal axis.

7.4 A Polymer in a Slit between Two Impenetrable Substrates

In this section we consider a two-dimensional directed polymer in a slit between two walls. The lower wall is at height 0, the upper wall at height $N \in \mathbb{N}$. The polymer follows a ballot path (see Fig. 7.12). Each visit to the lower wall comes with an adsorption energy σ, each visit to the upper wall with an adsorption energy ξ, both of which are taken to be non-negative.

7.4.1 Model

We fix $N \in \mathbb{N}$ and replace the set of paths and the Hamiltonian in (7.1–7.2) by

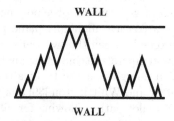

Fig. 7.12. A directed polymer in a slit.

$$\mathcal{W}_{n,N} = \big\{ w = (i, w_i)_{i=0}^n : w_0 = 0, w_{i+1} - w_i = \pm 1 \ \forall 0 \le i < n,$$
$$0 \le w_i \le N \ \forall 0 \le i \le n \big\}, \qquad (7.69)$$
$$H_{n,N}(w) = -\sigma K_n(w) - \xi K_{n,N}(w),$$

with $\sigma, \xi \in [0, \infty)$ and

$$K_n(w) = \sum_{i=1}^n 1_{\{w_i = 0\}}, \quad K_{n,N}(w) = \sum_{i=1}^n 1_{\{w_i = N\}}, \qquad w \in \mathcal{W}_{n,N}. \quad (7.70)$$

The path measure replacing (7.3) becomes

$$P_{n,N}^{\sigma,\xi}(w) = \frac{1}{Z_{n,N}^{\sigma,\xi}} e^{\sigma K_n(w) + \xi K_{n,N}(w)}, \qquad w \in \mathcal{W}_{n,N}, \qquad (7.71)$$

where we drop the reference measure P_n, which is harmless because P_n is the uniform distribution on ballot paths of length n. The main objects of interest are the finite width free energy

$$f_N(\sigma, \xi) = \lim_{n \to \infty} \frac{1}{n} \log Z_{n,N}^{\sigma,\xi} \qquad (7.72)$$

and its infinite width counterpart

$$f(\sigma, \xi) = \lim_{N \to \infty} f_N(\sigma, \xi). \qquad (7.73)$$

The motivation for this model is as follows. Consider a *colloidal dispersion* consisting of particles on which polymers are adsorbed. When two colloidal particles approach one another, the adsorbed polymers become confined to the space between the particles. The resulting loss of entropy induces a repulsive force between the particles, preventing them from clustering. This is called *steric stabilization*. On the other hand, when the adsorbed polymers bridge the space between the two particles and become adsorbed on both, the result is an additional attractive force, which may or may not exceed the repulsive force. In the former case the particles cluster. This is called *sensitized flocculation*. If the colloidal particles are large compared to the adsorbed polymers, then it is fair to approximate the space between the two particles by a slit between two parallel planes. Colloidal dispersions are used e.g. in paints and in pharmaceuticals. For more background, see Napper [247].

Pioneering work was done by DiMarzio and Rubin [91], who considered a three-dimensional random walk model of a polymer between two parallel walls with a short-range attractive monomer-plane potential. They showed that there is a repulsive force between the two walls at high temperature and an attractive force at low temperature, but their analysis is restricted to the case where the adsorption energies on the two walls are the same ($\sigma = \xi$). Brak, Owczarek, Rechnitzer and Whittington [39] analyzed a two-dimensional directed model, based on ballot paths, allowing for different adsorption energies on the two walls ($\sigma \ne \xi$). They arrived at a complete description of the phase diagram and the force diagram. We will describe their work below.

7.4.2 Generating Function

Let $\mathcal{W}_{n,N}^0$ be the subset of paths in $\mathcal{W}_{n,N}$ that end at the lower wall, i.e., $w_n = 0$ (see Fig. 7.12). Abbreviate $a = e^\sigma$ and $b = e^\xi$. Define the generating function

$$G_N(z; a, b) = \sum_{n=0}^{\infty} Z_{n,N}^0 \, z^n, \qquad z \in \mathbb{C}, \, N \in \mathbb{N}, \tag{7.74}$$

where

$$Z_{n,N}^0 = \sum_{w \in \mathcal{W}_{n,N}^0} a^{l(w)} b^{u(w)} \tag{7.75}$$

is the partition sum for paths ending at the lower wall, with $l(w)$ and $u(w)$ denoting the number of visits by w to the lower, respectively, upper wall. Our task is to identify the singularity $z_N(a, b)$ of $z \mapsto G_N(z, a, b)$ that lies closest to the origin on $(0, \infty)$, since

$$f_N(a, b) = \lim_{n \to \infty} \frac{1}{n} \log Z_{n,N}^0 = -\log z_N(a, b). \tag{7.76}$$

(It is trivial to show that $Z_{n,N}^0 / Z_{n,N} \leq 1$ decays at most polynomially fast in n for fixed N, implying the first equality in (7.76).)

The following theorem gives an explicit formula for $G_N(z; a, b)$.

Theorem 7.8. *Let* $q = (1 - 2z^2 - \sqrt{1 - 4z^2})/2z^2$. *Then, for all* $z \in \mathbb{C}$ *and* $N \in \mathbb{N}$,

$$G_N(z; a, b) = \frac{(1+q)[(1+q-bq) + (1+q-b)q^N]}{(1+q-aq)(1+q-bq) - (1+q-a)(1+q-b)q^N}. \tag{7.77}$$

Proof. The argument given here is a bit sketchy. For details we refer to [39]. The computation of $G_N(z; a, b)$ proceeds via induction on N. A functional recursion relation is obtained by building up the paths in a slit of width $N+1$ from the paths in a slit of width N after replacing each hit of the top wall by a zig-zag path (with increments $\nearrow$ and $\searrow$ alternating) of an arbitrary length. The generating function of a zig-zag path equals $1/(1 - bz^2)$, and so we get the recursion

$$G_{N+1}(z; a, b) = G_N(z; a, 1/(1 - bz^2)), \qquad N \in \mathbb{N}, \tag{7.78}$$

which is to be solved subject to the initial condition $G_1(z; a, b) = 1/(1-abz^2)$, the generating function for zig-zag paths in a slit of width 1.

By induction on N it can be shown that the solution of (7.78) is of the form

$$G_N(z; a, b) = \frac{P_N(z; 0, b)}{P_N(z; a, b)} \tag{7.79}$$

with $P_N(z; a, b)$ a polynomial in z. Inserting (7.79) into (7.78), we get the recursion

$$P_{N+1}(z; a, b) = (1 - bz^2) P_N(z; a, 1/(1 - bz^2)), \qquad N \in \mathbb{N}, \qquad (7.80)$$

which is to be solved subject to the initial condition $P_1(z; a, b) = 1/(1 - abz^2)$. In terms of the variable q defined by $z = \sqrt{q}/(1 + q)$, the solution reads (with a little help from MAPLE)

$$P_N(z; a, b) = \frac{(1 + q - aq)(1 + q - bq) - (1 + q - a)(1 + q - b)q^N}{(1 - q)(1 + q)^{N+1}}. \qquad (7.81)$$

Substituting (7.81) into (7.79), we get the claim. □

The dominant singularity of the r.h.s. of (7.77) is $q_N(a, b)$, defined to be the zero of the denominator with the smallest value of $|z|$, solving

$$q^N = \frac{(1 + q - aq)(1 + q - bq)}{(1 + q - a)(1 + q - b)}. \qquad (7.82)$$

A closed form solution is possible only for special values of (a, b), which we list next:

$$\begin{aligned}
(a, b) = (1, 1) \qquad & q_N = e^{2\pi i/(N+1)} \quad & f_N(1, 1) = \log(2\cos[\pi/(N+2)]), \\
(a, b) = (2, 1) \qquad & q_N = e^{2\pi i/(2N+2)} \quad & f_N(2, 1) = \log(2\cos[\pi/(2N+2)]), \\
ab = a + b, \, a > 2 \quad & q_N = 1/(a - 1) \quad & f_N(a, b) = \log(a/\sqrt{a - 1}),
\end{aligned}$$
$$(7.83)$$

and similar formulas when $(a, b) = (1, 2)$ and $ab = a + b$, $b > 2$. For other values of (a, b) the solution can be found numerically (see [39]). Note that the curve $ab = a + b$ is special: along this curve the free energy is *independent* of N, and is equal to the free energy for wetting (one wall).

7.4.3 Effective Force

The *effective force* exerted by the polymer on the two walls at width N is

$$\phi_N(a, b) = f_{N+1}(a, b) - f_N(a, b). \qquad (7.84)$$

We thus find that for all $N \in \mathbb{N}$ the effective force $\phi_N(a, b)$ is zero on the curve $ab = a + b$, strictly negative above the curve (attraction) and strictly positive below the curve (repulsion). The force tends to zero as $N \to \infty$. The rate at which this happens depends on (a, b). As shown in [39], the decay can be *exponential* (short-range force) or *polynomial* (long-range force) in N. The phase diagram is drawn in Fig. 7.13, the force diagram in Fig. 7.14. Both these figures are valid in the limit $n \to \infty$ followed by $N \to \infty$. Note that these limits may in general not be interchanged.

In [39], the phase transitions in Fig. 7.13 are shown to be second order along the horizontal and vertical critical lines, and first order along the oblique critical line. Across the latter, the densities of visits by the polymer to the

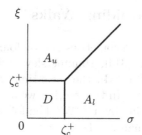

Fig. 7.13. Phase diagram for a directed polymer in a slit. There are three phases: D (= desorbed), A_l (= adsorbed on lower wall), A_u (= adsorbed on upper wall); $\zeta_c^+ = \log 2$ is the critical value for wetting in halfspace.

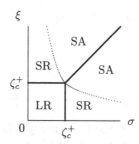

Fig. 7.14. Force diagram for a directed polymer in a slit. There are three types of forces: LR (= long-range repulsion), SR (= short-range repulsion), SA (= short-range attraction). On the dotted curve the force is zero.

walls make a jump. Remarkably, on this critical line the force turns out to be twice as large (for large N) as near the line, so that there is also a discontinuity in the strength of the force.

In terms of colloidal dispersions, Fig. 7.14 shows that long-range repulsive forces promote steric stabilization (entropy dominates), while short-range attractive forces promote sensitized flocculation (energy dominates). The special curve $ab = a + b$ is not a phase boundary.

Owczarek, Prellberg and Rechnitzer [260] provide a complete finite-size scaling analysis, i.e., they show how the free energy and the force scale as the length n of the polymer and the width N of the slit tend to infinity simultaneously. Brak, Iliev, Rechnitzer and Whittington [37] carry out the same analysis as in Brak, Owczarek, Rechnitzer and Whittington [39], but for generalized ballot paths rather than ballot paths. The formulas are more complicated, but the results are similar. They use the so-called "wasp-waist"-method, which leads to a system of difference equations that is somewhat easier to analyze than the functional equations encountered in Section 7.4.2. The method is flexible enough to allow for generalization to partially directed self-avoiding walks in higher dimension.

7.5 Adsorption of Self-avoiding Walks

Early studies of SAWs in confined geometries can be found in Guttmann and Whittington [139], Whittington [314], Hammersley and Whittington [144], and Whittington and Soteros [317]. In the meantime, a vast combinatorial. literature exists on such problems. In this section we ask the question: What happens when we consider wetting by a SAW? The following result is due to Hammersley, Torrie and Whittington [142].

Consider SAWs on $\mathbb{Z}^d$, $d \geq 2$, that start at the origin and are confined to the halfspace $\{x_1 \geq 0\}$. Let $c_n^+(v)$, $n, v \in \mathbb{N}$, be the number of halfspace SAWs with $n+1$ vertices that have $v+1$ vertices in the hyperplane $\{x_1 = 0\}$. Consider the partition sum

$$Z_n^+(\zeta) = \sum_{v \in \mathbb{N}} c_n^+(v)\, e^{\zeta v}, \qquad n \in \mathbb{N}, \zeta \in \mathbb{R}. \tag{7.85}$$

The associated free energy is

$$f^+(\zeta) = \lim_{n \to \infty} \frac{1}{n} \log Z_n^+(\zeta), \qquad \zeta \in \mathbb{R}. \tag{7.86}$$

As shown in [142], the existence of the free energy follows from a standard subadditivity argument. The proof is similar to Part 1 of the proof of Theorem 6.5, and therefore we skip the details.

Theorem 7.9. *For every $d \geq 2$, the function $\zeta \mapsto f^+(\zeta)$ is non-analytic at a critical value $\zeta_c^+ \in (0, \log\{\mu(d)/\mu(d-1)\})$, where $\mu(d)$ is the connective constant of SAW in $\mathbb{Z}^d$ defined in (2.19) (with $\mu(1) = 1$).*

Proof. First, for $\zeta \leq 0$ we may estimate

$$Z_n^+(\zeta) \leq Z_n^+(0) = \sum_{v \in \mathbb{N}} c_n^+(v) = c_n^+, \tag{7.87}$$

with c_n^+ the number of halfspace SAWs. It is known (see Whittington [313]) that

$$\lim_{n \to \infty} \frac{1}{n} \log c_n^+ = \log \mu(d). \tag{7.88}$$

Next, taking only the single term with $v = 0$ in (7.85), we have

$$Z_n^+(\zeta) \geq c_n^+(0). \tag{7.89}$$

By translating all halfspace SAWs one unit in the positive x_1-direction and adding an edge from the origin to the new starting point, we see that $c_n^+(0) = c_{n-1}^+$. Together with (7.87–7.89), this yields

$$f^+(\zeta) = \log \mu(d) \qquad \forall \zeta \leq 0. \tag{7.90}$$

To control $f^+(\zeta)$ for $\zeta \geq 0$, we note that a lower bound is obtained by restricting (7.85) to halfspace SAW's that lie entirely in the hyperplane $\{x_1 = 0\}$, i.e.,

$$Z_n^+(\zeta) \geq c_n^+(n) \, e^{\zeta n}, \qquad \zeta \in \mathbb{R}. \tag{7.91}$$

Since $c_n^+(n)$ is the number of SAWs in $\mathbb{Z}^{d-1}$ with $n+1$ vertices, we thus obtain

$$f^+(\zeta) \geq \log \mu(d-1) + \zeta, \qquad \zeta \in \mathbb{R}. \tag{7.92}$$

By combining (7.90) and (7.92), we see that $\zeta \mapsto f^+(\zeta)$ must be non-analytic at some critical value $\zeta_c^+ \in [0, \log\{\mu(d)/\mu(d-1)\}]$.

Considerable extra work is needed to show that ζ_c^+ does not lie on the edges of this interval, in particular, $\zeta_c^+ > 0$ as in the directed walk model studied in Section 7.2. For details we refer the reader to Hammersley, Torrie and Whittington [142]. Janse van Rensburg [187] gives an easier proof of $\zeta_c^+ > 0$ based on a comparison with wetting of self-avoiding polygons. This proof in fact yields the lower bound $\zeta_c^+ \geq [8(d-1)]^{-1} \log \sqrt{1 + [\mu(d)]^{-2}}$. $\square$

Janse van Rensburg and Rechnitzer [195] estimate ζ_c^+ for SAWs in $d = 2, 3$ with the help of Monte Carlo simulation. This yields $\zeta_c^+(2) \approx 0.56$ and $\zeta_c^+(3) \approx 0.29$. The phase transition clearly comes out as second order.

The same result as in Theorem 7.9 applies to the pinned SAW, i.e., when we drop the hard wall constraint. Also here $\zeta_c \in [0, \log\{\mu(d)/\mu(d-1)\}]$. It is believed that $\zeta_c = 0$, but a proof is not known. For both models the phase transition is expected to be second order. Proofs are missing, but simulations support the claim.

7.6 Extensions

(1) It is possible to include the boundary cases $a = 0$ and $a = \infty$ in (7.5). In the former case (where $b(\cdot)$ is marginally summable) the phase transition is infinite order, in the latter case (where the tail of $b(\cdot)$ is smaller than polynomial) the behavior is similar to that when $a \in (1, \infty)$. See Giacomin [116], Chapter 2, for details. If the random walk S is transient, i.e., $B(1) = \sum_{k \in \mathbb{N}} b(k) < 1$, then the same theory as in Section 7.1 can be used after normalization of $b(\cdot)$. In this case $\zeta_c = \log[1/B(1)] > 0$ (compare with (7.39)). Thus, the results equally well apply to a $(1 + d)$-dimensional pinning model, where the substrate is a d-dimensional hyperplane and the path measure P is that of a d-dimensional random walk, which for $d \geq 3$ is always transient (Spitzer [284], Section 8).

(2) Deuschel, Giacomin and Zambotti [89] and Caravenna, Giacomin and Zambotti [56] show that the path measure P_n^+, the analogue of P_n in (7.3) for the wetting model, in the limit as $n \to \infty$ converges weakly to a path measure P_∞^+ on the set of infinite directed non-negative paths. There are three relevant regimes (see Fig. 7.15): $\zeta < \zeta_c^+$, $\zeta = \zeta_c^+$ and $\zeta > \zeta_c^+$. In these regimes, P_∞^+ is the law of a transient, null-recurrent, respectively, positive recurrent Markov

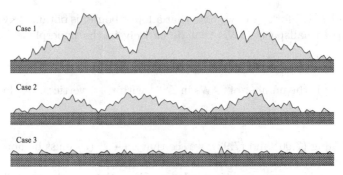

Fig. 7.15. A visual representation of the three regimes for the wetted polymer (courtesy of J.-D. Deuschel, G. Giacomin and L. Zambotti). Case 1: $\zeta < \zeta_c^+$; Case 2: $\zeta = \zeta_c^+$; Case 3: $\zeta > \zeta_c^+$.

process, whose transition probabilities can be written down explicitly. In the subcritical regime the scaling limit of P_∞^+ is Brownian meander, i.e., Brownian motion conditioned to stay positive. In the critical regime the intersection set under P_∞^+ turns out to be a regenerative set process with a *fractal structure*: it has "Lévy exponent" equal to $1 \wedge \zeta$. In the supercritical regime, finally, P_∞^+ has exponentially decaying correlations, and the distribution of the length of an excursion away from the wall has an exponential tail. (Similar results holds for the pinning model.)

In [89] continuous increments of the paths are considered, with an interaction Hamiltonian that sums a general potential at the values of the increments and sums the interactions with the wall. An explicit formula is derived for ζ_c^+ in terms of the partition sum of the model without interaction with the wall. In [56] the results are refined and also discrete increments of the path are allowed. In all cases, $\zeta_c^+ > 0$, i.e., the wetting transition occurs at a strictly positive value of the adsorption energy. Sohier [281] identifies the finite-size scaling limits of the path measure, i.e., when the polymer is close to criticality and has a length that is of the order of the correlation length.

(3) The results in Section 7.3 have been extended to SAWs (Orlandini, Tesi and Whittington [256]). For the wetting version the force-temperature diagram looks as drawn in Fig. 7.16, which is re-entrant. The reason is that, unlike a ballot path, a SAW has a positive entropy when it is adsorbed. This entropy gradually gets lost as more and more of the polymer desorbes when an increasing force is applied to it. Consequently, at low temperature the critical force needs to increase to counterbalance this loss. Simulations can be found in Krawczyk, Prellberg, Owczarek and Rechnitzer [219].

(4) The results in Section 7.4 have been extended to SAWs. It is shown in Janse van Rensburg, Orlandini and Whittington [194] that the same phase diagram as in Fig. 7.13 holds, with ζ_c^+ the critical value for the halfspace SAW. Most features of the phase diagram are proved rigorously, some are

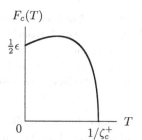

Fig. 7.16. Plot of $T \mapsto F_c(T)$ for SAW.

conjectured. In Janse van Rensburg, Orlandini, Owczarek, Rechnitzer and Whittington [192] and Martin, Orlandini, Owczarek, Rechnitzer and Whittington [237] these features are substantiated via exact enumeration and Monte Carlo simulation.

(5) Janse van Rensburg [189], [190], [191], and Janse van Rensburg and Ye [197], study directed polymers in two-dimensional wedges constrained to begin and end on the same side of the wedge, compute the free energy, and find the repulsive entropic force the polymer exerts on the wedge as the angle of the wedge is varied. In all cases the force diverges logarithmically with the inverse of the angle as the latter tends to zero, a property that appears to be robust. The problem is delicate, because without the constraint on the endpoints the free energy does not depend on the wedge angle (Hammersley and Whittington [144]) and the force is zero. In [189] an attractive interaction with one side of the wedge is added and the critical value for the wetting transition is determined.

(6) Whittington [315] considers wetting of a copolymer performing a partially directed self-avoiding walk and carrying two types of monomers, occurring in alternating order, of which only one type interacts with the interface. The critical threshold for adsorption is computed, and turns out to be strictly less than twice the critical threshold for the homopolymer. Moghaddam, Vrbová and Whittington [241] look at more general forms of periodicity and compare the associated critical thresholds. They also do a Monte Carlo study when the copolymer is a SAW, and find that the numerical differences with directed models tend to be small.

(7) Caravenna and Pétrélis [59], [60] consider a multi-interface medium, consisting of an infinite array of layers with width $L_n \in 2\mathbb{N}$, at which pinning occurs with energy $-\zeta$. The polymer follows a ballot path and for $\zeta > 0$ (attracting case) displays three regimes, depending on how fast L_n grows with n: (1) $L_n - C(\zeta) \log n \to -\infty$: infinitely many interfaces are visited; (2) $L_n - C(\zeta) \log n$ bounded: finitely many interfaces are visited; (3) $L_n - C(\zeta) \log n \to \infty$: only one interface is visited. Here, $C(\zeta)$ is a certain explicit function of ζ. As L_n increases from 1 to $C(\zeta) \log n$, the scaling of the

vertical displacement of the right endpoint of the polymer decreases smoothly from order $\sqrt{n}$ to order $\log n$. As L_n increases beyond $C(\zeta)\log n$, the vertical displacement jumps from order $\log n$ to order 1. For $\zeta < 0$ (repulsive case) there are three regimes also: (1) $L_n \ll n^{1/3}$: infinitely many interfaces are visited; (2) $L_n \asymp n^{1/3}$: finitely many interfaces are visited; (3) $L_n \gg n^{1/3}$: only one interface is visited.

(8) Vrbová and Whittington [306] and Janse van Rensburg [187] consider self-avoiding polygons with an interaction Hamiltonian that has both self-attraction (energy $-\gamma$ for monomers that are self-touching) and attraction to a linear wall (energy $-\zeta$ for monomers on the wall), which is a combination of the models studied in Chapter 6 and Section 7.2. Based on earlier work in the physics literature (Foster [107], Foster and Yeomans [109], Foster, Orlandini and Tesi [108]) and on simulations (Vrobová and Whittington [307], [308]), the phase diagram in the (γ, ζ)-plane for $d = 3$ is believed to have the shape drawn in Fig. 7.17. In [306] it is proved that there is an adsorption transition (between the (DE+DC)-phase and the (AE+AC)-phase) for all values of γ, with $\zeta_c^+(\gamma) \in [0, \log(\mu(3)/\mu(2)) + 2\gamma]$ for $\gamma \geq 0$. It is also proved that if there is a collapse transition in the desorbed phase (between the DE-phase and the DC-phase), $then$ the associated critical curve is linear, i.e., the critical threshold γ_c for collapse does not depend on ζ (a similar result was found in [107] and [109] for a directed version of the model). In [187] it is proved that $\zeta_c^+(\gamma) > 0$ for all $\gamma \in \mathbb{R}$. There is strong evidence that the phase transitions DE/AE, DE/DC and AE/AC are all second order. The phase transitions DC/AC and DC/AE appear to be first order. Simulations further suggest that in $d = 2$ the AC-phase is absent. This has been proved for directed models (Foster and Yeomans [109]). More refined simulations appear in Krawczyk, Owczarek, Prellberg and Rechnitzer [220]. For a recent overview, see Owczarek and Whittington [261].

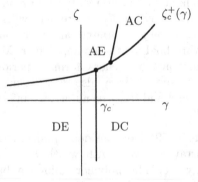

Fig. 7.17. Conjectured phase diagram for adsorption and collapse of self-avoiding polygons in $d = 3$. There are four phases: DE (= desorbed-extended), DC (= desorbed-collapsed), AE (= adsorbed-extended) and AC (= adsorbed-collapsed). Only the critical curve between DE+DC and AE+AC has been shown to exist. Note the occurrence of two tricritical points.

Similar behavior is expected for walks instead of polygons. However, for walks the free energy has only been shown to exists for $\zeta \in \mathbb{R}$ and $\gamma \leq 0$, i.e., for repulsive self-touchings, in which case it is equal to that for polygons (see [306]). It should be equal for all $\zeta, \gamma \in \mathbb{R}$.

(9) Velenik [305] provides an overview of pinning and wetting for interfaces, which are higher-dimensional analogues of polymers. The interaction Hamiltonian for these interfaces is rather general, allowing for a flexible description of a number of associated physical phenomena, such as the "roughening transition".

(10) Caputo, Martinelli and Toninelli [49] study the pinned SRW-polymer in the adsorbed phase subject to a Metropolis dynamics w.r.t. the Hamiltonian defined in (7.1–7.2) in which two successive increments $\nearrow\searrow$ can be replaced by $\searrow\nearrow$, and vice versa. The average transition time between a typical desorbed and a typical adsorbed configuration is computed. Dynamic models of polymers are important and interesting, but are mathematically largely unexplored. They fall outside the scope of the present monograph.

7.7 Challenges

(1) Prove that the phase transitions in the pinning and the wetting model of SAW, as described in Section 7.5, are second order. Prove that $\zeta_c = 0$. (It is already known that $\zeta_c^+ > 0$.) Show that $\zeta \mapsto f(\zeta)$ and $\zeta \mapsto f^+(\zeta)$ are infinitely differentiable on (ζ_c, ∞), respectively, (ζ_c^+, ∞).

(2) Prove the qualitative properties of the force-temperature diagram in Fig. 7.16 from Section 7.6, Extension (3).

(3) Prove the qualitative properties of the phase diagram in Fig. 7.17 from Section 7.6, Extension (8). In particular, show that there is a phase transition between AE and AC and between DE and DC. Determine the shape of the critical curve between AE and AC, and show that this curve does not meet the critical curve between DE+DC and AE+AC at the same point as the vertical critical curve between DE and DC does (i.e., show that there are two tricritical points as drawn). Try to determine the order of the phase transitions. All these are hard challenges.

(4) Extend the analysis in Sections 7.1–7.3 to the situation where the substrate has a finite width, i.e., in (7.2) the indicator $1_{\{w_i=0\}}$ is replaced by $1_{\{0 \leq w_i \leq N\}}$ for some $N \in \mathbb{N}$. It is conjectured in Caravenna, Giacomin and Zambotti [56] that the scaling limits of the path measure are the same for all $N \in \mathbb{N}$, both for pinning and wetting.

(5) What can be said about the multi-interface medium in Section 7.6, Extension (7) when the widths of the layers are random, i.e., when $\mathbb{Z}^2$ is sliced

into rows of random sizes $H_k L_n$, $k \in \mathbb{Z}$, with $(H_k)_{k \in \mathbb{Z}}$ i.i.d. random variables taking values in $[C_1, C_2]$ for some $0 < C_1 < C_2 < \infty$?

(6) For all tractable models of wetting based on SRW it is known that the desorption transition is second order in the temperature without a force and first order in the temperature with a force, no matter how small (see Orlandini, Tesi and Whittington [254]). Investigate to what extent this property is robust.

(7) Prove the robustness of the property that "the force is proportional to the logarithm of the inverse of the wedge angle" as described in Section 7.6, Extension (5).

Part B

Polymers in Random Environment

OUTLINE:

In Part B we look at five models of polymers in random environment.

- In Chapter 8, we consider a polymer with positive and negative *charges*, arranged randomly along the chain, resulting in repulsion and attraction at the locations where the polymer intersects itself. This is a model for a polymer in an *electrolyte*: the charges on the polymer are surrounded by an ionic atmosphere that screens the charges so that they are felt "on site" only. Our main result will be that the *annealed* charged polymer is *subdiffusive* in any $d \geq 1$. We obtain the full scaling behavior. The annealed model is appropriate for polymers consisting of *amphoteric* monomers, i.e., monomers containing both acidic and basic functional groups, since such monomers may change their charge as the pH of the solution in which they are immersed is varied.

- In Chapter 9, we look at a copolymer, consisting of a random concatenation of *hydrophobic* and *hydrophilic* monomers, interacting with two solvents, *oil* and *water*, separated by a *linear* interface. If the interaction between monomers and solvents is strong and unbiased enough, then the polymer *localizes* near the interface, otherwise it *delocalizes* away from the interface. Our main result is a detailed analysis of the *quenched* critical curve separating the two phases, the order of the phase transition, and the path behavior in each of the two phases.

- In Chapter 10, we extend the results of Chapter 9 to a model where the linear interface is replaced by a *percolation-type* interface. This is a primitive model for a copolymer in an *emulsion* consisting of *mesoscopically* large droplets of oil floating in water. Various forms of (de)localization are possible, depending on whether the oil droplets percolate or not.

- In Chapter 11, the polymer interacts with a linear substrate composed of different types of atoms or molecules, occurring in a random order. Each time the polymer hits the substrate it picks up a reward or a penalty according to the type of atom or molecule it encounters. If the reward to hit the substrate is strong enough, then the polymer *localizes*, otherwise it *delocalizes*. Both the *pinning* and the *wetting* version of the model are of interest. Our main result is a detailed analysis of the *quenched* critical curve separating the two phases, the order of the phase transition, and the path behavior in each of the two phases. Moreover, we investigate when the quenched and the annealed critical curve are different, referred to as *relevant disorder*, and when they coincide, referred to as *irrelevant disorder*. The wetting version of the model can be used to describe the *denaturation transition* of DNA, where the two strands detach with increasing temperature, as well as the *unzipping transition* of DNA under the action of a force.

- In Chapter 12, finally, we consider a polymer interacting with a *random potential* field. An example is a hydrophobic homopolymer living in a *microemulsion* consisting of *microscopically* small droplets of oil and water. Depending on the dimension and on the strength of the interaction, the polymer can be in a phase of *weak disorder* or a phase of *strong disorder*, where the polymer is diffusive, respectively, superdiffusive. Our main result is a classification of these two phases. Another application is a polymer in a *gel* with different sizes of pockets.

Except in Chapter 8, we will consider only *directed paths*, i.e., paths that cannot backtrack (already used in Section 6.2 and Chapter 7). This restriction is necessary in order to make the models mathematically tractable.

The presence of the random environment adds on a considerable layer of complication, and necessitates the development of new ideas and techniques. Some of these are successful in elucidating the behavior of the polymer, others only partially resolve key questions.

Polymers in random environment is a challenging and captivating area, with problems that are driven by specific applications. Many of these problems are "easy to state but hard to solve", and frequently cannot be tackled at full power with non-rigorous methods (such as mean-field approximations and replica computations), in the best of the tradition of mathematical physics.

WARNING: In Part B we use the symbol ω to denote the random environment. Be careful to distinguish this from the symbol w that is used to denote the polymer path.

8

Charged Polymers

In this chapter we consider a model introduced in Kantor and Kardar [203], where each monomer carries a *random charge*, and each self-intersection of the polymer is rewarded when the two charges of the associated monomers have opposite sign and is penalized when they have the same sign. This model is a variation on the weakly self-avoiding walk described in Chapters 3 and 4, with a random self-interaction driven by the charges.

We will focus on the *annealed* path measure, of the type defined in (1.5). We will show that the annealed charged polymer is in a *collapsed phase*, irrespective of its overall charge distribution, and is *subdiffusive* with a scaling limit that can be computed explicitly, namely, Brownian motion conditioned to stay inside a finite ball. The free energy will be different for neutral and for non-neutral charged polymers, even though the scaling limit is the same. Once more *local times* will prove to be useful. In particular, the large deviation behavior of the local times of SRW will play a crucial role in the identification of the scaling limit.

DNA and proteins are polyelectrolytes, with each monomer in a specific charged state that depends on the pH of the solution in which they are immersed. Our annealed analysis applies to polymers consisting of *amphoteric* monomers, i.e., monomers containing both acidic and basic functional groups. An amphoteric monomer can be positively charged at low pH and negatively charged at high pH. In equilibrium the individual monomers do not retain their charge: the charges fluctuate in time, with the overall charge of the polymer being more or less constant. For polyelectrolytes with amphoteric monomers there is a particular value of the pH, called the *iso-electric point*, at which the overall charge is zero, separating the regimes of positive and negative overall charge.

In Section 8.1 we define the model, in Section 8.2 we compute the annealed free energy, in Section 8.3 we prove the subdiffusive scaling, while in Section 8.4 we make a link with the parabolic Anderson model.

F. den Hollander, *Random Polymers*,
Lecture Notes in Mathematics 1974, DOI: 10.1007/978-3-642-00333-2_8,
© Springer-Verlag Berlin Heidelberg 2009, Reprint by Springer-Verlag Berlin Heidelberg 2012

8.1 A Polymer with Screened Random Charges

Let

$$\omega = (\omega_i)_{i\in\mathbb{N}_0} \text{ is i.i.d. with } \mathbb{P}_p(\omega_0 = +1) = p \text{ and } \mathbb{P}_p(\omega_0 = -1) = 1 - p, \quad (8.1)$$

with $p \in (0,1)$. This random sequence labels the charges of the monomers along the polymer (see Fig. 8.1), with p the density of the positive charges and $1 - p$ the density of the negative charges. Throughout the sequel, $\mathbb{P}_p$ denotes the law of ω.

Our set of paths and Hamiltonian are

$$\mathcal{W}_n = \{w = (w_i)_{i=0}^n \in (\mathbb{Z}^d)^{n+1} : w_0 = 0, \|w_{i+1} - w_i\| = 1 \ \forall 0 \le i < n\},$$
$$H_n^\omega(w) = \beta I_n^\omega(w),$$
$$(8.2)$$

where $\beta \in (0,\infty)$ and (compare with (3.2))

$$I_n^\omega(w) = \sum_{\substack{i,j=0 \\ i<j}}^n \omega_i\omega_j \, 1_{\{w_i=w_j\}}. \quad (8.3)$$

For fixed ω, the quenched path measure is

$$P_n^{\beta,\omega}(w) = \frac{1}{Z_n^{\beta,\omega}} e^{-\beta I_n^\omega(w)} P_n(w), \qquad w \in \mathcal{W}_n, \quad (8.4)$$

where we recall that P_n is the projection onto $\mathcal{W}_n$ of the law P of SRW. What this path measure does is reward self-intersections when the charges are opposite ($\omega_i\omega_j = -1$) and penalize self-intersections when the charges are the same ($\omega_i\omega_j = 1$). Note that the interaction between the charges is felt "on site" only, which is why we speak of *screened random charges*. Think of this as modeling a polymer in an electrolyte. Each charge gets surrounded by an ionic atmosphere, which reduces the spatial range of the charge-charge interaction (see Section 8.6, Challenge (4), for other options).

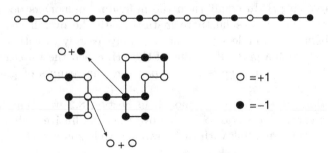

Fig. 8.1. A polymer carrying random charges. The path may or may not be self-avoiding. The charges only interact at self-intersections.

The cases $p = 0$ and $p = 1$ correspond to the weakly self-repellent random walk. Indeed, if $\omega \equiv +1$ or $\omega \equiv -1$, then all the charges are the same and there are only penalties.

We will be focusing on the *annealed model* with path measure (recall (1.5))

$$\mathbb{P}_n^{\beta,p}(w) = \frac{1}{\mathbb{Z}_n^{\beta,p}} \left[\int e^{-\beta I_n^\omega(w)} \mathbb{P}_p(d\omega) \right] P_n(w), \qquad w \in \mathcal{W}_n. \qquad (8.5)$$

This describes the situation in which the charges can vary along the polymer and can take part in the equilibration. Intuitively, we expect:

(a) $p = \frac{1}{2}$: The total charge is on average *neutral*, the repulsion and the attraction balance out, and the polymer behaves qualitatively like SRW, i.e., has a diffusive scaling.

(b) $p \neq \frac{1}{2}$: The total charge is on average *non-neutral*, but makes a large deviation under the annealed measure so that it becomes neutral. The price for this large deviation is an entropy factor that shows up in the free energy. Apart from this factor, the polymer behaves as if $p = \frac{1}{2}$.

In Sections 8.2–8.3 we will show that the above intuition is correct, except that actually the attraction wins from the repulsion as soon as the total charge is neutral, resulting in the polymer being *subdiffusive*.

8.2 Scaling of the Free Energy

8.2.1 Variational Characterization

The annealed partition sum equals

$$\mathbb{Z}_n^{\beta,p} = \mathbb{E}_p \left(Z_n^{\beta,\omega} \right) \quad \text{with} \quad Z_n^{\beta,\omega} = E \left(e^{-\beta I_n^\omega(w)} \right). \qquad (8.6)$$

(Note that $\log \mathbb{Z}_n^{\beta,p}$ is different from $\mathbb{E}_p(\log Z_n^{\beta,\omega})$, which is the average of the quenched free energy.) The following result is due to Biskup and König [21] for $p = \frac{1}{2}$. Note that β plays no role in the scaling behavior.

Theorem 8.1. *For every $d \geq 1$, $\beta \in (0,\infty)$ and $p \in (0,1)$,*

$$\lim_{n \to \infty} \frac{(\alpha_n)^2}{n} \left\{ \log \mathbb{Z}_n^{\beta,p} - n \frac{1}{2} \log[4p(1-p)] \right\} = -\chi \qquad (8.7)$$

with $\alpha_n = (n/\log n)^{1/(d+2)}$ and $\chi \in (0,\infty)$ given by (8.26) below.

Proof. Let (similarly as in (3.8))

$$\widehat{I}_n^\omega = \sum_{i,j=0}^n \omega_i \omega_j \, 1\{w_i = w_j\}. \qquad (8.8)$$

Then $\widehat{I}_n^\omega = 2I_n^\omega + (n+1)$. Hence, in (8.5) we may as well put $\widehat{I}_n^\omega$ in the exponential weight factor (which only changes β to 2β). We will do so and henceforth drop the overscript.

We begin by carrying out the first expectation in (8.6). To that end, define

$$G^{\beta,p}(l) = \log \mathbb{E}_p\left(\exp\left[-\beta\left(\sum_{k=1}^{l}\omega_k\right)^2\right]\right), \qquad l \in \mathbb{N}_0. \tag{8.9}$$

Then, writing

$$I_n^\omega(w) = \sum_{i,j=0}^{n} \omega_i\omega_j \sum_{x\in\mathbb{Z}^d} 1\{w_i = x\}1\{w_j = x\} = \sum_{x\in\mathbb{Z}^d}\left(\sum_{i=0}^{n}\omega_i 1\{w_i = x\}\right)^2, \tag{8.10}$$

using Fubini's theorem to interchange expectations, noting that ω is i.i.d., and using (8.9), we may rewrite (8.6) as

$$\begin{aligned}
\mathbb{Z}_n^{\beta,p} &= E\left(\mathbb{E}_p\left(e^{-\beta I_n^\omega(w)}\right)\right) \\
&= E\left(\mathbb{E}_p\left(\prod_{x\in\mathbb{Z}^d}\exp\left[-\beta\left(\sum_{k=1}^{\ell_n(x)}\omega_{\tau_k(x)}\right)^2\right]\right)\right) \\
&= E\left(\exp\left[\sum_{x\in\mathbb{Z}^d} G^{\beta,p}(\ell_n(x))\right]\right),
\end{aligned} \tag{8.11}$$

where $\tau_k(x)$ is the time of the k-th visit of w to site x and $\ell_n(x)$ is the local time of w at site x defined in (3.9).

The following lemma allows us to reduce the proof to the case $p = \frac{1}{2}$, and shows that $G^{\beta,\frac{1}{2}}(l)$ tends to $-\infty$ at a very slow rate, which will serve us later on.

Lemma 8.2. *For every $\beta \in (0,1)$ and $p \in (0,1)$,*

$$G^{\beta,p}(l) = A(p)l - \tfrac{1}{2}\log l + O(1) \qquad as\ l \to \infty \tag{8.12}$$

with $A(p) = \frac{1}{2}\log[4p(1-p)] \leq 0$.

Proof. Abbreviate $\Sigma_l = \sum_{k=0}^{l}\omega_k$, $l \in \mathbb{N}$. A change of measure from $\mathbb{P}_p$ to $\mathbb{P}_{\frac{1}{2}}$ gives

$$G^{\beta,p}(l) = Al + G_*^{\beta,p}(l) \tag{8.13}$$

with

$$G_*^{\beta,p}(l) = \log\mathbb{E}_{\frac{1}{2}}\left(\exp\left[-\beta(\Sigma_l)^2 + B\Sigma_l\right]\right), \tag{8.14}$$

where $A = A(p) = \frac{1}{2}\log[4p(1-p)]$ and $B = B(p) = \frac{1}{2}\log[p/(1-p)]$. To prove (8.12), we pick $N \in \mathbb{N}$ large and estimate

$$e^{G_*^{\beta,p}(l)} \geq e^{-\beta N^2 - |B|N} \, \mathbb{P}_{\frac{1}{2}}(|\Sigma_l| \leq N),$$
$$e^{G_*^{\beta,p}(l)} \leq e^{-\beta N^2 + |B|N} + e^{B^2/4\beta} \, \mathbb{P}_{\frac{1}{2}}(|\Sigma_l| \leq N). \tag{8.15}$$

Here, we use that $-\beta x^2 + Bx$ is maximal at $x = B/2\beta$, where it assumes the value $B^2/4\beta$, and is non-increasing in $|x|$ for $|x|$ large enough. By the local central limit theorem we have, for fixed N,

$$\mathbb{P}_{\frac{1}{2}}(|\Sigma_l| \leq N) \sim C(N)\, l^{-1/2} \quad \text{as } l \to \infty \tag{8.16}$$

for some $C(N) \in (0, \infty)$. Substituting this into (8.15) and letting $l \to \infty$ followed by $N \to \infty$, we obtain

$$G_*^{\beta,p}(l) = -\tfrac{1}{2} \log l + O(1) \quad \text{as } l \to \infty, \tag{8.17}$$

with the dependence on β and p sitting in the error term only. Combining (8.13) and (8.17), we get the claim in (8.12). Note that the above argument shows that for large l the term $B\Sigma_l$ in (8.14) is negligible, implying that $G_*^{\beta,p}(l) = G^{\beta,\frac{1}{2}}(l) + O(1)$ as $l \to \infty$. $\square$

Because $\sum_{x \in \mathbb{Z}^d} \ell_n(x) = n+1$, it follows from (8.11) and (8.13) that

$$Z_n^{\beta,p} = e^{A(n+1)} \, E \left(\exp \left[\sum_{x \in \mathbb{Z}^d} G_*^{\beta,p}(\ell_n(x)) \right] \right). \tag{8.18}$$

The frontfactor in (8.18) explains the shift of the free energy in (8.7). Thus, it remains to identify the scaling behavior of the expectation in (8.18) with the help of (8.17).

8.2.2 Heuristics

For all β and p,

$$[G^{\beta,p}(l+1) - A] \geq [G^{\beta,p}(l) - A] + [G^{\beta,p}(1) - A] \quad \forall l \in \mathbb{N}_0. \tag{8.19}$$

Indeed, from (8.9) we have

$$e^{G^{\beta,p}(l+1)} = E_p \left(e^{-\beta(\Sigma_{l+1})^2} \right) = E_p \left(e^{-\beta(\Sigma_l)^2} e^{-\beta} \left[p e^{-2\beta \Sigma_l} + (1-p) e^{2\beta \Sigma_l} \right] \right)$$
$$\geq e^{G^{\beta,p}(l)} e^{-\beta} [4p(1-p)]^{1/2} = e^{G^{\beta,p}(l)} e^{G^{\beta,p}(1)} e^A, \tag{8.20}$$

where we use that $pe^{-2\beta x} + (1-p)e^{2\beta x} \geq [4p(1-p)]^{1/2}$ for all $\beta, x \in \mathbb{R}$. Because $G^{\beta,p}$ is negative, we may refer to (8.19) as sublinearity. This property is inherited by $G_*^{\beta,p}$ defined in (8.13). A consequence of this sublinearity is that the local times in (8.18), which are subject to the restriction $\sum_{x \in \mathbb{Z}^d} \ell_n(x) = n+1$, have a tendency to "pile up" rather than to "spread out". This explains

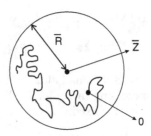

Fig. 8.2. A random walk starting at 0 conditioned to stay inside the ball with radius $R_n \sim \bar{R}\alpha_n$ and center $Z_n \sim \bar{Z}\alpha_n$.

why the polymer has a tendency to be subdiffusive. The following argument, which makes this intuition precise, is sketchy. We will indicate later how to fill in the details.

The maximal contribution to $\mathbb{Z}_n^{\beta,p}$ comes from a strategy where the polymer stays inside $B_{R_n}(0) \subset \mathbb{R}^d$, the closed ball of radius R_n (to be determined later) centered at 0, and distributes its local times evenly over this ball (see Fig. 8.2). The cost for staying inside $B_{R_n}(0)$ up to time n is

$$\exp\left[-\frac{\lambda_d\, n}{(R_n)^2}\,[1+o(1)]\right] \qquad \text{as } n \to \infty, \tag{8.21}$$

where λ_d is the principal Dirichlet eigenvalue of $-\Delta_{\mathbb{R}^d}/2d$ on $B_1(0)$, with $\Delta_{\mathbb{R}^d}$ the continuous Laplacian. Conditionally on the polymer staying inside $B_{R_n}(0)$ up to time n, the local time per site is $(n/\omega_d(R_n)^d)[1+o(1)]$, with $\omega_d = |B_1(0)|$. Since, by (8.17), $l \mapsto G_*^{\beta,p}(l)$ is almost constant for large l, the contribution of this strategy to (8.18) is

$$\exp\left[-\frac{1}{2}\omega_d(R_n)^d \log\left(\frac{n}{\omega_d(R_n)^d}\right)[1+o(1)]\right] \qquad \text{as } n \to \infty. \tag{8.22}$$

The two exponentials in (8.21–8.22) are of the same order when we choose R_n such that

$$\frac{n}{(R_n)^2} \asymp (R_n)^d \log\left(\frac{n}{(R_n)^d}\right), \tag{8.23}$$

i.e., $R_n \asymp \alpha_n = (n/\log n)^{1/(d+2)}$ (where $\asymp$ means that the ratio of the two sides is bounded above and below by strictly positive and finite constants). Therefore, putting $R_n = \bar{R}\alpha_n$, we get (8.7) with

$$\chi = \inf_{R\in(0,\infty)} \left[\frac{\lambda_d}{R^2} + \frac{1}{d+2}\omega_d\, R^d\right], \tag{8.24}$$

where the scaled radius R is to be optimized over. The infimum is achieved at $R = \bar{R}$ with

$$\bar{R} = \left[\frac{2(d+2)\lambda_d}{d\omega_d}\right]^{1/(d+2)}, \tag{8.25}$$

and equals

$$\chi = \left[\frac{\omega_d}{2}\right]^{2/(d+2)} \left[\frac{(d+2)\lambda_d}{d}\right]^{d/(d+2)}. \tag{8.26}$$

8.2.3 Large Deviations

The above argument can be made rigorous with the help of large deviation techniques. For background on what follows next, we refer the reader to the Saint-Flour lectures by Bolthausen [28], Sections 2.5–2.6. First, the strategy of staying inside a large ball and distributing the local times evenly clearly provides a *lower bound* for the free energy. Second, the fact that no other strategy does better, is proved in three steps:

(1) *Compactification*: The random walk is "wrapped around" a large box containing $B_{R_n}(0)$, i.e., is periodized w.r.t. this box. By the sublinearity of $l \mapsto G_*^{\beta,p}(l)$, this wrapping increases the contribution to the expectation in (8.18), thus providing an *upper bound* for the free energy.

(2) *Scaling*: For the wrapped random walk, after scaling the large box down to a finite box, the scaled local times are close to those of a "mollified" Brownian motion. The latter satisfy an LDP, with a rate function given by the standard Donsker-Varadhan rate function for the empirical measure of Brownian motion.

(3) *Optimization*: By optimizing over the profile of the scaled local times, we get (8.7) with χ given by a functional variational problem involving empirical measures on a finite box. Here it is important that, because $l \mapsto G_*^{\beta,p}(l)$ is almost constant for large l, only the *order* of the local times is relevant, not the precise form of their spatial profile. Accordingly, the functional variational problem for χ reduces to the scalar variational problem stated in (8.24), in which only λ_d and ω_d enter. The fact that the empirical measures take their support in a ball is due to radial symmetry in combination with a standard isoperimetric inequality.

The full argument requires a number of smoothing and approximation techniques. It is easy to show that the optimal path spends a "negligible" amount of time outside the ball $B_{R_n}(0)$. It is harder to show that the optimal path never leaves the ball. For the latter we refer to Bolthausen [27]. □

8.3 Subdiffusive Behavior

The following theorem shows that the annealed charged polymer is in the *collapsed phase*, with subdiffusive scaling, irrespective of its overall charge.

Theorem 8.3. *For every $d \geq 1$, $\beta \in (0, \infty)$ and $p \in (0, 1)$, under the path measure $\mathbb{P}_n^{\beta,p}$,*

$$\left(\frac{1}{\alpha_n} S_{\lfloor nt \rfloor}\right)_{0 \le t \le 1} \Longrightarrow (\Theta_t)_{0 \le t \le 1} \qquad as \ n \to \infty \qquad (8.27)$$

with $(\Theta_t)_{t \ge 0}$ *Brownian motion on* $\mathbb{R}^d$ *conditioned on not leaving the ball with radius* $\bar{R}$ *given by (8.25) and with a randomly shifted center* $\bar{Z}$.

Proof. The proof of Theorem 8.1 shows that the annealed charged polymer up to time n lives in a ball of radius R_n (see Fig. 8.2). The profile of its scaled local times is given by the eigenfunction associated with λ_d. Its scaled endpoint is distributed as the endpoint of a Brownian motion on a ball with radius 1, with a drift towards the center of this ball according to this eigenfunction. The center of the ball is not 0, but is randomly shifted according to that same eigenfunction. We again refer to Bolthausen [27] for details; the behavior is the same as for the range of SRW. □

8.4 Parabolic Anderson Equation

An interesting connection is the following. Let $S = (S_t)_{t \ge 0}$ be continuous-time SRW on $\mathbb{Z}^d$ jumping at rate 1, and let $(\ell_t)_{t \ge 0}$ be its associated local time process defined by $\ell_t(x) = \int_0^t 1\{S_t = x\} dt$, $x \in \mathbb{Z}^d$, $t \ge 0$. Define

$$Z_t^{\beta,p} = E\left(\exp\left[\sum_{x \in \mathbb{Z}^d} G^{\beta,p}(\ell_t(x))\right]\right), \qquad t \ge 0, \qquad (8.28)$$

which is the continuous-time analogue of (8.11). Then

$$Z_t^{\beta,p} = \mathbb{E}(u(0,t)), \qquad (8.29)$$

with $u \colon \mathbb{Z}^d \times [0,\infty) \to \mathbb{R}$ the solution of the *parabolic Anderson equation*

$$\begin{cases} \frac{\partial u}{\partial t}(x,t) = \Delta_{\mathbb{Z}^d} u(x,t) + \xi(x) u(x,t), & x \in \mathbb{Z}^d, \ t \ge 0, \\ u(x,0) = 1, & x \in \mathbb{Z}^d, \end{cases} \qquad (8.30)$$

where $\Delta_{\mathbb{Z}^d}$ is the discrete Laplacian and $\xi = \{\xi(x) \colon x \in \mathbb{Z}^d\}$ is an i.i.d. field of non-positive random variables whose probability law is given by the moment generating function

$$\mathbb{E}\left(e^{l\xi(0)}\right) = e^{G^{\beta,p}(l)}, \qquad l \in \mathbb{N}_0, \qquad (8.31)$$

where we write $\mathbb{E}$ to denote expectation w.r.t. ξ. The relation between (8.28) and (8.29–8.31) is an immediate consequence of the *Feynman-Kac representation* of the solution of (8.30) given ξ, which reads

$$u(x,t) = E\left(\exp\left[\int_0^t \xi(S_s) ds\right]\right) = E\left(\exp\left[\sum_{x \in \mathbb{Z}^d} \xi(x) \ell_t(x)\right]\right). \qquad (8.32)$$

Indeed, if (8.32) is substituted into (8.29), Fubini's theorem is used to inter-change the two expectations, and (8.31) is used afterwards, then the result is precisely (8.28). In Biskup and König [21] the annealed Lyapunov exponents of (8.30), i.e., the growth rates of the successive moments of $u(0, t)$, are com-puted for several choices of $\xi(0)$ whose moment generating function satisfies an appropriate scaling assumption. The quantity of interest in Theorem 8.1 corresponds to the first Lyapunov exponent for the special case where $\xi(0)$ is given by (8.31).

Remark: The moment generating function in (8.31) actually is not completely monotone as a function of l and therefore does not properly define the random variable $\xi(0)$. However, this can be repaired by perturbing $G^{\beta,p}(l)$ slightly, without affecting its asymptotics in (8.17).

The parabolic Anderson equation is used to model the evolution of reac-tant particles in the presence of a catalyst. For an overview, see Gärtner and König [113].

8.5 Extensions

(1) The argument in Sections 8.2–8.3 is easily extended to random charges that take values in $\mathbb{R}$, provided they have a finite moment generating function.

(2) Chen [66] studies the annealed scaling behavior of $I_n^\omega(w)$ defined in (8.3) for random charges that have a symmetric distribution with a finite moment generating function (so that the total charge is on average neutral). A central limit theorem is derived, a law of the iterated logarithm, as well as moder-ate and large deviation estimates. The behavior is different in $d = 1$, $d = 2$ and $d \geq 3$. The moderate and large deviation estimates are further refined in Asselah [11]. Chen [67] contains an extensive study of the large deviation prop-erties of random walk intersections. This is useful also for the self-repellent polymer studied in Chapters 3–4.

(3) For the quenched model, with path measure $P_n^{\beta,\omega}$ given by (8.4), the two main questions are: (i) Is the free energy self-averaging in ω? (ii) Is there a phase transition from a collapsed state to an extended state at some $\beta_c \in (0, \infty)$? Both these questions remain unresolved. For $d = 1$, partial results have been obtained in a different direction, as we describe in Extensions (4), (5) and (6) below.

(4) Derrida, Griffiths and Higgs [86] and Derrida and Higgs [88] consider the case where the steps of the random walk are drawn from $\{0, 1\}$ rather than $\{-1, +1\}$, which makes the model a bit more tractable, both analytically and numerically. In [86] the charge disorder is binary, and numerical evidence is found for the free energy to be self-averaging and to exhibit a "freezing transition" at a critical threshold $\beta_c \in (0, \infty)$, i.e., the quenched charged

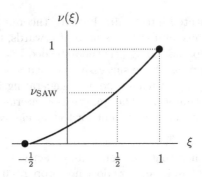

Fig. 8.3. Qualitative shape of $\xi \mapsto \nu(\xi)$ for the one-sided charged polymer conditioned on the overall charge.

polymer is ballistic when $\beta < \beta_c$ and subballistic when $\beta > \beta_c$. In the latter regime it seems that the end-to-end distance scales like n^ν, with $\nu = \nu(\beta)$ depending on β. In this regime, long and rare stretches of the polymer that are globally neutral find it energetically favorable to collapse onto single sites. Numerical simulation gives that $\beta_c \geq 0.48$. In [88] the charge disorder is standard normal and the total charge $\sum_{i=1}^n \omega_i$ is *conditioned* to grow like n^ξ, $\xi \in [-\frac{1}{2}, 1]$. It is found numerically that the end-to-end distance scales like n^ν, with $\nu = \nu(\xi)$ depending on ξ and growing roughly linearly from $\nu(-\frac{1}{2}) = 0$ to $\nu(1) = 1$, with $\nu(\frac{1}{2}) \approx 0.574$ (see Fig. 8.3). The latter is the exponent for the quenched charged polymer when the charges are typical.

(5) Martinez and Petritis [238] choose the charge distribution to be given by the increments of a SRW *conditioned* to return to 0 after n steps, which guarantees that the total charge is neutral. Instead of looking at the free energy conditioned on the charges, they study the free energy *conditioned on the path and averaged over the disorder*, i.e.,

$$\frac{1}{n} \log \mathbb{E}\left(e^{-\beta I_n^\omega(w)} \mid w\right), \qquad w \in \mathcal{W}_n, \tag{8.33}$$

and show that this is self-averaging in w in the limit as $n \to \infty$, and has no phase transition in β.

(6) A continuum version of the model in [238] is studied in Buffet and Pulé [46]. Here, the path of the polymer is given by a Brownian motion $B = (B_t)_{t \in [0,1]}$, the charge distribution is given by a Brownian bridge $b = (b_t)_{t \in [0,1]}$, while the interaction is no longer "on site" but is given by a two-body potential $h \colon \mathbb{R} \to \mathbb{R}$ of the form

$$h(x) = \int_\mathbb{R} g(x - y)g(y)\,\mathrm{d}y, \qquad x \in \mathbb{R}, \tag{8.34}$$

for some $g \colon \mathbb{R} \to \mathbb{R}$ that is square integrable, so that h is even, positive definite, bounded and integrable. The interaction Hamiltonian is taken to be

$$H_t^{\beta,b}(B) = \frac{\beta t}{2} \int_0^1 db_u \int_0^1 db_v \, h\big(t(B_u - B_v)\big), \tag{8.35}$$

where t is the length of the polymer, and time is scaled down to $[0,1]$ for convenience: tB_u is the position of the polymer at time u, $\sqrt{t}b_u$ is the total charge carried by the polymer up to time u. The object of interest is the free energy conditioned on B and averaged over b, i.e.,

$$f(\beta; B) = \lim_{t \to \infty} \frac{1}{t} \log Z_t^\beta(B) \qquad B - a.s. \tag{8.36}$$

with

$$Z_t^\beta(B) = \mathbb{E}\left(e^{-H_t^{\beta,b}(B)} \mid B\right), \tag{8.37}$$

where $\mathbb{E}$ denotes expectation over b. The latter partition sum is computed in terms of a series involving the local times of B. It turns out that, in contrast with the discrete model in [238], the free energy is *not* self-averaging, i.e., $f(\beta; B)$ depends on B. However, $\beta \mapsto f(\beta; B)$ is shown to be analytic on $[0, \infty)$, which rules out a phase transition. The analysis indicates that there is a collapse transition at some $\beta_c = \beta_c(B) \in (-\infty, 0)$, which corresponds to the "unphysical" regime where charges of the same sign attract each other and charges of the opposite sign repel each other. For $\beta < \beta_c(B)$ the free energy diverges (because $\log Z_t^\beta(B)$ scales differently with t, similarly as in (6.7) for the polymer with self-intersections and self-touchings considered in Chapter 6). The analysis gives an explicit upper bound for $\beta_c(B)$, namely,

$$\beta_c(B) \leq -\left[L(B) \int_{\mathbb{R}} h(x)dx\right]^{-1}. \tag{8.38}$$

with $L(B)$ the maximal local time of B on $\mathbb{R}$.

(7) Almost nothing is known rigorously for SAW with random charges. The appropriate modifications of (8.2–8.3), relevant only for $d \geq 2$, read

$$\mathcal{W}_n = \big\{w = (w_i)_{i=0}^n \in (\mathbb{Z}^d)^{n+1}: \, w_0 = 0, \, \|w_{i+1} - w_i\| = 1 \, \forall \, 0 \leq i < n,$$
$$w_i \neq w_j \, \forall \, 0 \leq i < j \leq n\},$$
$$H_n^\omega(w) = \beta I_n^\omega(w), \tag{8.39}$$

and

$$I_n^\omega(w) = \frac{1}{2d} \sum_{\substack{i,j=0 \\ i<j}}^n \omega_i \omega_j \, 1_{\{|w_i - w_j|=1\}}. \tag{8.40}$$

For $d = 2$ and $d = 3$, a number of results have been obtained with the help of exact enumeration, series analysis and Monte Carlo techniques. Monari and Stella [242] argue that for $p = \frac{1}{2}$ there is a collapse transition as β *increases* from 0 through a critical value $\beta_c^+(\frac{1}{2}) \in (0, \infty)$. This transition seems to be in the same universality class as the θ-transition in homopolymers mentioned

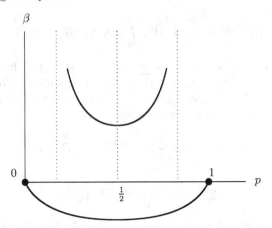

Fig. 8.4. Conjectured phase diagram for SAW with random charges. The top curve is $p \mapsto \beta_c^+(p)$, the lower curve is $p \mapsto \beta_c^-(p)$. The polymer is collapsed both above the top curve and below the bottom curve, and is extended elsewhere.

in Section 6.1.3. Kantor and Kardar [204], Grassberger and Hegger [127] and Golding and Kantor [126] find that, for $p > \frac{1}{2}$ but close to $\frac{1}{2}$, a similar collapse transition occurs, but at a value $\beta_c^+(p) \in (0, \infty)$ that increases with p, while for p close to 1 there is no collapse transition, i.e., $\beta_c^+(p) = \infty$, and the behavior is like SAW for all $\beta \in (0, \infty)$ (see Fig. 8.4). As β *decreases* from 0, there appears to be a collapse transition at a critical value $\beta_c^-(p) \in (-\infty, 0)$ for all $p \in [\frac{1}{2}, 1)$. This corresponds to the "unphysical" regime where charges of the same sign attract each other and charges of the opposite sign repel each other. In this regime, for $p = 1$ the model is the same as that of the SAW-version of the homopolymer in Section 6.1 with $\gamma = -\beta$ (recall (6.1–6.2)), and should be qualitatively similar for $p \in [\frac{1}{2}, 1)$ as well (see Fig. 8.4).

For further background we refer to the overview paper by Soteros and Whittington [283], Sections 2.3, 3.2.3 and 4.2.3. The "unphysical" regime mentioned in Extensions (6) and (7) can be reinterpreted as describing a copolymer consisting of hydrophobic and hydrophilic monomers, a topic that will be treated in Chapter 9.

8.6 Challenges

(1) Recall (1.3–1.5). Find the growth rate of the *quenched* free energy $\log Z_n^{\beta,\omega}$, the partition sum appearing in (8.4). Is this rate ω-a.s. constant or is it sample dependent? What about the *average quenched* free energy $\mathbb{E}_p(\log Z_n^{\beta,\omega})$?

(2) For $p = \frac{1}{2}$, find out whether or not the polymer is ω-a.s. subdiffusive under the quenched law $P_n^{\beta,\omega}$ as $n \to \infty$. The fluctuations of the charges in ω are expected to push the polymer farther apart.

(3) For $p \neq \frac{1}{2}$, show that, under the quenched law $P_n^{\beta,\omega}$ as $n \to \infty$, the charged polymer ω-a.s. has the same qualitative behavior as the soft polymer (which corresponds to the case $p = 0$ and $p = 1$). In particular, in $d = 1$ show that the charged polymer ω-a.s. is ballistic and that its speed is smaller than that of the soft polymer studied in Chapter 3.

(4) Investigate whether it is possible to deal with charges whose interaction extends beyond the "on site" interaction in (8.3), like a Coulomb potential (polynomial decay) or a Yukawa potential (exponential decay). A Yukawa potential arises from a Coulomb potential via screening of the charges.

9

Copolymers Near a Linear Selective Interface

A *copolymer* is a polymer consisting of different types of monomer. In this chapter we consider a two-dimensional *directed* copolymer, consisting of a random concatenation of *hydrophobic* and *hydrophilic* monomers, near a linear interface separating two immiscible solvents, *oil* and *water* (see Fig. 9.1). We will be interested in the *quenched* path measure (of the type defined in (1.3)). We will show that, as a function of the strength and the bias of the interaction between the monomers and the solvents, this model has a phase transition between a *localized* phase, where the copolymer stays near the interface, and a *delocalized* phase, where the copolymer wanders away from the interface. The critical curve separating the two phases has interesting properties, some of which remain to be clarified. The main techniques used are the *subadditive ergodic theorem*, the *method of excursions*, *large deviations* and *partial annealing estimates*.

In Section 9.1 we define the model, in Section 9.2 we study the free energy, in Section 9.3 we prove the existence of the phase transition, in Section 9.4 we identify the qualitative properties of the critical curve, while in Section 9.5 we describe the qualitative properties of the two phases.

In Chapter 10 we will turn to a model where the linear interface is replaced by a random interface, coming from large blocks of oil and water arranged in a percolation-type fashion.

The order of the monomers is determined by the *polymerization process* through which the copolymer is grown (see Section 1.1). Since the monomers cannot reconfigure themselves without some chemical reaction occurring, a copolymer is an example of a system with *quenched* disorder. Different copolymers typically have different orderings of monomers. Therefore, in order to determine the physical properties of a system of copolymers, an average must be taken over the possible monomer orderings. This is why it is natural to consider the monomers as being drawn *randomly* according to some appropriate probability distribution, i.i.d. in our case.

Copolymers near liquid-liquid interfaces are of interest due to their extensive application as surfactants, emulsifiers, and foaming or antifoaming agents.

F. den Hollander, *Random Polymers*,
Lecture Notes in Mathematics 1974, DOI: 10.1007/978-3-642-00333-2_9,
© Springer-Verlag Berlin Heidelberg 2009, Reprint by Springer-Verlag Berlin Heidelberg 2012

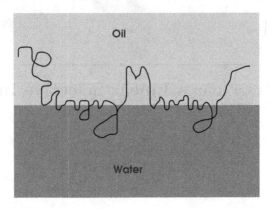

Fig. 9.1. An undirected copolymer near a linear interface.

Many fats contain stretches of hydrophobic and hydrophilic monomers, arranged "in some sort of erratic manner", and therefore are examples of random copolymers. The transition between a localized and a delocalized phase has been observed experimentally e.g. in neutron reflection studies of copolymers consisting of blocks of ethylene oxide and propylene oxide near a hexane-water interface (Phipps, Richardson, Cosgrove and Eaglesham [266]). Here, a thin layer of hexane, approximately 10^{-5} m thick, is spread on water. In the localized phase, the copolymer is found to stretch itself along the interface in a band of width approximately 20 Å.

Soteros and Whittington [283], Sections 2.2, 3.2.2 and 4.2.2, describes a number of rigorous, approximate, numerical and heuristic results for copolymers near linear interfaces. An earlier reference is Whittington [316]. Giacomin [116], Chapters 6–9, gives an overview of rigorous and numerical results based on the method of excursions.

9.1 A Copolymer Interacting with Two Solvents

For $n \in \mathbb{N}_0$, let

$$\mathcal{W}_n = \left\{ w = (i, w_i)_{i=0}^n : w_0 = 0, w_{i+1} - w_i = \pm 1 \ \forall \, 0 \leq i < n \right\} \tag{9.1}$$

denote the set of all n-step directed paths in $\mathbb{N}_0 \times \mathbb{Z}$ that start from the origin and at each step move either north-east or south-east (called ballot paths; recall Fig. 1.4). This is the set of configurations of the copolymer. An example of a path in $\mathcal{W}_n$ is drawn in Fig. 9.2. Let

$$\omega = (\omega_i)_{i \in \mathbb{N}} \text{ be i.i.d. with } \mathbb{P}(\omega_1 = +1) = \mathbb{P}(\omega_1 = -1) = \tfrac{1}{2}. \tag{9.2}$$

This random sequence labels the order of the monomers along the copolymer (see Fig. 9.2). Throughout the sequel, $\mathbb{P}$ denotes the law of ω. As Hamiltonian we pick, for fixed ω,

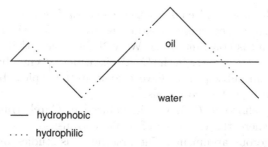

Fig. 9.2. A 14-step directed copolymer near a linear interface. Oil is above the interface, water is below. The drawn edges are hydrophobic monomers, the dashed edges are hydrophilic monomers.

$$H_n^\omega(w) = -\lambda \sum_{i=1}^{n} (\omega_i + h) \operatorname{sign}(w_{i-1}, w_i), \qquad w \in \mathcal{W}_n, \qquad (9.3)$$

with $\operatorname{sign}(w_{i-1}, w_i) = \pm 1$ depending on whether the edge (w_{i-1}, w_i) lies above or below the horizontal axis, and $\lambda, h \in \mathbb{R}$. Without loss of generality we may restrict the interaction parameters to the quadrant

$$\text{QUA} = \{(\lambda, h) \in [0, \infty)^2\}. \qquad (9.4)$$

The choice in (9.3) has the following interpretation. Think of $w \in \mathcal{W}_n$ as a directed copolymer on $\mathbb{N}_0 \times \mathbb{Z}$, consisting of n monomers represented by the edges in the path (rather than the sites). The lower halfplane is water, the upper halfplane is oil. The monomers are labeled by ω, with $\omega_i = +1$ meaning that monomer i is hydrophobic and $\omega_i = -1$ that it is hydrophilic. The term $\operatorname{sign}(w_{i-1}, w_i)$ equals $+1$ or -1 depending on whether monomer i lies in the oil or in the water. Thus, the Hamiltonian rewards matches and penalizes mismatches of the chemical affinities between the monomers and the solvents. The parameter λ is the *disorder strength*. The parameter h plays the role of an *disorder bias*: $h = 0$ corresponds to the hydrophobic and hydrophilic monomers interacting equally strongly, while $h = 1$ corresponds to the hydrophilic monomers not interacting at all. Only the regime $h \in [0, 1]$ is of interest (because for $h > 1$ both types of monomers prefer to be in the oil and the copolymer is always delocalized).

The law of the copolymer given ω is denoted by P_n^ω and is defined as in (1.3), i.e.,

$$P_n^\omega(w) = \frac{1}{Z_n^\omega} e^{-H_n^\omega(w)} P_n(w), \qquad w \in \mathcal{W}_n, \qquad (9.5)$$

where P_n is the law of the n-step directed random walk, which is the uniform distribution on $\mathcal{W}_n$. This law is the projection on $\mathcal{W}_n$ of the law P of the infinite directed walk whose vertical steps are SRW. Note that we have suppressed λ, h from the notation.

In what follows we will consider the *quenched free energy* $f(\lambda, h)$ of the copolymer, i.e., the free energy conditioned on ω. We will first show that $f(\lambda, h)$ exists and is constant ω-a.s. We will then show that $f(\lambda, h)$ is non-analytic along a *critical curve* in the (λ, h)-plane, derive a number of properties of this curve, and subsequently have a look at the typical behavior of the copolymer in each of the two phases.

The model defined in (9.3) was introduced in Garel, Huse, Leibler and Orland [111], where the existence of a phase transition was argued on the basis of non-rigorous arguments. The first rigorous studies were carried out in Sinai [274] and in Bolthausen and den Hollander [31]. The latter paper proved the existence of a phase transition, derived a number of properties of the critical curve, and raised a number of questions. Since then, several papers have appeared in which these questions have been settled and various aspects of the model have been further elucidated, leading not only to many interesting results, but also to challenging open problems. We will refer to these papers as we go along.

In Giacomin [116], Chapters 6–9, the more general situation is considered where the ω_i are $\mathbb{R}$-valued with a finite moment generating function and the $w_{i+1} - w_i$ are the increments of a random walk in the domain of attraction of a stable law. We will comment on this extension in Section 9.6. The special situation treated below (SRW and binary disorder) already collects the main ideas.

A key observation is the following. The energy of a path is a sum of contributions coming from its *successive excursions* away from the interface (see Fig. 9.2). What is relevant for the energy of these excursions is when they start and end, and whether they are above or below the interface. This simplifying feature, together with the directed nature of the path, is of great help when we want to do explicit calculations.

9.2 The Free Energy

Our starting point is the following self-averaging property, which is taken from Bolthausen and den Hollander [31].

Theorem 9.1. *For every* $\lambda, h \in \mathrm{QUA}$,

$$f(\lambda, h) = \lim_{n \to \infty} \frac{1}{n} \log Z_n^\omega \tag{9.6}$$

exists ω-*a.s. and in* $\mathbb{P}$-*mean, and is constant* ω-*a.s.*

Proof. The proof consists of two parts. In Lemma 9.2 we prove that the claim holds when the random walk is constrained to return to the origin at time $2n$. In Lemma 9.3 we show how to remove this constraint.

Fix $\lambda, h \in \mathrm{QUA}$. Abbreviate

$$\Delta_i = \mathrm{sign}(S_{i-1}, S_i) \tag{9.7}$$

and define

$$Z_{2n}^{\omega,0} = E\left(\exp\left[\lambda \sum_{i=1}^{2n}(\omega_i + h)\Delta_i\right] 1_{\{S_{2n}=0\}}\right), \tag{9.8}$$

where we recall that P is the law of SRW, $S = (S_i)_{i\in\mathbb{N}_0}$.

Lemma 9.2. $\lim_{n\to\infty} \frac{1}{2n} \log Z_{2n}^{\omega,0}$ exists ω-a.s. and in $\mathbb{P}$-mean, and is constant ω-a.s.

Proof. We need the following three properties:

(I) $Z_{2n}^{\omega,0} \geq Z_{2m}^{\omega,0} Z_{2n-2m}^{T^{2m}\omega,0}$ for all $0 \leq m \leq n$, with T the left-shift acting on ω as $(T\omega)_i = \omega_{i+1}$, $i \in \mathbb{N}$.
(II) $n \to \frac{1}{2n} \mathbb{E}(\log Z_{2n}^{\omega,0})$ is bounded from above.
(III) $\mathbb{P}(T\omega \in \cdot) = \mathbb{P}(\omega \in \cdot)$.

Property (I) follows from (9.8) by inserting an extra indicator $1_{\{S_{2m}=0\}}$ and using the Markov property of S at time $2m$. Property (II) holds because

$$\mathbb{E}\left(\log Z_{2n}^{\omega,0}\right) \leq \log \mathbb{E}\left(Z_{2n}^{\omega,0}\right)$$

$$= \log E\left((\cosh\lambda)^{2n} \exp\left[\lambda h \sum_{i=1}^{2n}\Delta_i\right] 1_{\{S_{2n}=0\}}\right) \tag{9.9}$$

$$\leq 2n(\log\cosh\lambda + \lambda h).$$

Property (III) is trivial. Thus, $\omega \mapsto (\log Z_{2n}^{\omega,0})_{n\in\mathbb{N}_0}$ is a *superadditive random process* (which is the analogue of a superadditive deterministic sequence mentioned in Section 1.3). It therefore follows from the *subadditive ergodic theorem* (Kingman [214]) that $\lim_{n\to\infty} \frac{1}{2n} \log Z_{2n}^{\omega,0}$ converges ω-a.s. and in $\mathbb{P}$-mean, and is measurable w.r.t. the tail sigma-field of ω. Since the latter is trivial, i.e., all events not depending on finitely many coordinates have probability 0 or 1, the limit is constant ω-a.s. $\square$

Our original partition sum at time $2n$ was

$$Z_{2n}^{\omega} = E\left(\exp\left[\lambda \sum_{i=1}^{2n}(\omega_i + h)\Delta_i\right]\right), \tag{9.10}$$

which is (9.8) but without the indicator. Thus, in order to prove Theorem 9.1 we must show that this indicator is harmless as $n \to \infty$. Since $|\log(Z_{2n}^{\omega}/Z_{2n+1}^{\omega})| \leq \lambda(1+h)$, it will suffice to consider time $2n$.

Lemma 9.3. There exists a $C < \infty$ such that $Z_{2n}^{\omega,0} \leq Z_{2n}^{\omega} \leq CnZ_{2n}^{\omega,0}$ for all $n \in \mathbb{N}$ and ω.

Proof. The lower bound is obvious. The upper bound is proved as follows (compare with the proof of Theorem 7.1). For $1 \leq k \leq n$, consider the events

$$
\begin{aligned}
A_{2n,2k}^+ &= \{S_i > 0 \text{ for } 2n - 2k + 1 \leq i \leq 2n\}, \\
B_{2n,2k}^+ &= \{S_i > 0 \text{ for } 2n - 2k + 1 \leq i < 2n, S_{2n} = 0\},
\end{aligned}
\tag{9.11}
$$

and similarly for $A_{2n,2k}^-, B_{2n,2k}^-$ when the excursion is below the interface. By conditioning on the last hitting time of 0 prior to time $2n$, we may write

$$
\begin{aligned}
Z_{2n}^\omega &= Z_{2n}^{\omega,0} + \sum_{k=1}^n Z_{2n-2k}^{\omega,0}\, E\Bigg(\exp\Bigg[\lambda \sum_{i=2n-2k+1}^{2n} (\omega_i + h)\Delta_i \Bigg] \\
&\qquad\qquad \times 1\{A_{2n,2k}^+ \cup A_{2n,2k}^-\} \,\Big|\, S_{2n-2k} = 0 \Bigg) \\
&= Z_{2n}^{\omega,0} + \sum_{k=1}^n Z_{2n-2k}^{\omega,0}\, \frac{a(2k)}{b(2k)}\, E\Bigg(\exp\Bigg[\lambda \sum_{i=2n-2k+1}^{2n} (\omega_i + h)\Delta_i \Bigg] \\
&\qquad\qquad \times 1\{B_{2n,2k}^+ \cup B_{2n,2k}^-\} \,\Big|\, S_{2n-2k} = 0 \Bigg).
\end{aligned}
\tag{9.12}
$$

The reason for the second equality in (9.12) is that $\Delta_i = +1$ for all $2n-2k+1 \leq i \leq 2n$ on the events $A_{2n,2k}^+, B_{2n,2k}^+$ and $\Delta_i = -1$ for all $2n - 2k + 1 \leq i \leq 2n$ on the events $A_{2n,2k}^-, B_{2n,2k}^-$ (ω is fixed). We have

$$
\begin{aligned}
P(A_{2n,2k}^+ | S_{2n-2k} = 0) &= P(A_{n,k}^- \mid S_{2n-2k} = 0) = a(2k), \\
P(B_{2n,2k}^+ | S_{2n-2k} = 0) &= P(B_{n,k}^- \mid S_{2n-2k} = 0) = b(2k),
\end{aligned}
\tag{9.13}
$$

where (compare with (7.4))

$$
\begin{aligned}
a(2k) &= P(S_i > 0 \text{ for } 1 \leq i \leq 2k \mid S_0 = 0), \\
b(2k) &= P(S_i > 0 \text{ for } 1 \leq i < 2k, S_{2k} = 0 \mid S_0 = 0).
\end{aligned}
\tag{9.14}
$$

Moreover, there exist $0 < C_1, C_2 < \infty$ such that $a(2k) \sim C_1/k^{1/2}$ and $b(2k) \sim C_2/k^{3/2}$ as $k \to \infty$ (see Spitzer [284], Section 1). Hence $a(2k)/b(2k) \leq Ck$, $k \in \mathbb{N}$, for some $C < \infty$. Finally, without the factor $a(2k)/b(2k)$ the last sum in (9.12) is precisely $Z_{2n}^{\omega,0}$. Hence we get

$$
Z_{2n}^\omega \leq (1 + Cn)\, Z_{2n}^{\omega,0},
\tag{9.15}
$$

which proves the claim.   □

Lemmas 9.2–9.3 complete the proof of Theorem 9.1.   □

9.3 The Critical Curve

Now that we have proved the existence of the quenched free energy, we proceed to study its properties, in particular, we proceed to look for a phase transition in QUA. In Section 9.3.1 we define the localized and the delocalized phases, while in Section 9.3.2 we prove the existence of a non-trivial critical curve separating the two. Sections 9.4–9.5 will be devoted to establishing the qualitative properties of the critical curve and the two phases.

Grosberg, Izrailev and Nechaev [133] and Sinai and Spohn [276] study the annealed version of the model, in which Z_n^ω is averaged over ω. The free energy and the critical curve can in this case be computed exactly, but they provide little information on what the quenched model does. Nevertheless, in Section 9.4.1 we will use the annealed model to obtain an upper bound on the quenched critical curve.

9.3.1 The Localized and Delocalized Phases

The quenched free energy $f(\lambda, h)$ is continuous, nondecreasing and convex in each variable (convexity follows from Hölder's inequality, as shown in Section 1.3). Moreover, we have

$$f(\lambda, h) \geq \lambda h \qquad \forall (\lambda, h) \in \text{QUA}. \qquad (9.16)$$

Indeed, since $P(\Delta_i = +1 \ \forall 1 \leq i \leq n) \sim C/n^{1/2}$ for some $C > 0$ as $n \to \infty$ (see Spitzer [284], Section 7), it follows from (9.3–9.5) that

$$
\begin{aligned}
Z_n^\omega &= E\left(\exp\left[\lambda \sum_{i=1}^n (\omega_i + h)\Delta_i\right]\right) \\
&\geq \left(\exp\left[\lambda \sum_{i=1}^n (\omega_i + h)\Delta_i\right] 1_{\{\Delta_i = +1 \ \forall 1 \leq i \leq n\}}\right) \\
&= \exp\left[\lambda \sum_{i=1}^n (\omega_i + h)\right] P(\Delta_i = +1 \ \forall 1 \leq i \leq n) \\
&= \exp[\lambda h n + o(n) = O(\log n)] \qquad \omega - a.s.,
\end{aligned}
\qquad (9.17)
$$

where in the last line we use the strong law of large numbers for ω. Thus, we see that the lower bound in (9.16) corresponds to the strategy where the copolymer wanders away from the interface in the upward direction. This leads us to the following definition (see Fig. 9.3).

Definition 9.4. *We say that the copolymer is:*
(i) *localized if $f(\lambda, h) > \lambda h$,*
(ii) *delocalized if $f(\lambda, h) = \lambda h$.*
We write

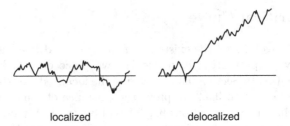

localized   delocalized

Fig. 9.3. Expected path behavior in the two phases.

$$\mathcal{L} = \{(\lambda, h) \in \text{QUA}: f(\lambda, h) > \lambda h\},$$
$$\mathcal{D} = \{(\lambda, h) \in \text{QUA}: f(\lambda, h) = \lambda h\}, \tag{9.18}$$

to denote the localized, respectively, the delocalized region in QUA.

In case (i), the copolymer is able to beat on an exponential scale the trivial strategy of moving upward. It is intuitively clear that this is only possible by crossing the interface at a positive frequency, but a proof of the latter requires work. In case (ii), the copolymer is not able to beat the trivial strategy on an exponential scale. In principle it could do better on a smaller scale, but it actually does not, which also requires work. Path properties are dealt with in Section 9.5.1.

Albeverio and Zhou [1] prove that if $\lambda > 0$ and $h = 0$, then $\log Z_n^\omega$ satisfies a law of large numbers and a central limit theorem (as a random variable in ω). This result readily extends to the entire localization regime.

9.3.2 Existence of a Non-trivial Critical Curve

The following theorem, taken from Bolthausen and den Hollander [31], shows that $\mathcal{L}$ and $\mathcal{D}$ are separated by a non-trivial critical curve (see Fig. 9.4).

Theorem 9.5. *For every* $\lambda \in [0, \infty)$ *there exists an* $h_c(\lambda) \in [0, 1)$ *such that the copolymer is*

$$\begin{array}{ll} localized & if \ 0 \leq h < h_c(\lambda), \\ delocalized & if \ h \geq h_c(\lambda). \end{array} \tag{9.19}$$

Moreover, $\lambda \mapsto h_c(\lambda)$ *is continuous and strictly increasing on* $[0, \infty)$, *with* $h_c(0) = 0$ *and* $\lim_{\lambda \to \infty} h_c(\lambda) = 1$.

Proof. Let

$$g_n^\omega(\lambda, h) = \frac{1}{n} \log E\left(\exp\left[\lambda \sum_{i=1}^n (\omega_i + h)(\Delta_i - 1)\right]\right). \tag{9.20}$$

Then, because $\lambda \sum_{i=1}^n (\omega_i + h) = \lambda h + o(n)$ ω-a.s., Theorem 9.1 says that

$$\lim_{n \to \infty} g_n^\omega(\lambda, h) = g(\lambda, h) \qquad \omega - a.s. \tag{9.21}$$

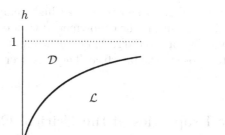

Fig. 9.4. Plot of $\lambda \mapsto h_c(\lambda)$.

with

$$g(\lambda, h) = f(\lambda, h) - \lambda h \tag{9.22}$$

the *excess free energy*. In terms of g, (9.18) becomes

$$\mathcal{L} = \{(\lambda, h) \in \text{QUA}\colon g(\lambda, h) > 0\},$$
$$\mathcal{D} = \{(\lambda, h) \in \text{QUA}\colon g(\lambda, h) = 0\}. \tag{9.23}$$

For $\lambda \in [0, \infty)$, define

$$h_c(\lambda) = \inf\{h \in [0, \infty)\colon (\lambda, h) \in \mathcal{D}\}. \tag{9.24}$$

To study this function, it is expedient to change variables by putting

$$\theta = \lambda h, \qquad \theta_c(\lambda) = \lambda h_c(\lambda), \qquad \bar{g}(\lambda, \theta) = g(\lambda, h). \tag{9.25}$$

Because λ and θ appear linearly in the Hamiltonian in (9.3), we know that $(\lambda, \theta) \mapsto \bar{g}(\lambda, \theta)$ is convex (see Section 1.3). Moreover, since $\bar{g} \geq 0$, we have $\mathcal{D} = \{(\lambda, \theta)\colon \bar{g}(\lambda, \theta) \leq 0\}$, i.e., $\mathcal{D}$ is a level set of the function $\bar{g}$. Since $\bar{g}$ is a convex function, it follows that $\mathcal{D}$ is a convex set, implying in turn that $\mathcal{L}$ and $\mathcal{D}$ are separated by a *single* critical curve

$$\theta_c(\lambda) = \inf\{\theta \in [0, \infty)\colon (\lambda, \theta) \in \mathcal{D}\} \tag{9.26}$$

that is itself a convex function.

We know that $h_c(0) = 0$. Now, Theorems 9.6–9.7 below will imply that

$$0 < \liminf_{\lambda \downarrow 0} \frac{1}{\lambda} h_c(\lambda) \leq \limsup_{\lambda \downarrow 0} \frac{1}{\lambda} h_c(\lambda) < \infty. \tag{9.27}$$

Moreover, Theorem 9.7 will imply that $\lim_{\lambda \to \infty} h_c(\lambda) = 1$. All that therefore remains to be done is to show that h_c is strictly increasing on $[0, \infty)$. To that end, note that $\theta_c(0) = 0$, while (9.27) shows that $\theta_c(\lambda)$ is of order λ^2 near 0. Consequently, $\lambda \mapsto \theta_c(\lambda)/\lambda$ is strictly increasing $[0, \infty)$ (with limiting value 0 at $\lambda = 0$). $\square$

The critical curve in Fig. 9.4 shows that at high temperature (small λ) *entropic* effects dominate, causing the copolymer to wander away from the interface, while at low temperature (large λ) *energetic* effects dominate, causing the copolymer to stay close to the interface. The crossover value of λ depends on the value of h.

9.4 Qualitative Properties of the Critical Curve

In Section 9.4.1 we derive an upper bound on the critical curve, in Section 9.4.2 a lower bound, while in Section 9.4.3 we indicate that it has a (positive and finite) slope at 0.

9.4.1 Upper Bound

The next result, also taken from Bolthausen and den Hollander [31], provides an upper bound on h_c. This bound is the critical curve for the *annealed* model.

Theorem 9.6. $h_c(\lambda) \leq (2\lambda)^{-1} \log \cosh(2\lambda)$ *for all* $\lambda \in (0, \infty)$.

Proof. Estimate (recall (9.20–9.22))

$$g(\lambda, h) = \lim_{n \to \infty} \frac{1}{n} \mathbb{E}\left(\log E\left(\exp\left[\lambda \sum_{i=1}^{n} (\omega_i + h)(\Delta_i - 1) \right] \right) \right)$$

$$\leq \lim_{n \to \infty} \frac{1}{n} \log E\left(\mathbb{E}\left(\exp\left[\lambda \sum_{i=1}^{n} (\omega_i + h)(\Delta_i - 1) \right] \right) \right) \qquad (9.28)$$

$$= \lim_{n \to \infty} \frac{1}{n} \log E\left(\prod_{i=1}^{n} \left[\tfrac{1}{2} e^{-2\lambda(1+h)} + \tfrac{1}{2} e^{-2\lambda(-1+h)} \right]^{1_{\{\Delta_i = -1\}}} \right).$$

The first equality comes from the fact that Kingman's subadditive ergodic theorem holds in $\mathbb{P}$-mean. The inequality follows from Jensen's inequality and Fubini's theorem. The second equality uses (9.2). The right-hand side is ≤ 0 as soon as the term between square brackets is ≤ 1. Consequently,

$$(2\lambda)^{-1} \log \cosh(2\lambda) > h \quad \Longrightarrow \quad g(\lambda, h) = 0, \qquad (9.29)$$

which yields the desired upper bound on $h_c(\lambda)$. $\square$

Note that the proof of Theorem 9.6 is a *partial annealing estimate* in disguise, because to transform (9.18) into (9.23) we have used the law of large numbers for ω.

9.4.2 Lower Bound

The following counterpart of Theorem 9.6, due to Bodineau and Giacomin [22], provides a lower bound on $h_c(\lambda)$.

Theorem 9.7. $h_c(\lambda) \geq (\frac{4}{3}\lambda)^{-1} \log \cosh(\frac{4}{3}\lambda)$ *for all* $\lambda \in (0, \infty)$.

Proof. As we will see, the lower bound comes from strategies where the copolymer dips below the interface during rare long stretches in ω where the empirical mean is sufficiently biased.

We begin with some notation. Pick $l \in 2\mathbb{N}$. For $j \in \mathbb{N}$, let

$$I_j = ((j-1)l, jl] \cap \mathbb{N}, \qquad \Omega_j = \sum_{i \in I_j} \omega_i. \qquad (9.30)$$

Pick $\delta \in (0, 1]$. Define recursively

$$i_0^\omega = 0, \qquad i_j^\omega = \inf\{k \geq i_{j-1}^\omega + 2: \Omega_k \leq -\delta l\}, \ j \in \mathbb{N}, \qquad (9.31)$$

and abbreviate $\tau_j^\omega = i_j^\omega - i_{j-1}^\omega - 1$, $j \in \mathbb{N}$. (In (9.31) we skip at least 2 to make sure that $\tau_j^\omega \geq 1$, which is needed below.) For $n \in \mathbb{N}$, put

$$t_{n,l,\delta}^\omega = \sup\{j \in \mathbb{N}: i_j^\omega \leq \lfloor n/l \rfloor\}, \qquad (9.32)$$

and consider the set of paths (see Fig. 9.5)

$$\mathcal{W}_{n,l,\delta}^\omega = \Big\{w \in \mathcal{W}_n: \ w_i \leq 0 \text{ for } i \in \cup_{j=1}^{t_{n,l,\delta}^\omega} I_{i_j^\omega},$$
$$w_i \geq 0 \text{ for } i \in \{(0, n] \cap \mathbb{N}\} \setminus \cup_{j=1}^{t_{n,l,\delta}^\omega} I_{i_j^\omega}\Big\}. \qquad (9.33)$$

By restricting the path to $\mathcal{W}_{n,l,\delta}^\omega$, we find that the quantity in (9.20) can be bounded from below as

$$g_n^\omega(\lambda, h) = \frac{1}{n} \log E\left(\exp\left[\lambda \sum_{i=1}^n (\omega_i + h)(\Delta_i - 1)\right]\right)$$

$$\geq \frac{1}{n} \log E\left(\exp\left[\lambda \sum_{i=1}^n (\omega_i + h)(\Delta_i - 1)\right] 1_{\mathcal{W}_{n,l,\delta}^\omega}\right)$$

$$\geq \frac{1}{n} \log\left\{\left(\prod_{j=1}^{t_{n,l,\delta}^\omega} b(\tau_j^\omega l)\, b(l)\, e^{-2\lambda(-\delta + h)l}\right) a\left(n - i_{t_{n,l,\delta}^\omega} l\right)\right\}$$

$$\geq \frac{1}{n} \sum_{j=1}^{t_{n,l,\delta}^\omega} \log b(\tau_j^\omega l) + \frac{t_{n,l,\delta}^\omega}{n} [\log b(l) + 2\lambda(\delta - h)l] + \frac{1}{n} \log a(n),$$

$$(9.34)$$

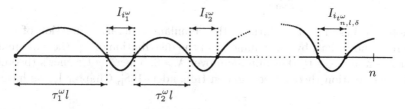

Fig. 9.5. A path in $\mathcal{W}_{n,l,\delta}^\omega$.

where $a(k)$ and $b(k)$ are defined in (9.13), the second inequality uses that $\Omega_j \leq -\delta l$ for $j = 1, \ldots, t_{n,l,\delta}^\omega$, and the third inequality uses that $k \mapsto a(k)$ is non-increasing. We need to compute the ω-a.s. limit of the right-hand side of (9.34) as $n \to \infty$.

The last term vanishes as $n \to \infty$, because $a(n) \asymp n^{-1/2}$. Furthermore, since there exists a $C > 0$ such that $b(k) \geq Ck^{-3/2}$ for $k \in 2\mathbb{N}$, we have

$$
\frac{1}{n} \sum_{j=1}^{t_{n,l,\delta}^\omega} \log b(\tau_j^\omega l) \geq \frac{1}{n} \sum_{j=1}^{t_{n,l,\delta}^\omega} \log \left[C(\tau_j^\omega l)^{-3/2} \right]
$$

$$
\geq -\frac{3 t_{n,l,\delta}^\omega}{2n} \log \left[\frac{\sum_{j=1}^{t_{n,l,\delta}^\omega} \tau_j^\omega l}{t_{n,l,\delta}^\omega} \right] + \frac{t_{n,l,\delta}^\omega}{n} \log C,
$$

(9.35)

where the second inequality uses Jensen's inequality. Moreover, applying the ergodic theorem to $(\Omega_j)_{j \in \mathbb{N}}$ while recalling (9.31–9.32), we have

$$
\lim_{n \to \infty} \frac{t_{n,l,\delta}^\omega}{n} = p_{l,\delta} \qquad \omega - \text{a.s.},
$$

(9.36)

where

$$
p_{l,\delta} = \frac{1}{l} q_{l,\delta}(1 - q_{l,\delta}) \text{ with } q_{l,\delta} = \mathbb{P}(\Omega_1 \leq -\delta l).
$$

(9.37)

Since $\sum_{j=1}^{t_{n,l,\delta}^\omega} \tau_j^\omega l \leq n - t_{n,l,\delta}^\omega l$, it follows from (9.36) that

$$
\limsup_{n \to \infty} \frac{\sum_{j=1}^{t_{n,l,\delta}^\omega} \tau_j^\omega l}{t_{n,l,\delta}^\omega} \leq \lim_{n \to \infty} \frac{n - t_{n,l,\delta}^\omega l}{t_{n,l,\delta}} = p_{l,\delta}^{-1} - l \qquad \omega - \text{a.s.}
$$

(9.38)

Combining (9.34–9.38) and recalling (9.21), we arrive at

$$
g(\lambda, h) \geq -\tfrac{3}{2} p_{l,\delta} \log \left(p_{l,\delta}^{-1} - l \right) + p_{l,\delta} \left[2 \log C - \tfrac{3}{2} \log l + 2\lambda(\delta - h)l \right].
$$

(9.39)

This inequality is valid for all $l \in 2\mathbb{N}$ and $\delta \in (0, 1]$.

Next, let

$$
\Sigma(\delta) = \sup_{\lambda > 0} \left[\lambda\delta - \log \mathbb{E}\left(e^{-\lambda\omega_1} \right) \right], \qquad \delta \in (0, 1],
$$

(9.40)

denote the Legendre transform of the cumulant generating function of $-\omega_1$, where we note that, by the symmetry of the distribution of ω_1, the supremum over $\lambda \in \mathbb{R}$ reduces to the supremum over $\lambda > 0$. Then, by Cramér's theorem of large deviation theory (see e.g. den Hollander [168], Chapter I), we have

$$
\lim_{l \to \infty} \frac{1}{l} \log q_{l,\delta} = -\Sigma(\delta) \qquad \forall \delta \in (0, 1].
$$

(9.41)

Hence, letting $l \to \infty$ in (9.39) and using (9.37), we obtain

$$\exists \, \delta \in (0,1]: \quad -\tfrac{3}{2}\Sigma(\delta) + 2\lambda(\delta - h) > 0 \quad \Longrightarrow \quad g(\lambda, h) > 0. \qquad (9.42)$$

Taking the inverse Legendre transform in (9.40), we have

$$\sup_{\delta \in (0,1]} \left[\tfrac{4}{3}\lambda(\delta - h) - \Sigma(\delta) \right] = -\tfrac{4}{3}\lambda h + \log \mathbb{E}\left(e^{-\tfrac{4}{3}\lambda \omega_1} \right) \qquad (9.43)$$

$$= -\tfrac{4}{3}\lambda h + \log \cosh\left(\tfrac{4}{3}\lambda\right), \qquad \lambda > 0.$$

Combining (9.42–9.43), we get

$$\left(\tfrac{4}{3}\lambda\right)^{-1} \log \cosh\left(\tfrac{4}{3}\lambda\right) > h \quad \Longrightarrow \quad g(\lambda, h) > 0, \qquad (9.44)$$

which yields the desired lower bound on $h_c(\lambda)$. $\quad\square$

9.4.3 Weak Interaction Limit

The upper and lower bounds in Theorems 9.6–9.7 are sketched in Fig. 9.6. Numerical work in Caravenna, Giacomin and Gubinelli [55] indicates that $\lambda \mapsto h_c(\lambda)$ lies somewhere halfway between these bounds (see also Garel and Monthus [244]).

The following weak interaction limit is proved in Bolthausen and den Hollander [31].

Theorem 9.8. *There exists a $K_c \in (0, \infty)$ such that*

$$\lim_{\lambda \downarrow 0} \frac{1}{\lambda} h_c(\lambda) = K_c. \qquad (9.45)$$

Proof. The idea behind the proof is that, as $\lambda, h \downarrow 0$, the excursions away from the interface become longer and longer (i.e., entropy gradually takes over from energy). As a result, both w and ω can be approximated by Brownian motions. In essence, (9.45) follows from the scaling property

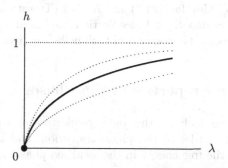

Fig. 9.6. Upper and lower bounds on $\lambda \mapsto h_c(\lambda)$.

$$\lim_{a \downarrow 0} a^{-2} f(a\lambda, ah) = \widetilde{f}(\lambda, h), \qquad \lambda, h \geq 0, \tag{9.46}$$

where $\widetilde{f}(\lambda, h)$ is the free energy of a space-time continuous version of the copolymer model, with Hamiltonian

$$H_t^b(B) = -\lambda \int_0^t (\mathrm{d}b_s + h \, \mathrm{d}s) \, \mathrm{sign}(B_s) \tag{9.47}$$

replacing (9.3), and with path measure given by the Radon-Nikodym derivative

$$\frac{\mathrm{d}P_t^b}{\mathrm{d}P}(B) = \frac{1}{Z_t^b} \, \mathrm{e}^{-H_t^b(B)} \tag{9.48}$$

replacing (9.5). Here, $B = (B_s)_{s \geq 0}$ is a path drawn from Wiener space, replacing (9.1), P is the Wiener measure (i.e., the law of standard Brownian motion on Wiener space), and $(b_s)_{s \geq 0}$ is a standard Brownian motion, replacing (9.2), playing the role of the quenched randomness. The proof of (9.46) is based on a *coarse-graining* argument. Due to the presence of exponential weight factors, (9.46) is a much more delicate property than the standard invariance principle relating simple random walk and Brownian motion. Moreover, (9.45) follows from (9.46) only after the latter has been shown to be "stable against perturbations" in λ, h. For details we refer to [31].

In the continuum model, the quenched critical curve turns out to be *linear* with slope K_c. The value of K_c is not known. $\square$

The proof of (9.45) shows that the same scaling property holds for the model in which the h-dependence sits in the probability law of ω rather than in the Hamiltonian, i.e., $\mathbb{P}(\omega_i = \pm 1) = \frac{1}{2}(1 \pm h)$ and $H_n^\omega(w) = \lambda \sum_{i=1}^n \omega_i \Delta_i(w)$ instead of (9.2–9.3). This describes a copolymer where the monomers occur with different densities but interact equally strongly. Alternatively, we could allow for more general ω, assuming values in $\mathbb{R}$ according to a symmetric distribution with a finite exponential moment (see Section 9.6). Thus, K_c has a certain degree of *universality*. See also Section 9.6, Extension (1).

The bounds in Theorems 9.6–9.7 give $K_c \in [\frac{2}{3}, 1]$. There have been various papers in the literature arguing in favor of $K_c = \frac{2}{3}$ (Stepanow, Sommer and Erukhimovich [285], Monthus [243]) and $K_c = 1$ (Trovato and Maritan [298]). Toninelli [295] proves that $K_c < 1$ (see Section 9.6, Extension (2)). The numerical work in Caravenna, Giacomin and Gubinelli [55] gives $K_c \in [0.82, 0.84]$.

9.5 Qualitative Properties of the Phases

In Section 9.5.1 we look at the path properties in the two phases, in Section 9.5.2 at the order of the phase transition, and in Section 9.5.3 at the smoothness of the free energy in the localized phase.

9.5.1 Path Properties

We next state two theorems identifying the path behavior in the two phases. The first is due to Biskup and den Hollander [20], the second to Giacomin and Toninelli [120]. These theorems confirm the naive view put forward in Section 9.3 when we defined $\mathcal{L}$ and $\mathcal{D}$ (recall Fig. 9.3).

Theorem 9.9. *ω-a.s. under P_n^ω as $n \to \infty$:*
(a) *If $(\lambda, h) \in \mathcal{L}$, then the path intersects the interface with a strictly positive density, while the lengths and the heights of its excursions away from the interface are exponentially tight.*
(b) *If $(\lambda, h) \in \mathrm{int}(\mathcal{D})$, then the path intersects the interface with a zero density.*

Theorem 9.10. *ω-a.s. under P_n^ω as $n \to \infty$, if $(\lambda, h) \in \mathrm{int}(\mathcal{D})$, then the path intersects the interface $O(\log n)$ times.*

The idea behind Theorem 9.9 is that in the localized regime, where the excess free energy $g = g(\lambda, h)$ defined in (9.22) is strictly positive, the probability for the copolymer to be away from the interface during a time l is roughly e^{-gl} as $l \to \infty$. This is because the copolymer contributes an amount gl to the free energy when it stays near the interface, but contributes 0 when it moves away. The idea behind Theorem 9.10 is that strictly inside the delocalized regime, where the excess free energy is zero, the probability for the copolymer to return to the interface a large number of times is small. This idea is exploited with the help of concentration inequalities.

It is believed that strictly inside the delocalized regime the number of intersections with the interface is in fact $O(1)$. This has only been proved deep inside the delocalized phase, namely, above the annealed upper bound in Theorem 9.6, where ideas similar to those that went into the proof of Theorem 7.3 can be exploited (Giacomin and Toninelli [120]). See Section 9.7, Challenge (3).

9.5.2 Order of the Phase Transition

The next theorem, due to Giacomin and Toninelli [123], shows that the phase transition is *at least* of second order.

Theorem 9.11. *For every $\lambda \in (0, \infty)$,*

$$0 \leq g(\lambda, h) = O\left([h_c(\lambda) - h]^2\right) \qquad as\ h \uparrow h_c(\lambda). \tag{9.49}$$

Proof. We give the proof for the case where $\omega = (\omega_i)_{i \in \mathbb{N}}$ is an i.i.d. sequence of standard normal random variables, rather than binary random variables as in (9.2). At the end of the proof we indicate how to adapt the argument.

Recall the definitions in the proof of Theorem 9.7. Consider the set of paths (see Fig. 9.7)

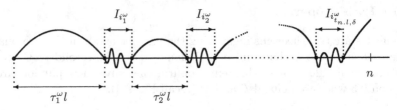

Fig. 9.7. A path in $\widehat{\mathcal{W}}^\omega_{n,l,\delta}$.

$$\widehat{\mathcal{W}}^\omega_{n,l,\delta} = \Big\{ w \in \mathcal{W}_n \colon w_i = 0 \text{ for } i \in \cup_{j=1}^{t^\omega_{n,l,\delta}} \partial I_{i^\omega_j},$$
$$w_i \geq 0 \text{ for } i \in \{(0,n] \cap \mathbb{N}\} \setminus \cup_{j=1}^{t^\omega_{n,l,\delta}} I_{i^\omega_j} \Big\} \tag{9.50}$$

with $\partial I_j = \{(j-1)l, jl\}, j \in \mathbb{N}$. By restricting the path to $\widehat{\mathcal{W}}^\omega_{n,l,\delta}$, we get

$$\widetilde{Z}^\omega_n \geq \Big(\prod_{j=1}^{t^\omega_{n,l,\delta}} b(\tau^\omega_j l) \, \widetilde{Z}_l^{T^{i^\omega_j l} \omega} \Big) \, a\Big(n - i_{t^\omega_{n,l,\delta}} l \Big), \tag{9.51}$$

where T is the left-shift acting on ω, and the tilde is used to indicate that the partition sums are taken with $\Delta_i - 1$ instead of Δ_i, as in (9.20) and (9.34). Letting $n \to \infty$, using that $k \mapsto a(k)$ is non-increasing with $\lim_{n\to\infty} \frac{1}{n} \log a(n) = 0$, and noting that the convergence in (9.21) holds in $\mathbb{P}$-mean as well, we obtain

$$g(\lambda, h) \geq \liminf_{n\to\infty} \frac{1}{n} \mathbb{E}\Big(\sum_{j=1}^{t^\omega_{n,l,\delta}} \log b(\tau^\omega_j l) \Big) + \liminf_{n\to\infty} \frac{1}{n} \mathbb{E}\Big(\sum_{j=1}^{t^\omega_{n,l,\delta}} \log \widetilde{Z}_l^{T^{i^\omega_j l} \omega} \Big). \tag{9.52}$$

The first term in the right-hand side of (9.52) was computed in (9.35–9.38) and equals $-\frac{3}{2} p_{l,\delta} \log(p_{l,\delta}^{-1} - l)$. The expectation under the limit in the second term equals (recall (9.31–9.32))

$$\frac{1}{n} \mathbb{E}\Big(\sum_{j=1}^{t^\omega_{n,l,\delta}} \log \widetilde{Z}_l^{T^{i^\omega_j l} \omega} \Big) = \mathbb{E}\Big(\frac{t^\omega_{n,l,\delta}}{n} \Big) \mathbb{E}\Big(\log \widetilde{Z}^\omega_l \mid \Omega_1 \leq -\delta l \Big). \tag{9.53}$$

Conditioning l i.i.d. normal random variables with mean 0 and variance 1 to have sum equal to $-\delta l$ is the same as taking l i.i.d. normal random variables with mean $-\delta$ and variance 1. Hence, the effect of the conditioning in (9.53) is that, in (9.20), ω_i becomes $\omega_i - \delta$, so that

$$\mathbb{E}\Big(\log \widetilde{Z}^\omega_l(\lambda, h) \mid \Omega_1 \leq -\delta l \Big) = \mathbb{E}\Big(\log \widetilde{Z}^\omega_l(\lambda, h - \delta) \Big), \tag{9.54}$$

where we add the parameters λ, h as arguments to exhibit the shift in h. Inserting (9.54) into (9.53), picking $h = h_c(\lambda)$ in (9.52), noting that $g(\lambda, h_c(\lambda)) = 0$, and recalling (9.36), we arrive at

$$0 \geq -\tfrac{3}{2}\, p_{l,\delta} \log\left(p_{l,\delta}^{-1} - l\right) + p_{l,\delta}\, \mathbb{E}\left(\log \widetilde{Z}_l^\omega(\lambda, h_c(\lambda) - \delta)\right). \tag{9.55}$$

This inequality is valid for all $l \in 2\mathbb{N}$ and $\delta \in (0,1]$.

Dividing (9.55) by l, letting $l \to \infty$ and using (9.41), we obtain

$$0 \geq -\tfrac{3}{2}\Sigma(\delta) + g\left(\lambda, h_c(\lambda) - \delta\right), \tag{9.56}$$

where Σ is the Cramér rate function for standard normal random variables, i.e., $\Sigma(\delta) = \tfrac{1}{2}\delta^2$. Hence we conclude that

$$g\left(\lambda, h_c(\lambda) - \delta\right) \leq \tfrac{3}{4}\delta^2, \tag{9.57}$$

which proves the claim.

It is easy to extend the proof to binary ω. All that is needed is to show that (9.54) holds asymptotically as $l \to \infty$ and to use that $\Sigma(\delta) \sim \tfrac{1}{2}\delta^2$ as $\delta \downarrow 0$. $\square$

9.5.3 Smoothness of the Free Energy in the Localized Phase

We conclude with a theorem by Giacomin and Toninelli [124] showing that the free energy is smooth throughout the localized phase. Consequently, our critical curve is the *only* location where a phase transition of *finite order* occurs.

Theorem 9.12. $(\lambda, h) \mapsto f(\lambda, h)$ *is infinitely differentiable on* $\mathcal{L}$.

Proof. The main idea behind the proof is that, for $(\lambda, h) \in \mathcal{L}$, the correlation between any pair of events that depend on the values of the path w of the copolymer in finite and disjoint subsets of $\{1, \ldots, n\}$ decays exponentially with the distance between these sets. This is formulated in the following lemma.

Lemma 9.13. *Let $\mathcal{K}$ be an arbitrary compact subset of $\mathcal{L}$. Then there exist $0 < c_1, c_2 < \infty$ (depending on $\mathcal{K}$) such that, for all $(\lambda, h) \in \mathcal{K}$, all $n \in \mathbb{N}$, all integers $1 \leq a_1 < b_1 < a_2 < b_2 \leq n$, and all pairs of events A and B that are measurable w.r.t. the sigma-fields generated by $(w_{a_1}, \ldots, w_{b_1})$, respectively, $(w_{a_2}, \ldots, w_{b_2})$,*

$$\mathbb{E}\left(\left|P_n^\omega(A \cap B) - P_n^\omega(A)\, P_n^\omega(B)\right|\right) \leq c_1 e^{-c_2(a_2 - b_1)}. \tag{9.58}$$

Proof. The proof of Lemma 9.13 is based on the following lemma. For $n \in \mathbb{N}$, let $P_n^{\omega, \otimes 2}$ be the joint law of two independent copies w^1 and w^2 of w of length n.

Lemma 9.14. *There exist $0 < c_1, c_2 < \infty$ (depending on $\mathcal{K}$) such that for all $(\lambda, h) \in \mathcal{K}$, all $n \in \mathbb{N}$, and all integers $1 \leq a < b \leq n$,*

$$\mathbb{E}\left(P_n^{\omega, \otimes 2}(E_{a,b})\right) \leq c_1 e^{-c_2(b-a)}, \tag{9.59}$$

where $E_{a,b} = \{\nexists\, i \in \{a+1, \ldots, b-1\}\colon w_i^1 = w_i^2 = 0\}$.

Proof. The main idea is that, for $(\lambda, h) \in \mathcal{L}$, the lengths of the excursions of the copolymer away from the interface are exponentially tight. This forces w^1 and w^2 to hit the interface in a set of sites with a strictly positive density. Therefore the probability that w^1 and w^2 hit the interface at the same site at least once in $\{a+1, \ldots, b-1\}$ tends to 1 exponentially fast as $b - a \to \infty$. For details we refer the reader to Giacomin and Toninelli [124]. □

Now note that the l.h.s. of (9.58) is equal to

$$\mathbb{E}\left(\left|E_n^{\omega, \otimes 2}\left[\{1_A(w^1)1_B(w^1) - 1_A(w^1)1_B(w^2)\} 1_E(w^1, w^2)\right]\right|\right). \tag{9.60}$$

Together with (9.59), this completes the proof of Lemma 9.13. □

We use Lemma 9.13 to complete the proof of Theorem 9.12. This can be done by appealing to the theorem of Arzela-Ascoli, according to which it is enough to prove that for every $k_1, k_2 \in \mathbb{N}$ the partial derivative

$$\frac{\partial^{k_1 + k_2}}{\partial \lambda^{k_1} \partial h^{k_2}} \left(\frac{1}{n} \mathbb{E} \log Z_n^\omega\right) \tag{9.61}$$

is bounded from above, uniformly in $n \in \mathbb{N}$ and $(\lambda, h) \in \mathcal{K}$.

In order to achieve the latter, we define, for $k \in \mathbb{N}$ and for $\{f_1, \ldots, f_k\}$ any family of bounded functions on the path space $\mathcal{W}_n$ whose supports are finite,

$$E_n^\omega[f_1(w); \ldots; f_k(w)] = \sum_{P \in \mathcal{P}} (-1)^{|P|-1}(|P|-1)! \prod_{p=1}^{|P|} E_n^\omega\left[\prod_{l \in P_p} f_l(w)\right], \tag{9.62}$$

where $\mathcal{P}$ denotes the set of all partitions $P = (P_p)_{p \in |P|}$ of $\{1, \ldots, k\}$, with $|P|$ the number of sets in the partition P. We need the following lemma for this quantity.

Lemma 9.15. *For all $k \in \mathbb{N}$ there exist $0 < c_1, c_2 < \infty$ (depending on k and $\mathcal{K}$) such that, for all $(\lambda, h) \in \mathcal{K}$,*

$$\mathbb{E}\left[\left|E_n^\omega[f_1; \ldots; f_k]\right|\right] \leq c_1 \|f_1\|_\infty \cdots \|f_k\|_\infty \, e^{-c_2(|\mathcal{I}| - [|\mathcal{S}(f_1)| + \cdots + |\mathcal{S}(f_k)|])}, \tag{9.63}$$

where $\mathcal{S}(f_1), \ldots, \mathcal{S}(f_k)$ are the supports of $f_1, \ldots, f_k$, and $\mathcal{I}$ is the smallest interval containing $\mathcal{S}(f_1), \ldots, \mathcal{S}(f_k)$.

Proof. The proof uses induction on k. We skip the details, noting only that the case $k = 1$ is trivial, while the case $k = 2$ is a direct consequence of Lemma 9.13 after we write bounded functions with finite support as linear combinations of indicators of events. □

The point of the notation introduced in (9.62) is that (9.61) can be written in the form

$$\frac{1}{n} \sum_{1 \le i_1, \ldots, i_{k_1} \le n} \sum_{1 \le j_1, \ldots, j_{k_2} \le n} \mathbb{E}\Big[(\omega_{i_1} + h) \ldots (\omega_{i_{k_1}} + h)$$

$$\times E_n^\omega \big[\Delta_{i_1}(w); \ldots; \Delta_{i_{k_1}}(w); \Delta_{j_1}(w); \ldots; \Delta_{j_{k_2}}(w)\big]\Big]. \tag{9.64}$$

Since the ω_i's are bounded, the proof will therefore be complete once we show that

$$\frac{1}{n} \sum_{1 \le i_1, \ldots, i_k \le n} \mathbb{E}\Big[\big|E_n^\omega [\Delta_{i_1}(w); \ldots; \Delta_{i_k}(w)]\big|\Big], \qquad k = k_1 + k_2, \tag{9.65}$$

is bounded uniformly in $n \in \mathbb{N}$ and $(\lambda, h) \in \mathcal{K}$. For this we use Lemma 9.15, to bound (9.65) from above by

$$\frac{1}{n} \sum_{m=1}^{n-1} \sum_{(i_1, \ldots, i_k) \in T_n^m} c_1 e^{-c_2(m-k)}, \tag{9.66}$$

where

$$T_n^m = \big\{(i_1, \ldots, i_k) : 1 \le i_1, \ldots, i_k \le n, \\ \max\{i_1, \ldots, i_k\} - \min\{i_1, \ldots, i_k\} = m\big\}. \tag{9.67}$$

Since $|T_n^m| \le nm^k$, (9.66) is at most $c_1 e^{c_2 k} \sum_{m \in \mathbb{N}} m^k e^{-c_2 m} < \infty$. □

9.6 Extensions

(1) With the exception of Theorem 9.11, all the results described in Sections 9.2–9.5 extend to the situation where the ω_i are $\mathbb{R}$-valued with a finite moment generating function. For instance, define

$$h^*(\lambda) = \frac{1}{2\lambda} \log \mathbb{E}\big(e^{2\lambda\omega_1}\big), \qquad \lambda \in (0, \infty), \tag{9.68}$$

and assume that h^* is finite on $(0, \infty)$. Then the proofs of Theorems 9.6–9.7 show that

$$h^*(\tfrac{2}{3}\lambda) \le h_c(\lambda) \le h^*(\lambda), \qquad \lambda \in (0, \infty). \tag{9.69}$$

Moreover, if the random walk is replaced by a renewal process whose return times to the interface have distribution $P(S_i > 0 \ \forall 0 < i < n, S_n = 0) \sim n^{-1-a}L(n)$ as $n \to \infty$ for some $a > 0$ and some function L that is slowly varying at infinity (as in (7.5)), then the same proofs yield

$$h^*(\tfrac{1}{1+a}\lambda) \le h_c(\lambda) \le h^*(\lambda), \qquad \lambda \in (0, \infty). \tag{9.70}$$

We refer the reader to Giacomin [116], Chapters 6–8, for a full account of these extensions. The reason why the generalization from random walks to renewal processes works is precisely that the Hamiltonian in (9.3) decomposes into contributions coming from single excursions away from the interface. Therefore only the asymptotics of the law of these excursions matters. It also allows for the inclusion of $(1 + d)$-dimensional excursions away from a d-dimensional flat interface with $d \geq 2$.

Theorem 9.11 has been extended to bounded disorder, and also to continuous disorder subject to a mild entropy condition that is satisfied e.g. for Gaussian disorder (see Giacomin and Toninelli [122], [123]).

(2) Toninelli [294] shows that the upper bound in Theorem 9.6 is strict for un-bounded disorder and large λ. This result is extended by Bodineau, Giacomin, Lacoin and Toninelli [23] to arbitrary disorder (subject to h^* in (9.68) being finite) and arbitrary λ. The proofs are based on *fractional moment estimates* of the partition sum. Toninelli [295] further refines the latter technique to show that $K_c < 1$ for arbitrary disorder, thereby ruling out $K_c = 1$ (recall the discussion at the end of Section 9.4.3). Bodineau, Giacomin, Lacoin and Toninelli [23] also show that the lower bound in Theorem 9.7 is strict for arbitrary disorder and small λ, at least for a large subclass of excursion re-turn time distributions. The proof is based on finding appropriate *localization strategies*, in the spirit of the computation in Section 9.4.2.

(3) The difficulty behind improving the upper bound is that the typical length of the excursions diverges as the critical curve is approached from below. Consequently, any attempt to do a higher order partial annealing in the hope to improve the first order partial annealing argument in (9.28) is doomed to fail. This fact was observed in Orlandini, Rechnitzer and Whittington [253] and in Iliev, Rechnitzer and Whittington [182], and was subsequently proved for a large class of models in Caravenna and Giacomin [54]. In the latter paper, the setting is an arbitrary Hamiltonian $w \mapsto H_n^\omega(w)$ with the property that there exists a sequence $(D_n)_{n \in \mathbb{N}}$ of subsets of $\mathbb{Z}^d$ such that

$$\lim_{n \to \infty} \frac{1}{n} \log P(w_i \in D_i \ \forall 1 \leq i \leq n) = 0,$$

$$w_i \in D_i \ \forall 1 \leq i \leq n \implies H_n^\omega(w) = 0 \ \forall \omega, \tag{9.71}$$

where P is the reference path measure. Given an arbitrary local, bounded and measurable function $\omega \mapsto F(\omega)$ with $\mathbb{E}(F(\omega)) = 0$, it is shown

$$\liminf_{n \to \infty} \frac{1}{n} \log \mathbb{E}\big(E\big(\exp[-H_n^\omega(S)]\big)\big) > 0$$

$$\implies \quad \liminf_{n \to \infty} \frac{1}{n} \log \mathbb{E}\Big(E\Big(\exp\Big[-H_n^\omega(S) - \sum_{i=1}^n F(T^i\omega)\Big]\Big)\Big) > 0, \tag{9.72}$$

where T is the left-shift acting on ω. What this says is that if the *annealed* free energy is strictly positive, then it remains strictly positive after adding the empirical average of a centered local function of the disorder. In other words, the two annealed free energies have the same critical curve. Note that the centering implies that the *quenched* free energies associated with H and $H+F$ are the same: the term $\sum_{i=1}^{n} F(T^i \omega)$ does not depend on S. Thus, in order to improve an annealed estimate of a critical curve one has to resort to adding non-local functions of the disorder, which are computationally unattractive.

(4) Caravenna, Giacomin and Gubinelli [55] carry out a detailed numerical analysis of the critical curve, with full statistical control on the errors. They exploit the superadditivity property noted in the proof of Theorem 9.1, which implies that

$$g(\lambda, h) = \sup_{n \in \mathbb{N}} \frac{1}{2n} \log \mathbb{E}(\log Z_{2n}^{*,\omega}). \tag{9.73}$$

Hence, for any given (λ, h), if $\mathbb{E}(\log Z_{2n}^{*,\omega})$ for some finite n, then $(\lambda, h) \in \mathcal{L}$. This leads to a sharp lower bound for the critical curve. It is much harder to get a decent upper bound. Computations are carried out up to $n = 2 \times 10^8$, and concentration inequalities are used to estimate the expectation $\mathbb{E}$ with only relatively few samples of ω. Giacomin and Sohier (private communication) have pushed the computation up to $n = 10^{12}$, thereby improving the lower bound for the critical curve even further.

Interestingly, the results show that to a remarkable degree of accuracy

$$h_c(\lambda) \approx h^*(K_c \lambda), \qquad \lambda \in (0, \infty), \tag{9.74}$$

with K_c the constant in (9.45), both for binary and standard normal disorder. Nevertheless, Bodineau, Giacomin, Lacoin and Toninelli [23] prove that equality in (9.74) cannot hold for all λ, by showing that the critical curve depends on the fine details of the excursion return time distribution and not only on the universal constant K_c.

(5) The restriction to the quadrant in (9.4) can be trivially removed. Indeed, because $f(\lambda, h)$ is antisymmetric under the transformations $\lambda \to -\lambda$ and $h \to -h$ (recall (9.2–9.3)), the full phase diagram consists of four critical curves, one in each quadrant of $\mathbb{R}^2$, which are images of the critical curve in the first quadrant (drawn in Fig. 9.4) under reflection in the horizontal and the vertical axes. For instance, below the critical curve in the fourth quadrant (i.e., for $h \leq -h_c(\lambda)$ and $\lambda \geq 0$) the copolymer is delocalized into the water rather than into the oil.

A further observation is the following. By subtracting $\lambda n + \lambda h \sum_{i=1}^{n} \omega_i$ from the Hamiltonian H_n^ω in (9.3), we obtain a new Hamiltonian

$$\widehat{H}_n^\omega(w) = -2\lambda(1+h) \sum_{i=1}^{n} \frac{1+\omega_i}{2} \frac{1+\Delta_i}{2} - 2\lambda(1-h) \sum_{i=1}^{n} \frac{1-\omega_i}{2} \frac{1-\Delta_i}{2}, \tag{9.75}$$

where we recall (9.7). In terms of the reparameterization $\alpha = 2\lambda(1+h)$ and $\beta = 2\lambda(1-h)$, this reads

$$\widehat{H}_n^\omega(w) = -\alpha \sum_{i=1}^n 1_{\{\omega_i = \Delta_i = 1\}} - \beta \sum_{i=1}^n 1_{\{\omega_i = \Delta_i = -1\}}, \qquad (9.76)$$

in which energy $-\alpha$ is assigned to the hydrophobic monomers in the oil, $-\beta$ to the hydrophilic monomers in the water, and 0 to the other two combinations. By the strong law of large numbers for ω, we have $H_n^\omega - \widehat{H}_n^\omega = \lambda n + o(n)$ ω-a.s. Therefore the quenched free energies associated with H_n^ω and $\widehat{H}_n^\omega$ differ by a term $\log \lambda$, which allows for a direct link between the respective phase diagrams. The form in (9.76) is used in a number of papers (see Extensions (7–10) below), and also in Chapter 10, where we look at a copolymer near a random selective interface.

In the (α, β)-model the quenched critical curve $\alpha \mapsto \beta_c(\alpha)$ takes the form in Fig. 9.8. This curve is continuous, strictly increasing and concave as a function of α, with a finite asymptote as $\alpha \to \infty$, and with a curvature as $\alpha \downarrow 0$ that tends to K_c, the universal constant in Theorem 9.8. The former comes from the bound $1 - h_c(\lambda) \le C/\lambda$, $\lambda \to \infty$, $C < \infty$, implied by Theorem 9.7. The latter comes from the observation that, by (9.45), as $\lambda, \alpha \downarrow 0$,

$$\alpha - \beta_c(\alpha) = 4\lambda h_c(\lambda) \sim 4 K_c \lambda^2 = \tfrac{1}{4} K_c \big(\alpha + \beta_c(\alpha)\big)^2 \sim \tfrac{1}{4} K_c (2\alpha)^2 = K_c \alpha^2. \qquad (9.77)$$

(6) Bolthausen and Giacomin [29] look at the version of the model where the disorder is *periodic* (earlier work can be found in Grosberg, Izrailev and Nechaev [133]). For this case the free energy can be expressed in terms of a variational formula. However, it turns out to be delicate to deal with large periods, since computations quickly become prohibitive. In particular, the fact that random disorder arises from periodic disorder as the period tends to infinity seems hard to implement when probing the fine details of the critical curve. Caravenna, Giacomin and Zambotti [57], [58] look at the path properties for periodic disorder, showing that under the law P_∞^ω, i.e., the weak limit

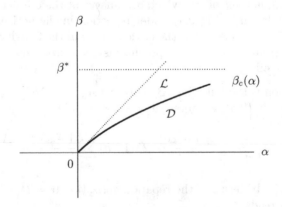

Fig. 9.8. Plot of $\alpha \mapsto \beta_c(\alpha)$.

of the law P_n^ω, the path is transient in the interior of $\mathcal{D}$, null-recurrent at the boundary $\partial \mathcal{D}$, and positive recurrent in $\mathcal{L}$. In each of the three cases they identify the scaling limit of the path.

(7) Den Hollander and Wüthrich [173] consider a version of the copoly-mer model with an *infinite array* of linear interfaces stacked on top of each other, equally spaced at distance L_n with n the length of the copolymer. If $\lim_{n\to\infty} L_n = \infty$, then this model has the same free energy as the single inter-face model, because the copolymer "sees only one interface at a time". Under the assumption that $\lim_{n\to\infty} L_n/\log n = 0$ and $\lim_{n\to\infty} L_n/\log\log n = \infty$, they show that the copolymer hops between neighboring interfaces on a time scale of order $\exp[\chi L_n]$, for some $\chi = \chi(\lambda, h) > 0$, when $(\lambda, h) \in \mathcal{L}$. The reason is that in the localized phase the excursions away from the interface are exponentially unlikely in their length and height (recall Theorem 9.9(a)). See also Extension (7) in Section 7.6.

(8) Martin, Causo and Whittington [236] consider a version of the single interface model in which the path is self-avoiding, the sites (rather than the edges) in the path represent the monomers, while the Hamiltonian is given by (9.76). They identify the qualitative properties of the phase diagram in the (α, β)-plane with the help of exact enumeration methods, series analysis techniques and rigorous bounds. It turns out that there are two critical curves (which are images of each other w.r.t. the line $\{\alpha = \beta\}$ when both monomer types occur with density $\frac{1}{2}$). Both curves lie in the first and in the third quadrant, with horizontal and vertical asymptotes, and meet each other at the origin. Below the lower curve the copolymer is delocalized into the water, above the upper curve it is delocalized into the oil, while in between the two curves it is localized near the interface. Madras and Whittington [233], building on earlier work by Maritan, Riva and Trovato [235], prove that inside the first quadrant the lower curve lies strictly inside the first octant except at the origin, and is a continuous, non-decreasing and concave function of α (and thus is similar to the one in Fig. 9.8, showing that the SAW-restriction does not alter the qualitative properties of the phase diagram). Causo and Whittington [64] analyze the phase diagram with the help of Monte Carlo techniques, and their results suggest that the phase transition is second order in the third quadrant but higher order in the first quadrant. If the latter were true, then this would imply that the origin is a tricritical point, i.e., a point where three phases meet. The results in [235], [236] and [233] apply to SAW's in dimensions $d \geq 2$.

(9) James, Soteros and Whittington [198], [199] generalize the SAW-version of the (α, β)-model by adding to the Hamiltonian an energy $-\gamma$ for each monomer that lies at the interface, irrespective of its type. It turns out that the value of γ affects the phase diagram (see Fig. 9.9). Indeed, as later shown in Madras and Whittington [233], for $\gamma < 0$ the phase diagram is qualitatively like that for $\gamma = 0$, but not for $\gamma > 0$. Indeed, there are *two critical values* $0 \leq \gamma_1 \leq \gamma_2 < \infty$ such that for $\gamma \in (\gamma_1, \gamma_2)$ the two critical curves are separated,

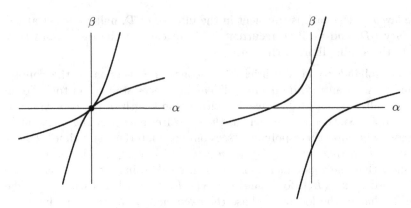

Fig. 9.9. Phase diagrams for the (α, β, γ)-model. The first curve is for $\gamma \in (-\infty, \gamma_1]$, the second curve for $\gamma \in (\gamma_1, \gamma_2)$. For $\gamma \in [\gamma_2, \infty)$ both curves have moved out to infinity.

i.e., they pass through the second and the fourth quadrant (resulting in a single localized phase), while for $\gamma \in [\gamma_2, \infty)$ the two critical curves have "moved out to infinity", i.e., the system is localized for all values of (α, λ). It is shown that $\gamma_2 < \infty$. It is believed that $\gamma_1 = 0$ and $\gamma_2 > 0$, but this remains open. For $\alpha = \beta = 0$ the model reduces to the homogeneous pinning model described in Section 7.1. Also the results in [198] and [199] apply to SAW's in dimensions $d \geq 2$. Exact enumeration is done for $d = 3$.

Habibzadah, Iliev, Saguia and Whittington [140] analyze the (α, β, γ)-model for generalized ballot paths rather than ballot paths (recall Fig. 1.4). The qualitative properties of the phase diagram are the same as for self-avoiding paths.

(10) Iliev, Orlandini and Whittington [181] study the effect of a force in the (α, β)-model with $\alpha < 0$ and $\beta > 0$, i.e., when the copolymer is delocalized below the interface. The force is applied to the endpoint of the copolymer and is perpendicular to the interface in the upward direction (in the spirit of the models considered in Section 7.3). Both for ballot paths and for generalized ballot paths the quenched free energy is computed in the first order Morita approximation, i.e., Z_n^ω is averaged over ω conditioned on $n^{-1} \sum_{i=1}^n \omega_i = 0$. Even though the Morita free energy is only an upper bound for the quenched free energy, it is argued that both free energies lead to the same critical force, which can therefore be computed explicitly. The critical curve in the force-temperature diagram is strictly increasing and so the phase transition is *not re-entrant* (compare with Section 7.3).

The situation is different when $\beta \geq \alpha > 0$. In that case the phase transition is *re-entrant* for $\beta > \alpha$, with the critical curve in the force-temperature diagram hitting 0 at some finite temperature, and *not re-entrant* for $\beta = \alpha$, with the critical curve being strictly increasing. This fits with the observations made in Extension (4), where the critical curve in the phase diagram of the (α, β)-plane without force was found to touch the line $\{\alpha = \beta\}$ (see Fig. 9.8). It is argued that the critical curve in the force-temperature diagram in the Morita approximation is an upper bound for the one of the quenched model.

(11) For the directed version of the (α, β, γ)-model, Orlandini, Tesi and Whittington [254] and Orlandini, Rechnitzer and Whittington [253] analyze the phase diagram in the Morita approximation, i.e., they compute the annealed free energy subject to restrictions on the first and second moments of ω. This gives only limited rigorous information on the properties of the critical curves, but it provides considerable insight into the mechanisms driving the phase transition. The analysis was subsequently extended by Iliev, Rechnitzer and Whittington [182], who found evidence that the lower critical curve is non-analytic at the origin when $\gamma = 0$, at a point in the third quadrant when $\gamma < 0$, and at a point in the first quadrant when $\gamma > 0$. The nature of these tricritical points remains unclear.

(12) Brazhnyi and Stepanow [40] consider a model of a random copolymer in $\mathbb{R}^3$ where the Hamiltonian includes elasticity, self-repulsion and self-attraction, thereby combining aspects from Chapters 3–6 and 9. Obviously, such models display highly complex behavior, but they are worthwhile to investigate when the aim is to model particular experimental situations. With a combination of analytical and numerical techniques it is shown that localization is disfavored by self-repulsion, favored by self-attraction, and can be re-entrant as a function of the densities of the different monomer types.

9.7 Challenges

(1) Improve Theorem 9.11 by finding out whether the phase transition is second order or higher order. Numerical analysis seems to indicate that the order is higher (Trovato and Maritan [298], Causo and Whittington [64], Caravenna, Giacomin and Gubinelli [55]). Monthus [243] suggests that the order is infinite.

(2) We know from Theorem 9.12 that the quenched free energy is infinitely differentiable on $\mathcal{L}$. Find out whether or not it is analytic. At points where this fails the system is said to have a Griffiths-McCoy singularity (Griffiths [132], McCoy [239]). Such singularities are known to occur is some disordered systems, e.g. the low-temperature dilute Ising model in zero magnetic field. The singular behavior is due to the occurrence of rare but arbitrarily large regions where the disorder is such that locally the system is almost at a phase transition. Is the critical curve, i.e., the boundary of $\mathcal{L}$, analytic?

(3) Improve Theorem 9.10 by showing that the path intersects the interface only finitely often. This has been proved by Giacomin and Toninelli [120] deep inside $\mathcal{D}$, namely, for (λ, h) on or above the annealed critical curve that is the upper bound in Theorem 9.6.

(4) Identify the analogue of the weak interaction limit in (9.46–9.48) when the reference random walk is not SRW (see Section 9.6, Extension (1)).

(5) The global properties of the copolymer do not depend on the fact that the monomer types are drawn in an i.i.d. rather than a stationary ergodic fashion, but the local properties do. Since in real polymers the order of the monomer types is at best Markov (as a result of the underlying polymerization process), it is interesting to try and extend the results in Sections 9.2–9.5 to the Markov setting.

(6) Investigate whether or not the critical curve for the SAW-version of the (α, β)-model defined in Section 9.6, Extension (8), has a positive and finite curvature as $\alpha \downarrow 0$. Is this curvature equal to K_c, as in the directed version of the (α, β)-model, or not? Prove that $\gamma_1 = 0$ in the SAW-version of the (α, β, γ)-model defined in Section 9.6, Extension (9).

10

Copolymers Near a Random Selective Interface

In this chapter we consider a different version of the directed copolymer model analyzed in Chapter 9, namely, one where the linear interface is replaced by a *random interface*. In particular, rather than putting the oil and the water in two halfplanes, we place them in large square blocks in a random percolation-type fashion. This is *a crude model of a copolymer in an emulsion*, consisting of oil droplets floating in water (see Fig. 10.1). We will see that this model exhibits a remarkably rich critical behavior, with *two phases* in a supercritical percolation regime and *four phases* in a subcritical percolation regime. *Large deviations* and *partial annealing estimates* will again play a central role, together with *coarse-graining*, entropy estimates and variational calculus.

The random interface model was introduced in den Hollander and Whittington [172] and the crude properties of the phase diagram were derived there. Finer details of the phase diagram were subsequently derived in den Hollander and Pétrélis [170], [171]. The text below follows these three papers.

One application of the model is the stabilization of milk by a protein called casein. This is a copolymer that adsorbs at the fat-water interface, wrapping itself around the tiny fat droplets that float in the water, thereby preventing them to coagulate. Another application is ink, which consists of carbon particles that are coated, for instance, with albumin (from egg white) or gum arabic (a polysaccharide obtained from the Acacia tree), with the same stabilizing effect.

In Section 10.1 we define the model. In Section 10.2 we derive a *variational formula* for the quenched free energy in terms of constituent quenched free energies at the block level and an underlying set of frequencies at which the copolymer can visit the blocks. In Section 10.3 we use the results from Section 10.2 to study the phase diagram in the supercritical percolation regime. In Section 10.4 we turn our attention to the subcritical percolation regime.

F. den Hollander, *Random Polymers*,
Lecture Notes in Mathematics 1974, DOI: 10.1007/978-3-642-00333-2_10,
© Springer-Verlag Berlin Heidelberg 2009, Reprint by Springer-Verlag Berlin Heidelberg 2012

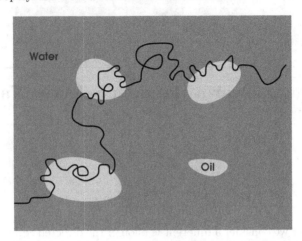

Fig. 10.1. An undirected copolymer in an emulsion.

10.1 A Copolymer Diagonally Crossing Blocks

Each positive integer is randomly labeled A or B, with probability $\frac{1}{2}$ each, independently for different integers. The resulting labeling is denoted by

$$\omega = \{\omega_i \colon i \in \mathbb{N}\} \in \{A, B\}^{\mathbb{N}} \tag{10.1}$$

and represents the *randomness of the copolymer*, with A denoting a hydrophobic monomer and B a hydrophilic monomer. Fix $p \in (0, 1)$ and $L_n \in \mathbb{N}$. Partition $\mathbb{R}^2$ into square blocks of size L_n:

$$\mathbb{R}^2 = \bigcup_{x \in \mathbb{Z}^2} \Lambda_{L_n}(x), \qquad \Lambda_{L_n}(x) = x L_n + (0, L_n]^2. \tag{10.2}$$

Each block is randomly labeled A or B, with probability p, respectively, $1-p$, independently for different blocks, with A denoting oil and B denoting water. The resulting labeling is denoted by

$$\Omega = \{\Omega(x) \colon x \in \mathbb{Z}^2\} \in \{A, B\}^{\mathbb{Z}^2} \tag{10.3}$$

and represents the *randomness of the emulsion* (see Fig. 10.2).

Let

- $\mathcal{W}_n$ = the set of n-step *directed self-avoiding paths* starting at the origin and being allowed to move *upwards, downwards and to the right* (recall Fig. 1.4):

$$\mathcal{W}_n = \{w = (w_i)_{i=0}^n \colon w_0 = 0,\; w_{i+1} - w_i \in \{\uparrow, \downarrow, \rightarrow\}\; \forall\, 0 \le i < n, \\ w_i \ne w_j\; \forall\, 0 \le i < j \le n\}. \tag{10.4}$$

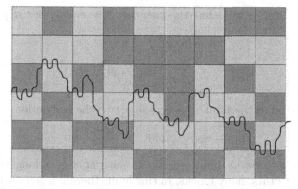

Fig. 10.2. A directed self-avoiding path crossing blocks of oil and water diagonally. The light-shaded blocks are oil, the dark-shaded blocks are water. Each block is L_n lattice spacings wide in both directions. The path carries hydrophobic and hydrophilic monomers on the lattice scale, which are not indicated.

- $\mathcal{W}_{n,L_n}$ = the subset of $\mathcal{W}_n$ consisting of those paths that enter blocks at a corner, exit blocks at one of the two corners *diagonally opposite* the one where it entered, and in between *stay confined* to the two blocks that are seen upon entering (see Fig. 10.2).

In other words, after the path reaches a site xL_n for some $x \in \mathbb{Z}^2$, it must make a step to the right, it must subsequently stay confined to the pair of blocks labeled x and $x + (0, -1)$, and it must exit this pair of blocks either at site $xL_n + (L_n, L_n)$ or at site $xL_n + (L_n, -L_n)$. This restriction – which is put in to make the model mathematically tractable – is unphysical. Nonetheless, as we will see in Sections 10.3–10.4, the model has physically very relevant behavior.

Given ω, Ω and n, with each path $w \in \mathcal{W}_{n,L_n}$ we associate an energy given by the Hamiltonian

$$H_{n,L_n}^{\omega,\Omega}(w) = -\sum_{i=1}^{n} \left(\alpha\, 1_{\{\omega_i = \Omega_{(w_{i-1},w_i)}^{L_n} = A\}} + \beta\, 1_{\{\omega_i = \Omega_{(w_{i-1},w_i)}^{L_n} = B\}} \right), \quad (10.5)$$

where (w_{i-1}, w_i) denotes the i-th step of the path and $\Omega_{(w_{i-1},w_i)}^{L_n}$ denotes the label of the block that the i-th step lies in. What this Hamiltonian does is count the number of AA-matches and BB-matches and assign them energy $-\alpha$ and $-\beta$, respectively, where $\alpha, \beta \in \mathbb{R}$. Note that, as in (9.3), the interaction is assigned to edges rather than to vertices, i.e., we identify the monomers with the steps of the path. We will see later that without loss of generality we may restrict the interaction parameters to the cone

$$\text{CONE} = \{(\alpha, \beta) \in \mathbb{R}^2 : \alpha \geq |\beta|\}. \quad (10.6)$$

Given ω, Ω and n, we define the *quenched free energy per step* as

$$f_{n,L_n}^{\omega,\Omega} = \frac{1}{n} \log Z_{n,L_n}^{\omega,\Omega},$$

$$Z_{n,L_n}^{\omega,\Omega} = \sum_{w \in \mathcal{W}_{n,L_n}} \exp\left[-H_{n,L_n}^{\omega,\Omega}(w)\right]. \tag{10.7}$$

We are interested in the limit $n \to \infty$ subject to the restriction

$$L_n \to \infty \qquad \text{and} \qquad \frac{1}{n}L_n \to 0. \tag{10.8}$$

This is a *coarse-graining* limit where the path spends a long time in each single block yet visits many blocks. In this limit, there is a separation between a *polymer scale* and an *emulsion scale*.

In Theorem 10.1 below we will see that

$$\lim_{n \to \infty} f_{n,L_n}^{\omega,\Omega} = f = f(\alpha, \beta; p) \qquad \text{exists } \omega, \Omega - a.s., \tag{10.9}$$

is finite and non-random, and can be expressed as a variational problem involving the free energies of the polymer in each of the four possible pairs of adjacent blocks it may encounter and the frequencies at which the polymer visits each of these pairs of blocks on the coarse-grained block scale. This variational problem will be the starting point of our analysis.

10.2 Preparations

This section contains some preparatory definitions, lemmas and theorems that are needed to formulate our main results in Sections 10.3–10.4 for the phase diagram. We state the variational formula for the free energy, compute some path entropies, and identify the block free energies in terms of a *linear interface free energy*. Proofs of the lemmas and theorems stated below are elementary and can be found in den Hollander and Whittington [172].

10.2.1 Variational Formula for the Free Energy

We begin with two sets of definitions.

I. For $L \in \mathbb{N}$ and $a \geq 2$ (with aL integer), let $\mathcal{W}_{aL,L}$ denote the set of aL-step directed self-avoiding paths starting at $(0,0)$, ending at (L,L), and in between not leaving the two adjacent blocks of size L labeled $(0,0)$ and $(-1,0)$. For $k, l \in \{A, B\}$, let

$$\psi_{kl}^{\omega}(aL, L) = \frac{1}{aL} \log Z_{aL,L}^{\omega},$$

$$Z_{aL,L}^{\omega} = \sum_{w \in \mathcal{W}_{aL,L}} \exp\left[-H_{aL,L}^{\omega,\Omega}(w)\right] \text{ when } \Omega(0,0) = k \text{ and } \Omega(0,-1) = l,$$

$$\tag{10.10}$$

denote the free energy per step in a kl-block when the number of steps inside the block is a times the size of the block. Let

$$\lim_{L\to\infty} \psi_{kl}^\omega(aL, L) = \psi_{kl}(a) = \psi_{kl}(\alpha, \beta; a).$$ (10.11)

Note here that k labels the type of the block that is diagonally crossed, while l labels the type of the block that appears as its neighbor at the starting corner (see Fig. 10.3). We will see in Section 10.2.3 that the limit exists ω-a.s. and is non-random, and that ψ_{AA} and ψ_{BB} take on a simple form whereas ψ_{AB} and ψ_{BA} do not.

II. Let $\mathcal{W}$ denote the class of all *coarse-grained paths* $W = \{W_j : j \in \mathbb{N}\}$ that step diagonally from corner to corner (see Fig. 10.4, where each dashed line with arrow denotes a single step of W). For $n \in \mathbb{N}$, $W \in \mathcal{W}$ and $k, l \in \{A, B\}$, let

$$\rho_{kl}^\Omega(W, n) = \frac{1}{n} \sum_{j=1}^{n} 1_{E_{kl}^\Omega(j)}$$ (10.12)

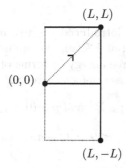

(L, L)

$(0, 0)$

$(L, -L)$

Fig. 10.3. Two neighboring blocks. The dashed line with the arrow indicates that the coarse-grained path makes a step diagonally upwards.

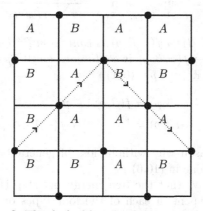

Fig. 10.4. W sampling Ω. The dashed lines with arrows indicate the steps of W.

with $E_{kl}^{\Omega}(j)$ the event that (W_{j-1}, W_j) diagonally crosses a k-block in Ω having an l-block in Ω as its neighbor at the starting corner. Abbreviate

$$\rho^{\Omega}(W, n) = \left(\rho_{kl}^{\Omega}(W, n)\right)_{k,l \in \{A,B\}},\tag{10.13}$$

which is a 2×2 matrix with nonnegative elements that sum up to 1. Let $\mathcal{R}^{\Omega}(W)$ denote the set of all limits points of the sequence $\{\rho^{\Omega}(W, n) \colon n \in \mathbb{N}\}$, and put

$$\mathcal{R}^{\Omega} = \text{the closure of the set } \bigcup_{W \in \mathcal{W}} \mathcal{R}^{\Omega}(W).\tag{10.14}$$

Clearly, $\mathcal{R}^{\Omega}$ exists for all Ω. Moreover, since Ω has a trivial sigma-field at infinity (i.e., all events not depending on finitely many coordinates of Ω have probability 0 or 1) and $\mathcal{R}^{\Omega}$ is measurable with respect to this sigma-field, we have

$$\mathcal{R}^{\Omega} = \mathcal{R}(p) \qquad \Omega - a.s.\tag{10.15}$$

for some *non-random closed* set $\mathcal{R}(p)$. This set, which depends on the parameter p controlling Ω, is the set of all possible limit points of the frequencies at which the four pairs of adjacent blocks can be seen along an infinite coarse-grained path.

With I and II above we have introduced the two main quantities of the model, namely, ψ_{kl}, $k, l \in \{A, B\}$, and $\mathcal{R}(p)$. We are now ready to formulate the key theorem expressing the free energy in terms of these quantities. Let $\mathcal{A}$ be the set of 2×2 matrices whose elements are ≥ 2.

Theorem 10.1. (a) *For all $(\alpha, \beta) \in \mathbb{R}^2$ and $p \in (0, 1)$,*

$$\lim_{n \to \infty} f_{n,L_n}^{\omega,\Omega} = f = f(\alpha, \beta; p) \qquad exists \ \omega, \Omega - a.s. \ and \ in \ mean,\tag{10.16}$$

is finite and non-random, and is given by

$$f = \sup_{(a_{kl}) \in \mathcal{A}} \sup_{(\rho_{kl}) \in \mathcal{R}(p)} \frac{\sum_{k,l} \rho_{kl} a_{kl} \psi_{kl}(a_{kl})}{\sum_{k,l} \rho_{kl} a_{kl}}.\tag{10.17}$$

(b) *The function $(\alpha, \beta) \mapsto f(\alpha, \beta; p)$ is convex on $\mathbb{R}^2$ for all $p \in (0, 1)$.*
(c) *The function $p \mapsto f(\alpha, \beta; p)$ is continuous on $(0, 1)$ for all $(\alpha, \beta) \in \mathbb{R}^2$.*
(d) *For all $(\alpha, \beta) \in \mathbb{R}^2$ and $p \in (0, 1)$,*

$$\begin{aligned}
f(\alpha, \beta; p) &= f(\beta, \alpha; 1 - p),\\
f(\alpha, \beta; p) &= \tfrac{1}{2}(\alpha + \beta) + f(-\beta, -\alpha; p).
\end{aligned}\tag{10.18}$$

(Part (d) is the reason why without loss of generality we may restrict the parameters to the cone in (10.6).)

Theorem 10.1 says that the free energy per step is obtained by keeping track of the times spent in each of the four types of blocks, summing the free energies of the four types of blocks given these times, and afterwards optimizing over these times and over the coarse-grained random walk. The latter carries no entropy because of (10.8). Thus, we see that *self-averaging* occurs both at the *polymer scale* and at the *emulsion scale*.

Theorem 10.1 is the starting point for our analysis of the phase diagram in Sections 10.3–10.4. What we need to do is collect the necessary information on the key ingredients of (10.17). This is done in Sections 10.2.2–10.2.3. We will see that, to get the qualitative properties of the phase diagram, only very little information is needed about $\mathcal{R}(p)$.

The behavior of f as a function of (α, β) is different for $p \geq p_c$ and $p < p_c$, where $p_c \approx 0.64$ is *the critical percolation density for directed bond percolation on the square lattice*. The reason is that the coarse-grained paths W, which determine the set $\mathcal{R}(p)$, sample Ω just like paths in directed bond percolation on the square lattice rotated by 45 degrees sample the percolation configuration (see Fig. 10.4). We will see in Sections 10.3–10.4 that, consequently, the phase diagrams in the supercritical and subcritical regimes are different.

10.2.2 Path Entropies

We next state two lemmas that identify the entropy of a path diagonally crossing a block or running along the interface in a block (the numbers appearing in these lemmas are model specific). We need these lemmas in Section 10.2.3. Their proofs are straightforward.

Let

$$\text{DOM} = \{(a,b): a \geq 1 + b, b \geq 0\}. \tag{10.19}$$

For $(a,b) \in \text{DOM}$, let $N_L(a,b)$ denote the number of aL-step self-avoiding directed paths from $(0,0)$ to (bL, L) whose vertical displacement stays within $(-L, L]$ (aL and bL are integer). Let

$$\kappa(a,b) = \lim_{L \to \infty} \frac{1}{aL} \log N_L(a,b). \tag{10.20}$$

Lemma 10.2. (i) $\kappa(a,b)$ exists and is finite for all $(a,b) \in \text{DOM}$.
(ii) $(a,b) \mapsto a\kappa(a,b)$ *is continuous and strictly concave on* DOM *and analytic on the interior of* DOM.
(iii) *For all* $a \geq 2$,

$$a\kappa(a,1) = \log 2 + \tfrac{1}{2}\left[a \log a - (a-2) \log(a-2)\right]. \tag{10.21}$$

(iv) $\sup_{a \geq 2} \kappa(a,1) = \kappa(a^*,1) = \tfrac{1}{2} \log 5$ *with unique maximizer* $a^* = \tfrac{5}{2}$.
(v) $(\frac{\partial}{\partial a}\kappa)(a^*,1) = 0$ *and* $a^*(\frac{\partial}{\partial b}\kappa)(a^*,1) = \tfrac{1}{2} \log \tfrac{9}{5}$.

For $\mu \geq 1$, let $\hat{N}_L(\mu)$ denote the number of μL-step self-avoiding paths from $(0,0)$ to $(L,0)$ with no restriction on the vertical displacement (μL is integer). Let

$$\hat{\kappa}(\mu) = \lim_{L \to \infty} \frac{1}{\mu L} \log \hat{N}_L(\mu). \tag{10.22}$$

162   10 Copolymers Near a Random Selective Interface

Lemma 10.3. (i) $\hat{\kappa}(\mu)$ *exists and is finite for all* $\mu \geq 1$.
(ii) $\mu \mapsto \mu\hat{\kappa}(\mu)$ *is continuous and strictly concave on* $[1,\infty)$ *and analytic on* $(1,\infty)$.
(iii) $\hat{\kappa}(1) = 0$ *and* $\mu\hat{\kappa}(\mu) \sim \log\mu$ *as* $\mu \to \infty$.
(iv) $\sup_{\mu \geq 1} \mu[\hat{\kappa}(\mu) - \frac{1}{2}\log 5] < \frac{1}{2}\log\frac{9}{5}$.

10.2.3 Free Energies per Pair of Blocks

In this section we identify the block free energies defined in (10.11). Because AA-blocks and BB-blocks have no interface, both ψ_{AA} and ψ_{BB} can be computed explicitly.

Lemma 10.4. *For all* $(\alpha,\beta) \in \mathbb{R}^2$ *and* $a \geq 2$, ω-a.s. *and in mean,*

$$\psi_{AA}(a) = \tfrac{1}{2}\alpha + \kappa(a,1) \qquad and \qquad \psi_{BB}(a) = \tfrac{1}{2}\beta + \kappa(a,1). \qquad (10.23)$$

To compute $\psi_{AB}(a)$ and $\psi_{BA}(a)$, we first consider the free energy per step when the path moves in the vicinity of a *single linear interface* $\mathcal{I}$ separating a solvent A in the upper halfplane from a solvent B in the lower halfplane including the interface itself (see Fig. 10.5). To that end, for $c \geq b > 0$, let $\mathcal{W}_{cL,bL}$ denote the set of cL-step directed self-avoiding paths starting at $(0,0)$ and ending at $(bL,0)$. Define

$$\psi_L^{\omega,\mathcal{I}}(c,b) = \frac{1}{cL} \log Z_{cL,bL}^{\omega,\mathcal{I}} \qquad (10.24)$$

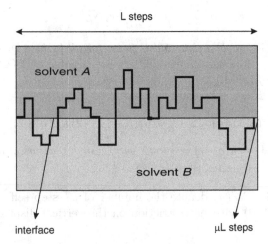

Fig. 10.5. Illustration of (10.24–10.25) for $c = \mu$ and $b = 1$.

with

$$Z_{cL,bL}^{\omega,\mathcal{I}} = \sum_{w \in \mathcal{W}_{cL,bL}} \exp\left[-H_{cL}^{\omega,\mathcal{I}}(w)\right],$$

$$H_{cL}^{\omega,\mathcal{I}}(w) \tag{10.25}$$

$$= -\sum_{i=1}^{cL} \left(\alpha\, 1_{\{\omega_i=A,(w_{i-1},w_i)>0\}} + \beta\, 1_{\{\omega_i=B,(w_{i-1},w_i)\leq 0\}} \right),$$

where $(w_{i-1}, w_i) > 0$ means that the i-th step lies in the upper halfplane and $(w_{i-1}, w_i) \leq 0$ means that the i-th step lies in the lower halfplane or in the interface.

Lemma 10.5. *For all $(\alpha, \beta) \in \mathbb{R}^2$ and $c \geq b > 0$,*

$$\lim_{L\to\infty} \psi_L^{\omega,\mathcal{I}}(c,b) = \phi^{\mathcal{I}}(c/b) = \phi^{\mathcal{I}}(\alpha,\beta;c/b) \tag{10.26}$$

exists ω-a.s. and in mean, and is non-random.

Lemma 10.6. *For all $(\alpha, \beta) \in \mathbb{R}^2$ and $a \geq 2$,*

$$a\psi_{AB}(a) = a\psi_{AB}(\alpha,\beta;a)$$
$$= \sup_{\substack{0\leq b\leq 1,\, c\geq b \\ a-c\geq 2-b}} \left\{ c\phi^{\mathcal{I}}(c/b) + (a-c)[\tfrac{1}{2}\alpha + \kappa(a-c,1-b)] \right\}. \tag{10.27}$$

Lemma 10.7. *Let $k, l \in \{A, B\}$.*
(i) For all $(\alpha, \beta) \in \mathbb{R}^2$, $a \mapsto a\psi_{kl}(\alpha, \beta; a)$ is continuous and concave on $[2, \infty)$.
(ii) For all $a \in [2, \infty)$, $\alpha \mapsto \psi_{kl}(\alpha, \beta; a)$ and $\beta \mapsto \psi_{kl}(\alpha, \beta; a)$ are continuous and non-decreasing on $\mathbb{R}$.

Lemma 10.5 is an immediate consequence of the subadditive ergodic theorem. Indeed, $\phi^{\mathcal{I}}$ is a *single interface free energy* and can be treated as in Chapter 9. The idea behind Lemma 10.6 is that the polymer follows the AB-interface over a distance bL during cL steps and then wanders away from the AB-interface to the diagonally opposite corner over a distance $(1-b)L$ during $(a-c)L$ steps. The optimal strategy is obtained by maximizing over b and c (see Fig. 10.6). In den Hollander and Pétrélis [170] it is shown that the supremum in (10.27) is uniquely attained and that $a \mapsto \psi_{kl}$, $k, l \in \{A, B\}$, is strictly concave.

With (10.23) and (10.27) we have identified the key ingredients in the variational formula for the free energy in (10.17) in terms of the single interface free energy $\phi^{\mathcal{I}}$. This constitutes a *major simplification*, in view of the methods and techniques available from Chapter 9.

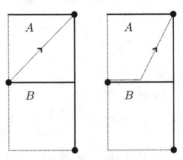

Fig. 10.6. Two possible strategies inside an AB-block: The path can either move straight across or move along the interface for awhile and then move across. Both strategies correspond to a coarse-grained step diagonally upwards as in Fig. 10.3. The dotted lines represent the motion of the path on the block level. On a microscopic scale the path is erratic, which is however not indicated.

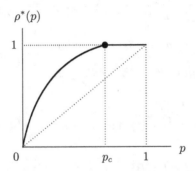

Fig. 10.7. Plot of $p \mapsto \rho^*(p)$.

10.2.4 Percolation

Let

$$\rho^*(p) = \sup_{(\rho_{kl}) \in \mathcal{R}(p)} [\rho_{AA} + \rho_{AB}], \qquad p \in (0,1). \tag{10.28}$$

This is the maximal frequency of A-blocks crossed by an infinite coarse-grained path (recall (10.12–10.15)). The graph of $p \mapsto \rho^*(p)$ is sketched in Fig. 10.7.

The elements of $\mathcal{R}(p)$ are matrices

$$\begin{pmatrix} \rho_{AA} & \rho_{AB} \\ \rho_{BA} & \rho_{BB} \end{pmatrix} \tag{10.29}$$

whose elements are non-negative and sum up to 1. It is easy to see that $p \mapsto \mathcal{R}(p)$ is continuous in the Hausdorff metric. Moreover, for $p \geq p_c$, $\mathcal{R}(p)$ contains matrices of the form

$$\begin{pmatrix} 1 - \gamma & \gamma \\ 0 & 0 \end{pmatrix} \qquad \text{for } \gamma \text{ in a closed subset of } (0,1). \tag{10.30}$$

The reason is that, when $p > p_c$, the A-blocks percolate and W may eventually either stay inside the infinite A-cluster or move along its boundary.

We have now gathered enough information on the ingredients of the variational formula in Theorem 10.1 for the free energy to be able to start addressing the phase diagram. In Section 10.3 we consider the supercritical regime $p \geq p_c$, where the oil blocks percolate, in Section 10.4 the subcritical regime, where the oil blocks do not percolate. These regimes display completely different behavior, since CONE defined in (10.6) favors the oil blocks over the water blocks.

10.3 Phase Diagram in the Supercritical Regime

The phase diagram is relatively simple in the supercritical regime. This is because the *oil blocks percolate*, and so the coarse-grained path can choose between moving into the oil or running along the interface between the oil and the water (see Fig. 10.8). We may therefore expect the behavior to be *qualitatively similar to that for a single linear interface.* In Section 10.4 we will see that the phase diagram is considerably more complex in the subcritical regime. This is because when the oil does not percolate there is no (!) infinite interface available.

In Section 10.3.1 we derive a criterion for localization in terms of the single interface free energy $\phi^{\mathcal{I}}$ defined in Lemma 10.5. In Section 10.3.3 we use this criterion to state a theorem identifying the qualitative properties of the critical curve. In Section 10.3.3 we give the proof of this theorem. Section 10.3.4 states finer details of the critical curve, the proofs of which are technical and are omitted.

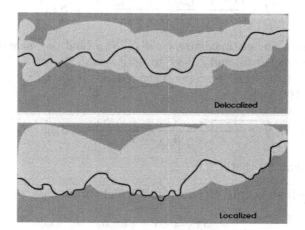

Fig. 10.8. Two possible strategies when the oil percolates.

10.3.1 Free Energy in the Two Phases

The following result gives a preliminary identification of the two phases.

Theorem 10.8. *Let $p \geq p_c$. Then*

$$\begin{aligned}
\mathcal{D} &= \{(\alpha, \beta) \in \text{CONE}: S_{AB} = S_{AA}\}, \\
\mathcal{L} &= \{(\alpha, \beta) \in \text{CONE}: S_{AB} > S_{AA}\},
\end{aligned} \tag{10.31}$$

where

$$S_{kl} = \sup_{a \geq 2} \psi_{kl}(a), \qquad k, l \in \{A, B\}. \tag{10.32}$$

Proof. The proof uses (10.30) and Theorem 10.1(a), in combination with the inequalities

$$S_{BB} \leq S_{AA}, \qquad S_{BA} \leq S_{AB}, \tag{10.33}$$

which hold because $\beta \leq \alpha$. We prove that for $p \geq p_c$:

$$\begin{aligned}
\text{(i)} \quad & S_{AB} = S_{AA} \Longrightarrow f = S_{AA}. \\
\text{(ii)} \quad & S_{AB} > S_{AA} \Longrightarrow f > S_{AA}.
\end{aligned} \tag{10.34}$$

(i) Suppose that $S_{AB} = S_{AA}$. Then, because $\psi_{kl}(a) \leq S_{kl}$ for all $a \geq 2$ and $k, l \in \{A, B\}$ by (10.32), Theorem 10.1(a) and (10.33) yield

$$f \leq \sup_{(a_{kl}) \in \mathcal{A}} \sup_{(\rho_{kl}) \in \mathcal{R}(p)} \frac{\sum_{k,l} \rho_{kl} a_{kl} S_{kl}}{\sum_{k,l} \rho_{kl} a_{kl}} \leq \sup_{k,l} S_{kl} = S_{AB} = S_{AA}. \tag{10.35}$$

On the other hand, (10.30) and Theorem 10.1(a) yield

$$f \geq \frac{(1-\gamma)\bar{a}_{AA} S_{AA} + \gamma \bar{a}_{AB} S_{AB}}{(1-\gamma)\bar{a}_{AA} + \gamma \bar{a}_{AB}} = S_{AA}, \tag{10.36}$$

where $\bar{a}_{AA}, \bar{a}_{AB}$ are any maximizers of S_{AA}, S_{AB} (and the value of γ is irrelevant). Combine (10.35) and (10.36) to get $f = S_{AA}$.

(ii) Suppose that $S_{AB} > S_{AA}$. Then (10.30) with $0 < \gamma < 1$ and Theorem 10.1(a) yield

$$f \geq \frac{(1-\gamma)\bar{a}_{AA} S_{AA} + \gamma \bar{a}_{AB} S_{AB}}{(1-\gamma)\bar{a}_{AA} + \gamma \bar{a}_{AB}} > S_{AA}. \tag{10.37}$$

Here it is important that $\gamma > 0$. $\square$

It follows from Lemma 10.2(iv), (10.23) and (10.32) that

$$S_{AA} = \tfrac{1}{2}\alpha + \tfrac{1}{2}\log 5. \tag{10.38}$$

Theorem 10.8, together with (10.34) and (10.38), yields:

Theorem 10.9. *Let $p \geq p_c$. Then $(\alpha, \beta) \mapsto f(\alpha, \beta; p)$ is non-analytic along the curve in* CONE *separating the two regions*

$$\mathcal{D} = \text{delocalized phase} = \left\{ (\alpha, \beta) \in \text{CONE: } f(\alpha, \beta; p) = \tfrac{1}{2}\alpha + \tfrac{1}{2}\log 5 \right\},$$
$$\mathcal{L} = \text{localized phase} = \left\{ (\alpha, \beta) \in \text{CONE: } f(\alpha, \beta; p) > \tfrac{1}{2}\alpha + \tfrac{1}{2}\log 5 \right\},$$
$$(10.39)$$

where $\lim_{n \to \infty} \frac{1}{n} \log |\mathcal{W}_{n, L_n}| = \tfrac{1}{2}\log 5$ *is the entropy per step of the walk in a single block subject to (10.8).*

The intuition behind Theorem 10.9 is as follows. Suppose that $p > p_c$. Then the A-blocks percolate. Therefore the polymer has the option of moving to the infinite cluster of A-blocks and staying in that infinite cluster forever, thus seeing only AA-blocks. In doing so, it loses an entropy of at most $o(n/L_n) = o(n)$ on the coarse-grained scale, it gains an energy $\tfrac{1}{2}\alpha n + o(n)$ on the lattice scale (because only half of its monomers are matched), and it gains an entropy $(\tfrac{1}{2}\log 5)n + o(n)$ on the lattice scale (see the top half of Fig. 10.8). Alternatively, the path has the option of following the boundary of the infinite cluster (at least part of the time), during which it sees AB-blocks and (when $\beta \geq 0$) gains more energy by matching more than half of its monomers (see the bottom half of Fig. 10.8). Consequently,

$$f(\alpha, \beta; p) \geq \tfrac{1}{2}\alpha + \tfrac{1}{2}\log 5. \qquad (10.40)$$

The boundary between the two regimes in (10.39) corresponds to the crossover where one option takes over from the other. Note that the critical curve does *not* depend on p. Because $p \mapsto f(\alpha, \beta; p)$ is continuous (by Theorem 10.1(c) in Section 10.2.1), the same critical curve occurs at $p = p_c$. Note further that f does not depend on p in the delocalized phase, but does in the localized phase.

10.3.2 Criterion for Localization

The key result identifying the critical curve in the supercritical regime is the following criterion in terms of the free energy of the *single linear interface*.

Proposition 10.10. *Let $p \geq p_c$. Then $(\alpha, \beta) \in \mathcal{L}$ if and only if*

$$\sup_{\mu \geq 1} \mu[\phi^{\mathcal{I}}(\alpha, \beta; \mu) - \tfrac{1}{2}\alpha - \tfrac{1}{2}\log 5] > \tfrac{1}{2}\log \tfrac{9}{5}. \qquad (10.41)$$

Proof. Combining (10.38) with Lemma 10.6, together with the reparameterization $\mu = c/b$ and $\nu = (a - c)/b$, we get

$$S_{AB} - S_{AA} = \sup_{\mu \geq 1, \nu \geq 1} \frac{\mu[\phi^{\mathcal{I}}(\mu) - S_{AA}] - \nu[\tfrac{1}{2}\log 5 - u(\nu)]}{\mu + \nu} \qquad (10.42)$$

with

$$u(\nu) = \sup_{\frac{2}{\nu+1} \leq b \leq 1} \kappa(b\nu, 1 - b), \qquad \nu \geq 1. \tag{10.43}$$

Abbreviate $v(\nu) = \nu[\frac{1}{2} \log 5 - u(\nu)]$. Below we will show that

$$\begin{aligned}&\text{(i)} \ \ v(\nu) > \tfrac{1}{2} \log \tfrac{9}{5} \ \ \text{for all } \nu \geq 1,\\&\text{(ii)} \ \lim_{\nu \to \infty} v(\nu) = \tfrac{1}{2} \log \tfrac{9}{5}.\end{aligned} \tag{10.44}$$

This will imply the claim as follows. If $\mu[\phi^{\mathcal{I}}(\mu) - S_{AA}] \leq \frac{1}{2} \log \frac{9}{5}$ for all μ, then by (i) the numerator in (10.42) is strictly negative for all μ and ν, and so by (ii) the supremum is taken at $\nu = \infty$, resulting in $S_{AB} - S_{AA} = 0$. On the other hand, if $\mu[\phi^{\mathcal{I}}(\mu) - S_{AA}] > \frac{1}{2} \log \frac{9}{5}$ for some μ, then, for that μ, by (i) and (ii) the numerator is strictly positive for ν large enough, resulting in $S_{AB} - S_{AA} > 0$.

To prove (10.44), we will need the following inequality. Abbreviate $\chi(a, b) = a\kappa(a, b)$. Then by Lemma 10.2(ii) we have, for all $(s, t) \neq (u, v)$ in DOM,

$$\begin{aligned}\chi(s, t) - \chi(u, v) &= \int_0^1 dw \, \frac{\partial}{\partial w} \chi(u + w(s - u), v + w(t - v))\\&> \left[\frac{\partial}{\partial w} \chi(u + w(s - u), v + w(t - v))\right]_{w=1} \tag{10.45}\\&= (s - u) \left(\frac{\partial}{\partial a} \chi\right)(s, t) + (t - v) \left(\frac{\partial}{\partial b} \chi\right)(s, t).\end{aligned}$$

To prove (10.44)(i), put $b = a/\nu$ in (10.43) and use Lemma 10.2(iv–v) to rewrite the statement in (10.44)(i) as

$$\kappa\left(a, 1 - \frac{a}{\nu}\right) < \kappa(a^*, 1) - \frac{a^*}{\nu} \left(\frac{\partial}{\partial b} \kappa\right)(a^*, 1) \ \text{ for all } \nu \geq 1 \text{ and } \frac{2\nu}{\nu + 1} \leq a \leq \nu. \tag{10.46}$$

But this inequality follows from (10.45) by picking $s = a^*$, $t = 1$, $u = a$, $v = 1 - \frac{a}{\nu}$, canceling a term $a^*\kappa(a^*, 1)$ on both sides, using that $(\frac{\partial}{\partial a}\kappa)(a^*, 1) = 0$, and afterwards canceling a common factor a on both sides.

To prove (10.44)(ii), we argue as follows. Picking $b = \frac{a^*}{\nu}$ in (10.43), we get from Lemma 10.2(iv) that

$$v(\nu) \leq \nu \left[\kappa(a^*, 1) - \kappa\left(a^*, 1 - \frac{a^*}{\nu}\right)\right]. \tag{10.47}$$

Letting $\nu \to \infty$, we get from Lemma 10.2(v) that

$$\limsup_{\nu \to \infty} v(\nu) \leq a^* \left(\frac{\partial}{\partial b} \kappa\right)(a^*, 1) = \tfrac{1}{2} \log \tfrac{9}{5}. \tag{10.48}$$

Combine this with (10.44)(i) to get (10.44)(ii). $\quad\square$

Proposition 10.10 says that localization occurs if and only if the free energy per step for the single linear interface exceeds the free energy per step for an AA-block by a certain positive amount. This excess is needed to *compensate* for the loss of entropy that occurs when the path runs along the interface for awhile before moving upwards from the interface to end at the diagonally opposite corner (recall Fig. 10.3). The constants $\frac{1}{2}\log 5$ and $\frac{1}{2}\log\frac{9}{5}$ are special to our model.

10.3.3 Qualitative Properties of the Critical Curve

• Statement of Qualitative Properties

With the help of Proposition 10.10, we get the following theorem identifying the qualitative properties of the supercritical phase diagram (see Fig. 10.9). The proof is deferred to Section 10.3.3.

Theorem 10.11. *Let* $p \geq p_c$.
(a) *For every* $\alpha \geq 0$ *there exists a* $\beta_c(\alpha) \in [0, \alpha]$ *such that the copolymer is*

$$\begin{aligned} \text{delocalized if } &-\alpha \leq \beta \leq \beta_c(\alpha), \\ \text{localized} \quad \text{if } &\beta_c(\alpha) < \beta \leq \alpha. \end{aligned} \tag{10.49}$$

(b) *The function* $\alpha \mapsto \beta_c(\alpha)$ *is independent of* p, *continuous, non-decreasing and concave on* $[0,\infty)$. *There exists an* $\alpha^* \in (0,\infty)$ *such that*

$$\begin{aligned} \beta_c(\alpha) &= \alpha \text{ if } \alpha \leq \alpha^*, \\ \beta_c(\alpha) &< \alpha \text{ if } \alpha > \alpha^*, \end{aligned} \tag{10.50}$$

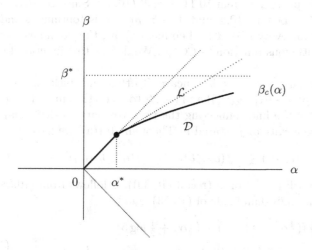

Fig. 10.9. Plot of $\alpha \mapsto \beta_c(\alpha)$ for $p \geq p_c$. Compare with Fig. 9.8.

and

$$\lim_{\alpha \downarrow \alpha^*} \frac{\alpha - \beta_c(\alpha)}{\alpha - \alpha^*} \in [0, 1). \tag{10.51}$$

Moreover, there exists a $\beta^* \in [\alpha^*, \infty)$ such that

$$\lim_{\alpha \to \infty} \beta_c(\alpha) = \beta^*. \tag{10.52}$$

The intuition behind Theorem 10.11 is as follows. Pick a point (α, β) inside $\mathcal{D}$. Then the polymer spends almost all of its time deep inside the A-blocks. Increase β while keeping α fixed. Then there will be a larger energetic advantage for the polymer to move some of its monomers from the A-blocks to the B-blocks by *crossing the interface inside the AB-block pairs*. There is some entropy loss associated with doing so, but if β is large enough, then the energy advantage will dominate, so that AB-localization sets in. The value at which this happens depends on α and is strictly positive. Since the entropy loss is finite, for α large enough the energy-entropy competition plays out not only below the diagonal, but also below a horizontal asymptote. On the other hand, for α small enough the loss of entropy dominates the energetic advantage, which is why the critical curve has a piece that lies on the diagonal. The larger the value of α the larger the value of β where AB-localization sets in. This explains why the critical curve moves to the right and up. The trivial inequality $\phi^{\mathcal{I}}(\mu) \leq \alpha + \hat{\kappa}(\mu)$ and the gap in Lemma 10.3(iv) are responsible for the linear piece of the critical curve. At the end of this linear piece the critical curve has a slope discontinuity, because the linear interface is already strictly inside its localized region.

• Proof of Qualitative Properties

In this section we prove Theorem 10.11. Recall (10.32). Since $S_{AA}(\alpha, \beta)$ does not depend on β by Lemma 10.4, and $\beta \mapsto S_{AB}(\alpha, \beta)$ is continuous and non-decreasing on $\mathbb{R}$ for every $\alpha \in \mathbb{R}$ by Lemma 10.7(ii), the boundary between $\mathcal{D}$ and $\mathcal{L}$ is a continuous function in CONE. We denote this function by $\alpha \mapsto \beta_c(\alpha)$.

We first show that the curve is concave. To that end, pick any $\alpha_1 < \alpha_2$, and consider the points $(\alpha_1, \beta_c(\alpha_1))$ and $(\alpha_2, \beta_c(\alpha_2))$ on the curve. Let (α_3, β_3) be the midpoint of the line connecting the two. We want to show that $\beta_3 \leq \beta_c(\alpha_3)$. By the convexity of f, stated in Theorem 10.1(a), we have

$$f(\alpha_3, \beta_3) \leq \tfrac{1}{2}[f(\alpha_1, \beta_c(\alpha_1)) + f(\alpha_2, \beta_c(\alpha_2))]. \tag{10.53}$$

Since the curve itself is part of $\mathcal{D}$ (recall (10.39)), it follows from (10.38) and (10.34)(i) that the right-hand side of (10.53) equals

$$\tfrac{1}{2}\left[\left(\tfrac{1}{2}\alpha_1 + \tfrac{1}{2}\log 5\right) + \left(\tfrac{1}{2}\alpha_2 + \tfrac{1}{2}\log 5\right)\right]$$
$$= \tfrac{1}{2}\left(\frac{\alpha_1 + \alpha_2}{2}\right) + \tfrac{1}{2}\log 5 = \tfrac{1}{2}\alpha_3 + \tfrac{1}{2}\log 5. \tag{10.54}$$

Thus, $f(\alpha_3, \beta_3) \leq \frac{1}{2}\alpha_3 + \frac{1}{2}\log 5$. But, by (10.34), the reverse inequality is true always, and so equality holds. Consequently, $(\alpha_3, \beta_3) \in \mathcal{D}$, which proves the claim that $\beta_3 \leq \beta_c(\alpha_3)$.

The concavity in combination with the lower bound in part (i) of the following lemma show that the curve is non-decreasing. This lemma settles most of Theorem 10.11.

Lemma 10.12. *Let $p \geq p_c$.*
(i) $\beta_c(\alpha) \geq \log(2 - e^{-\alpha})$ *for all $\alpha \geq 0$.*
(ii) $\beta_c(\alpha) < 8\log 3$ *for all $\alpha \geq 0$.*
(iii) $\beta_c(\alpha) = \alpha$ *for all $0 \leq \alpha \leq \alpha_0$, where $\alpha_0 \approx 0.125$ is implicitly defined by*

$$\sup_{\mu \geq 1} \mu\left[\hat{\kappa}(\mu) + \frac{1}{2}\alpha_0 - \frac{1}{2}\log 5\right] = \frac{1}{2}\log\frac{9}{5}. \tag{10.55}$$

Proof. (i) We have, recalling (10.24–10.25),

$$\phi^{\mathcal{I}}(\mu) = \lim_{L \to \infty} \frac{1}{L}\log Z_L^{\omega,\mathcal{I}}(\mu) \quad \omega - a.s. \tag{10.56}$$

with

$$Z_L^{\omega,\mathcal{I}}(\mu) = \sum_{w \in \mathcal{W}_{\mu L, L}} \exp\left[-H_{\mu L}^{\omega,\mathcal{I}}(w)\right]$$

$$H_{\mu L}^{\omega,\mathcal{I}}(w) = -\sum_{i=1}^{\mu L}\left(\alpha 1_{\{\omega_i = A, (w_{i-1}, w_i) > 0\}} + \beta 1_{\{\omega_i = B, (w_{i-1}, w_i) \leq 0\}}\right). \tag{10.57}$$

We will derive an upper bound on $\phi^{\mathcal{I}}(\mu)$ by doing a so-called *first-order partial annealing estimate*, similar to the one we did in Section 9.4. This estimate is also referred to as a first-order Morita approximation (see Orlandini, Rechnitzer and Whittington [253] for a general account of Morita approximations). The estimate consists in writing

$$H_{\mu L}^{\omega,\mathcal{I}}(w) = -\sum_{i=1}^{\mu L}\alpha 1_{\{\omega_i = A\}}$$

$$-\sum_{i=1}^{\mu L} 1_{\{(w_{i-1}, w_i) \leq 0\}}\left(-\alpha 1_{\{\omega_i = A\}} + \beta 1_{\{\omega_i = B\}}\right), \tag{10.58}$$

using that the first term is $-\mu L\frac{1}{2}\alpha[1 + o(1)]$ ω-a.s. as $L \to \infty$ and is independent of w, substituting the latter into (10.57), and performing the expectation over ω. This gives

$$\mathbb{E}\left(\log Z_L^{\omega,\mathcal{I}}(\mu)\right)$$

$$\leq \mu L\frac{1}{2}\alpha[1 + o(1)] + \log \sum_{w \in \mathcal{W}_{\mu L, L}}\prod_{i=1}^{\mu L} 1_{\{(w_{i-1}, w_i) \leq 0\}}\left\langle e^{-\alpha 1_{\{\omega_i = A\}} + \beta 1_{\{\omega_i = B\}}}\right\rangle$$

$$\leq \mu L\frac{1}{2}\alpha[1 + o(1)] + \mu L[\hat{\kappa}(\mu) + o(1)] + \mu L\log\left(\frac{1}{2}e^{-\alpha} + \frac{1}{2}e^{\beta}\right), \tag{10.59}$$

where we use Jensen's inequality as well as the i.i.d. property of ω for the first inequality, and we use Lemma 10.3(i) for the second inequality. Consequently,

$$\phi^{\mathcal{I}}(\mu) = \lim_{L\to\infty} \frac{1}{\mu L} \mathbb{E}\left(\log Z_L^{\omega,\mathcal{I}}(\mu)\right) \leq \tfrac{1}{2}\alpha + \hat{\kappa}(\mu) + \log\left(\tfrac{1}{2}e^{-\alpha} + \tfrac{1}{2}e^{\beta}\right). \quad (10.60)$$

Suppose that

$$\log\left(\tfrac{1}{2}e^{-\alpha} + \tfrac{1}{2}e^{\beta}\right) \leq 0. \quad (10.61)$$

Then substitution of (10.60) into (10.42) gives

$$S_{AB} - S_{AA} \leq \sup_{\mu\geq 1,\, \nu\geq 1} \frac{\mu[\hat{\kappa}(\mu) - \tfrac{1}{2}\log 5] - \nu[\tfrac{1}{2}\log 5 - f(\nu)]}{\mu + \nu}. \quad (10.62)$$

But the right-hand side is the same as $S_{AB} - S_{AA}$ when $\alpha = \beta = 0$ (as can be seen from (10.42) because $\phi^{\mathcal{I}}(\mu) = \hat{\kappa}(\mu)$ when $\alpha = \beta = 0$), and therefore is equal to 0. Hence, recalling (10.14), we find that (10.61) implies that $(\alpha, \beta) \in \mathcal{D}$. Consequently,

$$\log\left(\tfrac{1}{2}e^{-\alpha} + \tfrac{1}{2}e^{\beta_c(\alpha)}\right) \geq 0 \qquad \text{for all } \alpha \geq 0, \quad (10.63)$$

which gives the lower bound that is claimed.

(ii) We will show that there exists a $\mu_0 > 1$ such that

$$\mu_0[\phi^{\mathcal{I}}(\mu_0) - S_{AA}] > \tfrac{1}{2}\log\tfrac{9}{5} \qquad \text{for all } \alpha \geq 0 \text{ when } \beta \geq 8\log 3. \quad (10.64)$$

This will prove the claim via Proposition 10.10.

Consider the polymer along the single infinite interface $\mathcal{I}$. Fix ω. In ω, look for the strings of B's that are followed by a string of at least three A's. Call these B-strings "good", and call all other B-strings "bad". Let $w(\omega)$ be the path that starts at $(0,0)$, steps to $(0,1)$ and proceeds as follows. Each time a good B-string comes up, the path moves down from height 1 to height 0 during the step that carries the A just preceding the good B-string, moves at height 0 during the steps that carry the B's inside the string, moves up from height 0 to height 1 during the step that carries the first A after the string, and moves at height 1 during the step that carries the second A after the string. The third A can be used to either move from height 1 to height 0 in case the next good B-string comes up immediately, or to move at height 1 in case it is not. When a bad B-string comes up, the path stays at height 1.

Along $w(\omega)$, we have that all the A's lie in the upper halfplane, all the bad B-strings lie in the upper halfplane, while all the good B-strings lie in the interface. Asymptotically, the good B-strings contain $\tfrac{1}{4}$-th of the B's. Hence $\tfrac{1}{8}$-th of the steps carry a B that is in a good B-string. Moreover, the number of steps between heights 0 and 1 is $\tfrac{1}{8}$ times the number of steps at heights 0 and 1 (because the average length of a good B-string is 2), and so $w(\omega)$ travels a distance L in time $\tfrac{9}{8}L$ for L large, which corresponds to $\mu = \mu_0 = \tfrac{9}{8}$. Thus, the value of the Hamiltonian in (10.25) at the path $w(\omega)$ equals

$$H_{\mu_0 L}^{\omega, \mathcal{I}}(w(\omega)) = -\mu_0 L \left(\tfrac{1}{2}\alpha + \tfrac{1}{8}\beta\right)[1 + o(1)] \qquad \omega - a.s. \text{ as } L \to \infty. \quad (10.65)$$

Therefore, recalling (10.26), we have

$$\phi^{\mathcal{I}}(\mu_0) \geq \tfrac{1}{2}\alpha + \tfrac{1}{8}\beta. \quad (10.66)$$

Via (10.38) this gives

$$\mu_0[\phi^{\mathcal{I}}(\mu_0) - S_{AA}] \geq \mu_0 \left(\tfrac{1}{8}\beta - \tfrac{1}{2}\log 5\right). \quad (10.67)$$

Consequently, the inequality in (10.64) holds as soon as

$$\tfrac{1}{8}\beta > \tfrac{8}{9}\log 3 + \tfrac{1}{18}\log 5. \quad (10.68)$$

Since the right-hand side is strictly smaller than $\log 3$, this proves the claim.

(iii) Pick $\alpha = \beta$. Then $\phi^{\mathcal{I}}(\mu) \leq \alpha + \hat{\kappa}(\mu)$. If $\alpha \in [0, \alpha_0]$, with α_0 given by (10.55), then this bound in combination with (10.38) and Proposition 10.10 gives $S_{AB} = S_{AA}$. Thus, $\{(\alpha, \alpha): \alpha \in [0, \alpha_0]\} \subset \mathcal{D}$. $\square$

Lemma 10.12, together with the concavity of $\alpha \mapsto \beta_c(\alpha)$ (shown prior to Lemma 10.12), proves Theorem 10.11, except for the slope discontinuity stated in (10.51). But the latter follows from the fact that if the piece of β_c on $[\alpha^*, \infty)$ is "analytically continued" outside CONE, then it hits the vertical axis at a strictly positive value, namely, α_0. Indeed, for $\alpha = 0$ we have $\phi^{\mathcal{I}}(\mu) = \tfrac{1}{2}\beta + \hat{\kappa}(\mu)$, because there is zero exponential cost for the path to stay in the lower halfplane (recall (10.24–10.26)). Consequently, the criterion for delocalization in Proposition 10.10, $S_{AB} = S_{AA}$, reduces to

$$\sup_{\mu \geq 1} \mu \left[\hat{\kappa}(\mu) + \tfrac{1}{2}\beta - \tfrac{1}{2}\log 5\right] \leq \tfrac{1}{2}\log \tfrac{9}{5}. \quad (10.69)$$

This inequality holds precisely when $\beta \leq \alpha_0$ with α_0 defined in (10.55).

10.3.4 Finer Details of the Critical Curve

In den Hollander and Pétrélis [170] the following three theorems are proved, which complete the analysis of the phase diagram in Fig. 10.9.

Theorem 10.13. *Let $p \geq p_c$. Then $\alpha \mapsto \beta_c(\alpha)$ is strictly increasing on $[0, \infty)$.*

Theorem 10.14. *Let $p \geq p_c$. Then for every $\alpha \in (\alpha^*, \infty)$ there exist $0 < C_1 < C_2 < \infty$ and $\delta_0 > 0$ (depending on p and α) such that*

$$C_1 \delta^2 \leq f(\alpha, \beta_c(\alpha) + \delta; p) - f(\alpha, \beta_c(\alpha); p) \leq C_2 \delta^2 \qquad \forall \delta \in (0, \delta_0]. \quad (10.70)$$

Theorem 10.15. *Let $p \geq p_c$. Then $(\alpha, \beta) \mapsto f(\alpha, \beta; p)$ is infinitely differentiable throughout $\mathcal{L}$.*

Theorem 10.13 implies that the critical curve never reaches the horizontal asymptote, which in turn implies that $\alpha^* < \beta^*$ and that the slope in (10.51) is > 0. Theorem 10.14 shows that the phase transition along the critical curve in Fig. 10.9 is *second order off the diagonal*. In contrast, we know that the phase transition is *first order on the diagonal*. Indeed, the free energy equals $\frac{1}{2}\alpha + \frac{1}{2}\log 5$ on and below the diagonal segment between $(0,0)$ and (α^*, α^*), and equals $\frac{1}{2}\beta + \frac{1}{2}\log 5$ on and above this segment as is evident from interchanging α and β. Theorem 10.15 tells us that the critical curve in Fig. 10.9 is the *only* location in CONE where a phase transition of *finite* order occurs.

The proofs of Theorems 10.13–10.15 are technical. They rely on *perturbation arguments*, in combination with *exponential tightness of the excursions away from the interface* inside the localized phase, and are hard because we have two variational formulas to fight with: (10.17) and (10.27). For details we refer to [170].

Theorems 10.14–10.15 are the analogues of Theorems 9.11–9.12 for the single flat infinite interface. For the latter model, Theorem 9.11 says that the phase transition is *at least of second order*, i.e., only the quadratic upper bound is proved. As mentioned in Section 9.7, Challenge (1), numerical simulation seems to indicate that the order is higher, contrary to what we have here.

The mechanisms behind the phase transition in the two models are different. While for the single interface model the copolymer makes long excursions away from the interface and dips below the interface during a fraction of time that is at most of order δ^2, in our emulsion model the copolymer runs along the interface during a fraction of time that is of order δ, and in doing so stays close to the interface. Moreover, because near the critical curve for the emulsion model the single interface model is already strictly inside its localized phase, there is a variation of order δ in the single interface free energy as a constituent part of the emulsion free energy. Thus, the δ^2 in the emulsion model is the product of two factors δ, one coming from the time spent running along the interface and one coming from the variation of the constituent single interface free energy away from its critical curve.

The proof of Theorem 10.15 uses some of the ingredients of the proof of Theorem 9.12 in Giacomin and Toninelli [122]. However, in the emulsion model there is an extra complication, namely, the speed per step to move one unit of space forward may vary (because steps are up, down and to the right), while in the single interface model this is fixed at one (because steps are north-east and south-east). Consequently, the infinite differentiability with respect to this speed variable needs to be controlled too. This is handled by taking the Legendre transform w.r.t. this variable. Along the way we need to prove *uniqueness of maximizers* and *non-degeneracy of the Jacobian matrix at these maximizers*, in order to be able to apply implicit function theorems. For details we refer to den Hollander and Pétrélis [170].

10.4 Phase Diagram in the Subcritical Regime

In the subcritical regime the phase diagram is *much more complex* than in
the supcritical regime. The reason is that *the oil does not percolate,* and so the
copolymer no longer has the option of moving into the oil (in case it prefers
to delocalize) nor of running along the interface between the oil and the water
(in case it prefers to localize). Instead, it has to every now and then cross
blocks of water, even though it prefers the oil. Below we list the main results,
skipping the mathematics.

It turns out that there are *four phases,* separated by three critical curves,
meeting at *two tricritical points,* all of which depend on p. The phase diagram
is sketched in Fig. 10.10. For details on the derivation, we refer to den Hollander and Pétrélis [171]. The copolymer has the following behavior in the four
phases of Fig. 10.10, as illustrated in Figs. 10.11–10.12:

- $\mathcal{D}_1$: *fully delocalized* into A-blocks and B-blocks, but never inside both
  blocks of a neighboring pair.
- $\mathcal{D}_2$: *fully delocalized* into A-blocks and B-blocks, and inside both blocks of
  a neighboring BA-pair.
- $\mathcal{L}_1$: *partially localized* near the interface in BA-pairs.
- $\mathcal{L}_2$: *partially localized* near the interface in both BA-pairs and AB-pairs.

This is to be compared with the much simpler behavior in the two phases of
Fig. 10.9, as given by Fig. 10.8.

The intuition behind the phase diagram is as follows. In $\mathcal{D}_1$ and $\mathcal{D}_2$, β is not
large enough to induce localization. In both types of block pairs, the reward
for running along the interface is too small compared to the loss of entropy
that comes with having to cross the block at a steeper angle. In $\mathcal{D}_1$, where α

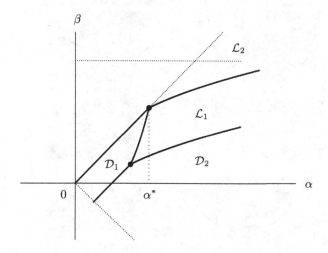

Fig. 10.10. Plot of the phase diagram for $p < p_c$. There are four phases, separated
by three critical curves, meeting at two tricritical points.

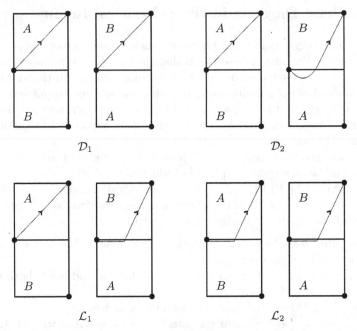

Fig. 10.11. Behavior of the copolymer, inside the four block pairs containing oil and water, for each of the four phases in Fig. 10.10.

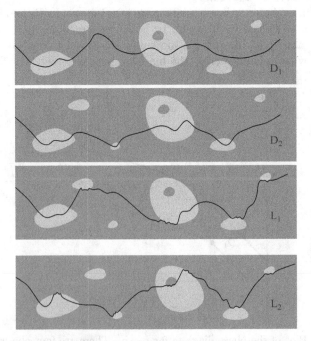

Fig. 10.12. Illustration of the four phases in Fig. 10.10 according to Fig. 10.11. Compare with Fig. 10.8.

and β are both small, the copolymer stays on one side of the interface in both types of block pairs. In $\mathcal{D}_2$, where α is larger, when the copolymer diagonally crosses a water block (which it has to do every now and then because the oil does not percolate), it dips into the oil block before doing the crossing. Since β is small, it still has no interest to localize. In $\mathcal{L}_1$ and $\mathcal{L}_2$, β is large enough to induce localization. In $\mathcal{L}_1$, where β is moderate, localization occurs in those block pairs where the copolymer crosses the water rather than the oil. This is because $\alpha > \beta$, making it more advantageous to localize away from water than away from oil. In $\mathcal{L}_2$, where β is large, localization occurs in both types of block pairs.

Note that the boundary between $\mathcal{D}_1$ and $\mathcal{D}_2$ is linear, as drawn in Fig. 10.10. This is because in $\mathcal{D}_1$ and $\mathcal{D}_2$ the free energy is a function of $\alpha - \beta$ only. This boundary extends above the horizontal because no localization can occur when $\beta \leq 0$. In $\mathcal{L}_1$ and $\mathcal{L}_2$ the free energy is no longer a function of $\alpha - \beta$. Note that there are *two tricritical points*, one that depends on p and one that does not.

Very little is known so far about the fine details of the four critical curves in the subcritical regime. The reason is that in none of the four phases does the free energy take on a simple form (contrary to what we saw in the supercritical regime, where the free energy is simple in the delocalized phase). In particular, in the subcritical regime there is no simple criterion like Proposition 10.10 to characterize the phases. In den Hollander and Pétrélis [171] it is shown that the phase transition between $\mathcal{D}_1$ and $\mathcal{D}_2$ has order ≤ 2, between $\mathcal{D}_1$ and $\mathcal{L}_1$ has order in $(1, 2]$, and between $\mathcal{D}_2$ and $\mathcal{L}_1$ has order at most that of the single linear interface (which has order ≥ 2 by the analogue of Theorem 9.11). It is further shown that the free energy is infinitely differentiable in the interior of $\mathcal{D}_1$ and $\mathcal{D}_2$. The same is believed to be true for $\mathcal{L}_1$ and $\mathcal{L}_2$, but a proof is missing.

It was argued in den Hollander and Whittington [172] that *the phase diagram is discontinuous at $p = p_c$*. Indeed, none of the three critical curves in the subcritical phase diagram in Fig. 10.10 converges to the critical curve in the supercritical phase diagram in Fig. 10.9. This is because *percolation or no percolation of the oil completely changes the character of the phase transition(s)*.

10.5 Extensions

(1) It is straightforward to extend the model to more than two types of monomers and solvents, say, $A, B, C, \dots$, and to put parameters $\alpha, \beta, \gamma, \dots$ in the Hamiltonian in (10.5) to reward the different matches. All that changes in the variational formula for the quenched free energy in (10.17) is that the sum over kl runs over all possible pairs drawn from $\{A, B, C, \dots\}$. There are now more than two regimes of percolation, but the behavior is qualitatively similar.

(2) It is possible to relax the corner restriction in the definition of the path space $\mathcal{W}_{n,L_n}$ defined below (10.4), by allowing the copolymer to exit at the corner straight across the corner where it enters a pair of adjacent blocks. The result is that in (10.17) an additional block pair free energy ψ_{kl}^* is needed that records the contribution to f from such straight crossings. Also this free energy can be expressed in terms of crossing entropies and the linear interface free energy $\phi^{\mathcal{I}}$, just like the free energies ψ_{kl} were in Lemmas 10.4 and 10.6. The coarse-grained path W defined in Section 10.2.1 will now be able to make three rather than two steps, and the link with oriented percolation is therefore somewhat different. The resulting phase diagram becomes even richer. In particular, in the supercritical regime a second critical curve appears that is associated with the phase transition of the single linear interface.

10.6 Challenges

(1) Prove the analogues of Theorems 9.9–9.10 for a single block in the supercritical regime. This ought to be straightforward. Use these analogues to prove that the typical path indeed behaves as sketched in Figures 10.8 and 10.11.

(2) In the subcritical regime, the first and the second critical curve are understood only qualitatively. For instance, it has not been shown that these curves are monotone in α and p. For the third critical curve (between $\mathcal{L}_1$ and $\mathcal{L}_2$) the situation is worse: we do not even have a rigorous proof that the free energy is non-analytic at this curve, for the simple reason that it is not given by an explicit expression on either side of the curve. The challenge is to provide such a proof and thereby establish the existence of the phase transition.

(3) We know from Theorem 10.15 that in the supercritical regime the quenched free energy is infinitely differentiable on $\mathcal{L}$. Find out whether or not it is analytic. This is the analogue of Challenge (2) in Section 9.7. In the subcritical regime infinite differentiability has only been proved on the interior of $\mathcal{D}_1$ and $\mathcal{D}_2$ (see the end of Section 10.4), and the same challenge applies here.

(4) What happens when the block sizes become random? For instance, we may slice $\mathbb{Z}^2$ into rows and columns of random sizes $H_k L_n$, $k \in \mathbb{Z}$, and $V_k L_n$, $k \in \mathbb{Z}$, with $(H_k)_{k \in \mathbb{Z}}$ and $(V_k)_{k \in \mathbb{Z}}$ i.i.d. random variables taking values in $[C_1, C_2]$ for some $0 < C_1 < C_2 < \infty$. In the resulting blocks we may, as before, choose to place oil with probability p and water with probability $1 - p$, and ask what happens in the limit (10.8). We expect a qualitatively similar phase diagram, but the order of the phase transition is likely to depend on the law of H_1 and V_1 near C_1 and C_2. The challenge is to prove or disprove this. Compare with Section 7.7, Challenge (5).

(5) Investigate to what extent the behavior in the subcritical phase is qualitatively similar to that of the polymer in a random potential treated in Chapter 12. In both models the path has to hunt for "energetically favorable" locations in a random environment that are "not too far way".

(6) Attempt to remove the corner restriction in the definition of the path space $\mathcal{W}_{n,L_n}$ below (10.4) altogether. In the supercritical regime, does the phase diagram still have only one critical curve?

11

Random Pinning and Wetting of Polymers

In this chapter we look at *pinning* and *wetting* of a polymer by a *random substrate*. To that end, we think of the substrate as being composed of different types of atoms or molecules, occurring in a *random order*, and each time the polymer visits the substrate it picks up a reward or a penalty according to the type it encounters (see Fig. 11.1). Alternatively, we may think of a random copolymer consisting of different types of monomers, which interact differently with a homogeneous substrate. We will again be focusing on the phase transition between a *localized phase* and a *delocalized phase*. The results to be described below are an extension of those obtained for the homogeneous pinning model treated in Chapter 7.

The random pinning model is close in nature to the model of a copolymer near a linear selective interface described in Chapter 9. Therefore we will be able to re-use many of the ideas and techniques that were exposed in Chapter 9, allowing for a shortening of the proofs. In fact, in some sense the random pinning model is mathematically easier than the copolymer model. On the other hand, it has distinctive properties. One such property, which we will describe below, is that in some cases the disorder is *relevant*, meaning that the quenched critical curve is distinct from the annealed critical curve, while in other cases the disorder is *irrelevant*, meaning that the two critical curves coincide. This dichotomy does not occur for the copolymer model, where the disorder is always relevant. Another distinctive property is that the weak interaction limit of the random pinning model is trivial, in the sense that the disorder is not felt, whereas we saw that this is not the case for the copolymer model.

In Sections 11.1–11.5 we define the random pinning model, the quenched free energy and the quenched critical curve, and identify the qualitative properties of the critical curve and the two phases. This part runs parallel to Sections 9.1–9.5 for the copolymer near a linear selective interface. In Section 11.6 we focus on the issue of relevant vs. irrelevant disorder. In Section 11.7 we take a brief look at the extension of the results in Sections 11.1–11.6 to wetting. The wetting analogue may be used to describe the *denaturation transition*

F. den Hollander, *Random Polymers*,
Lecture Notes in Mathematics 1974, DOI: 10.1007/978-3-642-00333-2_11,
© Springer-Verlag Berlin Heidelberg 2009, Reprint by Springer-Verlag Berlin Heidelberg 2012

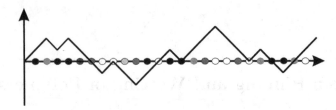

Fig. 11.1. A directed homopolymer near a linear interface with random disorder. Different shades of white, grey and black represent different values of the disorder.

of DNA with increasing temperature. This is done via the so-called Poland-Sheraga model, which views the two strands of DNA as polymers adsorbed onto each other. In Section 11.8 we consider what happens in the presence of a force, both for pinning and wetting, thus providing an extension of the results obtained in Section 7.3 for the homopolymer. The wetting version may be used to describe the *unzipping* of DNA under the action of a force: one strand of the DNA is pulled off the other strand as if it were a polymer adsorbed onto a substrate. Since the binding energies in an AT-pair and a CG-pair are different, and the order in which these pairs occur is more or less random, DNA can be modeled as random wetting with *binary disorder*.

The main techniques in the present chapter are the *subadditive ergodic theorem*, the *method of excursions* and *partial annealing estimates*.

11.1 A Polymer Near a Linear Penetrable Random Substrate: Pinning

For $n \in \mathbb{N}_0$, let

$$\mathcal{W}_n = \{ w = (i, w_i)_{i=0}^n \colon w_0 = 0, \ w_i \in \mathbb{Z} \ \forall \, 0 \le i \le n \}, \qquad (11.1)$$

denote the set of all n-step directed paths in $\mathbb{N}_0 \times \mathbb{Z}$ starting from the origin. This set labels the configurations of the polymer where, as in (7.1), we allow for arbitrary increments. Let

$$\omega = (\omega_i)_{i \in \mathbb{N}} \text{ be i.i.d. with } \mathbb{P}(\omega_1 = +1) = \mathbb{P}(\omega_1 = -1) = \tfrac{1}{2}. \qquad (11.2)$$

This random sequence labels the order of the species along the interface $\mathbb{N} \times \{0\}$. Throughout the sequel, $\mathbb{P}$ denotes the law of ω. (In Section 11.9, Extension (3), we will comment on the extension to more general disorder.) As Hamiltonian we pick, for fixed ω,

$$H_n^\omega(w) = -\sum_{i=1}^n (\lambda \omega_i - h) \, 1_{\{w_i = 0\}}, \qquad w \in \mathcal{W}_n, \qquad (11.3)$$

with $(\lambda, h) \in \mathrm{QUA}$ (recall (9.4)). Here, λ is the *disorder strength* and h is the *disorder bias*. Thus, a visit of the polymer to the substrate at time i contributes an energy $-(\lambda\omega_i - h)$. Note that, compared to (9.3), the parameter λ appears in front of the disorder rather than in front of the sum, and h appears with the opposite sign. This is the notation used in most of the literature.

The law of the copolymer given ω is denoted by P_n^ω and is defined as in (9.5), i.e.,

$$P_n^\omega(w) = \frac{1}{Z_n^\omega} e^{-H_n^\omega(w)} P_n(w), \qquad w \in \mathcal{W}_n, \tag{11.4}$$

where, as in Section 7.1, P_n is the projection onto $\mathcal{W}_n$ of the path measure P of a directed irreducible random walk $S = (S_i)_{i\in\mathbb{N}_0}$. We will make the same assumptions on S as in (7.4–7.6), namely, that $b(k) = P(S_k = 0, S_i \neq 0 \,\forall\, 0 < i < k)$ satisfies $\sum_{k\in\mathbb{N}} b(k) = 1$ and

$$b(k) \sim k^{-1-a} L(k) \qquad \text{as } k \to \infty \text{ through } \Lambda \tag{11.5}$$

for some $a \in (0,\infty)$ and some slowly varying function L, where $\Lambda = \{k \in \mathbb{N}: b(k) > 0\}$. As in Chapter 7, the fact that under the law P the successive visits of S to 0 form a *renewal process* will drive the computations, and the proofs will be valid for general renewals not necessarily arising from random walk as long as (11.5) is in force.

Comparing (11.1–11.3) with (7.1–7.2), we see that in the random pinning model the random sequence $(\lambda\omega_i - h)_{i\in\mathbb{N}}$ takes over the role of the parameter ζ in the homogeneous pinning model.

11.2 The Free Energy

Our starting point is the following analogue of Theorem 9.1.

Theorem 11.1. *For every* $\lambda, h \in \text{QUA}$,

$$f(\lambda, h) = \lim_{n\to\infty} \frac{1}{n} \log Z_n^\omega \tag{11.6}$$

exists ω-*a.s. and in* $\mathbb{P}$-*mean, and is constant* ω-*a.s.*

Proof. The proof of Theorem 9.1 can be carried over almost verbatim. Indeed, consider the constrained partition sum

$$Z_{2n}^{\omega,0} = E\left(\exp\left[\sum_{i=1}^{2n}(\lambda\omega_i - h)\,1_{\{S_i=0\}}\right]1_{\{S_{2n}=0\}}\right). \tag{11.7}$$

Then $\omega \mapsto (\frac{1}{2n}\log Z_{2n}^\omega)_{n\in\mathbb{N}_0}$ is a superadditive random process, and so the analogue of Lemma 9.2 holds. Moreover, the same estimate as in Lemma 9.3 applies, showing that $Z_{2n}^{\omega,0}$ and Z_{2n}^ω differ by at most a polynomial factor. Finally, $|\log(Z_{2n}^\omega/Z_{2n+1}^\omega)| \leq \lambda + h$, which removes the restriction to even n. $\square$

490

11.3 The Critical Curve

11.3.1 The Localized and Delocalized Phases

The same argument as in (9.16–9.17) shows that

$$f(\lambda, h) \geq 0 \qquad \forall\, (\lambda, h) \in \mathrm{QUA}, \qquad (11.8)$$

with the lower bound corresponding to the polymer wandering away from the interface in either the upward or the downward direction. This leads us to the following analogue of Definition 9.4.

Definition 11.2. *We say that the polymer is:*
(i) *localized if $f(\lambda, h) > 0$,*
(ii) *delocalized if $f(\lambda, h) = 0$.*
We write

$$\begin{aligned}
\mathcal{L} &= \{(\lambda, h) \in \mathrm{QUA}\colon\ f(\lambda, h) > 0\}, \\
\mathcal{D} &= \{(\lambda, h) \in \mathrm{QUA}\colon\ f(\lambda, h) = 0\},
\end{aligned} \qquad (11.9)$$

to denote the localized and the delocalized region in QUA.

11.3.2 Existence of a Non-trivial Critical Curve

The quenched free energy $f(\lambda, h)$ is continuous and convex in (λ, h) (recall Section 1.3), and non-increasing in h for every λ. Moreover, $f(0, h)$ is the free energy of the homogeneous pinning model (with $\zeta = -h$).

The analogue of Theorem 9.5 reads as follows (see Fig. 11.2).

Theorem 11.3. *For every $\lambda \in [0, \infty)$ there exists an $h_c(\lambda) \in [0, \infty)$ such that the polymer is*

$$\begin{aligned}
&\text{localized} && \text{if } 0 \leq h < h_c(\lambda), \\
&\text{delocalized} && \text{if } h \geq h_c(\lambda).
\end{aligned} \qquad (11.10)$$

Moreover, $\lambda \mapsto h_c(\lambda)$ is continuous, strictly increasing and convex on $[0, \infty)$, with $h_c(0) = 0$ and $\lim_{\lambda \to \infty} h_c(\lambda)/\lambda = 1$.

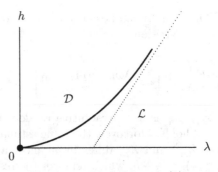

Fig. 11.2. Plot of $\lambda \mapsto h_c(\lambda)$.

Proof. The fact that $\mathcal{L}$ and $\mathcal{D}$ are separated by a single critical curve $\lambda \mapsto h_c(\lambda)$, defined by $h_c(\lambda) = \inf\{h \in \mathbb{R}: f(\lambda, h) = 0\}$, is immediate from the fact that $h \mapsto f(\lambda, h)$ is non-increasing for every $\lambda \in [0, \infty)$. Since the level sets of a convex function are convex, $\mathcal{D} = \{(\lambda, h) \in \text{QUA}: f(\lambda, h) \leq 0\}$ is a convex set and h_c is a convex function.

Since the critical threshold for pinning in the homogeneous model is 0, we have that $h_c(0) = 0$. Theorem 11.5 below states that $h_c(\lambda) > 0$ for $\lambda > 0$. Together with the convexity this yields that $\lambda \mapsto h_c(\lambda)$ is strictly increasing on $[0, \infty)$. Theorem 11.4 below implies that $h_c(\lambda)/\lambda \leq 1$. It therefore remains to show that $\liminf_{\lambda \to \infty} h_c(\lambda)/\lambda \geq 1$.

To that end, note that

$$Z_n^\omega \geq E\left(e^{(\lambda - h)|I_n^\omega|} 1_{\{S \in \mathcal{W}_n^\omega\}}\right), \tag{11.11}$$

where

$$\begin{aligned} I_n^\omega &= \{i \in \{1, \ldots, n\}: \omega_i = 1\}, \\ \mathcal{W}_n^\omega &= \{w \in \mathcal{W}_n: w_i = 0 \;\forall i \in I_n^\omega, \; w_i \neq 0 \;\forall i \in \{1, \ldots, n\}\backslash I_n^\omega\}. \end{aligned} \tag{11.12}$$

We have $\lim_{n \to \infty} \frac{1}{n} I_n^\omega = \frac{1}{2}$ ω-a.s. Moreover, $\lim_{n \to \infty} \frac{1}{n} \log P(\mathcal{W}_n^\omega) = \Sigma$ with $\Sigma = \sum_{k \in \Lambda} 2^{-(k+1)} \log b(k)$. It therefore follows from (11.6) and (11.11) that

$$f(\lambda, h) \geq \tfrac{1}{2}(\lambda - h) + \Sigma. \tag{11.13}$$

But $\Sigma \in (-\infty, 0)$ by (11.5). Therefore, for any $\delta > 0$, we have $f(\lambda, h) > 0$ when $\lambda > -2\Sigma/\delta$ and $h \leq (1 - \delta)\lambda$. Consequently, $\liminf_{\lambda \to \infty} h_c(\lambda)/\lambda \geq 1 - \delta$. Let $\delta \downarrow 0$ to get the claim. $\square$

The mean value of the disorder is $\mathbb{E}(\lambda \omega_1 - h) = -h$. Thus, we see from Fig. 11.2 that for the random pinning model localization may even occur for *moderately negative* mean values of the disorder, contrary to what we found in Section 7.2 for the homogeneous pinning model, where localization occurs only for $\zeta > 0$. In other words, even a globally repulsive random interface can pin the polymer: all that the polymer needs to do is to hit the positive values of the disorder and avoid the negative values as much as possible.

11.4 Qualitative Properties of the Critical Curve

11.4.1 Upper Bound

The analogue of Theorem 9.6 yields the following upper bound for h_c. This upper bound is the critical curve for the *annealed* model.

Theorem 11.4. $h_c(\lambda) \leq h_c^{\text{ann}}(\lambda)$ with $h_c^{\text{ann}}(\lambda) = \log\cosh(\lambda)$ *for all* $\lambda \geq 0$.

Proof. Compute, as in (9.28),

$$
\begin{aligned}
f(\lambda, h) &= \lim_{n \to \infty} \frac{1}{n} \mathbb{E}\left(\log E\left(\exp\left[\sum_{i=1}^{n}(\lambda\omega_i - h)1_{\{S_i=0\}}\right]\right)\right) \\
&\leq \lim_{n \to \infty} \frac{1}{n} \log E\left(\mathbb{E}\left(\exp\left[\sum_{i=1}^{n}(\lambda\omega_i - h)1_{\{S_i=0\}}\right]\right)\right) \qquad (11.14) \\
&= \lim_{n \to \infty} \frac{1}{n} \log E\left(\prod_{i=1}^{n}\left[\tfrac{1}{2}e^{\lambda-h} + \tfrac{1}{2}e^{-\lambda-h}\right]^{1_{\{S_i=0\}}}\right).
\end{aligned}
$$

The right-hand side is ≤ 0 as soon as the term between square brackets is ≤ 1. Consequently,

$$
h > \log\cosh(\lambda) \quad \Longrightarrow \quad f(\lambda, h) = 0, \qquad (11.15)
$$

which yields the desired upper bound on $h_c(\lambda)$. $\square$

It is natural to wonder whether it is possible to improve the upper bound in Theorem 11.4 by doing partial annealing estimates, i.e., by doing a first-order or higher-order Morita approximation (see e.g. Orlandini, Rechnitzer and Whittington [253]). The answer is no. As already noted in Extension (3) in Section 9.6, Caravenna and Giacomin [54] have shown that such approximations do not lead to an improvement. In fact, we will see in Section 11.2 that the upper bound in Theorem 11.4 is sharp in some cases.

11.4.2 Lower Bound

There is no analogue of Theorem 9.6. Indeed, this lower bound comes from strategies where the polymer dips below the interface during rare long stretches in ω where the empirical mean is sufficiently biased (see Fig. 9.5). Such strategies no longer work for the random pinning model, because the path only picks up a reward or a penalty *at* the interface. The following lower bound for h_c is proved in Alexander and Sidoravicius [6].

Theorem 11.5. $h_c(\lambda) > 0$ *for* $\lambda > 0$.

Proof. We follow the proof given in Giacomin [116], Section 5.2. This proof has the advantage that it quantifies the lower bound. In what follows, we first restrict ourselves to the subclass of excursion laws $b(\cdot)$ for which $b(1), b(2) > 0$ (recall (7.7)). At the end of the proof we indicate how to remove this restriction.

The key idea is to rewrite Z_n^ω as an expectation w.r.t. P of an exponential weight factor in which the disorder has a strictly positive expectation under $\mathbb{P}$. Once that is done, a homogeneous localization strategy will yield the desired result.

To formulate the rewrite, let

$$\chi = \frac{b(1)^2}{b(1)^2 + b(2)}, \qquad \widehat{\omega}_i = \widehat{\omega}_i(\lambda, h) = \log\left[\chi\, e^{\lambda\omega_i - h} + (1 - \chi)\right], \quad i \in \mathbb{N}.$$
(11.16)

Let $e(\lambda, h) = \mathbb{E}(\widehat{\omega}_1)$ and note that, because $\chi \in (0,1)$,

$$\exists\, \delta > 0 \colon \forall\, \lambda > 0\ \exists\, h_0 = h_0(\lambda, \delta) > 0 \colon \qquad e(\lambda, h) > \delta\ \forall\, h \le h_0.$$
(11.17)

For $i \in \mathbb{N}$, abbreviate $\delta_i = 1_{\{S_i = 0\}}$ and $\widehat{\delta}_i = 1_{\{S_{i-1} = S_{i+1} = 0\}} 1_{\{i\,\text{odd}\}}$.

Lemma 11.6. *For* $n \in 2\mathbb{N}$,

$$Z_n^\omega = E\left(\exp\left[\sum_{i=1}^{n}(\lambda\omega_i - h)\delta_i(1 - \widehat{\delta}_i) + \sum_{i=1}^{n}\widehat{\omega}_i\widehat{\delta}_i\right]\right).$$
(11.18)

Proof. Fix $n \in 2\mathbb{N}$. Let

$$\mathcal{I}_n = \{i \in \{1, 3, \ldots, n-3, n-1\}\colon S_{i-1} = S_{i+1} = 0\}.$$
(11.19)

Number the points in $\{0\} \cup (\cup_{i \in \mathcal{I}_n}\{i-1, i+1\}) \cup \{n\}$ as $0 = n_1 \le n_2 \le \ldots n_{k-1} \le n_k = n$ with $k = 2|\mathcal{I}_n| + 2$ (which includes repetitions). Note that $n_{2j-1} \le n_{2j}$ and $n_{2j+1} = n_{2j} + 2$ for $j = 1, \ldots, |\mathcal{I}_n|$, and that $\mathcal{I}_n = \{n_{2j} + 1\colon j = 1, \ldots, |\mathcal{I}_n|\}$. Further note that all the n_j's are even, while all the elements of $\mathcal{I}_n$ are odd.

Partition the set of possible excursions according to the values taken by $\mathcal{I}_n$. Then, for $A \subset \{1, 3, \ldots, n-3, n-1\}$, we have

$$E\left(\exp\left[\sum_{i=1}^{n}(\lambda\omega_i - h)\delta_i\right] 1_{\{\mathcal{I}_n = A\}}\right)$$

$$= \left(\prod_{j=1}^{|A|}\widehat{Z}_{n_{2j-1},n_{2j}}^{\omega,0}\left[b(1)^2\, e^{\lambda\omega_{n_{2j}+1} - h} + b(2)\right] e^{\lambda\omega_{n_{2j}+1} - h}\right)\widehat{Z}_{n_{2|A|+1},n}^{\omega},$$
(11.20)

where

$$\widehat{Z}_{u,v}^\omega = E\left(\exp\left[\sum_{i=u+1}^{v}(\lambda\omega_i - h)\delta_i\right] 1_{\{\widehat{\delta}_{u+1} = \cdots = \widehat{\delta}_{v-1} = 0\}}\right), \quad u, v \in \mathbb{N}_0,\ u \le v,$$
(11.21)

and $\widehat{Z}_{u,v}^{\omega,0}$ is the same with an extra indicator $1_{\{\delta_n = 1\}}$ under the expectation. The term between square brackets in the r.h.s. of (11.20) comes from the fact that the sites in A may or may not contain a renewal time.

By (11.16), we have

$$b(1)^2\, e^{\lambda\omega_{n_{2j}+1} - h} + b(2) = [b(1)^2 + b(2)]\, e^{\widehat{\omega}_{n_{2j}+1}}.$$
(11.22)

494

Substituting this into (11.20) and summing over A, we obtain the relation in (11.18). Indeed, the effect of the substitution in (11.22) is that, when $\widehat{\delta}_i = 1$, the contribution of the disorder at site $n_{2j} + 1$ changes from $e^{\lambda \omega_{n_{2j}+1} - h}$ to $e^{\widehat{\omega}_{n_{2j}+1}}$. $\square$

Put $M_b = \sum_{k \in \mathbb{N}} k b(k)$. For the remainder of the proof we distinguish two cases: (I) $M_b < \infty$ and (II) $M_b = \infty$.

(I) Estimate, using Jensen's inequality and Lemma 11.6,

$$\frac{1}{n} \mathbb{E}(\log Z_n^\omega) \geq \frac{1}{n} \mathbb{E}\left(E\left[\sum_{i=1}^n (\lambda \omega_i - h)\delta_i(1 - \widehat{\delta}_i) + \sum_{i=1}^n \widehat{\omega}_i \widehat{\delta}_i \right] \right)$$
$$= -h \frac{1}{n} \sum_{i=1}^n E(\delta_i) + h \frac{1}{n} \sum_{i=1}^n E(\delta_i \widehat{\delta}_i) + e(\lambda, h) \frac{1}{n} \sum_{i=1}^n E(\widehat{\delta}_i). \tag{11.23}$$

By the standard renewal theorem, we have

$$\lim_{n \to \infty} \frac{1}{n} \sum_{i=1}^n E(\delta_i) = \frac{1}{M_b}. \tag{11.24}$$

Since

$$E(\widehat{\delta}_{2i-1}) = P(S_{2i-2} = S_{2i} = 0) = E(\delta_{2i-2})\left[b(1)^2 + b(2)\right],$$
$$E(\delta_{2i-1}\widehat{\delta}_{2i-1}) = P(S_{2i-2} = S_{2i-1} = S_{2i} = 0) = E(\delta_{2i-2})\, b(1)^2, \tag{11.25}$$

the statement in (11.24) implies that

$$\lim_{n \to \infty} \frac{1}{n} \sum_{i=1}^n E(\widehat{\delta}_i) = \frac{b(1)^2 + b(2)}{2M_b}, \qquad \lim_{n \to \infty} \frac{1}{n} \sum_{i=1}^n E(\delta_i \widehat{\delta}_i) = \frac{b(1)^2}{2M_b}, \tag{11.26}$$

where the factor $\frac{1}{2}$ arises because $\widehat{\delta}_i = 0$ when i is even. Therefore, letting $n \to \infty$ in (11.23) and using (11.24–11.26), we obtain

$$f(\lambda, h) = \lim_{n \to \infty} \frac{1}{n} \mathbb{E}(\log Z_n^\omega)$$
$$\geq \frac{1}{M_b}\left[e(\lambda, h) \frac{b(1)^2 + b(2)}{2} + h\left(\frac{b(1)^2}{2} - 1\right)\right]. \tag{11.27}$$

This in turn yields

$$e(\lambda, h) > h \frac{2 - b(1)^2}{b(1)^2 + b(2)} \quad \Longrightarrow \quad (\lambda, h) \in \mathcal{L}, \tag{11.28}$$

which proves the claim because of (11.17).

(II) The lower bound in (11.28) is useless when $M_b = \infty$. However, we may repeat the argument in (I) after tilting $b(\cdot)$. To that end, define, for $r > 0$ (compare with (7.18)),

$$b^r(k) = \frac{e^{-rk}b(k)}{B(r)}, \qquad k \in \mathbb{N}, \tag{11.29}$$

where $B(r) = \sum_{k \in \mathbb{N}} e^{-rk}b(k) < \infty$. Write P^r for the path measure corresponding to the excursion law $b^r(\cdot)$. We will exploit the inequality

$$\log \int e^X \, d\mu = \log \int e^{X - \log(d\nu/d\mu)} \, d\nu \geq \int X \, d\nu - R(\nu|\mu)$$

$$\forall \mu, \nu, X : \ \mu \ll \nu, \ \nu \ll \mu, \int |X| d\nu < \infty, \tag{11.30}$$

where $R(\nu|\mu) = \int \log(d\nu/d\mu) \, d\nu \in [0, \infty]$ denotes the relative entropy of μ w.r.t. ν. Using (11.30) with $\mu = P$, $\nu = P^r$ and X the r.h.s. of (11.18), we may again use Lemma 11.6 to estimate

$$\mathbb{E}(\log Z_n^\omega) \geq \mathbb{E}\left(E_r \left[\sum_{i=1}^n (\lambda \omega_i - h)\delta_i(1 - \widehat{\delta}_i) + \sum_{i=1}^n \widehat{\omega}_i \widehat{\delta}_i \right] \right) - R(P_n^r | P_n). \tag{11.31}$$

Therefore, letting $n \to \infty$, inserting the relation

$$\lim_{n \to \infty} \frac{1}{n} R(P_n^r | P_n) = R(r) \ \text{ with } \ R(r) = -r + \frac{[-\log B(r)]}{M_{b^r}} \tag{11.32}$$

(see Giacomin [116], Appendix B1; note that $M_{b^r} = \mathrm{d}[-\log B(r)]/dr)$, and using the analogues of (11.24–11.26) for P^r, we obtain

$$f(\lambda, h) \geq \frac{1}{M_{b^r}} \left[e(\lambda, h) \frac{b^r(1)^2 + b^r(2)}{2} + h \left(\frac{b^r(1)^2}{2} - 1 \right) \right] - R(r). \tag{11.33}$$

Now let $r \downarrow 0$, note that $R(r) = O(r)$, and use that, by (11.5), $M_{b^r} = O(r^{-1+a})$, $a > 0$, up to a slowly varying correction. Then (11.33) once again yields (11.28). This completes the proof of Theorem 11.5 when $b(1), b(2) > 0$.

The condition $b(1), b(2) > 0$ is less restrictive than it seems. For instance, for SRW we have $\Lambda = \{k \in \mathbb{N} : b(k) > 0\} = 2\mathbb{N}$, and so the polymer does not see the disorder on the odd-numbered sites. The random pinning problem therefore is the same as for excursions with length distribution $\bar{b}(\cdot)$ given by $\bar{b}(k) = b(2k)$, $k \in \mathbb{N}$, for which $\bar{b}(1), \bar{b}(2) > 0$.

We finish by explaining what to do when $b(1) = 0$ and $b(2), b(3) > 0$. The sketch given below easily generalizes to arbitrary Λ.

Let $\lambda_1, \lambda_2, \lambda_3, h \geq 0$, and define

$$H_n^\omega(w) = -\sum_{i=1}^n \left(\lambda_{1+(i-1)(\text{mod } 3)}\omega_i - h \right) 1_{\{w_i = 0\}}, \qquad w \in \mathcal{W}_n. \tag{11.34}$$

This Hamiltonian gives weight λ_j to the disorder at the sites in $j + 3\mathbb{N}_0$, $j = 1, 2, 3$. Let $f(\lambda_1, \lambda_2, \lambda_3, h)$ denote the associated quenched free energy. For every $h \geq 0$, this function is convex and symmetric in $\lambda_1, \lambda_2, \lambda_3$, and hence

$$f(\lambda_1, \lambda_2, \lambda_3, h) \geq f(0, 0, \lambda_3, h). \tag{11.35}$$

This in turn implies that $h_c(\lambda_1, \lambda_2, \lambda_3) \geq h_c(0, 0, \lambda_3)$ for the critical surface separating the localized phase from the delocalized phase (compare with Fig. 11.2). Picking $\lambda_1 = \lambda_2 = \lambda_3 = \lambda$ to recover the original Hamiltonian in (11.3), we find that $h_c(\lambda) \geq h_c(0, 0, \lambda)$, and so it suffices to prove that $h_c(0, 0, \lambda) > 0$ for $\lambda > 0$.

The Hamiltonian in (11.34) with $\lambda_1 = \lambda_2 = 0$ has disorder only at the sites in $3\mathbb{N}$. We can therefore repeat the argument given above for the case $b(1), b(2) > 0$, replacing the even sites by $6\mathbb{N}$, the odd sites by $3 + 6\mathbb{N}$, $b(1)^2$ by $b(3)^2$, and $b(2)$ by $b(2)^3 e^{-2h}$. Indeed, $b(2)^3 e^{-2h}$ is the contribution to the partition sum coming from 3 excursions of length 2 that hit the substrate at distances 2 and 4 but miss the substrate at distance 3. Consequently, in (11.16) we get $\chi = \chi(h) = b(3)^2 / [b(3)^2 + b(2)^3 e^{-2h}]$, but (11.17) continues to hold, and the rest of the argument is the same. □

The proof of Theorem 11.5 yields that (see (11.16–11.17) and (11.28))

$$h_c(\lambda) \geq c_1 \lambda^2, \qquad \lambda \in [0, c_2]. \tag{11.36}$$

Together with the upper bound in Theorem 11.4 this implies that

$$h_c(\lambda) \asymp \lambda^2, \qquad \lambda \downarrow 0. \tag{11.37}$$

Pétrélis [264] gives an alternative proof of Theorem 11.5 for the case of SRW by proving (11.36).

11.4.3 Weak Interaction Limit

Pétrélis [265] proves for SRW that

$$\lim_{a \downarrow 0} a^{-2} f(a\lambda, ah) = \widetilde{f}(h) \qquad \forall \lambda \geq 0, \tag{11.38}$$

where $\widetilde{f}(h)$ is the free energy of the space-time continuous Hamiltonian $H_t(B) = -hL_t(B)$, where $B = (B_s)_{s \geq 0}$ is standard Brownian motion. Thus, the disorder *vanishes* in the weak interaction limit, which is in sharp contrast with what we found in Section 9.4.3 for the copolymer near a selective interface. Since $\widetilde{f}(h) > 0$ for all $h > 0$ (which is the Brownian analogue of what we found in Theorem 7.2 for the homopolymer), it follows from (11.38) that $\lim_{\lambda \downarrow 0} h_c(\lambda)/\lambda = 0$, which is in agreement with (11.37). In Section 11.6 we will see that, actually, $\lim_{\lambda \downarrow 0} h_c(\lambda)/\lambda^2 = \lim_{\lambda \downarrow 0} h_c^{\mathrm{ann}}(\lambda)/\lambda^2 = \frac{1}{2}$.

11.5 Qualitative Properties of the Phases

Theorems 9.9–9.12 carry over to the random pinning model. We refer to Giacomin and Toninelli [120], [122], [123], [124] for proofs. Similarly as in in Chapter 9, the localization strategies that are used to obtain lower bounds on the free energies are those where the polymer hits the substrate at stretches carrying an atypically high density of +1's in ω. This allows for a comparison of the free energy at different value of the parameters, which is needed to carry out the proper perturbation arguments close to the critical curve.

The fact that the phase transition is at least of second order shows that the disorder has a tendency of *smoothing*. Indeed, for the homogeneous pinning model we saw in Section 7.1.3 that the order of the phase transition depends on the tail exponent a, namely, the order is ≥ 2 when $a > \frac{1}{2}$ and < 2 when $a < \frac{1}{2}$. The smoothing is in accordance with what is called the *Harris criterion* (Harris [149]):

- "Arbitrary weak disorder modifies the nature of a phase transition when the order of the phase transition in the non-disordered system is < 2".

11.6 Relevant versus Irrelevant Disorder

In this section we address the question:

- When is the upper bound in Theorem 11.4 sharp?

We will see that the answer depends on the excursion length distribution, i.e., on the exponent a and the slowly varying function $L(\cdot)$ in (11.5), and possibly also on λ.

- The disorder is said to be *relevant* when $h_c(\lambda) < h_c^{\mathrm{ann}}(\lambda)$ and *irrelevant* when $h_c(\lambda) = h_c^{\mathrm{ann}}(\lambda)$.

Relevant disorder means that the ability of the quenched polymer to mimic the behavior of the annealed polymer breaks down close to the annealed critical curve. Theorem 11.7 below summarizes results derived in Alexander [5], Toninelli [293], [294], Giacomin and Toninelli [125], Derrida, Giacomin, Lacoin and Toninelli [85], Alexander and Zygouras [7], [8], and Giacomin, Lacoin and Toninelli [119]. The exponent $a = \frac{1}{2}$ turns out to be critical, in accordance with the Harris criterion stated above (see Fig. 11.3), and the exponent $a = 0$ is included as well.

Theorem 11.7. *Subject to* (11.5), *the disorder is:*
- *relevant*
(1) *for all* $\lambda > 0$ *when* $a \in (\frac{1}{2}, \infty)$;
(2) *for all* $\lambda > 0$ *when* $a = \frac{1}{2}$ *and* $L(\infty) = 0$ *or* $\lim_{K \to \infty}[L(K)]^{-1}$ $\sum_{k=1}^{K} k^{-1}[L(k)]^{-2} = \infty$;

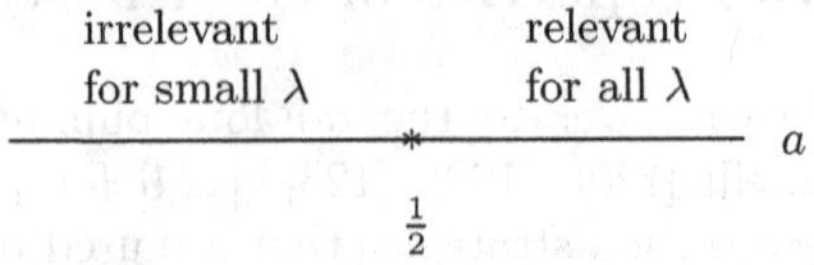

	irrelevant	relevant	
	for small λ	for all λ	

$$\frac{}{\qquad\qquad\qquad * \qquad\qquad\qquad} \quad a$$

$$\frac{1}{2}$$

Fig. 11.3. Relevant vs. irrelevant disorder as summarized in Theorem 11.7. At the threshold value $a = \frac{1}{2}$ the behavior is delicate (marginally relevant vs. irrelevant disorder). At $a = 0$ the disorder is irrelevant for all λ.

- *irrelevant*
(3) *for small $\lambda > 0$ when $a = \frac{1}{2}$ and $\sum_{k \in \mathbb{N}} k^{-1}[L(k)]^{-2} < \infty$;*
(4) *for small $\lambda > 0$ when $a \in \left(0, \frac{1}{2}\right)$;*
(5) *for all $\lambda > 0$ when $a = 0$.*

Proof. A complete proof would take up far too much space. We give a brief sketch of the underlying heuristics, explaining why the case $a = \frac{1}{2}$ is critical. The following argument is taken from Toninelli [296]. For simplicity the disorder is taken to be standard Gaussian instead of binary. In that case, $f^{\mathrm{ann}}(\lambda, h) = f^{\mathrm{homo}}(\frac{1}{2}\lambda^2 - h)$, with $f^{\mathrm{homo}}(h)$ the free energy of the homopolymer with parameter h (i.e., (7.1–7.2) with $\zeta = h$), implying that $h_c^{\mathrm{ann}}(\lambda) = \frac{1}{2}\lambda^2$.

We begin by writing down the identity

$$f_n(\lambda, h) = \frac{1}{n}\,\mathbb{E}(\log Z_n^\omega) = f_n^{\mathrm{homo}}(\Delta) + R_n(\lambda, \Delta), \qquad (11.39)$$

where $\Delta = \frac{1}{2}\lambda^2 - h$,

$$R_n(\lambda, \Delta) = \frac{1}{n}\log \mathbb{E}\left(E_n^\Delta\left(\exp\left[\sum_{i=1}^n (\lambda\omega_i - \tfrac{1}{2}\lambda^2)\,1_{\{S_i=0\}}\right]\right)\right), \qquad (11.40)$$

and E_n^Δ denotes expectation w.r.t. the path measure P_n^Δ of the homopolymer of length n with parameter Δ (i.e., (7.3) with $\zeta = \Delta$). Note that *irrelevance* of the disorder for small λ amounts to the second term in the r.h.s. of (11.39) being *much smaller* than the first term. Formally expanding the r.h.s. of (11.40) in λ for n, Δ fixed, we get

$$R_n(\lambda, \Delta) = -\frac{1}{2}\lambda^2 \frac{1}{n}\sum_{i=1}^n \left[P_n^\Delta(S_i = 0)\right]^2 + O_\Delta(\lambda^3), \qquad \lambda \downarrow 0, \qquad (11.41)$$

where the notation O_Δ means that the error term may depend on Δ. Since

$$\lim_{n\to\infty} \frac{1}{n}\sum_{i=1}^n P_n^\Delta(S_i = 0) = \frac{\partial f^{\mathrm{homo}}(\Delta)}{\partial \Delta} \qquad (11.42)$$

and $P_n^\Delta(S_i = 0)$ is essentially independent of i as long as $1 \ll i \ll n$, we obtain from (11.39–11.42) that

$$f\left(\lambda, \tfrac{1}{2}\lambda^2 - \Delta\right) = f^{\text{homo}}(\Delta) - \frac{1}{2}\lambda^2 \left[\frac{\partial f^{\text{homo}}(\Delta)}{\partial \Delta}\right]^2 + O_\Delta(\lambda^3), \quad \lambda \downarrow 0. \quad (11.43)$$

What this relation does is approximate the quenched free energy close to the annealed critical curve in terms of the free energy of the homopolymer. By Theorem 7.4, we have

$$f^{\text{homo}}(\Delta) \sim \Delta^{1 \vee (1/a)} L^*(1/\Delta), \quad \frac{\partial f^{\text{homo}}}{\partial \Delta}(\Delta) \sim \Delta^{0 \vee (1-a)/a} L^{**}(1/\Delta), \quad \Delta \downarrow 0, \quad (11.44)$$

with L^* and L^{**} slowly varying functions. Inserting this into (11.43), we see that the second term in (11.43) is much smaller than the first term if and only if $1/a < 2(1-a)/a$, i.e., $a \in (0, \tfrac{1}{2})$. Therefore, this is the range of a-values for which the disorder is irrelevant.

It is straightforward to extend the above argument to arbitrary disorder with finite moment generating function (see Section 11.9, Extension (3)). The l.h.s. of (11.43) then reads $f(\lambda, h_c^{\text{ann}}(\lambda) - \Delta)$. See [296] for details.

With the help of interpolation and replica coupling techniques, it is shown in Giacomin and Toninelli [125] that for cases (3) and (4) the expansion in (11.43) holds true, even in the sharper form

$$f\left(\lambda, h_c^{\text{ann}}(\lambda) - \Delta\right) = f^{\text{homo}}(\Delta) - \frac{1}{2}\lambda^2 \left[\frac{\partial f^{\text{homo}}(\Delta)}{\partial \Delta}\right]^2 [1 + O(1)], \quad \lambda, \Delta \downarrow 0. \quad (11.45)$$

From this it is deduced that $f(\lambda, h_c^{\text{ann}}(\lambda) - \Delta)$ and $f^{\text{homo}}(\Delta)$ are comparable when $\lambda, \Delta \downarrow 0$, implying that $h_c^{\text{ann}}(\lambda) = h_c(\lambda)$ for λ sufficiently small.

Cases (1) and (2) are dealt with in Derrida, Giacomin, Lacoin and Toninelli [85], and in Alexander and Zygouras [7]. The main idea in [85] is the following. Choose h close to $h_c(\lambda)$ such that the *annealed system is localized*, i.e., $h < h_c^{\text{ann}}(\lambda)$. Consider the quenched system with n of the order of the correlation length in the annealed system with $n = \infty$, and show that $\mathbb{E}([Z_n^\omega]^\gamma)$ is small for some appropriate $\gamma \in (0, 1)$ depending on a. After that show that, as n is increased, this fractional moment does not grow appreciably. The latter implies that the *quenched system is delocalized*, i.e., $h \geq h_c(\lambda)$, and so it follows that $h_c^{\text{ann}}(\lambda) > h_c(\lambda)$.

Case (5) is settled in Alexander and Zygouras [8]. $\square$

Not surprisingly, the critical exponent $a = \tfrac{1}{2}$ is the most delicate. The physics literature contains conflicting statements. For instance, Forgacs, Luck, Nieuwenhuizen and Orland [104] claim irrelevance for small $\lambda > 0$ (for Gaussian disorder and ballot paths), Derrida, Hakim and Vannimenus [87] claim relevance for all $\lambda > 0$ (for arbitrary mean zero disorder and generalized ballot paths), while Gangardt and Nechaev [110] predict irrelevance for all $\lambda > 0$ (for

binary disorder and ballot paths). Ballot paths and generalized ballot paths
correspond to $a = \frac{1}{2}$ and $L(\infty) \in (0, \infty)$. Case (2) of Theorem 11.7 therefore
shows that in all these examples the disorder is actually relevant.

For the cases of relevant disorder, bounds on the gap $h_c^{\text{ann}}(\lambda) - h_c(\lambda)$ have
been derived in the papers cited prior to Theorem 11.7. As $\lambda \downarrow 0$, this gap
decays like

$$h_c^{\text{ann}}(\lambda) - h_c(\lambda) \asymp \begin{cases} \lambda^2, & \text{if } a \in (1, \infty), \\ \lambda^2 \psi(1/\lambda), & \text{if } a = 1, \\ \lambda^{2a/(2a-1)}, & \text{if } a \in (\frac{1}{2}, 1), \end{cases} \tag{11.46}$$

with $\psi(\cdot)$ slowly varying and vanishing at infinity when $L(\infty) \in (0, \infty)$.

Partial results are known for $a = \frac{1}{2}$. For instance, when $\sum_{k \in \mathbb{N}} k^{-1}[L(k)]^{-2}$
$= \infty$ the gap decays faster than any polynomial, which implies that the dis-
order can at most be *marginally relevant*, a situation where standard per-
turbative arguments cannot work. When $L(\infty) \in (0, \infty)$, the gap lies between $\exp[-\lambda^{-4}]$ and $\exp[-\lambda^{-2}]$ for $\lambda > 0$ small enough, modulo constants
in the exponent. When $L(k) = O([\log k]^{-\frac{1}{2}-\theta})$, $k \to \infty$, $\theta > 0$, the gap lies
above $\exp[-\lambda^{-\theta'}]$ for all $\theta' < \theta$ and $\lambda > 0$ small enough. Both cases correspond
to marginal relevance.

The more the stretches where the quenched polymer hits the interface look
like the typical disorder, the closer $h_c(\lambda)/h_c^{\text{ann}}(\lambda)$ is to 1. Thus, a moderate
gap means that the depinning transition is fairly uniform until close to the
critical curve, while a small gap means that it is fairly uniform all the way
to the critical curve. The above mentioned results on the gap imply that
$\lim_{\lambda \downarrow 0} h_c(\lambda)/h_c^{\text{ann}}(\lambda) = 1$ when $a \in [0, 1]$ and < 1 when $a \in (1, \infty)$.

In the physics literature, relevant disorder is used both for a change in
the critical curve *and* a change in the order of the phase transition. As we
see from the above, the two come together, except possibly in the marginally
relevant case.

11.7 A Polymer Near a Linear Impenetrable Random Substrate: Wetting

Essentially all the results stated in Sections 11.1–11.6 carry over from pinning
to wetting. There is the same parallel here as between Sections 7.1 and 7.2. For
details see e.g. the review paper by Toninelli [296]. Garel and Monthus [112]
provides extensive numerics for the critical behavior in the wetting model.

DNA is a string of adenine-thymine and cytosine-guanine base pairs forming a double helix. A and T share two hydrogen bonds, C and G share three.
If we think of the two strands as performing random walks in two-dimensional
space subject to the restriction that they do not cross each other, then the
distance between the two strands is a random walk in the presence of a wall.
This representation of DNA is called the Poland-Sheraga model (see reference [268] and Fig. 11.4). The localized phase $\mathcal{L}$ corresponds to the bounded

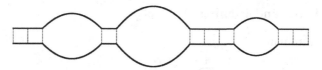

Fig. 11.4. Schematic representation of the two strands of DNA in the Poland-Sheraga model. The dotted lines are the interacting base pairs, the loops are the denaturated segments without interaction.

phase of DNA, where the two strands are attached, the delocalized phase $\mathcal{D}$ corresponds to the denaturated phase, where the two strands are detached. Since the order of the base pairs in DNA is irregular and their binding energies are different, we may think of DNA as a wetted copolymer with binary disorder. The order of the base pairs will of course not be i.i.d., but the comparison with the wetting model is reasonable for a qualitative description. If we want to allow for mismatches between the bases and assign them zero binding energy, then we have ternary disorder. Upon heating, the hydrogen bonds that keep the base pairs together can break and the two strands can separate, either partially or completely. This is called *denaturation*. See Cule and Hwa [79] and Kafri, Mukamel and Peliti [202] for background.

It is not realistic to presume that the two strands can be modeled as SRWs. However, the theory of wetting allows for a general excursion law, as long as it satisfies the regularity property in (11.5). Hence, we may attempt to pick an excursion law that approximates the true spatial behavior of DNA strands, one that takes into account for instance the self-avoidance *within* the denaturated segments. There is an extended literature on this subject (see e.g. Kafrie, Mukamel and Peliti [202], Richard and Guttmann [270]). A typical range of values for the tail exponent a of the excursion law used for DNA is $a \in [1.1, 1.2]$. In [202] it is claimed that the denaturation transition is first-order. This indicates that the Poland-Sheraga model has its limitations, because we know from Section 11.5 that the presence of disorder is smoothing the denaturation transition to second order or higher. Perhaps the DNA molecules that are used in experiments are not long enough ($n \leq 10^6$) for the asymptotic analysis ($n \to \infty$) to apply, or the self-avoidance *between* the different denaturated segments plays a role.

11.8 Pulling a Polymer off a Substrate by a Force

The results in Section 7.3, pertaining to what happens when a force is applied to a homopolymer that is pinning or wetting a substrate, can be extended to the disordered setting. What is described below is taken from Giacomin and Toninelli [121].

11.8.1 Force and Pinning

Our Hamiltonian is

$$H_n^\omega = -\lambda \sum_{i=1}^n \omega_i \, 1_{\{w_i=0\}} - \phi w_n, \qquad w \in \mathcal{W}_n, \tag{11.47}$$

with $\lambda, \phi > 0$ and $\mathcal{W}_n$ the set of paths in (11.1) (compare with (7.48)). It is not hard to prove that the quenched free energy

$$f(\lambda, \phi) = \lim_{n\to\infty} \frac{1}{n} \log Z_n^\omega \tag{11.48}$$

exists ω-a.s. and is self-averaging. The following is the analogue of Theorem 7.6, where we restricted ourselves to the special case where the reference random walk can only make steps of size ≤ 1, namely, $P(S_1 = -1) = P(S_1 = 1) = \frac{1}{2}p$ and $P(S_1 = 0) = 1 - p$ for some $p \in (0, 1)$, as in (7.50).

Theorem 11.8. *For every $\lambda, \phi > 0$, the free energy exists and is given by*

$$f(\lambda, \phi) = f(\lambda) \vee g(\phi), \tag{11.49}$$

with $f(\lambda)$ the free energy of the pinned copolymer without force, and

$$g(\phi) = \log \big[p \cosh(\phi) + (1 - p) \big]. \tag{11.50}$$

Proof. The proof is similar to that of Theorem 7.6, with $f(\lambda)$ taking over the role of the free energy of the homopolymer. For details we refer to [121]. ∎

The critical force is given by the formula

$$\phi_c(\lambda) = g^{-1}(f(\lambda)) \tag{11.51}$$

with g^{-1} the inverse of g. As for homopolymers in Section 7.3, we may put $\lambda = 1/T$, $F = \phi/T$, $F_c(T) = \phi_c(1/T)/T$ as in (7.62), and attempt to plot $T \mapsto F_c(T)$. Since we do not have a closed form expression for $f(\lambda)$, this cannot be done in full detail. However, the qualitative properties of the force-temperature diagram can be deduced by looking at the asymptotics of $f(\lambda)$ for $\lambda \downarrow 0$, respectively, $\lambda \to \infty$. Our main point of interest here is to see whether the force-diagram is *re-entrant*, i.e., whether $T \mapsto F_c(T)$ goes through a minimum or a maximum, at least for some values of p. Precisely when re-entrant behavior occurs depends on the fine details of the model, as we already saw in Section 7.3 for the homopolymer.

It is shown in [121] that

$$F_c(T) = \begin{cases} 0, & \text{as } T \to \infty, \\ \frac{1}{2} + T\left[\frac{1}{2}\sum_{k\in\mathbb{N}} 2^{-k} \log b_p(k) - \log(\frac{p}{2})\right] + o(T), & \text{as } T \downarrow 0, \end{cases} \tag{11.52}$$

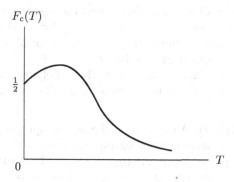

Fig. 11.5. Plot of $T \mapsto F_c(T)$ for p small.

with $b_p(\cdot)$ the excursion length distribution of the random walk with parameter p. Numerically this gives re-entrant behavior as soon as $p < 0.8006$ (see Fig. 11.5). The second line in (11.52) comes from the fact that, in the strong interaction limit, the free energy is dominated by paths that visit the substrate precisely at those i with $\omega_i = 1$ (compare with (11.11–11.13)). Note that the first line gives limit 0 instead of $1/p$ found for the homopolymer in (7.64). This is because in the weak interaction limit the free energy $f(\lambda)$ for the copolymer vanishes faster than the free energy $f^{\text{homo}}(\zeta)$ for the homopolymer, namely, $f(\lambda) = O(\lambda^4)$, $\lambda \downarrow 0$, compared to $f^{\text{homo}}(\zeta) \asymp \zeta^2$, $\zeta \downarrow 0$. Indeed, this follows from the annealed bound $f(\lambda) \leq f^{\text{homo}}(\log \cosh(\lambda))$ (recall (11.14)).

11.8.2 Force and Wetting

The extension to wetting is straightforward. The analogue of Theorem 7.7 can be guessed by looking at Theorem 11.8: simply replace $f(\lambda)$ by $f^+(\lambda)$. It is again possible to determine when the phase transition is re-entrant. In [121] it is shown that

$$
F_c^+(T) = \begin{cases} 0, & \text{as } T \to \infty, \\ \frac{1}{2} + T \left[\frac{1}{2} \sum_{k \in \mathbb{N}} 2^{-k} \log b_p^+(k) - \log(\frac{p}{2}) \right] + o(T), & \text{as } T \downarrow 0, \end{cases}
$$

(11.53)

with $b_p^+(1) = b_p(1)$ and $b_p^+(k) = \frac{1}{2} b_p(k)$, $k \in \mathbb{N}\backslash\{1\}$. Numerically this gives re-entrant behavior as soon as $p < 0.7162$.

Iliev, Orlandini and Whittington [181] compute the free energy of the wetted polymer with force in the first-order Morita approximation, i.e., by averaging Z_n^ω over ω conditioned on $n^{-1} \sum_{i=1}^n \omega_i = 0$. Even though this does not give the correct quenched free energy, it is argued that the force-temperature diagram is qualitatively similar. An argument is given why the quenched critical force is bounded from above by the critical force in the Morita approximation, which is explicitly calculable. Both ballot paths ($p = 1$) and generalized ballot paths ($p = \frac{2}{3}$) are considered, and the disorder is taken to be binary with $\mathbb{P}(\omega_1 = 1) = \rho$ and $\mathbb{P}(\omega_1 = 0) = 1 - \rho$ for some $\rho \in [0,1]$. The force-temperature diagram is found to be re-entrant for all $\rho \in (0,1)$. The reason is that at zero temperature the 1-monomers lie on the substrate, while the 0-monomers may or may not. Hence, the copolymer has a strictly positive entropy, implying that the force-temperature diagram has a strictly positive slope at $T = 0$.

The two strands of DNA can be pulled apart by applying a force to the end of one strand (e.g. with optical tweezers), while anchoring the end of the other strand to some physical support (e.g. a glass slide). In this way an *unzipping transition* can be observed. For background, see Lubensky and Nelson [229], Marenduzzo, Trovato and Maritan [234], who argue in favor of a re-entrant force-temperature phase diagram. Danilowicz, Kafri, Conroy, Coljee, Weeks and Prentiss [80] describe an experiment with double-stranded lambda-phage DNA at temperatures ranging from 15 to 50 degrees Celcius. Only in the range from 24 to 35 degrees Celcius do the measurements agree with the predictions of the theory based on the Poland-Sheraga model.

11.9 Extensions

(1) Similarly as noted in Section 7.6, Extension (1), for the homopolymer pinning model, all results for the random pinning model carry over when the reference random walk S is transient, by normalization of $b(\cdot)$. Alexander and Sidoravicius [6] show that the lower bound in Theorem 11.5 extends to an arbitrary excursion length distribution and an arbitrary disorder under only minor regularity conditions. The main result of [6] is that *disorder strictly enhances pinning*, i.e., $h_c(\lambda) > h_c^{\text{homo}}(\lambda)$ for all $\lambda > 0$, the latter being the critical curve of the homopolymer whose Hamiltonian is obtained from (11.3) after replacing the disorder by its average value. Moreover, it is shown that $\lim_{\lambda \to \infty} h_c(\lambda)/\lambda = \infty$ when the disorder is unbounded, and that $h_c(\lambda) = \infty$ for all $\lambda > 0$ when the disorder does not have a finite exponential moment and the tail of the excursion length distribution is bounded from below by a power law with exponent < -1.

(2) Alexander [4] looks at the order of the phase transition when the disorder is standard Gaussian and the excursions are either recurrent or transient with a length distribution that has an exponential tail. It turns out that in the transient case the disorder is relevant for all $\lambda > 0$, and both the quenched and the annealed free energy have a phase transition of order < 2. In the recurrent case the behavior depends in a subtle way on the subexponential prefactors modulating the exponential decay, and various scenarios are possible. This, once more, shows that the issue of relevance vs. irrelevance is highly delicate.

(3) With the exception of the analogue of Theorem 9.11, i.e., the fact that the phase transition is second order or higher, all the results described in

Sections 11.1–11.8 extend to the situation where the ω_i's are $\mathbb{R}$-valued with a moment generating function that is finite on the positive halfline. For details we refer the reader to Giacomin [116], Chapter 5. In most of the papers cited prior to Theorem 11.7 proofs are written out for Gaussian disorder only, and it is stated "somewhat loosely" that proofs carry over to arbitrary disorder. In principle this is straightforward, although the critical exponent $a = \frac{1}{2}$ (marginally relevant vs. irrelevant disorder) requires attention. As in Section 9.6, Extension (1), the analogue of Theorem 9.11 has been proved for bounded disorder and for continuous disorder subject to a mild entropy condition (Giacomin and Toninelli [122], [123]).

It is shown in Toninelli [294] that any disorder that is *unbounded from above* is relevant for λ large enough, no matter what the choice of $a \in (0, \infty)$ and $L(\cdot)$ is. This supplements parts (1) and (2) of Theorem 11.7.

(4) Bolthausen, Caravenna and de Tilière [30] consider the limit when the disorder is strong but has a low density (referred to as a reduced wetting model). They take as Hamiltonian $H_n(w) = \lambda \sum_{i=1}^{n} \omega_i \, 1_{\{w_i=0\}}$, with the ω_i's i.i.d. $\{0,1\}$-valued such that $\mathbb{P}(\omega_1 = 1) = e^{-c\lambda}$ and $\mathbb{P}(\omega_1 = 0) = 1 - e^{-c\lambda}$ for some $c \in (0, \infty)$, and let $\lambda \to \infty$. They find that for ballot paths (SRW with $a = \frac{1}{2}$) the value $c = \frac{2}{3}$ is the critical threshold for localization, in the sense that $f_c(\lambda)$, the quenched free energy, is > 0 for λ large when $c < \frac{2}{3}$ and $= 0$ for λ large when $c > \frac{2}{3}$. The proof is based on a renormalization argument, showing that the main contribution to the free energy comes from those paths that hit the substrate precisely at those i with $\omega_i = 1$. Toninelli [294] gives a simpler proof of the same result based on fractional moment estimates of the partition sum, and shows that the critical threshold is $c = 1/(1+a)$ when the excursion length distribution has exponent $a \in (0, \infty)$.

(5) Toninelli [291] obtains finite-size estimates for the depinning transition. It is found that, for any $a \in (0, \infty)$ and $\lambda > 0$, if $h = h_n$ varies with the length n of the polymer such that $\lim_{n\to\infty} n^\chi [h_n - h_c(\lambda)] = c$ with $\chi \geq 0$ and $c \in \mathbb{R}$, then the number of visits of the polymer to the substrate is $O(n^{\frac{2}{3}} \log n)$ for $\chi \geq \frac{1}{3}$, $O(n^{2\chi} \log n)$ for $0 \leq \chi < \frac{1}{3}$ and $c > 0$, and $O(n^{1-\chi})$ for $0 \leq \chi < \frac{1}{3}$ and $c < 0$.

(6) Toninelli [291], [292] studies the correlation length in the pinning and wetting model in the localized phase. Let

$$\mathrm{Cov}_l^\omega(k) = P_\infty^\omega(S_l = S_{l+k} = 0) - P_\infty^\omega(S_l = 0) \, P_\infty^\omega(S_{l+k} = 0), \qquad (11.54)$$

where P_∞^ω is the weak limit of the path measure P_n^ω as $n \to \infty$. For the wetting model with SRW, it is shown that, uniformly in $l \in \mathbb{N}$,

$$\lim_{k\to\infty} \frac{1}{k} \log \mathrm{Cov}_l^\omega(k) = -f(\lambda, h),$$

$$\lim_{k\to\infty} \frac{1}{k} \log \mathbb{E}\big(\mathrm{Cov}_l^\omega(k)\big) = -\mu(\lambda, h), \qquad (11.55)$$

where

$$\mu(\lambda, h) = -\lim_{n\to\infty} \frac{1}{n} \log \mathbb{E}(1/Z_n^\omega) \qquad \omega - a.s. \tag{11.56}$$

For wetting with a random walk different from SRW, and for pinning, only upper bounds are derived. What (11.55) says is that the *quenched* and the *average quenched* correlation length are controlled by the free energy, respectively, the "inverted free energy". It is shown in Giacomin and Toninelli [124] that $\mu(\lambda, h) < f(\lambda, h)$ throughout the localized phase, and in Giacomin and Toninelli [125] that $\mu(\lambda, h) \asymp f(\lambda, h)$ as $h \uparrow h_c(\lambda)$ for every $\lambda > 0$ when the disorder is irrelevant (recall cases (3) and (4) in Theorem 11.7).

The quantity $\mu(\lambda, h)$ plays a central role in the estimate of the largest gap between two successive hits of the substrate. Namely, it is shown in [124] that, in the localized phase,

$$\lim_{n\to\infty} P_n^\omega \left(\left| \frac{\mathrm{maxgap}_n(T)}{\log n} - \frac{1}{\mu(\lambda, h)} \right| > \epsilon \right) = 0 \quad \forall \epsilon > 0 \quad \text{in } \mathbb{P}\text{-probability.} \tag{11.57}$$

This is to be compared with what we found for the homopolymer in the proof of Theorem 7.3(a).

(7) Caravenna, Giacomin and Zambotti [57], [58] derive the scaling properties of the path measure for periodic disorder. The results are similar to those mentioned in Section 7.6, Extension (2).

(8) Cheliotis and den Hollander [65] derive a *variational formula* for the critical curve $\lambda \mapsto h_c(\lambda)$ in Theorem 11.3. The derivation is based on a quenched large deviation principle obtained by Birkner, Greven and den Hollander [17] for the empirical process of "words read off from a random letter sequence by an independent renewal process". The variational formula leads to a *necessary and sufficient* criterion for relevant disorder in terms of positivity of a certain relative entropy. From this criterion it is deduced that, for any choice of the disorder distribution and the excursion length distribution, there is a *critical* value $\lambda_c \in [0, \infty]$ such that the disorder is irrelevant for $\lambda \in [0, \lambda_c]$ and relevant for $\lambda \in (\lambda_c, \infty)$. Moreover, upper and lower bounds are derived for λ_c.

(9) Janvresse, de la Rue and Velenik [196] provide a necessary and sufficient condition for localization on a *single sample* of the environment. The Hamiltonian is (11.3) with $\lambda > 0$ and $h = 0$, and it is shown that

$$\liminf_{n\to\infty} E_n^\omega \left(\frac{1}{n} \sum_{i=1}^n \omega_i \, 1_{\{w_i=0\}} \right) > 0 \quad \Longleftrightarrow \quad \liminf_{n\to\infty} \frac{1}{n} \sum_{i=1}^n \omega_i > 0. \tag{11.58}$$

(10) Pétrélis [264] looks at the model with Hamiltonian

$$H_n^\omega(w) = \lambda \sum_{i=1}^{n} (1 + s\eta_i)\, 1_{\{w_i = 0\}} + h \sum_{i=1}^{n} \text{sign}(w_{i-1}, w_i), \tag{11.59}$$

with $\lambda, h, s \geq 0$ and $\eta = (\eta_i)_{i \in \mathbb{N}}$ an i.i.d. sequence such that η_1 has zero mean. This model describes the random pinning of a hydrophobic homopolymer at an interface between oil and water in which random droplets of a third solvent are present. For ballot paths (SRW) it is shown that the quenched free energy $f(\lambda, h, s)$ exists, and has a localized phase with $f(\lambda, h, s) > h$ and a delocalized phase with $f(\lambda, h, s) = h$. The critical curve $\lambda \mapsto h_c(\lambda, s)$ is finite, strictly increasing and convex on $[0, \lambda_c(s))$, for some $\lambda_c(s) \in (0, \infty)$, and is infinite on $(\lambda_c(s), \infty)$. It is not known whether $h_c(\lambda_c(s), s)$ is finite or not. Trivial bounds are

$$h_c(\lambda, 0) \leq h_c(\lambda, s) \leq h_c \left(\lambda + \log \mathbb{E}(e^{\lambda s \eta_1}), 0 \right), \tag{11.60}$$

with $h_c(\lambda, 0) = -\frac{1}{4} \log[1 - 4(1 - e^{-\lambda})^2]$ the critical curve for the homopolymer (see Chapter 7). This in turn yields bounds on $\lambda_c(s)$, namely, $\bar{\lambda}_c(s) \leq \lambda_c(s) \leq \log 2$ with $\bar{\lambda}_c(s)$ the solution of the equation $\lambda + \log \mathbb{E}(e^{\lambda s \eta_1}) = \log 2$. In [264] it is shown that if η_1 has finite variance, then there exist $c_1, c_2 > 0$ such that

$$h_c(\lambda, s) \geq -\frac{1}{4} \log \left[1 - 4 \left(1 - e^{-\lambda - c_2 s^2 \lambda^2} \right)^2 \right], \tag{11.61}$$

$$s \in [0, c_1], \ \lambda \in [0, \log 2 - c_2 s^2 \lambda^2).$$

Together with the upper bound in (11.60), this implies that $h_c(\lambda, s) \asymp \lambda^2$, $\lambda \downarrow 0$. The bounds in (11.60–11.61) are drawn in Fig. 11.6.

(11) Pétrélis [265] studies a combination of the models considered in Chapters 9 and 11 when the interface has a finite width, namely,

$h_c(\lambda, s)$

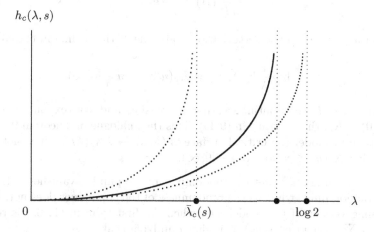

Fig. 11.6. The thick dotted lines are the upper and lower bound in (11.60). The thick drawn line is the lower bound in (11.61). The critical curve $\lambda \mapsto h_c(\lambda, s)$ lies somewhere in between the top two lines.

$$H_n^{\omega,\eta}(w) = -\lambda \sum_{i=1}^n (\omega_i + h)\,\mathrm{sign}(w_{i-1}, w_i) - \bar\lambda \sum_{i=1}^n \sum_{j=-M}^{M} (\eta[M]_i)_j\, 1_{\{w_i=j\}},$$

$$(11.62)$$

where $\lambda, \bar\lambda, h \geq 0$, $\omega = (\omega_i)_{i\in\mathbb{N}}$ is an i.i.d. sequence of random variables with a symmetric distribution whose variance equals 1, $M \in \mathbb{N}$, and $\eta[M] = (\eta[M]_i)_{i\in\mathbb{N}}$ is an i.i.d. sequence of random $(2M+1)$-vectors, independent of ω, whose components have a finite moment generating function. It is shown for ballot paths (SRW) that the quenched free energy $f(\lambda, \bar\lambda, h)$ exists, is non-analytic at a critical surface $(\lambda, \bar\lambda) \mapsto h_c(\lambda, \bar\lambda)$, and satisfies a weak scaling property similar to (9.46–9.48), namely,

$$\lim_{a\downarrow 0} a^{-2} f(a\lambda, a\bar\lambda, ah) = \widetilde{f}(\lambda, \bar\lambda\Sigma, h) \tag{11.63}$$

with

$$\Sigma = \sum_{j=-M}^{M} \mathbb{E}\big((\eta[M]_1)_j\big) \tag{11.64}$$

the average of the sum of the components of $\eta[M]_1$, $f(\lambda, \bar\lambda, h)$ the free energy of the space-time continuous Hamiltonian

$$H_t^b(B) = -\lambda \int_0^t (\mathrm{d}b_s + h\,\mathrm{d}s)\,\mathrm{sign}(B_s) - \bar\lambda L_t(B), \tag{11.65}$$

$B = (B_s)_{s\geq 0}$ and $b = (b_s)_{s\geq 0}$ two independent standard Brownian motions, and $L_t(B)$ the local time of B at 0. The path measure is

$$\frac{\mathrm{d}P_t^b}{\mathrm{d}P}(B) = \frac{1}{Z_t^b}\,\mathrm{e}^{-H_t^b(B)}. \tag{11.66}$$

After a "stable against perturbations" argument, the scaling in (11.63) yields that

$$\lim_{\lambda\downarrow 0} \frac{1}{\lambda} h_c(\lambda, u\lambda) = K_c(u), \qquad u \in [0, \infty), \tag{11.67}$$

with $u \mapsto K_c(u)$ continuous, non-decreasing and convex on $[0, \infty)$, and $K_c(0) = K_c$, the constant in (9.45). This the analogue of Theorem 9.8. In the continuum model the critical surface is $(\lambda, \bar\lambda) \mapsto \lambda K_c(\bar\lambda/\lambda)$. It is not known whether $K_c(u) < \infty$ for all $u \in [0, \infty)$.

We see from (11.65) that the randomness of the pinning vanishes in the weak interaction limit, whereas the randomness of the copolymer does not. For the pinning version of the model, i.e., when the first term in (11.62) is replaced by $-h\sum_{i=1} \mathrm{sign}(w_{i-1}, w_i)$, it is shown in [265] that

$$\widetilde{f}(0, \bar\lambda, h) = h \vee \tfrac{1}{2}\left(\lambda^2 + \tfrac{h^2}{\lambda^2}\right) \tag{11.68}$$

and

$$\lim_{\bar{\lambda}\downarrow 0} \frac{1}{\bar{\lambda}^2} h_c(0, \bar{\lambda}) = \Sigma^2. \qquad (11.69)$$

This describes the weak scaling limit of a homopolymer consisting of hydrophobic monomers near an oil-water interface of finite width with random pinning.

(12) Wüthrich [322] considers the modification of the pinning model in which the Hamiltonian is (compare with (11.3))

$$H_n^{\omega,\eta}(w) = -\lambda \sum_{i=1}^{n} (\omega_i + h) \, 1_{\{w_i=0\}} \, 1_{\{\eta_i=1\}}, \qquad w \in \mathcal{W}_n, \qquad (11.70)$$

with $(\lambda, h) \in \mathrm{QUA}$, $\omega = (\omega_i)_{i \in \mathbb{N}}$ as in (11.2), and $\eta = (\eta_i)_{i \in \mathbb{N}}$ a second i.i.d. sequence with $\mathbb{P}(\eta_1 = 1) = p$ and $\mathbb{P}(\eta_1 = -1) = 1 - p$, $p \in (0, 1)$. This describes a copolymer consisting of hydrophobic and hydrophilic monomers (labeled by ω) in a medium consisting of water in the plane and droplets of oil located at random positions on the horizontal axis (labeled by η). The $(1 + d)$-version of the directed model is studied. It is proved that there is a critical line $\lambda \mapsto h_c^p(\lambda)$, separating a localized phase from a delocalized phase, which is continuous and non-decreasing on $(0, \infty)$. It is further shown that the distance between the endpoint of the copolymer and the interface is of order 1 in the localized phase and of order $\sqrt{n}$ deep inside the delocalized phase.

(13) Lacoin [225] considers pinning on a class of hierarchical lattices. Pick $b, s \in \mathbb{N}\backslash\{1\}$, and let D_0 be the graph with two vertices, A and B, connected by a single edge. Define D_n, $n \in \mathbb{N}_0$, iteratively as follows: to get D_{n+1}, replace each edge in D_n by b branches in parallel, each with s edges in series. For $n \in \mathbb{N}_0$, the path space $\mathcal{W}_n$ consists of all directed paths from A to B, of length s^n. The interface is chosen to be a reference path $w^* \in \mathcal{W}_n$, and the Hamiltonian is

$$H_n^{\omega}(w) = \sum_{i=1}^{s^n-1} (\lambda \omega_i - h) \, 1_{\{w_i=w_i^*\}}, \qquad w \in \mathcal{W}_n. \qquad (11.71)$$

For the case $b \in (1, s)$ (the case $b \in (s, \infty)$ can be dealt with via duality), the following is shown: (1) $b < \sqrt{s}$: the disorder is relevant for all $\lambda > 0$, with gap

$$h_c^{\mathrm{ann}}(\lambda) - h_c(\lambda) \asymp \lambda^{2a/(2a-1)}, \qquad \lambda \downarrow 0, \qquad (11.72)$$

where $a = \log(s/b)/\log s \in (\frac{1}{2}, 1)$; (2) $b > \sqrt{s}$, the disorder is irrelevant for small $\lambda > 0$; (3) $b = \sqrt{s}$: the disorder is marginally relevant for small $\lambda > 0$, with a gap that lies between $\exp[-\lambda^{-4}]$ and $\exp[-\lambda^{-2}]$, modulo constants in the exponent. Giacomin, Lacoin and Toninelli [118], [119] consider a different class of hierarchical lattices, with bond rather than site disorder, and obtain similar results. Earlier work on hierarchical pinning appeared in Derrida and Griffiths [83], and Derrida, Hakim and Vannimenus [87].

(14) Cranston, Hryniv and Molchanov [78] consider a continuous-time SRW on the complete graph K_n (with n vertices and $\binom{n}{2}$ edges) jumping at rate 1. Pinning takes place at a single site, say 0. For the homopolymer, the Hamiltonian up to time t reads $H_t[K_n](w) = -\lambda \int_0^t ds\, 1_{\{w_s=0\}}$, and the free energy for $n, t \to \infty$ such that $t = o(n)$ equals $f(\lambda) = 0 \vee (\lambda - 1)$. For the copolymer, the Hamiltonian up to time t reads $H_t^b[K_n](w) = -\lambda \int_0^t (db_s + h\,ds)\, 1_{\{w_s=0\}}$, with $(b_s)_{s\geq 0}$ a standard Brownian motion, and the quenched free energy (which is self-averaging) is shown to be explicitly computable with critical curve $h_c(\lambda) = 1/\lambda$.

(15) Velenik [305] contains an overview of pinning of higher-dimensional interfaces by disorder on a surface.

11.10 Challenges

(1) Improve the lower bound in Theorem 11.5. Is there a nice formula comparable with that in Theorem 9.7 for the copolymer near a selective interface?

(2) Complete the classification of relevant vs. irrelevant disorder in Section 11.6, i.e., for $a = \frac{1}{2}$ find the necessary and sufficient condition on the slowly varying function $L(\cdot)$ in (11.5) that separates relevant from irrelevant disorder.

(3) Find out whether the phase transition in the pinning model is second order or higher order when the disorder is relevant (recall Section 11.5). Numerical analysis seems to indicate that the order is higher. This is the analogue of Section 9.7, Challenge (1).

(4) Find out whether or not $(\lambda, h) \mapsto f(\lambda, h)$ is analytic on $\mathcal{L}$. Is $\lambda \mapsto h_c(\lambda)$ analytic for those cases where it differs from $h_c^{\mathrm{ann}}(\lambda)$ for all $\lambda \in (0, \infty)$? This is the analogue of Section 9.7, Challenge (2).

(5) Investigate to what extent the relations in (11.55) for wetting carry over when the reference random walk is not SRW, in particular, for $(\lambda, h) \in \mathcal{L}$ close to the quenched critical curve. It is shown in Giacomin [117] that they do not hold in general for pinning. Clarify the relation between $\mu(\lambda, h)$ and $f(\lambda, h)$ in the case of relevant disorder.

(6) For the model described in Section 11.9, Extension (11), with $M = 1$, is the critical surface $(\lambda, \bar{\lambda}) \mapsto h_c(\lambda, \bar{\lambda})$ different for $\bar{\lambda} \neq 0$ than for $\bar{\lambda} = 0$, i.e., does arbitrarily small disorder modify the critical behavior, as it does for the copolymer model studied in Chapter 9?

(7) Extend the analysis to disorder that is Markov. This is the analogue of Section 9.7, Challenge (5).

(8) Determine the fine details of the critical curve in the droplet model in Section 11.9, Extension (12), using the techniques from Chapter 9.

12

Polymers in a Random Potential

In Chapters 9–10 we studied a copolymer in the vicinity of a linear, respectively, a random selective interface between two solvents. The application we had in mind was a copolymer consisting of a random concatenation of hydrophobic and hydrophilic monomers, living in a medium consisting of oil and water located in two halfspaces, respectively, in *mesoscopic* droplets arranged in a percolation-type fashion. In Chapter 10 it was important that the droplets were large compared to the size of the monomers, so that a coarse-graining technique w.r.t. the random environment could be put to use. In the present chapter we look at what happens when the oil and water droplets are *microscopic*, i.e., have the same size as the monomers.

The model we will look at can be thought of as describing a homopolymer living on $\mathbb{N} \times \mathbb{Z}^d$, with $\mathbb{N}$ playing the role of time and $\mathbb{Z}^d$, $d \geq 1$, playing the role of space. The disorder is associated with the sites in $\mathbb{N} \times \mathbb{Z}^d$ (see Fig. 12.1). This model is commonly referred to as the *directed polymer in random environment*, but after what we saw in Chapters 9–10 this name is no longer distinctive.

In Section 12.1 we define the model and record a few basic properties, most notably, we identify an underlying *martingale* that plays a key role in the analysis. In Section 12.2 we state three theorems. The first theorem shows that there is a unique critical temperature separating a phase of *weak disorder* from a phase of *strong disorder*, in which the behavior of the polymer is diffusive, respectively, superdiffusive. The second and third theorem give sufficient conditions for the occurrence of the two phases in terms of the parameters in the model. In Sections 12.3–12.6 we give the proofs of these theorems. In Section 12.7 we address the question to what extent the sufficient condition for weak disorder can be sharpened. This leads to an interesting characterization of the critical temperature, and also reveals a link with the random pinning model studied in Chapter 11. Part of what is written below is drawn from the overview given in Comets, Shiga and Yoshida [70].

The one-dimensional version of the model made its first appearance in the physics literature, in work by Huse and Henley [176], Huse, Henley

F. den Hollander, *Random Polymers*,
Lecture Notes in Mathematics 1974, DOI: 10.1007/978-3-642-00333-2_12,
© Springer-Verlag Berlin Heidelberg 2009, Reprint by Springer-Verlag Berlin Heidelberg 2012

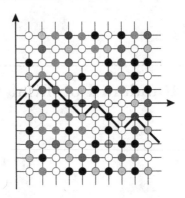

Fig. 12.1. A directed homopolymer in random environment. Different shades of white, grey and black represent different values of the disorder, e.g. microscopic droplets of oil and water of different sizes.

and Fisher [177], Kardar [205], Kardar and Nelson [206], and Kardar and Zhang [207], where it was used to describe roughening of domain walls in the two-dimensional Ising model with random impurities. Since then it has been used to describe a variety of different growth phenomena, including the formation of magnetic domains in spin-glasses, turbulence in viscous incompressible fluids, and the motion of fronts in forest fires and in bacterial colonies on Petri dishes. The zero-temperature limit of the model, where the polymer is controlled by the lowest energies it can find in the disorder, corresponds to oriented first-passage percolation. This is why the directed polymer in random environment is sometimes referred to as "oriented first-passage percolation at positive temperature".

12.1 A Homopolymer in a Micro-emulsion

The set of paths is

$$\mathcal{W}_n = \{w = (w_i)_{i=0}^n \in (\mathbb{Z}^d)^{n+1} \colon w_0 = 0, \ \|w_{i+1} - w_i\| = 1 \ \forall \, 0 \leq i < n\}. \tag{12.1}$$

The random environment

$$\omega = \{\omega(i, x) \colon i \in \mathbb{N}, \ x \in \mathbb{Z}^d\} \tag{12.2}$$

consists of an i.i.d. field of $\mathbb{R}$-valued non-degenerate random variables with cumulant generating function

$$c(\beta) = \log \mathbb{E}(e^{\beta \omega(1,0)}), \qquad \beta \in [0, \infty), \tag{12.3}$$

which is assumed to be finite. We write $\mathbb{P}$ to denote the law of ω. The Hamiltonian is

$$H_n^{\beta,\omega}(w) = -\beta \sum_{i=1}^{n} \omega(i, w_i), \qquad w \in \mathcal{W}_n, \tag{12.4}$$

where β plays the role of the *strength of the disorder*. The *quenched path measure* is

$$P_n^{\beta,\omega}(w) = \frac{1}{Z_n^{\beta,\omega}} \, e^{-H_n^{\beta,\omega}(w)} \, P_n(w), \qquad w \in \mathcal{W}_n, \tag{12.5}$$

where P_n is the projection onto $\mathcal{W}_n$ of the law P of SRW.

The interpretation of (12.1–12.5) is that of a directed polymer in a random potential that picks up random penalties or rewards depending on the sign and the value of the potential at the sites that it visits (see Fig. 12.1). For binary disorder the model can be used to describe a hydrophobic homopolymer in a random micro-emulsion of oil and water droplets located at the sites in $\mathbb{N} \times \mathbb{Z}^d$, with $\omega(i, x) = +1$ denoting the presence of an oil droplet at site (i, x) and $\omega(i, x) = -1$ the presence of a water droplet. Alternatively, we may think of a copolymer with hydrophobic and hydrophilic monomers, in which case $\omega(i, x) = +1$ stands for a match hydrophobic-oil or hydrophilic-water and $\omega(i, x) = -1$ for a mismatch hydrophilic-oil or hydrophobic-water. Yet another application is that of a polymer in a *gel* with different sizes of pockets. This is important because DNA molecules can be separated from one another by *electrophoresis* in a gel.

Note that $\beta \mapsto c(\beta)$ defined in (12.3) is strictly increasing and strictly convex on $[0, \infty)$ with $c(0) = 0$. Key examples of disorder are standard Gaussian with $c(\beta) = \frac{1}{2}\beta^2$ and p-Bernoulli with $c(\beta) = \log[pe^{\beta} + (1-p)e^{-\beta}]$, $p \in (0, 1)$.

12.2 A Dichotomy: Weak and Strong Disorder

12.2.1 Key Martingale

The existence of the *quenched* free energy

$$f(\beta) = \lim_{n \to \infty} \frac{1}{n} \log Z_n^{\beta,\omega} \qquad \omega - a.s. \tag{12.6}$$

is proved in Carmona and Hu [61] (Gaussian disorder) and Comets, Shiga and Yoshida [69] (general disorder), by combining a subadditivity argument with a concentration inequality. This technique is by now standard, and we refer the reader to Chapters 9–11. The *annealed* free energy equals

$$f^{\mathrm{ann}}(\beta) = \lim_{n \to \infty} \frac{1}{n} \log \mathbb{E}(Z_n^{\beta,\omega}) = c(\beta), \tag{12.7}$$

since $\mathbb{E}(Z_n^{\beta,\omega}) = E(\mathbb{E}(\exp[\beta \sum_{i=1}^{n} \omega(i, S_i)])) = e^{nc(\beta)}$ for all $n \in \mathbb{N}_0$.

The key object in the analysis of the model defined in Section 12.1 is the following quantity:

$$M_n^{\beta,\omega} = \frac{Z_n^{\beta,\omega}}{\mathbb{E}(Z_n^{\beta,\omega})} = e^{-nc(\beta)}\, Z_n^{\beta,\omega}, \qquad n \in \mathbb{N}_0. \tag{12.8}$$

The reason is that $(M_n^{\beta,\omega})_{n\in\mathbb{N}_0}$ is a *martingale* w.r.t. the filtration generated by ω, i.e., $(\mathcal{F}_{(0,n]}^{\omega})_{n\in\mathbb{N}_0}$ with $\mathcal{F}_{(0,n]}^{\omega} = \sigma[\omega(i,x)\colon 1 \leq i \leq n,\, x \in \mathbb{Z}^d]$ the sigma-algebra of the random environment up to time n. This fact is easily verified by writing

$$M_n^{\beta,\omega} = E\left(\prod_{i=1}^{n} e^{\beta\omega(i,S_i)-c(\beta)}\right), \qquad M_0^{\beta,\omega} = 1, \tag{12.9}$$

and recalling (12.3). Note that $\mathbb{E}(M_n^{\beta,\omega}) = 1$ and $M_n^{\beta,\omega} > 0$ for all $n \in \mathbb{N}_0$.

By the martingale convergence theorem (Durrett [96], Section 4.2), we have

$$M^{\beta,\omega} = \lim_{n\to\infty} M_n^{\beta,\omega} \quad \text{exists } \omega - a.s.} \tag{12.10}$$

Moreover, since the event $\{\omega\colon M^{\beta,\omega} > 0\}$ is measurable w.r.t. the tail sigma-algebra $\cap_{n\in\mathbb{N}_0}\sigma[\omega(i,x)\colon i > n,\, x \in \mathbb{Z}^d]$, it follows from the Kolmogorov zero-one law (Durrett [96], Section 1.8) that the following dichotomy holds:

$$\begin{aligned} \text{(WD)}: \quad & \mathbb{P}(M^{\beta,\omega} > 0) = 1, \\ \text{(SD)}: \quad & \mathbb{P}(M^{\beta,\omega} = 0) = 1. \end{aligned} \tag{12.11}$$

We will see in what follows that the first case characterizes *weak disorder*, for which the behavior of the polymer is diffusive in the $\mathbb{Z}^d$-direction, while the second case characterizes *strong disorder*, for which the behavior is superdiffusive (see Fig. 12.2 for an impression of the typical path behavior). In Sections 12.4–12.5 we will derive *sufficient conditions* on d, β and $c(\cdot)$ for each of the two cases. Note that if $\beta = 0$, then the polymer performs SRW in the $\mathbb{Z}^d$-direction, which fits with (WD) because $M_n^{0,\omega} = 1$ for all $n \in \mathbb{N}_0$.

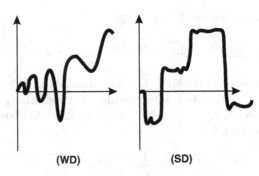

(WD)   **(SD)**

Fig. 12.2. Typical path behavior in the phases of weak disorder (WD) and strong disorder (SD).

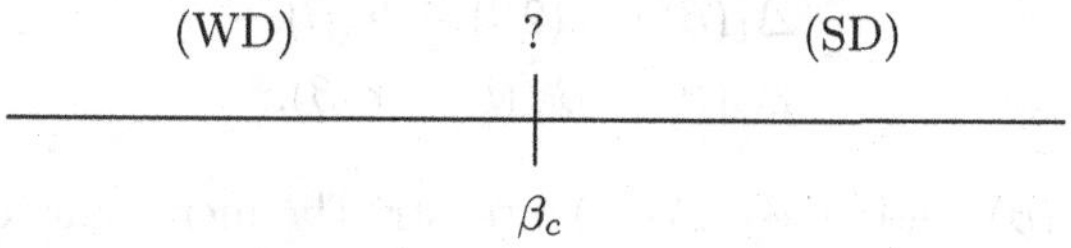

(WD)   ?   (SD)

β_c

Fig. 12.3. Separation of the two phases.

12.2.2 Separation of the Two Phases

The following theorem is due to Comets and Yoshida [74], and shows that
(WD) and (SD) are separated by a unique critical value of the interaction
strength β (see Fig. 12.3).

Theorem 12.1. *For any choice of the disorder distribution, $\beta \mapsto \mathbb{E}(\sqrt{M^{\beta,\omega}})$
is non-increasing on $[0,\infty)$. Consequently, there exists a $\beta_c \in [0,\infty]$ such that*

$$\beta \in [0, \beta_c) \implies \text{(WD)},$$
$$\beta \in (\beta_c, \infty) \implies \text{(SD)}. \tag{12.12}$$

Theorem 12.1 will be proved in Section 12.3. It is not known what happens
at the critical value (see Section 12.9, Challenge (2)).

It follows from (12.6–12.12) that

$$f(\beta) = f^{\text{ann}}(\beta) \qquad \forall \beta \in [0, \beta_c], \tag{12.13}$$

where the equality at $\beta = \beta_c$ can be added because both free energies are
convex and hence continuous in β. It is expected that $f(\beta) < f^{\text{ann}}(\beta)$ for
all $\beta \in (\beta_c, \infty)$, but this has not yet been proved in full generality. See
Section 12.8, Extension (2).

12.2.3 Characterization of the Two Phases

In this section we state two results for the path behavior in (WD), respectively,
(SD). These results are formulated in Theorems 12.2–12.3 below and will be
proved in Sections 12.5–12.6. They also yield a lower and an upper bound
on β_c.

We need some more notation. Let

$$\pi_d = (P \times P')(\exists n \in \mathbb{N} : S_n = S'_n) \tag{12.14}$$

denote the collision probability of two independent copies of the random
walk S. Since S is SRW, $S - S'$ is SRW at twice the speed, and so we have
$\pi_d = F_d$ with $F_d = P(\exists n \in \mathbb{N} : S_n = 0)$ the probability of return to the
origin, for which we saw in Section 2.1 that $F_d = 1$ in $d = 1, 2$ and $F_d < 1$ in
$d \geq 3$. For $\beta \in [0, \infty)$, define

$$\Delta_1(\beta) = c(2\beta) - 2c(\beta),$$
$$\Delta_2(\beta) = \beta c'(\beta) - c(\beta). \tag{12.15}$$

Both $\beta \mapsto \Delta_1(\beta)$ and $\beta \mapsto \Delta_2(\beta)$ are strictly increasing on $[0, \infty)$ with $\Delta_1(0) = \Delta_2(0) = 0$. From the representations

$$\Delta_1(\beta) = \Delta_2(\beta) + \int_\beta^{2\beta} c''(v)(2\beta - v)\, dv,$$
$$\Delta_2(\beta) = \int_0^\beta c''(v)v\, dv, \tag{12.16}$$

we see that $\Delta_1 > \Delta_2$ on $(0, \infty)$. Furthermore, define

$$\mathrm{max}_n^{\beta,\omega} = \max_{x \in \mathbb{Z}^d} P_{n-1}^{\beta,\omega}(S_n = x), \qquad n \in \mathbb{N}, \tag{12.17}$$

where the r.h.s. concerns the n-step polymer weighted according to the Hamiltonian up to time $n-1$ rather than time n, so that the last step is like a step of the random walk (this is done for later convenience). The quantity in (12.17) is a measure of how "localized" the polymer is at time n in the given random environment ω.

Theorem 12.2. *Suppose that*

(I) $d \geq 3$ *and* $\Delta_1(\beta) < \log(1/\pi_d)$.

Then

$$\lim_{n\to\infty} \frac{1}{n} E_n^{\beta,\omega}(\|S_n\|^2) = 1 \qquad \omega - a.s. \tag{12.18}$$

and

$$\lim_{n\to\infty} \mathrm{max}_n^{\beta,\omega} = 0 \qquad \omega - a.s. \tag{12.19}$$

Theorem 12.3. *Suppose that*

(II) $d = 1, 2$ *and* $\beta > 0$ *or* $d \geq 3$ *and* $\Delta_2(\beta) > \log(2d)$.

Then there exists a $c = c(d, \beta) > 0$ *such that*

$$\limsup_{n\to\infty} \mathrm{max}_n^{\beta,\omega} \geq c \qquad \omega - a.s. \tag{12.20}$$

The first half of Theorem 12.2 is due to Imbrie and Spencer [183], Bolthausen [24], Sinai [275] and Song and Zhou [282] (see also Olsen and Song [250] and Kifer [213]). The second half of Theorem 12.2, as well as Theorem 12.3, are due to Carmona and Hu [61] and Comets, Shiga and Yoshida [69]. The proofs in [183] and [275] are based on expansion techniques, the proofs in [24] and [282] on martingale arguments.

12.2.4 Diffusive versus Non-diffusive Behavior

The statements in (12.18–12.19) are typical for *diffusive* behavior. What is noteworthy is that the diffusion constant is *not renormalized* by the disorder. This is markedly different from what we found e.g. for the soft polymer in Chapter 4, Theorem 4.1, where the self-repellence – no matter how weak – renormalizes the diffusion constant by a factor > 1.

The statement in (12.20), which indicates *non-diffusive* behavior, suggests that the endpoint of the polymer is concentrated in a small region around a "most favorable site" (which depends on ω). Possibly the whole polymer is concentrated inside a "most favorable corridor". It has been conjectured that, throughout (SD), ω-a.s. as $n \to \infty$

$$E_n^{\beta,\omega}(\|S_n\|^2) \approx n^{2\nu(d)}, \qquad \log Z_n^{\beta,\omega} - \mathbb{E}(\log Z_n^{\beta,\omega}) \approx n^{\chi(d)}, \qquad (12.21)$$

where the symbol $\approx$ is to be interpreted in the following sense:

$$\nu(d) = \inf\left\{\nu > 0\colon \mathbb{E}\left(\lim_{n\to\infty} P_n^{\beta,\omega}\left(\max_{0\le i\le n}\|S_i\| \le n^\nu\right)\right) = 1\right\},$$
$$\chi(d) = \sup\left\{\chi > 0\colon \liminf_{n\to\infty} \mathbb{E}\left([\log Z_n^{\beta,\omega} - \mathbb{E}(\log Z_n^{\beta,\omega})]^2\right) \ge n^{2\chi}\right\}. \qquad (12.22)$$

Both exponents are believed not to depend on β, throughout (SD), and to satisfy $\chi(d) = 2\nu(d) - 1$, with $\nu(1) = \frac{2}{3}$ and $\nu(2) \in (\frac{1}{2}, \frac{2}{3})$. Partial results towards (12.21) have been obtained by Piza [267] (for SRW and general disorder), Petermann [262] and Mejane [240] (for Gaussian random walk and Gaussian disorder), and Carmona and Hu [62] (for SRW and Gaussian disorder). See Section 12.7, Extension (7), for more details. There are deep links between (12.21) and results for fluctuations of surfaces in first-passage percolation and the Ising model, and fluctuations of spectra of random matrices. For an overview we refer the reader to Krug and Spohn [221].

12.2.5 Bounds on the Critical Temperature

The conditions in (I) and (II) are mutually exclusive because $\Delta_1 > \Delta_2$ on $(0, \infty)$ and $\pi_d > 1/2d$. In fact, there is a gap between (I) and (II), so that Theorems 12.2–12.3 do not cover the full parameter regime. Note that (I) holds for *all* $c(\cdot)$ when β is sufficiently small, while (II) holds for *many* $c(\cdot)$ when β is sufficiently large. Indeed, there are examples of disorder distributions for which (I) holds for all $\beta \in [0, \infty)$ and (II) is empty, for instance, for the p-Bernoulli distribution with $p \ge \pi_d$, because $\Delta_1(\infty) = \Delta_2(\infty) = \log(1/p)$. Only when $p < \pi_d$ is (II) non-empty. For the standard Gaussian distribution we have $\Delta_1(\beta) = \beta^2$ and $\Delta_2(\beta) = \frac{1}{2}\beta^2$, and (II) is non-empty.

In view of the above observations, it is natural to introduce two critical values,

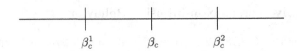

Fig. 12.4. Bounds on the critical temperature. Depending on the disorder distribution, either $0 < \beta_c^1 < \beta_c^2 < \infty$ or $0 < \beta_c^1 < \beta_c^2 = \infty$ or $\beta_c^1 = \beta_c^2 = \infty$.

$$\beta_c^1 = \sup \left\{ \beta \in [0, \infty) \colon \ \Delta_1(\beta) < \log(1/\pi_d) \right\},$$
$$\beta_c^2 = \inf \left\{ \beta \in [0, \infty) \colon \ \Delta_2(\beta) > \log(2d) \right\}, \tag{12.23}$$

which satisfy

$$0 < \beta_c^1 \leq \beta_c^2 \leq \infty. \tag{12.24}$$

To ensure that $\beta_c^2 < \infty$, so that (II) is non-empty, it suffices that $\Delta_2(\infty) = \infty$, which by (12.16) amounts to $\int_0^\infty c''(v)v \, dv = \infty$. The latter holds, for instance, when the disorder is: (1) unbounded; (2) bounded with no mass at the supremum of its support and a regularly varying density near this supremum. Indeed, in case (1) we have $\infty = c'(\infty) = c'(0) + \int_0^\infty c''(v) \, dv$, while in case (2) an easy computation gives $c''(v) \asymp 1/v^2$ as $v \to \infty$. An example of (2) is the uniform distribution on $[0, 1]$, for which $\Delta_1(\beta) \sim \Delta_2(\beta) \sim \log \beta$ as $\beta \to \infty$.

Theorems 12.2 and 12.3 will be proved in Sections 12.5 and 12.6, respectively. The proofs will show that

$$\text{(I)} \Longrightarrow \text{(WD)}, \qquad \text{(II)} \Longrightarrow \text{(SD)}. \tag{12.25}$$

Recalling Theorem 12.1 and (12.23), we thus have (see Fig. 12.4)

$$\beta_c \in [\beta_c^1, \beta_c^2]. \tag{12.26}$$

Note that, because $\Delta_1 > \Delta_2$ on $(0, \infty)$, if $\beta_c^2 < \infty$, then $\beta_c^1 < \beta_c^2$.

12.3 Proof of Uniqueness of the Critical Temperature

In this section we prove Theorem 12.1.

Proof. Abbreviate $f(x) = \sqrt{x}$, $x \in [0, \infty)$. We will show that, for all $n \in \mathbb{N}_0$ and $\beta \in \mathbb{R}$,

$$\mathbb{E}\left(f(M_n^{\beta, \omega}) \right) < \infty, \qquad \mathbb{E}\left(\frac{\partial}{\partial \beta} f(M_n^{\beta, \omega}) \right) < \infty, \tag{12.27}$$

and

$$\frac{\partial}{\partial \beta} \mathbb{E}\left(f(M_n^{\beta, \omega}) \right) = \mathbb{E}\left(\frac{\partial}{\partial \beta} f(M_n^{\beta, \omega}) \right) \leq 0. \tag{12.28}$$

Since $\mathbb{E}(M_n^{\beta, \omega}) = 1$ for all $n \in \mathbb{N}_0$ and $\beta \in \mathbb{R}$, we have

$$\lim_{n\to\infty} \mathbb{E}\left(f(M_n^{\beta,\omega})\right) = \mathbb{E}\left(f(M^{\beta,\omega})\right). \tag{12.29}$$

Combining (12.28–12.29), we get the claim in Theorem 12.1.

To prove (12.27–12.28), we fix any $\beta_0 \in (0,\infty)$ and first show that, for all $n \in \mathbb{N}$ and $p \in [1,\infty)$,

$$\sup_{0\le\beta\le\beta_0} M_n^{\beta,\omega}, \quad \sup_{0\le\beta\le\beta_0} [M_n^{\beta,\omega}]^{-1}, \quad \sup_{0\le\beta\le\beta_0} \left|\frac{\partial}{\partial\beta} M_n^{\beta,\omega}\right| \in L^p(\mathbb{P}). \tag{12.30}$$

Indeed, recalling (12.4), abbreviating

$$e_n^{\beta,\omega}(S) = e^{\beta I_n^\omega(S) - nc(\beta)}, \qquad I_n^\omega(S) = \sum_{i=1}^n \omega(i, S_i), \tag{12.31}$$

and using Fubini's theorem, we have

$$\begin{aligned}
\mathbb{E}\left([M_n^{\beta,\omega}]^p\right) &= \mathbb{E}\left(\left[E\left(e_n^{\beta,\omega}(S)\right)\right]^p\right) \\
&\le \mathbb{E}\left(E\left([e_n^{\beta,\omega}(S)]^p\right)\right) = \mathbb{E}\left(E\left(e^{p\beta I_n^\omega(S) - pnc(\beta)}\right)\right) = e^{n[c(p\beta) - pc(\beta)]}
\end{aligned} \tag{12.32}$$

and

$$\begin{aligned}
\mathbb{E}\left([M_n^{\beta,\omega}]^{-p}\right) &= \mathbb{E}\left(\left[E\left(e_n^{\beta,\omega}(S)\right)\right]^{-p}\right) \\
&\le \mathbb{E}\left(E\left(\left[e_n^{\beta,\omega}(S)\right]^{-p}\right)\right) = e^{n[c(-p\beta) + pc(\beta)]},
\end{aligned} \tag{12.33}$$

both of which are finite. A similar estimate can be done for the third quantity in (12.30), because

$$\frac{\partial}{\partial\beta} M_n^{\beta,\omega} = \frac{\partial}{\partial\beta} E\left(e_n^{\beta,\omega}(S)\right) = E\left([I_n^\omega(S) - nc'(\beta)] e_n^{\beta,\omega}(S)\right). \tag{12.34}$$

Next, since $f'(x) = \frac{1}{2\sqrt{x}} \le \frac{1}{2}(x + x^{-1})$, $x \in (0,\infty)$, we have

$$\left|\frac{\partial}{\partial\beta} f(M_n^{\beta,\omega})\right| = \left|f'(M_N^{\beta,\omega})\frac{\partial}{\partial\beta} M_n^{\beta,\omega}\right| \le \left|\frac{1}{2}\left(M_n^{\beta,\omega} + [M_n^{\beta,\omega}]^{-1}\right)\frac{\partial}{\partial\beta} M_n^{\beta,\omega}\right|. \tag{12.35}$$

Together with (12.30) and Hölder's inequality this yields that, for all $n \in \mathbb{N}_0$,

$$\sup_{0\le\beta\le\beta_0} \left|\frac{\partial}{\partial\beta} f(M_n^{\beta,\omega})\right| \in L^1(\mathbb{P}). \tag{12.36}$$

We can now complete the proof (12.27–12.28). To that end we write

$$f(M_n^{\beta_0,\omega}) = f(M_n^{0,\omega}) + \int_0^{\beta_0} d\beta \frac{\partial}{\partial\beta} f(M_n^{\beta,\omega}). \tag{12.37}$$

The claim in (12.27) and the equality in (12.28) follow by combining (12.37) with (12.30) and using Fubini's theorem. To get the inequality in (12.28), we recall (12.34) to write

$$
\begin{aligned}
\mathbb{E}\left(\frac{\partial}{\partial\beta} f(M_n^{\beta,\omega})\right) &= \mathbb{E}\left(f'(M_n^{\beta,\omega}) \frac{\partial}{\partial\beta} M_n^{\beta,\omega}\right) \\
&= E\left(\mathbb{E}\left(f'(M_n^{\beta,\omega}) \left[I_n^\omega(S) - nc'(\beta)\right] e_n^{\beta,\omega}(S)\right)\right).
\end{aligned}
\tag{12.38}
$$

For fixed S, $e_n^{\beta,\omega}(S)\mathbb{P}(d\omega)$ is a product measure. Since $\omega \mapsto [I_n^\omega(S) - nc'(\beta)]$ is non-decreasing and $\omega \mapsto f'(M_n^{\beta,\omega})$ is non-increasing (because $f' \leq 0$ and $\omega \mapsto M_n^{\beta,\omega}$ is non-decreasing), it therefore follows from the FKG-inequality (Fortuin, Kasteleyn and Ginibre [106]) that

$$
\begin{aligned}
&\mathbb{E}\left(f'(M_n^{\beta,\omega}) \left[I_n^\omega(S) - nc'(\beta)\right] e_n^{\beta,\omega}(S)\right) \\
&\leq \mathbb{E}\left(f'(M_n^{\beta,\omega}) e_n^{\beta,\omega}(S)\right) \mathbb{E}\left(\left[I_n^\omega(S) - nc'(\beta)\right] e_n^{\beta,\omega}(S)\right).
\end{aligned}
\tag{12.39}
$$

However, for all S,

$$
\begin{aligned}
\mathbb{E}\left(\left[I_n^\omega(S) - nc'(\beta)\right] e_n^{\beta,\omega}(S)\right) &= \mathbb{E}\left(\frac{\partial}{\partial\beta} e_n^{\beta,\omega}(S)\right) \\
&= \frac{\partial}{\partial\beta} \mathbb{E}\left(e_n^{\beta,\omega}(S)\right) = \frac{\partial}{\partial\beta} 1 = 0,
\end{aligned}
\tag{12.40}
$$

and so the inequality in (12.28) follows by combining (12.38–12.40).   □

12.4 Martingale Estimates

We first note that, subject to (I),

$$
\lim_{n\to\infty} M_n^{\beta,\omega} = M^{\beta,\omega} \quad \text{exists } \omega - a.s. \text{ and in } L^2(\mathbb{P}).
\tag{12.41}
$$

Indeed,

$$
\begin{aligned}
\mathbb{E}([M_n^{\beta,\omega}]^2) &= (E \times E')\left(\prod_{i=1}^n \mathbb{E}\left(e^{\beta[\omega(i,S_i)+\omega(i,S_i')]-2c(\beta)}\right)\right) \\
&= (E \times E')\left(\prod_{i=1}^n \left[1_{\{S_i \neq S_i'\}} + e^{c(2\beta)-2c(\beta)} 1_{\{S_i=S_i'\}}\right]\right) \\
&= (E \times E')\left(e^{[c(2\beta)-2c(\beta)] V_n(S,S')}\right),
\end{aligned}
\tag{12.42}
$$

where S, S' are two independent copies of the random walk, and $V_n(S,S') = \sum_{i=1}^n 1_{\{S_i=S_i'\}}$ denotes their *collision local time* up to time n. Put $V(S,S') =$

$\lim_{n\to\infty} V_n(S, S')$. Since $(P \times P')(V(S, S') = k) = (1 - \pi_d)(\pi_d)^k$, $k \in \mathbb{N}$, we see that

$$(I) \quad \Longleftrightarrow \quad \sup_{n\in\mathbb{N}} \mathbb{E}([M_n^{\beta,\omega}]^2) < \infty. \tag{12.43}$$

The claim in (12.41) therefore follows from the martingale convergence theorem.

The proofs of Theorems 12.2–12.3 given in Sections 12.5–12.6 rely on two technical lemmas, which are extensions of (12.41). These lemmas, which are taken from Comets, Shiga and Yoshida [70], will be stated and proved in Sections 12.4.1–12.4.2 below. The first lemma is a generalization of an earlier result by Bolthausen [24] and Song and Zhou [282].

12.4.1 First Estimate

Let $\phi: \mathbb{N}_0 \times \mathbb{Z}^d \to \mathbb{R}$ be such that

(C1) $|\phi(n, x)| \leq C_1 + C_2 n^{q/2} + C_3 \|x\|^q$ for all $(n, x) \in \mathbb{N}_0 \times \mathbb{Z}^d$ and some $C_1, C_2, C_3 \in [0, \infty)$, $q \in [0, \infty)$,

(C2) $(\phi(n, S_n))_{n\in\mathbb{N}_0}$ is a martingale w.r.t. the filtration generated by S, i.e., $(\mathcal{F}_{(0,n]}^S)_{n\in\mathbb{N}_0}$ with $\mathcal{F}_{(0,n]}^S = \sigma[S_i: 0 \leq i \leq n]$ the sigma-algebra of SRW up to time n.

Define

$$M_n^{\beta,\omega}(\phi) = E\left(\phi(n, S_n) \prod_{i=1}^{n} e^{\beta\omega(i, S_i) - c(\beta)}\right), \qquad M_0^{\beta,\omega}(\phi) = \phi(0, 0), \tag{12.44}$$

and note that, by (C2), $(M_n^{\beta,\omega}(\phi))_{n\in\mathbb{N}_0}$ is a martingale w.r.t. the filtration generated by ω (compare with (12.9)).

Lemma 12.4. *Assume* (I) *and* (C1–C2). *If $q > 0$, then*

$$\lim_{n\to\infty} n^{-q/2} \max_{0\leq i\leq n} M_i^{\beta,\omega}(\phi) = 0 \qquad \omega - a.s. \tag{12.45}$$

Moreover, if $0 \leq q < (d-2)/2$, then

$$\lim_{n\to\infty} M_n^{\beta,\omega}(\phi) = M^{\beta,\omega}(\phi) \qquad \text{exists } \omega - a.s. \text{ and in } L^2(\mathbb{P}). \tag{12.46}$$

Proof. We already proved (12.46) for $q = 0$ in (12.41), and so we may assume that $q > 0$. The proof will be based on the following additional estimate.

Lemma 12.5. *Assume* (I) *and* (C1–C2). *Then*

$$\mathbb{E}\left([M_n^{\beta,\omega}(\phi)]^2\right) = O(\psi_n) \qquad \text{as } n \to \infty, \tag{12.47}$$

with

$$\psi_n = \begin{cases} 1, & \text{if } \theta < 0, \\ \log n, & \text{if } \theta = 0, \\ n^\theta, & \text{if } \theta > 0, \end{cases} \tag{12.48}$$

where $\theta = q - (d-2)/2$.

Proof. First, following a computation similar as in (12.42), we write

$$\mathbb{E}\big([M_n^{\beta,\omega}(\phi)]^2\big) = [\phi(0, S_0)]^2$$
$$+ \sum_{k=1}^{n} \left(e^{\Delta_1(\beta)} - 1\right)^k \sum_{1 \le i_1 < \cdots < i_k \le n} (E \times E') \left([\phi(i_k, S_{i_k})]^2 \chi_{i_1 \ldots i_k}(S, S')\right), \tag{12.49}$$

where $i_0 = 0$ and we abbreviate

$$\chi_{i_1 \ldots i_k}(S, S') = 1_{\left\{S_{i_1} = S'_{i_1}, \ldots, S_{i_k} = S'_{i_k}\right\}}. \tag{12.50}$$

By (C1), the first term in the r.h.s. of (12.49) is bounded from above by C_1^2, while the summand in the second term is bounded from above by

$$3(C_3)^2 A_{i_1,\ldots,i_k} + [3(C_1)^2 + 3(C_2)^2 (i_k)^q] B_{i_1,\ldots,i_k}, \tag{12.51}$$

with

$$\begin{aligned} A_{i_1,\ldots,i_k} &= (E \times E')\big(\|S_{i_k}\|^{2q} \chi_{i_1 \ldots i_k}(S, S')\big), \\ B_{i_1,\ldots,i_k} &= (E \times E')\big(\chi_{i_1 \ldots i_k}(S, S')\big). \end{aligned} \tag{12.52}$$

Next, the first quantity in (12.52) can be estimated as follows (C denotes a generic constant that may change from line to line):

$$A_{i_1,\ldots,i_k} \le k^{2q-1} \sum_{l=1}^{k} (E \times E')\big(\|S_{i_l} - S_{i_{l-1}}\|^{2q} \chi_{i_1 \ldots i_k}(S, S')\big)$$
$$= k^{2q-1} \sum_{l=1}^{k} \left\{ \prod_{m=1}^{l-1} (E \times E')\big(\chi_{i_m - i_{m-1}}(S, S')\big) \right\}$$
$$\times (E \times E')\big(\|S_{i_l - i_{l-1}}\|^{2q} \chi_{i_l - i_{l-1}}(S, S')\big)$$
$$\times \left\{ \prod_{m=l+1}^{k} (E \times E')\big(\chi_{i_m - i_{m-1}}(S, S')\big) \right\}$$
$$= k^{2q-1} \sum_{l=1}^{k} C(i_l - i_{l-1})^{q-d/2} \prod_{\substack{1 \le m \le k \\ m \ne l}} (E \times E')\big(\chi_{i_m - i_{m-1}}(S, S')\big). \tag{12.53}$$

Here, the first line uses the triangle inequality and Jensen's inequality (for $2q \ge 1$; see later for $2q < 1$), the second line uses the Markov property of S and S', while the third line uses the bound

$$(E \times E')(\|S_n\|^{2q} \chi_n) = \sum_{x \in \mathbb{Z}^d} E(\|S_n\|^{2q} 1_{\{S_n = x\}}) \, P'(S_n' = x)$$

$$\leq C \, n^{-d/2} \, E(\|S_n\|^{2q}) \leq C \, n^{q-(d/2)}, \qquad n \in \mathbb{N}. \tag{12.54}$$

Summing (12.53), we obtain

$$\sum_{1 \leq i_1 < \cdots < i_k \leq n} A_{i_1,\ldots,i_k} \leq C \, \psi_n \, k^{2q} \left(\frac{\pi_d}{1 - \pi_d} \right)^{k-1}, \tag{12.55}$$

where we use that $\sum_{j=1}^{n} j^{q-d/2} \leq C\psi_n$, $n \in \mathbb{N}$, and

$$\sum_{j=1}^{\infty} (E \times E')(\chi_j(S, S')) = \frac{\pi_d}{1 - \pi_d}, \tag{12.56}$$

noting that $\pi_d < 1$ when $d \geq 3$.

A similar bound holds for the second quantity in (12.52), namely,

$$\sum_{1 \leq i_1 < \cdots < i_k \leq n} (i_k)^q B_{i_1,\ldots,i_k} \leq C \, \psi_n \, k^{2q} \left(\frac{\pi_d}{1 - \pi_d} \right)^{k-1}, \tag{12.57}$$

as is easily checked after replacing everywhere $\|S_i\|^{2q}$ by i^q, $i \in \mathbb{N}$. Inserting (12.55–12.57) into (12.51) and the latter into (12.49), we arrive at

$$E([M_n^{\beta,\omega}(\phi)]^2) \leq C + C \, \psi_n \sum_{k=1}^{n} k^{2q} \left(e^{\Delta_1(\beta)} - 1 \right)^k \left(\frac{\pi_d}{1 - \pi_d} \right)^{k-1}, \tag{12.58}$$

which gives the claim because $(e^{\Delta_1(\beta)} - 1)\frac{\pi_d}{1-\pi_d} < 1$ precisely when $\Delta_1(\beta) < \log(1/\pi_d)$.

It remains to deal with the case $2q < 1$. We can keep (12.53) provided we drop the factor k^{2q-1}. This is harmless, because the geometric factor in (12.58) absorbs the polynomial factor anyway. $\square$

We are finally ready to give the proof of (12.45–12.46). To get (12.46), note that $q < (d-2)/2$ corresponds to $\theta < 0$, and so Lemma 12.5 gives that $\sup_{n \in \mathbb{N}} E([M_n^{\beta,\omega}(\phi)]^2) < \infty$. The claim therefore follows from the martingale convergence theorem. To get (12.45), abbreviate

$$\bar{M}_n^{\beta,\omega}(\phi) = \max_{0 \leq i \leq n} M_n^{\beta,\omega}(\phi). \tag{12.59}$$

Pick $\delta > 0$, put $m(n) = \lceil n^{1/\delta} \rceil$, and estimate

$$E\left(\bar{M}_{m(n)}^{\beta,\omega}(\phi) > m(n)^\delta \sqrt{\psi_{m(n)}} \right) \leq [m(n)^{2\delta} \psi_{m(n)}]^{-1} \, E\left([\bar{M}_{m(n)}^{\beta,\omega}(\phi)]^2 \right)$$

$$\leq [m(n)^{2\delta} \psi_{m(n)}]^{-1} \, 4 \, E\left([M_{m(n)}^{\beta,\omega}(\phi)]^2 \right)$$

$$\leq C m^{-2\delta} \leq C n^{-2}, \tag{12.60}$$

where the second inequality follows from Doob's inequality for martingales (Durrett [96], Section 4.4), and the third inequality uses Lemma 12.5. Therefore, by the Borel-Cantelli lemma, the events

$$\left\{ \bar{M}^{\beta,\omega}_{m(n)}(\phi) > m(n)^\delta \sqrt{\psi_{m(n)}} \right\}, \qquad n \in \mathbb{N}, \tag{12.61}$$

occur finitely often ω-a.s. Since $n \mapsto \bar{M}^{\beta,\omega}_n(\phi)$ is non-decreasing and $n \mapsto m(n)$ grows polynomially, the latter implies that also the events $\{\bar{M}^{\beta,\omega}_n(\phi) > n^\delta \sqrt{\psi_n}\}$, $n \in \mathbb{N}$, occur finitely often ω-a.s. However, by (12.48), we have $n^\delta \sqrt{\psi_n} = o(n^{q/2})$ for δ small enough, because $q > 0$ and $\theta < q$ (the latter because (I) requires that $d \geq 3$). This proves the claim.   □

12.4.2 Second Estimate

Let

$$I^{\beta,\omega}_n = \sum_{x \in \mathbb{Z}^d} [P^{\beta,\omega}_{n-1}(S_n = x)]^2, \qquad n \in \mathbb{N}. \tag{12.62}$$

Lemma 12.6. *If $\beta \neq 0$, then $\{M^{\beta,\omega} = 0\} = \{\sum_{n \in \mathbb{N}} I^{\beta,\omega}_n = \infty\}$ ω-a.s. If, moreover, $\mathbb{P}(M^{\beta,\omega} = 0) = 1$, then there exist $c_1, c_2 \in (0, \infty)$ such that*

$$\exp\left[-c_1 \sum_{i=1}^n I^{\beta,\omega}_i \right] \leq M^{\beta,\omega}_n \leq \exp\left[-c_2 \sum_{i=1}^n I^{\beta,\omega}_i \right] \qquad \omega - a.s. \tag{12.63}$$

Proof. It suffices to prove that the following implications hold ω-a.s.:

$$M^{\beta,\omega} = 0 \implies \sum_{n \in \mathbb{N}} I^{\beta,\omega}_n = \infty \implies (12.63). \tag{12.64}$$

The proof of (12.64) is based on Doob's decomposition theorem (Durrett [96], Section 4.2). Let $X_n = -\log M^{\beta,\omega}_n$, $n \in \mathbb{N}_0$ ($X_0 = 0$), where for ease of notation we suppress β, ω from the notation. Split X_n into a martingale part and an associated compensator part,

$$X_n = A_n + B_n, \qquad n \in \mathbb{N}_0 \ (A_0 = B_0 = 0), \tag{12.65}$$

defined by

$$\begin{aligned} \Delta A_n &= A_n - A_{n-1} = -\log(1 + \delta_n) + \mathbb{E}\left(\log(1 + \delta_n) \mid \mathcal{F}^\omega_{(0,n-1]} \right), \\ \Delta B_n &= B_n - B_{n-1} = -\mathbb{E}\left(\log(1 + \delta_n) \mid \mathcal{F}^\omega_{(0,n-1]} \right), \end{aligned} \tag{12.66}$$

where

$$\delta_n = E^{\beta,\omega}_{n-1}\left(e^{\beta\omega(n,S_n) - c(\beta)} \right) - 1, \tag{12.67}$$

and $\mathcal{F}^\omega_{(0,n-1]}$ is the sigma-algebra generated by ω up to time $n-1$ (recall Section 12.2.1). The split in (12.65–12.67) arises from $\Delta X_n = X_n - X_{n-1} = \Delta A_n + \Delta B_n$ and the relation

$$\frac{M^{\beta,\omega}_n}{M^{\beta,\omega}_{n-1}} = 1 + \delta_n, \tag{12.68}$$

the latter being immediate from (12.4–12.5) and (12.8–12.9). Note that $n \mapsto A_n$ is a martingale w.r.t. the filtration generated by ω, and that (12.66) determines A_n and B_n recursively since $A_0 = B_0 = 0$.

With the help of the relations

$$\delta_n = \sum_{x \in \mathbb{Z}^d} P^{\beta,\omega}_{n-1}(S_n = x) \left[e^{\beta\omega(n,x)-c(\beta)} - 1 \right],$$
$$\mathbb{E}\left(\delta_n \mid \mathcal{F}^\omega_{(0,n-1]} \right) = 0, \tag{12.69}$$

it can be easily shown (see [70] for details) that there is a $c \in (0, \infty)$ such that

$$\Delta\langle A\rangle_n \le c I^{\beta,\omega}_n, \qquad c^{-1} I^{\beta,\omega}_n \le \Delta B_n \le c I^{\beta,\omega}_n. \tag{12.70}$$

Here, $n \mapsto \langle A\rangle_n$ is the compensator of $n \mapsto (X_n)^2$, determined recursively from

$$\Delta\langle A\rangle_n = \mathbb{E}\left([\Delta A_n]^2 \mid \mathcal{F}^\omega_{(0,n-1]} \right), \qquad n \in \mathbb{N} \ (\langle A\rangle_0 = 0). \tag{12.71}$$

It follows from (12.70) that ω-a.s.

$$\sum_{n \in \mathbb{N}} I^{\beta,\omega}_n < \infty \quad \Longrightarrow \quad \langle A\rangle \in \mathbb{R}, \ B \in \mathbb{R} \quad \Longrightarrow \quad A \in \mathbb{R}, \ B \in \mathbb{R}$$
$$\Longrightarrow \quad X \in \mathbb{R} \quad \Longrightarrow \quad M^{\beta,\omega} > 0, \tag{12.72}$$

where $X, A, B, \langle A\rangle$ denote the limits of $X_n, A_n, B_n, \langle A\rangle_n$ as $n \to \infty$. Note that $n \mapsto \langle A\rangle_n$ and $n \mapsto B_n$ are non-decreasing (the latter because of Jensen's inequality in combination with the second lines in (12.66) and (12.69)), and that $-\log M^{\beta,\omega} = X = A + B$. The second implication in (12.72) uses that A exists and is finite a.s. on the set $\{\langle A\rangle < \infty\}$ (Durrett [96], Section 4.4). This settles the first implication in (12.64).

To get the second implication in (12.64), we argue that

$$B = \infty \quad \Longrightarrow \quad \lim_{n\to\infty} \frac{-\log M^{\beta,\omega}_n}{B_n} = \lim_{n\to\infty} \frac{X_n}{B_n} = 1. \tag{12.73}$$

Indeed, either $\langle A\rangle < \infty$, in which case $A \in \mathbb{R}$ and $\lim_{n\to\infty} A_n/B_n = A/B = 0$, or $\langle A\rangle = \infty$, in which case we write $A_n/B_n = (A_n/\langle A\rangle_n)(\langle A\rangle_n/B_n)$ and use that the first factor tends to zero while the second factor is bounded away from infinity by (12.70), to again get $\lim_{n\to\infty} A_n/B_n = 0$. Combining (12.73) and (12.70), we arrive at the second implication in (12.64). $\square$

12.5 The Weak Disorder Phase

In this section we prove Theorem 12.2. The proof relies on Lemmas 12.4 and 12.6.

- Proof of the first implication in (12.25):

Proof. Pick $\phi \equiv 1$ ($q = 0$) in (12.44). Then $M_n^{\beta,\omega}(\phi) = M_n^{\beta,\omega}$, and so (12.41) yields that $\mathbb{E}(M^{\beta,\omega}) = \lim_{n\to\infty} \mathbb{E}(M_n^{\beta,\omega}) = 1$. By the zero-one law for the event $\{M^{\beta,\omega} > 0\}$, we therefore have (WD). $\square$

- Proof of (12.18):

Proof. Pick $\phi(n,x) = \|x\|^2 - n$ ($q = 2$) in (12.44), and write

$$E_n^{\beta,\omega}(\|S_n\|^2) - n = M_n^{\beta,\omega}(\phi)/M_n^{\beta,\omega}. \tag{12.74}$$

By (12.45) in Lemma 12.4, we have $\lim_{n\to\infty} n^{-1} M_n^{\beta,\omega}(\phi) = 0$ ω-a.s., while we know from (WD) proved above that $M_n^{\beta,\omega}$ is ω-a.s. bounded away from 0 uniformly in n. $\square$

- Proof of (12.19):

Proof. Recalling (12.17) and (12.62), we have the sandwich

$$[\max_n^{\beta,\omega}]^2 \le I_n^{\beta,\omega} \le \max_n^{\beta,\omega}. \tag{12.75}$$

It follows from Lemma 12.6 and (WD) proved above that $\sum_{n\in\mathbb{N}} I_n^{\beta,\omega} < \infty$ ω-a.s. Therefore

$$\sum_{n\in\mathbb{N}} [\max_n^{\beta,\omega}]^2 < \infty \qquad \omega - a.s., \tag{12.76}$$

which implies the claim. $\square$

12.6 The Strong Disorder Phase

In this section we prove Theorem 12.3. The proof relies on Lemma 12.6.

- Proof of the second implication in (12.25):

Proof. Pick $\gamma \in (0,1)$. Recalling (12.9) and using the estimate $(u + v)^\gamma \le u^\gamma + v^\gamma$, $u, v \ge 0$, we have

$$
\begin{aligned}
[M_n^{\beta,\omega}]^\gamma &= \left[E\left(\prod_{i=1}^n e^{\beta\omega(i,S_i)-c(\beta)} \right) \right]^\gamma \\
&= \left[\sum_{\substack{x\in\mathbb{Z}^d \\ \|x\|=1}} \tfrac{1}{2d} e^{\beta\omega(1,x)-c(\beta)} E\left(\prod_{j=1}^{n-1} e^{\beta\omega(j+1,x+S_j)-c(\beta)} \right) \right]^\gamma \\
&\le \sum_{\substack{x\in\mathbb{Z}^d \\ \|x\|=1}} \left[\tfrac{1}{2d} e^{\beta\omega(1,x)-c(\beta)} \right]^\gamma \left[E\left(\prod_{j=1}^{n-1} e^{\beta\omega(j+1,x+S_j)-c(\beta)} \right) \right]^\gamma .
\end{aligned}
\tag{12.77}
$$

Since the last expectation has the same law as $M_{n-1}^{\beta,\omega}$, we get

$$\mathbb{E}([M_n^{\beta,\omega}]^\gamma) \leq r(\gamma)\,\mathbb{E}([M_{n-1}^{\beta,\omega}]^\gamma), \tag{12.78}$$

where

$$r(\gamma) = (2d)^{1-\gamma}\,\mathbb{E}\left(e^{\beta\omega(1,x)-c(\beta)}\right)^\gamma. \tag{12.79}$$

Next, $\gamma \mapsto \log r(\gamma)$ is convex and infinitely differentiable, with $r(0) = 2d$ and $r(1) = 1$. The inequality $\Delta_2(\beta) > \log(2d)$ in condition (II) is equivalent to the statement that $[\log r]'(1) > 0$, and hence guarantees that $r(\gamma) < 1$ for some $\gamma \in (0,1)$. Choosing this γ in the above estimates, we conclude from (12.78) that there exists a $c \in (0,\infty)$ such that

$$\limsup_{n\to\infty} n^{-1} \log M_n^{\beta,\omega} \leq -c \qquad \omega - a.s. \tag{12.80}$$

Hence we have proved (SD) under the second half of condition (II). In a similar manner it can be shown that for every $\beta > 0$ there exists a $c \in (0,\infty)$ such that

$$
\begin{aligned}
d = 1: \quad & \limsup_{n\to\infty} n^{-\frac{1}{3}} \log M_n^{\beta,\omega} \leq -c \qquad \omega - a.s., \\
d = 2: \quad & \limsup_{n\to\infty} (\log n)^{-\frac{1}{2}} \log M_n^{\beta,\omega} \leq -c \qquad \omega - a.s.
\end{aligned}
\tag{12.81}
$$

This implies that (SD) also holds under the first half of condition (II). For details we refer to [69]. □

• Proof of (12.20):

Proof. With the help of (12.62), (12.80) and Lemma 12.6, we estimate

$$\limsup_{n\to\infty} \max_{x\in\mathbb{Z}^d} P_{n-1}^{\beta,\omega}(S_n = x) \geq \limsup_{n\to\infty} \frac{1}{n} \sum_{i=1}^{n} I_i^{\beta,\omega}$$
$$\geq -\liminf_{n\to\infty} \frac{1}{c_1 n} \log M_n^{\beta,\omega} \geq c/c_1 > 0, \tag{12.82}$$

which proves the claim. □

12.7 Beyond Second Moments

Recall (12.43), which says that (I) holds if and only if $\sup_{n\in\mathbb{N}} \mathbb{E}([M_n^{\beta,\omega}]^2) < \infty$, implying (WD) as stated in (12.25). Higher moments are of no use to improve (I). Indeed, Coyle [77] shows that $\lim_{n\to\infty} \mathbb{E}([M_n^{\beta,\omega}]^\alpha)$ diverges for all $\alpha \geq \alpha_0$ for some $\alpha_0 = \alpha_0(d,\beta) \in (0,\infty)$ when $d \geq 3$ and $\beta \in (0,\infty)$. The way forward is to consider lower moments. Clearly, to get (WD) it is sufficient that

$$\sup_{n\in\mathbb{N}} \mathbb{E}([M_n^{\beta,\omega}]^\alpha) < \infty \qquad \text{for some } \alpha \in (1,2], \tag{12.83}$$

and so we are looking for a condition on d and β under which the latter is true. Monthus and Garel [245] conjecture that the second moment estimate is sharp, i.e., $\beta_c^1 = \beta_c$ in (12.26). However, we shall see that this is wrong.

12.7.1 Fractional Moment Estimates

Evans and Derrida [97], using convexity arguments, derive a sufficient condition for (12.83), namely,

$$c(\alpha\beta) - \alpha c(\beta) < \log[1/\rho(\alpha)] \tag{12.84}$$

with

$$\rho(\alpha) = \sum_{n\in\mathbb{N}}\sum_{x\in\mathbb{Z}^d} \left[(P\times P')(S_i \neq S_i' \text{ for } 0 < i < n, S_n = S_n' = x)\right]^{\frac{1}{2}\alpha}. \tag{12.85}$$

Together with the definitions

$$\beta_c(\alpha) = \sup\left\{\beta \in [0,\infty): c(\alpha\beta) - \alpha c(\beta) < \log[1/\rho(\alpha)]\right\}, \qquad \alpha \in (1,2], \tag{12.86}$$

and

$$\bar\beta_c = \sup_{\alpha\in(1,2]} \beta_c(\alpha), \tag{12.87}$$

this yields

$$\beta < \bar\beta_c \quad\Longrightarrow\quad \text{(WD)}. \tag{12.88}$$

Note that $\rho(2) = \pi_d$, so that for $\alpha = 2$ the criterion in (12.84) reduces to (I), i.e., from (12.23) we have

$$\beta_c(2) = \beta_c^1. \tag{12.89}$$

Consequently, $\bar\beta_c \geq \beta_c^1$. Thus, the task is to compute $\bar\beta_c$ for different choices of d and the disorder distribution and to see in which cases $\bar\beta_c > \beta_c(2)$. In [97] this task is taken up numerically for standard Gaussian disorder, and it is shown that $\bar\beta_c > \beta_c^1$ for d sufficiently large.

Camanes and Carmona [48] carry out a systematic study of $\rho(\alpha)$ when $d \geq 3$, showing that there is an $\alpha_0 \in [1,2)$ such that $\alpha \mapsto \rho(\alpha)$ is infinite on $(1,\alpha_0]$, is continuous and non-increasing on $(\alpha_0, 2]$, and satisfies $\lim_{\alpha\downarrow\alpha_0} \rho(\alpha) = \infty$. Since $\rho(2) = \pi_d < 1$, it follows that there is an $\alpha_1 \in (\alpha_0, 2)$ such that $\rho(\alpha_1) = 1$, so that $\alpha \mapsto \beta_c(\alpha)$ is defined on $[\alpha_1, 2]$ and satisfies $\beta_c(\alpha_1) = 0$. They then show that, subject to a certain entropy condition on the random walk and the disorder, $\alpha \mapsto \beta_c(\alpha)$ goes through a maximum at some $\bar\alpha \in (\alpha_1, 2)$, which therefore is the optimal value for the estimate. Based on this analysis, they show that $\bar\beta_c = \beta_c(\bar\alpha) > \beta_c^1$ for standard Gaussian disorder in $d \geq 5$, binomial disorder in $d \geq 4$ with small mean, and Poisson disorder in $d \geq 3$ with small mean (see Fig. 12.5).

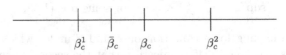

Fig. 12.5. Bounds on the critical temperature.

12.7.2 Size-biasing

Birkner [16] carries the analysis a crucial step further by proving the following result.

Theorem 12.7. *Let*

$$\theta^* = \sup\big\{\theta \geq 1\colon E\big(\theta^{V(S,S')}\big) < \infty \ S' - a.s.\big\} \qquad (12.90)$$

with $V(S,S') = \sum_{n\in\mathbb{N}} 1_{\{S_n = S'_n\}}$ *the collision local time of two independent SRWs, both starting at* 0. *(Note that* θ^* *is non-random because the tail sigma-algebra of* S' *is trivial.) Define*

$$\beta_c^* = \sup\big\{\beta \in [0,\infty)\colon c(2\beta) - 2c(\beta) < \log\theta^*\big\}. \qquad (12.91)$$

Then

$$\beta < \beta_c^* \quad \Longrightarrow \quad \text{(WD)}. \qquad (12.92)$$

Proof. The proof is based on a size-biasing argument, which runs as follows. Define (suppress from β from the notation)

$$e = \{e(i,x)\}_{i\in\mathbb{N},x\in\mathbb{Z}^d}, \qquad e(i,x) = e^{\beta\omega(i,x)-c(\beta)}, \qquad (12.93)$$

which is an i.i.d. $(0,\infty)$-valued random field. Define a size-biased version of this random field, written

$$\hat{e} = \{\hat{e}(i,x)\}_{i\in\mathbb{N},x\in\mathbb{Z}^d} \qquad (12.94)$$

and independent of e, such that

$$\mathbb{P}\big(\hat{e}(1,0) \in \cdot\big) = \mathbb{E}\big(e(1,0)\,1_{\{e(1,0)\in\,\cdot\}}\big) \qquad (12.95)$$

(no normalization is needed because $\mathbb{E}(e(1,0)) = 1$). Given S', put

$$\hat{e}_{S'} = \{e_{S'}(i,x)\}_{i\in\mathbb{N},x\in\mathbb{Z}^d}, \qquad \hat{e}_{S'}(i,x) = 1_{\{S'_i \neq x\}}\,e(i,x) + 1_{\{S'_i = x\}}\,\hat{e}(i,x), \qquad (12.96)$$

and define

$$\hat{M}_n^{e,\hat{e},S'} = E\left(\prod_{i=1}^n \hat{e}_{S'}(i,S_i)\right). \qquad (12.97)$$

This is a size-biased version of the martingale defined in (12.9), which in the present notation is

$$M_n^e = E\left(\prod_{i=1}^n e(i,S_i)\right). \qquad (12.98)$$

The point of (12.97) is that for any bounded function $f\colon [0,\infty) \to \mathbb{R}$,

$$\mathbb{E}\big(M_n^e\,f(M_n^e)\big) = (\mathbb{E} \times \hat{\mathbb{E}} \times E')\big(f(\hat{M}_n^{e,\hat{e},S'})\big), \qquad (12.99)$$

where $\mathbb{E}, \hat{\mathbb{E}}, E, E'$ denote expectation w.r.t. $e, \hat{e}, S, S'$, respectively. Indeed, this identity follows from the following computation:

$$
\begin{aligned}
\mathbb{E}\big(M_n^e f(M_n^e)\big) &= \mathbb{E}\left[E'\left(\prod_{i=1}^n e(i, S_i') \right) f\left(E\left(\prod_{i=1}^n e(i, S_i) \right) \right) \right] \\
&= E'\left(\mathbb{E}\left[\left(\prod_{i=1}^n e(i, S_i') \right) f\left(E\left(\prod_{i=1}^n e(i, S_i) \right) \right) \right] \right) \\
&= E'\left((\mathbb{E} \times \hat{\mathbb{E}})\left[f\left(E\left(\prod_{i=1}^n \hat{e}_{S'}(i, S_i) \right) \right) \right] \right) \\
&= (\mathbb{E} \times \hat{\mathbb{E}} \times E')\big(f(\hat{M}_n^{e,\hat{e},S'}) \big),
\end{aligned}
\tag{12.100}
$$

where the third equality uses (12.95–12.96).

Next, it follows from (12.99) that

$$(M_n^e)_{n \in \mathbb{N}_0} \text{ is uniformly integrable} \quad \Longleftrightarrow \quad (\hat{M}_n^{e,\hat{e},S'})_{n \in \mathbb{N}_0} \text{ is tight.} \tag{12.101}$$

However,

$$(\mathbb{E} \times \hat{\mathbb{E}})\big(\hat{M}_n^{e,\hat{e},S'}\big) = E\left(\theta^{\sum_{i=1}^n 1_{\{S_i = S_i'\}}} \right) \le E\big(\theta^{V(S,S')}\big), \qquad \theta = e^{c(2\beta) - 2c(\beta)}, \tag{12.102}$$

and hence $E(\theta^{V(S,S')}) < \infty$ S'-a.s. is enough to ensure the r.h.s. of (12.101). This completes the proof because the l.h.s. of (12.101) is equivalent to (WD). $\square$

Since $(E \times E')(\theta^{V(S,S')}) < \infty$ when $\theta < 1/\pi_d$, we see from (12.90) that $\theta^* \ge 1/\pi_d$, so that $\beta_c^* \ge \beta_c^1$. It was proved in Birkner, Greven and den Hollander [18] that $\theta^* > \pi_d$ in $d \ge 5$, implying that $\beta_c^* > \beta_c^1$ (see Fig. 12.6). Note that this gap holds *irrespective* of the distribution of the disorder, in contrast with the gap that was obtained by Camanes and Carmona [48] with the help of fractional moment estimates (see Section 12.7.1). In [18] it is conjectured that the gap persists in $d = 3, 4$. For $d = 4$ this conjecture is settled in Birkner and Sun [19].

Theorem 12.7 is optimal, in the sense that the size-biasing is a fractional moment estimate of order α with α "infinitesimally close" to 1. The only point where something is lost in the argument is (12.102), where an expectation is

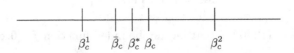

Fig. 12.6. Bounds on the critical temperature.

taken w.r.t. to the disorder. Presumably there exists a $\beta_c^{**} > \beta_c^*$ such that the r.h.s. of (12.101) holds for $\beta < \beta_c^{**}$. If so, then $\beta_c^{**} = \beta_c$, the critical threshold between (WD) and (SD) found in Theorem 12.1.

12.7.3 Relation with Random Pinning

As noted in Birkner and Sun [19], there is an interesting connection between the gap $\beta_c^* > \beta_c(2)$ obtained in Section 12.7.2 and the question of relevant vs. irrelevant disorder in the random pinning model discussed in Section 11.6.

Return to (12.96–12.97). Given S', put

$$\check{e}_{S'} = \{\check{e}_{S'}(i,x)\}_{i\in\mathbb{N}, x\in\mathbb{Z}^d}, \qquad \check{e}_{S'}(i,x) = 1_{\{S_i'\neq x\}}\, e(i,x) + 1_{\{S_i'=x\}}\, \hat{e}(i,0),$$
(12.103)

and define

$$\check{M}_n^{e,\check{e},S'} = E\left(\prod_{i=1}^n \check{e}_{S'}(i,S_i)\right).$$
(12.104)

Then, because the disorder is i.i.d., the distribution of $\check{M}_n^{e,\check{e},S'}$ as a function of $e, \check{e}, S'$ is the same as the distribution of $M_n^{e,\hat{e},S'}$ as a function of $e, \hat{e}, S'$. Therefore, tightness of $(M_n^{e,\hat{e},S'})_{n\in\mathbb{N}_0}$ is equivalent to tightness of $(\check{M}_n^{e,\check{e},S'})_{n\in\mathbb{N}_0}$. Consequently, recalling (12.101), we have

$$\sup_{n\in\mathbb{N}_0}(\mathbb{E}\times E')\left(\check{M}_n^{e,\check{e},S'}\right) < \infty \ a.s. \quad\Longrightarrow\quad (WD).$$
(12.105)

Now, an easy computation gives

$$(\mathbb{E}\times E')\left(\check{M}_n^{e,\check{e},S'}\right) = (\mathbb{E}\times E')\left(\exp\left[\sum_{i=1}^n (\beta\check{\omega}_i - c(\beta))\, 1_{\{S_i=S_i'\}}\right]\right)$$
$$= \tilde{E}\left(\exp\left[\sum_{i=1}^n (\beta\check{\omega}_i - c(\beta))\, 1_{\{\tilde{S}_i=0\}}\right]\right),$$
(12.106)

where $\check{\omega} = (\check{\omega}_i)_{i\in\mathbb{N}}$ is i.i.d. with moment generating function

$$\check{\mathbb{E}}\left(e^{\beta\check{\omega}_1}\right) = e^{c(2\beta)-c(\beta)},$$
(12.107)

and $\tilde{S} = (\tilde{S}_i)_{i\in\mathbb{N}_0}$ is the random walk whose excursion lengths away the origin have the same distribution as the first collision time of two independent copies of the random walk $S = (S_i)_{i\in\mathbb{N}_0}$. Recalling (11.3–11.4), we see that the r.h.s. of (12.106) is the partition sum of the random pinning model with parameters $(\lambda, h) = (\beta, c(\beta))$, disorder $\check{\omega}$ and reference random walk $\tilde{S}$.

Let $\tilde{f}(\beta)$ denote the quenched free energy associated with the above random pinning model. The annealed free energy $\tilde{f}^{\mathrm{ann}}(\beta)$ is the free energy of the homopolymer with parameter $\zeta = c(2\beta) - 2c(\beta)$, which is > 0 if and only if $c(2\beta) - 2c(\beta) > \log(1/\pi_d)$ (recall Section 7.6, Extension (1)), i.e., if and only if $\beta > \beta_c(2)$. It follows from the results in Section 11.6 that, for all $\tilde{S}$ whose excursion length distribution is such that the disorder is relevant in the random pinning model (which includes SRW), there exists a $\beta > \beta_c(2)$, sufficiently close to $\beta_c(2)$, such that $\tilde{f}(\beta) = 0 < \tilde{f}^{\mathrm{ann}}(\beta)$. This in turn implies that $\beta_c > \beta_c(2)$.

What the above considerations show is that the directed polymer in random environment with disorder ω and reference random walk S has a gap $\beta_c > \beta_c(2)$ as soon as the random pinning model with *biased disorder* $\breve{\omega}$ and reference random walk given by the *difference random walk* $\tilde{S}$ has relevant disorder. Note that if ω has finite moment generating function, then so has $\breve{\omega}$. Thus, we can use the results in Section 11.6, which only depend on the type of random walk $\tilde{S}$.

12.8 Extensions

(1) Essentially all the results described in Sections 12.1–12.7 carry over from SRW to an arbitrary reference random walk, with π_d in (12.14) as the key quantity for the weak disorder regime. The diffusive scaling in (12.18) only holds when S has zero mean and finite variance, but can be adapted accordingly. It is natural to conjecture that $\beta_c > \beta_c^1$ for all transient $\tilde{S}$ with excursion length exponent $a \in (0, \infty)$. Indeed, presumably something is lost by taking the expectation w.r.t. the disorder ω in (12.105), so that the results for the random pinning model do not give the sharp criterion for the gap.

(2) It is proved in Comets and Yoshida [74] that for arbitrary disorder with finite moment generating function there exists a $\tilde{\beta}_c \in [0, \infty]$ such that

$$f(\beta) \begin{cases} = f^{\mathrm{ann}}(\beta), & \text{if } \beta \in [0, \tilde{\beta}_c], \\ < f^{\mathrm{ann}}(\beta), & \text{if } \beta \in (\tilde{\beta}_c, \infty). \end{cases} \tag{12.108}$$

This implies that $\tilde{\beta}_c$ is a point of non-analyticity of the quenched free energy, satisfying $\tilde{\beta}_c \geq \beta_c$ because of (12.13). It is proved in Comets and Vargas [71] that $\tilde{\beta}_c = 0$ for arbitrary disorder and $d = 1$, and in Lacoin [226] that $\tilde{\beta}_c = 0$ for Gaussian disorder and $d = 2$. We saw in Theorem 12.3 that $\beta_c = 0$ in $d = 1, 2$. It is expected that $\tilde{\beta}_c = \beta_c$ in general.

Note that, by (12.6–12.8),

$$f^{\mathrm{ann}}(\beta) - f(\beta) = - \lim_{n \to \infty} \frac{1}{n} \log M_n^{\beta, \omega} \qquad \omega - a.s. \tag{12.109}$$

Thus, the second half of Lemma 12.6 shows that, for all $\beta \in (\tilde{\beta}_c, \infty)$, $\liminf_{n \to \infty} \frac{1}{n} \sum_{i=1}^{n} I_i^{\beta, \omega} \geq c > 0$ ω-a.s. In combination with the sandwich in (12.75), the latter implies the localization property in (12.20), which therefore holds beyond condition (II) in Theorem 12.3.

(3) As shown in Imbrie and Spencer [183], Bolthausen [24], Sinai [275], Song and Zhou [282], under condition (I) in Theorem 12.2 the CLT holds for the endpoint of the polymer. Comets and Yoshida [74] prove that, in fact, $d \geq 3$ and (WD) imply the CLT. Conlon and Olsen [75] and Coyle [76] explain that the reason why the diffusion constant is not renormalized is that the disorder is *uncorrelated in time*. As a result, under the annealed path measure the polymer evolves as SRW. As soon as time-correlation is added, the diffusion constant in general becomes renormalized, similarly as for random walk in random environment (see the Saint-Flour lectures by Zeitouni [323]).

(4) Song and Zhou [282] and Albeverio and Zhou [2] derive a rate of convergence result for (12.18). In particular, they show that, subject to (I) and for any $\delta > 0$,

$$\mathbb{E}\big(E_n^{\beta,\omega}(\|S_n\|^2 - n)\big) \sim \begin{cases} O\big(n^{\frac{6-d}{4}+\delta}\big), & \text{if } d = 3, 4, 5, \\ O\big([\log n]^{1+\delta}\big), & \text{if } d = 6, \\ O(1), & \text{if } d \geq 7. \end{cases} \tag{12.110}$$

Partial results had been obtained earlier in [183], [24].

(5) Carmona and Hu [61], [62] (Gaussian disorder) and Comets, Shiga and Yoshida [69] (general disorder) derive large deviation bounds for the endpoint of the polymer.

(6) Essentially all results in Sections 12.1–12.5 carry over from SRW to an arbitrary irreducible random walk. The gap $\beta_c^* > \beta_c^1$ is proved in [18] for any symmetric strongly transient random walk with a regularly varying tail at infinity, and is conjectured to be present for any symmetric transient random walk.

(7) It is observed numerically in Huse and Henley [176] that in $d = 1$

$$E_n^{\beta,\omega}(\|S_n\|^2) \asymp n^{\frac{4}{3}} \quad \text{as } n \to \infty \quad \omega - a.s. \tag{12.111}$$

An intuitive explanation of this superdiffusive scaling is given in Huse, Henley and Fisher [177], Kardar and Nelson [206], and Hwa and Fisher [180], based on the idea that in low dimension even for small β the disorder is "effectively strong at large times and large distances" (see also Kardar [205] and Kardar and Zhang [207]). There still is no proof available. For a Gaussian version of the model, in which SRW is replaced by a random walk in $\mathbb{R}$ with Gaussian steps and the disorder is a Gaussian random field in $\mathbb{N} \times \mathbb{R}^d$ whose covariance function is a two-sided exponential, Petermann [262] shows that the exponent $\nu(1)$ in (12.21) satisfies $\nu(1) \geq \frac{3}{5}$. Mejane [240] complements this result by showing that $\nu(d) \leq \frac{3}{4}$ for $d \geq 1$ as soon as the covariance function is bounded and integrable, extending earlier work by Piza [267] for SRW and bounded disorder.

(8) Comets and Yoshida [72], [73], [74] study a continuous space-time version of the model in which the reference process is Brownian motion and the

random environment is a *Poisson point process* with constant intensity. The Hamiltonian at time t is chosen to be $-\beta$ times the number of points in the 1-neighborhood of the path up to time t. Essentially all the results described in Sections 12.2–12.4 carry over, including the bound $\nu(d) \leq \frac{3}{4}$ for $d \geq 1$ mentioned in Extension (7). Stochastic analysis techniques allow for more elegant and flexible proofs. This model is interesting both for $\beta > 0$ (attractive points) and $\beta < 0$ (repulsive points), and has a close link with the work on Brownian motion in a random environment consisting of finite balls whose centers are located on a Poisson point process, as studied in the monograph by Sznitman [288] and in Wüthrich [318], [319], [320], [321].

(9) Sinai [275] proves a local central limit theorem for the discrete model, showing that, under condition (I) and conditionally on the event $\{S_n = x\}$ with $\|x\| = o(\sqrt{n})$, the polymer feels the disorder only near 0 and x, and in between behaves like a SRW conditioned to hit x at time n. Vargas [304] extends this result to the continuous model with Poisson point process disorder mentioned in Extension (8).

(10) Moriarty and O'Connell [246] consider a continuous version of the directed polymer in random environment given by the partition sum

$$Z_N^{\beta,N}(B) = \int dt_1 \cdots \int dt_{N-1} \, 1_{\{0 \leq t_1 < \cdots < t_{N-1} \leq N\}}$$
$$\times \exp\left[\beta \sum_{i=1}^{N} [B_i(t_i) - B_i(t_{i-1})]\right], \tag{12.112}$$

where $N \in \mathbb{N}$, $\beta \in \mathbb{R}$, $t_0 = 0$, $t_N = N$, and $B = (B_m)_{m \in \mathbb{N}}$ is a sequence of independent standard Brownian motions on $\mathbb{R}$. This models a directed polymer of length N visiting N layers successively, running along each layer for some time and picking up a Brownian increment. It is shown that, for every $\beta \neq 0$,

$$f(\beta) = \lim_{N \to \infty} \frac{1}{N} \log Z_N^{\beta,N}(B) = -\mathcal{L}[-\psi](-\beta^2) - \log(\beta^2) \quad B-a.s., \tag{12.113}$$

where $\psi = \Gamma'/\Gamma$ with Γ the Gamma-function, and $\mathcal{L}[-\psi]$ is the Legendre transform of $-\psi$. This settles a conjecture put forward in O'Connell and Yor [249]. It is further shown that $\beta \mapsto f(\beta)$ is analytic and strictly convex on $\mathbb{R}$ (with $f(0) = 1$). Hence, there is no phase transition as a function of β. The reason for the latter is that the system is in the strong disorder phase for every $\beta \neq 0$. Indeed, the annealed free energy is $f^{\mathrm{ann}}(\beta) = \frac{1}{2}\beta^2 + 1$, because

$$\mathbb{E}(Z_N^{\beta,N}(B)) = \int dt_1 \cdots \int dt_{N-1} \, 1_{\{0 \leq t_1 < \cdots < t_{N-1} \leq N\}}$$
$$\times \mathbb{E}\left(e^{\beta[B_1(N) - B_1(0)]}\right) = \frac{N^{N-1}}{(N-1)!} e^{\frac{1}{2}\beta^2 N}, \tag{12.114}$$

and it is easy to check that $f(\beta) < f^{\mathrm{ann}}(\beta)$ for all $\beta \neq 0$.

It is immediate from (12.112) that

$$\lim_{\beta \to \infty} \frac{1}{\beta} \log Z_N^{\beta,N}(B) = \sup_{0 \le t_1 < \cdots < t_{N-1} \le N} \sum_{i=1}^{N} [B_i(t_i) - B_i(t_{i-1})]. \qquad (12.115)$$

Let us abbreviate the r.h.s. by $M_N^N(B)$. Then it follows from the work of Baryshnikov [13] and Gravner, Tracy and Widom [128] on spectra of random matrices and related growth models that $N^{-1/3}[M_N^N(B) - 2N]$ converges in distribution as $N \to \infty$ to the Tracy-Widom law (see Tracy and Widom [297]). This fits well with the conjecture that the fluctuation exponent for the partition sum of the directed polymer in random environment equals $\chi(1) = \frac{1}{3}$ (recall Section 12.2.4).

(11) Buffet, Patrick and Pulé [45] consider the directed polymer in random environment on a binary tree, extending earlier work by Derrida and Spohn [84]. Paths can only move forward in the tree, and the reference measure on path space is taken to be the counting measure. Since $\mathbb{E}(Z_n^{\beta,\omega}) = [2e^{c(\beta)}]^n$, the annealed free energy equals $f^{\mathrm{ann}}(\beta) = c(\beta) + \log 2$. The quenched free energy is self-averaging and satisfies

$$f(\beta) = \begin{cases} f^{\mathrm{ann}}(\beta), & \text{if } \beta \in [0, \beta_c], \\ f^{\mathrm{ann}}(\beta_c), & \text{if } \beta \in (\beta_c, \infty), \end{cases} \qquad (12.116)$$

where $\beta_c \in (0, \infty]$ is the critical threshold separating (WD) and (SD). This threshold is shown to solve the equation $g'(\beta) = 0$ with $g(\beta) = \beta^{-1} f^{\mathrm{ann}}(\beta)$. The function $\beta \mapsto g(\beta)$ is strictly decreasing on $(0, \beta_c)$ and strictly increasing on (β_c, ∞). It is plotted in Fig. 12.7 for the case where the disorder distribution is exponential on the negative axis with mean $-\lambda^{-1}$, $\lambda > 0$, in which case $c(\beta) = \log[\lambda/(\beta+\lambda)]$ and β_c solves the equation $\log[2\lambda/(\beta+\lambda)] = -\beta/(\beta+\lambda)$. For standard Gaussian disorder $c(\beta) = \frac{1}{2}\beta^2$ and $\beta_c = \sqrt{2 \log 2}$.

On the tree, (12.83) holds whenever $g'(\beta) < 0$, i.e., for $\beta \in (0, \beta_c)$. The computations are easy because after two paths separate they never meet again.

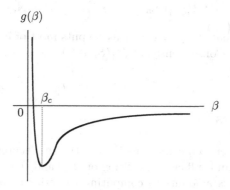

Fig. 12.7. Plot of $\beta \mapsto g(\beta)$ on the binary tree for a disorder distribution that is exponential on the negative axis.

(12) Ioffe and Velenik [185] study what happens for *undirected* paths. They consider the following partition sum:

$$Z^{\lambda,\beta,\omega}(N) = \sum_{w \in \mathcal{W}(N)} e^{-\lambda|w| - \beta \sum_{i=1}^{|w|} \omega(w_i)}, \qquad \lambda > \log(2d), \ \beta > 0, \quad (12.117)$$

where $\omega = \{\omega(x) \colon x \in \mathbb{Z}^{d+1}\}$ are i.i.d. $[0,\infty)$-valued random variables whose support contains 0, $\mathcal{W}(N)$ is the set of nearest-neighbor paths on $\mathbb{Z}^{d+1}$ of arbitrary length starting at 0 and ending at the hyperplane at distance N from 0 perpendicular to the 1-st direction, $|w|$ is the length of $w \in \mathcal{W}(N)$, and w_i is the site that w visits at time i. They show that if $d \geq 3$, then for every $\lambda > \log(2d)$ and $\beta \in [0, \beta_0)$ for some $\beta_0 = \beta_0(d, \lambda) \in (0, \infty)$,

$$\lim_{N \to \infty} M^{\lambda,\beta,\omega}(N) = \lim_{N \to \infty} \frac{Z^{\lambda,\beta,\omega}(N)}{\mathbb{E}(Z^{\lambda,\beta,\omega}(N))} \qquad (12.118)$$

exists and is strictly positive ω-a.s., and the path obeys diffusive scaling with a *renormalized* diffusion constant. This is precisely the analogue of Theorem 12.2, but now for undirected paths. The proofs are hard, because the martingale property no longer is available, and are based on an elaborate coarse-graining argument.

(13) Vargas [303] extends (SD) to disorder without finite moment generating function, where the martingale method breaks down. Vargas [304] proves that if β and the disorder distribution are such that $c(\beta) = \infty$, then for every $\delta \in (0,1)$ there exists an $\epsilon(\delta) > 0$ such that

$$\liminf_{n \to \infty} \mathbb{E}\left(\frac{1}{n} \sum_{m=1}^{n} P_{m-1}^{\beta,\omega}\left(S_m \in A_m^{\beta,\omega,\epsilon(\delta)}\right) \right) \geq \delta, \qquad (12.119)$$

where

$$A_m^{\beta,\omega,\epsilon} = \{x \in \mathbb{Z}^d \colon P_{m-1}^{\beta,\omega}(S_m = x) > \epsilon\}, \qquad m \in \mathbb{N}, \ \epsilon > 0. \qquad (12.120)$$

What this says is that the polymer measure puts most of its mass on ϵ-atoms for ϵ small enough. Consequently, $(P_{n-1}^{\beta,\omega}(S_n \in \cdot))_{n \in \mathbb{N}}$ is *asymptotically purely atomic*.

12.9 Challenges

(1) Close the gap between conditions (I) and (II) in Theorems 12.2–12.3, i.e., find the necessary and sufficient condition on d, β and $c(\cdot)$ under which (WD) and (SD) hold. This amounts to computing the critical threshold $\beta_c^{**} = \beta_c$ mentioned at the end of Section 12.6.

(2) Prove that $\tilde{\beta}_c = \beta_c$ in general, as already indicated in Section 12.8, Extension (2). Determine what happens at the critical threshold β_c. Is this part of (WD) or (SD)?

(3) Prove that $\beta_c > \beta_c^1$ as soon as the random walk S is such that the difference random walk $\tilde{S}$ is transient with excursion length exponent $a \in (0, \infty)$.

(4) Clarify the relation between $\bar{\beta}_c$ in (12.87) and β_c^* in (12.91). Is it the case that $\beta_c^* > \bar{\beta}_c$?

(5) Prove (12.111) and compute $\nu(1)$ and $\nu(2)$. Is there a scaling limit for the path that is sample-independent and, if so, then what is this scaling limit?

(6) Prove that in (SD) the path is confined to a "most favorable corridor" (recall the remarks made below Theorems 12.2–12.3). Show that in $d = 1$ the width of this corridor is of order 1 and its typical fluctuations are of order $n^{2/3}$ (see Section 12.2.4).

(7) Investigate what happens when the polymer lives on a randomly branching tree or on a supercritical percolation cluster, both carrying an i.i.d. random environment.

References

1. S. Albeverio and X.Y. Zhou, Free energy and some sample path properties of a random walk with random potential, J. Stat. Phys. 83 (1996) 573–622.
2. S. Albeverio and X.Y. Zhou, A martingale approach to directed polymers in a random environment, J. Theoret. Probab. 9 (1996) 171–189.
3. D. Aldous, Self-intersections of 1-dimensional random walks, Probab. Theory Relat. Fields 72 (1986) 559–587.
4. K.S. Alexander, Ivy on the ceiling: first-order polymer depinning transitions with quenched disorder, Markov Proc. Relat. Fields 13 (2007) 663–680.
5. K.S. Alexander, The effect of disorder on polymer depinning transitions, Commun. Math. Phys. 279 (2008) 117–146.
6. K.S. Alexander and V. Sidoravicius, Pinning of polymers and interfaces by random potentials, Ann. Appl. Probab. 16 (2006) 636–669.
7. K.S. Alexander and N. Zygouras, Quenched and annealed critical points in polymer pinning models, preprint 2008.
8. K.S. Alexander and N. Zygouras, Equality of critical points for polymer depinning transitions with loop exponent one, preprint 2008.
9. S.E. Alm and S. Janson, Random self-avoiding walks on one-dimensional lattices, Commun. Statist. Stochastic Models 6 (1990) 169–212.
10. S. Asmussen, *Applied Probability and Queues* (2nd. ed.), Applications of Mathematics, Vol. 51, Springer, New York, 2003.
11. A. Asselah, Annealed large deviation estimates for the energy of a polymer, work in progress.
12. M.N. Barber and B.W. Ninham, *Random and Restricted Walks: Theory and Applications*, Gordon and Breach, New York, 1970.
13. Yu. Baryshnikov, GUEs ans queues, Probab. Theory Relat. Fields 119 (2001) 256–274.
14. P. Billingsley, *Probability and Measure*, John Wiley & Sons, New York, 1968.
15. N.H. Bingham, C.M. Goldie and J.L. Teugels, *Regular Variation*, Cambridge University Press, Cambridge, 1987.
16. M. Birkner, A condition for weak disorder for directed polymers in random environment, Elect. Comm. Prob. 9 (2004) 22–25.
17. M. Birkner, A. Greven and F. den Hollander, Quenched large deviation principle for words in a letter sequence, EURANDOM Report 2008–034, to appear in Probab. Theory Relat. Fields.

234   References

18. M. Birkner, A. Greven and F. den Hollander, Collision local time of transient random walks and intermediate phases in interacting stochastic systems, preprint 2008.
19. M. Birkner and R. Sun, Annealed vs quenched critical points for a random walk pinning model, preprint 2008.
20. M. Biskup and F. den Hollander, A heteropolymer near a linear interface, Ann. Appl. Probab. 9 (1999) 668–687.
21. M. Biskup and W. König, Long-time tails in the parabolic Anderson model with bounded potential, Ann. Probab. 29 (2001) 636–682.
22. T. Bodineau and G. Giacomin, On the localization transition of random copolymers near selective interfaces, J. Stat. Phys. 117 (2004) 801–818.
23. T. Bodineau, G. Giacomin, H. Lacoin and F.L. Toninelli, Copolymers at selective interfaces: new bounds on the phase diagram, J. Stat. Phys. 132 (2008) 603–626.
24. E. Bolthausen, A note on diffusion of directed polymers in a random environment, Commun. Math. Phys. 123 (1989) 529–534.
25. E. Bolthausen, On self-repellent one-dimensional random walks, Probab. Theory Relat. Fields 86 (1990) 423–441.
26. E. Bolthausen, On the construction of the three-dimensional polymer measure, Probab. Theory Relat. Fields 97 (1993) 81–101.
27. E. Bolthausen, Localization of a two-dimensional random walk with an attractive path interaction, Ann. Probab. 22 (1994) 875–918.
28. E. Bolthausen, *Large Deviations and Interacting Random Walks*, Lecture Notes in Mathematics 1781, Springer, Berlin, 2002, pp. 1–124.
29. E. Bolthausen and G. Giacomin, Periodic copolymers at selective interfaces: a large deviations approach, Ann. Appl. Probab. 15 (2005) 963–983.
30. E. Bolthausen, F. Caravenna and B. de Tilière, The quenched critical point of a diluted polymer model, to appear in Stoch. Proc. Appl.
31. E. Bolthausen and F. den Hollander, Localization transition for a polymer near an interface, Ann. Probab. 25 (1997) 1334–1366.
32. E. Bolthausen and C. Ritzmann, A central limit theorem for convolution equations and weakly self-avoiding walks, preprint 2007, unpublished.
33. E. Bolthausen and U. Schmock, On self-attracting random walks, in: *Stochastic Analysis* (eds. M.C. Cranston and M.A. Pinsky), Proc. Sympos. Pure Math., Vol. 57, American Mathematical Society, Providence, RI, 1995, pp. 23–44.
34. E. Bolthausen and U. Schmock, On self-attracting d-dimensional random walks, Ann. Probab. 25 (1997) 531–572.
35. R. Brak, J.W. Essam and A.L. Owczarek, New results for directed vesicles and chains near an attractive wall, J. Stat. Phys. 93 (1998) 155–193.
36. R. Brak, A.J. Guttmann and S.G. Whittington, A collapse transition in a directed walk model, J. Phys. A: Math. Gen. 25 (1992) 2437–2446.
37. R. Brak, G.K. Iliev, A. Rechnitzer and S.G. Whittington, Motzkin path models of long chain polymers in slits, J. Phys. A: Math. Gen. 40 (2007) 4415–4437.
38. R. Brak, A.L. Owczarek and T. Prellberg, A scaling theory of the collapse transition in geometric cluster models of polymers and vesicles, J. Phys. A: Math. Gen. 26 (1993) 4565–4579.
39. R. Brak, A.L. Owczarek, A. Rechnitzer and S.G. Whittington, A directed model of a long chain polymer in a slit with attractive walls, J. Phys. A: Math. Gen. 38 (2005) 4309–4325.

40. V.A. Brazhnyi and S. Stepanow, Adsorption of a random heteropolymer with random self-interactions onto an interface, Eur. Phys. J. B27 (2002) 355–362.

41. D.C. Brydges and J.Z. Imbrie, End-to-end distance from the Green's function for a hierarchical self-avoiding walk in four dimensions, Commun. Math. Phys. 239 (2003) 523–547.

42. D.C. Brydges and G. Slade, A collapse transition for self-attracting walks, Resenhas do Instituto da Matemática e Estatística da Universidade de São Paulo, 1 (1994) 363–372.

43. D.C. Brydges and G. Slade, The diffusive phase of a model of self-interacting walks, Probab. Theory Relat. Fields 103 (1995) 285–315.

44. D.C. Brydges and T. Spencer, Self-avoiding walk in 5 or more dimensions. Commun. Math. Phys. 97 (1985) 125–148.

45. E. Buffet, A. Patrick and J.V. Pulé, Directed polymers on trees: a martingale approach, J. Phys. A: Math. Gen. 26 (1993) 1823–1834.

46. E. Buffet and J.V. Pulé, A model of continuous polymers with random charges, J. Math. Phys. 38 (1997) 5143–5152.

47. T.W. Burkhardt, Localization-delocalization transition in a solid-on-solid model with a pinning potential, J. Phys. A: Math. Gen. 14 (1981) L63–L68.

48. A. Camanes and P. Carmona, Directed polymers, critical temperature and uniform integrability, preprint 2007, submitted to Markov Proc. Relat. Fields.

49. P. Caputo, F. Martinelli and F.L. Toninelli, On the approach to equilibrium for a polymer with adsorption and repulsion, Elect. J. Probab. 13 (2008) 213–258.

50. S. Caracciolo, G. Parisi and A. Pelissetto, Random walks with short-range interaction and mean-field behavior, J. Stat. Phys. 77 (1994) 519–543.

51. F. Caravenna, *Random Walk Models and Probabilistic Techniques for Inhomogeneous Polymer Chains*, Ph.D. Thesis, University of Milan Bicocca and University of Paris 7, October 21, 2005.

52. F. Caravenna and J.-D. Deuschel, Pinning and wetting transition for (1+1)-dimensional fields with Laplacian interaction, to appear in Ann. Probab.

53. F. Caravenna and J.-D. Deuschel, Scaling limits of (1+1)-dimensional pinning models with Laplacian interaction, preprint 2008.

54. F. Caravenna and G. Giacomin, On constrained annealed bounds for pinning and wetting models, Elect. Comm. Probab. 10 (2005) 179–189.

55. F. Caravenna, G. Giacomin and M. Gubinelli, A numerical approach to copolymers at selective interfaces, J. Stat. Phys. 122 (2006) 799–832.

56. F. Caravenna, G. Giacomin and L. Zambotti, Sharp asymptotic behavior for wetting models in (1+1)-dimension, Elect. J. Probab. 11 (2006) 345–362.

57. F. Caravenna, G. Giacomin and L. Zambotti, A renewal theory approach to periodic copolymers with adsorption, Ann. Appl. Probab. 17 (2007) 1362–1398.

58. F. Caravenna, G. Giacomin and L. Zambotti, Infinite volume limits of polymer chains with periodic charges, Markov Proc. Relat. Fields 13 (2007) 679–730.

59. F. Caravenna and N. Pétrélis, A polymer in a multi-interface medium, preprint 2008, submitted to Ann. Appl. Prob.

60. F. Caravenna and N. Pétrélis, Depinning of a polymer at an infinity of interfaces, preprint 2008.

61. P. Carmona and Y. Hu, On the partition function of a directed polymer in a Gaussian random environment, Probab. Theory Relat. Fields 124 (2002) 431–457.

62. P. Carmona and Y. Hu, Fluctuation exponents and large deviations for directed polymers in a random environment, Stoch. Proc. Appl. 112 (2004) 285–308.

236   References

63. M.C.T.P. Carvalho and V. Privman, Directed walk models of polymers at interfaces, J. Phys. A: Math. Gen. 21 (1988) L1033–L1037.

64. M.S. Causo and S.G. Whittington, A Monte Carlo investigation of the localization transition in random copolymers at an interface, J. Phys. A: Math. Gen. 36 (2003) L189–L195.

65. D. Cheliotis and F. den Hollander, Variational characterization of the critical curve for pinning of random polymers, work in progress.

66. X. Chen, Limit law for the energy of a charged polymer, preprint 2007, to appear in Ann. I. H. Poincaré P & S.

67. X. Chen, *Random Walk Intersections: Large Deviations and Some Related Topics*, to appear in Mathematical Surveys and Monographs, American Mathematical Society.

68. N. Clisby, R. Liang and G. Slade, Self-avoiding walk enumeration via the lace expansion. J. Phys. A: Math. Theor. 40 (2007) 10973–11017.

69. F. Comets, T. Shiga and N. Yoshida, Directed polymers in random environment: Path localization and strong disorder, Bernoulli 9 (2003) 705–723.

70. F. Comets, T. Shiga and N. Yoshida, Probabilistic analysis of directed polymers in a random environment: a review, in: *Stochastic Analysis on Large Scale Systems*, Adv. Stud. Pure Math. 39 (2004), pp. 115–142.

71. F. Comets and V. Vargas, Majorizing multiplicative cascades for directed polymers in random media, Alea 2 (2006) 267–277.

72. F. Comets and N. Yoshida, Brownian directed polymers in random environment, Commun. Math. Phys. 254 (2004) 257–287.

73. F. Comets and N. Yoshida, Some new results on Brownian directed polymers in random environment, RIMS Kokyuroku 1386 (2004) 50–66.

74. F. Comets and N. Yoshida, Directed polymers in random environment are diffusive at weak disorder, Ann. Probab. 34 (2006) 1746–1770.

75. J.G. Conlon and P.A. Olsen, A Brownian motion version of the directed polymer problem, J. Stat. Phys. 84 (1996) 415–454.

76. L.N. Coyle, A continuous time version of random walks in a random potential, Stoch. Proc. Appl. 64 (1996) 209–235.

77. L.N. Coyle, Infinite moments of the partition function for random walks in a random potential, J. Math. Phys. 39 (1998) 2019–2034.

78. M. Cranston, O. Hryniv and S. Molchanov, Homo- and hetero-polymers in the mean-field approximation, preprint 2008.

79. D. Cule and T. Hwa, Denaturation of heterogenous DNA, Phys. Rev. Lett. 79 (1997) 2375–2378.

80. C. Danilowicz, Y. Kafri, R.S. Conroy, V.W. Coljee, J. Weeks and M. Prentiss, Measurement of the phase diagram of DNA unzipping in the temperature-force plane, Phys. Rev. Lett. 93 (2004) 078101.

81. K. De'Bell and T. Lookman, Surface phase transitions in polymer systems, Rev. Mod. Phys. 65 (1993) 87–113.

82. A. Dembo and O. Zeitouni, *Large Deviations Techniques and Applications* (2nd. ed.), Springer, New York, 1998.

83. B. Derrida and R.B. Griffiths, Directed polymers on disordered hierarchical lattices, Europhys. Lett. 8 (1989) 111–116.

84. B. Derrida and H. Spohn, Polymers on disordered trees, spin glasses, and traveling waves, J. Stat. Phys. 51 (1988) 817–840.

85. B. Derrida, G. Giacomin, H. Lacoin and F.L. Toninelli, Fractional moment bounds and disorder relevance for pinning models, preprint 2007, to appear in Commun. Math. Phys.

86. B. Derrida, R.B. Griffiths and P.G. Higgs, A model of directed random walks with random self-interactions, Europhys. Lett. 18 (1992) 361–366.

87. B. Derrida, V. Hakim and J. Vannimenus, Effect of disorder on two-dimensional wetting, J. Stat. Phys. 66 (1992) 1189–1213.

88. B. Derrida and P.G. Higgs, Low-temperature properties of directed random walks with random self-intersections, J. Phys. A: Math. Gen. 27 (1994) 5485–5493.

89. J.-D. Deuschel, G. Giacomin and L. Zambotti, Scaling limits of equilibrium wetting models in (1+1)-dimension, Probab. Theory Relat. Fields 132 (2005) 471–500.

90. J.-D. Deuschel and W. Stroock, *Large Deviations*, Academic Press, London, 1989.

91. E.A. DiMarzio and R.J. Rubin, Adsorption of a chain polymer between two plates, J. Chem. Phys. 55 (1971) 4318–4336.

92. M. Doi and S.F. Edwards, *The Theory of Polymer Dynamics*, International Series of Monographs on Physics, Vol. 73, Clarendon Press, Oxford, 1986.

93. K. Dušek (ed.), *Polymer Networks*, Advances in Polymer Sciences 44, Springer, Berlin, 1982.

94. N. Dunford and J.T. Schwartz, *Linear Operators, Part II: Spectral Theory*, Interscience Publishers, New York, 1963.

95. B. Duplantier and H. Saleur, Exact tricritical exponents for polymers at the FTHETA point in two dimensions, Phys. Rev. Lett. 59 (1987) 539–542.

96. R. Durrett, *Probability: Theory and Examples* (2nd. ed.), Duxbury Press, Belmont, 1996.

97. M.R. Evans and B. Derrida, Improved bounds for the transition temperature of directed polymers in a finite-dimensional random medium, J. Stat. Phys. 69 (1992) 427–437.

98. M.E. Fisher, Walks, walls, wetting, and melting, J. Stat. Phys. 34 (1984) 667–729.

99. G.J. Fleer, M.A. Cohen Stuart, T. Cosgrove and J.M.H.M. Scheutjens and B. Vincent, *Polymers at Interfaces*, Chapmann and Hall, London, 1993.

100. P.J. Flory, *Principles of Polymer Chemistry*, Cornell University Press, Ithaca, 1949.

101. P.J. Flory, The configuration of a real polymer chain, J. Chem. Phys. 17 (1949) 303–310.

102. P. Flory, *The Statistical Mechanics of Chain Molecules*, Interscience Publishers, New York, 1969.

103. P.J. Flory, Spatial configuration of macromolecular chains, Nobel Lecture, 1974. (http://nobelprize.org/nobel_prizes/chemistry/laureates/1974/).

104. G. Forgacs, J.M. Luck, Th.M. Nieuwenhuizen and H. Orland, Wetting of a disordered substrate: Exact critical behavior in two dimensions, Phys. Rev. Lett. 57 (1986) 2184–2187.

105. G. Forgacs, V. Privman and H.L. Frisch, Adsorption-desorption transition of polymer chains interacting with surfaces, J. Chem. Phys. 90 (1989) 3339–3345.

106. C.M. Fortuin, P.W. Kasteleyn and J. Ginibre, Correlation inequalities on some partially ordered sets, Commun. Math. Phys. 22 (1971) 89–103.

238   References

107. D.P. Foster, Exact evaluation of the collapse phase boundary for two-dimensional directed polymers, J. Phys. A: Math. Gen. 23 (1990) L1135–L1138.

108. D.P. Foster, E. Orlandini and M.C. Tesi, Surface critical exponents for models of polymer collapse and adsorption: the universality of the θ and θ' points, J. Phys. A: Math. Gen. 25 (1992) L1211–L1217.

109. D.P. Foster and J. Yeomans, Competition between self-attraction and adsorption in directed self-avoiding polymers, Physica A 177 (1991) 443–452.

110. D.M. Gangardt and S.K. Nechaev, Wetting transition on a one-dimensional disorder, preprint 2007.

111. T. Garel, D.A. Huse, S. Leibler and H. Orland, Localization transition of random chains at interfaces, Europhys. Lett. 8 (1989) 9–13.

112. T. Garel and C. Monthus, Two-dimensional wetting with binary disorder: a numerical study of the loop statistics, Eur. Phys. J. B46 (2005) 117–125.

113. J. Gärtner and W. König, The parabolic Anderson model. In: *Interacting Stochastic Systems* (J.-D. Deuschel and A. Greven, eds.). Springer, Berlin, 2005, pp. 153–179.

114. P.-G. de Gennes, *Scaling Concepts in Polymer Physics*, Cornell University Press, Ithaca, 1979.

115. P.-G. de Gennes, Soft matter, (http://nobelprize.org/nobel_prizes/physics/laureates/1991/).

116. G. Giacomin, *Random Polymer Models*, Imperial College Press, London, 2007.

117. G. Giacomin, Renewal convergence rates and correlation decay for homogeneous pinning models, Elect. J. Probab. 13 (2008) 513–529.

118. G. Giacomin, H. Lacoin and F.L. Toninelli, Hierarchical pinning models, quadratic maps and quenched disorder, preprint 2007.

119. G. Giacomin, H. Lacoin and F.L. Toninelli, Marginal relevance of disorder for pinning models, preprint 2008.

120. G. Giacomin and F.L. Toninelli, Estimates on path delocalization for copolymers at selective interfaces, Probab. Theory Relat. Fields 133 (2005) 464–482.

121. G. Giacomin and F.L. Toninelli, Force-induced depinning of directed polymers, J. Phys. A: Math. Gen. 40 (2007) 5261–5275.

122. G. Giacomin and F.L. Toninelli, Smoothing of depinning transitions for directed polymers with quenched disorder, Phys. Rev. Lett. 96 (2006) 070602.

123. G. Giacomin and F.L. Toninelli, Smoothing effect of quenched disorder on polymer depinning transitions, Commun. Math. Phys. 266 (2006) 1–16.

124. G. Giacomin and F.L. Toninelli, The localized phase of disordered copolymers with adsorption, Alea 1 (2006) 149–180.

125. G. Giacomin and F.L. Toninelli, On the irrelevant disorder regime of pinning models, preprint 2007, submitted to Ann. Probab.

126. I. Golding and Y. Kantor, Two-dimensional polymers with random short-range interactions, Phys. Rev. E 56 (1997) R1318–R1321.

127. P. Grassberger and R. Hegger, On the collapse of random copolymers, Europhys. Lett. 31 (1995) 351–356.

128. J. Gravner, C.A. Tracy and H. Widom, Limit theorems for height fluctuations in a class of discrete space and time growth models, J. Stat. Phys. 102 (2001) 1085–1132.

129. C.T. Greenwood and E.A. Milne, *Natural High Polymers*, Contemporary Science Paperbacks 18, Oliver & Boyd, Edinburgh, 1968.

130. A. Greven and F. den Hollander, Variational characterization of the speed of a one-dimensional self-repellent random walk, Ann. Appl. Probab. 3 (1993) 1067–1099.

131. P. Griffin, Accelerating beyond the third dimension: Returning to the origin in simple random walk, Math. Scientist 15 (1990) 24–35.

132. R.B. Griffiths, Nonanalytic behavior above critical point in a random Ising ferromagnet, Phys. Rev. Lett. 23 (1969) 17–19.

133. A. Grosberg, S. Izrailev and S. Nechaev, Phase transition in a heteropolymer chain at a selective interface, Phys. Rev. E 50 (1994) 1912–1921.

134. R. Guida and J. Zinn-Justin, Critical exponents of the N-vector model, J. Phys. A: Math. Gen. 31 (1998) 8103–8121.

135. N. Gunari, A.C. Balazs and G.C. Walker, Force-induced globule-coil transition in single polystyrene chains in water, J. Am. Chem. Soc. 129 (2007) 10046–10047.

136. N. Gunari and G.C. Walker, Nanomechanical fingerprints of individual blocks of a diblock copolymer chain, Langmuir 24 (2008) 5197–5201.

137. A.J. Guttmann, Asymptotic analysis of power-series expansions, in: *Phase Transitions and Critical Phenomena* (eds. C. Domb and J.L. Lebowitz), Vol. 11, Academic Press, New York, 1989.

138. A.J. Guttmann (ed.), *Polygons, Polyominoes and Polyhedra*, Springer, Berlin, in press.

139. A.J. Guttman and S.G. Whittington, Self-avoiding walks in a slab of finite thickness: A model of steric stabilisation, J. Phys. A: Math. Gen. 11 (1978) L107–L110.

140. N. Habibzadah, G.K. Iliev, A. Saguia and S.G. Whittington, Some Motzkin path models of random and periodic copolymers, J. Phys.: Conf. Ser. 42 (2006) 111–123.

141. J.M. Hammersley, On the rate of convergence to the connective constant of the hypercubical lattice, Quart. J. Math. Oxford 12 (1961) 250–256.

142. J.M. Hammersley, G.M. Torrie and S.G. Whittington, Self-avoiding walks interacting with a surface, J. Phys. A: Math. Gen. 15 (1982) 539–571.

143. J.M. Hammersley and D.J.A. Welsh, Further results on the rate of convergence to the connective constant of the hypercubical lattice, Quart. J. Math. Oxford 13 (1962) 108–110.

144. J.M. Hammersley and S.G. Whittington, Self-avoiding walks in wedges, J. Phys. A: Math. Gen. 18 (1985) 101–111.

145. T. Hara, Decay of correlations in nearest-neighbour self-avoiding walk, percolation, lattice trees and animals, Ann. Probab. 36 (2008) 530–593.

146. T. Hara, R. van der Hofstad and G. Slade, Critical two-point functions and the lace expansion for spread-out high-dimensional percolation and related models, Ann. Probab. 31 (2003) 349–408.

147. T. Hara and G. Slade, Self-avoiding walk in five or more dimensions, I. The critical behaviour, Commun. Math. Phys. 147 (1992) 101–136.

148. T. Hara and G. Slade, The lace expansion for self-avoiding walk in five or more dimensions, Rev. Math. Phys. 4 (1992) 235–327.

149. A.B. Harris, Effect of random defects on the critical behaviour of Ising models, J. Phys. C 7 (1974) 1671–1692.

150. B.J. Haupt, T.J. Senden and E.M. Sevick, AFM evidence of Rayleigh instability in single polymer chains, Langmuir 18 (2002) 2174–2182.

240   References

151. M. Heydenreich, Long-range self-avoiding walk converges to α-stable processes, EURANDOM Report 2008–038.

152. M. Heydenreich, R. van der Hofstad and A. Sakai, Mean-field behavior for long- and finite-range Ising model, percolation and self-avoiding walk, J. Stat. Phys. 132 (2008) 1001–1049.

153. R. van der Hofstad, The constants in the central limit theorem for the one-dimensional Edwards model, J. Stat. Phys. 90 (1998) 1295–1310.

154. R. van der Hofstad, *One-Dimensional Random Polymers*, CWI Tract 123, Stichting Mathematisch Centrum, Amsterdam, 1998.

155. R. van der Hofstad, The lace expansion approach to ballistic behaviour for one-dimensional weakly self-avoiding walk, Probab. Theory Relat. Fields 119 (2001) 311–349.

156. R. van der Hofstad, Spread-out oriented percolation and related models above the upper critical dimension: induction and superprocesses, Ensaios Mathemáticos, Vol. 9, Sociedade Brasileira de Matemática, 2005, pp. 91–181.

157. R. van der Hofstad and F. den Hollander, Scaling for a random polymer, Commun. Math. Phys. 169 (1995) 397–440.

158. R. van der Hofstad, F. den Hollander and W. König, Central limit theorem for a weakly interacting random polymer, Markov Proc. Relat. Fields 3 (1997) 1–63.

159. R. van der Hofstad, F. den Hollander and W. König, Central limit theorem for the Edwards model, Ann. Probab. 25 (1997) 573–597.

160. R. van der Hofstad, F. den Hollander and W. König, Large deviations for the one-dimensional Edwards model, Ann. Probab. 31 (2003) 2003–2039.

161. R. van der Hofstad, F. den Hollander and W. König, Weak interaction limits for one-dimensional random polymers, Probab. Theory Relat. Fields 125 (2003) 483–521.

162. R. van der Hofstad, F. den Hollander and G. Slade, A new inductive approach to the lace expansion, Probab. Theory Relat. Fields 111 (1998) 253–286.

163. R. van der Hofstad and A. Klenke, Self-attractive random polymers, Ann. Appl. Prob. 11 (2001) 1079–1115.

164. R. van der Hofstad, A. Klenke and W. König, The critical attractive random polymer in dimension one, J. Stat. Phys. 106 (2002) 477–520.

165. R. van der Hofstad and W. König, A survey of one-dimensional random polymers, J. Stat. Phys. 103 (2001) 915–944.

166. R. van der Hofstad and G. Slade, A generalised inductive approach to the lace expansion, Probab. Theory Relat. Fields 122 (2002) 389–430.

167. R. van der Hofstad and G. Slade, The lace expansion on a tree with applications to networks of self-avoiding walks, Adv. Appl. Math. 30 (2003) 471–528.

168. F. den Hollander, *Large Deviations*, American Mathematical Society, Fields Institute Monographs 14, Providence, RI, 2000.

169. F. den Hollander and N. Pétrélis, A mathematical model for a copolymer in an emulsion, EURANDOM Report 2007–032, to appear in J. Math. Chem.

170. F. den Hollander and N. Pétrélis, On the localized phase of a copolymer in an emulsion: supercritical percolation regime, EURANDOM Report 2007–048, to appear in Commun. Math. Phys.

171. F. den Hollander and N. Pétrélis, On the localized phase of a copolymer in an emulsion: subcritical percolation regime, EURANDOM Report 2008–031, to appear in J. Stat. Phys.

172. F. den Hollander and S.G. Whittington, Localization transition for a copolymer in an emulsion, Theor. Prob. Appl. 51 (2006) 193–240.

173. F. den Hollander and M. Wüthrich, Diffusion of a heteropolymer in a multi-interface medium, J. Stat. Phys. 114 (2004) 849–889.

174. M. Holmes, A.A. Járai, A. Sakai and G. Slade, High-dimensional graphical networks of self-avoiding walks, Canad. J. Math. 56 (2004) 77–114.

175. B.D. Hughes, *Random Walks and Random Environments*, Vols. 1 and 2, Oxford Science Publications, Clarendon Press, Oxford, 1996.

176. D.A. Huse and C.L. Henley, Pinning and roughening of domain walls in Ising systems due to random impurities, Phys. Rev. Lett. 54 (1985) 2708–2711.

177. D.A. Huse, C.L. Henley and D. Fisher, [Roughening by impurities at finite temperatures] Huse, Henley, and Fisher respond, Phys. Rev. Lett. 55 (1985) 2094.

178. T. Hwa, Disorder-induced depinning transition, Phys. Rev. B 51 (1995) 455–469.

179. T. Hwa and D. Cule, Polymer adsorption on disordered substrates, Phys. Rev. Lett. 79 (1997) 4930.

180. T. Hwa and D.S. Fisher, Anomalous fluctuations of directed polymers in random media, Phys. Rev. B 49 (1994) 3136–3154.

181. G. Iliev, E. Orlandini and S.G. Whittington, Adsorption and localization of random copolymers subject to a force: The Morita apprximation, Eur. Phys. J. B 40 (2004) 63–71.

182. G. Iliev, A. Rechnitzer and S.G. Whittington, Random copolymer localization in the Morita approximation, J. Phys. A: Math. Gen. 38 (2005) 1209–1223.

183. J.Z. Imbrie and T. Spencer, Diffusion of directed polymers in a random environment, J. Stat. Phys. 52 (1988) 609–626.

184. D. Ioffe and Y. Velenik, Ballistic phase for self-interacting random walks, in: *Analysis and Stochastics of Growth Processes and Interface Models* (eds. P. Mörters, R. Moser, H. Schwetlick, M. Penrose and J. Zimmer), Oxford University Press, Oxford, 2008.

185. D. Ioffe and Y. Velenik, Diffusivity of semi-directed polymers at weak disorder, manuscript in preparation.

186. Y. Isozaki and N. Yoshida, Weakly pinned random walk on the wall: pathwise descriptions of the phase transition, Stoch. Proc. Appl. 96 (2001) 261–284.

187. E.J. Janse van Rensburg, Collapsing and adsorbing polygons, J. Phys. A: Math. Gen. 31 (1998) 8295–8306.

188. E.J. Janse van Rensburg, *The Statistical Mechanics of Interacting Walks, Polygons, Animals and Vesicles*, Oxford University Press, Oxford, 2000.

189. E.J. Janse van Rensburg, Adsorbing bargraph in a q-wedge, J. Phys. A: Math. Gen. 38 (2005) 8505–8525 (corrigendum: J. Phys. A: Math. Gen. 38 (2006) 8505–8525).

190. E.J. Janse van Rensburg, Forces in Motzkin paths in a wedge, J. Phys. A: Math. Gen. 39 (2006) 1581–1608.

191. E.J. Janse van Rensburg, Moments of directed paths in a wedge, J. Phys.: Conf. Ser. 42 (2006) 147–162.

192. E.J. Janse van Rensburg, E. Orlandini, A.L. Owczarek, A. Rechnitzer and S.G. Whittington, Self-avoiding walks in a slab with attractive walls, J. Phys. A: Math. Gen. 38 (2005) L823–L828.

242   References

193. E.J. Janse van Rensburg, E. Orlandini, M.C. Tesi and S.G. Whittington, Self-averaging in random self-attracting polygons, J. Phys. A: Math. Gen. 34 (2001) L37–L44.

194. E.J. Janse van Rensburg, E. Orlandini and S.G. Whittington, Self-avoiding walks in a slab: rigorous results, J. Phys. A: Math. Gen. 39 (2006) 13869–13902.

195. E.J. Janse van Rensburg and A. Rechnitzer, Multiple Markov chain Monte Carlo study of adsorbing self-avoiding walks in two and three dimensions, J. Phys. A: Math. Gen. 37 (2004) 6875–6898.

196. E.J. Janvresse, T. de la Rue and Y. Velenik, Pinning by a sparse potential, Stoch. Proc. Appl. 115 (2005) 1323–1331.

197. E.J. Janse van Rensburg and L. Ye, Forces in square lattice directed paths in wedges, J. Phys. A: Math. Gen. 38 (2005) 8493–8503.

198. E.W. James, C.E. Soteros and S.G. Whittington, Localization of a random copolymer at an interface: an exact enumeration study, J. Phys. A: Math. Gen. 36 (2003) 11575–11584.

199. E.W. James, C.E. Soteros and S.G. Whittington, Localization of a random copolymer at an interface: an untethrered self-avoiding walk model, J. Phys. A: Math. Gen. 36 (2003) 1–14.

200. I. Jensen, Enumeration of self-avoiding walks on the square lattice, J. Phys. A: Math. Gen. 37 (2004) 5503–5524.

201. I. Jensen, homepage (www.ms.unimelb.edu.au/~iwan).

202. Y. Kafri, D. Mukamel and L. Peliti, Why is the DNA denaturation transition first order?, Phys. Rev. Lett. 85 (2000) 4988-4991.

203. Y. Kantor and M. Kardar, Polymers with random self-interactions, Europhys. Lett. 14 (1991) 421–426.

204. Y. Kantor and M. Kardar, Collapse of randomly self-interacting polymers, Europhys. Lett. 28 (1994) 169–174.

205. M. Kardar, Roughening by impurities at finite temperatures, Phys. Rev. Lett. 55 (1985) 2923.

206. M. Kardar and D.R. Nelson, Commensurate-incommensurate transitions with quenched random impurities, Phys. Rev. Lett. 55 (1985) 1157–1160.

207. M. Kardar and Y.C. Zhang, Scaling of directed polymers in random media, Phys. Rev. Lett. 58 (1987) 2087–2090.

208. T. Kennedy, Ballistic behavior in a 1D weakly self-avoiding walk with decaying energy penalty, J. Stat. Phys. 77 (1994) 565–579.

209. T. Kennedy, Monte Carlo tests of Stochastic Loewner Evolution predictions for the 2D self-avoiding walk, Phys. Rev. Lett. 88 (2002) 130601.

210. H. Kesten, Ratio theorems for random walk II, J. d'Analyse Math. 11 (1963) 323–379.

211. H. Kesten, On the number of self-avoiding walks, J. Math. Phys. 4 (1963) 960–969.

212. H. Kesten and F. Spitzer, Ratio theorems for random walks, J. d'Analyse Math. 11 (1963) 285–322.

213. Y. Kifer, The Burgers equation with a random force and a general model for directed polymers in random environments, Probab. Theory Relat. Fields 108 (1997) 29–65.

214. J.F.C. Kingman, Subadditive ergodic theory, Ann. Probab. 6 (1973) 883–909.

215. F.B. Knight, Random walks and a sojourn density process of Brownian motion, Transactions of the AMS 109 (1963) 56–86.

216. W. König, The drift of a one-dimensional self-avoiding random walk, Probab. Theory Relat. Fields 96 (1993) 521–543.

217. W. König, The drift of a one-dimensional self-repellent random walk with bounded increments, Probab. Theory Relat. Fields 100 (1994) 513–544.

218. W. König, A central limit theorem for a one-dimensional polymer measure, Ann. Probab. 24 (1996) 1012–1035.

219. J. Krawczyk, T. Prellberg, A.L. Owczarek and A. Rechnitzer, Stretching of a chain polymer adsorbed at a surface, J. Stat. Mech. Theor. Exp. (2004) P10004.

220. J. Krawczyk, A.L. Owczarek, T. Prellberg and A. Rechnitzer, Pulling absorbing and collapsing polymers from a surface, J. Stat. Mech. Theor. Exp. (2005) P05008.

221. H. Krug and H. Spohn, Kinetic roughening of growing surfaces, in: *Solids far from Equilibrium* (ed. C. Godrèche), Cambridge University Press, Cambridge, 1991.

222. S. Kumar, I. Jensen, J. Jacobsen and A.J. Guttmann, Role of conformational entropy in force-induced biopolymer unfolding, Phys. Rev. Lett. 98 (2007) 128101.

223. S. Kusuoka, On the path property of Edwards' model for long polymer chains in three dimensions, Proc. Bielefeld Conf. on *Infinite Dimensional Analysis and Stochastic Processes* (ed. S. Albeverio), Res. Notes Math. 124 (Pitman Advanced Publishing Program IX), Boston, Pitman, 1985, pp. 48–65.

224. S. Kusuoka, Asymptotics of polymer measures in one dimension, Proc. Bielefeld Conf. on *Infinite Dimensional Analysis and Stochastic Processes* (ed. S. Albeverio), Res. Notes Math. 124 (Pitman Advanced Publishing Program IX), Boston, Pitman, 1985, pp. 66–82.

225. H. Lacoin, Hierarchical pinning model with site disorder: Disorder is marginally relevant, preprint 2008.

226. H. Lacoin, New bounds for the free energy of directed polymer in dimension $1 + 1$ and $1 + 2$, preprint 2008.

227. G.F. Lawler, O. Schramm and W. Werner, On the scaling limit of planar self-avoiding walks, Proc. Sympos. Pure Math., Vol. 72, Part 2, American Mathematical Society, Providence, RI, 2002, pp. 339–364.

228. J.M.J. van Leeuwen and H.J. Hilhorst, Pinning of a rough interface by an external potential, Phys. A 107 (1981) 319–329.

229. D.K. Lubensky and D.R. Nelson, Pulling pinned polymers and unzipping DNA, Phys. Rev. Lett. 85 (2000) 1572–1575.

230. N. Madras and G. Slade, *The Self-Avoiding Walk*, Birkhäuser, Boston, 1993.

231. N. Madras and A.D. Sokal, The pivot algorithm: A highly efficient Monte Carlo method for the self-avoiding walk, J. Stat. Phys. 50 (1988) 109–186.

232. N. Madras and S.G. Whittington, Self-averaging in finite random copolymers, J. Phys. A: Math. Gen. 35 (2002) L427–L431.

233. N. Madras and S.G. Whittington, Localization of a random copolymer at an interface, J. Phys. A: Math. Gen. 36 (2003) 923–938.

234. D. Marenduzzo, A. Trovato and A. Maritan, Phase diagram of force-induced DNA unzipping in exactly solvable models, Phys. Rev. E 64 (2001) 031901.

235. A. Maritan, M.P. Riva and A. Trovato, Heteropolymers in a solvent at an interface, J. Phys. A: Math. Gen. 32 (1999) L275–L280.

244   References

236. R. Martin, M.S. Causo and S.G. Whittington, Localization transition for a randomly coloured self-avoiding walk at an interface, J. Phys. A: Math. Gen. 33 (2000) 7903–7918.

237. R. Martin, E. Orlandini, A.L. Owczarek, A. Rechnitzer and S.G. Whittington, Exact enumeration and Monte Carlo results for self-avoiding walks in a slab, J. Phys. A: Math. Gen. 40 (2007) 7509–7521.

238. S. Martinez and D. Petritis, Thermodynamics of a Brownian bridge polymer model in a random environment, J. Phys. A29 (1996) 1267–1279.

239. B. McCoy, Incompleteness of the critical exponent description for ferromagnetic systems containing random impurities, Phys. Rev. Lett. 23 (1969) 383–386.

240. O. Mejane, Upper bound of a volume exponent for directed polymers in a random environment, Ann. I. H. Poincaré P & R 40 (2004) 299–308.

241. M. S. Moghaddam, T. Vrbová and S.G. Whittington, Adsorption of periodic copolymers at a planar interface, J. Phys. A: Math. Gen. 33 (2000) 4573–4584.

242. P. Monari and A.L. Stella, θ-point universality of random polyampholytes with screened interactions, Phys. Rev. E 59 (1999) 1887–1892.

243. C. Monthus, On the localization of random heteropolymers at the interface between two selective solvents, Eur. Phys. J. B13 (2000) 111–130.

244. C. Monthus and T. Garel, Delocalization transition of the selective interface model: distribution of pseudo-critical temperatures, J. Stat. Mech. (2005) P12011.

245. C. Monthus and T. Garel, Freezing transition of the directed polymer in a 1+d random medium: Location of the critical temperature and unusual critical properties, Phys. Rev. E 74 (2006) 011101.

246. J. Moriarty and N. O'Connell, On the free energy of a directed polymer in a Brownian environment, Markov Proc. Relat. Fields 13 (2007) 251–266.

247. D.H. Napper, *Polymeric Stabilization of Colloidal Dispersions*, Academic Press, London, 1983.

248. B. Nienhuis, Exact critical exponents of the $O(n)$ models in two dimensions, Phys. Rev. Lett. 49 (1982) 1062–1065.

249. N. O'Connell and M. Yor, Brownian analogues of Burke's theorem, Stoch. Proc. Appl. 96 (2001) 285–304.

250. R. Olsen and R. Song, Diffusion of directed polymers in a strong random environment, J. Stat. Phys. 83 (1996) 727–738.

251. Y. Oono, On the divergence of the perturbation series for the excluded-volume problem in polymers, J. Phys. Soc. Japan 39 (1975) 25–29.

252. Y. Oono, On the divergence of the perturbation series for the excluded-volume problem in polymers, II. Collapse of a single chain in poor solvents, J. Phys. Soc. Japan 41 (1976) 787–793.

253. E. Orlandini, A. Rechnitzer and S.G. Whittington, Random copolymers and the Morita approximation: polymer adsorption and polymer localization, J. Phys. A: Math. Gen. 35 (2002) 7729–7751.

254. E. Orlandini, M.C. Tesi and S.G. Whittington, A self-avoiding walk model of random copolymer adsorption, J. Phys. A: Math. Gen. 32 (1999) 469–477.

255. E. Orlandini, M.C. Tesi and S.G. Whittington, Self-averaging in models of random copolymer collapse, J. Phys. A: Math. Gen. 33 (2000) 259–266.

256. E. Orlandini, M.C. Tesi and S.G. Whittington, Adsorption of a directed polymer subject to an elongational force, J. Phys. A: Math. Gen. 37 (2004) 1535–1543.

257. E. Orlandini and S.G. Whittington, Statistical topology of closed curves: Some applications in polymer physics, Rev. Mod. Phys. 79 (2007) 611–642.

258. A.L. Owczarek and T. Prellberg, Exact solution of semi-flexible and super-flexible interacting partially directed walks, J. Stat. Mech. (2007) P11010.

259. A.L. Owczarek, T. Prellberg and R. Brak, The tricritical behavior of self-interacting partially directed walks, J. Stat. Phys. 72 (1993) 737–772.

260. A.L. Owczarek, T. Prellberg and A. Rechnitzer, Finite-size scaling functions for directed polymers confined between attracting walls, J. Phys. A: Math. Theor. 41 (2008) 035002.

261. A.L. Owczarek and S.G. Whittington, Interacting lattice polygons, in: Polygons, Polynominoes and Polyhedra (ed. A.J. Guttmann), Springer, Berlin, in press, pp. 305–319.

262. M. Petermann, Superdiffusivity of directed polymers in random environment, Ph.D. thesis, University of Zürich, 2000.

263. N. Pétrélis, Localisation d'un Polymère en Interaction avec une Interface, Ph.D. Thesis, University of Rouen, France, February 2, 2006.

264. N. Pétrélis, Polymer pinning at an interface, Stoch. Proc. Appl. 116 (2006) 1600–1621.

265. N. Pétrélis, Polymer pinning: weak coupling limits, to appear in Ann. I. H. Poincaré.

266. J.S. Phipps, R.M. Richardson, T. Cosgrove and A. Eaglesham, Neutron reflection studies of copolymers at the hexane/water interface, Langmuir 9 (1993) 3530–3537.

267. M.S.T. Piza, Directed polymers in a random environment: some results on fluctuations, J. Stat. Phys. 89 (1997) 581–603.

268. D. Poland and H.A. Sheraga (eds.), Theory of Helix-Coil Transitions in Biopolymers, Academic Press, New York, 1970.

269. V. Privman, G. Forgacs and H.L. Frisch, New solvable model of polymer chain adsorption at a surface, Phys. Rev. B 37 (1988) 9897–9900.

270. C. Richard and A.J. Guttmann, Poland-Sheraga models and the DNA denaturation transition, J. Stat. Phys. 115 (2004) 943–965.

271. B. Roynette, P. Vallois and M. Yor, Limiting laws associated with Brownian motion perturbed by normalized exponential weights, C. R. Math. Acad. Sci. Paris 337 (2003) 667–673.

272. R.J. Rubin, Random-walk model of chain-polymer adsorption at a surface, J. Chem. Phys. 43 (1965) 2392–2407.

273. F. Seno and A.L. Stella, θ point of a linear polymer in 2 dimensions: a renormalization group analysis of Monte Carlo enumerations, J. Physique 49 (1988) 739–748.

274. Ya.G. Sinai, A random walk with random potential, Theor. Prob. Appl. 38 (1993) 382–385.

275. Ya.G. Sinai, A remark concerning random walks with random potentials, Fund. Math. 147 (1995) 173–180.

276. Ya.G. Sinai and H. Spohn, Remarks on the delocalization transition for heteropolymers, in: Topics in Statistical and Theoretical Physics (eds. R.L. Dobrushin et al.), AMS Transl. 177, Providence, RI, 1996, pp. 219–223.

277. G. Slade, The diffusion of self-avoiding walk in high dimensions, Commun. Math. Phys. 110 (1987) 661–683.

278. G. Slade, Convergence of self-avoiding walk to Brownian motion in high dimensions, J. Phys. A 21 (1988) L417–L420.

552

246   References

279. G. Slade, The scaling limit of self-avoiding random walk in high dimensions, Ann. Proab. 17 (1989) 91–107.
280. G. Slade, *The Lace Expansion and its Applications*, Lecture Notes in Mathematics 1879, Springer, Berlin, 2006.
281. J. Sohier, Finite size scaling for homogeneous pinning models, preprint 2008.
282. R. Song and X.Y. Zhou, A remark on diffusion of directed polymers in random environments, J. Stat. Phys. 85 (1996) 277–289.
283. C.E. Soteros and S.G. Whittington, The statistical mechanics of random copolymers, J. Phys. A: Math. Gen. 37 (2004) R279–R325.
284. F. Spitzer, *Principles of Random Walk* (2nd. ed.), Springer, New York, 1976.
285. S. Stepanow, J.-U. Sommer and I.Ya. Erukhimovich, Localization transition of random copolymers at interfaces, Phys. Rev. Lett. 81 (1998) 4412–4415.
286. S.F. Sun, C.-C. Chou and R.A. Nash, Viscosity study of the collapsed state of polystyrene, J. Chem. Phys. 93 (1990) 7508–7509.
287. S.-T. Sun, I. Nishio, G. Swislow and T. Tanaka, The coil-globule transition: radius of gyration of polystyrene in cyclohexane, J. Chem. Phys. 73 (1980) 5971–5975.
288. A.-S. Sznitman, *Brownian Motion, Obstacles and Random Media*, Springer, Berlin, 1998.
289. M.C. Tesi, E.J. Janse van Rensburg, E. Orlandini and S.G. Whittington, Monte Carlo study of the interacting self-avoiding walk model in three dimensions, J. Stat. Phys. 82 (1996) 155–181.
290. M.C. Tesi, E.J. Janse van Rensburg, E. Orlandini and S.G. Whittington, Interacting self-avoiding walks and polygons in three dimensions, J. Phys. A: Math. Gen. 29 (1996) 2451–2463.
291. F.L. Toninelli, Critical properties and finite-size estimates for the depinning transition of directed random polymers, J. Stat. Phys. 126 (2007) 1025–1044.
292. F.L. Toninelli, Correlation lengths for random polymer models and for some renewal sequences, Elect. J. Probab. 12 (2007) 613–636.
293. F.L. Toninelli, A replica-coupling approach to disordered pinning models, Commun. Math. Phys. 280 (2008) 389–401.
294. F.L. Toninelli, Disordered pinning models and copolymers: beyond annealed bounds, Ann. Appl. Probab. 18 (2008) 1569–1587.
295. F.L. Toninelli, Coarse graining, fractional moments and the critical slope of random polymers, preprint 2008, to appear in Elect. J. Probab.
296. F.L. Toninelli, Localization transition in disordered pinning models. Effect of randomness on the critical properties, in: *Proceedings of the 5th Prague Summer School on Mathematical Statistical Mechanics* (September 2006) (ed. R. Kotecký), Springer, Berlin, to appear.
297. C.A. Tracy and H. Widom, Level-spacing distribution and the Airy kernel, Commun. Math. Phys. 159 (1994) 151–174.
298. A. Trovato and A. Maritan, A variational approach to the localization transition of heteropolymers at interfaces, Europhys. Lett. 46 (1999) 301–306.
299. D. Ueltschi, A self-avoiding walk with attractive interactions, Probab. Theory Relat. Fields 124 (2002) 189–203.
300. C. Vanderzande, *Lattice Models of Polymers*, Cambridge Lecture Notes in Physics 11, Cambridge University Press, 1998.
301. S.R.S. Varadhan, Appendix to K. Symanzik, Euclidean quantum field theory, in: *Local Quantum Theory* (ed. R. Jost), Academic Press, 1969.

302. V. Vargas, *Polymères dirigés en milieu aléatoire et champs multifractaux*, Ph.D. Thesis, University of Paris 7, France, November 23, 2006.

303. V. Vargas, A local limit theorem for directed polymers in random media: the continuous and the discrete case, Ann. Inst. H. Poincaré. Probab. Statist. 42 (2006) 521–534.

304. V. Vargas, Strong localization and macroscopic atoms for directed polymers, Probab. Theory Relat. Fields 138 (2007) 391–410.

305. Y. Velenik, Localization and delocalization of random interfaces, Probability Surveys 3 (2006) 112–169.

306. T. Vrbová and S.G. Whittington, Adsorption and collapse of self-avoiding walks and polygons in three dimensions, J. Phys. A: Math. Gen. 29 (1996) 6253–6264.

307. T. Vrbová and S.G. Whittington, Adsorption and collapse of self-avoiding walks in three dimensions: A Monte Carlo study, J. Phys. A: Math. Gen. 31 (1998) 3989–3998.

308. T. Vrbová and S.G. Whittington, Adsorption and collapse of self-avoiding walks at a defect line, J. Phys. A: Math. Gen. 31 (1998) 7031–7041.

309. W. Werner, *Random Planar Curves and Schramm-Loewner Evolutions*, Lecture Notes in Mathematics 1840, Springer, Berlin, 2004, pp. 107–195.

310. J. Westwater, On Edwards' model for polymer chains, Commun. Math. Phys. 72 (1980) 131–174.

311. J. Westwater, On Edwards' model for polymer chains. III. Borel summability, Comm. Math. Phys. 84 (1982) 459–470.

312. J. Westwater, On Edwards' model for polymer chains, in: *Trends and Developments in the Eighties* (S. Albeverio and P. Blanchard, eds.), Bielefeld Encounters in Math. Phys. 4/5, World Scientific, Singapore, 1984.

313. S.G. Whittington, Self-avoiding walks terminally attached to an interface, J. Chem. Phys. 63 (1975) 779–785.

314. S.G. Whittington, Self-avoiding walks with geometrical constraints, J. Stat. Phys. 30 (1983) 449–456.

315. S.G. Whittington, A directed walk model of copolymer adsorption, J. Phys. A: Math. Gen. 31 (1998) 8797–8803.

316. S.G. Whittington, Random copolymers, Physica A 314 (2002) 214–219.

317. S.G. Whittington and C.E. Soteros, Polymers in slabs, slits and pores, Isr. J. Chem. 31 (1991) 127–133.

318. M.V. Wüthrich, Scaling identity for crossing Brownian motion in a Poissonian potential, Probab. Theory Relat. Fields 112 (1998) 299–319.

319. M.V. Wüthrich, Superdiffusive behavior of two-dimensional Brownian motion in a Poissonian potential, Ann. Probab. 26 (1998) 1000–1015.

320. M.V. Wüthrich, Fluctuation results for Brownian motion in a Poissonian potential, Ann. Inst. H. Poincaré Probab. Statist. 34 (1998) 279–308.

321. M.V. Wüthrich, Numerical bounds for critical exponents of crossing Brownian motion, Proc. Amer. Math. Soc. 130 (2002) 217–225.

322. M.V. Wüthrich, A heteropolymer in a medium with random droplets, Ann. Appl. Probab. 3 (2006) 1653–1670.

323. O. Zeitouni, *Random Walks in Random Environment*, Lecture Notes in Mathematics 1837, Springer, Berlin, 2004, pp. 189–312.

Index

List of Participants

37th Probability Summer School, Saint-Flour, France

July 8-21, 2007

Lecturers

Jérôme BUZZI	École Polytechnique, Palaiseau, France
Frank den HOLLANDER	Leiden Univ., The Netherlands
Jonathan C. MATTINGLY	Duke Univ., Durham, USA

Participants

Jean-Baptiste BARDET	Univ. Rennes 1, F
Anne-Laure BASDEVANT	Univ. Pierre et Marie Curie, Paris, F
Mireia BESALU	Univ. Barcelona, Spain
Michel BONNEFONT	Univ. Paul Sabatier, Toulouse, F
Pavel BUBAK	Univ. Warwick, Coventry, UK
Alain CAMANES	Univ. Nantes, F
Francesco CARAVENNA	Univ. Padova, Italy
Pavel CHIGANSKY	Univ. du Maine, Le Mans, F
Alina CRUDU	Univ. Rennes 1, F
Sébastien DARSES	Univ. Pierre et Marie Curie, Paris, F
Omar de la CRUZ	Univ. Chicago, USA
Latifa DEBBI	Univ. Setif, Algérie
François DELARUE	Univ. Denis Diderot, Paris, F
Maxime DELEBASSEE	Univ. Paul Sabatier, Toulouse, F
Hacène DJELLOUT	Univ. Blaise Pascal, Clermont-Ferrand, F

254   List of Participants

Leif DOERING	TU Berlin, Germany
Pierre FOUGERES	Univ. Paris 10, F
Antoine GERBAUD	Institut Fourier, Grenoble, F
Ricardo GOMEZ	Univ. Nacional Autónoma México
Dan GOREAC	Univ. Brest, F
Ludovic GOUDENEGE	Univ. Rennes 1, F
Erika HAUSENBLAS	Univ. Salzburg, Austria
Martin HUTZENTHALER	Goethe Univ., Frankfurt, Germany
Barbara JASIULIS	Univ. Wroclaw, Poland
Wouter KAGER	EURANDOM, Eindhoven, The Netherlands
Tamas KOI	Budapest Univ. Technology, Hungary
Maxime LAGOUGE	Univ. Denis Diderot, Paris, F
Matthieu LERASLE	INSA Toulouse, F
Shuyan LIU	Univ. Lille 1, F
Eva LOECHERBACH	Univ. Paris 12, F
Grégory MAILLARD	École Polytechnique Fédérale Lausanne, CH
Florent MALRIEU	Univ. Rennes 1, F
Charles MANSON	Univ. Warwick, Coventry, UK
Scott McKINLEY	Duke Univ., Durham, USA
Oana MOCIOALCA	Kent State Univ., USA
Peter NANDORI	Budapest Univ. Technology, Hungary
Francesca R. NARDI	Eindhoven, The Netherlands
Eulalia NUALART	Univ. Paris 13, F
Alberto OHASHI	Univ. Campinas, Brazil
Nicolas PETRELIS	EURANDOM, Eindhoven, The Netherlands
Jean PICARD	Univ. Blaise Pascal, Clermont-Ferrand, F
Mikael ROGER	Univ. Rennes 1, F
Marco ROMITO	Univ. Firenze, Italy
Jean-Pierre ROZELOT	Observatoire Côte d'Azur, F
E. SAINT LOUBERT BIE	Univ. Blaise Pascal, Clermont-Ferrand, F
Christian SELINGER	Univ. Luxembourg
Laurent SERLET	Univ. Blaise Pascal, Clermont-Ferrand, F
Arvind SINGH	Univ. Pierre et Marie Curie, Paris, F
Julien SOHIER	Univ. Denis Diderot, Paris, F
Cristian SPITONI	Leiden Univ., The Netherlands
Mitja STADJE	ORFE, Princeton, USA

Ramon Van HANDEL   California Inst. Techn., Pasadena, USA
Andrea WATKINS   Duke Univ., Durham, USA
Hendrik WEBER   Univ. Bonn, Germany
Radoslaw WIECZOREK   Polish Academy Sciences, Katowice, Poland
Lorenzo ZAMBOTTI   Univ. Pierre et Marie Curie, Paris, F

Programme of the School

Main Lectures

Jérôme Buzzi	Hyperbolicity through entropies
Frank den Hollander	Random polymers
Jonathan Mattingly	Ergodicity of dissipative SPDEs

Short Lectures

Pavel Bubak	Asymptotic strong Feller property for degenerate diffusions
Alain Camanes	Directed polymers in random environment and uniform integrability
Francesco Caravenna	The quenched critical point of a diluted disordered polymer model
Sébastien Darses	Stochastic derivatives
Latifa Debbi	On deterministic and stochastic fractional partial differential equations
François Delarue	Stochastic analysis of a numerical scheme for transport
Maxime Delebassee	Uniqueness in graph percolation
Pierre Fougères	Markovian type semilinear problems associated with some (infinite dimensional) functional inequalities
Ricardo Gómez	Almost isomorphisms of Markov shifts
Erika Hausenblas	SPDEs driven by Poisson random measures
Wouter Kager	Patterns and ratio limit theorems for Markovian random fields

258   Programme of the School

Shuyan Liu	Estimation of parameters of stable distributions on convex cones
Grégory Maillard	Intermittency on catalysts
Oana Mocioalca	Space regularity of stochastic heat equations driven by irregular Gaussian processes
Francesca Romana Nardi	Ideal gas approximation for a two-dimensional rarefied gas under Kawasaki dynamics
Eulalia Nualart	Hitting probabilities for systems of fractional stochastic heat equations on the circle
Alberto Ohashi	Finite dimensional invariant manifolds for SPDEs driven by fractional Brownian motion
Nicolas Pétrélis	Copolymer in an emulsion
Mikaël Roger	A central limit theorem for some (random) compositions of automorphisms of the torus
Marco Romito	Markov solutions for the stochastic Navier-Stokes equations in dimension three
Jean-Pierre Rozelot	A story of the sun oblateness: from Princeton, 1996 to Pic du Midi, 2006
Laurent Serlet	Can super-Brownian excursion become old, big or heavy?
Arvind Singh	The speed of a cookie random walk
Julien Sohier	The homogeneous polymer pinning model near criticality
Cristian Spitoni	Metastability for reversible probabilistic cellular automata with self-interaction
Ramon Van Handel	Robustness and approximations in nonlinear filtering
Radoslaw Wieczorek	Fragmentation-coagulation processes as limits of individual based models